ASTRONOMY

Fourth Edition

Tom Bullock

West Valley College

DEDICATION

This book is affectionately dedicated to my Mother, who not only brought me into this world, but nourished my curiosity and interest in knowing, understanding, and appreciating it.

About the cover:

The title of this book is partially taken from the occurrence of a total solar eclipse, a photograph of which is on the cover. As Moon slips between Earth and Sun, the latter is gradually eclipsed. At the moment Sun first becomes completely covered by Moon, the event is called Second Contact. It is a very emotional event, as it suddenly turns nighttime and stars and planets become visible in the sky. Such an eclipse is one of the most fascinating astronomical phenomena observable from Earth with the naked eye. For an explanation of the complete process, refer to Chapter 3 in the Book.

Text Design: Tom Bullock
Cover Design: Tom Bullock
Cover Photograph: Total solar eclipse taken from Atacama Desert, Chile, November 3, 1994. One-second exposure of eclipsed Sun with Hasselblad camera and 280-mm lens riding piggyback on a Celestron Super-8 Telescope, on high-speed Ektachrome film. Later, a 25-second double exposure of background sky and full moonlit landscape at Death Valley, California. *Tom Bullock*

Copyright © 1987, 1989, 1994, 1996 by Kendall/Hunt Publishing Company

ISBN 0-7872-2428-6

All rights reserved. No part of this publication may be reproduced, stored in a retrieval system, or transmitted, in any form or by any means, electronic, mechanical, photocopying, recording, or otherwise, without the prior written permission of the copyright owner.

Printed in the United States of America

10 9 8 7 6 5 4 3 2 1

CONTENTS.

PART I.—PHYSICAL ASTRONOMY.

	PAGE
CHAP. I.—General Properties of Matter	15
Sec. 1. Gravitation	17
Div. 1. Centre of Gravity, 19; Div. 2. Specific Gravity and Density	21
Sec. 2. General Laws of Motion	21
Sec. 3. Compound Motion	22
Div. 1. Curvilinear Motion, 24; Div. 2. Centrifugal and Centripetal Forces	26
II.—Angular Measurement	27

PART II.—DESCRIPTIVE ASTRONOMY.

CHAP. I.—Solar System	29
II.—The Sun	30
III.—Interior Planets	34
Sec. 1. Mercury	35
Sec. 2. Venus	37
IV.—The Earth	42
V.—Exterior Planets	51
Sec. 1. Mars	53
Sec. 2. Asteroids	56
Sec. 3. Jupiter	58
Sec. 4. Saturn	61
Sec. 5. Uranus	63
Sec. 6. Neptune	63
VI.—Satellites	65
Sec. 1. The Moon	66
Sec. 2. Jupiter's Satellites	87
Sec. 3. Rings and Moons of Saturn	90
Sec. 4. Moons of Uranus	93
Sec. 5. Moons of Neptune	94
VII.—Motion of the Earth	96
Sec. 1. Diurnal Motions of the Earth	96
Sec. 2. Annual Motion of the Earth	100
Sec. 3. Seasons	102
VIII.—Time	108
IX.—Ecliptic and Zodiac	114
X.—Tides	117
XI.—Eclipses	123
Sec. 1. Solar Eclipses	125
Sec. 2. Lunar Eclipses	131
Sec. 3. Transits	134
Div. 1. Transits of Mercury, 136; Div. 2. Transits of Venus	136
Sec. 4. Occultations	138
XII.—Comets	140
XIII.—Systems of Astronomy	155
Sec. 1. Ptolemaic System	155
Sec. 2. Egyptian System	156
Sec. 3. Tychonic System	156
Sec. 4. Copernican System	157

PART III.—SIDEREAL ASTRONOMY.

CHAP. I.—Fixed Stars	159
Sec. 1. Apparent Motions and Positions of the Stars	161
Div. 1. Precession, 164; Div. 2. Nutation, 165; Div. 3. Refraction, 166; Div. 4. Aberration, 169; Div. 5. Proper Motion, 172; Div. 6. Parallax	174
CHAP. II.—Distances of the Fixed Stars	177
III.—Milky Way	180
IV.—Magnitude of the Stars	182
V.—Appearance of the Stars	184
Sec. 1. Variable and Periodic Stars	184
Sec. 2. Temporary Stars	185
Sec. 3. Compound Stars	187
Div. 1. Double Stars, 187; Div. 2. Multiple Stars	189
Sec. 4. Clusters and Nebulæ	190
Div. 1. Double Nebulæ, 193; Div. 2. Nubeculæ or Magellanic Clouds	194
VI.—Meteors	195
Sec. 1. Meteorites	196
Sec. 2. Zodiacal Light	198
VII.—Constellations	198
Sec. 1. Zodiacal Constellations	200
Sec. 2. Northern Constellations	212
Sec. 3. Southern Constellations	240

PART IV.—PRACTICAL ASTRONOMY.

CHAP. I.—General Properties of Light	260
Sec. 1. Refraction of Light	261
Sec. 2. Reflection of Light	264
II.—The Observatory	267
III.—Telescopes	268
Sec. 1. Refracting Telescopes	268
Sec. 2. Achromatic Telescopes	269
Sec. 3. Equatorial Telescopes	270
Sec. 4. Transit Instruments	270
IV.—Micrometer	274
V.—Instrumental Adjustments	275
Sec. 1. The Vernier	275
Sec. 2. The Level	276
Sec. 3. Plumb-Line	277
Sec. 4. Artificial Horizon	277
Sec. 5. Collimator	278
Sec. 6. Transit Clock	279
VI.—Graduated Circles	281
Sec. 1. Mural Circle	281
Sec. 2. Transit Circle	284
Sec. 3. Altitude and Azimuth Instrument	287

PART V.—TREATISE ON THE GLOBES.

CHAP. I.—Problems on the Terrestrial Globe	287
II.—Problems on the Celestial Globe	307

PART VI.—HISTORY OF ASTRONOMY.

CHAP. I.—Sec. 1. From the Earliest Times to the Christian Era	320
Sec. 2. Astronomical Instruments in use from the Earliest Times to the Christian Era	327
II.—Sec. 1. From the Christian Era to the year 1600	329
Sec. 2. Astronomical Instruments invented from the Christian Era to the year 1600	335
III.—Sec. 1. From the beginning of the Seventeenth to the end of the Eighteenth Century	336
Sec. 2. Astronomical Instruments in use from the beginning of the Seventeenth to the end of the Eighteenth Century	341
IV.—History of Astronomy from the beginning to the middle of the Nineteenth Century	343
NOTES	345
ASTRONOMICAL DICTIONARY	411

Figure T of C–1 *Compare this Table of Contents of an Astronomy textbook dated 1855 with that of this Book. Notice especially that the interest in the universe as observed at the time was a description of objects: their appearances and motions. An understanding of the origin and evolution of celestial objects awaited the development of astrophysics, and the application of such fields as chemistry to the Heavens.*

Overview of Book Contents

PREFACE *Page* **vii**

SECTION 1 *Page* **1**   OBSERVING THE SKY

Chapter 1	Organization of the Universe	**2**
Chapter 2	Mapping the Sky We Observe	**18**
Chapter 3	Observing Motions of Objects in Sky	**34**
Chapter 4	Birth and Growth of Modern Science	**64**
Chapter 5	Telescopes and Radiation	**90**

SECTION 2 *Page* **121** EVOLUTION OF THE STARS

Chapter 6	Atoms, Matter, and Radiation	**122**
Chapter 7	Sun, Our Nearest Star	**140**
Chapter 8	Characteristics and Properties of Stars	**164**
Chapter 9	The Birth of Stars	**190**
Chapter 10	The Death of Stars	**210**
Chapter 11	Pulsars, Neutron Stars, Black Holes	**224**

SECTION 3 *Page* **251** EVOLUTION OF GALAXIES

Chapter 12	The Milky Way, Our Home in Space	**252**
Chapter 13	Galaxies Beyond the Milky Way	**276**
Chapter 14	Active Core Galaxies and Quasars	**296**
Chapter 15	The Origin of the Universe	**310**

SECTION 4 *Page* **335** EVOLUTION OF PLANETS

Chapter 16	The Origin of the Solar System	**336**
Chapter 17	Exploring the Earth-like Planets	**376**
Chapter 18	Exploring Jupiter and Saturn	**424**
Chapter 19	Exploring Uranus, Neptune, and Pluto	**450**

SECTION 5 *Page* **467** EVOLUTION OF LIFE IN THE UNIVERSE

Chapter 20	Is Anyone Out There?	**468**
Chapter 21	The Future of Life in the Universe	**492**

GLOSSARY/INDEX *Page* **502**

SEASONAL STAR CHARTS **Appendix**

Detailed Book Contents

PREFACE *Page* **vii**
- The Book's theme **xi**
- For the learner **xii**
- My goal **xii**
- The drama of extraterrestrials **xiii**
- The drama of evolution **xiii**
- Methods of astronomy **xiv**
- Life in the universe **xv**
- Style of presentation **xv**
- Second Contact **xvii**
- Chapter contents **xviii**
- A Final Word **xxi**

SECTION 1 *Page* **1**   OBSERVING THE SKY

Chapter 1   Organization of the Universe **2**
- Scale of things in the universe **4**
- Brief tour of the universe **10**
- General characteristics of things **12**
- Learning Objectives/New Terms **17**

Chapter 2   Mapping the Sky We Observe **18**
- The Location of Sun in space **19**
- Ancient concepts of the sky **23**
- Mapping the sky **27**
- Learning Objectives/New Terms **32**

Chapter 3   Observing Motions of Objects in Sky **34**
- Observing the stars' movements **35**
- Observing Sun's motions **39**
- Observing movements of the planets **42**
- Observing Moon's movements **44**
- Observing movement of the equinoxes **44**
- Eclipses **49**
- Ocean tides **58**
- Astronomy vs astrology **60**
- Scientific objections to astrology **60**
- Learning Objectives/New Terms **63**

Chapter 4   Birth and Growth of Modern Science **64**
- Worldview **65**
- The birth of modern science **68**
- Galileo enters the drama **75**
- Opposition to Galileo and Copernicus **76**
- Science in northern Europe **81**
- What is science? **87**
- Learning Objectives/New Terms **89**

Chapter 5	Telescopes and Radiation **90**	
	Electromagnetic Energy **91**	
	Measuring radiation at Earth's surface **97**	
	Telescopes **99**	
	New developments in telescope design **106**	
	Operating a telescope **108**	
	Radio telescopes **112**	
	Telescopes in Space **114**	
	Learning Objectives/New Terms **119**	

SECTION 2 *Page* **121** *EVOLUTION OF STARS*

Chapter 6	Atoms, Matter, and Radiation **122**
	Atomic structure **123**
	Atoms and sunlight **126**
	Temperature **130**
	Conclusions drawn from spectra **135**
	Motions of objects in space **137**
	Learning Objectives/New Terms **139**
Chapter 7	Sun, Our Nearest Star **140**
	Observing Sun **141**
	Detailed studies of Sun **144**
	Solar behavior: Babcock Model **153**
	Sun-Earth relationships **153**
	Journey Into Sun **156**
	Learning Objectives/New Terms **162**
Chapter 8	Characteristics and Properties of Stars **164**
	Determining brightnesses of stars **165**
	Separating temperature/luminosity **170**
	Binary stars: determining their masses **172**
	Interpretation of the H–R diagram **179**
	Comparison of stars **183**
	Extraterrestrials? **186**
	Learning Objectives/New Terms **186**
Chapter 9	The Birth of Stars **190**
	Locating Young Stars **191**
	Interiors of birthing stars **197**
	Learning Objectives/New Terms **208**
Chapter 10	The Death of Stars **210**
	Interiors of Dying Stars **212**
	Red giant to white dwarf **219**
	Learning Objectives/New Terms **223**

Chapter 11	Pulsars, Neutron Stars, Black Holes **224**
	Supernovae **225**
	Theoretical basis for neutron stars **227**
	Detection of supernovae **233**
	Cause of supernovae **235**
	Novae **236**
	Summary of Star Death **238**
	Black Holes—Theoretical Basis **239**
	Black Holes—Observational evidence **246**
	Learning Objectives/New Terms **250**

SECTION 3 *Page* **251** *EVOLUTION/GALAXIES*

Chapter 12	The Milky Way, Our Home in Space **252**
	Evolution of the universe **253**
	Discovery of our galaxy **254**
	Determining Distances to objects **256**
	Properties of star clusters **260**
	Shape of the Milky Way galaxy **265**
	Evolution of the Milky Way **268**
	Learning Objectives/New Terms **274**
Chapter 13	Galaxies Beyond the Milky Way **276**
	Observing Galaxies **277**
	Distances to galaxies **282**
	Classifying galaxies **285**
	Galactic rotation **291**
	Evolution of galaxies **294**
	Formation of galaxies **294**
	Learning Objectives/New Terms **295**
Chapter 14	Active Core Galaxies and Quasars **296**
	Peculiar galaxies **297**
	The quasar controversy **300**
	Characteristics of quasars **301**
	Problems in explaining quasars **304**
	The birth of galaxies **313**
	Learning Objectives/New Terms **313**
Chapter 15	The Origin of the Universe **314**
	Size and extent of the universe **315**
	Curvature of space **317**
	Origin of the Universe **319**
	Test of cosmological theories **322**
	Conditions after the Big Bang **323**
	Mapping past history of the universe **325**
	Current Research in cosmology **328**
	Assumptions of cosmologists **331**
	Future of the universe **332**
	Learning Objectives/New Terms **333**

SECTION 4 Page 335 EVOLUTION OF PLANETS

Chapter 16   The Origin of the Solar System **336**
Exploration of the Solar System **337**
Solar system survey—contents **338**
Solar system survey—appearances **340**
Solar system survey—distances **341**
Solar system survey—chemistry **342**
Solar system debris—comets **344**
Solar system debris—meteoroids **353**
Solar system debris—asteroids **357**
Asteroids and dinosaurs **364**
Tunguska event **366**
Meteorites **366**
Extrasolar planetary systems **370**
The search for planetary systems **371**
Learning Objectives/New Terms **376**

Chapter 17   Exploring the Earth-like Planets **378**
General factors in planet building **379**
Planet Earth **381**
Future of Earth's surface **387**
The Earth as a Planet **387**
Conditions favorable to atmospheres **389**
Moon **390**
Origin of Moon **393**
Venus **394**
Mercury **403**
Mars **405**
Tiny satellites of Mars **416**
Formation of the solar system **416**
Learning Objectives/New Terms **422**

Chapter 18   Exploring the Giants Jupiter, Saturn **424**
Voyager Mission **425**
Jupiter **426**
Satellites of the outer solar system **434**
Jupiter's family of satellites **435**
Saturn **440**
Satellites of Saturn **444**
Learning Objectives/New Terms **449**

Chapter 19   Exploring Uranus, Neptune, and Pluto **450**
General features of Uranus **451**
Voyager 2 and Uranus finally meet **453**
Satellites of Uranus **456**
Neptune **458**
Satellites of Neptune **461**
Pluto **463**
Beyond Pluto **465**
Future of *Voyager* satellites **465**
Learning Objectives/New Terms **466**

SECTION 5 Page 467 EVOLUTION OF LIFE IN THE UNIVERSE

Chapter 20   Is Anyone Out There? **468**
"When" will we conduct the search? **470**
"Why" should we conduct the search? **474**
"Who" should we search for? **474**
"How" should we conduct the search? **476**
"What" sort of signal do we search for? **480**
Effects of contact **484**
Probability of successful contact **485**
Attempts to contact others **489**
Learning Objectives/New Terms **491**

Chapter 21   The Future of Life in the Universe **492**
Space station **496**
Astronomy in the Future **500**
Learning Objectives/New Terms **501**

GLOSSARY/INDEX Page 502

SEASONAL STAR CHARTS Appendix

Preface

What is Astronomy?

Figure Preface–1 A small family of galaxies, called Stephen's Quintet, is but a small sample of the contents of space. Each galaxy contains about 100 billion stars, and there are about 100 billion galaxies in the known universe. It is the origin, structure and evolution of stars and galaxies that is a central theme of astronomy. (Lick Observatory)

CENTRAL THEME: What motivates people to study the universe of planets, stars and galaxies, and how can this Book be used to understand, appreciate and enjoy the discoveries of modern astronomers?

Early in 1992, astronomers surprised the public and the scientific community alike with the announcement that they had obtained convincing evidence of the manner in which the entire universe in which we live began some 20 billion years ago. To some, it was not only an important scientific discovery, but one that had personal importance – a spiritual discovery. As reported in the news media, one of the codiscoverers stated that "It was like looking into the face of God."

This so-called *Big Bang* theory is not new. It has been around for several decades. But it has always been just that – a theory. Supporting evidence accumulated over the years had been circumstantial, and far from conclusive. The scientific community was strongly divided on the issue. What the astronomers now seemed to be saying was that their observations provided the evidence needed to make it a scientific truth rather than a scientific theory.

Less controversial in the field of astronomy is the observation that stars and their planets form out of clouds of gas and dust that inhabit space – that they live for a while, and that they die by ejecting clouds of gas and dust back out into space. Thus new stars and planets form out of the recycled remains of stars that have already died. Since our own star – Sun – is an average star by whatever criteria used, astronomers believe that it too was born in this fashion and will die in the same fashion as other stars. Furthermore, if we adopt the consensus of evolutionary biologists, life on Earth evolved out of the raw materials present on our planet at the time of its formation.

What Are We Doing Here? Just for the sake of argument, let's tentatively adopt the above as being true: the Big Bang scenario for the origin of the universe, and chemical/biological evolution for the origin of life on Earth. Now this is a very interesting and even exciting conclusion for humans to ponder and accept. But it also presents us with some challenges and difficulties. For example, what is the purpose of human beings in the universe if we are here as the result of random chemical processes in a vast ocean of chemical arrangements? And where do we go from here? Should we proceed to venture out into the cosmos to populate the distant solar systems with our kind – rather like a stellar Noah's Ark? In light of these possible truths, perhaps the most important task is that of redefining what it means to be human.

What is the goal of human existence if we are simply by-products of the random interactions between chemicals in the chill vacuum of space or on the surface of an glob of chemicals we call Earth? What is the purpose of life? What are we here for? What is the correct strategy for a happy life? One need only read the morning newspaper or watch the evening news to realize that the world is experiencing considerable pain over our inability to agree on answers to such questions. One might even argue that it is because so many people adopt the scientific view of the origin of life that we are in such turmoil in the first place!

So it seems to me that the focus of a book claiming to survey the universe and its contents should be not on what we know (or claim to know) to be true, but on how we know (or claim to know) something is true. In other words, it is the process of gathering evidence to support our theories that is of greatest importance in science. What evidence do you need to be convinced that the universe has taken 20 billion years to evolve into you? And of course there is that other nagging question that attaches to any attempt to define our origins – what does it mean? That is certainly not an easy question to answer, and is certainly not one that can be answered with the scientific method.

Let me stimulate your thinking. If we got here by the process of cosmic evolution: by-products of matter behaving according the laws of physics, chemistry, and biology – then it seems likely that similar events occur elsewhere, on planets orbiting other stars. Are there as many extraterrestrial civilizations as there are stars in the sky? If so, have They had any influence on our own evolution in the past? What would be the human consequences if we ever make contact with Them? Should we prepare for that eventuality? And if that should come to pass – for better or worse – will our beliefs about what it is to be human change in a drastic way?

If you are even slightly nervous about establishing contact with "beings" out there, let me hasten to add that we started a program for doing that over 30 years ago. It is a little late to begin worrying about it. Signals have already been sent out and are making their way toward what might be distant civilizations. And it hasn't been just the United States, either. Just in case you are interested, here is a partial listing of past and current attempts to establish contact:

- 1960   USA   2 stars
- 1963   USSR   1 quasar
- 1966   Australia   1 galaxy
- 1968-69   USSR   12 stars
- 1970-72   France   10 nearest stars
- 1972-76   USA   674 stars
- 1972-present USSR   All-sky search
- 1974-76   Canada   70 stars
- 1974   USA   1 star cluster
- 1975-79   Netherlands   50 star fields
- 1977   USA   200 stars
- 1977-present Germany   3 stars
- 1978   USA   185 stars
- 1978   USA/Australia   25 star clusters
- 1979   Australia   Close Sun-like stars
- 1979-81   USA   200 stars
- 1981   Netherlands   Center of Milky Way
- 1982   Canada   Center of Milky Way
- 1985-95   USA   Project *META*
- 1992-94   USA   NASA's *HRMS*
- 1990-present Argentina   Project *META II*
- 1994-present USA (Arecibo)   Project *SERENDIP*
- 1995-present USA   Project *Phoenix*
- 1995-present USA   Project *BETA*

What Are You Doing Here? I can think of several reasons why you have chosen to learn about astronomy:

- You think that it will be fun.

- You were given a telescope for your birthday and you want to know how to use it and what there is to look at in the sky with it.

- You want to expand your Worldview — your belief system by which to live — by including the contents and behavior of the physical world around you.

- You like to learn about new discoveries in science.

- You have chosen to be a teacher, and that is one of the subjects that you will teach.

- Your close friend is interested in astronomy, and you want to be able to share it with him/her.

- You have chosen to graduate or obtain a degree from a college or university, and one of the requirements is to take a class in the physical sciences. You have chosen astronomy because:
  - One of the above.
  - It worked in well with your schedule.
  - Nothing else seemed exciting.
  - It seemed like the easiest way to fulfill the requirement.

What Am I Doing Here? I have written this book to fill the need for an introductory textbook of the science of Astronomy that emphasizes the process of inquiry by people into the workings of nature. Its intended audience is the college student who has chosen not to major in science, but who wants to understand how scientists go about their work — especially how they manage to piece together theories of how the planets, galaxies, the entire universe, and even you and I came to exist in time and space.

You may point out that the choice to pick up this book was not exactly of your own choosing. Perhaps you are required by your college or university to complete a class in a physical science such as Astronomy in order to graduate, obtain a credential of some type, or move up the academic ladder. That's not entirely surprising — science is an activity in which we are all participants as well as subjects. We DO science at the same time that we are affected by others DOING science. So it is reasonable that colleges devoted to producing well-informed citizens require each of its graduates to take at least one survey course in one of the sciences.

Being a well-informed citizen means that you help shape the uses of science in our democratic society. After all, some of your tax monies go toward supporting projects that scientists convince Congress to fund. Learning the methods of science in one subject (such as astronomy) allows you to understand the process of science in other subjects (like biology) as well. In other words, the process of science is applied in the same manner by all of the scientific disciplines. The terms are different, of course. But the approach is the same.

Mostly what I want as an outcome of your reading of the Book is an appreciation for the universe in which we live. It is our home. It is ours to enjoy. Personally, I think that the best way to enjoy anything is to participate. So I believe that you will enjoy learning the subject if you participate in outdoor activities during which the topics discussed in the text can be viewed *directly*. Thus throughout the Book I suggest various sky-viewing activities intended to enhance your pleasure of learning Astronomy. These usually include a map of the portion of the sky in which the activity is to be accomplished. In other words, I would like the Book to be user-friendly in that it relates to the real sky that is observable to anyone.

What can you do with Astronomy? The subject of astronomy is one to be enjoyed outside of the classroom or confines of the home. It can be the focus of outdoor as well as indoor activities. Specifically, it can considerably enhance any adventure to a location where a dark nighttime sky is available. There are also celestial events unique to various locations around the world, such as the *auroras*. Here are a few of the experiences you might want to try during your reading of this Book:

- Observe the Northern Lights (Chapter 7)
- Observe eclipses (Chapter 3)
- Visit a Planetarium show
- Observe a "green flash" (Chapter 5)
- Participate in a "star party" conducted by amateur astronomers
- Visit an observatory (Chapter 5)

For those of you who want to "go with the flow" and see where the subject matter of astronomy takes you, be aware that it can easily influence the development of your *"World View."* That is, we are not born into this world with a set of well-identified "truths" by which to live out our lives, or even by which to run the planet. The advance of civilization is a history of the conflicts that result from two opposing forces operating within each of us — the need to lead others and the need to be led by others. Once we find a "truth" by which we want to live out our own lives, we feel compelled to share it with others and even yearn for them to adopt it as their own "truth." On the other end, we feel the need to "fit in," to be like others, to belong, to be liked by others. Kings, emperors, and presidents have attained and maintained power over others not by the sword alone, but also by attracting the consensus of followers who needed to "belong."

Rulers then and now often claim a divine right to their

rulership, citing religious writings or even omens as the source of their "truth." Wars are fought to maintain the claimed divine rulership, the soldiers believe it because it is "their" ruler. The way out of this treadmill of blind obedience to authority is certainly the road toward rulership by consensus — what we in the West call democracy. Rulers are still needed to guide the process by which consensus is established, especially as the needs of members of society increase in number and the needs begin to overlap one another, competing for attention and resources.

It is not an accident that the overthrow of absolute rulership and establishment of the earliest democracies and the rapid rise of science occurred hand in hand. Science, as a method of inquiry, works best when it is non-directed — when it arises simply from native curiosity. Democracy is an attribute of modern science as well as our political system. There are no "kings" or "rulers" in science. There are certainly people we refer to as "experts" in science. This is a label applied to anyone who by hard work and insight have convinced a large number of scientists that their experiments and observations reveal an accurate description of how nature works. But these experts have no authority over anyone. Every scientist is free to listen to or ignore what they have to say. In fact, science moves forward by virtue of the fact that some scientists do NOT listen to (or at least do not believe) the experts.

There is also a method of consensus-building, a procedure by which a scientist's description of a "truth" gets aired. The accepted method of announcing a new theory or discovery to the scientific community at large is through the British magazine *Nature*. As the proposed theory or discovery is analyzed and discussed at conventions, meetings, or in individual laboratories around the world, it accumulates a list of "acceptances" and/or "rejections." It either fits in with a larger body of findings or can be duplicated by other scientists, or it conflicts with other previously accepted discoveries or cannot be verified by anyone else. In any case, it receives its "day in court" through the mechanisms by which scientists share their ideas with one another.

Anyone who attempts to bypass the accepted method of proposing a new theory or discovery is therefore suspect, as if he/she wants to announce a "truth" without going through the process of giving other scientists a chance to give it their "seal of approval." This very thing happened a few years ago when some scientists chose to announce a revolutionary discovery of "cold fusion" in a bottle. Basically, they claimed that they had detected the occurrence of the process by which Sun shines in a bottle under controlled conditions. If true, it would have signaled the discovery of an energy source that would have transformed human civilization. It has yet to be duplicated by anyone since.

The Language of Science In this Book, you are asked to learn new terms and add vocabulary to your speech while learning astronomy. The new terms are **boldface** when used for the first time in the text. They are listed at the conclusion of each Chapter so that you can test the efficiency of your learning at the conclusion of reading each Chapter. They are defined in the Glossary at the end of the Book. Humans use language to express and convey thoughts, ideas, and feelings. Scientists do as well. But in addition to using words and numbers, they use such tools as symbols, diagrams, graphs and formulae. To learn astronomy is to learn the special language of astronomy. This may seem tedious at first, but in the long run it makes the task of learning astronomy much easier. Take, for example, the matter of just how our Sun manages to generate the energy that sustains life on Earth.

I could, in a few paragraphs, describe the solar energy process in terms that you could understand without having had courses in physics or chemistry. But I can summarize the process more simply by writing the symbolic expression: $E = mc^2$. In doing so, I have reduced a lot of words into a few symbols. There are several advantages for doing so. Two of the most important are that it saves space and it allows for understanding the process without regard to the native language of the reader. In other words, scientists around the world share a common language of scientific notation. For example, the symbol "**c**" in the expression $E = mc^2$ is reserved for the *speed of light* in every scientific laboratory in the world. The "**E**" stands for the energy being produced within Sun, and "**m**" stands for the amount of mass that is consumed as energy is produced.

Mathematics is, without a doubt, the "language of nature." You will learn in Chapter 20 that scientists are so convinced of this that mathematical expressions are assumed to be the language by which extraterrestrial civilizations will first contact us. Although mathematics is treated very lightly throughout the Book, keep in mind that behind each of the concepts of astronomy is a set of equations and/or calculations that lead to the concept that I discuss in words. That is the meaning behind my frequent use of the phrase *"According to calculations, astronomers have concluded that...."* Therefore, my task is to convey in words what is more easily expressed as mathematical formulae.

The language of science is precise. There is no interpretation of words. They are adopted for the purpose of expressing the behavior of something in space. They do not have value connected to them. Consequently scientists do not argue over the use of words. Think about the last conversation you had with a friend. What you probably remember best is not so much the actual words that he/she used, but the "meaning" of the subject discussed. The object of words — in other words — is to convey meaning to another person. Once you have conveyed the meaning, you can forget the words that were used.

Reflect on the last disagreement you had with a friend. Did you eventually argue over the meaning of a word that one of you used? Such words as "God" and "love" and "care" and "mad" and "afraid" and "angry" are loaded with value because of the different levels of importance we attach to them. Scientists may argue over whether an object found wandering in the solar system is an asteroid or a comet, but fundamentally they are not arguing over the use of the word, but whether the observational data best fits the definition of comet or asteroid.

That is another convincing argument for using symbols in science — universal agreement. After all, the *object of science is to understand how nature works*, and nature is not debatable. It works, regardless of how confused we might be in attempting to understand all of its aspects. Nature is simple, so by concentrating a number of descriptive words into a single symbol we get closer to understanding how it behaves — we are "hearing" it speak to us.

The Book's Theme

The overall theme of this Book is **curiosity**. *The end product of that curiosity is* **knowledge**. *The method/process by which that curiosity leads to knowledge is* **science**.

Curiosity is the driving force behind much of human behavior. It reveals itself every time we ask "WHY?" or "HOW?" This book is the result of my own curiosities — "How can I write a book about Astronomy that stimulates students toward FEELING the excitement of viewing and UNDERSTANDING the nighttime sky?" "How can I write a book that stimulates students toward refining their beliefs about their role in the universe and the meaning of life?" Some people fear curiosity because it "...kills the cat." I will not address myself to that question. That is for the reader to determine for himself/herself. It is sufficient to feel joy while satisfying my curiosities about the universe, and marveling that others experience the same joy. One of the great joys of life is to observe the curiosities of children.

Knowledge speaks for itself. It requires no lengthy explanation. By itself, knowledge is neutral — it is neither good nor bad. It is how we act upon that knowledge that creates a basis for goods or bads. Again, it is not my task to determine whether or not our accumulated knowledge about the universe was worth the time, energy or money. It is sufficient for me to experience the joy of possessing what little knowledge I have, and in many cases acting upon it.

Science is a method of inquiry. The word "science" comes from the Latin word "*scientia*", which means "to know." That is not to say that knowledge can come only from the methods of science. Knowledge comes from a variety of methods. But we cannot deny that it is the progress of science that most influences our daily lives. Perhaps I should say that it is our choosing to use the products of science that most influences our daily lives. I need cite but one example to illustrate the difference between knowledge and the use of knowledge.

Early in the Twentieth Century, scientists developed the means to investigate the structure of the matter of which all material objects are made. In the process, they discovered that the atom was composed of smaller particles — neutrons, protons and electrons. Although this knowledge allowed scientists to explain the manner in which Sun generates the energy by which we live, it allowed scientists to develop the atomic bomb, and eventually the hydrogen bomb.

Science versus Technology Although the distinction between science and technology is often a blurry and gray one, we generally consider them to be quite separate forms of human activity. Many countries of the world adopt and use the modern technology that springs from science without themselves actually adopting the methods of science. Science means knowing, or the activity associated with arriving at the state of knowing. On the other hand, the word "technology" comes from the Greek "*techne*", which means "art" or "skill." Technology is the skill of making things using knowledge (science). Thus one could loosely define **science** as knowledge we possess of how the universe operates, whereas **technology** is the application of that knowledge in reorganizing that same universe for our own benefit — including doing more science.

This example may clarify the distinction between science and technology. Chemists doing science are curious about how chemicals can be arranged in various ways to form new chemicals. By examining the structure of those new chemicals, they are able to learn something about the inner workings of the original chemicals. Now a technologist, learning of the producing of those new chemicals, might experiment further with them in an effort to use them in an industrial process for making a product that you and I will purchase — a sexier brand of toothpaste, for example. A portion of the purchase price goes toward paying the salary of the technologist, since he/she is working for a company that makes products for our benefit (?). But the original chemist had no goal other than to understand the workings of nature (chemicals) itself.

We often refer to such work as *pure* research, inasmuch as there is no product, only further knowledge. Of course, that chemist must be supported by someone, so governments use some of our tax monies to pay for the research (usually through colleges and universities). At the same time, they well know that out of such pure research comes new understandings of the workings of nature, which in turn some technologist will use to create new products that will create new jobs that will provide more tax monies to support more research, and so on.

The laws of physics allow us the opportunity to construct space habitats in orbit around Earth. In that unique environment of weightlessness and perfect vacuum, new products will be made and returned to Earth to enrich our lives. But at the same time, in those same habitats, scientists will be learning new things about how the universe works. In what direction that new knowledge takes us cannot be foreseen. It may result in our going even further out into space, or withdrawing back to our blue planet.

For the Learner

Astronomy is the study and description of the universe and its components — stars, galaxies, and planets. Since it is conducted in the minds of people, it is a piece of art. Like a painting, drawing, or sculpture, it expresses an idea and/or attempts to elicit a feeling or emotion about something. In this case, astronomy attempts to convey in words, diagrams, and/or mathematical formulae an accurate representation of the things that are out there in the universe, and the manner in which they are behaving.

Unlike a piece of art, however, astronomy is accomplished by a multitude of artists, not just a single person. When we observe a piece of art, we are often times as interested in technique as in content; that is, we are interested in just how it is that the artist was able to construct the piece. So it is with astronomy. And because all people who engage in the work of astronomy have agreed to apply a certain set of rules while working on the piece of art, they have also decided to call themselves scientists in order to emphasize the fact that they do use a technique common to all. Of course we call that technique the **scientific method**.

It is important to me to stress the manner in which this technique operates. Especially is it important to note that the piece of art itself (astronomy) is <u>not</u> the actual thing it attempts to describe. The thing is the thing, the description is the description. So when you read of a description of the manner in which Sun shines, for example, you are not to assume that description <u>is</u> the manner in which Sun shines. Sun is going to shine regardless of what attempts we make to explain its behavior. But I would also emphasize that it is the goal of scientists to make the description as accurate as possible so as to most completely explain as much as possible about Sun.

From what I have said already, you might correctly wonder if scientists have yet painted an accurate picture of how Sun shines. Well, let me put it this way. After completing the first draft of the painting (may I begin calling scientific paintings **theories**?) in the 1930s and 1940s, scientists (with the encouragement of the President of the United States, who was anxious to end the Second World War) built an artificial sun using that first draft. That was the hydrogen bomb.

In other words, the building of the bomb was based upon processes that we <u>thought</u> were operating inside of Sun, but only by making a model of those processes could we be sure that we were correct. The fact that it was successful (?) strongly suggests that our theory was correct. True, there are lots of things about Sun that we don't yet understand, but the method of inquiry seems to work.

My Goal

Astronomy is much more than what can be seen in the sky. And that is what this book is about. I will begin with and then enlarge upon what you already know and have experienced regarding the sky. What I will share with you are the conclusions reached by people after they asked questions about the objects in the sky and then conducted experiments and observations in order to answer those questions. In essence, scientists do what you and I do all the time — observe, question, formulate conclusions about what they observe, and then test their conclusions in future experiences.

For the astronomer, this process requires the use of sophisticated instruments and high-tech hardware and software. The accuracy of the conclusion is still dependent upon the imagination and creativity of the human being, and that is the emphasis I would like to be obvious throughout the book. What separates the Nobel Prize winner from the average scientist is the use of the imagination in piecing together the diverse set of observations into a concise theory that works. By "works," of course, I mean a theory that is able to explain all of the observations and experiments related to that particular field of study.

My major goal in this book is to personalize the subject of astronomy for you, to make it user-friendly, to make it relevant to your life, to make it change your life in the way you think, feel, and perhaps even act. I say that because that is what it has done for me and for many other people around me over the years I have been teaching astronomy. I would like the sky to be 3-dimensional to you: for it to be more than a mere canopy of white dots above your head. I would like you to be able to visualize the stars and other objects as having features, and even personalities in a sense.

As much as possible, therefore, I will attempt to relate the material in the book to an actual experience you can have outdoors in the presence of the "real" universe. The universe is, after all, mostly out "there," not "in here." So I hope that you will take the time and effort to go outside and look at some of the examples of objects being discussed in the book.

Self-Discovery Although the specific focus here is astronomy, you will be introduced to many other scientific disciplines as well — chemistry, physics, geology, meteorology, atomic physics, and even a little biology and sociology. The subject of astronomy is the study of the universe. And everything that is, is in the universe. So anything you can think of is a legitimate topic for astronomy. Although I won't go too far afield, it will be necessary to borrow knowledge from adjacent fields of study in order to more clearly explain the behavior of objects in space.

What particularly fascinates me about the subject of astronomy is its tendency to explain my own connectedness with the universe. Since I, too, belong to the universe, I am a legitimate topic for study. Isn't it interesting, for example, that I am even capable of studying the universe and making some sense out of it? Isn't it amazing that I am capable of analyzing the light just now arriving from the Great Andromeda Galaxy, located 2 million light-years away, and be able to explain the behavior of the objects of which it is composed?

I am able, in a very real sense, to communicate with an object that is located far away. Admittedly, it is probably a one-way communication, but isn't even that an amazing realization? How is it, out of all the things that inhabit the universe, that it is only we humans that appear to possess the skills necessary to understand and explain the intricate workings of another little corner of the universe?

Thinking So this book is also about thinking. I could just as easily have titled it "The Joy of Thinking", because, in a sense, most astronomical research has no practical benefit or application. The goal is knowledge about how the universe works. Obtaining that knowledge requires the making of instruments which creates jobs, but the knowledge is saleable only if one writes a book about the knowledge. So what will you do with the knowledge you obtain herein? My hope is that it will enlarge your horizon of understanding, stimulate your thinking, and motivate you toward further exploration, further discovery, and further understanding.

When, on the evening news, you see a photograph of a galaxy recently discovered, I hope you *understand* what you are seeing and that which is beyond the photograph itself — the distances, the movements, and the processes that are taking place therein. You do this all the time, I'm sure. If someone mentions the name of your favorite car or flavor of ice cream, or you see a photograph of your favorite vacation resort, you go immediately beyond the name or the photograph to a broader association established through experience. So I want to encourage you to experience astronomy as much as possible as you read the Book.

The Drama of Extraterrestrials

I have observed that we humans are attracted to drama — we can't easily pass up bold headlines in the newspaper, an intriguing spy thriller, a provocative love story, or a daily soap opera on TV. Our curiosities lead us to wonder why something happened, who committed the crime, who fell in love, or who cheated who. Novels, plays, operas, movies, and other forms of entertainment tap that remarkable human appetite.

And so it is with this Book. It is not a collection of separate facts. It is a synthesis of ideas collected over the centuries by a variety of scientists and other thinkers, weaved together to form a dramatic tapestry reflecting our current thinking about the universe in which we live.

I choose to use the **SETI Program** (the **S**earch for **ExtraT**errestrial **I**ntelligence) as a central theme. Since 1959, many countries of the world have been engaged in SETI programs in the attempt to establish contact with distant civilizations. It is questionable, of course, whether such attempts stand a reasonable chance of success. But what is of interest to us is the question — Why is it that the astronomical community is so convinced of the existence of life beyond Earth that several counties are willing to devote time and money and resources to contact that life?

In order to answer that question, I would need to tell you what it is that astronomers and other scientists know about the universe beyond Earth (**extraterrestrial** = outside of Earth) that convinces them that there is probably life out there. But that is a course in astronomy, which of course is what this Book is about.

Each topic in the book relates to this central theme or question. At any time during your reading, you should be able to associate a particular topic with the central theme. At the conclusion, you will have a clear understanding of the justification for the SETI programs. Don't get me wrong — I'm not attempting to convince you that there are "beings" out there. My interest is in explaining why it is that the scientific community has arrived at that conclusion. Not every scientist believes that the Program has value. There will always be debate between doing (and paying for) things that are "practical" and doing things simply to fulfill our curiosities and stimulate our intellects.

Science proceeds by consensus — the majority rules. But often times the majority view is wrong. New observations and/or experiments are added to the existing ones and new conclusions are reached. So you're on your own as regards the conclusion you reach — you may end up opposing the prevailing viewpoint. Again, my interest is that you have a basis for your conclusion.

Many of our theories in astronomy are controversial — like a political candidate, there is not a consensus as to the correctness of a theory. In science, this is healthy. It means that the method of scientific inquiry is alive and well. After all, new observations of the universe are constantly being made, with corresponding new interpretations of what they suggest about the universe. So I will be emphasizing the controversies that have and currently are shaping the painting we call astronomy. The drama of astronomy will be highlighted by many of the current controversies.

The Drama of Evolution

I could just as easily have entitled this book "Evolution" rather than the title I chose. For many people, the word "evolution" has strong emotional associations, raising thoughts of Darwin's controversial theories and images of apes becoming men and women. But in a broad sense, of course, the word "evolution" means change — the form, content, and process by which things (and even ideas) change over time. That essentially is what the study of modern astronomy is all about — How are the objects in space evolving over time? Or, to put it in more simplified terms, what are the life/death cycles of the objects occupying space — their birth, life, and death?

Now, to the casual observer of science, this may seem like a radical approach to a subject that is usually associated with arrangements of stars in the sky, beautiful photographs of distant planets returned to Earth via satellites launched by the inhabitants months or years prior to receipt of the photos, or even to the expected events on Earth caused by a passing comet. But such is the current conclusion of scientists who make space their area of interest. The universe around us consists of objects that have had a birth from preexisting objects, and indeed that will change into other objects as time allows chemical reactions to run their course.

Perhaps the best way to illustrate the evolution of the discipline of astronomy is to look at the list of topics in the table of contents of an astronomy textbook published in 1842, and then compare them with the topics to be covered in this Book (Diagram T of C-1). What should be obvious is that our current interest is in the manner in which astronomical objects evolve , whereas a century ago the interest was in the mere descriptions and positions of celestial objects. That is why I chose to use it on the Title Page as an introduction to this Book. I would like you to appreciate how far we have come.

Before the 1920s, astronomers were restricted mostly to the task of organizing the solar system and the sky in a geometrical way — where things are located, how they move, and what their positions will be at some future date. In order to go beyond that limited scope to a study of changes within the objects themselves, the behavior of matter at the atomic level needed to be discovered.

The great discoveries in atomic physics in the early part of this century allowed for the opening up of entirely new branches of astronomy, so that today there are astronomers who spend most of their time not at the telescope, but in the atomic physics laboratory trying to squeeze out more information about the behavior of matter. The motivation for doing this often leads to completely unexpected things — for example, the discovery of objects in space whose behavior just cannot be explained using our present framework of knowledge regarding atoms and molecules.

Even though I deplore the sensationalism inherent in most TV programming and newspapers, it does tell me something about we humans. We are attracted to and intrigued by drama, mystery and controversy. In the news media, the emphasis is generally on the dramatic aspects of human behavior, very often involving violence and social misbehavior. I would also like to capture your appetite for drama, mystery and controversy, but of a different sort. That is the drama surrounding the manner in which nature works, and especially the way in which humans attempt to explain the manner in which nature works.

I discuss the various topics in modern astronomy according to a basic question — a basic mystery — about the universe. Often times, the answering of that question or the solving of that mystery involves controversy. Different scientists will interpret the evidence in different ways. That is exciting too.

Methods of Astronomy

Before we explore the details of how we are surrounded by change, I feel that it is very necessary to develop within you — the reader — a confidence in the conclusions of modern astronomers. You see, astronomy as a scientific discipline is somewhat different from other sciences in that, with two exceptions, it does not have available to it samples of the actual objects it claims to know so much about. How can we claim that stars are born, live, and die if we do not have some ""star-stuff"" in the laboratory to study?

A geologist can obtain samples of rocks to support theories of how regions of Earth's surface came to be; a botanist can pluck samples of plants to study under the microscope in hopes of learning their growth and behavior. But with the exception of meteorites, telling us something about the asteroid belt, and moon rocks, obtained with a $24-billion program to land a few men on Moon, astronomers must learn about objects in space by studying the light and other types of radiation coming from the objects.

So unless I can establish at the very offset your confidence in the tools and instruments that are used to study that light and radiation, you are certainly going to wonder about the certainty of the conclusions and theories being offered by them. The first few chapters are dedicated to that goal of developing your awareness and appreciation for the astronomical tools and instruments used in dissecting the composition and behavior of stars, galaxies, and planets.

Stars The next logical topic to discuss is that of stars — their birth, life, and death cycle. But why stars? Why not those objects with which we are most familiar — Earth and moon, and perhaps the other planets to which we have launched (and in two cases on which we have actually landed) satellites? Because stars are the fundamental building blocks of the universe, or to put it another way, most of the material substance of the universe is contained within stars. So logically we should begin our discussion of evolution with those things that are most commonplace.

To put it another way, let us consider our local experience — Earth and Sun. Sun is a star, similar in all respects to the other stars seen in the nighttime sky. It contains 99.37% of all the matter in our solar system, being about one million times larger than our planet — Earth. But most of the remaining matter, that 0.63%, is contained in Jupiter, which is about 1,000 times larger than Earth.

So it is obvious that planets do not make up much of the material composition of the universe, and therefore will not be given undue emphasis at the beginning of a textbook in astronomy. Of course, since our star, Sun, is the nearest star to us, it would seem reasonable that we know and understand it best, so we will actually begin our discussion of the life cycle of stars with a discussion of our sun.

Galaxies It turns out that stars, like people, prefer not to live alone in space. The force of gravity has acted to pull them into giant families called galaxies. Our sun is located in a family (galaxy) called the Milky Way, and it appears that our sun has over 100 billion brother and sister stars in that family. But how did those stars gather together to form that family, and how is it able to retain its shape as a giant pinwheel? Those are the topics of Chapters 10 through 12 — the life cycle of the galaxies; their birth, life, and (perhaps) eventual fate.

Universe The galaxies are islands of stars, occupying space as far as our telescopes are capable of seeing. Thus we can think of the universe as the total collection of galaxies, the space in which they are embedded, and the time that has elapsed since their making. So the next topic will be the evolution of the universe as a whole — its birth, how it came to exist in the form we see today, and its possible fate. In dealing with the topic of the origin of the entire universe, therefore, we will be treating space and time as entities.

When I speak of the origin of the universe, we speak simultaneously of the origin of space and time as well as of matter. Thus was the great insight of Albert Einstein, who taught us that space and time are measurable quantities just as is matter, and therefore are capable of evolutionary change.

Having thus covered the conclusions of modern astronomers as regards the evolution of stars, galaxies, and the universe itself, I will return to our more local experience and consider the origin of the planets and debris orbiting Sun. After all, even though Earth is rather minute compared to Sun, there does appear to be some interesting features about it that set it apart from stars and galaxies. So we will ask ourselves how it is that such a diversity of planets, moons, and debris was left circling Sun as it formed, and the extent to which we can expect such events to take place around other stars.

Life in the Universe

So what else can be included under the topic of evolution? Stars, galaxies, the universe itself, and planets with associated debris — it would appear that such a list includes just about all there is. But what about life itself? Should it be placed in a separate category, to be covered only in textbooks related to biology? The details of the structure and behavior of life forms and their origins on Earth are normally covered in that separate discipline called biology, but that is only for convenience.

By the time we have gotten through the topics of stars, galaxies, the universe, and planets, a general pattern of evolution will have emerged. The slow, gradual change of the material substances of the universe behaving by the laws of physics and chemistry evolve into new material substances with new characteristics and behavior patterns.

It will be tempting, therefore, to explain the origin of life on Earth by using the same patterns of evolution. Because time and space prevent a full, detailed coverage of this topic, I will only touch upon the general patterns discovered by evolutionary biologists, and the features of the scientific theory of the origin of life on Earth.

Regardless of how convinced you are of the theory of evolution, you might want to consider another interesting consequence of the pattern of evolutionary change in the universe. If life developed and flourished on Earth through evolutionary changes, then it is conceivable that similar events have occurred elsewhere in space, perhaps on the surfaces of planets around each and every star in the sky!

And if we have developed the means by which to announce ourselves to the rest of the universe via radio communication, have other civilizations of life forms developed communication techniques and at this very moment be sending out signals in the hope of contacting someone? Presumedly they would not closely resemble us in appearance, but can't we assume that they would have the capacity for understanding and even changing their environment?

The fact that we have not yet received any signal that would indicate the presence of any such "beings" out there does not discourage us from our conclusion that they may well be there. The assumptions upon which we base our means of searching and the very content and form of the expected language are no doubt faulty, but of course if we never try because we don't know what the correct assumptions are, we will never get anywhere. Absence of evidence is not evidence of absence.

Style of Presentation

Such is the content of this textbook — the evolution of objects in *space*. But I would like to stress the process by which the topics of evolution are handled in the book. As each of the topics of evolution is presented, I will pose three questions to be considered:

- <u>What</u> do we know (or claim to know)?
- <u>How</u> do we know?
- <u>What</u> does it mean?

What Do We Know? Science is commonly thought to be a collection of facts or truths, arrived at by scientists acting as detectives trying to understand how things work. Such is the way that science courses are commonly taught — as a collection of facts or truths, the "What do we know?"s. But the methods of science do not allow for the universe to be so easily categorized in terms of "truths" or "facts."

In fact, scientists work on the *assumption* that there is no such thing as truth that can be determined by the scientific method. Every scientific statement has some level of believability or acceptability, usually based upon the consensus of all scientists working in the subject area.

To that extent, the scientific method is a democratic system based upon the votes of participating scientists. There is no authority-no king, no president, no emperor who can claim to have a monopoly over what is right or truth. The consensus of scientists about a particular theory today may well change so that what is accepted as being true today is found to be false tomorrow.

This process of inquiry, of course, goes on continuously, so that scientific knowledge itself evolves. Every conclusion in science, and therefore every statement in this book, at some point in time falls somewhere along a line between the extremes of certainty and uncertainty.

Let me illustrate with an example. Frequently I am asked if astronomers really have discovered black holes in space. The expression "black hole", by the way, has become such a commonly-used one that it is used in reference to anything that is all take and no give – the Internal Revenue Service, for example.

Anyway, my response is that the best way to explain the phenomena we observe in the vicinity of a few stars is to theorize the presence of black holes. That does not prove that they exist. At the moment – with the observations and data that are available – it is our best guess.

The difference between guess and theory is a matter of how much data, both observational and theoretical, is available. To the astronomer who happened to catch a random event on a photograph, it would be a guess. But to the theoretician who has had a chance to study the photograph and compare it with a other photographs and related data, it is a theory.

When the scientific community has had sufficient time to study it further and integrate it into the network of related objects, it is accepted as scientific truth. That doesn't mean it will never change — new observations may eventually reveal a fatal flaw. Reflecting on the trial of Galileo during the Renaissance, we will see how an incorrect theory survived for over 1200 years!

How Do We Know? The focus of any book on science, it seems to me, should therefore not be on the "What do we know?"s, but on the "How do we know?"s. Let me provide a few examples. Few of you would argue about the shape of Earth, the accepted one being generally that of a ball or sphere. If asked to offer evidence of your belief, few of you would be in a position to offer experimental evidence or convincing proof. In fact, there is an organization based in New England called the Flat Earth Society which believes and teaches the theory that Earth is indeed flat.

Well, I am not trying to convince you that Earth may be flat. Personally, I am not willing to spend much time researching the evidence of Earth's roundness to convince myself. In other words, on a scale of 0 to 10 from uncertainty to certainty, I would place the theory of Earth's shape somewhere around 9.99. There is very little evidence to doubt its roundness. But the subject of the existence of black holes is another matter.

If one were to take a vote amongst astronomers today as to their theory of the death of our star, Sun, you would find only a 91% agreement, say. And if you surveyed their opinion as to the Big Bang theory for the origin of the universe, you would find a 46% agreement, perhaps. The point of this discussion is that scientific theories are more or less incomplete, and therefore to some extent not entirely believable. So a major emphasis in this book is a discussion of the *evidence that supports a particular theory, not just a description of that theory.*

This process requires a flexible mind and a tolerance for ambiguity. In many cases the evidence is unclear and apparently conflicting, as if both suspects in a mystery novel have motives for committing the crime. It should be obvious by now that the excitement of science is not in the end product itself, the theory. The excitement of science is in the process of getting there – the collecting of clues and observations which, when pieced together, allow a picture to unfold before our eyes.

What Does It Mean? I suspect that very few of you, if any at all, are using this textbook as a basis for your going on to become an astronomer or even engage in a scientific discipline in general. So why do educational institutions require students to take a course such as this? Of what value is the learning of a science like astronomy if it is not used on a day-to-day basis?

I hope to inject such value in the process of developing the topics mentioned above through the answering of the third question, "What does it mean?" It is difficult to imagine anyone knowing even the barest of the conclusions of modern astronomy, and at the same time not having reflected upon his/her purpose or meaning in the universe.

After all, most of the universe is *out there*, above our heads, and *down there*, below our feet. How can we assign any importance to ourselves at all in light of our smallness in terms of matter, space, and time? Let me offer one way of approaching that question. I suspect that Sun is unaware of the existence of the next nearest star, *Alpha Centauri*. But I am aware of it, as well as the 100 billion other stars that make up the Milky Way galaxy. Isn't it interesting (and unexplainable) that one part of the universe (my consciousness) is capable of communicating with (being aware of) another part of the universe! Perhaps my significance lies in my ability to know how insignificant I am in terms of matter, space, and time. Or, as Albert Einstein pondered:

"The most incomprehensible thing about the universe is that it is comprehensible."

Example of Style Let me offer an example of how I use the three questions "What do we know?", "How do we know?", and "What does it mean?" in the Book.

- *What do we know?* The Milky Way galaxy contains some 100 billion stars.
  How do we know? We take photographs of the sky and count the number of tiny dots that appear in the photographs.

- *What do we know?* There are some 10 billion galaxies in the known universe.
  How do we know? We take photographs of the depths of space, and count the number of galaxies that appear as smudges in the photographs. Using simple multiplication, we can then estimate the number of stars in the known universe — 100 billion times 10 billion equals 1000 billion billion.

A mathematician calculated the size of a container necessary to hold that number of grains of sand. To his surprise he found that number of grains of sand is approximately the number occupying all the beaches on Earth! There are — in other words — as many stars in the known universe as there are grains of sand on all the beaches of Earth. The next time you spend a day at the beach, pick up a handful of sand and let it shift through your fingers, and ponder the significance of what I have said. One of those grains, perhaps one caught beneath your fingernail, represents our star, Sun.

- *What does it mean?* Whatever you conclude from that experience, and of course each of us will have our unique and personal conclusions, I suspect that you will be a changed person. Not drastically, necessarily, but changed in the way you see the world (and life) around you. That is what I am referring to when I mention the "What does it mean?" of astronomy. Hopefully you will encounter many of them while reading this Book.

That is what any book or course of instruction is intended to do — to change you. Above the entrance to any classroom should be a sign that says — *"You are here not to worship what is known, but to question it."* So if you don't go away after reading this Book a changed person, you didn't pay close enough attention.

Someone once said that children enter school as question marks (?) and leave as periods (.). I do not want you to be one of those periods — period. I would like you to enjoy the act of learning just as you enjoy a good meal, or a walk on the beach. Learning is most enjoyable, of course, if you are able to relate the experience to your own life. Well, you are in the universe! You are a part of it! So wouldn't anyone want to know about their own home?

I would like you to arrive at the final page of this Book with an appreciation of astronomy as an evolving method of obtaining knowledge and understanding about the universe, and not just an awareness of what that knowledge is. Or, to put it another way, "The thrill is in the chase, not in the kill." I hope that you enjoy your **Second Contact** with astronomy.

For most of you, this is not the first contact you have had with the subject of astronomy. That is one reason I have titled this book **Second Contact**. You have, no doubt, spent evenings outdoors gazing at the star-studded sky. Or you have had occasion to attend a presentation at a local planetarium in which an artificial (but quite accurate) sky was projected onto the dome over your head.

At the very least, you've been aware of the changing position of Sun in the sky, the varying lengths of day and night during the year, and the changing shape of Moon in the sky. Perhaps you're fortunate enough to have a neighbor who occasionally sets up a telescope in the backyard, and who has enticed you over in order to see the rings of Saturn, the four Galilean moons of Jupiter, or the craters on Moon.

As a child, you may have heard or read stories about the constellations in the sky, reflections from an earlier age when people offered different explanations as to how the sky was formed in the first place, and why the stars were placed in the patterns we still see today. Some of the names of those patterns are so familiar to us that they have been used as logos for marketing purposes. Such words as "Orion," "Polaris," "Vega," "Quasar," and "Pulsar" are taken from astronomical dictionaries.

Second Contact

Let me explain why I have chosen the title for this book. The dictionary defines drama as "A series of actions or events that involve the quality of conflict."

On July 11th of 1991, thousands of people from around the world converged on the Big Island of Hawaii and on the southern tip of Baja California to witness one of nature's most spectacular displays — a Total <u>Solar Eclipse</u>. For the most part, these adventurers were not astronomers who had a particular interest in studying events in space. They were ordinary people from all walks of life — people like you and I. Those in Hawaii witnessed Sun's sudden disappearance for a maximum of 4 minutes and 11 seconds, while observers in Mexico had the slightly longer experience of 6 minutes and 56 seconds.

Those may not seem like sufficiently long enough periods of time to justify the time and expense of traveling so far, and yet because of this eclipse's unusual length it was billed by astronomers and the media alike as the "Eclipse of the Century." Astronomers refer to the precise moment when Moon just begins to move in front of Sun to block out sunlight streaming toward Earth as *First Contact*. At the precise moment when Moon completely blocks out Sun and all of the sunlight headed toward Earth, observers appropriately located on Earth find themselves completely in the dark. It becomes nighttime in the middle of the day. That dramatic moment is called *Second Contact*.

The same expression – *Second Contact* – is used to refer to that moment during a Total <u>Lunar</u> Eclipse when Moon just slips completely within Earth's shadow cast into space. For reasons explained in Chapter 3, Moon at that dramatic moment is a reddish-orange color. There is an orange ball hanging in the sky! So those moments of Second Contact are the types of dramas that I associate with astronomy, and therefore I wanted the reader to get caught up in the excitement as well.

That is one reason I chose to title the book **Second Contact**. The second reason has to do with the reason that I wrote the Book in the first place. Learning astronomy has changed my life. Astronomy has become incorporated into just about every aspect of my life – above and beyond the fact that I earn my living by teaching classes in it. It is a prominent component of the design and furnishing of my home. It is incorporated into my travel and vacation plans. It is an integral part of my life philosophy. And so I know that it can and will become the same for other people as well.

After all, most of the universe is above our heads – out there even beyond the stars in the sky. Doesn't everyone have a natural yearning for knowing their location in the universe? So in offering this Book as a guide toward understanding the universe in which we find ourselves, I hope to build upon your previous experience with astronomy, even if that is limited to an occasional evening under the stars during a camping adventure or outdoor picnic or a viewing of the latest Star Trek movie. This Book will be your *Second Contact* with astronomy.

The third reason for the title of the Book is a way of expressing appreciation to Carl Sagan for the manner in which I had <u>my</u> *Second Contact* with astronomy. When I first began teaching, I sensed that my students did not appear very excited about the subject. That was probably due to the fact that I wasn't very excited about the subject either. The textbook that I had chosen to use in my first classes was a traditional one, describing each of the planets, explaining the characteristics of the stars, and theorizing about the origin of the universe.

But there seemed to be something missing. There was no obvious drama. There was no drama that connected me to the universe outside of myself. The book was asking my students (and me as well) to understand the universe out there but omitted any reference to my <u>participation</u> in the universe. In other words, it was difficult to find any relevance of astronomy to my everyday life. Then I was introduced to the writings of Carl Sagan, and in particular to his book *The Search for Intelligent Life in the Universe*. This textbook had a theme, a drama, a unifying concept for my classes. It was a simple yet thought-provoking theme: "Is it possible that we are not the only conscious beings in the entire universe?" "And if so, is it possible that we can establish contact with *them* and learn from *them*?" "And if we can (and do), what might the consequences be?"

I have held to that theme ever since, believing that in order to have an comprehensive judgement on the matter of life elsewhere, one should also have a comprehensive understanding of the nature and behavior of the contents of the universe. The reason the word Contact enters this discussion of the central theme of my Book is that Carl Sagan went on to write a bestselling book titled *Contact*. In it, he portrays the possible implications of establishing contact with ET's (extraterrestrials).

Just as a TV documentary is not the event itself, but only someone's representation or interpretation of it, so a book about astronomy is not astronomy. To have the experience of astronomy yourself, you begin by going outside and experiencing the real sky rather than reading about it or listening to someone talk about it. That doesn't mean that reading a book about or listening to a discussion about astronomy is a waste of time.

This change in my approach to teaching astronomy is reflected in my attempt to have the reader of this book personalize the subject for his/her own life. Look for ways of integrating the topics of astronomy into your own life, whether it be looking up at the sky more frequently, watching a science documentary on television, or enriching your belief system. I have attempted to make the book user-friendly, to make it engaging enough so that you will look forward to reading it and learning something new and exciting about the universe. Perhaps you'll make your *Third Contact* with astronomy a visit to northern Chile on November 3rd of 1994, or Europe on August 11 of 1999, to witness a total solar eclipse yourself!

Chapter Contents

You are about to scan the Table of Contents of this Book out of curiosity of what you can expect as you read it. Although I have chosen specific titles for each Chapter, I'd like to make sure that you know in advance that there is a flow to the progress of the Chapters. Not only does one Chapter build upon the previous one, but there is an emerging theme as the Chapters become components in a structure, much like bricks make up a house. Here is how I describe the theme that evolves out of the Chapters:

Chapter 1: We take a tour of the universe, from Earth to the most distant reaches of space. While doing so, we learn how the objects that occupy space are organized and how they behave. In the end, we have a working vocabulary with which to understand and discuss the various topics that make up the subject of astronomy.

Chapter 2: We return to Earth, since it from here that astronomers launch their studies of objects in space. We study how the sky is organized and mapped for the purpose of further study, adding names of some popular objects to our vocabulary.

Chapter 3: Sun, Moon, planets and stars we observe in the sky move with respect to the horizon and also with respect to the map of the sky we learned. They move according to well-defined laws of motion that allow for prediction of their future positions and behavior. Using the vocabulary we learned in the first two Chapters, we learn how these motions are observed from Earth's surface, and how they are explained by fundamental laws of nature — the laws of physics.

Chapter 4: The implication of Chapters 1-3 is that astronomers can explain the behaviors of objects in space using the scientific method, in spite of the fact that those objects are located at vast distances from Earth. Before being introduced to further claims about the universe, we sit in judgement at the trial of Galileo to compare contrasting methods of obtaining knowledge about the universe, and, in the process, observe the birth of modern science. Only if we accept the scientific method can we accept the conclusions of modern astronomer.

Chapter 5: Modern science wins out over the methods of the ancients, so we feel confident that technological gadgets like telescopes image objects located billions of light-years away, and that high-speed computers can explain their origins and behaviors. We examine how modern astronomers use these tools to obtain information from space, so that we can have confidence in their discoveries.

Chapter 6: Telescopes, ground-based and in orbit around Earth and Sun, are the basic tools by which information is gathered from space. But they merely collect radiation which is then recorded by various types of detectors. What does the radiation have to do with the objects that sent it? We learn about the connection between radiation and matter — how the chemistry of the universe is encoded in beams of radiation.

Chapter 7: We now apply our knowledge of telescopes and their ability to collect and analyze radiation to the nearest star, our Sun — which is the gateway to our understanding of more distant stars. We observe at close range the various features of Sun, their behaviors over time, learn how Sun shines, and how all of these matters influence life on Earth.

Chapter 8: Having learned the characteristics and behavior of a single star, we now apply similar methods to the study of other stars. In the spirit of the scientific method, we categorize stars according to the various characteristics determined by telescopes and attached instruments, and then look for patterns that might reveal the evolution of a star — its birth, life, and death.

Chapter 9: Surveying the vast numbers of stars in the sky, and knowing how stars generate their energies, astronomers conclude that stars form out of the basic ingredients that drift between the stars — clouds of gas and dust. We explore these regions of space, examining them just as we do the stars themselves, and see evidence that convinces us that indeed the stars evolve from the basic chemicals that pervade the universe.

Chapter 10: If stars form from clouds of chemicals, and "burn" those chemicals in the act of shining, then it would seem that eventually stars must run out of "fuel" and die. We apply telescopes to the study of those sites in space that reveal what appears to be the debris of stellar death. In surveying the different kinds of sites, we learn the different types of death that a star can have, and what characteristics the star had that determines the particular type of death.

Chapter 11: Although it appears that most stars die in rather quiet fashion, some explode violently. In the process, they not only create remains in the form of black holes that defy logic and laws of physics as currently understood, but scatter gas and dust out into space that eventually forms into new stars. Death leads to new life. We look at evidence for the bizarre black holes, and the remains of exploded stars that is participating in the birth of new generations of stars.

Chapter 12: Most of the events discussed in the previous Chapters occur within a large family of stars and clouds of gas and dust called the Milky Way galaxy. Sun and its host of planets are members of this organization of 200 billion stars. There is organization to the placement of stars, the study of which provides clues to its origin some 15 billion years ago. We now learn how the process of star birth, life, and death led to the Milky Way's present structure and behavior.

Chapter 13: Astronomers estimate that there are as many galaxies outside of the boundaries of the Milky Way as there are stars within it. They are the fundamental building blocks of the large-scale universe. Photographic studies reveal that they come in a variety of shapes, sizes, and stellar compositions. They, too, appear to have formed in a manner similar to our galaxy. We categorize galaxies in order to look for patterns that assist in explaining their origin, much as we did for stars in Chapter 8. We learn that galaxies, like stars, have a birth, life, and even death.

Chapter 14: Those galaxies that we observe to be most distant from us are also observed farthest back in time. They appear as they were when the universe itself was very young. They appear and behave very differently than the galaxies close to us, so we call them active galaxies. We compare their characteristics with those of the nearby galaxies, concluding, in the process, that the universe of galaxies that we observe in our immediate vicinity may have formed out of the remnants of these strange objects.

Chapter 15: Our study of the characteristics and behavior of the quasars leads to the conclusion that they were the first objects to form after the event that brought the universe into existence, the Big Bang. The evidence for a "beginning" of the universe piles up year after year. We critically analyze the observations that support this conclusion, and learn how astronomers explain the chain of events that followed the actual creation event and that led to the objects that we observe in the universe today. Finally, we look at an alternative theory, the Steady State theory, in order to avoid having to accept a "beginning" for the universe.

Chapter 16: Despite the fact that stars and clouds of gas and dust make up 99.9% of the universe, that remaining 0.01% includes some rather interesting sideshow events — like planets and you and I. We have one solar system to study in detail in hopes of understanding its origin and formation. We study the characteristics and behaviors of the various categories of objects that orbit Sun, planets as well as the debris such as asteroids and comets, and again look for patterns that help us explain how they came to be.

Chapter 17: The small, rocky planets close to Sun are significantly different from the large, gaseous planets located in the deep-freeze portion of the solar system. We examine these Earth-like planets in close detail, again looking for features that assist us in explaining a common origin for all of them. At the same time, in anticipation of Chapter 21, we look within those same features for possibilities of a human presence on any of them sometime in the future.

Chapter 18: Using long-distance space probes, we have rediscovered the planets Jupiter and Saturn and their satellites. The tens of thousands of photographs taken by the probes reveal strange new worlds for us to explain. We examine the photographs and other data carefully for clues to their behavior as well as their formation.

Chapter 19: The frozen worlds of Uranus and Neptune were almost complete strangers to us prior to the arrival of the satellite *Voyager 2* in 1986. Likewise, the distant world of Pluto was not well understood until just recently. We examine the photographs and other data for clues to their properties, behaviors, and origins.

Chapter 20: Having fit the theory of the formation of our solar system into the general framework of the theories of stellar and galactic formation after the Big Bang, we naturally wonder if planetary formation occurs around every star. And if planets exist around other stars, is it possible that intelligence has evolved on the surfaces of some of them? And if so, can we assume that we have enough in common to allow us to communicate? We explore the attempts have been made and are currently being made to establish contact, and critically examine the assumptions upon those searches are based.

Chapter 21: Regardless of whether or not there is anyone out there with whom to share our knowledge, we humans seem intent on making excursions into space, perhaps even to take up permanent residency on the surfaces of Moon and planets on the way to visiting the distant stars. We examine the proposals to cut our dependency on Earth and to take giant strides into the cosmos. Our search for meaning in the universe takes on new forms as we do so.

A Final Word

I am going to say some things that many a physicist would argue about. I am not always going to be technically correct. It is far more important to me to convey the concept than to be absolutely technically accurate. This will be very obvious, for example, as I discuss the atomic structure of matter and the behavior of atoms. Words are like brush strokes on canvas. Together they make up the painting, an attempt to convey a feeling or thought to the viewer. So will I use words in this book. Learn them well so that you can see the picture I am trying to paint. A Zen Buddhist saying goes something like this:

> *A fishing pole is designed to catch a fish.*
> *Once the fish is caught,*
> *you may throw away the pole.*
> *A bow is designed to shoot an arrow.*
> *Once the arrow has been shot,*
> *you may throw away the bow.*
> *Words are designed to convey a meaning.*
> *Once you have got the meaning,*
> *you may throw away the words.*
> *So where is that person who has forgotten words*
> *so that I might have a word with him.*

I trust that you will be one of those persons who gets the meaning. It will make your appreciation of astronomy that much greater.

Section 1

Observing the Sky

Chapter 1

Organization of the Universe

Figure 1–1 This photograph of planet Earth was taken by the Apollo 13 crew during its journey home. The most visible landmass of Baja, California, is at the very center of the photograph. (NASA)

CENTRAL THEME: What is the architecture of the universe from the small to the large, and what language do astronomers use to describe its components?

I love to travel to places I've never been before. There is always a sense of excitement and amazement while I am surrounded by strange landscapes and people who dress differently, eat differently, talk differently, act differently, and believe differently. To more fully appreciate the experience, I prepare for the trip by learning some of the vocabulary, by listening to some of the music, and especially by reading about the history and culture of the region. I learn some of the cultural "language" so that I can better understand what I am experiencing. Astronomy is also a foreign culture. Since we will visit some very new and different environments in the course of reading this Book, I want to start off by developing a "language" or vocabulary with which we share thoughts and ideas about the universe.

It doesn't seem obvious when you look into the sky at night that the universe is very organized, but it certainly would be apparent if you were capable of high-speed flight out into the cosmos. In fact, if it were not for the fact that the universe operates according to some rather fundamental laws of nature, there really wouldn't be a field of study called astronomy in the first place. Actually, it is really not much different "out there" than it is right around us. And throughout the book I will associate objects in space – their makeup and behavior – with objects on Earth, in hopes that it turns the unfamiliar into the familiar. I do the same as we "travel" into the objects with which we are familiar to look at them afresh from the microscopic point of view.

The subject matter of astronomy covers the full range of objects in the universe – from particles smaller than the atom, to objects that are billions of miles in diameter, from events that last for only a billionth of a second to those that last for tens of billions of years. Perhaps that is one reason that astronomy fascinates so many people – it seems limitless in the scope of its study. That is a rather interesting idea, actually, because one of the topics of astronomy, *cosmology* by name, is the *study of the origin of the universe*. So in a very real sense, anything in the universe falls within the scope of the field of astronomy because anything that is in the universe presumedly originated with the universe. Cosmologists must explain everything in the universe according to whatever theory they propose. If they can't, then the theory is weakened as a result. The more things they observe that are explainable by the proposed theory, the stronger (more acceptable/believable) the theory is.

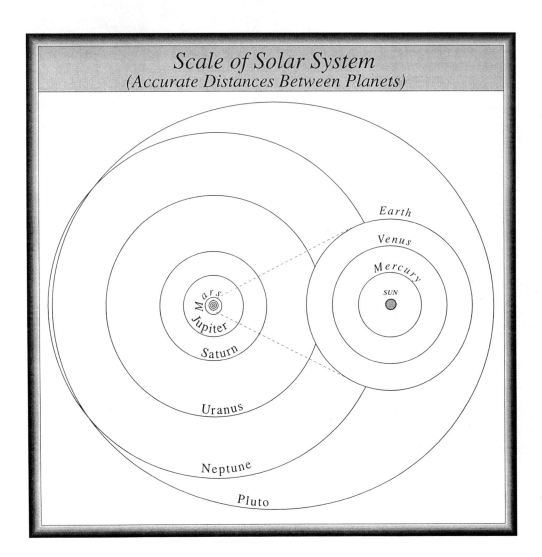

Figure 1-2 A scale drawing of our solar system reveals the immense amount of space between the planets, and the grouping of the planets into two categories: those relatively close to Sun, and those at great distances. Notice that Pluto's orbit crosses that of Neptune, so that Pluto is occasionally closer to Sun than Neptune.

Scale of Things in the Universe

Things in the Universe The most obvious things to be seen in the sky with the naked eye are Moon, Sun, planets, and stars. Moon orbits Earth. Earth and eight other planets orbit Sun — and the stars are scattered throughout the sky. Sun is our star — the nearest star to us. That's about it for night-to-viewing. There are occasional visitors to the sky, such as meteors and comets and "guest stars," but their short-lived appearances don't allow us to study them for long periods (like years) of time.

Groupings of Things in the Universe So Moon orbits around Earth while Earth orbits around Sun, allowing us to specify one obvious grouping — moons group around planets (Saturn has at least 18 moons), and planets group around stars (at least around our star). The grouping of our nine planets (and other minor objects) around Sun is called the **solar system** (Figure 1-2). Now look up at any two stars in the sky. Statistically, one of those two dots of light is in itself a combination of two dots — the result of two stars going around one another. How do we know that to be true? It is simple. We look at what appears to be a single dot through a telescope, which magnifies the star so as to separate it into two stars.

Figure 1–4 Many stars are bound by gravity into groups or clusters. This cluster, called the Pleiades *or Seven Sisters, is visible to the naked eye.* (Lick Observatory)

It is the same principle as in watching a distant car approaching at night with headlights glaring. At a great distance, a single light appears. But as the car gets closer, our eyes make out each of the headlights. Astronomers have, using this technique, determined that at least 50 percent all the stars we see in the sky at night are actually star-pairs, or what are referred to as **binary stars**. To be strictly accurate, at least one-half are *multiple stars*, inasmuch as some consist of more than two stars (Figure 1-3). For example, the middle star (dot) in the handle of the *Big Dipper* — a group of stars I assume you know — is in fact the combined light of four stars.

If the thought has occurred to you that Sun must be unusual because it does not have a star around which to orbit, you might ponder this for a moment — there is some evidence that we do have another star in our solar system. We'll consider the evidence for that theory in Chapter 16. But more importantly, there is good reason to consider the planet Jupiter as if it is the other member of a binary pair. Chemically, Jupiter is almost identical to Sun, and contains two-and-a-half times more material than the rest of the solar system (less Sun) put together! To an extraterrestrial, our solar system would appear to comprise Sun, Jupiter, and some debris. Jupiter is the star that failed — it just doesn't have enough material to make it shine like Sun.

Figure 1–3 Each dot in this photograph is a star. But close examination of these single dots suggests that in at least one half of the cases, each is the result of two or more stars in orbit around one another. (Lick Observatory)

You might also be wondering if there are similar conditions around other stars — do any of those dots in the sky consist of the combined light of a star and planets? I would be getting far ahead of the story if I elaborated on the subject here. Let me just say that recent observations provide convincing evidence that *planetary systems* do exist elsewhere, but the method used to detect them requires astronomical techniques that are more fully explained in Chapter 16. But I'm glad you thought of it.

You do not need to use a telescope to observe stars assembling into families or **star clusters**. An excellent example of a small grouping of stars that you have probably seen on a dark, clear winter evening is the *Pleiades*, or *Seven Sisters*. It is commonly mistakened for the "Little Dipper." But it is not. Seen through a telescope or binoculars, it is actually a rather large cluster of stars — one hundred or so in number (Figure 1-4). Between the stars are vast but extremely thin clouds of gas and dust, called **nebulae** (Figure 1-5). Gas and dust clouds don't account for very much of the material in our galaxy (less than one percent), but because of their great sizes in space they are easily observed through telescopes. All things considered, then, most of our galaxy consists of stars. So astronomers consider stars to be the basic building blocks of the

Figure 1–6 *The famous* Andromeda galaxy, *designated* M31 *in the* Messier Catalog, *is one of the nearest galaxies to the Milky Way. Because of its great size and number of stars, it is visible to the naked eye if observed from a dark location away from city lights.* (Lick Observatory)

universe. All of the stars and clusters of stars we see in the nighttime sky are members of a large group of stars that orbit around a common center. We call such a large grouping of stars a **galaxy**. Our galaxy was given the name the **Milky Way** because of its appearance in the sky. That is, most of the stars in the sky appear to lie along a "path of white" across the sky. Altogether, the Milky Way galaxy contains approximately one-hundred billion (100,000,000,000) stars, including our star — Sun. In the interest of conserving space when writing large numbers, mathematicians prefer to use **powers of ten** instead of writing a bunch of zero's. So we express the number of stars in the Milky Way as 10^{11}. The number 11 written as a superscript is the number of zero's that follow the figure 1. The average distance of Earth from Sun is 9.3×10^7 miles. Sun is located near the outer edge of the Milky Way galaxy.

Even an amateur's telescope allows us to observe the next grouping of things in the universe — other galaxies beyond the boundaries of the Milky Way. The best example of a galaxy located beyond the boundaries of the Milky Way is the **Andromeda galaxy** (Figure 1-6), noted for its naked-eye appearance as a smudge of light (at least when viewed from a dark location). There are hundreds, thousands, even millions of other galaxies beyond the borders

Figure 1–5 *When viewed through a telescope, many of the dots in the sky are revealed as immense clouds of gas and dust called nebulae. This particular one is called the* Trifid Nebula. (Lick Observatory)

CHAPTER 1 / Organization of the Universe

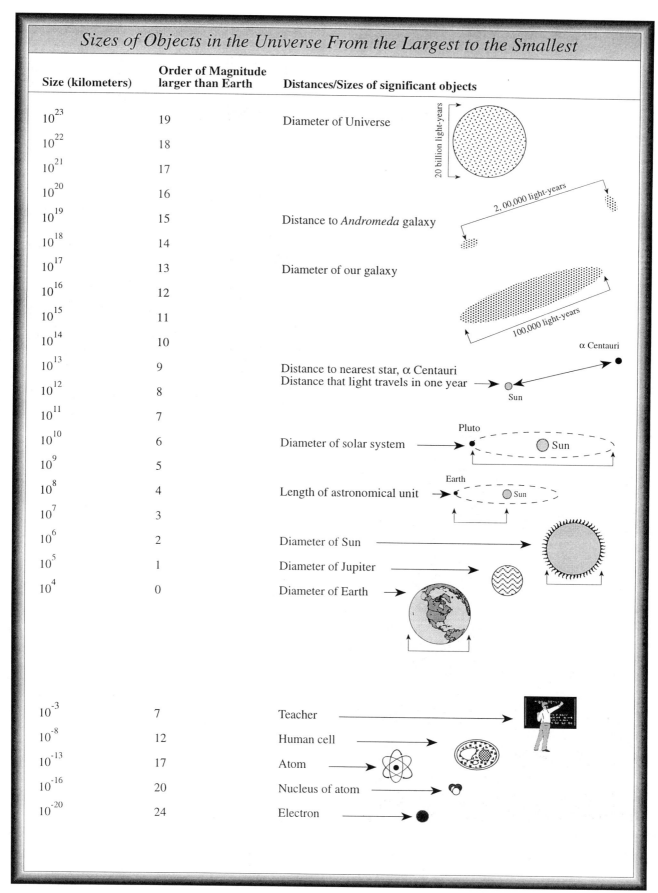

Figure 1–7 The range of the sizes and distances of the contents of the universe, from the very small to the very large.

of our own galaxy. To give you an idea of their distribution in space, there are more galaxies to be observed by telescope in the "bowl" of the familiar Big Dipper constellation than the number of stars in the entire sky visible to the human eye.

And galaxies too are organized into larger collections of galaxies. The Milky Way — our galaxy — is one of twenty or so galaxies that are members of what is called the **Local Group** of galaxies. The Milky Way is located near the outer edge of this **poor cluster of galaxies**. Beyond our small cluster are larger groups of galaxies, called **rich clusters of galaxies** (Figure 1-8). The name of the nearest rich cluster is taken from the pattern of stars that lies in the direction to the center of the cluster — the *Virgo* cluster.

And our very largest telescopes allow astronomers to map the distribution of thousands of clusters of galaxies, maps which then reveal the largest organization of all — **superclusters of galaxies**. At this scale, the galaxies appear to be distributed throughout the universe in the pattern of soap bubbles in a bubble bath. That is, from the largest perspective available to modern astronomers, galaxies appear to be distributed in space as if they are located where soap bubbles touch one another in a mass of bubbles. We don't yet know how to explain that feature of the universe.

Distances to Things No one needs to be reminded that the distances between things in the universe are immense. They are so immense, in fact, that the very word *astronomical* is now used in referring to something that has a large number associated with it — the "*astronomically* huge Federal debt," for example. Because of that fact, astronomers find it rather cumbersome to use common units of measurement such as feet and miles when referring to objects *outside of the solar system*. They could, but it would require using up lots of paper space in writing about the things in space. For example, using even a large unit of measurement like the mile to express the distance to the nearest star beyond Sun requires fourteen spaces ($\sim 25{,}000{,}000{,}000{,}000 = 2.5 \times 10^{13}$ miles)!

Rather, astronomers prefer to use the unit of measurement called the **light-year** when referring to distances to (or sizes of) stars, nebulae, galaxies, clusters of galaxies, and so forth. The *light-year* is simply the *distance that light travels in one year*. Since the speed of light is a constant (186,000 miles per second), and the year is a fixed number determined by the time that Earth requires to orbit Sun once, the light-year is also a fixed number (186,000 miles per second, times 60 seconds in a minute, times 60 minutes in an hour, times 24 hours in a day, times 365.25 days in a year — equals 5,870,000,000,000 *miles in a light-year*).

By the way, although extraterrestrials would find that the speed of light (using their own units of measurement, of course) is the same as what we find it to be here on Earth, they would not use the same unit of measurement as our "year." So we do not use the expression light-year in any of our messages sent out to contact "them." Their units of measurement of time are undoubtedly different than our unit, since ours are based upon the speed of rotation of Earth (hour, minute, second) and Earth's orbit around Sun (year). Refer to Figure 1-7 to see how distances are expressed using these units of measurement.

Actually, even distances within the solar system can be expressed in units based on the speed of light. We merely subdivide the year into its components of days, hours, minutes, or seconds. For example, since it requires 8 minutes for sunlight to make the 93,000,000-mile journey to Earth once it leaves the surface of Sun, we say that Sun is 8 *light-minutes* from Earth. Likewise, Moon is about 1 *light-second* from Earth. When we launch a probe into the outer portions of the solar system to photograph and study planets located there, we must consider the time required for signals to be sent to (and returned from) the probe. These travel times can be on the order of hours for some of the planets. Thus we can say that the planet Saturn is occasionally located 2 *light-hours* from Earth. The diameter of the solar system is about one-half light-day across. In other words, light requires about 12 hours to go from one edge of Pluto's orbit across to the other edge.

Figure 1-8 Each smudge in this photograph is a galaxy. This is a cluster of galaxies known as the Hercules cluster, named after the constelllation in which it is located. (NASA/STScI)

But astronomers also use another unit of measurement in expressing distances within the solar system that is not based upon a unit of time. It is the average distance of Earth from Sun (the 93 million miles mentioned above), and is called the **astronomical unit**, or **AU** for short. So we use a characteristic of Earth as a basis for expressing distances between objects in the solar system. For example, the planet Saturn averages 10 AU's away from Sun — it is 10 times farther away from Sun than Earth is.

Sizes, Masses, Densities of Things The units we use for expressing distances between things in space are also used for expressing **sizes** of things in space. For objects within the solar system, the unit of the *mile* or *kilometer* (1 mile = 1.6 kilometer) is sufficient — the diameter of Earth is about 8,000 miles, the diameter of Jupiter is about 88,000 miles, the diameter of Sun is about 800,000 miles, and so on. There are characteristics other than that of diameter of expressing the size of an object, of course. We can just as easily use the *radius* (one-half the diameter), *circumference* (distance around), or even *volume*.

Astronomers frequently use the value of a characteristic of one object as a yardstick for expressing the value of that same characteristic of another object. For example, we say that it would require approximately 1,000 Earths to fill up the volume of Jupiter, and 1,000 Jupiters to fill up the volume of Sun (that means it would require one-million Earths to fill up the volume of Sun!) (Figure 1-9). Since in astronomy we encounter objects that dwarf even the Sun in size, we often use one of Sun's characteristics as a unit measurement for expressing that same characteristic for another object. We frequently compare other stars to Sun according to radius, diameter, or volume (Figure 1-10). The star *Betelgeuse*, for example, has a diameter that is 400 times that of Sun — which means that it would require 64 million (6.4×10^7) Suns to fill up the volume of *Betelgeuse* (Figure 1-11).

The **mass** of a thing in space is an expression of the total amount of material that is contained within that thing. Do not confuse mass with weight. Certainly it is possible on Earth to measure the amount of material in my body by weighing myself on a scale. But how would I measure it in the "weightlessness" of space? I possess mass wherever I go, but I lose weight once I leave the confines of Earth and remain in space. Astronomers do have methods of determining the masses of things in space, and those methods will be presented in Chapter 8. For the moment, I am merely attempting to develop a vocabulary of commonly-used words so that you can appreciate a short tour of the variety of things that inhabit space.

As far as a unit of measurement for mass is concerned, astronomers pretty much use a unit with which you are familiar – *kilogram* (1 kg = 2.2 pounds). But again, rather than thinking in terms of a pure number, we usually express the mass of an object in terms of the mass of something with which we are familiar (but haven't necessarily weighed with

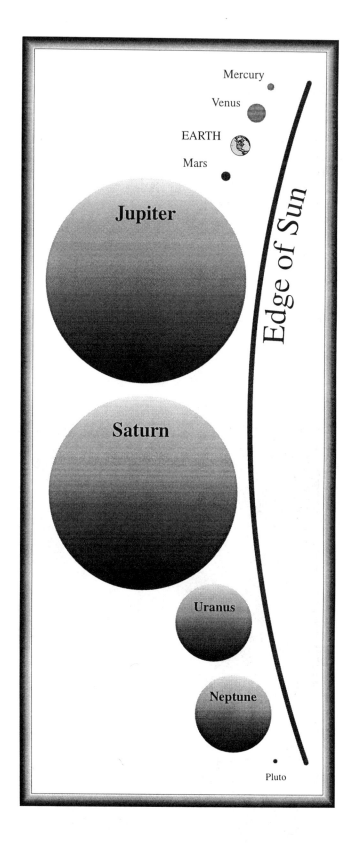

Figure 1-9 *Relative sizes of the major components of the solar system. Approximately one-thousand Earths could fit inside Jupiter, but approximately one-thousand Jupiters could fit inside of Sun.*

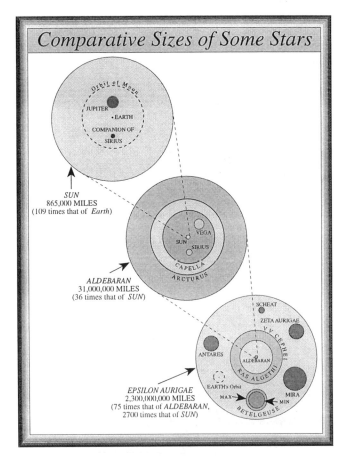

Figure 1–10 In calculating the sizes of stars, astronomers arrive at the conclusion that Sun is but an average star.

a scale). So it doesn't mean much to anyone to express the mass of Earth as being equal to 6×10^{24} kg. But you can easily relate to the statement that Earth is 81 times more massive than Moon — that Sun is 330,000 times more massive than Earth. Can you easily relate to the statement that star *HD 698* (a designation in a particular catalogue of stars) is 113 times more massive than Sun?

Having the sizes and masses of things at our disposal, we are now able to calculate another important characteristic of things in space — **density**. Density is simply *mass divided by volume* — mass per unit volume. It is an expression of how much matter is contained within a given volume. Two objects whose volumes are identical, but whose masses are different, have different densities. Likewise, two objects whose masses are identical but whose volumes are different, have different densities. The unit used in expressing density depends upon the units used in expressing mass and volume. Using Earth as an example, if we express its mass in kilograms and its volume in cubic kilometers, and reduce it to a scale comparable to human activity, we determine its density to be about 5.5 grams (abbreviated g) per cubic centimeter (abbreviated cm^3). Density of Earth is 5.5 g/cm^3. The densities of some materials with which you are familiar are:

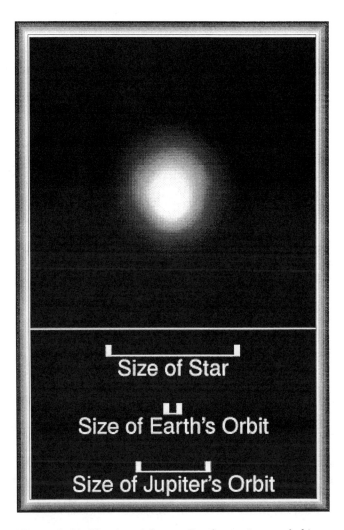

Figure 1–11 The size of the star Betelgeuse is revealed in the Hubble Space Telescope's first–ever photograph of the actual surface area of another star. Its size compared to distances within the solar system suggests that Sun is but a minor–sized star. (NASA/STScI)

Material	Density (g/cm3)
Gold	19.3
Lead	11.4
Iron	7.9
Earth (bulk)	5.6
Rock (typical)	2.5
Water	1.0
Wood (typical)	0.8
Insulating foam	0.1
Silica gel	0.02

Again, for our purposes, the numbers are not as important as being able to relate them to something with which we are familiar so that we can compare Earth's density with other objects in space. Water has a density of 1 g/cm^3. Since you are familiar with the consistency of water

CHAPTER 1 / Organization of the Universe

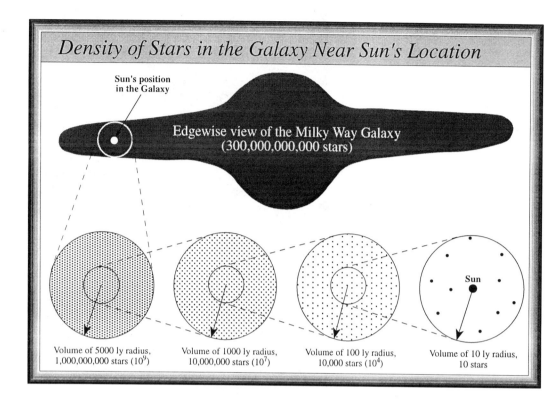

Figure 1-12 The number of stars within increasing volumes of space centered on Sun in the Milky Way galaxy. Despite the vast number of stars in the galaxy, however, the average spacing between stars is immense.

and Earth's rocks, you should be able to relate to these two numbers and their associated densities. And what if I were to inform you that Sun's density is 1.4 g/cm³. And that Saturn's density is 0.9 g/cm³! Or that objects that we call *neutron stars* have densities of 10^{15} g/cm³! Are you getting a sense of comparison between those things with which you are familiar and some things in space with which you are becoming familiar?

We must be careful when seeing the value for the density of an object in space that we interpret it correctly. While it is valid to say that the density of Earth is 5.5 times greater than that of water, that is only an average for the overall Earth. Because of the great pressures that exist at the center of Earth, the density of material there is much greater than that of material found on its surface. The same is true of Sun, and Jupiter, and most everything else in space.

Astronomers also use density to express the number of objects that occupy a given volume of space, such as the number of stars within different regions of the Milky Way galaxy (Figure 1-12). In this case, the use of density is an expression of the amount of distance between stars, providing a hint as to the vast amounts of space between stars in the universe.

Figure 1-13 We begin our tour of the universe by recalling our location on and relative size to Earth. The comparison in this figure is obviously not to scale.

10   SECTION 1 / Observing the Sky

Brief Tour of the Universe

Using the vocabulary and concepts introduced so far, let's take a short tour through the universe that we'll be exploring in more detail in the Chapters to come. The best way to do that, I believe, is to continue to scale the distances and sizes of things we encounter to dimensions that we can relate to. I begin by assuming that you are thoroughly aware of your size with respect to Earth, and your approximate location on its surface. There you are in Figure 1-13, exaggerated in size so as to make your location obvious.

Stepping further out into space so that we can observe Earth and its nearest neighbor to scale, we observe Earth-Moon system in Figure 1-14. At a distance of approximately 240,000 miles from Earth, Moon is 30 times farther away from Earth than Earth is in diameter. Moon has about one-fourth the diameter of Earth. Lining the planets up to scale according to size (but not distance) next to Sun reveals what I have already suggested — the minuteness of the planets compared to Sun, especially our planet Earth (look at Figure 1-9 again).

Moving out to a point that allows us to see the solar system in its entirety (as far as planets are concerned), we lose the ability to see a size for any of the planets — or even Sun itself. To put it in another perspective, imagine scaling the solar system down such that Sun is represented by a golf ball ($1^5/_8$ inches). Most of the planets are then represented by grains of sand, while Jupiter is a grain of rice. Earth (as a grain of sand) is located some 16 feet from Sun (as a golf ball). Jupiter is 80 feet from Sun, and Pluto is 640 feet.

That's our solar system — a central golf ball surrounded out to 640 feet by some specks of sand! Proceeding out to an even greater distance allows us to see Sun with respect to the nearest stars around it. Of course, the sizes of the stars themselves vanish at this scale. But if we again use golf balls to represent the average sizes of the nearest stars, we find that the average distance between them is about 800 miles!

In other words, the average distance between Sun and the next nearest star is represented by two golf balls — one located in San Francisco and one located in Vancouver, Canada! If we use golf balls to represent the distribution of stars in the Milky Way galaxy, we would be unable to fit all of the balls on Earth's surface (assuming an average spacing of 800 miles from one another). We need a spherical surface whose diameter is about 20 million miles in order to distribute the golf balls to represent the stars in the galaxy.

The total number of brother and sister stars we have in our galaxy is between 100 and 200 billion (10^{11}). Figure 1-12 illustrates the increasing number of stars in increasing volumes of space as we proceed out to the point at which we can observe the entire Milky Way and its contents. Of course, time and distance have not yet allowed us to view our Galaxy from outside, but we have plenty of photographs of other galaxies obtained from Earth. That shown in Figure 1-6 is one of the nearest to the Milky Way, the *Andromeda* galaxy. It contains somewhat more stars than does the Milky Way — some 200 to 400 billion. It, like ours, is about 100,000 light-years in diameter, and about 15,000 light-years thick at Sun's position in the Galaxy.

Looking at a photograph of a cluster of galaxies (Figure 1-8) reminds us of looking at a photograph of stars in the Milky Way galaxy. Each "smudge" on the photograph is a galaxy. In order to construct an appropriate scale of the distribution of galaxies in such a cluster, we'll reduce the average size of a galaxy to a model 12 inches (one foot) in diameter. At that scale, the nearby Andromeda galaxy is located about 20 feet away from the model representing our galaxy.

If we make a scaled model of the universe presently available to our telescopes, we'll have to distribute 10 billion or so of those 12-inch models of galaxies at 20-foot intervals. Furthermore, we will distribute them with respect to the manner in which galaxies appear to be distributed in space (the soap bubble analogy mentioned earlier). Look carefully at the plot of galaxies in Figure 1-15. You should notice a vague yet suggestive pattern as I have described. At the same time, you will gain an appreciation of the vast number of stars in the universe. There is an average of 100 billion (10^{11}) stars in each galaxy, and there are approximately 10 billion galaxies in the known universe. Simple arithmetic tells you that the total number of stars in the known universe approximates 1,000 billion billion (10^{21}).

Even if you are familiar with our Nation's debt, the number 10^{21} is probably too huge and abstract to relate to anything you've experienced in your everyday life. But I have a comparison that you can relate to. From a distance

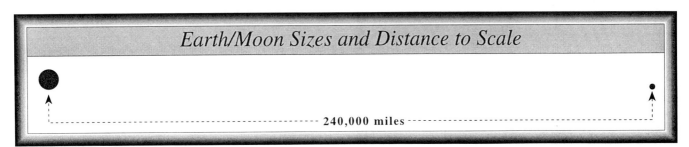

Figure 1–14 *An accurate scale of Earth and Moon according to their respective sizes and approximate distance from one another.*

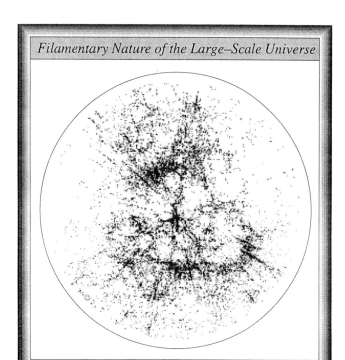

Figure 1–15 *Infinite care went into the plottting of 1,065 galaxies in one small region of the sky. The pattern of distribution is obvious but not well understood. It appears as if galaxies form along the intersections of bubbles in a bubble–bath mixture. (Harvard–Smithsonian Center for Astrophysics)*

a beach — like a galaxy of stars — looks like a smooth, flat plane. From up close, we realize it is made up of individual grains of sand, just as a close examination of a galaxy through a telescope allows us to see that it is made up of individual stars.

Although I wouldn't want to do it myself, it is a rather straightforward task to count the number of grains of sand required to fill an empty milk carton. And a little basic arithmetic allows us to estimate the number of milk cartons needed to empty a beach of all of its grains of sands. Finally, knowledge of geography allows us to estimate the number of sand-bearing beaches on Earth. Try it. You will arrive at the estimate that there are 10^{21} grains of sand on all of the world's beaches! And that is the estimated number of stars in the universe!

General Characteristics of Things

Aside from the properties of *size*, *mass*, and *density* that I defined earlier, objects in space have other attributes that are important to the astronomer who is intent on understanding their behavior. For example, the fact that we know that objects exist out there suggests a mechanism of communication. The fact that smaller objects are organized into larger organizations suggests a mechanism of force or energy that holds the group together. To understand these more subtle properties, we now take a tour deeper into the objects themselves.

Sources of Energy of Things The manner in which things in the universe communicate their presence and/or behavior to one another is through the mechanism of **energy**. Think about it for a moment — even in our day-to-day activities, we are only aware of events going on around us by virtue of our being able to detect (and properly interpret with our brains) energies like sound, light, touch, smell. Things out there in space also communicate their presence and behaviors by sending energies out into space toward other things. There are a given number of known types of energy that travel from one thing to another, and a given number of known processes that generate those energies.

The dominant process by which energy is transmitted is the one that is least obvious, simply because we are surrounded by it. The force that holds you and I to Earth, that holds Moon in orbit around Earth, that holds Earth (and other planets) in orbit around Sun, is also responsible for holding stars together in clusters, stars and clusters together in galaxies, and galaxies together in superclusters — **gravity**. Gravity is the "glue" that holds the universe together. It is the force that dominates the universe around us. It is certainly a source of energy — drop a rock on your foot if you want proof that energy can be released by gravity.

You might insist that when you drop the rock on your foot, the damage to your foot is caused not by the energy of gravity but by the motion of the rock itself. That is true. But it is gravity that is responsible for giving the energy to the rock in the first place. To clarify the distinction between movement and gravity, try another experiment — throw the same rock horizontally against a window in your house. Do you agree that the rock has energy? But this time it is not gravitational energy — it is energy associated with **motion**. Something that moves has energy. That motion may be obtained from the release of gravitational energy, but not everything that moves obtains its energy of motion from gravity.

So just where does the energy that is needed to put the rock in motion toward the window come from if not from gravity? Your arm, you say? Yes, but ultimately it comes from the release of **chemical** energy stored in the chemical bonds of *ATP* (adenosinetriphosphate) in your muscles, placed there as a consequence of your eating a hearty breakfast of oatmeal. Your body removes the energy stored in the chemical bonds of the oatmeal and transfers it to the molecules of ATP. In other words, when chemicals combine together to form new chemicals, there can be a release of energy. Burning wood is another familiar example of the release of chemical energy, in this case in the form of light and heat.

Let's return to the oatmeal — how does the energy get stored in the oatmeal in the first place, before it is extracted in your stomach? Sun. Yes, the process of photosynthesis causes chemicals in the oats to absorb sunlight which in turn creates chemical bonds which in turn store energy that which in turn is released in your stomach and stored in ATP. And the sunlight—how did it get started? Deep within the interior of Sun, particles of matter fuse together to form new particles, and in the process release energy that eventually streams out into space as "sunlight" (or more precisely, *radiation*).

This process of energy creation is called **nuclear fusion**, and is presumably the method by which all stars shine. Don't confuse what occurs inside of Sun with that occurs (hopefully never again!) in an atomic bomb. The latter process is nuclear fission – the breaking up or "splitting" of the atom, something that occurs throughout the universe, but not in such a dramatic way as an explosion.

The natural – the atomic bomb is unnatural – fission of atoms occurs in the process called **radioactive decay**. This source of energy is important in understanding events associated with planets in the solar system, including Earth. The energy generated by radioactive decay can take the form of light, heat, or the motions of subatomic particles acting on matter in the vicinity. Earth's interior is kept molten by the release of heat by radioactive decay, for example.

What is most important to astronomers as regards sources and types of energies in the universe is that since certain types of energies are released by certain types of processes, to know what type of energy is detected is to know what process is going on. And knowing what process is going on allows astronomers to go forward and backward in time to predict past and future events.

Compositions of Things Just as matter in the large-scale universe is highly organized by the force of gravity, matter at the atomic level is highly organized by a force that gives rise to the chemical energy previously discussed. For the moment, let's just say that the atoms that participate in chemical reactions and make up the things in space (including you and I and Earth) are held together by a chemical force. To understand the behavior of these things, astronomers need to understand atoms – every astronomer is a chemist. Stars may be the basic building blocks of the universe, but they are still made of atoms. And it is the behavior of those atoms that is ultimately responsible for the sum-total behavior of each star.

Chemists have determined that everything in the large-scale universe is composed of a combination of one or more of 92 **chemical elements**. You refer to them each time you mention oxygen, carbon, iron, gold, mercury, and so on. The gas oxygen is made up of billions of atoms of oxygen, in other words. Different things in space are made up of different combinations and/or ratios of those 92 elements (Figure 1-16).

Sun, for example, is made up mostly of the element *hydrogen*. One of the most important discoveries in as-

List of Chemical Elements and Atomic Numbers		
1 hydrogen	32 germanium	63 europium
2 helium	33 arsenic	64 gadolinium
3 lithium	34 selenium	65 terbium
4 beryllium	35 bromine	66 dysprosium
5 boron	36 krypton	67 holmium
6 carbon	37 rubidium	68 erbium
7 nitrogen	38 strontium	69 thulium
8 oxygen	39 yttrium	70 ytterbium
9 fluorine	40 zirconium	71 lutetium
10 neon	41 niobium	72 hafnium
11 sodium	42 molybdeum	73 tantalum
12 magnesium	43 technetium	74 tungsten
13 aluminum	44 ruthenium	75 rhenium
14 silicon	45 rhodium	76 osmium
15 phosphorus	46 palladium	77 iridium
16 sulfur	47 silver	78 platinum
17 chlorine	48 cadmium	79 gold
18 argon	49 indium	80 mercury
19 potassium	50 tin	81 thallium
20 calcium	51 antimony	82 lead
21 scandium	52 tellurium	83 bismuth
22 titanium	53 iodine	84 polonium
23 vanadium	54 xenon	85 astatine
24 chromium	55 cesuyn	86 radon
25 manganese	56 barium	87 francium
26 iron	57 lanthanum	88 radium
27 cobalt	58 cerium	89 actinium
28 nickel	59 praseodymium	90 thorium
29 copper	60 neodymium	91 protactinium
30 zinc	61 promethium	92 uranium
31 gallium	62 samarium	

Figure 1-16 *The Periodic Chart of chemical elements, of which all objects in the universe are composed. The number represents the size, weight and simplicity of the atoms that make up the chemical. The smaller the number, the smaller, lighter and simpler the atom.*

tronomy in the past 100 years or so is that it is possible to measure the chemical compositions of objects in space. That provides us with the basic means by which to understand the behavior of those objects. Using Sun as a reference again, hydrogen placed under extreme pressure "ignites" or "burns." So we can say with a rather high degree of certainty that since Sun is made up mostly of hydrogen, that process must be responsible for the sunlight that allows us to survive on Earth.

To discuss atoms requires that we extend our units of size measurements in the opposite direction — to the very small. Just as writing zero's to the left of the decimal makes a number get larger and larger, zero's placed to the right of the decimal causes the number to get smaller and smaller. Using the *powers of ten* notation, a negative superscript after the 10 expresses the number of zero's to the right of the decimal point. Thus one-tenth is written 10^{-1}, one-millionth is written 10^{-6}, and $1/500 = 2 \times 10^{-3}$. Just as the large numbers used to express sizes of and distances between things in space can seem abstract and removed from everyday experiences, so too these small measurement seem

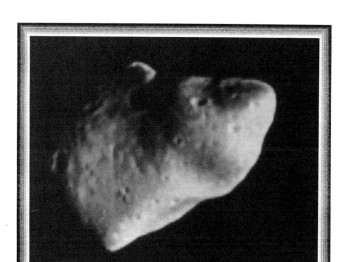

Figure 1–17 The asteroid Gaspra *was photographed by the Galileo satellite in November 1991 on its way to Jupiter. The oblong chunk of rock is 5 by 10 miles in size, and therefore lacks enough mass for its gravity to pull it into a spherical shape.* (NASA)

Temperature Conversion Chart			
Example	Fahrenheit	Celsius	Kelvin
Core of the Sun	27 million° F	15 million° C	15 million K
Surface temperature of the Sun	10,337° F	5,727° C	6,000 K
Maximum temperature of Venus	700° F	371° C	644 K
Boiling point of water	212° F	100° C	373 K
Normal body temperature	100° F	38° C	311 K
Normal room temperature	68° F	20° C	293 K
Freezing point of water	32° F	0° C	273 K
	0° F	-18° C	255 K
Minimum temperature on Earth	-100° F	-73° C	200 K
Polar ice cap of Mars	-198° F	-128° C	145 K
Average temperature on Saturn	-230° F	-146° C	127 K
Dark side of Mercury	-297° F	-183° C	90 K
Absolute zero: no molecular motion	-459° F	-273° C	0 K

Figure 1–18 Scientists prefer to express temperatures in Kelvins, since that scale is based on the lowest temperature that is possible in the universe. This conversion chart can be used to go from one scale to another.

unnecessary to anyone but a chemist. But I provide them simply that you will get a general grasp of the range of sizes of things that make up the universe. Figure 1-7 provides an overview of this range in terms of things with which you are familiar. You may want to refer back to it frequently as you read through the Chapters of this Book. It will provide you with a perspective of the organization of the universe.

Shapes of Things Looking beyond our limited Earthly view, we notice that the universe of things tends toward roundness. Although the basic building blocks — the stars — don't present that appearance to us, even through the largest telescope, there is every reason to believe that they too are round. But a close examination of the Andromeda galaxy (Figure 1-6), the planet Saturn (refer ahead to Figure 18-29), and the asteroid *Gaspra* (Figure 1-17) makes obvious the conclusion that there are exceptions to the roundness rule. There are two principles that explain the exceptions — *smallness* and *spinnedness*. Gravity operating within any object pulls equally in all directions, so a large mass of particles (atoms) is pulled by gravity into a spherical shape. Sun is a good example of a spherical object.

If the object is small, however, the force of gravity is not strong enough to pull it into a spherical shape, and it has a random shape. Averaging about nine miles in diameter, the asteroid *Gaspra* is a good example. In addition, if the object is not solid, the rate of spin causes it to "bulge" out along its equator and "flatten" at its poles. Saturn — being a large ball of liquid and gas — and the Andromeda galaxy — being a "fluid" of billions of stars — are good examples of this type of exception. As you will learn in Chapter 17, even Earth has an equatorial "bulge" as a result of its rate of spinnedness.

Temperatures of Things Mostly, the universe is extremely cold. That is, most of the volume of the universe consists of the vast and seemingly empty distances between stars and galaxies. Only when in close proximity to a star are you going to feel comfortably warm. Only if you are within 20 feet of those golf balls scattered at 800-mile intervals are you not going to require thermal undergarments. At the other extreme, if you somehow manage to survive a trip into the very center of Sun, you will experience temperatures that reach the 15-million-degree mark.

Although we in the United States have grown up using the temperature scale of *Fahrenheit*, it is not a very practical one when dealing with the range of temperatures in the large-scale universe. Neither does the Centigrade system that has made some inroads in our society. It is based upon the range of temperatures between the freezing and boiling points of water, but liquid water doesn't exist (as far as we know) beyond the boundaries of Earth's surface. Let me put it this way. **Temperature** is a measurement of the motion of the atoms that make up an object or substance. Evidence in the laboratory tells us that atoms cease to move at all when the temperature reaches minus 460 degrees Fahrenheit. We call that **Absolute Zero**. We have never been able to reach that temperature, but we have gotten very close to it in various experiments.

Since the space between stars is at about absolute zero, it is convenient to use a temperature scale that begins at that point and extends upward. This is the *Kelvin* system of temperature measurement, named after British physicist Lord Kelvin. In the Kelvin system, the degree symbol (°) is not used, and we merely say that a comfortable room temperature is about 300 Kelvins (300 K). That is part of the language adopted by scientist for their own convenience. The chart in Figure 1-18 will assist you in understanding the comparison between temperature scales.

Rate of Change of Things It won't surprise you to learn that the rates at which changes take place in different things are different. Chemical reactions occur at different rates, depending upon the chemicals involved and the conditions (e.g., *pressure* and *temperature*) within which the chemicals are located. Since astronomers are interested in the changes that take place within and around and between the things in space, they must spread their observations out over various time intervals in order to detect any changes that occur.

Those time intervals are as short as billionths of a second for events at the atomic level, to billions of years for events at the extreme distances in space. Well, obviously the human life span doesn't allow for observations at these latter extremes, and certainly instruments must be used to measure the changes that take place in extremely short time periods. The point is, however, that astronomy examines events that last extremely short periods of time as well as those that last longer than a lifetime.

Now for a mind-boggler. Astronomers observing events in space are allowed to see those events only as they were in the past. In a sense, astronomers study the universe as it used to be, not as it is. By and large, the manner in which astronomers obtain information about objects in space is based on the study of the radiation that comes from them. You <u>see</u> a star in the sky, and you know that it is there. Well, what you should be thinking is that "...it <u>was</u> there."

Radiation takes a finite period of time to go from one object to another, since it has a finite speed at which it travels. The longer it takes for the radiation to arrive from an object (the more distant it is from us), the further back in time we are observing it. You learned earlier that sunlight (a type of radiation) requires 8 minutes to travel to Earth. So when you look up at Sun, you are seeing it as it was 8 minutes ago (Figure 1-19). You have never seen (nor will ever see) Sun as it is a the moment, only as it was. When you look up at Moon, you see it as it <u>was</u> 1 second ago. When you look at the Andromeda galaxy through a telescope, you see it as it was 2 million years ago (it is 2 million light-years from us).

Although we often think of telescopes as instruments that "reach" out into the universe, that does not mean that they are getting closer to the objects being observed. They are — like the human eye — dependent upon receiving

Source of Light	Distance	Light Travel Time
Moon	1 light-*second*	1 *second*
Sun	8 light-*minutes*	8 *minutes*
Nearest Star	4.3 light-years	4.3 years
Average visible stars	300 light-years	300 years
Center of Milky Way	30,000 light-years	30,000 years
Magellanic clouds	200,000 light-years	200,000 years
Andromeda galaxy	2 million light-years	2 million years
Edge of Local Group	3 million light-years	3 million years
M 51 "Whirlpool" galaxy	10 million light-years	10 million years
Centaurus A radio galaxy	15 million light-years	15 million years
M 87 elliptical galaxy	40 million light-years	40 million years
Virgo cluster of galaxies	65 million light-years	65 million years
Coma cluster	350 million light-years	350 million years
Hydra cluster	2.5 billion light-years	2.5 billion years
Extremely distant galaxy 0902 + 34	10 billion light-years	10 billion years
Highly red-shifted quasar OQ172	17 billion light-years	17 billion years

Figure 1–19 The distance to an object in space is also an expression of the length of time that its light (radiation) took to go from that object to Earth. Therefore, it is also a measurement of how long ago the light (radiation) left the object being observed. This fact allows astronomers to study events in the past. This chart lists the distances and light travel times of some increasingly distant objects.

radiation from the objects through the front tube and into the eye or instrument. In principle, at least, this matter of always observing events in the past can be applied to our observing events on Earth.

After all, there is a small time interval between the moment light bounces off the teacher at the front of the classroom and the moment it enters your eye. You see him/her in the past — admittedly, not a significant time interval in the past, but one that is measurable at least. Light travels so fast that the time interval is too small to spark debate as to whether the teacher exists anymore or not. But the long travel times for radiation coming from distant galaxies must be considered when dealing with questions of past and present behavior.

Evolution of Things Some people avoid using the word **evolution** because they associate it with the theory that humans evolved from lower forms of life. But the context within which I use it throughout this book is that it means *change*. As we attempt to understand the behavior of things in space, we immediately arrive at the inevitable conclusion that everything in the universe is in a process of change — everything is evolving. We aren't able to observe profound changes in every object we study, simply because the human life span is terribly short compared to the times required for objects in space to go through enough change to be noticeable. But they do go through enough change for a pattern of behavior to be made clear to us.

That is immediately obvious when we look at Sun. It shines. It shines because hydrogen gas, under great pressure and temperature conditions, "burns." It will no doubt continue to do so long after automobiles are no longer used to get from one place to another. And what eventually will happen? Sun will cease to shine — it will "go out." That conclusion is not based upon an observation — it is simply a matter of our experience with the behavior of chemicals anywhere. When fuel is exhausted, rapid, radical change follows.

It is likewise with galaxies, since they consist mostly of stars. And of course we are eventually led by logic to ask the most fundamental of questions of all — where did the stars get the hydrogen gas in the first place? And what happens when all of the stars in the universe "go out?" Well, we don't have final answers to those questions yet, nor may we ever be so privileged to be able to explain our origins and destinies. But evidence for change is right in front of us, and human curiosity gradually scrapes away obstacles to understanding.

The excitement of this search for an understanding of origins is so powerful that much of modern astronomy revolves around it. In fact, after the first few Chapters of this book, I present details of the evidence that argues strongly for not only the evolution of stars and galaxies, but evolution of the entire universe from a creation event some 20 billion years ago. In the final few Chapters, I present details of the evidence that argues strongly for the evolution of the planets around Sun. With that as a springboard, you are invited to take a plunge into the pool of controversy about the origin of intelligent life both on Earth and elsewhere in space. Have a nice swim!

LEARNING OBJECTIVES: *Now that you have studied this Chapter, you should be able to:*

1. Name the groupings of objects in the universe, starting with the solar system and ending with clusters of galaxies.
2. Express the relative sizes of the objects in the universe, using Earth and Sun as rulers.
3. Explain how time and distance are related.
4. Explain the difference between weight, mass and density.
5. Describe the types of energy that operate in the universe, and the importance of each.
6. Express some temperatures of things with which you are familiar in the Kelvin scale.
7. Explain why objects in space can be spherical in shape or otherwise.
8. Define and use in a complete sentence each of the following **NEW TERMS**:

Absolute zero 14	**Milky Way** 5
Andromeda galaxy 5	**Motion energy** 12
Astronomical unit 8	**Nebulae** 5
Binary star 4	**Nuclear fusion** 12
Chemical element 13	**Poor cluster of galaxies** 7
Chemical energy 12	**Powers of ten** 5
Density 9	**Radioactive decay** 13
Energy 12	**Rich cluster of galaxies** 5
Evolution 16	**Size** 8
Galaxy 5	**Solar system** 4
Gravity 12	**Star cluster** 5
Light-year 7	**Supercluster of galaxies** 7
Local Group 7	**Temperature** 14
Mass 8	

Chapter 2

Mapping the Sky We Observe

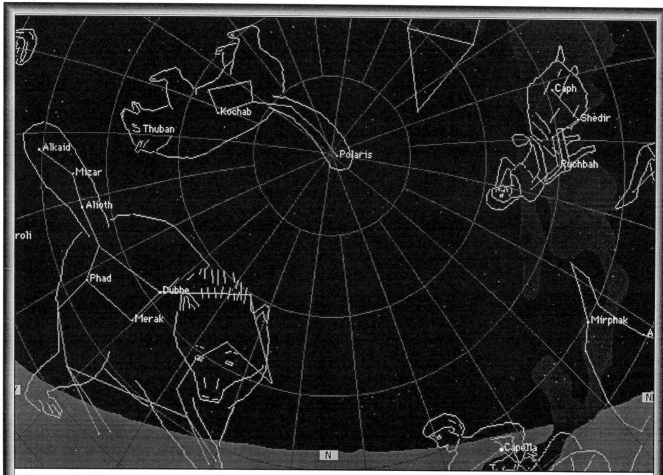

Figure 2–1 This sky map is an example of the power of modern home computers to create on the desktop the view of the sky as seen from any location on Earth at any time in history. This view of the northern sky is as seen from San Francisco early in the evening during early September of each year. The symbols and figures used on maps such as this are explained in this Chapter. (Carina Software)

CENTRAL THEME: What objects do we observe in the sky with our eyes, and how do astronomers organize them for further study?

You now have a general idea of the contents of the universe in which we live, and a short vocabulary with which to discuss your new-found knowledge and appreciation with another person. But you are probably not going to be one of those fortunate few who actually get to travel out there to encounter objects directly. Most of us are confined to planet Earth, from which we gaze outward and only imagine what it must be like out there. That shouldn't diminish our appreciation of the sky, however. I hope that you too find sufficient fulfillment just by using that incredible human faculty called the imagination to explore the cosmos from the comfort of our planet's surface.

The first chance you get, go outside at night, preferably away from any bright lights, and just look at the sky for awhile. Avoid thinking about anything other than the sky. Be conscious of any changes that occur. Be sensitive to any movements that may occur. You might try this on several occasions while reading this Chapter. It will help you to understand the concepts better, and it will for sure make you appreciate the subject of Astronomy better. It will at first seem rather haphazard and chaotic, even boring to look at for long. But as the mind begins to look for order, it notices that the stars seem to form patterns, the brightnesses of the stars seem to fall into a handy scale, and the colors of stars can be arranged into only a few groups. This is the starting point for studying space. My thesis for this Chapter is that:

The human mind has a need to make order out of chaos, and the mapping of the sky and the cataloging of its contents by characteristics of color and brightness is the foundation of the Astronomer's work.

Figure 2-2 *The galaxy seen in this photograph (NGC 5866) presents itself edgewise to us, in contrast to the galaxy M51 in Figure 13-15 that is face-on. It is obvious that stars appear more densely packed when seen edgewise than when seen face-on.* (Lick Observatory)

The Location of Sun in Space

Milky Way Galaxy We see Sun during the daytime, and the stars at night. As we have already learned, Sun is but an average star in a family of about 100 billion stars, the Milky Way galaxy. The only reason that Sun appears so differently to us is that it is the nearest star to us (only some 93 million miles), and sunlight scattered by Earth's atmosphere during the daytime prevents us from seeing the much dimmer stars.

The brightness of the daytime sky as a result of this scattering is greater than that of the brightest stars. From the surface of an airless body, however, one can observe both Sun and stars at the same time. Astronauts who travelled to Moon, for example, had that experience. Studies reveal that an observer in another galaxy looking at the Milky Way galaxy from an edgewise position would see it much like that of the galaxy shown in the photograph of Figure 2-2.

Sun, our star, is located some two-thirds of the way out from the center in what might be called the galactic suburbs. Sun, in addition to being average in size, temperature, age, and chemical composition, can neither claim to be strategically located in the Galaxy. From that vantage point out near the edge, we peer into different regions of the Galaxy when we scan the sky at night.

Milky Way Band Although the stars at night initially appear to be rather randomly distributed, you have no doubt noticed that a band of white runs across the sky. It is most noticeable when you are in a dark location, away from the light pollution of a large metropolitan area. This band of white, also called the *Milky Way*, is the result of the combined light of the millions of stars making up the Galaxy (Figure 2-3).

Because the Galaxy is flattened rather like two dinner plates put face to face, the stars are more concentrated in the sky when we view the Galaxy along the flattened portion, what we call the **galactic plane** (Figures 2-3 and 2-4). There are so many stars, in fact, that we are unable to see through the edgewise portion of the Galaxy to see what is beyond. In order to study or view objects outside of the galaxy (other galaxies, for example), astronomers must point their telescopes as far away from the band of the Milky Way as possible.

Figure 2–3 This 360–degree view of the dark nightime sky illustrates the flattened shape of our Galaxy. Looking at the band of white known as the Milky Way is viewing the Galaxy edgewise. The band consists of stars, luminous gas clouds, and dark dust clouds. The latter are seen as dark intrusions within the (generally) white band. (Lund Observatory, Sweden)

With a couple of exceptions, all of the objects we observe in the night sky are members of the Milky Way galaxy. When photographing another galaxy far away from our Galaxy, however, we necessarily capture in the photograph some of the stars within our own Galaxy that just happen to be in the way. Fortunately, the stars appear so small compared to the galaxies that they do little to interfere with the images of the galaxies.

Rotation of the Galaxy Galaxies are not static, motionless objects in space, but spin like giant pinwheels. When I speak of spinning, I mean that the stars orbit around the center, just as Earth orbits the center of the solar system (where Sun is located). You might wonder how stars orbit around something that has no single object at the center. It is easy to understand planets in the solar system orbiting Sun, for example. Well, the law of gravity simply says that an object can orbit around a single object or a collection of objects. In the case of a galaxy, a given star orbits around the center of gravity of the combined effect of all of the other billions of stars that are closer to the center than that star.

Of course, each star moves at a different rate around the center just as each planet in our solar system moves at a different rate around Sun. Just as our year is defined as the period required for Earth to orbit Sun once, a star's Galactic Year might be defined as the length of time required for that star to orbit the center of the galaxy once. Sun, for example, requires about 250 million years to complete one orbit around the center of the Milky Way galaxy. At a lesser distance from the center, fewer years are required; at a greater distance, more years are required. So when we speak of the rotation of a galaxy, we are actually referring to a large mass of stars each of which is orbiting around the center at a different rate.

The Proper Motion of Stars Let us consider the importance of this discovery. Imagine looking out along an imaginary line in a given direction into the Galaxy. The stars in that direction, along Line A in Figure 2-4, for example, are located at different distances from you, the viewer. Located at different distances from the center of the galaxy as well as from you, the stars are moving at different rates, just as planets move at different rates around Sun. But if that is true, shouldn't we see stars in the sky changing positions with respect to each other? Shouldn't the patterns of the constellations be changing from night to night?

If the stars were located close to us and at the same time moved fast enough, the patterns of stars that form the constellations would be forever changing. Whereas the rates of motions of stars around the center of the Milky Way are great enough (that of Sun is about 150 miles per second!), the distances to the stars is another matter. The distances of the stars from Earth, even the nearest ones, are so great that although their individual motions around the center of the galaxy are very great, the motions cannot be detected by the human eye.

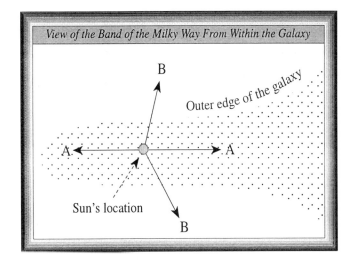

Figure 2–4 Since the Milky Way galaxy is flattened, we on Earth observe a greater concentration of stars when we look along the flattened portion (the band of the Milky Way) than when we look away from it (line A–A). To look outside the galaxy, we must therefore observe away from the band of the Milky Way, along the lines.

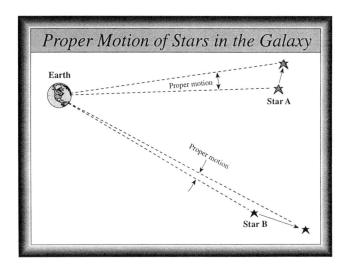

Figure 2–5 All stars in space are in motion with respect to one another, and in different directions. We on Earth observe their separate motions, called **proper motions**, projected onto the **celestial sphere**. Star A is moving slower than Star B, but its proper motion is less because it of the angle with which it is moving with respect to Earth.

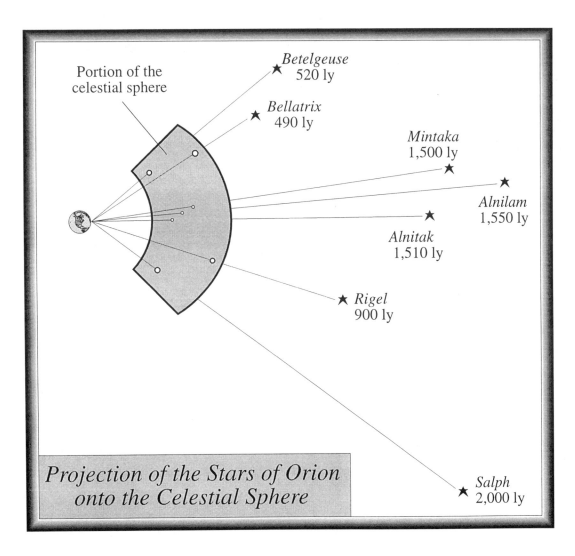

Figure 2–6 The stars that form the constellation of the Big Dipper (Ursa Major) are not grouped together in space, but rather are located at different distances from Earth. Their appearance as a "Dipper" is the result of our seeing them projected onto the imaginary celestial sphere.

CHAPTER 2 / Mapping the Sky We Observe

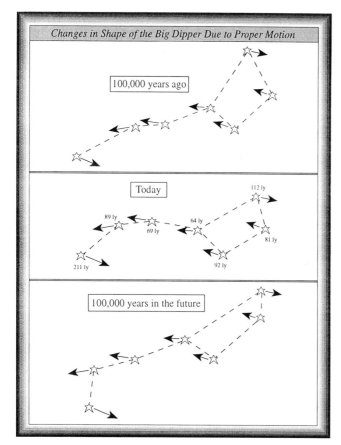

Figure 2–7 Because all of the stars in the Milky Way are moving at different speeds and in different directions, the "shapes" of the constellations are very slowly changing. The appearance of the Big Dipper as a "Dipper" is gradually disappearing.

Changing Shapes of Constellations The importance of proper motion can be made clear by using a couple of examples of constellations with which you may already be familiar. *Orion, the Hunter, is* a familiar constellation seen during the cold, clear nights of winter. Imagine being able to observe Earth from some distant location in space, and being able to observe the line of sight between Earthbound stargazers and the seven major stars forming Orion. The projection of the seven stars and their actual placements in space are shown in Figure 2-6.

Since the stars are located so far away from one another in space, they must be traveling at different rates and in different directions. Given enough time, the shape recognized today as *Orion* will eventually change to another shape, and perhaps require a new name. Admittedly, the time required for any group of stars to change shape sufficiently to be obvious to the human eye is enormous.

The group of seven stars that form the familiar pattern of the *Big Dipper* (Figure 2-1) will eventually change from its present recognizable shape of a water dipper to the ambiguous one shown in Figure 2-7. By measuring the rate of change over a period of a few years, it is easy to calculate the change that will take place over a long period of time — in this case, 100,000 years. But change it will. As I mention in a later Chapter, astrologers of the future will have to readjust their practice of assigning behavioral patterns to humans based upon the patterns of stars in the sky as those very patterns change. What appears today to be a pattern of a lion (*Leo*) may eventually look like a six-legged unicorn!

Distances The determination of distances to objects in space is an extremely important topic, and the accuracy of such measurements has important consequences throughout the field of Astronomy. A detailed survey of the methods astronomers use is reserved for a later Chapter. For the moment, however, here is an example of what I mean when I say that stars are very distant from us. Recall that the unit astronomers use to express distances beyond the confines of the solar system is based on the *speed of light* (186,000 miles per second) and the *time Earth requires to complete one orbit around Sun* (one year). It is not a unit of measurement that extraterrestrials are likely to use. "They" may well know the speed of light, but it is unlikely that They will know (let alone use) Earth's speed around Sun as a unit of measurement of distance.

The light-year is adopted (defined) as the distance that light travels in one year. It is equal to 6 trillion (6,000,000,000,000) miles! In one second, a ray of light circles Earth 8½ times. Light bouncing off Moon requires a little over one second to reach Earth. Sunlight leaving Sun's surface requires about 8 minutes to arrive at Earth's surface. The next nearest star to us is about 5 light-years away. 5 light-years is about the average spacing between stars in our Milky Way galaxy. It takes (on the average) about five years for the light of one star to get to its closest neighbor!

Astronomers routinely measure the motions using large telescopes, since telescopes have the ability to magnify not only size but motion as well. The measured motion of a star relative to the field of stars surrounding it is called the **proper motion** of that star. The proper motion of a star is one of the characteristics that astronomers go to great lengths to obtain. In Figure 2-5, the proper motions of two stars are shown in terms of angular change with respect to stars that haven't moved. Positions of objects and their motions are measured with respect to an imaginary sphere surrounding Earth called the **celestial sphere** (Figure 2-8).

Notice that the two stars move at different rates of speed in space, but they reveal different amounts of proper motion. In fact, the faster-traveling star shows the lesser amount of proper motion because of the angle of movement with respect to Earth. A star moving directly toward or away from Earth will not reveal any proper motion at all. It is rather like observing an extremely bright dot in the sky that appears fixed in the sky like the stars. Just about the time you decide to call 911 and report it as a UFO, it begins to move with respect to the stars. You then easily identify it as an airplane, realizing that it appeared stationary for a period of time because it was traveling directly toward you.

Ancient Concepts of the Sky

For most of human history the predominant belief was that the sky was permanent and unchanging and eternal. Things in the sky were fundamentally different than things on Earth, and behaved differently. Human senses provided the information for that conclusion. It was only as recently as 1916 that astronomers detected the very slight motions of stars relative to one another (*proper motion*), and had sufficient observational evidence to conclude that the universe out there is not fundamentally different from that which is close to and immediately surrounds us.

"Shooting" or "Falling" Stars Of course, you may argue, people long ago must have observed "falling" or "shooting" stars just as we do today (or tonight!). But they did not necessarily believe that those streaks of light in the sky were the result of stars "falling". In the first place, they had no basis for believing that the stars were "things" that could "fall." They were just as likely the rays of light pouring through holes punched in a large black sphere that surrounded Earth at a great distance!

Furthermore, even though a large number of streaks can be observed on any given night (at least when viewing away from the lights of the city), a careful observer will notice that there are never any stars missing! Ancient civilizations offered many theories to explain these phenomena that we today call **meteors**. Many believed they were sparks or cinders fired into the atmosphere, perhaps by volcanoes located some distance away.

Today we know that meteors are the result of dust-sized particles into which Earth runs as it orbits Sun. Friction with the atmosphere causes them to burn up. This might seem strange at first: How can a speck of dust cause such a bright streak of light in the sky? Well, it is not so much a matter of the size of the dust particle that matters. It is the fact that it is moving at a speed of about 20 miles per second when it encounters the molecules of air surrounding Earth. There is a lot of energy associated with something moving that fast, even if it is only a speck of dust. You wouldn't want to be hit by one. In fact, that is one reason why our astronauts must wear protective suits when

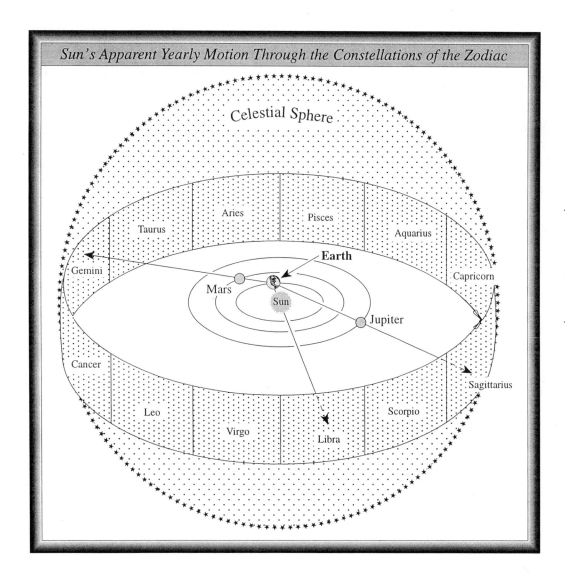

Figure 2–8 As Earth orbits Sun, the projection of Sun onto the celestial sphere makes Sun appear to move from West to East through the constellations of the Zodiac. This is the basis of the practice of astrology. Notice that since the solar system is mostly flattened like the galaxy, the planets also appear to move through the same constellations as they orbit Sun. In this particular case, Sun is "in" Libra, Jupiter is "in" Sagittarius, and Mars is "in" Gemini.

they "walk" in space. Above the protective layers of our atmosphere, any object is susceptible to such collisions. On the surface of Moon, for example, the rocks are finely pitted by such collisions.

Keep in mind that it is Earth that is moving, and therefore Earth that is running into the dust grains. In that sense, Earth's gravity is rather acting like a giant vacuum cleaner as Earth orbits Sun, sweeping up any dust particles that are caught within the solar system. So in spite of ancient people observing such streaks of light in the sky, they held onto the notion that the sky was eternal and unchanging.

Astrology You might be wondering what significance this change of thinking may have had, going from the ancient notion that the sky was unchanging and permanent to the modern notion that it is changing and impermanent. You will learn in Chapter 5 that the birth of modern science occurred when these very two different notions of the sky ended up on a collision course. But don't we retain today some of those ancient notions of the sky's permanence? If you ask a young child to point to where Heaven is, isn't it interesting that he/she will typically point up to the sky?

Have we perhaps inherited from our ancestors the notion that the permanence of the sky is associated with the permanence of Heaven? And there exists today another remnant of the ancient belief in the permanence of the sky that is even more widespread — astrology. **Astrology** is the field of study that professes to predict the future and to interpret the influence of celestial bodies on the lives and destinies of people.

Astrology is founded on an organization of the sky similar to that of the astronomer, but with a special emphasis on those fixed patterns in the sky called the **constellations of the zodiac**. It is through these that Sun, Moon, and planets move with time (Figure 2-8). This belief system originated before recorded history when people had every reason to believe the sky is unchanging and eternal. People must have felt helpless on Earth where such forces as earthquakes, lightning, tornadoes, and the like were

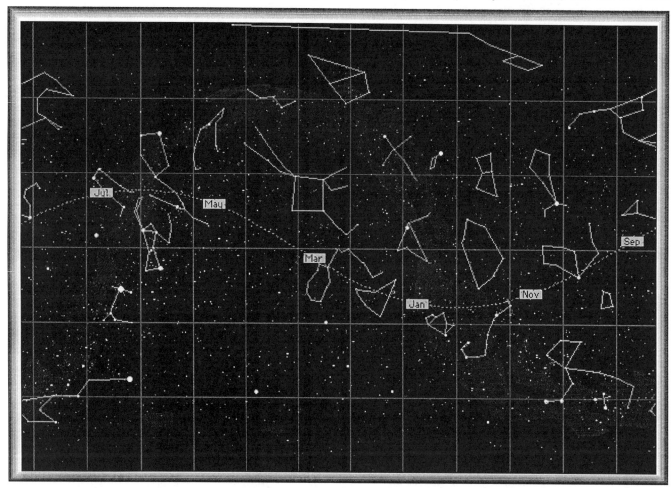

Figure 2–9 This is a map of that portion of the sky through which Sun appears to move as Earth orbits Sun. The dashed line is the path along which Sun moves, and some of the constellations of the Zodiac through which Sun moves are shown. According to astrology, a person born at the moment when Sun is in a particular constellation of the Zodiac is born "under" that particular "sign." The "official" boundaries between constellations are shown in Figure 2–13. (Carina Software)

beyond their control. They suspected that some type of control over their lives was coming from the direction of that permanent realm, the sky. Early peoples went to great lengths to organize the sky's contents and motions in an attempt to anticipate (and perhaps control) the effects those outside "forces" would have on them (Figure 2-9).

Astrology was (is) based on the assumption that forces outside of us can and do influence our lives. To study them is to have some degree of control over their supposed effects upon us. No longer do we believe that earthquakes occur when God sneezes or that the future can be seen in the entrails of birds laid out on a table top. But we are still susceptible to believing that the universe out there in some way affects our being, whether it be in patterns of stars in the sky, the appearance of the full Moon, or even stepping on a crack while walking along a sidewalk.

Origin of the Constellations So although the stars are moving at different rates and in different directions with respect to one another, naked-eye observations (viewing without the use of instruments such as telescopes, binoculars, etc.) by ancient and modern persons are incapable of detecting such motion. But why, you ask, do we still refer to patterns of stars in the sky? Why, if in fact the individual stars within a constellation are located at different distances from us and moving randomly with respect to one another, do we still talk about constellations? Well, the ancient observers had no way of knowing that constellations change shape, and it seemed reasonable to conclude that stars within a "pattern" in some way were associated with one another. But there were practical reasons as well for organizing the sky into constellations.

In order to conduct agriculture as a means of survival, it was absolutely essential for them to know the time of the year in order to know seasonal weather patterns. They needed to know when to expect the last killing frost, for example, so that seeds could be planted. And, of course, it was essential to know when to anticipate the first frost in the fall in order to harvest vulnerable crops before it hit and destroyed any crops left in the field.

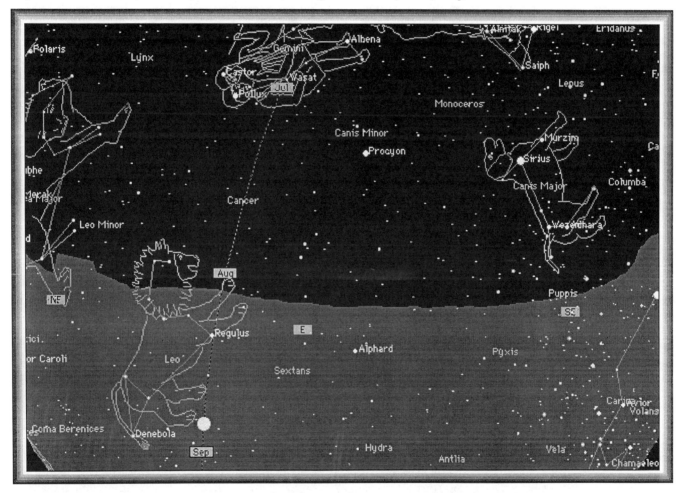

Figure 2–10 Ancient Egyptians observed a relationship between the rising of the life-sustaining waters of the Nile River and the first appearance of the bright star Sirius in the early morning sky. They henceforth assigned special "powers" to that brightest of stars. The diagram shows the eastern horizon as seen from Egypt just before sunrise. Sun's position on the ecliptic shows that it is late August. Shortly later in time, sunlight drowns out Sirius. A week earlier, Sun is closer to Sirius, making it impossible to see. (Carina Software)

Origin of Calendars Primitive peoples had neither the supermarkets to go to if their crops failed, nor calendars so familiar on our walls today for telling the time of the year. But they did have the sky as a calendar. That small wallet calendar you carry around with you is nothing more than a chart that tells you the position of Earth in its yearly orbit around Sun. So what you are actually carrying around with you is a small-scale model of the solar system.

The method of translating day and month into the position of Earth in space is one of the topics to be dealt with shortly. Suffice it to say that the position is known if one keeps track of the locations of the stars in the sky from night to night. It was not even necessary for them to understand why this was happening, why some seasons brought frost or even what frost was. They simply needed to associate certain weather conditions affecting their lives with the position of Sun with respect to patterns of stars in the sky. Curious minds work best on full stomachs.

An excellent example of the use of celestial objects for calendar keeping was the importance ancient Egyptians placed upon the brightest star in the sky, **Sirius**, visible during winter months (Figure 2-10). They noticed that the flooding of the banks of the Nile River, so necessary to the irrigation of the crops that line the river, occurred shortly after Sirius could first be seen in the morning twilight in the east. So they kept a close search for that event each year. No doubt there were those who believed that an actual relationship existed between Sirius and the Nile River, that the star caused the flooding just as some of us believe that the full Moon causes people to act strangely.

Another practical use to which knowledge of the positions of stars and constellations can be put is navigation. To get from one location on Earth's surface to another is relatively easy as long as one has easily-recognized landmarks or established trails. But on vast deserts or oceans that is not possible. Knowledge of star patterns allowed ancient mariners to sail from one land mass to another, in attempts to conduct trade, exchanging the products of one island for those of another. There is good evidence that such navigational skills allowed the Polynesians in the southern hemisphere to sail all the way to Hawaii to establish the culture that exists there today.

But why did they see patterns of a dragon, a bear, a warrior, a virgin in the sky? And why the stories that are associated with them? Well, imagine yourself trying to teach young children to recognize and remember some patterns of stars in the sky. Would you just sit the children down, draw lines between the dots that represent the stars, and require them to learn the patterns before they get any dessert? I suspect there is a more effective way. Besides, we are a naming species. We name everything. You could refer to your cat as "The Cat." Or you could call your two cats Cat 1 and Cat 2. But you'll probably use a name like Fluffy because it makes the cat more familiar in the literal sense.

I imagine a village elder, with children gathered around a fire, pointing with the tip of a glowing branch just removed from the fire, outlining a group of stars in the sky, telling a story of how that figure got into the sky. All the while, the children stare at the pattern, associating what they see with what they hear. I recall such stories told to me by my father in just that manner. I've never forgotten those stories, nor the star patterns. The abstract patterns became real through the use of visual imagery. For the ancients, the figures were no doubt selected with the intent of representing the animals with which they were familiar, or a brave warrior who had saved the village from attack by the members of a hostile village.

In the case of many constellations, the patterns were chosen to actually represent the seasonal changes that Earth experienced as it orbited Sun. For example, the constellation of *Libra*, patterned as a balance, represented that time of the year between summer and winter months, a balance between the growing and dormant seasons. It also represented that time of the year when days and nights are equal in length, the balance between light and dark.

Modern Constellations Today, astronomers officially recognize 88 constellations covering the entire sky. They include a menagerie of 14 men and women, 9 birds, two insects, 19 land animals, 10 water creatures, two centaurs, a head of hair, a serpent, a dragon, a flying horse, a river and 29 inanimate objects (such as a telescope, a reticle, and an air pump). The oldest description of the constellations as we know them comes from a poem, called Phaenomena, written about 270 BC by a Greek poet. One of the few groups of stars that just about every school child has at least heard about is the *Big Dipper*. Actually, the *Big Dipper* is not a constellation — it consists of seven stars within a larger group of stars that is a constellation. It is the Great Bear, or *Ursa Major*.

Even so, not every country sees the *Big Dipper* as a dipper. The British call it a Plough. The southern French call it a Saucepan. The Pawnee Indians see a stretcher on which a sick man is being carried. To the ancient Maya, it was a mythological parrot named Seven Macaw. To the early Egyptians, it was the thigh and leg of a bull. The ancient Chinese thought of it as a special chariot for the Emperor of the Heaven. For the Micmac Indians of Canada, as well as several North American Indian tribes, the bowl of the *Big Dipper* was a bear, and the three stars in the handle represented three hunters tracking the bear.

Not surprisingly, for a particular culture, the *Big Dipper* took on the shape of the familiar beasts and artifacts of the region in which the culture lived. In North Africa, people saw a camel. In the East Indies saw the pattern of stars as a shark or a canoe. The Sioux of central North America saw a pesky skunk in the pattern. And the Maya Indians of Mexico visualized an evil god Hunracan who was able to go anywhere at will to cause war and chaos. In the nineteenth century, the *Big Dipper* became the symbol of freedom for Southern runaway slaves who "followed the Drinking Gourd" to the northern states.

Figure 2–11 This piece of traditional "bark" art from the Aboriginals of Australia shows detailed representations of features in the sky. See the text for the story of the figures. (Tom Bullock)

As you read some of the stories of the constellations, you go away with the feeling that the story's content was likewise deliberately chosen. They tend to offer rules of behavior, or ethical and moral standards expected to be followed by village children. After all, schools as we know them today are recent developments. Thousands of years ago, when people were first organizing the sky, education was done within the context of the home, clan, or village. There is good reason to believe that many of the fairy stories that you were told or read as a child were written for the same purpose.

An intricate familiarity with the nighttime sky is obvious in an example of Australian Aboriginal art. Figure 2–11 illustrates a legend of two brothers who lost their lives while travelling in a canoe. Their floating bodies (in the central panel) became the starless parts of the Milky Way, and the capsized canoe turned into stars. The bodies are the dark shapes in the band of the Milky Way in the constellations of Serpens and Sagittarius. Modern astronomers know these regions as vast clouds of dust that block our view of the stars beyond. The canoe is a line formed by four stars located near *Antares*, the bright reddish star in the constellation of *Scorpius*.

The remaining area of the central panel is filled with stars of the Milky Way. The two outer panels with the wavy pattern represent the wake of the canoe in which the two brothers were travelling. They are the luminous parts of the Milky Way located near *Scorpius*, known to modern astronomers as glowing clouds of gas heated up by nearby hot stars. The upright figures of the two brothers can be seen in the top panel. The older brother stands on a rock on which he landed. This rock, represented by a black area underneath the figure, is another dust cloud located near the star designated *Theta Serpentis*.

Mapping the Sky

Just as geographers create maps of Earth's surface so that people can successfully go from one location to another, so astronomers map the sky so that people can properly locate and study objects in the sky (Figure 2–12).

Cataloguing the Stars The letters of the Greek alphabet are used to designate the *order of brightnesses of the stars in a constellation*. The brightest star in the constellation of *Canis Majoris* (the Big Dog) is therefore designated alpha (α) *Canis Majoris*. The second brightest star is designated beta (β) *Canis Majoris*, and so on. Astronomers use these designations on star maps and in their vocabulary when referring to stars. A scale of different sized dots is also used on star maps to represent the different brightnesses of stars.

In addition to inheriting the patterns of stars from ancient stargazers, we also inherited proper names for many of the brightest stars — notably from the Arabs. In the deserts of the dry, clean air of Saudi Arabia, ancient skywatchers spent considerable time studying, organizing, and mapping the sky. Thus α *Canis Majoris* is also known as *Sirius*, a star I mentioned earlier as an example of how people used stars for calendar-keeping. Figure 2–13 is a typical Star Chart (map) used by amateur astronomers, illustrating the use of sizes of dots to record stars according to brightness.

When you notice that the white dots on the chart are of different sizes, but do not be misled into thinking that those dots in any way represent the actual sizes of the stars. Even as seen through the most powerful telescopes on Earth, all stars (except Sun, and just recently *Betelgeuse*) appear as mere pinpoints of light. In actuality, most stars are quite enormous in size, but because they are so incredibly far away from us, their sizes are not seen or measurable directly. Their brightnesses can be easily measured, however, and that is what is indicated by the sizes of the dots.

★★

Brightnesses of Stars In order to measure the brightness of a star, astronomers attach a light-sensitive instrument to the telescope. It operates just like the light meter found in modern cameras, the **photometer**. It simply measures the amount of light coming from a distant object and entering the telescope, that amount indicating the object's brightness. Be careful when you use the expression "brightness". Remember that it in no way indicates how much light that object is actually emitting, but only how much we are receiving. The term **luminosity** is used to express the energy emitted by an object such as a star, but we don't need to discuss that characteristic of a star in great detail yet.

When expressing how bright an object is in the sky, in place of the somewhat vague term brightness, astronomers prefer to use the term **apparent visual magnitude** apparent because it is how the object appears to us, visual because the photometer measures visible radiation, and magnitude because a number is associated with that degree of brightness. A dim star located close to us and a very luminous star located far away from us can both appear to have the same brightness or apparent visual magnitude. On diagrams located throughout the book, you will see apparent visual magnitude abbreviated by the symbol (m_v).

Magnitude Scale 2,100 years ago, the Greek astronomer **Hipparchus** recorded on a chart the positions of all visible stars. To each dot he assigned a number between 1 and 6 to indicate the relative brightness of each star. He designated the very brightest stars as magnitude **1**, and the stars just barely visible magnitude **6**. The task must have been very time-consuming, inasmuch as some 3,000-4,000 stars are visible to the eye from a dark location when Moon is not in the sky.

Astronomers throughout the ages have used the magnitude scale developed by Hipparchus, but of course the development of telescopes allows us to see and catalog stars that are too dim to be seen with the naked eye. In order to assign an apparent visual magnitude number to those dimmer stars, the original six numbers used by Hipparchus had to be extended past the number 6 to larger numbers

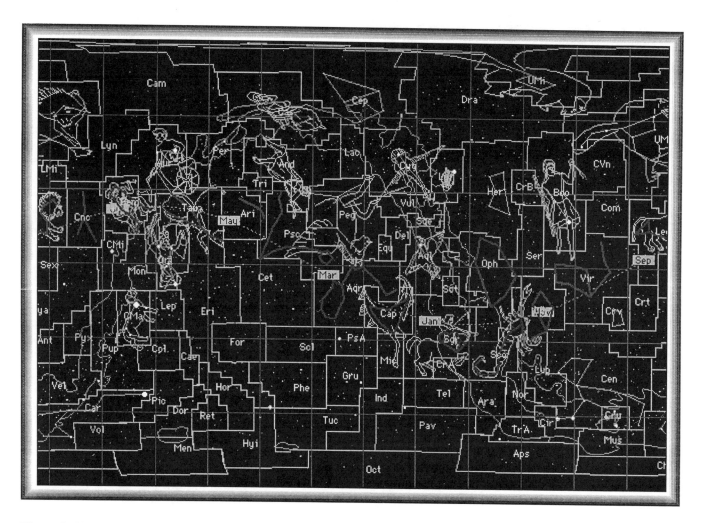

Figure 2–12 This star map of the sky includes the "official" boundaries adopted by astronomers throughout the world for the purpose of designating the location of an object on the celestial sphere. The abbreviations of the names of the constellations are also included. (Carina Software)

(but to represent dimmer stars!).

Radiation, both visible and invisible, consists of tiny bundles of energy called by scientists **photons**. The light-meter in a camera simply measures the number of photons reflected off an object. The device astronomers use to measure the amount of light coming from an object is called a **photometer**. So the apparent visual magnitude of a star is simply an expression of the number of photons that arrived at the detector, be it a telescope or the human eye.

When astronomers first used photometers to measure the number of photons arriving from the stars to which Hipparchus had assigned numbers 1 to 6, they found there were 100 times more photons recorded for magnitude 1 stars than for magnitude 6 stars. In other words, the magnitude 1 stars were 100 times brighter than the magnitude 6 stars. Therefore, the difference in brightness of two stars whose magnitudes differ by a factor of 1 (e.g., a 3 star and a 2 star) is approximately 2.5 times (Diagram to the right).

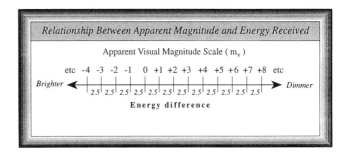

If we measure 6.25 times more photons arriving from star A than from star B, and we know that Star B's m_v is +8, then we can assign an m_v of +6 to star A. The calculation of m_v's is performed not just for stars, but for any object that appears in the sky (e.g., galaxies, planets, nebulae). Even the brightness of a temporary phenomenon such as a meteor is expressed as its m_v. For obvious reasons, the apparent visual magnitudes have been calculated for all of the objects in the sky, and are frequently included on star maps.

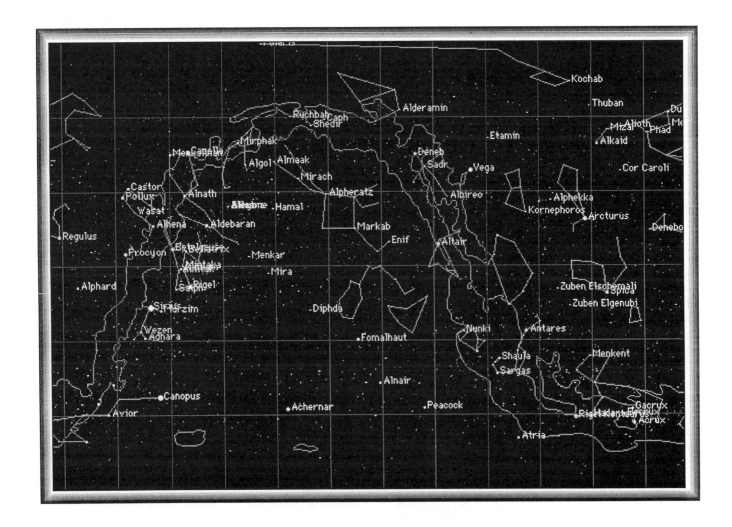

Figure 2–13 This star map includes the adopted names of some of the brightest stars in the sky. The sizes of the dots represent the relative brightnesses of those stars. (Carina Software)

CHAPTER 2 / Mapping the Sky We Observe

Boundaries Separating Constellations Up until 1922, the constellations inherited from ancient civilizations were only generally outlined in the sky. There was not an exact number of constellations, since many of the constellations of one civilization overlapped those of another civilization. Neither were there precise boundaries between adjacent constellations that were agreed upon by astronomers. By the 1920s, advances in telescopic observations made it necessary to establish boundaries in order for astronomers to share a common system of specifying the location of an object being studied.

In 1922, the astronomical community adopted 88 constellations for the entire sky, and by 1930 had established exact boundaries for each. The boundaries are sometimes shown on sky maps, although they do tend to clutter them up with detail that not everyone is going to use. The map in Figure 2-12, for example, shows the boundaries. No portion of the sky is left unclaimed by boundaries. If you point in any direction in the sky, you are pointing toward a given constellation. If you observe any object in the sky, even if it be an airplane temporarily in your field of view, it is in a constellation.

Using this system, astronomers now specify a particular star by to its order of brightness within the constellation's boundaries. To arrange the stars according to their brightnesses in the entire sky would be cumbersome. Breaking it up into small segments makes the task easier.

Maps, Magnitudes, and Coordinates On Earth's surface, we refer to the location of someone or something by referring to the two angles of latitude and longitude. **Latitude** is the angular distance of someone measured from Earth's equator northward or southward. The North Pole, for example, is at latitude 90 degrees North. Someone on the equator is located at latitude 0 degrees. **Longitude** is the angular distance of someone or something measured from the location of Greenwich, England. It is expressed as an angle between zero and 180 degrees East or West of Greenwich. Someone located in San Francisco, for example, is located at approximately Longitude 122 degrees West, Latitude 38 degrees North (Figure 2-14).

In order to specify the position of an object in the sky, astronomers use the two angles of **declination** and **right ascension**. The *declination* of an object is its angular distance measured north or south of an imaginary line in the sky called the **celestial equator**. The celestial equator is Earth's equator projected out into space and superimposed onto the celestial sphere, rotating along with it as Earth spins. It is the white line running horizontally across the center of the Chart in Figure 2-13. The white lines running parallel to the celestial equator are lines of declination, representing 30-degree intervals. The star *Fomalhaut*, lower center in the Figure, has a declination of about minus (because it is south of the celestial equator) 30 degrees.

There is no Greenwich in the sky from which to measure the angle that corresponds to longitude, so astronomers use a point on the celestial sphere called the **vernal equinox** from which to measure right ascension angles. The *right ascension* of an object is its angular distance along the celestial equator measured eastward from the vernal equinox. Astronomers choose to use that particular point from which to measure angles because of its historical significance. For the moment, think of it as Sun's location on the celestial sphere on March 21st, the time at which Sun is crossing the celestial equator from south to north (refer to Figure 2-9). The parallel white lines running perpendicular to the celestial equator in the Figure are lines of right ascension. For historical reasons, right ascension angles are measured in units of time: hours, minutes, and seconds. Hence the right ascension lines in Figure 2-9 (as well as Figure 2-13) are at 2-hour intervals. Since the line running through the vernal equinox represents the zero right ascension line, *Fomalhaut's* right ascension angle is approximately 23 hours.

Whenever the location of an object is expressed or recorded for astronomical purposes, it is in terms of right ascension and declination. When a telescope operator wants to point the telescope towards an object, he/she simply enters into the computer control panel the right ascension and declination angles for that particular object, and then pushes the search button. The telescope swings to the exact alignment for viewing or imaging the object.

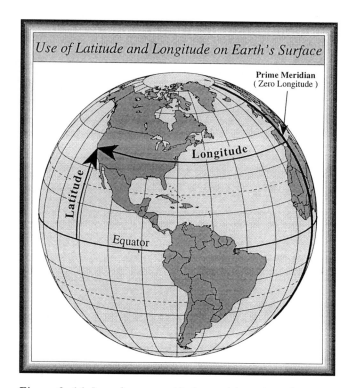

Figure 2–14 *In order to specify the position of an object (or person) on Earth, the two angles of Latitude and Longitude are used. Latitude is measured from Earth's equator. Longitude is measured from the prime meridian that runs through Greenwich, England.*

Telescope Power Expressed as m_v. An object's brightness is very important to an observer, especially an amateur astronomer who learns of a discovery made by professional astronomers using large telescopes. If the amateur wants to observe the object for any particular reason, he/she will need to know if it is bright enough to be detected by whatever telescope he/she is using. As you will learn in Chapter 5, a telescope is a light funnel. It is a device that collects photons and focuses them to a single point, the eyepiece where your eye or other detector is positioned. The larger the funnel, the more photons that are collected, and the brighter a given object will appear in the eyepiece.

So the size of the opening of a particular telescope determines the limit of just how dim of an object that the telescope is capable of detecting. The size of the opening determines how many photons enter the telescope to be detected, and the numerical value of m_v for an object is an expression of how many photons are detected from that object. The human eye requires a minimum number of photons to strike the retina before a message is sent to the brain to announce that something has been detected. If the telescope is capable of collecting and focusing to the eye enough photons from an object, the object is "seen." If that minimum number of photons is not focused onto the eye, the brain will not know the object exists. There are other types of detectors that astronomers use besides the human eye, but each has a particular minimum number of photons that must be collected before the object is recorded.

I can express this idea by using numbers. Assume that I am using a 6-inch telescope. Figure 2-15 tells me that I can see with it any object that has an m_v of +12 or brighter. I would not, for example, be able to use that telescope for observing the planet Pluto (which has an m_v of +15). To see Pluto, I would need to use at least a 14-inch telescope (Figure 2-15 again). For that reason, when you hear an amateur astronomer referring to his/her instrument, you will hear them talking about a 6-inch telescope, a 10-inch telescope, and so on. Since m_v is the symbol used to express the brightnesses of objects, an astronomer referring to a detailed star map can tell whether or not a given telescope is capable of observing an object of a given magnitude.

Since there are objects in the sky that are brighter than the brightest stars (can you think of any?), astronomers also extended the magnitude scale in the opposite direction, into the minus number range. This again uses the two-and-a-half-times-the-number-of-photons factor in assigning a number to those objects. Some of those bright objects, in addition to selected objects throughout the scale, are also shown in Figure 2-15.

It may require a little adjustment in your thinking to remember that those objects with negative numbers and small positive numbers are brighter than those with large positive numbers. It seems contrary to the manner in which we use numbers. But astronomers did not want to tamper with the original work of Hipparchus by rearranging the system to accommodate our usual way of thinking.

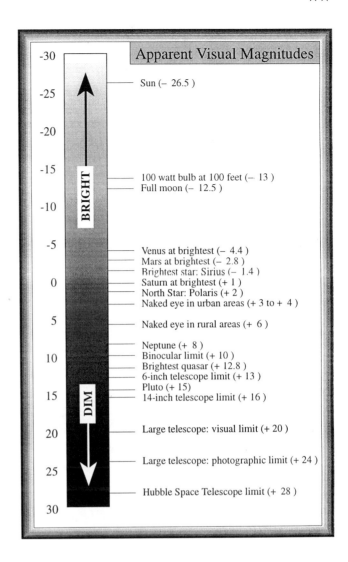

Figure 2–15 The range of brightnesses of objects that astronomers study with their instruments is expressed by the scale of apparent visual magnitudes. That range goes from the brightest (Sun is approximately –26) to the dimmest (very distant galaxies that are approximately +30).

★★

LEARNING OBJECTIVES: *Now that you have studied this Chapter, you should be able to:*

1. Describe Sun's position in the Milky Way galaxy.
2. Explain the two uses of the term Milky Way.
3. Describe how the rotation of the Milky Way galaxy causes the constellations to gradually change shape.
4. Explain how proper motion and the speed of a star in space are related.
5. Provide an example of a mistakened idea that ancient peoples had about the sky.
6. Describe the importance of the constellations of the Zodiac.
7. Provide an example of a practical application of the use of locating stars.
8. Explain how the magnitude scale is used to represent the brightness of objects in the sky, and what the difference between any two numbers represents.
9. Describe the contents of a Star Chart, and what basic information it contains.
10. Explain how the magnitude scale and the performance of a telescope are related.
12. Define and use in a complete sentence each of the following **NEW TERMS**:

Apparent visual magnitude (m_v) 28
Astrology 24
Celestial sphere 22
Celestial equator 30
Constellations of the zodiac 24
Declination 30
"Falling" star 23
Galactic plane 19
Hipparchus 28
Latitude 30

Longitude 30
Luminosity 28
Meteor 23
Photometer 28
Photons 29
Proper motion 22
Right Ascension 30
"Shooting" star 20
Sirius 26
Vernal equinox 30

Chapter 3

Observing Motions of Objects in the Sky

Figure 3–1 This photograph, taken from Latitude = 38 degrees North, illustrates a consequence of Earth's rotation. The streaks, called star trails, are the result of the stationary camera's shutter being left open during a long exposure. Can you calculate the length of the exposure? (Lick Observatory)

CENTRAL THEME: How do the objects we see in the sky appear to change with time, how do astronomers explain those changes, and how do any of those changes affect conditions on Earth?

So we have organized the sky into bounded sections called constellations, assigned names or letters to stars within those sections, and calculated numbers to assign to their brightnesses. This is done for the purpose of organizing the sky so that it is easier to keep track of the objects that we observe in the sky. There are also movements within the solar system, both short term and long term, that affect our attempt to organize the sky. You are familiar with some of them already — sunrise and sunset, for example. Try to relate as many of the movements as possible with your own experiences in looking at the sky. And supplement the reading by going outside and actually observing some of the features of the sky that I'm about to describe. There are five movements that are observable to the human eye — those of the **stars**, Sun, the **planets**, **Moon**, and the **Equinoxes**.

Observing the Stars' Movements

Rotation of the Celestial Sphere Let us spend an imaginary evening together watching the sky. Assuming that we are located somewhere in the United States, we notice over a period of time that the sky is moving. That is, stars in one direction appear to be getting higher and higher in the sky, and new stars appear to be coming up from the horizon. Those in the opposite direction appear to be getting lower and lower in the sky, eventually setting beneath the horizon. A photograph of this movement is shown on the Title page opposite. You can do this as well, by setting your camera on a tripod and holding the shutter open for as long as it is dark. Since the light of stars continues to fall onto the photographic film while the shutter is open, and the stars are moving all the while, the result will be the streaks you see in the photograph on the previous page.

Again outside, standing so that the setting stars are to our left (west) and the rising stars are to our right (east), we notice that there is a group of stars in front of us (in the north) that neither rise nor set. They appear to move around counterclockwise within a large circle, much like the hands of a huge clock (Figure 3-1). These are referred to as circumpolar stars, inasmuch as they are circling around a central star (can you locate it in Figure 3-1?). The reason these movements occur, of course, is because Earth is spinning, or what astronomers refer to as **rotation**. Days and nights are an obvious consequence of the rotation of Earth. This is quite a slow process, of course, so in order to make the above observations one must spend a few hours outdoors. For observers in the United States, the movements of stars with respect to the horizon will appear generally as shown in Figure 3-2. On your next adventure outdoors at night, you might enjoy verifying that it is so.

Turning back to observe the stars just rising in the east, we notice after an hour or so that the stars are not rising at right angles to the horizon, but rather at a fixed angle. In fact, the angle the setting stars make with respect to the horizon is exactly the same as the angle the rising stars make. The numerical value of that angle is determined by the location of the observer with respect to Earth's equator. The closer to the equator we are located, the steeper the angle — until at the equator itself, the stars rise and set at right angles to the horizon. An observer at either of the poles observes the stars moving parallel to the horizon, neither rising nor setting.

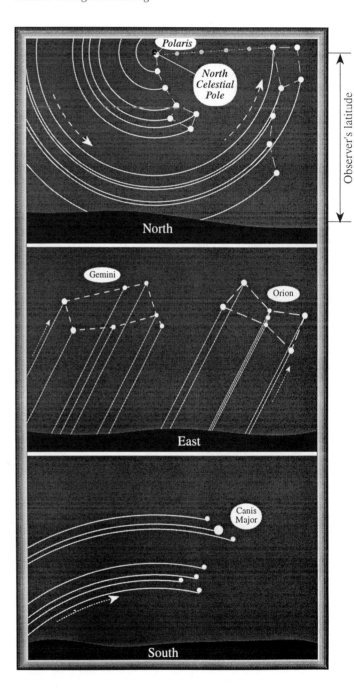

Figure 3–2 For an observer in the continental United States, the apparent motions of the stars relative to the horizon in three directions are shown. Can you predict what the motions are in the West?

CHAPTER 3 / Observing Motions of Objects in the Sky

★★★

The ancient observers detected the motion of stars with respect to the horizon as well. In fact, there is evidence that the early Polynesians used the varying rising and setting star angles as a navigation tool, and thereby located and settled the Hawaiian Islands. But they could neither sense any actual spinning of Earth nor offer any explanation how such a huge object like Earth could move, so they concluded that it was the sky that was moving — not heavy, massive Earth. The fact that the stars appeared neither to change in position nor change in brightness was further reason to arrive at that conclusion. They did not even consider the stars to be sizeable objects as we do now.

To them, the sky had somewhat of a mystical nature, consisting of small permanent dots attached in some manner to a huge rotating sphere surrounding Earth at some great distance. Even to us, it seems as if we are surrounded by a huge sphere in which the stars are embedded. This is the imaginary celestial sphere onto which we project the positions of objects and on which we therefore detect any movement between them. In fact, one could argue quite persuasively that stars were not even actual objects, not the large spheres of burning gases that we associate with them today. One could argue that surrounding Earth at some great distance is a huge, dark, spinning sphere, the surface of which has been randomly punched with holes of different sizes. Beyond this sphere there are burning fires, the light of which can be seen from Earth leaking through the holes. These are the stars.

Think about it as you gaze at the sky. Do you have any evidence to refute such a claim? I doubt it. It is not intuitively obvious that the stars have sizes and are actual objects. You and I think of them that way because we are enmeshed in the midst of cultural ideas that paint the scientific picture of stars being other Suns. We have accepted the evidence that others have provided.

Rotation of Earth I might challenge you in a similar matter as regards the movement of the stars. Do you have any evidence that Earth rotates? Again, I doubt that many nonscientists can produce such evidence. The motions of the stars, planets, Moon, and Sun can all quite satisfactorily be explained by assuming a stationary Earth and a host of objects obediently circling it. The most straightforward proof for Earth's rotation is that of the **Foucault Pendulum**, a demonstration that you might recall from a visit to a science museum. A heavy metal ball suspended by a strong wire swings back and forth along a constant line with respect to the stars. But as Earth rotates beneath the swinging ball, it appears as if the direction of the ball's swing keeps changing as it knocks over pegs or other objects placed in a circle around the pendulum. If Earth were not rotating, the ball would continue to swing back and forth along the same line. There is no other explanation for the behavior of the ball.

In a sense, neither Sun nor the stars rise in the east and set in the west. Earth, because it rotates from west to east, carries us toward those objects in the east and away from those in the west. It would be more accurate to say that Sun appears to get higher in the east as Earth spins towards it, or that Sun appears to get lower in the west as Earth spins away from it. But it doesn't really matter as long as the speaker knows what he/she means by it. Earth rotates around an imaginary axis that runs between the North and South Poles. This is Earth's **rotational axis**. All spinning objects have an axis of rotation.

We will learn more about the historical events that led to an overhaul of Earth-center theory in Chapter 4, resulting in the modern notion that Sun (and entire solar system) is located out in the boondocks of our galaxy. Suffice it to say that for most of human history the prevailing idea was that the entire universe centered on the motionless Earth. Everything moved around Earth. Even though we no longer seriously entertain this idea, however, we still refer to the huge sphere as the *celestial sphere*, and it is upon that imaginary sphere that we place objects for the purpose of mapping the sky.

The Celestial Poles Let us go back to that star around which the sky appeared to be moving in Figure 3-1. At the same time that Earth rotates, it orbits Sun. This latter motion is called **revolution**. Something going around something else is *revolving* around that something else. Something spinning about an axis is *rotating*. As Earth rotates and revolves around Sun, its rotational axis continuously points towards two imaginary points in the sky referred to as the **north** and **south celestial poles** (Figure 3-4). Because Earth is conveniently divided up into two equal segments, a northern half or northern *hemisphere*, and a southern half or southern *hemisphere*, the sky is

Figure 3–3 Facing toward the northeast from mid-latitudes in the United States, one observes stars rising at an angle. Can you determine how long the camera shutter was open during this photograph? (Tom Bullock)

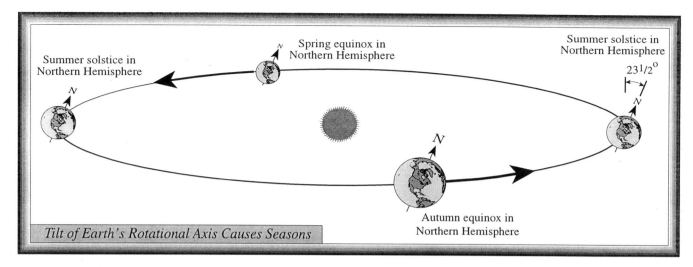

Figure 3–4 As Earth orbits Sun, its axis of rotation points to an imaginary point on the celestial sphere called the North Celestial Pole. Because Earth is tilted with respect to this orbital plane, we experience seasons.

correspondingly divided into two halves. These are the northern hemisphere of stars and the southern hemisphere of stars. They are joined by the line that is the projection of Earth's equator out into space referred to earlier as the celestial equator.

Now refer to Figure 3-5. The center of the northern hemisphere of Earth is, of course, the North Pole. Directly above the North Pole, on the celestial sphere, is the north celestial pole. Above the south pole, on the celestial sphere, is the south celestial pole. There is nothing located at those two points in the sky. They are only the extensions of Earth's rotational axis into space onto the celestial sphere. Located quite close to the north celestial pole, however, there is a magnitude 2 star named **Polaris** (α *Ursa Minor*, the brightest star in the constellation of the small bear).

Because *Polaris* is located so close to the north celestial pole, it is referred to as the **north star**. But you will learn shortly that *Polaris* hasn't always been located close to the north celestial pole — it is actually still getting closer to it. And after the year 2102, it will begin to slowly drift away from the north celestial pole. The movement is so slight, however, that you may find comfort in knowing that you can point *Polaris* out to your great-grandchildren and still refer to it as the north star.

Notice in both Figures 3-2 and 3-5 there is a region of the celestial sphere around the NCP that is always above the horizon as Earth rotates, and the stars within that region never set. This is called the **circumpolar** region, and the stars (and associated constellations) are *circumpolar* stars and constellations.

When astronomers speak of the direction *north*, they are referring to the direction of the north celestial pole — the point in the sky toward which Earth's rotational axis points. There is another direction *north* — magnetic north is the direction toward which the needle of a compass points. But that is based on Earth's magnetic field, which is not in alignment with the rotational axis.

Because of the geometry of Earth's position within the celestial sphere, the measured angle of *Polaris* (or more accurately, the north celestial pole) above the northern horizon is equal to the latitude of the observer (number of degrees above Earth's equator). In other words, your latitude on Earth is approximately equal to the angle of *Polaris* above the northern horizon (see Figure 3-5). If you

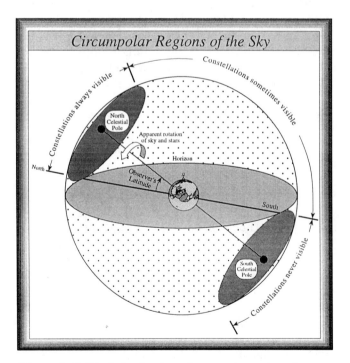

Figure 3–5 The latitude of an observer determines the number of circumpolar stars and constellations that are visible. By using geometry, one can show that a person's latitude is equal to the angle of the NCP above the northern horizon. This is very useful for navigating at sea.

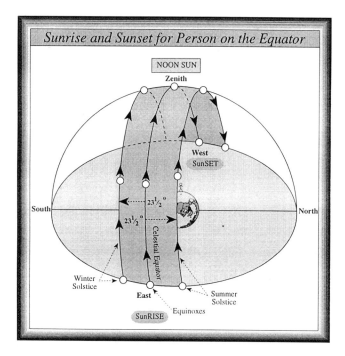

Figure 3-6 As seen from Earth's equator, all of the stars rise and set at right angles to the horizon. The celestial equator runs directly overhead, and the NCP and SCP lie directly on the horizon. The star Polaris therefore lies directly on the horizon as well.

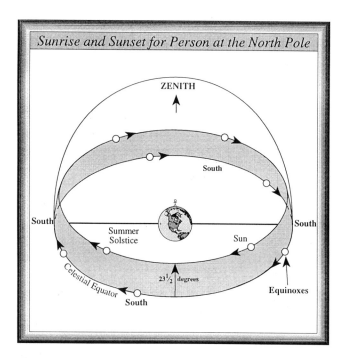

Figure 3-7 As seen from Earth's poles, none of the stars rise or set. The celestial equator runs directly along the horizon, and the NCP is at the zenith. The star Polaris therefore appears directly overhead as well.

observe *Polaris* directly on the horizon, you are located on the equator. If you observe it directly overhead — at a point called the **zenith** — you are located at the North Pole. So *Polaris* has and is very useful as a navigational aid. Using an instrument called a **sextant** to measure the angle between the horizon and *Polaris*, sailors can establish their latitude position on Earth.

Unfortunately, there is no corresponding south star — there is no bright star close to the south celestial pole. Historically, navigation in the southern hemisphere was limited by this fact. When you study a map of the early ocean voyages, you notice that few of the mariners ventured into the southern hemisphere. Those that did hugged the coastline quite carefully in order to find their way back home again.

Latitude Adjustment: the <u>Equator</u> Let us now take an imaginary trip to Earth's equator, to visit some exciting location such as the Galapagos Islands off the coast of Ecuador (Figure 3-6). As we travel southward and our latitude decreases — all the while observing the celestial sphere — we observe *Polaris* getting lower and lower in the sky, until it rests right on the horizon when we arrive at the equator. Looking at the remainder of the sky, we still observe the stars rising and setting, but they rise perpendicularly and set perpendicularly. Stars come up directly and set directly, not at an angle as they do where we live.

In addition, none of the stars appear circumpolar, and we can eventually observe all the stars in the southern and northern hemispheres. From a location in the United States, there is a group of stars circling the south celestial pole that never rises above the southern horizon. If you imagine watching Sun, Moon, and planets from the equator, you should be able to visualize that they — like the stars — rise and set directly. A consequence of this fact is that while at the equator, we experience equal days and nights: 12 hours between Sunrise and Sunset, and 12 hours between Sunset and Sunrise. The lengths of the days and nights do not vary as they do in the United States.

Latitude Adjustment: the <u>North Pole</u> Now to the North or South Pole (Figure 3-7). As we travel northward, Polaris gets higher and higher in the sky, until it reaches the zenith. We are now located at latitude 90 degrees North. The sky of stars now seems strange. The celestial sphere appears like a giant merry-go-round, with the stars neither rising nor setting. All stars are circumpolar. We can observe all of the stars in the northern hemisphere, but those in the southern hemisphere are forever concealed from our view. And again, if you visualize Sun's behavior (or that of Moon and planets) in the sky, it will go around and around and around, never getting lower or higher on a particular day. This is the "Land of the Midnight Sun."

Disregarding climatic considerations, can you offer any advantages to the placing of telescopic observatories at either the equator or the two poles?

Figure 3–8 Our modern time-keeping is based on Sun's movement across the sky relative to the N–S line called the meridian. Time "begins" at 12 noon when Sun crosses the meridian. During each hour of time, Earth rotates through 15 degrees. Hence Sun moves 15 degrees with respect to the meridian during each hour of time.

Observing Sun's Motions

Let us now consider the patterns of the Sun's movements in the sky. I say movements because there are two — one relative to the horizon — quite familiar to you — and one relative to the stars, which is not so obvious and which is actually rather difficult to observe. Keep in mind that motion is relative, and that a reference point, line, or plane must be implied or stated whenever motion is mentioned. For most human activity, the reference plane is implied to be Earth's surface — as when a police office cites me for going 65 m.p.h. in a 55 m.p.h. zone. But since astronomers deal with space outside Earth's surface, they are careful to state the reference point for any motion being considered.

Movement Relative to Horizon Sun rises generally in the east, climbs to some highest point directly to the south, and sets generally in the west. This motion is not due to Sun moving, of course: it is the result of Earth's rotation. Imagine observing Sun's movement if by magic Earth's atmosphere were to suddenly disappear. You would observe stars <u>and</u> Sun moving across the sky as one gigantic unit.

The air particles (molecules) scatter sunlight and cause the entire sky to be bright during the daytime, brighter than the brightest stars in the sky. If the sky is brighter than the stars beyond, then they cannot be observed. That, incidentally, is the reason that very few stars are observable from the vicinity of large metropolitan areas, except that it is the city lights being scattered, not sunlight. If Earth's atmosphere were to be removed, one could observe Sun and stars (and Moon and planets) moving as one gigantic unit.

Telling Time Someone once said that nature invented time so that everything wouldn't happen at once. After time "started," it became necessary to define a starting point, a point from which other events could be measured. Thus it was agreed that the starting point for "telling time" would be the point at which Sun reaches its highest point in the sky , what we call 12 noon. This is the basis for our time-keeping today. To be exact, our system of "telling time" is based on the position of Sun in the sky, not on the positions of hands on a clockface.

You refer to this system of time-keeping whenever you use the terms "AM" and "PM". In addition to being divided into northern and southern halves, the sky is also divided into an eastern half and western half, left and right halves as you look directly southward. These halves are joined by an imaginary line running north-south, running directly overhead through the zenith, a line that is called the **meridian** (Figure 3–8). When Sun is <u>before</u> the meridian during morning hours, it is AM (Latin, **a**nte-**m**eridian). When Sun is <u>after</u> the meridian, it is PM (Latin, **p**ost-**m**eridian). When Sun is <u>on</u> the meridian, it is 12 noon, and another "clock" day begins.

Since Earth rotates through 360 degrees in a period of 24 hours, Sun's position with respect to the meridian changes 15 degrees per hour. Therefore, we can anticipate that at 11 AM, Sun is located 15 degrees before the meridian. At 5 PM, Sun is located 75 degrees after the meridian. We could use this language in our communicating with other people about meetings, etc., but we'd probably not be very popular.

Movement Relative to Stars Instead of watching Sun move across the sky as Earth rotates, let us visualize the effects of Earth's movement (revolution) around Sun during a one-year period. To do this, we will perform another thought experiment, since we can't observe the effect directly. Imagine observing Sun in the sky during the daytime against the background of stars (i.e., on the celestial sphere) with Earth's atmosphere removed. We'll stop Earth's rotation, but allow Earth to continue to revolve around Sun. Earth does not appear to be moving, but Sun appears to move slowly with respect to the stars, moving from <u>west</u> to <u>east</u>, tracing out a line called the **ecliptic**.

★★★

In a one-year period, Sun traces out a path amongst the stars that completely circles Earth on the celestial sphere (refer back to Figures 2-8 and 2-9). The *ecliptic plane* is therefore defined as the *plane of Earth's orbit around Sun*, and the *ecliptic* is defined as the *apparent path that Sun traces on the celestial sphere* as Earth orbits Sun. Sun on a given day is at a particular point on the ecliptic. The ecliptic is broken up into 365+ segments, one for each day of the year.

The ecliptic is the basis of our calendar. The definition of "date" arises not from a number on a page on your calendar, but from the position of Sun at a point on the ecliptic. If you refer to a star chart (such as Figure 2-10), you will notice the ecliptic included on the map of the sky. The dates are placed along the ecliptic to indicate the positions of Sun throughout the year.

When you refer to a date of the year, you can visual in your mind Sun positioned amongst a group of stars, similar to looking at the hands of a clock and visualizing Sun in the sky at an angle with respect to the meridian (Figure 3-8). The fact that we no longer do this is simply evidence of our tendency to abstract nature's way of doing things, removing them from their original meaning. Can you think of other examples of this tendency?

Are you wondering why the word ecliptic is used to describe the path of Sun against the stars? Did you read the word as *eclipse* instead of ecliptic? If you did, it is understandable, since the association is intentional. It is only when *Moon is ON the ecliptic that an eclipse can occur.* As a brief introduction to the organization of the solar system, visualize it from the perspective of someone in space observing it edgewise (look ahead to Figure 3-13). Earth's orbital plane around Sun defines the ecliptic. But you notice that the entire solar system is flattened, very much like the galaxy in which it is contained.

We will see later that the common manner in which the solar system and Milky Way galaxy formed easily explains their flattened shapes, since the same laws of physics acted in both cases. Earth's Moon, even though it circles Earth rather than Sun, also lies close to this flattened plane, the ecliptic plane. This means that from our vantage point within the solar system, we always view the planets and Moon close to the ecliptic. Sun is, of course, on the ecliptic.

✴**Plotting Positions of Sun** Plotting the positions of Moon and planets on the celestial sphere is rather simple and straightforward. Armed with a map of the stars and the patterns they suggest, one need only draw in a dot on the map where Moon or a planet appears night after night. But since Sun cannot be seen at night, how did the ancients know the changing position of Sun against the background of stars? Their method was ingenious.

They waited until after sunset, just as the stars began to appear above the horizon where Sun had set. Being familiar with the patterns of stars adjacent to those appearing above the horizon, they could estimate just where Sun was relative to the pattern observed. Several nights later, they noticed that a slightly different set of stars would appear above the setting Sun, suggesting that Sun is moving "into" those seen on the previous night and drowning them out with its light. The sequence of observations would be similar to those of Figure 3-9 below.

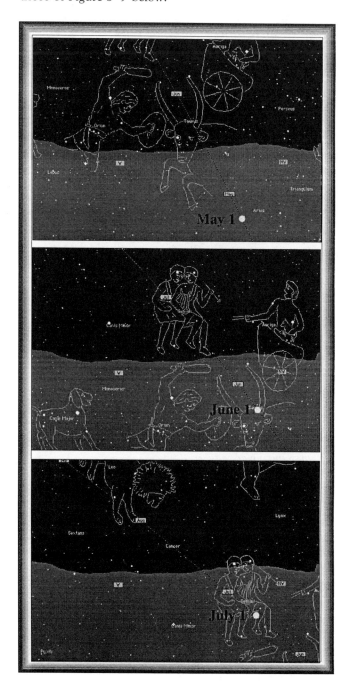

Figure 3-9 Ancient sky-watchers were well aware of the movement of Sun with respect to the stars. By observing the changing star patterns just after sunset, they deduced the changing position of Sun amongst the stars. The three views are at the same time of the evening on the dates of May 1st, June 1st, and July 1st. Notice the changing appearance of constellations just above the horizon just after sunset. (Carina Software)

Earth's Tilt and Seasons There are many interesting consequences of the fact that Earth's rotational axis is tilted 23.5° with respect to the plane of its orbit around Sun, not the least of which are the seasonal changes we experience. If it were not tilted, the ecliptic and celestial equator would be one and the same. But the ecliptic is angled so that Sun appears as far as 23.5° north of the celestial equator at the **summer solstice** (June 21), and 23.5° to the south of the celestial equator at the **winter solstice** (December 23). Between these two dates, Sun must necessarily cross the celestial equator – events we call the **spring (vernal) equinox**, when it crosses from south to north (March 21), and the **autumnal equinox** when it crosses from north to south (September 23).

This means that the lengths of day and night vary throughout the year, since during the 6-month interval between the spring and autumnal equinoxes Sun rises to the north of east, and sets to the north of west. During the 6 month interval between September 23 and March 21, Sun rises to the south of east and sets to the south of west. As shown in Figure 3-10, the result is that the length of the path of Sun across the sky varies throughout the year. When the path is longest (June 21), Sun is at its greatest distance north of the celestial equator, it reaches its highest point in the sky at 12 noon, the length of the day is longest, and the night is shortest.

When the path is shortest (December 23), Sun is at its greatest distance south of the celestial equator, it reaches its lowest point in the sky at 12 noon, the day is shortest, and the night is longest. Halfway between these two extremes, when Sun is crossing the celestial equator (March 21 and September 23), the days and nights are *equal* in length (hence the word **equinox**). These descriptions are for the northern hemisphere observer. I'll leave it to your imagination to work out, but if you are located somewhere in the southern hemisphere, the events occur at the opposite times of the year – the summer solstice occurs approximately December 23, the winter solstice approximately June 21, and so on.

A popular misconception is that seasons have something to do with Sun's distance from Earth. It is understandable that one would automatically assume that when it is hot during the middle of summer, Sun must be close. But in fact that is not the case – at least for those of us who live in the northern hemisphere. It turns out that Earth's orbit around Sun brings it *closest to Sun* on or about *January 4 each year!* This point on Earth's orbit is called **perihelion**. That means, of course, that when it is the summer solstice in the southern hemisphere, Sun is very close to perihelion.

Figure 3–10 (right) Sun's path in the sky varies during the year, so that it spends more time above the horizon during summer months (north of celestial equator) than during winter months (south of celestial equator). This results in varying amounts of sunlight and seasonal differences.

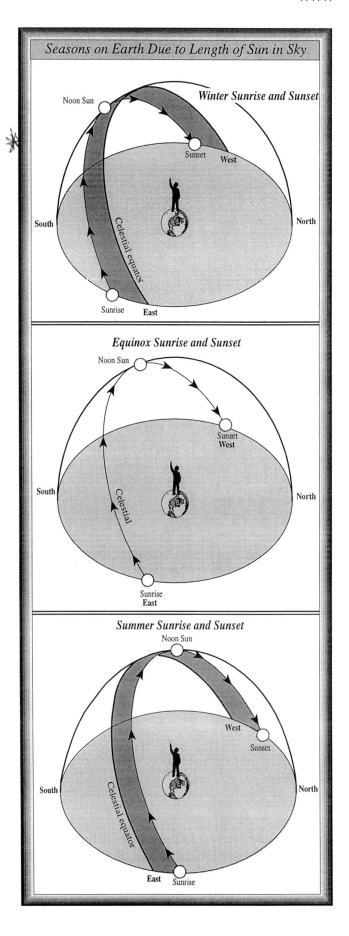

CHAPTER 3 / Observing Motions of Objects in the Sky

The general pattern of seasons is the result of two features, one of which I just mentioned. The longer Sun is in the sky (the longer the length of day), the more sunlight is collected at Earth's surface. The shorter the length of day, the less sunlight is collected. The second feature of Sun that influences seasons is its height in the sky during its daytime journey from horizon to horizon. When sunlight arrives at a low angle, a given amount of sunlight is dispersed over a larger area than when that same amount arrives at a more direct angle. Figure 3-11 illustrates the geometry of the extremes for a person living at 38 degrees North latitude. These two features determine the general pattern of seasons only. There are certainly factors such as elevation and proximity to oceans that influence local variations on this theme. Figure 3-12 illustrates the lengths of days and nights at different times of the year for observers at different locations on Earth. Notice the significance of the Tropics of Cancer and Capricorn as the extremes for the position of Sun directly overhead (at the zenith).

Seasons At the North Pole Now imagine rushing toward the North Pole on June 21st, this time concentrating not on the movement of stars, but on the movement of Sun. As we travel toward the pole, we notice Sun rising further and further to the north of east, and setting further and further to the north of west. By the time we arrive at latitude 66.5° North, we notice that Sun is skimming the horizon in the north at midnight and returning there the following night! Sun is not setting—you are within the arctic circle, the **Land of the Midnight Sun**.

Review Figure 3-7. At the North Pole itself, Sun remains above the horizon 24 hours a day between the spring and autumnal equinoxes, and is below the horizon 24 hours a day between September 23 and March 21. Your experience at the South Pole is similar, except at the opposite times of the year — Sun is continuously in the sky from September 23 to March 21. Try to imagine the conception that a civilization flourishing at the North Pole would develop to explain the nature of the universe, and Earth's relationship to it!

Constellations/Signs of the Zodiac If Sun appears to move along the ecliptic as Earth circles Sun, and every direction in the sky is within one of the 88 constellations, then necessarily Sun at any moment must be located within a particular constellation or on the boundary separating two adjacent constellations. As it moves along the ecliptic, Sun moves from within the boundaries of one constellation into another.

Ancient civilizations selected patterns of stars laying along the ecliptic and assigned them specific characteristics because they believed that the association between Sun and the star patterns behind them had both practical and spiritual significance. Practical, of course, because it allowed them to know the position of Sun relative to the stars, which is the basis of the calendar used for agricultural purposes.

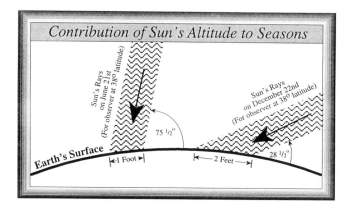

Figure 3-11 For an observer located at 38 degrees North latitude, the two extremes of sunlight occur at the solstices. The angle of sunlight determines how much will be absorbed by Earth's surface per unit area, and hence influences seasonal changes in temperatures.

Because they did not fully understand why seasons occurred, why objects moved in the sky the way they did (or even what they were!), they attached spiritual significance to the patterns they visualized in those stars.

Since we in the Western world are the inheritors of Greek/Roman traditions, we adopted their 12 constellations of the zodiac as those through which Sun travels during the year (Figure 2-8). As we will see shortly, Moon and planets also move through these same 12 constellations as they circle their parent bodies (since the solar system is flattened). At any particular moment, Sun, Moon, and each planet can be found in one of the constellations of the zodiac. Had we inherited the traditions of another civilization, the Chinese, for example, the ecliptic would be divided up not into 12 patterns of stars but 28.

Observing Movements of the Planets

People are curious. We do not care for uncertainty. We strive to understand the things we observe. Ancient observers saw a huge sphere of fixed stars in the sky, almost boring in its repetitive turning around Earth. They also observed seven objects (Moon, Sun, and 5 visible planets) that had motions very unlike those of the stars. These motions were repetitive, but only over long enough periods of time. That made them interesting, but difficult and challenging to explain.

The motions of Sun have already been discussed. I will discuss the motions of Moon later in this Chapter. The remaining five were the planets. The word "planet" is derived from the Greek word for "wandering" star. That is because a planet appears no different from a star, except that it moves relative to the background of distant stars. It "wanders" through the constellations of the zodiac. In summary, the planets are always found somewhere close to the ecliptic. This is apparent when we look at the flattened

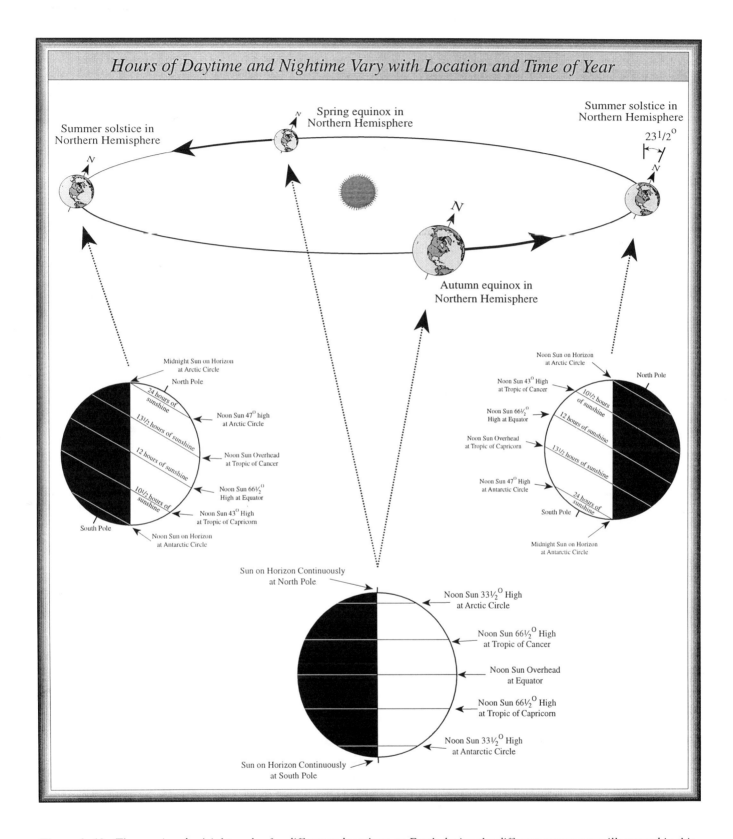

Figure 3–12 The varying day/night cycles for difference locations on Earth during the different seasons are illustrated in this diagram. You can use it in planning out summer/winter vacations to distant locations, especially those located to the extreme north and south latitudes.

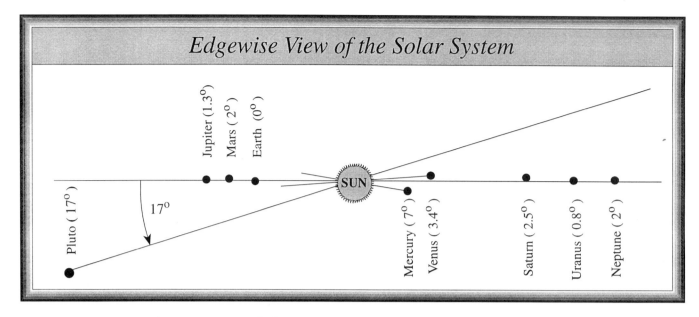

Figure 3–13 When the solar system is viewed edgewise, it appears to be flattened. Earth's orbit around Sun defines the ecliptic plane from which the orbits of other planets are measured. Except for Pluto, the planets generally appear on or near the ecliptic—i.e., in one of the constellations of the zodiac.

shape of the solar system from outside (Figure 3-13).

Incidentally, this allows for a convenient method of locating the ecliptic or the constellations of the zodiac in the sky: draw an imaginary line through any two or more planets, Sun, and/or Moon. Since these objects are always found on or close to the ecliptic, and the ecliptic runs through the constellations of the zodiac, the line you draw approximates the ecliptic.

Origin of the Week So important were these seven objects to the ancients that they believed that they must in some way influence events here on Earth. Some influences were obvious, like night/day cycles, the seasons, tides. So was it not possible that these same seven wandering objects might also affect our lives or behavior in some mysterious way? Are there reminders for us today of a possible connection between the cosmos "out there" and the cosmos "within"? Of course: seven, seven days in a week.

Each day is named after one of the 7 objects, and this is quite obvious when you look at the names of the days of the week in other languages (Figure 3-14). There is no movement of any object in the sky that occurs in the period of a week, as in the case of the day, month, and year. It is close to the period of time that Moon takes to move through one-fourth of its orbit around Earth, but not exactly.

Again, it is not difficult to understand why the ancients believed that celestial objects were responsible in some way for events on Earth. They observed some obvious connections (e.g., Sun with seasons), and simply extended the explanation to other events whose connections were not so obvious. Other examples? "Lunatic", from Moon, lunar. It was a common belief in ancient China that girls who looked directly at the full Moon would get pregnant!

Observing Moon's Movements

One of the motions of Moon in the sky is obvious to everyone. It — like Sun — rises in the east and sets in the West. This motion, as with Sun, we attribute to Earth's rotation. But there is something different about Moon. It changes shape from night to night, an obvious conclusion

Day	Ruling "Planet"	Anglo-Saxon Equivalent	Latin	French	Spanish
Sunday	Sun	-	Solis	Dimanche	Domingo
Monday	Moon	-	Lunae	Lundi	Lunes
Tuesday	Mars	Tiw	Martis	Mardi	Martes
Wesnesday	Mercury	Woden	Mercurii	Mercredi	Miercoles
Thursday	Jupiter (Jove)	Thor	Jovis	Jeudi	Jueves
Friday	Venus	Freya	Veneris	Vendredi	Viernes
Saturday	Saturn	Seterne	Saturni	Samedi	Sabado

Figure 3–14 The days of the week have an ancient origin, at a time when there were seven non–stellar objects in the sky that moved with respect to the stars. Because of that, they were associated with special "powers" or gods that were thought to influence events on Earth.

reached by any observer. What is probably less obvious is that while it is changing shape it is moving with respect to the stars (Figure 3-15).

Lunar Phases More careful observing reveals that there is a pattern to the changes in shape, or what we call the **phases** of Moon. Each night, Moon moves from west to east with respect to the background of stars. I am not referring to its motion relative to the horizon (which is the result of the rotation of Earth). Go outside on a night that Moon is out and approximate its position relative to some bright stars that are (hopefully) close by. Then return the following night at approximately the same time and note Moon's new position with respect to those same stars. It has moved!

Now take the experiment one step further. Note the day on which you made the second observation, and determine the number of days that go by until Moon returns back to that very position in the sky. You will find that it takes about 27.3 days. If you determine the time interval required to go from a particular phase back to that same phase again, it should be 29.5 days. The difference between these two time periods is simply due to the fact that Earth travels a certain distance around Sun in that same time interval, and Moon therefore has to travel slightly further in order to line up for the same phase again.

Look carefully at Figure 3-16. We refer to Moon's position in its (approximately) monthly orbit either by the name given to that particular phase or by the number of days it has been since the last new Moon. This reference to a number of days is referred to as the **age** of Moon. We define a new Moon as zero days old, a full Moon about 14.5 days old, and so on. In the diagram, keep in mind that you are standing on the rotating Earth observing Moon, which revolves around Earth much slower than Earth spins. While Moon is growing from 0 days (new) to 14.5 days old (full), we say that it is **waxing**. From full phase back to new Moon is declining or **waning**.

Since reflected sunlight is the only means by which we can observe Moon (and planets, too), the new Moon cannot be seen from Earth. During the full Moon, on the other hand, we can see the entire sunlit portion of Moon. The dark side of Moon is the half that is not lit up. The dark side of the new Moon is the side facing Earth. The dark side of the full Moon is the backside of Moon. Looking down from above Earth's north pole (Figure 3-16), the solar system's predominant motion is counterclockwise: Earth both revolves and rotates counterclockwise, Moon both revolves and rotates counterclockwise, all of the planets revolve (but not necessarily rotate) counterclockwise, and Sun rotates counterclockwise. Later we will see that this predominant motion goes a long way toward explaining the origin of the solar system.

Telling Time by Moon With these motions in mind, imagine yourself as the Earthbound person in Figure 3-16, rotating along with Earth. Is it obvious at what point in Earth's rotation you can see Sun just rising above the horizon? And also to observe sunset? So observing Moon at a given phase on a particular day, you should, in the same way, be able to predict the time of rising and setting for that particular phase of Moon. For example, the 1st quarter (waxing quarter) Moon rises at 12 noon, crosses your meridian at sunset, and sets at midnight. The waning gibbous Moon rises about 9 PM, and sets at 9 AM or so.

And it is also possible to work backwards: to tell the approximate time by observing the position of a certain phase of Moon in the sky. This technique requires your knowing that Sun's position relative to the meridian is the basis of clock time, and that Moon's phase indicates where Sun is located. Naturally, if Sun is up, you can approximate the time by estimating its position relative to the meridian. So Sun can be (and has been for centuries) used for telling time during the daytime, and, assuming it is in the sky, Moon can be used for telling time at night.

For example, if you observe the 3rd (waning) quarter

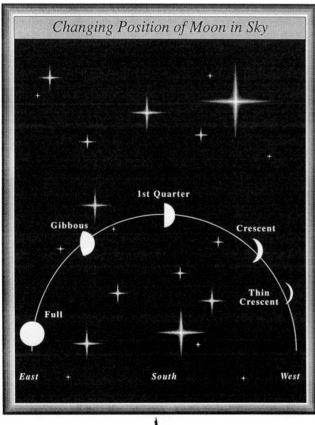

Figure 3-15 Observing Moon on successive nights <u>at the same time</u> reveals the sequence of lunar phases from thin crescent to full. Notice that Moon moves eastward with respect to the stars about 12 degrees each day.

Moon rising, it must be midnight. In Figure 3-16 the lit up side of Moon points toward Sun. If the right side is lit up, it is a growing (waxing) Moon, whereas the left side is lit up during a waning Moon. A little practice with this Figure allows you to develop a familiarity with the movements of Earth, Moon, and Sun that will, I'm sure, add to your enjoyment and appreciation of the sky.

Certainly you have noticed that the same face of Moon is always visible to Earthbound observers. Initially, one might conclude from this that Moon does not rotate. Well, it does, but it completes one rotation with each revolution. Moon rotates at the same rate that it revolves around Earth. This is also illustrated in Figure 3-16. The small circle on Moon's surface is an arbitrary fixed spot, and therefore rotates with Moon. Notice that it is always facing Earth.

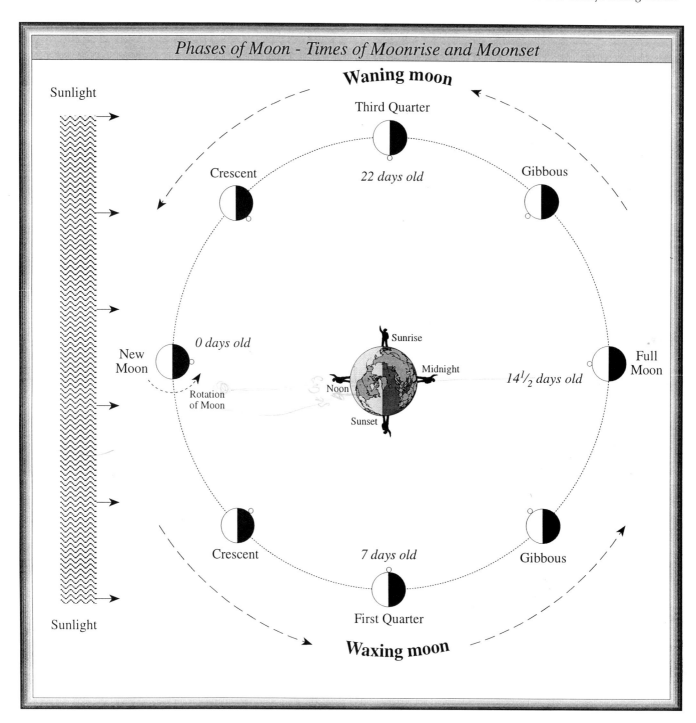

Figure 3–16 This representation of Moon's orbit around Earth in a 29–day period can be used to determine the time of moonrise and moonset, since time is location on Earth's surface. Or, as an alternative, you can determine your local time by knowing the location and phase of Moon in the sky.

Moon's Orbit An edgewise view of Moon's orbit around Earth (Figure 3-17) reveals that it is tipped about 5° with respect to the ecliptic. That is, Moon moves alternately from 5° above the ecliptic to 5° below the ecliptic. This suggests that during the 29.5-day cycle Moon requires to orbit Earth, it crosses the ecliptic twice, once when it crosses from below the ecliptic to above, and again when it crosses from above the ecliptic to below.

These two points on the ecliptic are called Moon's **nodes**. Moon is at a node twice each month. Moon spins as it circles Earth, rotating once for each orbit around Earth (i.e., its rotational period is equal to its period of revolution). Its rotation causes it to wobble slightly, the effect being that the nodes are not permanent points on the ecliptic, but gradually move in an 18-year cycle. In other words, Moon does not cross the ecliptic at the same points during each monthly cycle. The important point of this is that it is only when Moon is at or near a node that an eclipse can occur, because otherwise Moon will be either above or below Sun (in the case of a solar eclipse), or above or below Earth's shadow in space (in the case of a lunar eclipse). That is why it is named the ecliptic: only when Moon is at or is close to it can an eclipse occur (Figure 3-17).

But of course, that is not the only requirement for an eclipse. If it were, there would be two eclipses every month. In addition to Moon being at a node, Sun must be at one as well. If Sun is at the same node, there will be a solar eclipse (i.e., Moon is new). If Sun and Moon are at opposite nodes, there will be a lunar eclipse (i.e., Moon is full). What is the probability of these two conditions occurring simultaneously? Not very great — often enough to look forward to and to plan to observe, but not frequent enough to become mundane and commonplace. I will elaborate on the various types of eclipses and how you may observe them later in the Chapter, but there is one final movement in the sky that has rather interesting implications.

Observing Movement of the Equinoxes

Of all of the movements that can be observed in the sky, that which I am about to discuss is the most difficult to observe directly or indirectly. That is because the movement is so slow as to be detectable only over very long periods of time. But the implications of the movement are so important to astronomers as to warrant consideration. This is especially true for anyone who wants to understand why it is that astrology and astronomy have absolutely nothing in common except language.

Precession As you play with a top, you are aware of two motions of the top, that of spinning about an axis (rotation), and the other a wandering of the axis of the top itself. The top, in other words, does not remain exactly straight up and down, but changes angle with respect to the floor as it spins. If you trace the line that the extended axis of the top makes on the ceiling as it spins, you notice that it is in the shape of a circle (Figure 3-18). As the top slows down, the circle changes size, but, for Earth, let us ignore that fact. This wobbling of the top's axis is called **precession**.

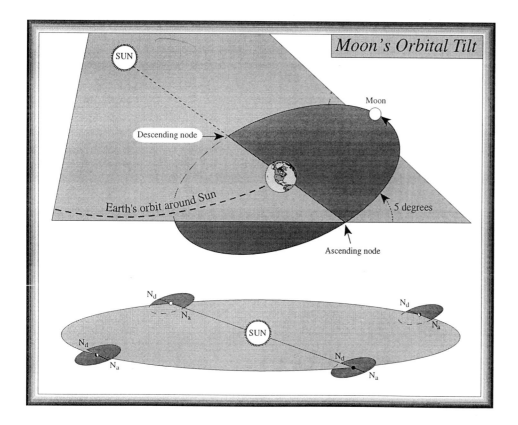

Figure 3–17 Moon's orbit around Earth is tilted by 5 degrees away from the ecliptic. As a consequence, eclipses can only occur when Moon is crossing the ecliptic—either on the ascending or the descending part of its orbit. But of course, another condition is necessary for an eclipse to occur—Sun must also be lined up with the Earth–Moon alignment. This latter condition occurs at both new and full phases of Moon.

Now Earth in going around Sun is behaving like a giant top. Its axis points, you have learned, toward imaginary points on the celestial sphere called the celestial poles. Close to the north celestial pole is the star Polaris. But Earth's axis precesses in the same manner as that of a top. The presence of Moon and Sun with respect to Earth's 23.5° tilt causes it to precess, and the effect is that the north and south celestial poles wander amongst the stars, tracing a complete circle in the sky once every 26,000 years (Figure 3-18).

An easy way to observe this is to go to a Planetarium and ask the operator to take you back in time by changing the direction in which the axis of the *starball* points. The starball is the spherical end of the instrument at the center of the room. Tiny holes in the ball allow for light from a bright source inside to shine onto the ceiling to create artificial "stars." Notice that the ball rotates around a shaft that attaches it to the upright instrument. To go backward or forward over long periods of time, the operator need only change the direction toward which the shaft points on the ceiling. This simulates a motion that Earth requires centuries to go through. After the operator makes that change, observe the motions of the stars as they rise and set.

Everything initially seems the same as what you currently observe in the sky. But if you look toward the north, you discover that *Polaris* is no longer stationary in the sky: it is moving in a small circle around the north celestial pole. It is coincidental that we have a North Star, a star close to the north celestial pole, at this time in history. For most of the 26,000-year precessional cycle, there is no star close to the north celestial pole, and consequently no North Star.

Figure 3-18 includes a map of the region of the sky through which the north celestial pole wanders during this long cycle, and the brighter stars that will be (or have been in the past) close to the north celestial pole. So although we are gradually losing *Polaris* as our North Star, we will gain another star called **Vega** in about 13,000 years. Of course, we eventually get *Polaris* back.

Equinoxes To the ancients, the four most important times of the year were the equinoxes and solstices, probably because they were used to time the cycles of growth (agriculture), and upon which life itself depended. The most important of these four was certainly March 21 — the *vernal equinox*. For it is then that Sun is arising from its long winter's journey beneath the celestial equator. Thus the vernal equinox was a reminder to all of the returning Sun and growing season. Perhaps we can't fully appreciate what this meant to them since we obtain our food from supermarkets and no longer have to depend upon the changing seasons for growing our own food.

Easter For those who have been raised within the Christian tradition, the most important celebration of the year is that of *Easter*. The day on which Easter falls is fixed by the Church, and is based on the spring equinox. It is always celebrated on the first Sunday after the first full Moon after the vernal equinox. We see in this close association a similar theme — the *arising of Sun* from beneath the celestial equator so that plant growth can take place again, and the *Resurrection of Christ* from the dead, offering hope for eternal life for those who believe in Him. You can carry this symbolic association as far as you like. The point is: the celebration of Easter is not based on the date that Christ is supposed to have died, but on an astronomical event in the sky that has to do with the behavior of Sun in the sky.

Astrological "Signs" When Hipparchus created his star map about 2,100 years ago, the vernal (spring) equinox was in the direction of the constellation of *Aries, the Ram*. That is where it is shown on his map. That is why the astrological column in the newspaper normally begins with *Aries*. It is the first constellation Sun is in after the most important day of the year. *Aries* is the first "sign" of the zodiacal year. In addition to causing the north celestial pole to wander amongst the stars, however, precession also causes the vernal equinox (intersection of ecliptic and celestial equator) to wander amongst the stars. Today, consequently, it is no longer located in *Aries* but in *Pisces, the Fish*. A person born on March 21st is no longer an *Aries*, but a *Pisces*!

Using the boundaries agreed upon by astronomers in 1930, it is easy to determine that Sun on March 21st has not been in the constellation of *Aries* since the year 50 BC (Figure 3-19 and 3-20). This affects not only those born on March 21st, but with the exception of a few born on the boundary between one sign and another, it affects everyone. *Each of us is actually one "sign" before the one the astrologers have been claiming.* I say this in the sense that Sun is not physically located in the constellation noted in the astrological column, but in the direction of the constellation prior to it. If you have been making decisions about your life based upon the morning newspaper, you might want to reconsider.

The Age of Aquarius Extending this idea backward and forward in time, you easily conclude that sometime before the time of Hipparchus, the vernal equinox was in *Taurus*, before that it was in *Gemini*, and so on. Since the precessional cycle is 26,000 years long, and there are 12 "signs" of the zodiac, the vernal equinox spends a little over 2,000 years in each "sign." Thus sometime in the near future it will slip out of the constellation of *Pisces* and into the adjacent one, *Aquarius, the Waterbearer*.

When exactly that occurs depends upon what boundaries separating the constellations you choose to use. Assuming you use the boundaries agreed upon by modern astronomers in 1930, the date that the vernal equinox enters the constellation of *Aquarius* is around the year 2000. At that moment, astrologers claim we will have entered the **Age of Aquarius.** So you see, astrologers do take account of the slippages due to precession, but not when they deal with the "sign" of a particular individual.

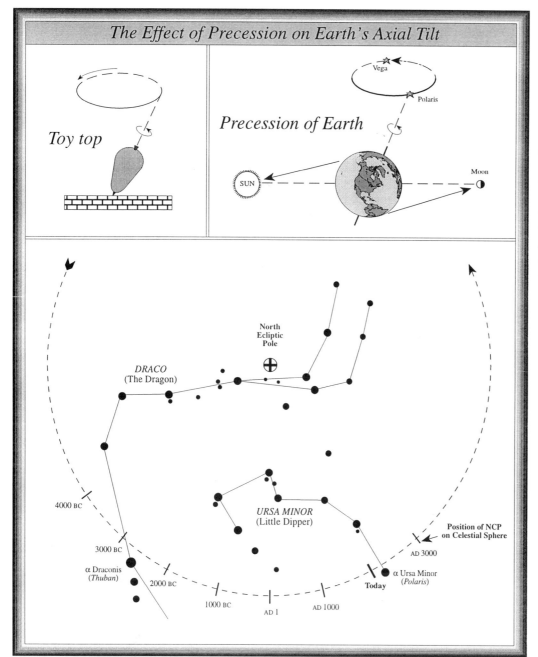

Figure 3–18 *A gravitational tug–of–war between tilted Earth and Sun and Moon causes a very slow wandering of Earth's axis of rotation. The NCP is the direction toward which this axis points, but that point wanders amongst the stars in a circular pattern. Consequently, the North Star Polaris will gradually move farther and farther away from it, and no longer be useful as a navigational guide.*

There are a few who do, but they are in the minority.

Just as a person exhibits the traits associated with his/her "sign", so humanity (so astrologers say) exhibits the traits associated with the constellation that the vernal equinox is in. For the past 2000 years or so, we have been collectively behaving like *Pisces* people. But when the vernal equinox soon enters *Aquarius*, we will collectively change our manner of behavior in accordance with the traits of that "sign." The consequences of this change (as suggested by astrologers) are reflected in the rock musical *Hair*, popular in the 1960s and 1970s.

Since I have already treated you to some of the lesser-known consequences of the motions of objects in the sky, there are others with which you are already familiar. Since they can be both practical and interesting to understand and observe, I'll explain them in some detail in hopes that you will integrate them into your life as well. Those are eclipses, tides, and astrology.

Eclipses

Eclipses of Sun and Moon were of much greater significance to ancient people than they are to us today, no doubt because they were considered (like comets) to have mythical significance because they occur only rarely, and occur in the inaccessible sky. Since events on Earth were thought to be influenced by celestial events, it was a wise court astronomer

★★★

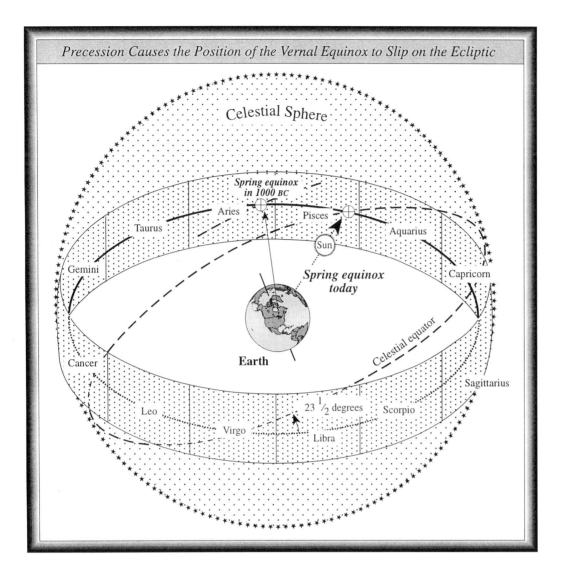

Figure 3–19 *As a consequence of Earth's precession, the position of Sun on the celestial sphere on a given date very slowly changes. Consequently, Sun is no longer in the constellation of Aries at the time of the spring equinox (at which time Sun crosses the celestial equator on its way northward). In addition, Sun on any particular date is not in the constellation noted in the astrology column of the newspapar.*

who worked out an accurate method of predicting eclipses in order to advise the king or emperor on matters of state interest, such as going to war or taking a wife!

As you will see, there is a great deal of regularity in the occurrences of eclipses, a fact that allows us to predict the exact dates and times that they can be observed. They can be exciting to observe, and I hope that by the time you read descriptions of their appearances, you'll want to rush out to make arrangements to see the next one.

Eclipses in General There are two types of eclipses, solar and lunar. Solar eclipses occur when Moon passes in front of Sun, when Moon is in the direction of Sun (i.e., during a new Moon). Lunar eclipses occur when Moon passes behind Earth, opposite Sun (i.e., during a full Moon), and enters the shadow Earth casts out into space (Figure 3-21). In order for there to be an exact alignment, however, it is also necessary that Moon be crossing or close to crossing the ecliptic (i.e., at or near a node).

Thus two conditions are necessary for an eclipse to occur (*new phase + node = solar eclipse, full phase + node = lunar eclipse*), which explains why they are so infrequent. The reason for this is that Moon does not circle Earth in the same plane that Earth circles Sun: Moon's monthly orbit around Earth is slightly tilted, so that Moon passes as much as 5 degrees above and below the line between Earth and Sun. It is only when Moon is close to that line that eclipses occur (Figures 3-22, 3-23, 3-24).

Solar Eclipses All objects in the solar system are illuminated by sunlight. They, like Earth, have a darkside (nighttime) and an illuminated side (daytime). But the shadow cast by the object is not confined to the surface — it extends out into space in the shape of a cone. It points away from Sun, of course. During a solar eclipse, this cone-shaped shadow — called the **umbra** — points directly toward Earth. It may or may not touch Earth, depending on how close Moon is to Earth at the time. If it reaches Earth, a **total eclipse** occurs. If it doesn't, an **annular eclipse** occurs (Figure 3-25).

Like Earth's orbit around Sun, Moon's orbit around Earth is not perfectly circular in shape, but elliptical

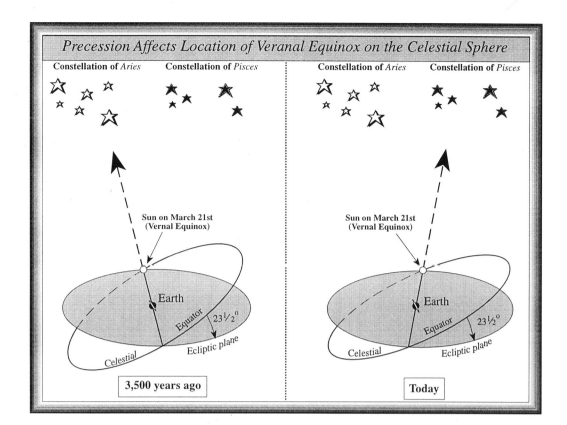

Figure 3–20 *Just as Sun at the time of the spring equinox has slowly slipped from the constellation of Aries into the constellation of Pisces over the past few thousand years, so will it eventually slip into the constellation of Aquarius. According to astrologers, humankind will at that time enter the "Age of Aquarius."*

(Kepler's 1st Law). During its 29.5-day orbit around Earth, Moon moves alternately closer to and further away from Earth. It is when Moon is closest to Earth, at a point called **perigee**, that Moon appears largest in the sky and that it will exactly block out Sun (If at the same time it is also at new phase and at a node. To summarize, when the new Moon is at a node and at perigee, its <u>apparent</u> size in the sky is identical to the <u>apparent</u> size of Sun and it completely blocks out Sun. This set of circumstances causes a region on Earth's surface to experience temporary nighttime.

If the new Moon is <u>close</u> to a node, but not exactly at it, only a portion of Sun is blocked out because Moon is not exactly lined up with Sun. So a **partial solar eclipse** will occur without there being either an annular or total solar eclipse. About 35% of the time this is the case. The annular eclipse, requiring two conditions, occurs about 37% of the time. Only 28% of all solar eclipses are total!

A list of solar eclipses for the remainder of this century is presented in Figure 3-26, just in case I have excited your curiosity enough so that you will want to observe one. Because the path of totality for a total solar eclipse is so small, any particular spot on Earth is treated to this event on average only once every 360 years! So don't wait for one to come to your hometown! As you see in Figure 3-26, the next one that is observable from somewhere in the United States is the year 2017!

Annular Solar Eclipse If the new Moon is at a node but not close to perigee, its apparent size is <u>less</u> than the apparent size of Sun, and sunlight leaks out from around Moon (Figure 3-25). The tip of the umbra (the region of darkness) does not reach Earth. During this type of eclipse,

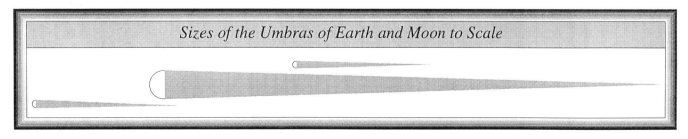

Figure 3–21 *A scale model of the Earth–Moon system according to sizes and distances and orbital inclination show why there is not an eclipse at every full or new moon. It is only when there is exact or near alignment that Moon's shadow reaches Earth (for a solar eclipse) or Moon moves through Earth's shadow (for a lunar eclipse).*

★★★

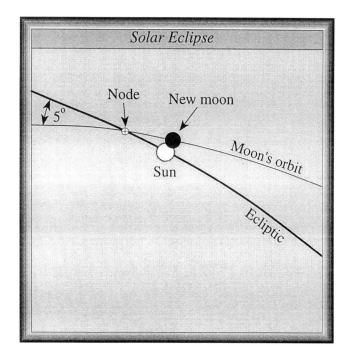

Figure 3–22 As viewed from the surface of Earth, a solar eclipse (left diagram) occurs when the new moon is at or near a node. A lunar eclipse (right diagram) occurs when the full moon is at or near a node.

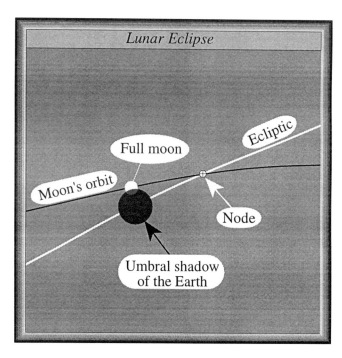

Figure 3–23 As viewed from the surface of Earth, a solar eclipse (left diagram) occurs when the new moon is at or near a node. A lunar eclipse (right diagram) occurs when the full moon is at or near a node.

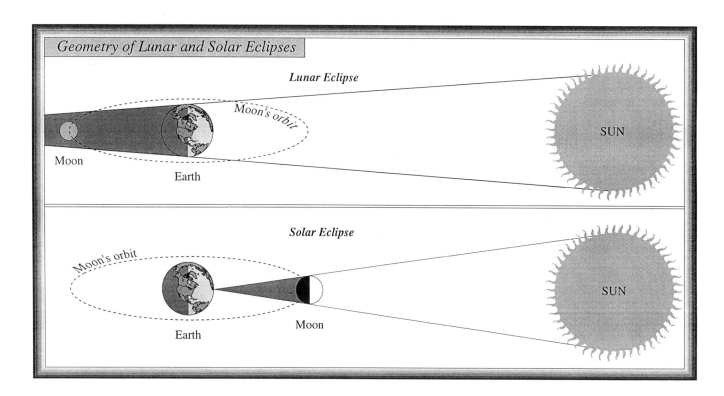

Figure 3–24 As viewed from space, a lunar eclipse (upper diagram) occurs when the full moon moves through Earth's shadow, whereas a solar eclipse (lower diagram) occurs when the new moon casts its shadow on Earth.

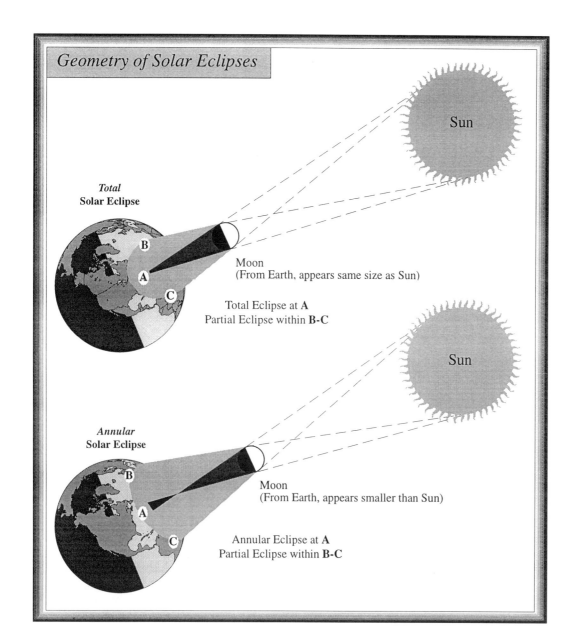

Figure 3–25 The difference between a total and annular solar eclipse arises from the distance of the new moon from Earth at the same time it is at a node. If it is at closest approach to Earth (perigee), a total solar eclipse will be visible from some location on Earth. If the new moon is farther away, its complete shadow will not fall on Earth and an annular eclipse will occur.

although much of Sun is blocked out (81% in Figure 3-27), the observer located directly beneath the tip of the umbra nonetheless experiences what seems to be normal daytime. You are not aware of what is happening unless you have been told in advance. And even then you need special equipment to actually observe what is happening. You need a telescope, binoculars, or telephoto lens with special filters attached in order to protect your eyes (and camera). Thus there is not much of an exodus of people to observe the narrow path of an annular solar eclipse.

Total Solar Eclipse One does not look through filters during the totality portion of a total solar eclipse, since sunlight is blocked out completely. You can observe totality with your naked eyes (refer to the photograph on the cover of the Book). Nevertheless, one must be extremely careful in observing such an event. You must wait until it is completely dark before looking directly at Sun, and you must also anticipate the exact moment when Moon passes Sun and daytime suddenly returns. If you are caught at that moment looking at Sun, severe eye damage can result, especially if you are observing through a telescope.

Before and after both total and annular eclipses there are partial eclipses: Sun appears to have a growing "bite" taken out of it as Moon passes in front of it. In other words, neither the total nor annular eclipses occur suddenly. Moon gradually approaches the moment when it covers the face of Sun. A partial eclipse is also observed by those located just outside the umbra during a total eclipse, or by those who are not located directly beneath the tip of the umbra during an annular eclipse. While those along a narrow path on Earth's surface are observing a total or annular solar eclipse, others outside of that path are observing a partial solar eclipse.

Total Solar Eclipses (A.D. 1994-2030)

Date	Duration of Totality (minutes)	Region(s) Where Totality is Visible
1994 Nov. 3	4.6	Chile, Brazil
1995 Oct. 24	2.4	Iran, India, Vietnam
1997 Mar. 9	2.8	Northeast Asia
1998 Feb. 26	4.4	Central America
1999 Aug. 11	2.6	Europe, India
2001 Jun. 21	4.9	Southern Africa
2002 Dec. 4	2.1	S. Africa, Australia
2003 Nov. 23	2.0	Antarctica
2005 Apr. 8	0.7	South Pacific
2006 Mar. 29	4.1	Africa, Russia
2008 Aug 1	2.4	Siberia, China
2009 Jul. 22	6.6	India, China, S. Pacific
2010 Jul. 11	5.3	South Pacific
2012 Nov. 13	4.0	Australia, S. Pacific
2013 Nov. 3	1.7	Africa
2015 Mar. 20	2.8	North Atlantic, Arctic
2016 Mar. 9	4.2	Southeast Asia, Pacific
2017 Aug. 21	2.7	United States
2019 Jul. 2	4.5	Pacific, South America
2020 Dec. 14	2.1	S. Pacific, S. America
2021 Dec. 4	1.9	Antarctica
2023 Apr. 20	1.3	Indian Ocean, Indonesia
2024 Apr. 8	4.5	S. Pacific, Mexico, East U.S.A.
2026 Aug. 12	2.3	Arctic, Greenland, N. Atlantic, Spain
2027 Aug. 2	6.4	N. Africa, Arabia, Indian Ocean
2028 Jul. 22	5.1	Indian Ocean, Australia, New Zealand
2030 Nov. 25	3.7	S. Africa, Indian Ocean, Australia

Figure 3–26 A list of total solar eclipses up until the year 2030 reveals the uniqueness of observing one from any particular location. One normally must travel to a distant location to observe the most spectacular type of eclipse. The only ones visible from the United States are highlited.

Observing a Total Solar Eclipse Since total solar eclipses are by far the most exciting type of eclipse to observe, I will elaborate on it a bit. Moon is continuously in orbit around Earth, just as Earth is continuously rotating. So when Moon's umbra reaches Earth during a total solar eclipse, the shadow projected onto Earth's surface travels at a speed of approximately 1,000 miles per hour. This shadow has the shape of an ellipse, and its size depends on the distances, and apparent sizes, of both Moon and Sun. At maximum (when Moon is exactly at perigee), it is about 170 miles wide. This is comparable in size to the shadow of a child's marble cast onto a tennis ball located 7 feet away.

As the shadow moves across Earth's surface, those fortunate enough to be within its path see Sun slowly covered by Moon. This partial phase lasts up to an hour or more, depending on where you are located with respect to the eclipse path. During most of this time, the dimming of Sunlight is hardly noticeable, and you can safely observe Sun only with proper viewing devices (such as a welder's helmet). Then, as Moon slowly slides exactly in front of Sun, the real fun begins! The sky becomes dark, and Moon's black disk looks almost like a hole in the sky, and visible around the edge of this black disk are usually several pinkish **solar prominences**, gigantic eruptions of hot gases that boil off Sun's surface in spectacular arch-shaped fashion (Figure 3–28).

At the instant just before Moon completely covers the disk of Sun, you may be treated to two bonuses. First, the last bit of sunlight that "leaks" around the edge of Moon may appear to break into "beads" as it passes through the valleys on Moon's cratered, uneven surface. This effect is known as **Bailey's Beads** after the English astronomer Francis Bailey, who first described them in 1836. They can be seen only for a few seconds, disappearing as Moon slips further into Sun's disk. Then, just as the last "bead" is visible, light from Sun's outer atmosphere begins to appear around the entire black disk of Moon, creating a glowing ring of light together with the last bright "bead" to create the **"Diamond Ring" Effect**. This too lasts only a brief moment before the entire disk of Sun is blocked by Moon and daytime gives way to nighttime.

Now it is dark, about as dark as that during a full Moon, but definitely more eerie and magical with this "black hole" in the sky. Nevertheless, the sky is dark enough for the brighter stars and planets to be seen, so that it is possible to observe the constellation that Sun is in at that particular time of the year. You experience darkness for as long as it takes the shadow to pass your particular location. Think about it. You are standing on Earth, and an elliptical-shaped shadow passes you. If you are located along the edge of the shadow, you will be in darkness for a shorter period of time than if you are located along the

Figure 3–27 The progression of an annular solar eclipse, from partial eclipse through annular and through partial again. Even during annularity, when most of Sun is blocked by Moon, the observer is in almost full daylight.

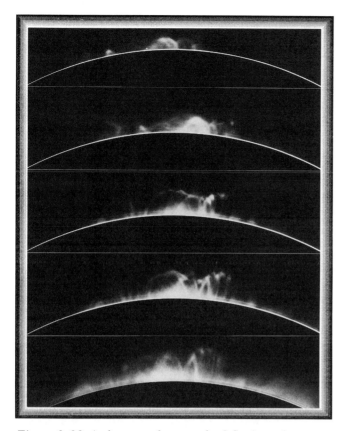

Figure 3-28 A close-up photograph of Sun's surface during a total solar eclipse reveals some of the violent activity along Sun's edge visible only at the time of such an eclipse. (Lick Observatory)

very centerline of the shadow. So it is that people who are intent on seeing such a spectacle, and travel halfway around the world to do so, are most likely to set up shop along that centerline.

The very longest that an observer can be in total darkness is about 7.5 minutes. Do I mean to say that people travel around the world to have a 7.5-minute experience! Yes, and you'll understand why if you ever make the effort to do it yourself (or you just happen to find yourself along the path by circumstance). Observers on either side of the path of the shadow — almost 2,200 miles — see only a partial eclipse. So you are thinking: what if you run fast in the same direction the shadow is moving? If you chase that 1,000 miles-per-hour shadow, you have a longer view of the eclipse, of course. Actually, some groups of people charter airplanes and chase the shadow for that reason. But it is expensive: usually only window seats and adjacent seats on one side of the airplane are occupied!

Sun during Totality The photograph on the Book's cover was taken at the total solar eclipse of 1994 in Chile, South America. Notice that although Sun is blocked out completely, there appears a bright crown of light around the dark circle of Moon. This is Sun's outer atmosphere, the **corona**, and it is the glow of these hot gases that light up

the landscape during a total solar eclipse. You see, Sun is not a solid object with a precise layer that we can call a surface. It has a fuzzy edge consisting of layers of hot gases at various temperatures and densities. What you and I call the "surface" as we look directly at Sun is called the **photosphere**. Just above that layer is a thinner but hotter layer called the **chromosphere**, and above that layer, extending for millions of miles out into space, is the thinner, even hotter region (corona) seen during the eclipse (Figure 3-29).

Neither the corona nor the chromosphere can be observed during the daytime because, being very thin, they emit less light than the light Earth's atmosphere scatters from the entire Sun. So the sky is brighter than the two layers. From Moon's surface, which of course has no atmosphere, it would be easier to observe these layers.

Another thing to get excited about during a total eclipse: if you look closely at the shape of the corona in the Book cover photograph, you notice a certain amount of structure in it, almost like streamers radiating out of Sun itself. As you will learn in Chapter 7, Sun is a huge, powerful magnet. Just as iron filings brought close to a magnet line up along its magnetic field, the hot atoms of gas making up the corona line up along Sun's magnetic field.

You will also learn that Sun goes through cycles of storm activity, and usually during any total solar eclipse some of these storms (prominences) can be observed. So, you see, there is much to be learned about Sun's cycle of activity during such an eclipse. Many professional as well as amateur astronomers carry considerable amounts of equipment to an eclipse site in order to gather important information.

It is rather unlikely that an astronomical observatory will just happen to lie along the eclipse path, so temporary ones are set up instead. The total eclipse of July 11, 1991,

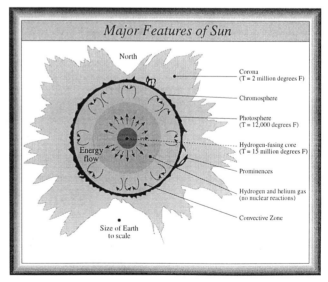

Figure 3-29 A cross-section diagram illustrates some of Sun's outer features visible during a total solar eclipse. We learn about them in detail in Chapter 7.

CHAPTER 3 / Observing Motions of Objects in the Sky   55

was an exception. The eclipse path went almost directly through the Big Island of Hawaii, at the center of which, atop 14,000-foot Mauna Kea, is one of the world's major astronomical sites. Important information about Sun was gained during that eclipse by the several telescopes located there. The map in Figure 3-30 will help you plan out your own excursion to witness one of these spectacular events.

Lunar Eclipses Generally speaking, lunar eclipses don't attract as much interest as solar eclipses. Perhaps that is because they can occur only at night, during which time people are usually in the comfort of their homes. But one type of lunar eclipse can be as interesting as that of a total solar eclipse, and can be seen by everyone on the nighttime side of Earth! A lunar eclipse can occur only during full Moon, since it is Moon passing through Earth's shadow on the side away from Sun that causes these eclipses. Moon must also be located at a node. However, just how close it is will determine which type of eclipse will occur. A list of some of the lunar eclipses to occur in the near future are shown in Figure 3-31.

Figure 3-33 illustrates the geometry of a lunar eclipse. It shows Earth's shadow in space from the side. Anything entering the inner region of the shadow (umbra) will be in darkness. Someone located anywhere within that complete shadow sees Earth entirely blocking Sun. An Earthbound observer claims someone in the umbra is in darkness. On the other hand, someone located anywhere within partial shadow (the **penumbra**) sees Earth blocking only a portion of Sun. To an observer on Earth, someone located there is in partial sunlight.

When the full Moon is at or very close to a node, it passes first through the penumbra, then through the umbra, and then out through the penumbra again. This is a **total lunar eclipse**. If the full Moon is only close to the node, it will just graze the umbra and cause only a **partial lunar eclipse**. If the full Moon is slightly farther away from the node, only a **penumbral lunar eclipse** will occur (Figure 3-32).

During the progress of a total lunar eclipse you can observe all three types of eclipses — as Moon gradually moves first through the penumbra on one side, into and through the umbra during totality, gradually out of the umbra into the penumbra on the other side of Earth's shadow, and then out of the penumbra into full sunlight. Whether Moon is at perigee or not is not important for lunar eclipses, since Moon never gets as far away from Earth as the tip of the umbra (Figure 3-32).

It seems logical that Moon should be invisible during the total part of a total lunar eclipse, but that is not the case. In fact, all observers on the nighttime side of Earth see a coppery-red Moon in the sky for the entire period during which Moon passes through the umbra. And that is spectacular! One need not travel far to observe this type of eclipse. But why does it appear red in the sky? Well, if it were not for the presence of Earth's atmosphere, Moon

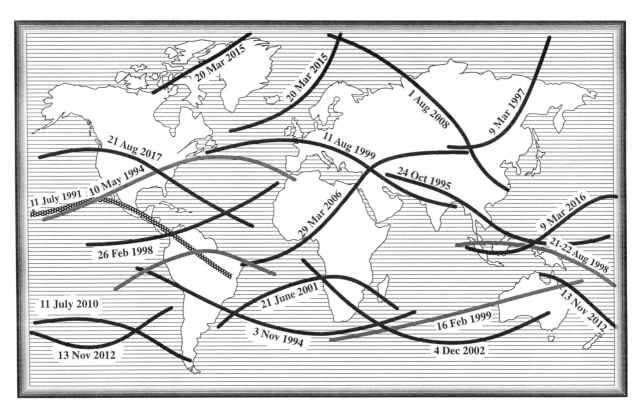

Figure 3-30 To round out your introduction to solar eclipses, this map of the more interesting future solar eclipses will assist you in your planning to observe one.

Dates/Times of Total Lunar Eclipses	
Date/Time (PST)	Length of Totality (Hours: Minutes)
1996, Apr. 4: 4 PM (**Total**)	1:26
1996, Sep. 27: 7 PM (**Total**)	1:10
1997, Mar. 24: 8:40 PM (**Partial**)	
2000, Jan. 21: 8:45 PM (**Total**)	1:16
2003, May 16: 7:41 PM (**Total**)	0:52
2004, Oct. 28: 7 PM (**Total**)	1:20
2005, Oct. 17: 4 am (**Partial**)	
2007, Aug. 28: 2:38 am (**Total**)	1:30
2008, Feb. 21: 7:27 PM (**Total**)	0:50
2010, Jun. 26: 3:40 am (**Partial**)	
2010, Dec. 21: 0:18 am (**Total**)	1:10

Figure 3–31 A list of future lunar eclipses. Lunar eclipse are much more easily seen, since everyone on the nighttime side of Earth is in a position to observe it.

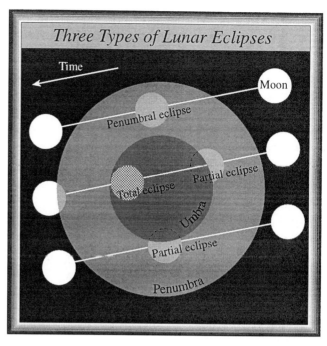

Figure 3–32 A cross–section of Earth's umbra and penumbra illustrates the geometry of the three types of lunar eclipses. How close Moon is to a node when it is Full determines whether it will be a penumbral, total, or partial eclipse. Notice that partial eclipses precede and follow a total lunar eclipse.

would disappear during a total lunar eclipse.

Earth's atmosphere acts like a miniature prism, breaking into the colors of the spectrum the light of Sun that leaks through. The red rays fall onto Moon (Figure 3-33). This is identical to the situation that occurs during sunsets. So there the full Moon is, glowing red against the black sky. At first, it seems not very distant, as if you are able to reach out and pluck it from the sky. Perhaps our brains, in determining distance, are confused by the change in color, and momentarily incorrectly interpret it as being close.

Moon Illusion This latter phenomenon is similar to an experience you have probably had yourself: your impression that the full Moon just rising in the eastern sky appears enormously larger than when it is higher in the sky. This is called the **Moon illusion**. Actually, if you take photographs of the same night's Moon both when it is close to the horizon and when it is high in the sky, you discover them to be identical in size.

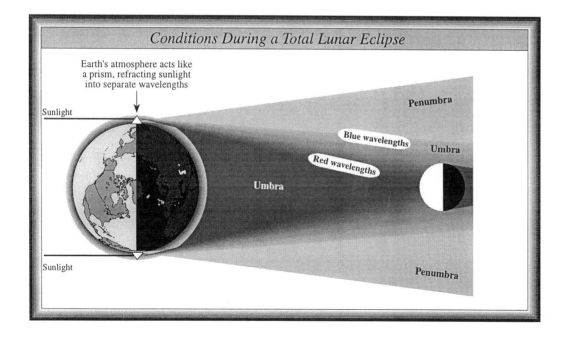

Figure 3–33 A cross–sectional view of Moon during the mid–point of a total lunar eclipse illustrates how sunlight is refracted by Earth's atmosphere so that only red wavelengths fall onto Moon. Notice the two types of shadows cast by Earth into space, and how Moon must move through the partial shadow (penumbra) prior to entering the total shadow (umbra).

★★★

Apparently the brain estimates sizes of things through comparison: in this case, by comparing the full Moon with objects it recognizes on the horizon near Moon. When Moon is high in the sky, there is nothing with which to compare Moon, so the brain says "small." There is a way, however, of temporarily confusing the brain and causing it to decide that the rising full Moon is not large after all. Look at the full Moon upside down! Apparently the brain isn't used to seeing the horizon upside down, and ignores the basis for comparison. The brain isn't always a reliable instrument, especially when dealing with things in the sky!

Ocean Tides

Now here is an interesting fact — eventually total solar eclipses will be a thing of the past, at least as far as Earth inhabitants are concerned. Moon is gradually moving away from Earth, so eventually the tip of Moon's umbra will no longer reach Earth. Annular eclipses will still occur, but not total eclipses. The reason for this leads us into the subject of ocean tides.

Most of you are familiar with tides, how they alternate between high and low. But are you familiar with the cycle of tides and its relationship to Moon's orbit around Earth? Actually, Sun is also a participant in the behavior of tides, but because it is so much more distant than Moon, its effect on tides is only one-third that of Moon. Sun is incredibly more massive than Moon, but it is also incredibly more distant. <u>Mass</u> and <u>distance</u> are the two variables determining the amount of influence one object has on another.

Gravity Earth's tides can be easily explained by using a simple yet eloquent principle of physics that expresses the amount of gravitational force between two objects. The following equation applies to concepts throughout the rest of the book, since it expresses the behavior of the basic "glue" of the universe, *gravity*:

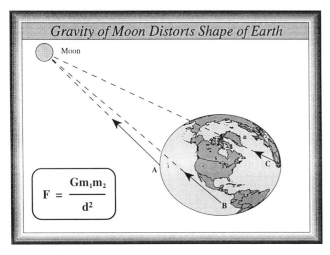

Figure 3–34 The law of gravity explains not only how Moon distorts the shape of Earth, but how ocean tides form.

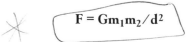

Gravity operates in a very definite way. This equation shows us how. The **F** represents the numerical value of the <u>F</u>orce of attraction between any two objects (1 and 2) in the universe whose <u>m</u>asses are m_1 and m_2. The **G** represents a <u>constant</u> of **G**ravity, a number built into the fabric of the universe. It might as well stand for **G**od, it is so constant. The **d** stands for the <u>distance</u> between the two masses. Expressed in words, the equation says:

The force of attraction between any two objects in the universe is directly proportional to the product of the masses of the two objects, and inversely proportional to the square of the distance between the objects.

The equation tells us is that the closer two objects are, the greater the force of attraction between them. And the more massive the objects are, the greater the attractive force. Simple. And this applies to every object that is made up of matter. You are not held on Earth because Earth is spinning — you are held here because you and Earth attract one another. If you place a scale between the two of you, the dial registers the force of attraction between the two of you.

Of course we call that force your **weight**. If you performed the same experiment on Moon's surface, you'd find that the force of attraction is less than that on Earth — one-sixth as much, to be exact. That is because in the equation above, the mass of Moon is less than that of Sun. The mass of your body, however, remains the same on both Moon and Earth.

It is quite easy to use the equation of gravity to explain how Moon "pulls" the oceans' waters up away from solid Earth towards it to form a tidal bulge on Earth's near side, pulls solid Earth away from oceans' waters to create a tidal bulge of water on Earth's far side, and even to stretch semisolid Earth slightly to create the slightly "squashed" shape of Earth (Figure 3-34).

Types of Tides Look at Figure 3-35. No matter where Moon is in its orbit around Earth, it pulls some water located beneath it away from Earth, forming one bulge. The equation tells us why: the water beneath Moon is closer to Moon than is Earth itself. Likewise, Moon pulls Earth away from the water on the far side of Earth to form the bulge on the far side. This is a little more difficult to conceptualize, but the equation tells us that it is so. You notice that when Sun, Moon, and Earth are lined up, at either full or new phases, the high tides are highest and low tides are lowest. These are called **spring tides**, not because they occur during the Spring season, but because the word comes from the German *springen* ("to rise up").

The equation tells us why this is so. At either new or full phases, Sun and Moon are aligned with Earth and their combined gravitational influences will be felt by the oceans. Sun and Moon are <u>both</u> attracting the water when they are lined up with Earth. Since there is only so much water in the oceans, if there are high bulges of water at one location, there must be low troughs of missing water somewhere else.

On the other hand, at the times of either quarter lunar phases, Moon and Sun are 90 degrees away from one another, and they compete, gravitationally-speaking, for the water. Moon wins out, and so the high tides are found beneath Moon. The not-so-low low tides are found beneath Sun. These are called **neap tides**. High neap tides are lower than spring high tides, and low neap tides are higher than low spring tides. During the two week interval between spring and neap tides, observers notice high tides gradually getting lower, and the lower tides gradually getting higher.

Now visualize this from the vantage point of a person standing on Earth's surface. This is how you experience the cycle of tides. Since Earth rotates faster than Moon revolves around Earth, you are taken alternately through two high tides and two low tides during every 24-hour period. The tidal bulges remain under Moon as Earth rotates beneath them. The water itself travels along with Earth's surface, but the bulges are fixed in position relative to Moon.

If you combine the tidal cycles in Figure 3-35 with what you learned about telling time by Moon (Figure 3-16), you can tell at what time, approximately, high and low tides will occur at the beach. To do so, you must observe Moon's phase and position in the sky. An easy way to visualize the method is to remember that when Moon is high in the sky, the water is high. When Moon is low (close to the horizon), the water is low.

Try this thought exercise. What phase of Moon and at what time of the day/night would you visit the beach in order to explore tide pools for marine life?

You may be surprised to learn that the same gravitational principle that causes ocean tides causes land tides as well. That is, when Moon is crossing your meridian high in the sky, it is pulling the crust of Earth up about one inch (and you along with it)! This has nothing whatsoever to do with earthquakes, by the way. Earth's crust behaves somewhat like asphalt, and adjusts to the slight distortions due to Moon's pull.

Extreme Tides If you combine what you just learned about tides with what you learned earlier about the shape of Moon's orbit around Earth, you can imagine that there is also a pattern in the strengths (heights) of spring tides. Spring high tides and spring low tides are not the same every month. Both Sun's and Moon's distances from Earth are additional factors that affect the tides. If Moon is at perigee at the same time it is new or full, the spring tides are unusually high and especially low. These are called **proxigean** tides. These tides are of special concern to people who live along coastlines, especially if they occur during the winter season during which time strong winds can add to the tides to cause flooding problems.

Slowing of Earth's Rotation As Earth rotates beneath the tidal bulges, there is a certain amount of friction between Earth's surface and the water. This is similar to the effect of spinning a cup of coffee. Notice that the coffee in contact with the side of the cup begins to move first, and gradually friction between the water molecules causes all of the coffee to spin along with the cup.

In the case of Earth (lowest portion of Figure 3-35), the tidal bulges are carried forward a slight amount by friction, so that instead of being precisely beneath Moon they are slightly out of alignment. An observer at the beach, for example, notices that the high tide arrives shortly after the full Moon has crossed the meridian, not when it is crossing. Looking at this misalignment, you notice that Moon is closer to one bulge than to the other. The equation tells us that Moon gravitationally pulls that nearer bulge with a stronger force than the more distant bulge. This differential attraction slows Earth's rotation, as if Moon were applying brakes to Earth.

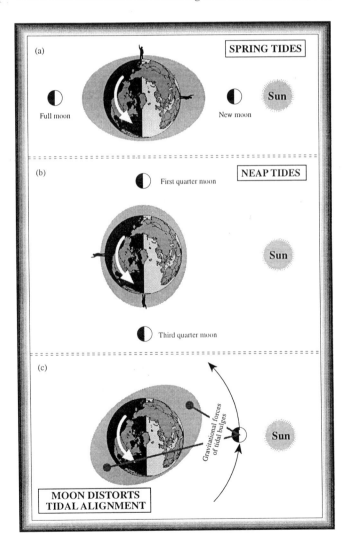

Figure 3-35 Spring tides occur during full and new moons, and neap tides occur at the two quarter phases.

★★★

A principle of physics (**conservation of angular momentum**) says that if something slows down and thereby loses energy, something else must absorb that same energy. Moon gets that energy lost by Earth as it slows down. Adding energy to Moon makes it speed up in its orbit around Earth, just as adding energy to a rocketship makes it leave and eventually circle Earth. This in turn causes Moon to move farther and farther away from Earth. And as the length of Earth's day gets greater and greater, the length of the month gets greater and greater.

Accurate measurements of Earth-Moon average distance reveals that this is actually happening, although the change is so small (1 inch per year), that it is not something of immediate concern. The length of the day is increasing by only about 1 second every 1,000 years. Although this is an extremely small amount, considered over the long time period during which Earth and Moon have existed, it is an important effect. According to a study of certain marine organisms that multiply according to the day-night cycle, the length of the day only 400 million years ago was about 22 hours. What do you think the length of the day was 5 billion years ago, when Earth and Moon had just formed?

Another consequence of Moon getting farther and farther away from Earth is that eventually total solar eclipses will not be visible from Earth's surface. Annular eclipses will still be observable, but when Moon moves so far away from Earth that the tip of the umbra no longer touches Earth, there will no longer be any expeditions to remote regions of Earth to witness that spectacular event. But don't worry, that won't be for another 50,000 years! The recession of Moon from Earth is quite slow.

Astronomy vs Astrology

As a method of pulling all of the concepts of this Chapter together, I now offer a practical application of what is known about the movements of objects in the sky. Since there has been considerable media attention given to the subject of astrology, I feel obliged to clarify the distinction between astronomy and astrology, and perhaps offer insight into the basis for the conflict between the two.

For centuries people attempted to make some sense of their existence, and to explain just how they are connected to the universe. The most obvious residual today of this tendency to connect *outer* and *inner* worlds is that of astrology: the belief and practice that alignments of celestial objects at the moment of birth influence each person and, in fact, that the changing alignments continue to influence all of us throughout our lives.

Newspaper Astrology The astrological column in the newspaper is a carry over from the distant past, at which time rational explanations for the forces of nature were unavailable. The astrological column is divided into 12 equal time intervals corresponding approximately to the intervals during which Sun is <u>supposedly</u> in each of the 12 constellations of the zodiac. So if you are born on October 2nd, as I was, then Sun is located *in* (the direction of) the constellation *Libra*, and your personality is thereby *influenced* by the traits associated with the "sign" of the *Scales*.

According to astrology, Sun enters *Libra* on September 23rd, the autumnal equinox, at which time not only are the days and nights *balanced* (equal), but the seasons are *balanced*. Sun is then about to move beneath the celestial equator for a six-month period, during which time cold and rain and (perhaps) snow will make the growing of most crops impossible. Thus a person born as a Libra tends to be a *balanced* person, always willing to listen to both sides of an argument, never going to one extreme or the other. We supposedly make good lawyers and teachers.

The professional astrologer does not generally care for the morning newspaper treatment of astrology, inasmuch as it reflects only Sun's placement in a constellation, which is only one-seventh of what is up there to do the influencing (actually 10, since we have only recently discovered Uranus, Neptune and Pluto). So if you decide to have your *chart* made by an astrologer, you have a map of the sky as it appeared at the moment of your birth (*not at conception*), together with the orientations of Sun, Moon, and planets with respect to one another. At this point, you are like a person with a cup of tea leaves. You need someone who can read the symbols (planets, Sun, Moon, or tea leaves) to tell you what they mean in terms of their influences upon you. Astrologers offer that service.

I compare this widely-practiced activity with the actual sky and the recent discoveries of astronomy. I do this, I would like to emphasize, not to be critical of the practice of astrology (poking fun is not the same as being critical!). I do it to make clear the distinctions between the two activities of astronomy and astrology, and how they differ in their approach to explaining the universe around us. You may disagree with the conclusions of astronomy and even be critical of its methods, but that is your choice. In any case, by comparing the two approaches you may learn something important — especially if you can identify the assumptions that scientists use in going about their work.

Astrology and astronomy evolved together, being inseparable when they sprang from the common need to organize the sky in order to conduct agriculture and explain the *mysteries* around them. But along the way, at some point in the evolution of human thought, the two became divorced and went their separate ways. The only thing they have in common now are memories of their common language — names of constellations and planets, the lines used in organizing the sky, and so on.

Scientific Objections to Astrology

- *Objection 1* The effect of precession invalidates any claim that Sun is physically located in the "sign" noted in the astrological column of the newspaper.

Explanation See the discussion just above.

- *Objection 2 Sun does not spend equal periods of time in each constellation, since each constellation is of a different size. So even if precession were not in effect, the astrological column would still be incorrect often.*

Explanation Looking at a rather typical drawing of the star patterns recorded by early peoples (Figure 3–36), it is quite obvious that they did not establish exact boundaries between constellations. When an object (like Sun) appeared somewhere between two patterns adopted as zodiacal constellations, it was a matter of opinion as to which it was actually in. It really didn't make a lot of difference anyway, since the ecliptic was divided equally into 12 segments without regard for the exact sizes of the star patterns.

If you make note of the sizes of the figures for *Libra* and *Virgo* in the same drawing, you notice that Virgo is much larger than *Libra*, and yet according to astrology Sun spends the same amount of time in each of the 12 constellations. Of course, this cannot be the case, inasmuch as Sun moves at a rather uniform rate along the ecliptic. This means that Sun, even to early observers, was not necessarily in the direction of the pattern of stars associated with that date.

- *Objection 3 There is no known force that accounts for celestial objects influencing humans on Earth.*

Explanation It is popular to cite the example of the tides as a possible explanation of celestial influences on humans. If Moon can cause tides on Earth, why not on the fluids within us? It is rather straightforward to apply the equation of gravity to that possibility. It turns out that the gravitational effect that the doctor had on you when you were delivered was greater than the combined effect of all celestial bodies! That is because they are all so very far away, and the doctor, although certainly not very massive, was very close.

- *Objection 4 There are actually 13 "signs" of the zodiac, not 12 as astrologers say.*

Explanation Astrologers have never adopted precise boundaries between constellations. If you accept the boundaries adopted by astronomers in 1930, there are actually 13 constellations that lie along the ecliptic (Figure 3–36). So people born between December 9 and 17 are born under the "sign" of *Ophiuchus*. This is not shown in the newspaper column.

- *Objection 5 Modern astrologers have integrated a few new discoveries into the chart-making scheme (Uranus, Neptune, and Pluto), but have left out all others.*

Explanation Early astrologers based their predictions on patterns of stars and the seven wanderers in the sky: Moon, Sun, and planets. As additional planets were discovered, they too were integrated into the chart-making scheme. But modern astronomers have discovered hundreds of asteroids, comets, pulsars, quasars, and nebulae: not to mention billions of galaxies that inhabit the known universe. Why are they not also included in astrologers' charts, if in fact

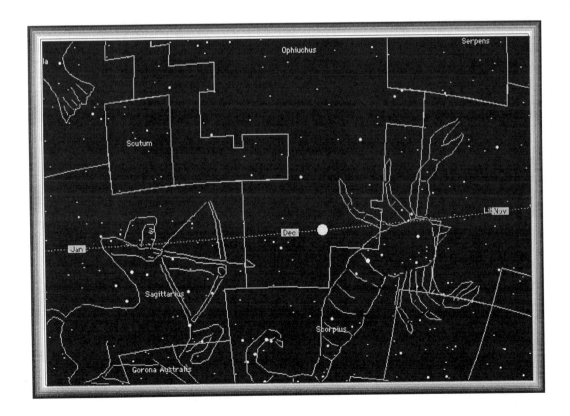

Figure 3–36 A portion of a modern star chart shows the location of Sun along the ecliptic during the middle of December. Using the boundaries between constellations recognized by astronomers, Sun is in the constellation of Ophiuchus during the period November 29th–December 18th. You will not find the name of that constellation on astrologers' charts. (Carina Software)

celestial bodies influence us in one way or another? Astrologers do not incorporate new scientific discoveries into their scheme, preferring a rigid and fixed system.

- *Objection 6 The patterns of the constellations, which presumedly influence us, are arbitrarily based only on the visible radiation coming from the stars.*

Explanation Early astrologers did not even know what stars were, let alone that they each give off different types of radiation in different amounts. So if our eyesights were tuned to infrared or ultraviolet wavelengths of radiation, the patterns of the constellations would appear entirely different to us. If the basic principle of astrology is that we are influenced in some way by celestial bodies, it seems strange that we are influenced only by the patterns formed by the visible wavelengths of light.

- *Objection 7 The stars in a given constellation are not fixed in the sky, and therefore the patterns of the constellations are gradually changing.*

Explanation If the characteristics of the figure in a constellation are also shared by the person born under that "sign," and the shape of the figure is gradually changing, then it is not clear what happens to the interpretation of the "signs" when the figures change. Astrology operates on the premise that there are 12 and only 12 "signs," and that their characteristics are permanent.

- *Objection 8 Different cultures visualize different figures for the stars, and therefore visualize entirely different constellations. This is significantly true for stars along the ecliptic which form constellations of the zodiac.*

Explanation Two children of different cultural backgrounds born at exactly the same moment in the same hospital have different "signs," and therefore have significantly different life patterns. Conventional astrology argues that their patterns should be similar.

- *Objection 9 The practice of astrology developed at a time when the only objects observed in the sky were stars, Sun, and Moon. Astrologers limit their charts, and therefore prediction of influences, to only those objects.*

Explanation Since 1930, astronomers have discovered billions of other galaxies, exploding stars, black holes, asteroids, thousands of comets, and so on and so on and so on. If humans are influenced by celestial objects, why do those newly-discovered objects not figure into the process?

- *Objection 10 The author has "identical" twin daughters who were born 6 minutes apart. I am not convinced that they have similar life patterns.*

Explanation There are no profound celestial changes within a 6-minute period of time. Therefore, identical twins have virtually the same astrological chart, and should therefore have similar life patterns.

Astrologers Respond It would be unfair of me to leave you at this point without offering some arguments by astrologers as to these observations and suggestions that they deal with a system which is no longer valid. Well, you see, the astrologer claims that the physical positions of objects in the sky really have nothing to do with what they are trying to teach in the first place.

The celestial objects are only symbols of a *reality* or *truth* that exists beyond the objects themselves, in the same way that tea leaves are symbols. Someone (the astrologer, the tea leaf reader) must look at the patterns and read the message contained therein. The astrologer simply uses some of the same terms that astronomers use. We differ in the definitions assigned to those terms.

So in the final analysis, one chooses to believe either those events which are shown to have some physical or numerical association with the rest of the universe (cause and effect), or those that invoke unknown forces operating behind symbols that nature provides in the form of physical objects and their movements (effect without cause, at least a quantitative cause). If astrology works for you in helping you understand yourself, furthering your success in achieving your goals, then you will no doubt use it as a tool. Only when it fails or prevents you from achieving your goals will you change direction.

Summary/Conclusion

Since earliest times people have been fascinated with the sky. It was something that not only allowed them to understand and predict the changing seasons and other cycles of nature, but it gave them a springboard from which to leap into a world view that integrated humans into a close identity with the universe. What has changed is the additional information that we have with which to interpret what we see.

But we still search for integration into the scheme of things. The movements of the stars, planets, and Moon allow us to organize the sky in such a way as to proceed in an orderly fashion. Being able to predict with great accuracy just when future eclipses will occur is a reminder to us of the laws of physics that govern the universe. We will insist on using such laws as we continue our evaluation of the possibility that life out there is also fascinated with the sky.

LEARNING OBJECTIVES: *Now that you have studied this Chapter, you should be able to:*

1. Describe the motions of Sun, Moon, planets, and stars as a result of Earth's rotation.
2. Describe the motions of Sun, Moon, and planets with respect to the zodiac as a result of Earth's revolution around Sun.
3. Describe the seasonal motion of Sun at Sunrise, noon, and Sunset relative to the horizon for an observer in mid-latitudes.
4. Describe the seasonal motion of Sun for an observer at the north pole and the equator.
5. State the astronomical event and/or cycle from which the day, week, month, and year are determined.
6. Describe the conditions necessary for the 6 types of eclipses to occur, where an observer must be located in order to witness each, and how each will appear.
7. Describe the phases of Moon in terms of the orientations of Earth, Moon, and Sun.
8. Determine the time of day or night by observing the Moon of a particular phase in a given direction in the sky.
9. Describe the orientation of Earth, Moon, and Sun necessary for spring and neap tides, and where a person must be to experience low and high tides of each type.
10. Explain how the ocean tides slow the rotation of Earth and cause Moon to recede from Earth.
11. Explain the difference between astronomy and astrology.
12. Explain the cause of precession, and the effects of precession as regards the positions of the vernal equinox and the north celestial pole on the celestial sphere.
13. Define and use in a complete sentence each of the following **NEW TERMS**:

Age of Aquarius 48
Age of Moon 45
Annular eclipse 50
Autumnal equinox 41
Bailey's beads 54
Celestial (North, South) pole 36
Chromosphere 55
Circumpolar 37
Conservation of angular momentum 59
Corona 55
Diamond ring effect 54
Ecliptic 39
Equinox 41
Foucault pendulum 36
Land of the Midnight Sun 42
Meridian 39
Moon illusion 57
Neap Tide 58
Node 47
North star 37
Partial solar (or lunar) eclipse 51
Penumbra 56
Penumbral lunar eclipse 56
Perigee 49

Perihelion 41
Phases 45
Photosphere 55
Polaris 37
Precession 47
Proxigean tides 59
Revolution 36
Rotation 35
Rotational axis 36
Sextant 38
Solar prominence 54
Spring (vernal) equinox 41
Spring Tide 58
Summer solstice 41
Total eclipse 50
Total lunar eclipse 56
Umbra 50
Vega 48
Waning Moon 45
Waxing Moon 45
Weight 58
Winter solstice 41
Zenith 38

Chapter 4

The Birth and Growth of Modern Science

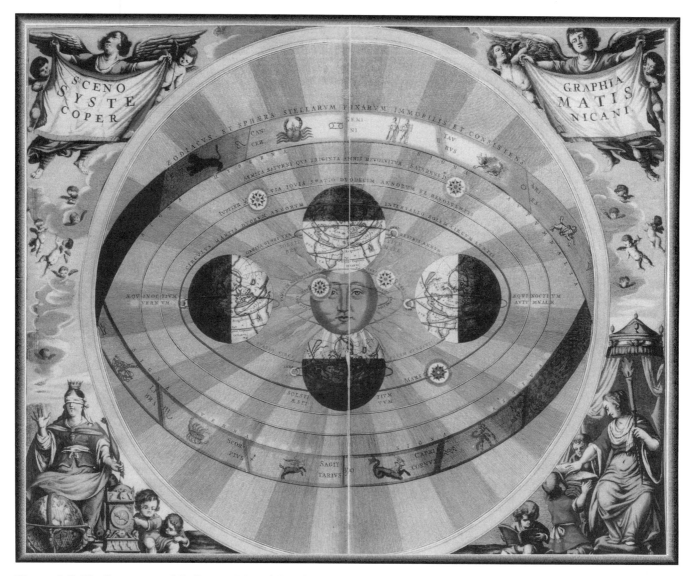

Figure 4–1 The Sun–centered (heliocentric) model of the universe is illustrated in this old manuscript. Notice that only five planets out to Saturn were known at the time. (Yerkes Observatory)

CENTRAL THEME: How did debate between the believers of Earth-centered and Sun-centered universe theories—resulting in the trial of Galileo—lead to the rise of modern science and its methods of obtaining "truth" about the universe?

It probably did not occur to you as you were reading the first few Chapters to question my claims that Earth spins, that Earth orbits around Sun, or even that Moon orbits Earth. These are such well-accepted doctrines of modern science, so embedded in our thinking, that few people argue for the contrary. There was a time, however, when we were quite young, that we thought differently. As a youngster, I thought Earth was quite stationary — that objects (like Sun) were actually moving in the sky. Try explaining to a 4- or 5-year-old child that Earth moves around Sun. You'll find what I mean. It is an abstraction to them. Their brains have not yet developed to the point of dealing with spatial concepts such as relative motion.

Scientific Models Neither, as you read the first few Chapters, did you question my use of diagrams to explain the behavior of some objects in space. Diagrams such as Figure 4-1 are what scientists call **models**, and are used to help us as well as themselves visualize and understand the positions and motions of objects we observe in the sky. Scientific models are not limited to diagrams, however. They take the form of mathematical equations, chemical formulae, geometrical shapes, and even actual craft work. It is important to keep in mind when relating to a scientific model that the model is not the thing in nature itself — it represents the thing in nature.

A young child has difficulty relating a model to the real thing. The model has one reality — the real thing another reality. The doll and model car are realities to the child, and in no way represent the real things. As science gets more and more complicated — dealing with more and more things that cannot be seen directly — it resorts more and more to models in the attempt to explain the universe around us. Accepting a model as an accurate representation of the real thing can then become a matter of serious debate. It was during the debate between two models explaining the motions of Sun, Moon, and planets that modern science had its accelerated beginnings.

There is an even larger issue at stake when debates occur between competing theories — exactly what criteria do we use in evaluating evidence before we are willing to accept and use it in explaining the universe and its contents?. In the Preface of the Book, I teased you with the proposal that perhaps there is a connection between blind acceptance of scientific theories on the one hand and social unrest and confusion about what we are doing here on the other hand. If that is so, then we might want to find an alternative to the scientific view of how life got here in the first place, and hope that it has something more specific to say about the meaning of life and what our role on the planet is.

I certainly don't have a "meaning of life" for you, but I can tell you how we came to adopt the scientific method in the first place, and how it evolved out of a system of belief that lacked an explanation for our existence. I'm certainly not going to recommend that we return to that ancient system of thought or practice. The point is that you and I are imbedded in a system of belief that is based on the scientific method whether we like it or not, and that by studying the origin of that method we can see its strengths and limitations. With that as a foundation, we are in a stronger position to evaluate our own criteria for accepting evidence for what we chose to believe.

Most early civilizations believed that Earth was stationary and that everything in the sky revolved around it. It is as if reasoning ability and the ability to deal with abstract concepts in the growing child is analogous to the growth of those same mental skills by civilizations in general. In other words, the mind of a young child today is representative of the general thinking that was possible in early civilizations. It is this change in thinking that I am interested in addressing in this Chapter. Why? Because if what I just said is true, then we are headed for significant change in our thinking about things in the future. If we want to be on the leading edge of major change in our thinking, it behooves us to be able to recognize the advanced warning signs so that we are prepared for consequences of the change. This leads to my major thesis for this Chapter:

> **Thesis**: *A person's belief system about life is influenced in part by his/her view about the physical world and its behavior. As that view changes, so also the belief system changes.*

A couple of examples can illustrate my point. I maintain that if I offered to you conclusive evidence that superior intelligent beings lived on a planet around the nearest star beyond Sun, your belief system about your life on Earth would change significantly. If scientists offered irrefutable evidence that primitive life forms were discovered in soil samples returned from Mars, your beliefs about life on Earth would be profoundly influenced. If scientists discovered the chemical that is responsible for the dying process, your beliefs about life on Earth would change.

Worldview

Every civilization, past and present, has a way of looking at the world around it. That view includes not only the external world *out there* but also the world *within* — and especially how one world affects the other world. In other words, civilizations through culture express beliefs not only about the relevance of stars and planets in the sky, but also about the nature and purpose of human existence, the nature of reality, the meaning of life, and so on. We refer to this set of beliefs about these worlds of inside and outside a **Worldview**.

It is a model of reality — different from a scientific model in that it is not limited to what is observable and

★★★★

measurable. The value of having a Worldview is that it helps people make sense of their own personal experiences in the world. It may not be spelled out in exact detail and distributed to all members of the larger society, but it is reflected in the culture — language, customs, art, music, literature, and so forth.

Over time, the Worldview evolves and changes, usually the result of the infusion of new ideas. At any particular time there are opposing views to the dominant one, of course, and often one of these gains enough acceptance to overthrow or replace the popular view. This change can be slow and gradual, or involve turmoil, revolt, and even violence. There seems to be a tendency to resist changes to the Worldview, to want to hold on to the old ways of thinking in spite of what seems like overwhelming evidence to the contrary. But what I emphasize here is that our search for a Worldview is an attempt to find a purpose for existence and that it necessarily includes our current understanding of the physical universe around us. Above all, it must allow us to see our lives as being meaningful.

Modern Science as a Worldview What I want to do in this Chapter is trace the origin of a major component of the current prevalent Worldview:

> *How it is that modern science as a form of inquiry has been so overwhelmingly accepted in our modern world? How did we get to the point of speaking in terms of atoms and molecules, space and time, force and momentum, energy and radiation?*

In answering these questions, I focus on the major changes that occurred in the *manner* in which we accept evidence for deciding whether a particular idea is true or not. The connection we make between the modern view of the physical world around us and our own personal *meaning of life* is not always clear or obvious, either to ourselves or others. After all, we are surrounded by the Worldview of our culture, and we are products of its teachings. A Zen Buddhist saying goes something like this:

> *"The fish is the last to know that it lives in water, simply because it is surrounded by it."*

So we are about to look at the history of modern science from our own prejudiced frames of minds, since modern science (as a thought process) has already gotten a foothold and has already conditioned your thinking (and mine). But let us try to clear as much prejudice from our minds as possible and imagine ourselves living at the time of the revolution that gave birth to modern science.

Early Cosmologies Figure 4-2 is a photograph of an old Chinese coin. The roundness of its shape is not surprising, but the square hole in the center seems odd. The early Chinese, like other civilizations, concluded that the uni-

Figure 4–2 Old Chinese coins reflect the cosmology of ancient Chinese cultures. The square represents the four corners of Earth, located at the center of the universe. The circular outside edge represents the celestial sphere that surrounds Earth. (Tom Bullock)

verse above their heads was eternal, unchanging, and predictable. The roundness of the coin symbolized their belief in this world of perfection, the smoothly turning celestial sphere.

Earth, on the other hand, has none of these attributes. There is only disease, war, famine, storm, and death. And yet it seems to be located at the center of the universe. Sun, Moon, planets, and even stars seem to be the objects moving, not the huge, sluggish Earth. The square hole of the coin represents Earth at the center of the universe, the four corners representing the four cardinal directions on Earth — north, south, east, west.

At the center of the sky is *Polaris*, the pole star, around which the sky turns. At the center of Earth is China. The Chinese believed, as did many civilizations, that for each event in the sky there is a corresponding event on Earth. Here are the two Chinese characters that spell the word "China":

66 SECTION 1 / Observing the Sky

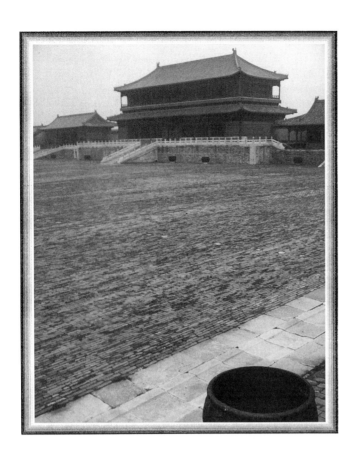

Figure 4–3 The Emperor's Palace in the Forbidden City in Beijing, China, was claimed to be located at the center of Earth. Since Earth was believed to be located at the center of the universe, the Emperor could claim that his rulership was divine. (Tom Bullock)

The characters translate as Middle *Kingdom*. Near the center of the capital of Beijing in China is the remarkable Forbidden City. It was constructed along a north-south line, as if it were a miniature Earth. At the center was the Emperor's Palace, a place so sacred that only the most noble members of his family were allowed to enter (Figure 4-3).

Polaris, around which all the stars in the sky appeared to rotate, was the equivalent of the Emperor's Palace. So if stars rotate around *Polaris*, and his Palace corresponds to *Polaris* in the sky, then people in China must rotate around (obey) the Emperor. He was considered to be God on Earth, to be strictly obeyed without reservation or hesitation. To look into his eyes directly was punishable by death. Such is the power that a ruler has (and some still do) exerted over the lives of others once they are led to believe that the power of the ruler derives from an arrangement of stars, writings considered "sacred," or any other number of claims. The coin represents a Worldview that explains the

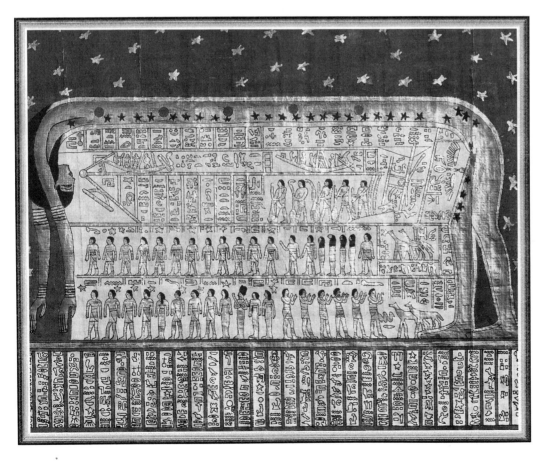

Figure 4–4 This painting on papyrus was copied from the ceiling of the tomb of Tutankhamun in the Valley of the Kings in Egypt. It represents the belief that the goddess Nut was responsible for holding up the sky. (Tom Bullock)

CHAPTER 4 / The Birth and Growth of Modern Science

origin and structure of the universe, and indeed the purpose of humans in it. The field of study which develops models with which to explain the origin and structure of the universe is **cosmology**.

In ancient Egypt, it was the goddess Nut who holds up the sky and within whose body the stars were thought to be embedded (Figure 4-4).

But least you conclude that only civilizations in "less-developed" countries adopted Worldviews of Earth-centered-ness, consider those societies in Europe living around the Mediterranean Sea. What does the word "Mediterranean" mean? Yes, *middle of Earth* ("Medi" = middle, "terranean" = Earth). Anyone living elsewhere – so it was believed – was a Barbarian. Jerusalem was thought to be the center of the universe by the early Christians. The Papago Indians of the American Southwest thought that a local mountain was the point around which the entire universe rotated. And the list goes on and on as we explore the roots of civilizations around the World.

Monuments to Earth Centered-ness Frequently, civilizations combined the spiritual with the practical. You are familiar with the arrangement of stones at *Stonehenge* in England. It is thought to represent a giant astronomical computer for predicting important times of the year for the purpose of agriculture and/or religious ceremonies. In the Yucatan Peninsula of Mexico lies the impressive stone pyramid of Chitza Itza, built by the mighty Mayan civilization that once thrived through much of the region. Each March 21st (Spring Equinox), thousands of people gather around the structure just before sunset to observe a light spectacle called "The Serpent Descending" (Figure 4-5).

Apparently the pyramid was constructed to make people take notice of or to celebrate the astronomical event during which Sun crosses the celestial equator on its way northward. As Sun sets lower and lower toward the horizon on that one day of the year, the stepped westward edge of the pyramid casts a shadow that resembles the sacred Mayan serpent onto the face of the stone walkway that goes up to the top of the structure. At the base of the steps is a stone heads of a serpent. So the entire effect is one of the appearance of a serpent descending from the top of the pyramid (at which location the Mayan priests performed their bloody sacrifices).

The Birth of Modern Science

This is not a comprehensive history of science throughout the centuries. I have chosen a particular interval of time and a particular event from within that interval that illustrates the upheavals that may occur during the transition from belief in one Worldview to adoption of a new one. At the same time, I illustrate the manner in which science works, and why it has become such a prominent part of our current view. The particular event I chose to use in my illustration is the trial of **Galileo**. Since he is generally regarded as being the Father of Modern Science, the choice is appropriate. He lived from 1564 to 1642, during which time Europe was going through monumental changes.

As you read this, visualize yourself a member of the jury that tried Galileo in 1633. Try to be as impartial as possible as you consider the charges against him, and to weigh carefully the evidence offered to support those charges and the evidence that Galileo offered on his own behalf. Try also to identify with the spirit of the time, to incorporate into your thinking the belief systems they had. To be able to do the latter, of course, you will have to be aware of your own preconceptions, those beliefs you already possess about the source of human knowledge. If you are able to identify with the mind set of the time, I shouldn't be surprised at all if you admit that you too would have voted with the Church to convict him.

Modern science has its origins in that historical period between the 14th and 16th centuries, a period we call the **Renaissance**. The word itself suggests some kind of upheaval in thought. In French, "*naissance*" means birth. So there was a **re**birth of sorts during this period. Of course, the revolution that resulted in the widespread use of science as a method of inquiry began in very small ways centuries prior to the Renaissance. One may even say that the revolution is still with us today, inasmuch as sizeable

Figure 4-5 In the Yucatan of Mexico is a Mayan pyramid that was apparently constructed in order to display a "descending serpent" with light from the setting Sun on the occasion of the two equinoxes. Thousands of people go there every year to witness the event. (Tom Bullock)

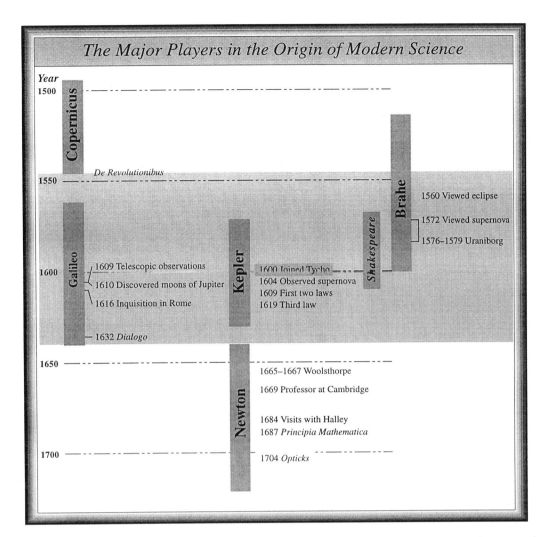

Figure 4-6 It was the 99-year period between the death of Copernicus (1543) and the birth of Newton (1642) that marked the transition between the ancient Earth-centered theory of Ptolemy and Aristotle and the Sun-centered theory of Copernicus. This also marked the transition between "natural philosophy" and modern science. The personalities that participated in this transition and some of the other historical events that were going on at the time are shown on this historical timeline.

portions of humanity have yet to adopt the methods of modern science. Figure 4-6 is a list of some of the personalities and events that occurred during the interval of the Galileo-Church confrontation. You may find it useful to refer back to it while reading the various episodes of the trial as you continue reading the Chapter.

Greek Astronomy The dominant Worldview held in Europe before and during Galileo's time had its roots in earlier Greek thought. Mostly it derived from the ideas of **Aristotle** almost 2,000 years earlier. Aristotle was a Greek thinker who opened up vast new areas of human knowledge in such fields as philosophy, history, politics, drama, and ethics, as well as astronomy. In his ideas of physics and chemistry and astronomy, of course, he was limited to knowledge obtained by his senses — mostly by his eyes. His observations of the sky caused him to conclude what most early thinkers concluded — the universe is centered on Earth, all things moving around Earth in what appear to be perfect circles.

To the Greeks, the circle was a symbol of perfection — the perfect shape. Since the stars, planets, Sun, and Moon do not show signs of change or decay, and move so effortlessly and with such precision across the sky, it was natural to assume they must dwell in a region of perfection and even obey perfect patterns of movement. The fact that Sun and full Moon both appear as perfect circles no doubt contributed to this belief.

In Chapter 16, we will find that the one object in the sky that obviously did not conform to the pattern of the perfect circle, the comet, was interpreted to be a messenger of doom and despair and destruction. This idea persists even today. To Aristotle, the predictability of the appearance and movements of objects in the sky could best be interpreted as being due to the fact that the heavens are not only made up of a different kind of substance than those found on Earth, but obeyed different laws than those working on Earth. It didn't strain his brain to conclude that events on Earth lacked the predictability, permanence, and smoothness of movement that events in the sky exhibited.

So Aristotle was not only dealing with the subject matter of astronomy, but those of physics, chemistry, geology, and others as well. His authority on these subjects was very great, no doubt because he effectively integrated them into a unified system that explained everything. So great, in fact, that the Catholic Church recognized them as the official Worldview up until the Renaissance, 2,000 years after his death! It took Galileo to question that great authority and prove evidence contradictory to his Worldview.

The Church did not adopt this Worldview all of a sudden. After all, the Christian Church did not even exist during Aristotle's life. So obviously Aristotle did not have anything to say about the truthfulness of Christianity. Only later, after Aristotle's death, did the Church slowly integrate the ideas of Aristotle into its theology. The ideas of Aristotle were easily tailored to suit the Church's need of evidence to support the idea of the existence of a Perfect and Powerful God Who lived in a Perfect Heaven far removed from the imperfect and ever-changing Earth.

The Church had, by the fifteenth century, unofficially adopted the philosophical system of Aristotle. Acceptance was not direct, but via the writings of the great Church thinker, **Saint Thomas Aquinas**. He had found that many of the Aristotelian concepts such as matter and form, nature, the process of knowledge and others were excellent vehicles for presenting the teachings of the Church. Although Saint Thomas himself was careful to warn readers against the temptation to dogmatize in philosophy, by the time of Galileo a stubborn dedication to the status quo had set in. Professors who ruled the universities and Churchmen who felt that all of Aristotle's philosophy had to be protected as the quasi-official doctrine of the Church resisted new ideas. By the time of Galileo, the Church (in southern Europe, at least) had established exclusive jurisdiction over the rites of passage of humans from the cradle to the grave.

The Ptolemaic System Actually, the astronomical system of Aristotle was not adopted by the Church directly, but through the work of **Ptolemy**. This mathematician lived and worked about 140 AD in Alexandria, Egypt, which at the time was a Greek colony. He developed a model of the universe based on the ideas of Aristotle, so the model is referred to as the **Ptolemaic system**. The basic idea of the *Ptolemaic system* is that Earth is located at the center of the universe, with all movements of Moon, Sun and planets

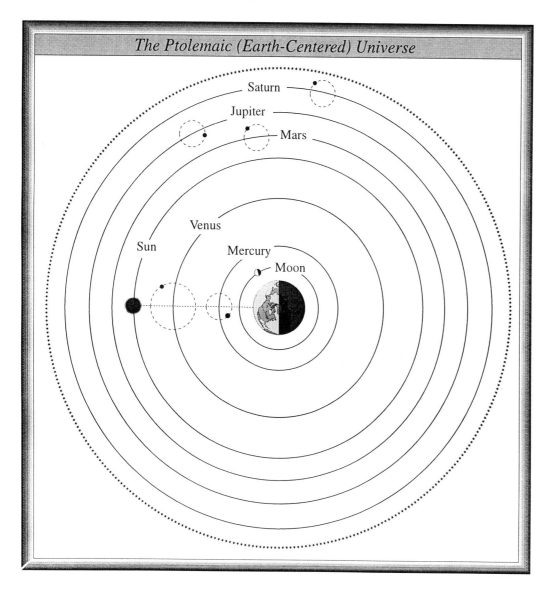

Figure 4–7 The Ptolemaic system was an Earth–centered theory that attempted to explain the motions of the seven "wandering" objects in the sky, as well as the sphere of fixed stars that moved as one complete unit. Notice that the five known planets were thought to move on small orbits (epicycles), the centers of which moved around Sun (deferent).

taking place around it. We also call this the **geocentric system**, since it places Earth at the center.

Figure 4-7 illustrates the Ptolemaic system as seen from the outside, a God's-eye view of the universe. You probably notice something odd about it, in addition to Earth being located at the center. The planets, instead of going directly around Earth, orbit around small orbital circles (**epicycles**), the centers of which go around Earth on a large orbit, the **deferent**. This complicated system was Ptolemy's way of explaining the motions of planets seen against the background of distant stars (i.e., on the celestial sphere).

Although Ptolemy did not know what planets actually were or in what manner they differed from stars, it was quite obvious that the five dots in the sky seemed to wander amongst the stars. Actually, the word "wanderer" is not exactly appropriate, since planets must move in very definite patterns if they belong to that realm considered to be Perfect and Orderly. These patterns could be best explained by assuming the behavior of the components as in Figure 4-8.

Retrograde Motion in the Ptolemaic System In the last Chapter, while discussing the motions of objects on the celestial sphere, I intentionally avoided describing the detailed motions of the planets. I generalized the apparent motions of the planets by suggesting that they move generally along the ecliptic, a fact that derives from the flattened shape of the solar system. But, in fact, if you trace the movement of a given planet from night to night, you find that it is quite complicated. It was the interpretation of these complicated motions that resulted in the debate between Galileo and the Church, the overthrow of the geocentric system, and the eventual adoption of the scientific method as a source of knowledge.

The planets move from west to east relative to the stars over long periods of time (days, weeks, months). (Note: do not confuse this motion with that of Sun during the day, and the stars, Moon, and planets during the night, which is from east to west. East-to-west movement is due to Earth's rotation.) While moving from west to east, a planet occasionally appears to slow down in its movement, reverse direction, move westward for a period of days, slow down again, reverse direction, and continue in its usual eastward direction. This **retrograde motion** occurs with predictable regularity.

Since both Moon and Sun appear to move smoothly along the ecliptic without exhibiting retrograde motion, in the Ptolemaic model they were not placed on an epicycle but directly on a deferent. But the planets Mercury and Venus reveal retrograde motion, since they are always found close to Sun as they move from one side of Sun to the other. So the model provided that the centers of the deferents of those two fast-moving planets are fixed along an imaginary line connected to Sun.

As the motions of Sun, Moon, and planets occur on

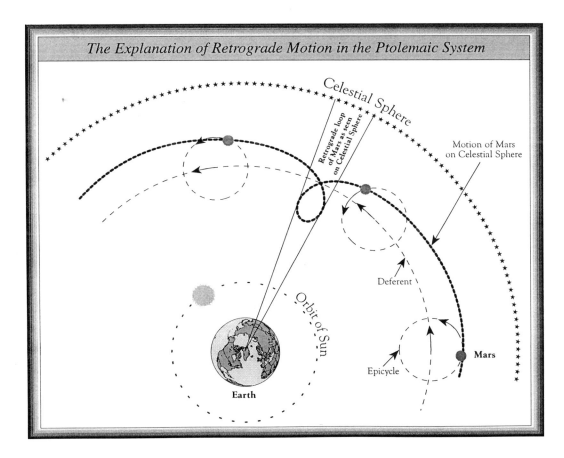

Figure 4–8 In order to explain the occasional movement of a planet in the reverse direction (from East to West), Ptolemy devised the model that placed planets on small orbits (epicycles) whose centers orbited Earth on a larger orbit (deferent). By varying the speed of the planet on the epicycle and the center of the epicycle on the deferernt, he was able to approximate the motions of the seven objects on the celestial sphere.

Figure 4–9 A page from an early manuscript illustrating the new Copernican universe. Earth is no longer at the center of the universe—it is rather one of six objects orbiting Sun. The stars, however, are still treated as a unit located on the celestial sphere. (Yerkes Observatory)

the celestial sphere, of course, the entire system of stars, Sun, Moon, and planets is rotating effortlessly around Earth once every 24 hours. It was only in observing and plotting the positions of the planets, Moon, and Sun from night to night that these gradual changes in position were (and still are) detected.

Measuring Positions of Planets During the time between Ptolemy and Galileo, more accurate devices for measuring the positions of objects amongst the stars were developed. This was especially due to the use of brass in making instruments that could be finely etched to measure angles to great precision. Thus it became obvious over the centuries that the Ptolemaic model as originally proposed only approximated the future positions of the planets. Modifications were occasionally made to the model so that its predictions of future positions conformed to the newer measurements.

This is, of course, a prerequisite for any good science. If a scientific model does not conform to real world measurements, then the model must be adjusted or discarded. In this case, for reasons that soon become apparent, the basic features were retained even as adjustments were made. In other words, Earth remained at the center of the universe, everything continued going around Earth, epicycles and deferents were perfect circles, and so forth. When the features of a theory (which of course the Ptolemaic model was) have to be continuously changed in order to adapt it to new observations, people (especially those making the new observations) begin having doubts about the validity of the theory itself. It is marriage of theory and observation that allows science to work so successfully.

Copernicus Comes Along The time was ripe for a change in the Worldview. New ideas were creeping into Europe from all over the world — literally. This was the age of exploration of the world globe. New lands, new goods, new peoples, new cultures, new ideas — Europe was alive with the excitement of these discoveries. As is commonly the case, it falls on one person to provide the seed that eventually precipitates the change to a new Worldview.

Copernicus was raised and educated in Poland by his uncle, who was an important bishop in the Church. His own life was a secluded one, but during his lifetime Europe entered the age of exploration, the renewal of learning, and the Reformation. Although destined from youth for a career as a religious administrator, he entered the University of Cracow in 1490, and was immediately exposed to mathematics and astronomy. These subjects remained amongst his greatest interests throughout his life.

When Copernicus carefully examined the Ptolemaic model in the early 1500s, he was disappointed in its inability to predict the future positions of the planets. So he began to search for another system or theory that could more exactly explain the movements of the bodies in the sky. It was the search for this more exact system that eventually led to the trial of Galileo. For the system that Copernicus proposed to replace the Ptolemaic system threatened the very foundation of the Church — or so the Church believed.

The Copernican system During his search for a better system, Copernicus was asking himself a simple question: *"How would the objects in the sky appear to move if I were standing on Sun rather than on Earth?"* To his surprise, while sketching out the Sun-centered solution, he was able to explain all observed motions in the sky and predict their future positions almost as accurately as the Ptolemaic model did. Figure 4-9 illustrates his new model: the Sun-centered or **heliocentric system**. It is frequently referred to as the **Copernican system** to make note of the person who first proposed it. It is, with some minor modifications, the model that we use today to explain the motions of objects in the sky.

His choice of Sun to be the center of the model is an obvious one — intuition told him that Sun's light and heat are absolutely essential for our survival. So it seemed reasonable that if it was so important to inhabitants on

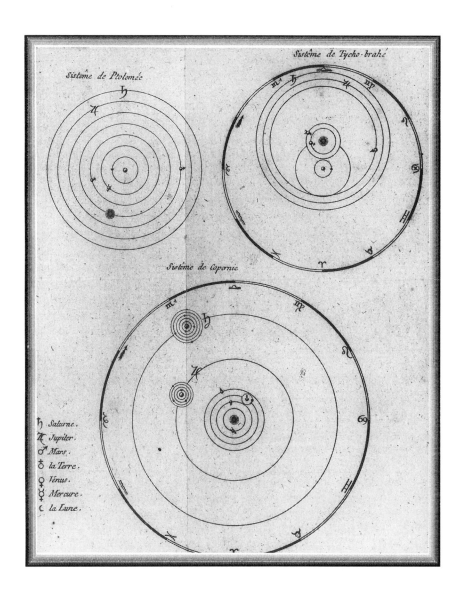

Figure 4–10 A page from an early manuscript illustrating the new Copernican universe (bottom diagram) that replaced the Ptolemaic or Earth-centered universe theory (upper–left diagram). The system proposed by Tycho Brahe, an astronomer whose life overlapped the lives of both Copernicus and Galileo, contained elements of both theories. Earth was still located at the center of the universe (upper–right diagram), but the planets orbited Sun instead of Earth. All three of these models attempted in their own way to explain the movements of objects in the sky. (Yerkes Observatory)

Earth, it might well be located at the center of the system of objects that life is not dependent on. But if the new model was no more accurate than the one it was replacing, how could he expect anyone to change from one to the other? Well, if you look at the two systems side by side, Figures 4-7 and 4-9, it is apparent that the Copernican system possesses a simplicity and elegance that the Ptolemaic model lacks — a beauty, if you will.

What does beauty have to do with science? If one assumes, as Copernicus did, that the universe is a creation of God, and that He is Beauty and Perfection, then He must have created the universe to reflect that Beauty and Perfection. Only on Earth is beauty and perfection in doubt. A system containing seven circles centered on the luminous Sun seemed to Copernicus more *beautiful* than a system of numerous epicycles and deferents, and therefore was more likely the kind of universe that God would create.

Copernicus did not intend to upset the Worldview of the time, nor did he want to dissolve the marriage the Church had nurtured between its own theology and the ideas of Aristotle. In fact, Copernicus was himself a canon in the Church. Nevertheless, being true to what he really believed, he composed his ideas of this revolutionary model in the form of a manuscript.

He was most likely aware of how the Church would respond to the publication of his heretical ideas, however, because the manuscript went to print just as he was on his deathbed. Tradition has it that the first copy was placed in his hands as he was dying in the year 1543. The title of the book was "*Concerning the Revolution of the Celestial Spheres,*" referring to the motion (revolution) of the planets around Sun, not Earth. This idea, as we will see shortly, was so radical and so threatening to the accepted Worldview of the Church that the word "revolution" in the title came to be used to refer to any radical idea or thought as well as the movement of one object around another.

For reasons that go beyond the scope of this short historical sketch, the book failed to cause a great stir either in the Church or among the public. It fell upon another person, Galileo, at a much later date, to attempt to use the book to convince the Church that the Copernican system was a more accurate description of the universe than the one held by the Church. Meanwhile, during the interval 1543 to 1609, events throughout Europe were providing additional ammunition for the eventual acceptance of the Copernican system.

Retrograde Motion in the Copernican System The central issue of dispute between the Ptolemaic and Copernican systems was over interpretation of the retrograde motion of the planets. It was the inaccuracy of the Ptolemaic System in explaining the retrograde motions of planets in the sky that provoked Copernicus to search for an alternative solution in the first place. The "circles upon circles" model was just not very accurate in predicting future positions of the planets. Just how did Copernicus explain the occasional backward movement of the planets?

By proposing that the planets revolve around Sun <u>at different rates</u>, Copernicus was able to explain what appeared to be the occasional backward motion of the planets. The system reveals how Earth occasionally passes up the slower-moving planets located farther away from Sun, and is in turn occasionally overtaken by those located closer to Sun. Hence it is simply a matter of relative motion, just as when you pass a car on the freeway it appears as if it is momentarily moving backwards relative to the sides of the freeway. Figure 4-11 illustrates this motion.

To fully appreciate this unfolding drama, it is important to keep in mind that offering an alternative model to explain retrograde motion does not in itself mean that the model is correct. The model follows directly from the Sun-centered theory itself, which of course was originally offered to replace the Earth-centered system because of the latter's inaccurate prediction of future planetary positions. Once adopted as the new theory, the Copernican system was no more accurate than the Ptolemaic system it replaced. By that time, however, new discoveries entered the courtroom as convincing pieces of evidence to support the system of Copernicus. That is where Galileo enters the stage.

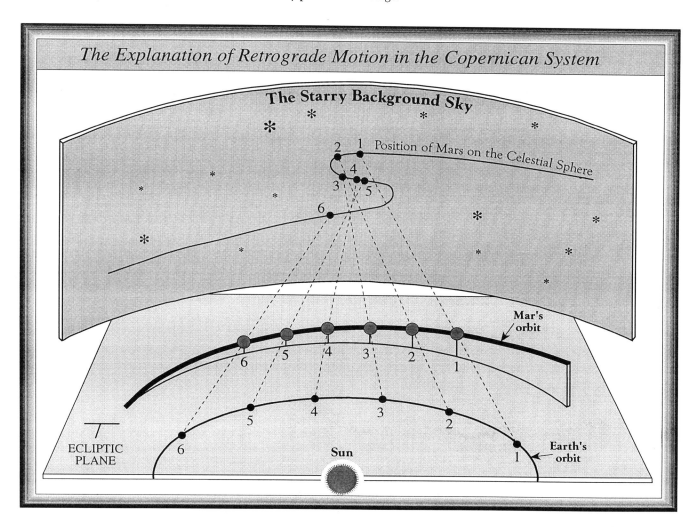

Figure 4-11 In the new Copernican system, the retrograde motion of planets is explained by this model of the theory. The planets orbit Sun at different speeds. When Earth overtakes a slower-moving planet like Mars, its appearance on the celestial sphere is backward-moving, from East to West. After passing it by, it appears to continue its normal West-to-East motion.

Galileo Enters the Drama

Galileo followed the arguments for the Sun-centered system presented in Copernicus's De revolutionibus, and became convinced of its correctness even before he first looked through his telescope. His telescopic discoveries simply reinforced his beliefs and convinced him that he had enough evidence to convince the Church as well. He attempted, in a sense, to defend the Copernican system in Court. To him, it was a matter of principle — one's beliefs about the behavior of the physical universe should be founded in reason and observation of the physical world, not on theological writings. Or to state it in his own words:

"The Church's role is to tell people how to go to Heaven, not to tell them how go the Heavens!"

The following is a short sketch of the events leading to the trial and condemnation of Galileo in his attempt to defend the Sun-centered system of Copernicus:

1564	Galileo born
1581	Begins studies at the University of Pisa
1586	Invents a hydrostatic balance
1589	Obtains a professorship of mathematics at the University of Pisa
1591	Resigns from Pisa after conflicts with Aristotelians
1597	Writes to Kepler that he has been a Copernican for several years
1608	Invention of the telescope by Hans Lippershey
1609	Constructs a telescope and begins observing the heavens
1610	Publishes the *Sidereus Nuncius*
1611	Visits Rome, Jesuit astronomers confirm his discoveries
1616	Theological Consultors of the Holy Office censure Copernican opinion as heretical, Galileo told not to hold or defend the theory, Copernicus's *De revolutionibus* placed on the list of Prohibited Books
1618	Appearance of a great comet stirs up debate about Copernican system
1624	Galileo returns to Rome to try to get the Copernican censure revoked. He is encouraged to write but told to stay within the limits of a hypothetical treatment
1630	Completes the *Dialogue on the Two Great World Systems* in which he intends to confirm the Copernican system
1632	The Dialogue is published, then halted, and Galileo is summoned to Rome
1633	The trial of Galileo is conducted, and he is required to retract his opinion on the Copernican system and his book is to be prohibited
1637	Galileo loses sight in both eyes, but continues to work on his new book, the *Two New Sciences*
1638	*Discourses on Two New Sciences* is published
1642	Galileo dies
1992	Galileo is officially reinstated by the Church

Galileo's Telescopic Discoveries You are still, I hope, imagining yourself as a juror listening to arguments for and against the heliocentric system of Copernicus. So far, it seems that its major asset is the "beauty" it possesses. But Galileo was the first to build a telescope for the purpose of exploring the sky, and what he discovered convinced Galileo that Copernicus was correct. His mind was prepared to accept what his eyes saw. The major discoveries, at odds with the teachings of the Church in the manner indicated, were:

- There are many more stars visible through the telescope than with the naked eye.
  Therefore, *The universe is much more vast than the Church teaches.*

- The dark markings on Moon appear to be due to a rough and scarred surface.
  Therefore, *Moon is not perfect, as the Church teaches.*

- The planet Venus is observed to go through a complete set of phases like Earth's Moon.
  Since, *In the Ptolemaic model, Venus never gets around to the far side of Sun to appear "full," the model must be revised or replaced.*

- The planet Jupiter is observed to have four Moons going around it.
  Since, *In the Ptolemaic model, Earth is the center of all motion, the model must be revised or replaced.*

- Sun is observed to have blemishes or dark markings on its surface.
  Therefore, *Sun is not perfect, as the Church teaches.*

The Church Responds The Church was not without its share of evidence to claim that the Ptolemaic system was alive and well. Take a short break to perform a simple experiment. Close one of your eyes while holding a finger about 12 inches in front of your face. Notice the point in the background against which your finger is held. Without moving your finger, close the open eye and open the other. Your finger appears to move against the background.

If you quickly blink your eyes, alternately, first one open and then the other, you notice your finger jumping back and forth at some angular distance. Continuing the experiment, as you blink off and on, gradually move your finger farther away from your face. The distance (angle) your finger appears to jump back and forth gets less and less as you move it farther away.

★★★★

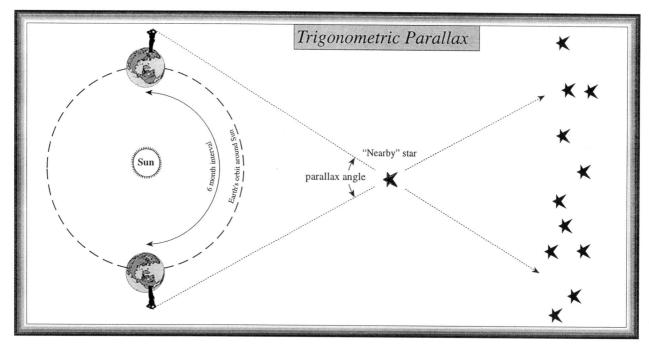

Figure 4-12 *The question of stellar parallax was a central issue during the Galileo–Church debate. If Earth orbits Sun instead if being fixed at the center of the universe, nearby stars should appear to move to and fro on the celestial sphere at 6-month intervals. Since no one had ever been able to observe such motion, it was difficult for Galileo to convince others that the Copernican theory was better than the Ptolemaic system which said that Earth was stationary.*

This effect is referred to as **parallax.** If the stars (whatever they might be) are located at different distances from Earth, one would expect the parallax effect to be detectable as Earth revolves around Sun (Figure 4-12). The nearer stars should appear to move back and forth with respect to the more distant stars during the one year that Earth takes to move from one side of its orbit around Sun to the other side and then back again. In the geocentric theory of Ptolemy, of course, the effect should not occur because Earth is assumed to be stationary in space. The Church scientists argued — reasonably, of course — that if indeed the Sun-centered system is a correct model of the universe, then parallax should be visible.

With the instruments available before and during the time of Galileo, no such parallax was detectable. Church scientists confronted Galileo with this fact, challenging him to explain how he could insist on the one hand that Earth orbits around Sun but on the other hand fail to detect any parallax. Galileo could only respond by suggesting that even the nearest stars are much too far away to reveal parallax — in the analogy I used earlier, your finger is too far away from your blinking eyes to appear to move back and forth.

Galileo calculated mathematically how far away "too far away" had to be in order for parallax not to detectable, and discovered that it was such a large number that even *he* must have wondered if perhaps he was wrong about Earth's movement in space. We know today that the very nearest star past Sun is about 30 trillion miles away (25,000,000,000,000 miles).

Unless you keep track of the Pentagon's annual budget, a number like this probably doesn't impress you. But to the person of Galileo's day, limited to small numbers in all aspects of daily life, a number like this seemed incomprehensible. Besides, the sky was thought to be the domain of God, and how absurd for Him to have removed Himself so far away from the very people He is attempting to serve! Heaven would be even *more* difficult to attain were it located so far away! The parallax of stars was not even discovered until 1838, long after the debate between Galileo and the Church was settled. By then, powerful telescopes had been developed to magnify not only the sizes of objects but also their motions and positions relative to other objects.

Opposition to Galileo and Copernicus

Publishing the Dialogue To be precise, the charge against Galileo, the reason he was called before the Inquisition, was that he had violated the instruction given to him by the Pope that he not teach the Copernican system. He was told that he could hold it as a *hypothesis*, but that he was not to portray it as a *truth*. Actually, he did not exactly teach the system. He wrote the fictional book *Dialogue Concerning the Two Chief World Systems* in which he pits a very learned and bright scientist defending the Copernican system against the unconvincing arguments of a somewhat slow and dim-witted person named *Simplicio*, who tries to defend the Ptolemaic system of the Church.

In the story, a third person acts as a moderator, asking the two to explain observations of the behavior of objects on Earth as well as in the sky. As he cleverly destroys each and every explanation offered by Simplicio, and offers much better explanations of his own, the scientist clearly makes Simplicio look like a fool, along with his theory of an Earth-centered universe.

So you think that the Church was being picky in calling the book heresy, just because it threatened the teachings of the Church? Well, consider a couple of additional facts. Galileo had been close friends with the Pope who eventually sent him before the Inquisition. A Cardinal at the time, he presented several objections to the Copernican system while he and Galileo walked and conversed in the gardens of the Vatican. Several of those very arguments appeared in the book as arguments coming from the mouth of Simplicio. When Pope Urban VIII read his very own arguments coming from the lips of such a fool, he was outraged. He felt personally insulted, as anyone would.

In addition, Galileo wrote the book in Italian rather than Latin, the language not only of the Church but academia as well. Common people were isolated from knowledge conveyed by writing, since official publications were written in Latin. Latin was taught only in the universities, and the Church determined who went go to the universities. Basically, the Church determined what knowledge was to be disseminated to people. By writing the book in Italian, Galileo made the book available to the guy on the street. It was an immediate best-seller.

Thought Control Certainly Galileo violated the spirit if not the letter of the instructions not to teach the heliocentric system of Copernicus. It is similar to the famous Scopes Trial of the 1930s, in which John Scopes was accused of teaching the theory of evolution in public school in Tennessee at a time when the prevailing Worldview was that man had been created in the image of God. John Scopes was on trial for teaching a subject that was banned from the public school system, but it was really the theory of evolution that was on trail.

Likewise, in the Galileo case, it was the heliocentric theory that was on trial, not Galileo. As a juror, you might feel that Galileo provided sufficient evidence on his behalf. Nevertheless, he was found guilty, and the Church clamped a tight lid on anyone supporting him or his ideas. He was not tortured or even imprisoned. He was placed under house arrest, but was free to conduct experiments and write. His most enduring book, *"The Two New Sciences"*, in which he develops the basic contents of a college physics course, was written during this time. But why did the Church go to such extremes to silence Galileo?

You are probably thinking that as a juror, you would have felt pressure from the Church to support its teaching that Earth is located at the center of the universe, regardless of Galileo's evidence to the contrary? Is that why the tribunal found him guilty, because they were afraid of what the Church would do to them? It is tempting to follow that line of reasoning, but you might first want to reflect on how thoroughly the Ptolemaic model was ingrained in their thinking after being around for 1,300 years. Consider how thoroughly intertwined were the theological ideas of the Church and the ideas of Aristotle dealing with the makeup and behavior of objects in the universe.

Trust in the Senses You see, we are so thoroughly embedded in our current Worldview of modern science, with its emphasis on instruments and devices for exploring and manipulating the invisible universe, that we fail to appreciate how mystical it must have seemed to those looking through a telescope for the first time. Why are you willing to believe what you see through a telescope, anyway? Or to put it more precisely, why are you willing to assume that what you see through the telescope actually exists *out there*, when it can't be seen with your eyes directly?

Well, you say, they could have pointed the telescope at an object on Earth and then verified its existence by going over and touching the object. But the sky was thought to be a different place. It was made of a different substance, made to be predictable, eternal, orderly. Just because telescopes work on Earth doesn't mean that they work on objects in the sky. *Especially if what you reportedly see conflicts with the established teaching of the Church.*

So what were Church scientists observing through the telescope? Perhaps what they saw were illusions, the work of the devil himself, leading people away from the Truth revealed by the Church? You and I think nothing of someone setting up an instrument in front of us and telling us that it will photograph the bones in our body, or the texture of a fly's wings, or even the structure of an atom itself. How is it that we learn to accept such claims of modern technology? Probably because we have seen it work so successfully in so many other situations. But Galileo was asking the Church to accept something quite new, a radical departure from any other technology available at the time.

Many Church scientists looked through Galileo's telescope and observed the same phenomena as Galileo. But it was easier for them to accept the possibility that they were being deceived than to assume that the teachings of the Church had been wrong for so many centuries. They needed more convincing evidence before they threw out a system that had worked for so long, especially a system that provided a meaning for man's existence. But just how did a "meaning of life" derive out of the geocentric system?

Macrocosm versus Microcosm The early Church combined the physics, chemistry, and astronomy of Aristotle with theology derived from the Bible and other early Christian writings to explain the purpose of man in the physical world. The overall theme was that the universe is focused on Earth, and man's presence on it. God created it for man, placing him at center stage so that he might marvel at His creation.

CHAPTER 4 / The Birth and Growth of Modern Science

★★★★

Everything on Earth, so Aristotle taught, is made up of a combination of four substances – *Earth* stuff, *water* stuff, *fire* stuff, and *air* stuff. A log of wood, for example, tossed into the fire, is broken down into its proportional substances: additional fire is released, air (smoke) is released, water escapes from the burning log, and a residue of Earth stuff (ash) remains. A rock does not break down in the same way because it consists almost entirely of Earth stuff.

Imagine performing a simple experiment during Galileo's trial. From some height above the ground, drop a rock and a leaf at the same time. You observe the rock hitting the ground first. A member of the tribunal would have remarked that the rock, consisting mostly of Earth stuff, belongs close to Earth, and therefore moves faster to get there. The leaf, on the other hand, has a higher percentage of water, air, and fire stuffs, and therefore is not in such a hurry to reach the ground. After all, when fire burns, where does the smoke (air) go? Up, of course. But why? Because it <u>belongs</u> there, responds the tribunal member. What evidence is there that air belongs up there, you ask? Because that's where air is – up there! Rocks are found on the ground because that's where they belong because they are made of Earth stuff.

Everything has its place in the universe. Those closest to the center of the universe (Earth) are the least permanent. Rocks are less permanent than water. Water is less permanent than air. Air, being most permanent, goes upward, because the sky is the most permanent thing of all. In order to explain the uniqueness of objects in the sky, placed beyond the confines of Earth, Aristotle proposed that they were formed of a fifth substance, **quintessence** (meaning "fifth substance"). It is infinitely light, is eternal, and is unchangeable. That is why objects in the sky appear fixed, permanent, and eternal.

Air, water, fire, and Earth does not come together to form wood. The natural state of affairs is for things to break down into lighter components, like smoke being released from wood. Observations of the behavior of matter on Earth led Aristotle to the conclusion that everything seeks a greater degree of permanence. Do you see how this idea could be applied to man within a theological context? Aristotle advanced his ideas of chemistry and physics primarily for the physical world, but the Church found it convenient to adapt the idea to include man as well.

This fusion of ideas did not happen overnight, but evolved over several centuries. It was an attempt to unify the world outside (**macrocosm**) with the world inside (**microcosm**). The Church taught that just as the external world is made of a hierarchy of substances, so also is man. And just as the four basic chemical substances are temporary and changeable, so also is man, since he is made up of the same substances as other objects in the external world.

Aristotle's Chemistry The sky is a constant reminder to each of us of everything's eventual destiny. And man himself must have within something that compares with the quintessence evident in the sky. The Church taught that he must also have the capacity to turn the temporary substances into the eternal: the quintessence. But how is this possible? How can earthly man, made of temporary substances, turn himself into the quintessence, the eternal substance?

Let me illustrate the importance of the concept of the quintessence in man and the sky. The eternal substance, the quintessence, is not found on Earth, since Earth is on the lowest rung of the ladder of permanence. But there is a substance that <u>appears</u> to have the property of permanence – gold. With the limited chemicals available to Aristotle and the early Church, gold could be neither tarnished nor combined with other substances. It was worthy of having the property of permanence, just like the heavenly bodies. Perhaps God provided for its presence on Earth so that we would be always reminded of His existence and the presence of Heaven above.

Is it any accident that the world economy is based on the gold standard, that the possession of gold has always been associated with power and wealth? Why did the Europeans rape and plunder the Americas in searching for gold? Why are wedding rings made of gold? Permanence. Marriage is a permanent arrangement, and the gold ring symbolizes that fact. It is also stated in the wedding vow.

Why are the chalices used during the most sacred portions of the Catholic Mass made of gold? Permanence. Catholics believe that the consecration occurring in the chalice involves the conversion of bread and wine into the body and blood of Christ. Only a substance that possesses the property of permanence is deemed worthy of being in contact with such a holy relic. To this day, our demand for gold reflects the ancient belief that it had properties like those of the sky.

The **alchemists**, ancestors of modern chemists, believed that if they could convert the base metals (combinations of Earth, water, fire, and air) into gold, they would achieve immortality (permanence). You might think they only wanted to increase their wealth by making gold, and no doubt for some that was true, but for the most part they wanted to achieve immortality. It was God, after all, who made Heaven and the quintessence of which it was made. By making gold out of Earthly substances, an alchemist would in effect be getting into the mind of God and performing something of which only He (being eternal Himself) was capable. In effect, he would be eternal because he would be God because he had made something that was eternal. The alchemists did not succeed, but out of their attempts evolved modern chemistry.

If objects on Earth are not made of the four primary substances, just what are they made of? Arrangements of atoms of carbon, nitrogen, hydrogen, and oxygen, of course. So say modern chemists. But have you ever seen a carbon atom? Or a nitrogen atom? We have adopted the modern myth (atoms) just as the ancients adopted the myth of Earth, water, fire and air. What will chemists of the 28th

Century say about the finer structure of matter?

Aristotle's Physics The physics of Aristotle had also been integrated into the Church's Worldview that prevailed at the time of Galileo. Try a simple experiment: push an object that is resting on the table in front of you. What happens? When you apply a force, it moves. When you cease pushing, the object returns to rest. It appears that the natural state of affairs for an object in the universe is rest, unless a force is acting on it. Forces acting on objects cause motion. Yes, one might observe that an arrow continues to move even after it leaves the force exerted by the bow. But as everyone knows (?), that is because the tip of the arrow is compressing the air as it flies along, causing the air to rush around to the rear of the arrow, which in turn pushes it along until friction stops it!

Now look up at the sky. Everything up there moves effortlessly, without any obvious force being applied to it. How can that be explained? Of course, God! The sky is the realm of Perfection, where perfectly light substance (quintessence) is easily moved by Unseen Hands. Perhaps God designed it that way, so that we would be constant reminded that there is a place of Perfection that awaits our deaths and towards which we must direct our behaviors.

You say that momentum keeps the arrow going? But what sort of mysterious force is that? Can you see momentum? How do you measure momentum? Do you see what I am saying? Whatever concept we use to explain an observation, there is at the root an abstraction: something mysterious, something still yet to be completely understood. In the matter at hand, you will find it easier to understand the resistance the Church offered to the new system of Copernicus if you remember that the theology of the Church and the astronomy/physics/chemistry/philosophy of Aristotle were inseparable.

The Church's Worldview offered a hierarchy to the universe from imperfection on Earth to Perfection in the Heavens. By virtue of our being human, we are less than perfect, but a hierarchy from imperfection to Perfection also exists within us. There is quintessence within us just as there is in the external world, and we are assured by the Church that by doing good works during our lives our quintessence will go to the place it belongs when we die: Heaven.

The Christian religion generally refers to the process of achieving permanence as salvation. The theology of Christianity is that imperfect man must follow a path of obedience in order to achieve permanence at death. Ordinary matter attains this state of perfection naturally. But man has a soul given to us by God, and therefore must make choices between Good and Evil. That was man's fall from grace, his rejection from the Garden of Evil.

The Church and Salvation It is obvious that we humans are not agreed on what the "good works" are that are necessary for achieving salvation. There doesn't seem to be an obvious formula for getting to Heaven. So what are we to do? Well, of course, we look to those who claim to know. But who that is himself/herself mortal (and therefore imperfect) claims to know such Truth, the Road to Salvation? Those who claim Perfection and Immortality, those who have been given the keys to Heaven: the Church.

The early Catholic Church claimed (and still does) that it was given the task of showing people the path to Heaven. The Church, founded by Jesus and headed by Saint Peter, according to the Biblical interpretation of the Church, was the only infallible source of knowledge. It was only through the Church that salvation could be achieved, because only it is permanent and infallible.

Prior to the Renaissance, to those living in Catholic Europe, the meaning of life was simple — one must live one's life in accordance with the teachings of the Church, the award for such "correct" living being salvation (entrance into Heaven after death). If one required a reminder that this is the goal/meaning of life, he/she needed look no further than the sky. The obvious permanence and predictability of bodies in the sky were daily (nightly?) reminders that there is a Heaven. The Church neatly weaved the physics, chemistry, and astronomy of Aristotle and its interpretation of the Bible into a beautiful and intricate Worldview that met the human need for meaning and purpose.

Galileo's Crime With that as background, it is easy to see why the Church resisted the introduction of the Sun-centered system. When Galileo arrived on the scene to challenge the Church's Worldview that placed Earth at the center of the universe, he failed to realize how carefully that view had been interwoven into the very theology that argued for the Church's existence in the first place.

In other words, if it could be argued that there is no hierarchy of substances in the macrocosm, that Earth is just another object circling Sun, then it can also be argued that there is no corresponding hierarchy within man himself. If that is so, then there is no state of perfection possible for man. There is no evidence in the physical world for believing in Heaven, a place for the saved. One is left only with the Church's reading of the Bible. The Church is no longer in the business of showing man how he might go to Heaven.

No wonder, then, that the Church went to such lengths to silence Galileo and other scientists who spoke out against its Worldview. It was not just a matter of holding onto a job because it needed money to live — it really did (and still does) believe in its mission. Churches even to day in general profess the same goal that the Church did at the time of Galileo: to show us the road to salvation. Churches feel defensive toward any claim that there is no Heaven after death.

Today, in Churches throughout the world, we participate in the same drama of striving toward Heaven, though few of us would argue that modern discoveries in as-

CHAPTER 4 / The Birth and Growth of Modern Science

tronomy conflict with Church dogma. What has developed over the centuries, of course, is a wider separation between theology (*how to go to Heaven*) and knowledge about the physical universe (*how go the heavens*).

In effect, Galileo upset the psychological security that derived from the neatly ordered interrelated hierarchies of astronomy, philosophy and theology. His opponents were afraid that if the geocentric concept fell, the whole construct of cosmology, the truth of the Scriptures and the anthropocentrism of creation would fall with it — a domino theory of sorts. The one thing he probably did wish posterity to learn from his personal tragedy was that no new opinion is wrong simply because it is new.

Effects of the Copernican Revolution Put in general terms, the Copernican Revolution was a conflict between two sources of knowledge — that stated by authority of the Church, and that obtained by reason and the senses. Are people capable of making sense of the universe around them through the use of instruments, reason, and experiment? Modern science says so. But the Church believed that when the senses argue again dogma of established authority, the senses must be wrong. Galileo started the ball rolling. That is why historians refer to the period of time following Galileo's death as the **Age of Reason**.

Man discovered that he was capable of reason so that he might be in charge of his own destiny. Man's purpose in the universe was no longer as clear, however. There is a price we pay for throwing out dogma and authority. In a sense, we are still living out the Copernican Revolution. We no longer place Earth at the center of the universe, but neither do we place man at the center with a well-defined purpose.

So what is our purpose now? What is the meaning of life? We've replaced the authority of the Church with trust in our senses and the gadgets of modern science. But science doesn't provide us knowledge about purpose or meaning since it does not proclaim to have access to Truth. We are left to our own devices in the search for Truth, or return to the authority offered by Churches or religion. The search can be a lonely one.

It would be easy to form an opinion that the Church did a terrible thing to Galileo, and that it is wrong to believe there is a connection between the physical world outside of man and the internal world of man. The connection, if there is one, just isn't as simple as that argued by the Church. That of the Church was founded in the incomplete and erroneous astronomy, physics, and chemistry of Aristotle. In fact, one can claim that with the success of the Copernican revolution, a long era of confusion about man's purpose in the universe began. That confusion is still with us today.

Take a survey amongst your friends. Ask them to explain their *meaning of life*. What kind of response do you get? In a very real sense, the success of the Copernican Revolution yanked man from the equation that linked him to the material universe. We have been groping for another link ever since. We no longer have a consensus as to what the purpose of human existence is. It can even be argued that the environmental crisis we face is a result of our failure to "link" ourselves to the physical universe.

The Church's Crime In 1758, the general ban of 1616 against Copernican writings was lifted. The special prohibitions were repealed in 1822 and the next edition of the Index of Forbidden Books, published in 1835, contained no trace of the infamous condemnation of heliocentric astronomy. It was always allowed at any time to write on the heliocentric system <u>as a hypothesis</u>.

It has also taken a long time for the Church to recover from the Galileo affair. It was obviously embarrassed by the trial of Galileo, and has been reluctant to reopen the question of his guilt or innocence. Galileo's book was not removed from the index of forbidden books until 1835. In 1979, Pope John Paul II announced that there are no irreconcilable differences between science and faith, and in 1980 appointed a commission of scientists, historians, and theologians to reexamine the evidence and the verdict in the trial of Galileo.

In 1984, the commission, in a preliminary report, acknowledged that the Church was wrong in silencing Galileo. And finally, in October 1992, the Pope officially and finally rehabilitated Galileo by accepting the commission's report that found that Galileo had been wrongfully condemned by the Roman Catholic Church. The pontiff added that faith and science are distinct but not necessarily opposite.

Conclusion of the Galileo Affair Galileo, despite the conviction at his trail, began a revolution in which reason and the senses were valid toward developing ideas about the nature of the physical world. Since the Church had made it clear in the conviction of Galileo that its authority was not to be questioned, however, this new science of inquiry moved to northern Europe where the Church's influence was slight. It was clear to learned people in southern Europe that they were not yet free to participate in the excitement of the Age of Reason. Not a slight consequence of this effect was that the Industrial Revolution began in northern Europe where people felt free to experiment with new ideas.

Lessons from the Galileo Affair On hindsight, there are a few important lessons that we can learn from the Galileo affair, especially as it relates to the practice of science in the modern world:

- What is the cogency of proof that is required of an independent thinker before he/she urges his/her views against established authority or tradition? The proofs (parallax and Foucault pendulum) needed to convince that Copernicus was correct had to wait more than a century after Galileo's death. The problem, however, is this: if a person's proofs must be so overwhelming that others will immediately accept them against established authority, then very few independent ideas will ever be proposed. Because most really new ideas, especially in science, require time-consuming research and thought by many people before rigorous proofs are found. As a practical matter, it is a practice in science to insist that "*Extraordinary claims require extraordinary evidence.*"

- The true originality of Galileo was his insistence that the book of nature is written only in mathematical language:

  "*Philosophy is written in this grand book, the universe, which stands continually open to our gaze. But the book cannot be understood unless one first learns to comprehend the language and read the letters in which it is composed. It is written in the language of mathematics, and its characters are triangles, circles, and other figures without which it is humanly impossible to understand a single word of it; without these one wanders about in a dark labyrinth.*"

- The aim of research is no longer knowledge of atoms and their motion "in themselves" separated from our experiential questioning. Rather, right from the beginning, we stand in the center of the controversy between nature and man, of which science is only a part. The familiar classification of the world into subject and object, inner and outer world, body and soul, somehow no longer applies and indeed leads to difficulties. In science, also, the object of research is no longer nature itself, but rather, nature exposed to our questioning and to this extent, we here also meet ourselves.

- What is the precise relationship between religious faith and modern science? Today it is not so much a question of religion "threatening" science as it is of a lingering fear that science will somehow undermine the doctrines of religion that are accepted on faith. What right does an individual have to challenge established authority when he/she feels he/she must? Those who wish to change the views of others cannot proceed to demolish what others believe without regard for their sensitivities and without something clearly superior to offer for their allegiance.

Science in Northern Europe

I have probably given you the impression that the "civilized" world outside of Rome was sitting around watching the drama between Galileo and the Church unfold. Of course not. The speed of communication and transportation did not allow for the rapid spread of information in those days. The authority of the Church was not very strong in northern Europe, and even before and during Galileo's life there were discoveries and observations being made that seriously challenged the Ptolemaic system.

In the north, there existed neither a religious nor intellectual climate that was resistant to change, since the influx of new ideas from around the world were themselves causing change. One can argue, therefore, that the revolution in Worldview that was precipitated by Galileo would have happened even if he had never confronted the Church. The change would have started from northern Europe and gradually spread to the south. Whether or not the Church would have confronted a later change in the same stubborn manner is anyone's guess.

Determination of Planetary Orbits Recall that the Ptolemaic model is based upon the belief that all objects move around Earth in perfect circles. That made for a very orderly system, but not a very predictable one. The purpose of a model is not just an accurate description of how the sky appears and how objects move in the sky, but also to predict the positions of objects in the future. The development of two devices spelled doom for the accuracy of the Ptolemaic system: the astrolabe and the clock.

The **astrolabe** is a device perfected by the Arabs for the purpose of determining the accurate positions of objects in the sky. They used this information in determining the direction to Mecca, the holy city toward which they faced while reciting their daily prayers. This was especially important as they spread across northern Africa and into Spain, getting further and further away from Mecca. The Europeans became aware of the astrolabe when they ejected the Arabs from Europe during the Crusades. The work of Ptolemy was uncovered at the same time.

The **clock** needs no introduction, but it may surprise you to learn that the early interest in the mechanical clock was to allow monks to be awakened for prayers during the night! Further refinement allowed for the accurate determination of time for the purpose of establishing the numerous feast days whose dates depended on astronomical data (the celebration of Easter is still fixed to an astronomical event). The feast day was of great concern to the priest, whose salvation depended upon its proper observance, but to the populace at large, whose life revolved around it. Sowing the seed, harvesting crops, market days, holidays: all were held on particular feast days.

With the discovery and gradual improvement of these two devices for measuring the positions of objects in the sky, by the time of Copernicus it was common knowledge that the Ptolemaic model could not predict the future position of a planet with any reasonable degree of accuracy. It was that fact that caused Copernicus to look for an alternate model for the heavens.

Although the placement of Sun at the center of the universe was a stroke of genius for Copernicus, his system was no more successful at planetary prediction than was his predecessor, Ptolemy. Recall that it was its "beauty" that impressed him, nothing else. Of course, Galileo had telescopic observations to convince himself of the superiority of the Copernican model over the Ptolemaic model. Both Copernicus and Galileo suffered from that same flaw in logic that Ptolemy had: motions in the sky must adhere to the perfect shape, the circle. Eventually, the work of two men led to the important conclusion that the circle may be a perfect shape, but that it is not to be found in the sky.

Astronomy goes Mathematical Planets move in elliptical orbits, not circular ones. This discovery resulted from the combined talents of two notable astronomers. The *technical skills* of a Danish nobleman, **Tycho Brahe** — who received financial backing from the King of Denmark to build instruments capable of very accurately measuring the positions of the planets as they move relative to the stars) — were incorporated into the *mathematical skills* of **Johannes Kepler** — who was able to fit the positions of the planets as recorded by Brahe into the mathematical shape of an ellipse. Kepler took the discovery of the elliptical shape of planetary orbits one step further — he discovered that the properties of elliptical orbits allowed him to explain other properties of the solar system not explained by the Ptolemaic model. He expressed these properties, now known to be valid for any system of orbiting objects, in the form of three laws of motion. These are called **Kepler's Laws of Planetary Motion**.

It had long been noticed that Sun does not move smoothly along the ecliptic. At some times of the year it appears to move faster than at other times. To account for this observation, Ptolemy had placed Earth off-center — but still retained Sun in circular orbit around it. Kepler made the concept more elegant in his **1st Law** — the **elliptical-orbit law**. Earth is located off-center at a point called the **focus** of the ellipse, but the orbit is no longer the perfect circle (see Figure 4-13).

Kepler then discovered mathematically that the varying speed of Sun relative to the stars — which is due to Earth's revolution around Sun — could be described by his **2nd Law**, or the **equal areas law**. In Figure 4-13, a planet moving for say a month's period of time in both intervals *a-b* and *c-d* sweeps out equal areas within the ellipse during those two intervals (count the number of squares in each area to verify the law). Of course, since the planet during the interval *a-b* is closer to Sun than when it is at *c-d*, it travels faster along the orbit and covers a larger portion of the orbit. But it sweeps out the same area as the slower moving portion of the orbit at *c-d*.

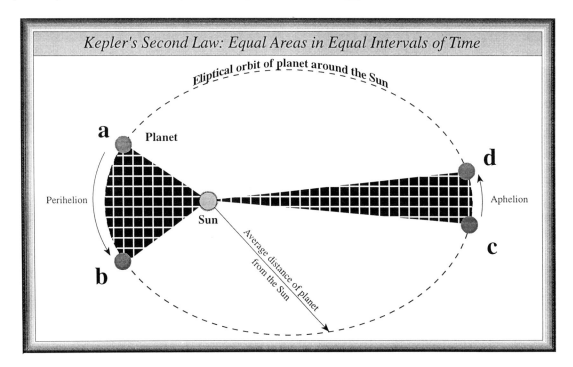

Figure 4-13 Kepler's 3 laws of planetary motion explained why it was that the future positions of the planets had been so difficult to predict accurately under both the Ptolemaic and original Copernican system. Planets orbit Sun in ellipcal orbits with Sun at one of the focii of the ellipse. In equal periods of time, a planet "sweeps out" equal areas. These laws essentially took the responsibility of moving planets out of the hands of God and put it into laws of physics. The language of Nature became mathematics.

Although it was well known that planets travel at different rates relative to the backdrop of the stars, it was not clear exactly what the rates had to do with distance from Earth (Ptolemaic model) or Sun (Copernican system). Kepler discovered that a mathematical relationship exists between the two variables. This is Kepler's **3rd Law**:

$$P^2 = A^3$$

where **P** is the time in years needed to circle Sun, and **A** is the distance from Sun in units of Earth's average distance from Sun (the astronomical unit – AU). One AU is approximately 93 million miles. The three laws explain (mathematically) the movements of objects in the solar system with great precision. In fact, all objects in space move along paths that derive from one of the four shapes obtained by cutting a cylinder in various ways (Figure 4-14). These shapes (circle, ellipse, parabola, hyperbola) are therefore called **conic sections**. Kepler was one of the first to recognize that nature is far from haphazard in the manner in which object behave. Even a tossed baseball conforms to one of the conic section shapes.

What you notice in these discoveries of Kepler is that they not only contradict the theory that objects in the sky move in perfect circles, but that they lead to the radical idea that mathematics can describe events in the sky. This seems innocent enough, unless you recall that the Church believed that imperfect man living on ever-changing Earth is unable to know the mind of God in explaining objects in the sky. Certainly not in the form of equations and numbers!

The reasoning of imperfect man cannot be valid for describing objects and motions in God's domain in the sky. The Church provided the model (from Ptolemy) for the general plan of the heavens, but the inner workings were the result of God's thinking, not man's. It was okay for man to explain events on Earth with mathematics, but to extend mathematics out into space presumed that man has a reasoning skill comparable to God Himself. Kepler published these discoveries between 1609 and 1618, just about the time Galileo was confronting the Church with his ideas.

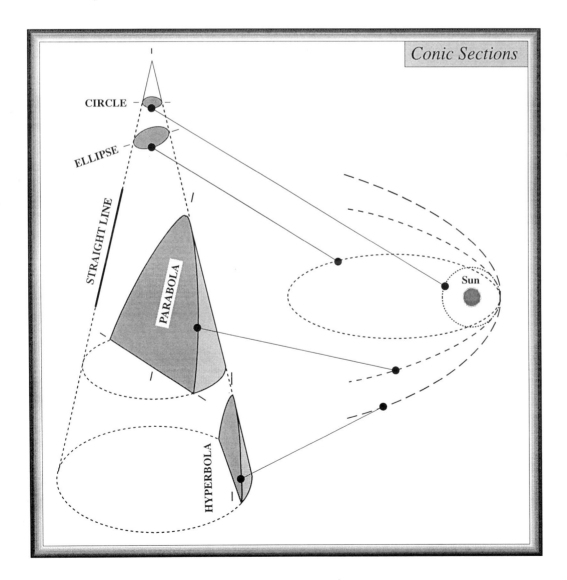

Figure 4–14 The orbits of all moving objects conform to one of the three shapes obtained by cutting a cone with flat planes. Cutting it at a slight angle creates an ellipse. Cutting it parallel to one side creates a parabola. Cutting it at right angles to the base creates a hyperbola. A circular orbit is a special circumstance of cutting the cone at right angles to the axis. A straight—line orbit is a special circumstance of cutting the cone parallel to one side right along the edge.

CHAPTER 4 / The Birth and Growth of Modern Science

Galileo knew of Kepler's discoveries, but rejected them because he could not bring himself to accept the possibility that the orbits of the planets did not conform to the perfect circle idea of the Greeks. Isn't that ironic! Galileo was brave and stubborn enough to confront the Church with a radical theory of his own, and was quite critical of the Church's unwillingness to be open-minded enough to accept it based upon his telescopic evidence. But at the same time he was unwilling to accept Kepler's evidence for the elliptical shape of planetary orbits because it led to a radical conclusion. This is rather common in the history of science as well as other area of human activity. We expect others to be open-minded in listening to and appreciating our Worldview, but find it difficult to do the same when listening to someone else's Worldview!

Newton's Mechanical Universe What Copernicus described as a geometric model of the sky, Kepler described mathematically. Yet there was another step necessary to complete the revolution in explaining the movements of objects in the sky. The hero of this part of the story is **Sir Isaac Newton**, who was born the same year that Galileo died. Galileo performed numerous experiments while attempting to explain the motions of objects on Earth, including the famous dropping of weights from the leaning tower of Pisa. These experiments led to the development of the branch of physics we call "mechanics." Newton carried the development to perfection.

The smooth and effortless movements of the eternal celestial objects had convinced Aristotle that the *natural state* of affairs for objects was perpetual motion. Earth, along with its imperfect inhabitants, had not yet reached that stage. Original sin committed in the Garden of Eden had cast man into the darkness of the lowest rung of the ladder of substances, Earth: heavy, immobile, and changeable. Galileo proposed a radically different idea. He theorized that perpetual motion on Earth is inhibited by friction, that an object would move forever after an initial force is applied. It is only because of an opposite force (friction) that makes it slows down.

How could anyone have thought otherwise, you ask? We have grown up with concepts like friction and gravity and momentum and centrifugal force, and cannot see the behavior of objects in any other way. These are not self-evident ideas, anymore than the rotation of Earth about an axis is. Prior to Galileo, Kepler, and Newton, the physics inherited from Aristotle explained motion by the applying of a force. If you give something a shove, it will move. When you remove the force, movement ceases. So how can one explain the motions of Sun, Moon, planets, and stars? The unseen hands of God, the Prime Mover, the Eternal Force.

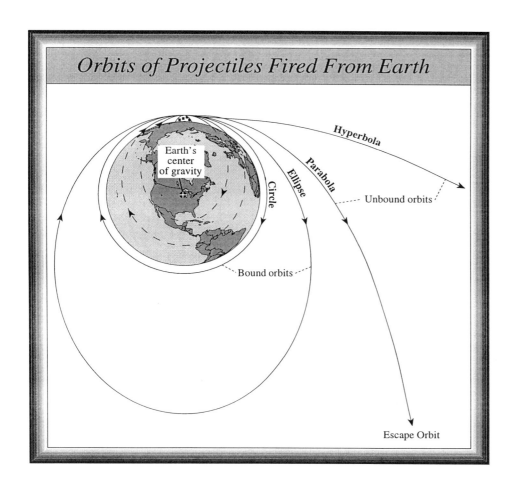

Figure 4–15 When an object is thrown or launched from Earth, it is conforming to one of the shapes derived from cutting the cone in the previous Figure 4–14. The center of Earth (as the gravitational center) is the focus of the orbit. The circle and ellipse are closed orbits (the object will eventually fall to Earth), whereas the parabola and hyperbola are open orbits (the object will leave Earth). To send a

Newton's Universal Law of Gravity Tradition tells us that while Newton was outside one day, he happened to glance up to see Moon in the sky, and at the same time see an apple fall from a tree. An idea flashed through his mind: perhaps there is a connection between these two events! They both fall toward Earth. But Moon, since it revolves around Earth, tries to fly away from it at the same time and at the same rate that it is falling toward Earth. So it never actually hits Earth Figure 4-15.

Thus was born the universal law of gravity, the same law we use to describe the behavior of Earth's tides. If the behavior of Moon can be explained by the equation, so reasoned Newton, all objects in the universe behave according to it as well. That is why the word universal is used in the description of the law. This conclusion, of course, rests firmly on the assumption that the behavior of objects on or near Earth can be explained by principles that operate throughout the universe. This is another way of saying that we assume that our particular region of space is commonplace and typical, and that what we find happening here can be found happening elsewhere as well. Whereas Kepler's genius showed us how the planets behave in going around Sun, Newton's genius shows us why they behave that way.

The laws of physics and chemistry that act within your reach are the same as those acting across the billions of light-years of space. We cannot prove that to be true; it is a matter of faith. But the fact that we can use those same laws to land men on Moon, send landers to other planets, and obtain photographs sent back from the distant planets is good enough reason to hold onto that faith. The laws continue to work for us. Only when observations of something in space don't conform to the laws will we look for new ones or develop a more generalized law that explains those new observations as well as all others.

Albert Einstein Sir Issac Newton's law of gravity, although successful in brokering the marriage between events on Earth and events in the sky, was not the final word as regards an accurate description of gravity. When it comes to the subject of black holes (see Chapter 11), we must abandon Newton's description of gravity as an attractive force and avoid saying that *"objects attract one another."* **Albert Einstein**, one of great scientists of this Century, revealed an entirely new universe to us in 1916 when he proposed that gravity, formerly thought to be a force of attraction, is actually the result of space being "warped" or "bent" in the vicinity of any object that has mass (Figure 4-16).

When a baseball falls toward Earth, it is because it is caught in space that is "warped" or "bent" by Earth's mass, and a property of another object with mass (the baseball) is to move in space to that place where space is most warped. You remain on Earth not because Earth attracts you, but because Earth bends the space around it. And you, occupying a portion of that space, move to that part that is most bent (Earth's surface). If it were not for the fact that Earth's crust is harder (more dense) than your body, you would go (be pulled, if you are still thinking of gravity as an attractive force) even further toward the center of Earth. Of course, you can always move away from the region of greatest warped-ness and venture out into less-bent space, but that requires adding energy to your body (as in a rocketship). Have you noticed that it even requires energy to climb up a ladder?

So what is it that causes matter to warp space? That is another ultimate question, a question of the origin of principles. Scientists do not know, and do not presume to be able to answer the question. Ever. That is not the domain of science. The goal of science is to explain the behavior of objects in the universe, the "how" of things. The scientist says that the warping of space by an object that has mass is a consequence of its property of "mass-ness." In other words, we have gotten to the point of explaining how mass warps space, but not to the point of knowing *why* the property of "mass-ness" causes warped-ness.

Science versus Religion Notice that what happened during the development of modern science was not just a more accurate description of the universe, but a change in the *cause* of events in the universe. For Ptolemy, objects in

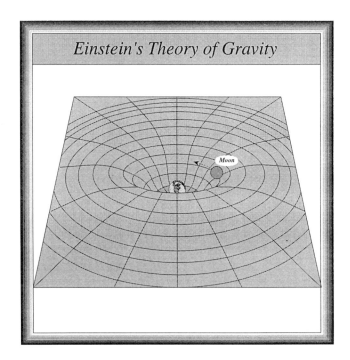

Figure 4-16 Albert Einstein's fame was established when he demonstrated that gravity, rather than being a force of attraction, is the result of matter "warping" or bending space. So Moon orbits Earth within this "warp." If Moon did not orbit Earth, it would fall toward Earth not because Earth pulls it, but because the space around Earth is "bend" and it is a property of objects made of matter to go to the least—bent space.

CHAPTER 4 / The Birth and Growth of Modern Science

the sky remained there because of the nature of their being eternal. The Church associated that property of eternalness with God, since the sky was His domain. Kepler described *how* those objects move in terms of the shapes of their orbits, using mathematical symbols that precisely described the movements. In his universal law of gravity, Newton explained *why* they moved as they did.

But from whence cometh this law of gravity and the mathematical formulae that describe the behavior of celestial objects? Well, you see, explanations for the ultimate origin of things keep getting pushed further and further into the background as new discoveries are made. We still search for ultimate causes. Modern science is unable to give complete explanations of the origin of things. And, it fact, science claims that by its very nature it is unable to do so. The Church did not need to fear the loss of God in peoples' lives. It is just that God is no longer necessary to explain the motions of the planets along in their orbits. One could argue, however, that He/She designed the laws by which they behave.

Where is God in Science? It is not because there is no *proof* of God's existence that scientists do not include the concept of God in their work, but because the very nature of the scientific process (the "how" of things) does not include a method of discovering the origin of principles (the "why" of things). Einstein spoke and wrote frequently about God in discussing his scientific ideas and theories, but he did not claim that God is the responsible agent for the workings of the universe.

I am often asked if I believe in God. You will not find mention of God in the typical Astronomy textbook. The general theme of textbooks describing the contents of space is that of human reason applying the laws of physics and chemistry to what we observe around us. For the scientist, the question is not whether there is a God, but rather whether it is necessary to invoke the use of God to understand and explain the universe.

At the very offset, of course, there is the problem of trying to agree on what "God" we are invoking. One person may believe that his/her "God" is needed, whereas

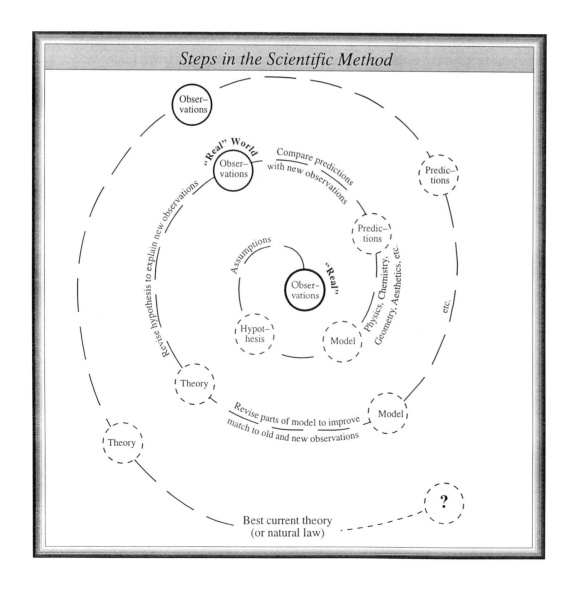

Figure 4–17 The scientific method is not unlike the manner in which humans make decisions about cause–effect relationships. We observe how things behave around us, and offer an explanation for how one thing affects another. We establish more general principles in hopes that our explanation can be applied to other events we observe. The scientist does the same thing, but uses in his/her explanation a wider range of laboratory or mathematical investigations than the average person would use.

another may not. Since scientists from around the world participate in the process of forming current theories about the universe, there is no consensus how a particular "God" fits into the picture. An individual scientist may use his/her personal "God" in the process of doing science, but the manner in which it is done is not included when reporting the results of the work to the larger community of scientist.

Even if we were to agree on a singular God, there would be intense disagreement on what instructions He/She issued us about the nature of the physical universe and how we are supposed to go about studying it. We fairly well agree that religions are useful (and perhaps necessary) for obtaining guidance in matters of morality and believe in God, but it is far from clear that any particular "God" has directed us not to explain the contents of space in terms of physics and chemistry.

What is Science?

Now that we have reviewed the events that gave rise to the scientific method, and witnessed the method by which it managed to replace the Earth-centered theory (and *theology*) with the Sun-centered theory (and perhaps theology), let us generalize about the process of science so we know what it is and what it isn't. What problems can it solve, and what problems is it not capable of handling?

Steps in the Scientific Method:

- A **hypothesis** is a single statement that requires testing. The claim that "All students love college" is a hypothesis that is subject to testing. The claim that "Earth spins about an axis" is a hypothesis that is subject to testing.

- A **model** is a description of a natural phenomenon. An Earth globe that rotates on a metal stand is a model, made, perhaps to test the hypothesis that Earth spins about an axis. It attempts to have us visualize what is invisible (i.e., we cannot observe Earth spinning). In that sense, a model is a metaphor — it is not the thing itself. It need not be a physical model, like a model of the solar system. It may be a series of mathematical equations. In many cases, it is impossible to construct a physical model because there is no physical analogy to the concept being expressed. For example, Einstein's famous relativity of time cannot be physically modeled.

    A scientific model attempts to explain what is observed in the "real" world, while at the same time predicting future behavior that is observable and measurable. It is this predictive aspect of a model that encourages scientific experimentation and observation — attempts to verify (or refute) the prediction. If the new observation does not match the prediction accurately, then the model either needs adjusting or an entirely new model is needed.

    Models must be verified by accurate observations. If two different models both explain the same set of observations with equal degrees of accuracy, other criteria are often times used to choose between them. *Aesthetics* is one of those criteria — the model that is simplest and most beautiful is generally preferred over one that is less so. The assumption is that nature is not complex, but behaves according to precise and simple principles. A model is never complete because there is always some degree of error in every observation. Therefore, all scientific models are tentative, some more so than others.

- A **theory** is a system of rules and principles that is capable of being applied to a wide variety of experiences in the "real" world, perhaps encompassing more than one model. A theory usually begins as a hypothesis, which, after extensive testing and application to a wide variety of circumstances, becomes more generalized in explaining the "real" world. Thus we refer to the theory of gravity. From our observations of the behavior of objects in space close to us, we have developed the theory of gravity. When we observe similar behavior in galaxies millions of light-years away, we evoke the theory of gravity to explain that behavior. The true test of a theory, then, is that it illuminate concrete features of publicly shared reality.

- A **natural law** is a theory that just about everyone accepts as being true, by virtue of it having been tested successfully over and over again. Please notice, however, that the "l" in law is always a small letter. There are never Absolutes in astronomy. Just about everyone accepted Earth-centered universe model for well over a thousand years before improved instruments were capable of showing errors in the predictions of the model.

Curiosity to Knowledge *The goal of science is knowledge.* Knowledge is obtained by sifting through evidence obtained by observation and experimentation. Curiosity Is the driving force behind the collecting of evidence. The pathway between curiosity and knowledge is science. But not all observations and experiments can be evaluated by science. There are a few assumptions that scientists use in evaluating evidence toward arriving at "truth:"

- The "*truth*" for which scientists search is a small "t" truth, not a capital "T" Truth. *There are no Absolutes or Truths in science.* There is simply a broad spectrum of knowledge we have about the universe, on one end of which are those things we don't know very well at all, and on the other end are the things that we think we know very well. As an example, we are reasonable certain that Earth is round and not flat. But we are not sure that there are black holes in space.

- *Humans are fallible.* No single scientist's observation or experiment can be trusted to establish a "truth" about something. Tentatively, it can. But the more people who make the same observation and/or achieve the same result with the same experiment, the more acceptable the truth is to the scientific community. The more radical the original proposed "truth" is, the more supporting evidence that is demanded. This is what is called *consensus* in science.

- *The universe operates on universal principles.* What happens once in a given set of circumstances will happen in the same identical manner under an identical set of circumstances. Actually, this assumption is implied in the previous paragraph. If a scientist performs an experiment under given conditions and obtains certain results, you and I should be able to obtain identical results under identical conditions. That is the basis for looking for *consensus*. In astronomy, we assume that the laws of physics and chemistry that operate on Earth, also operate in the far reaches of space and time.

- *Consensus in science is obtained by airing the conclusions of experimentation and observation before the entire scientific community.* This is usually accomplished by publication in the British journal *Nature*. Scientists doing research in the same or related subject of the article, will, if it has important implications in his/her field of work, respond to the article by doing further research to duplicate, extend, and/or refute the original research. Professional organization conferences, workshops, and informal discussions extend the contact between scientists doing work on the same and related fields. Gradually a consensus emerges.

LEARNING OBJECTIVES *Now that you have studied this Chapter, you should be able to:*

1. Describe the basic features of a scientific model, and the basic assumptions upon which it is based.
2. Describe the Earth-centered universe model in terms of the placement and motions of the observable objects in the sky.
3. Sketch the basic Ptolemaic model for the motions of the planets around the Earth, and how the use of epicycle and deferent explained retrograde motion.
4. State the assumptions upon which the Earth-centered theory was based.
5. State the strengths of the Ptolemaic model that led to its acceptance for over 1200 years.
6. Explain the features of the Church which, when added to the Ptolemaic model, resulted in the adoption of the model as a matter of Belief.
7. Describe the limitations of the Earth-centered theory that led to the proposing of the Copernican theory.
8. Describe the telescopic observations of Galileo, and how each contradicted the Earth-centered theory.
9. Explain why the Church was unable to accept the observations of Galileo, and state the basic roots of the Church that were being threatened.
10. Explain how annual stellar parallax was a major issue regarding the acceptance of the Sun-centered theory.
11. Explain how Brahe's contribution to astronomy illustrates one of the basic requirements for scientific model-building.
12. State Kepler's Three Laws of Planetary Motion, and explain how their discovery furthered the acceptance of the Copernican model.
13. Explain how Newton's explanation of natural motion in the universe furthered the acceptance of the Copernican model.
14. Explain how Einstein's model of gravity replaced that of Newton simply by challenging a couple of Newton's assumptions.
15. Contrast the general nature of scientific inquiry today with the nature of inquiry at the time of Aristotle and Ptolemy.
16. Define and use in a complete sentence each of the following **NEW TERMS**:

Age of Reason 80
Albert Einstein 85
Alchemist 78
Aristotle 69
Astrolabe 81
Clock 81
Conic sections 83
Copernicus 72
Copernican system 72
Cosmology 68
Deferent 71
Elliptical-orbit law 82
Epicycle 71
Equal areas law 82
Focus 82
Galileo 68
Geocentric system 71
Heliocentric system 72
Hypothesis 87

Johannes Kepler 82
Kepler's 3 laws of planetary motion 82
Macrocosm 78
Microcosm 78
Model 78
Natural Law 87
Parallax 76
Ptolemaic model/system 70
Ptolemy 70
Quintessence 78
Renaissance 68
Retrograde motion 71
Saint Thomas Aquinas 70
Scientific model 87
Sir Isaac Newton 84
Theory 87
Tycho Brahe 82
World-view 65

Chapter 5
Telescopes and Radiation

Figure 5–1 The Hubble Space Telescope (HST) *is deployed from the payload bay of the Space Shuttle Discovery in 1990. By placing the world-class telescope above the absorbing layer of Earth's atmosphere, astronomers obtain extremely detailed images of faint and distant objects in the universe.* (NASA)

CENTRAL THEME: How do astronomers collect radiation from celestial objects, and what do the different types of radiation tell us about those objects?

The Copernican revolution ushered in the methods of modern science, the trust in human senses and extensions of the human senses (instruments) to understand how the universe works. Other animals can run faster, can lift heavier loads, can detect a wider range of light-waves, can hear higher-pitched sounds, and even fly through the air. But humans have the ability to extend themselves beyond the range of the senses they were born with by building instruments and machines. The car, in that sense, is an extension of our legs. The computer is an extension of the brain. The telescope is an extension of the eyeball.

A robot space vehicle is not only an extension of the human eyeball, but the microwave link connecting it to Earth is an extension of the optic nerve connecting the eye to brain. In fact, just about every facet of modern astronomy is dependent upon these extensions, inasmuch as virtually all of our knowledge about the cosmos comes from our detection and analysis of radiation from distant objects. Astronomers do not study celestial objects: they study the radiation that presumedly comes from objects. I cannot prove that there are stars in space, but I can prove that I am detecting radiation from a point of light that I call a star.

Anyone who claims that stars, galaxies, and black holes exist is implying that there is a relationship between radiation and objects. If we see (receive radiation from) something, we presume it exists as an object. That, of course, is based upon our terrestrial experience, easily verified in the laboratory of Earth's environment. And if it works here, shouldn't it work with objects in space? Well, of course. But that was the outcome of the success of the Copernican revolution.

Actually, there are two portions of the cosmos that we *can* actually touch and study under microscopes here on Earth: meteorites and moon rocks. Meteorites are thought to come from the asteroid belt that circles Sun, and of course samples of Moon were returned to Earth as the successful outcome of the *Apollo* program. Information about everything else in the universe comes from our study of radiation.

Electromagnetic Energy

You are familiar with different types of light (red, blue, etc.) and radiation (X-ray, ultraviolet, etc.), but perhaps you didn't realize that all are members of a family of radiations called **electromagnetic radiation** (Figure 5-2). As family members, they have certain properties in common. The most important of these are *speed of travel, wavelength, frequency, and energy content*. It is the measurement of these properties that astronomers use in order to understand and explain the properties and behaviors of objects in space. Without exception, *all radiation travels at the same speed – the speed of light* (186,000 miles per second). To be strictly accurate, therefore, we should refer to it as the "speed of radiation" instead of the "speed of light." But the speed was determined and named before we even knew that other types of radiation existed, so it is now a part of the scientific vocabulary.

We humans are biologically equipped to detect (we are "sensitive" to) a very small percentage of the family members of radiation. An obvious one is **visible light**, that to which our eyes are tuned. But our bodies are also big eyeballs – in the sense that they detect (feel) radiation from Sun or fire or other sources of **heat**. *Heat* is also a type of radiation – one with lower energy content, longer wavelength, lower frequency. **Infrared radiation** is heat energy, so named because of the manner in which our skin senses it as heat. Other types of radiation may affect us, but they are not sensed as they interact with our bodies. Only later – after damage to our bodies becomes apparent – are we aware that the radiation was present.

Electromagnetic Spectrum Nature does not produce different *types* of radiation – *nature provides different objects and different processes that produce different amounts of radiation having different wavelengths, frequencies, and energy contents*. It is we humans who name and refer to intervals of wavelengths of radiation. We also make instruments capable of detecting particular intervals of wavelengths and name them according to the type of radiation they are designed (mostly) to detect. An X-ray instrument detects X-rays, a radio telescope detects radio waves, a gamma-ray instrument detects gamma rays. Figure 5-2 shows the entire range of wavelengths of radiation currently known, and the associated wavelengths in metric units. We call this range of radiation according to wavelength (or frequency or energy content) the **electromagnetic spectrum**.

Because the narrow range of visible wavelengths to which our eyes are tuned lies at approximately the center of the EM (for short) spectrum, we are tempted to attach special significance to visible radiation. However, we should avoid doing that. Although the study of the visible radiation that objects in the universe emit is certainly the foundation of astronomy, our ability to study other types has only recently been acquired. Prior to this century, astronomers knew only about what they saw with their eyes. As a result, discoveries in astronomy during the past few decades have been mind-boggling.

It is rather like your having learned to play a piano that had only ten keys, only to be introduced to a grand piano with a full set of 88 keys. Can you imagine how much more spectacular one of Brahms' piano concertos would sound! So we expect that there will be even more fascinating discoveries in store for us as we open new windows onto the universe. By "open new windows," I mean the building and use of instruments that tune into wavelengths of radiation that we have never searched for before. Let's now briefly explore the range of radiations available to us as we investigate the universe. Then we'll explore the behavior of radiation itself.

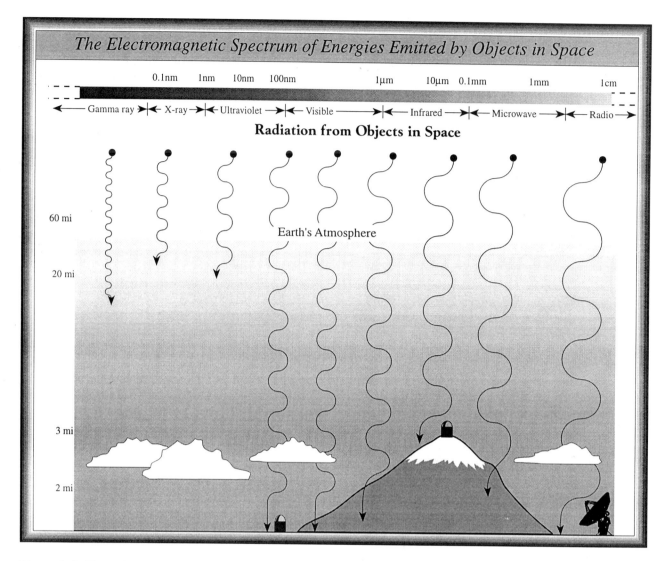

Figure 5–2 The entire range of wavelengths of electromagnetic energy emitted by objects in space attempt to penetrated to Earth's surface. But chemicals in the atmosphere selectively absorb out wavelengths in varying amounts so that much of the radiation never reaches the surface. The wavelengths are expressed in the metric system (meters) horizontally along the top of the diagram, and the heights above Earth's surface at which radiation is mostly absorbed are shown on the vertical scale.

Matter and Energy The contents of the universe can be conveniently divided up and placed into one of two categories — matter or energy. Matter is material substance, composed of atoms and molecules. The measurement of how much material (atoms and molecules) is in an object is referred to as the *mass* of the object. Planets, stars, and galaxies are made up of mass — as are our bodies. Radiation, on the other hand, is *energy*. Radiation travels between objects that have mass. Once energy is released (generated) by one object — such as a light bulb — it moves effortlessly through space at a constant speed (the speed of light) until it is absorbed by another object that has mass. Until it encounters another object, it continues to travel indefinitely through space.

I referred earlier to a bundle of energy as a *photon*. Radiation of any type is a stream of photons moving at constant velocity. Therefore, radiation is a stream of energy "bundles." As the photons travel along, they vibrate at a certain rate, very much like the strings of a guitar vibrate when they are stuck. If you look closely at vibrating strings, they look something like the illustration in Figure 5–3. The distance between the crests of two consecutive waves is called the **wavelength** of that vibration. For strings with identical lengths, there are more crests on the string that is vibrating rapidly than on the one vibrating slowly.

The **frequency** of a vibrating wave is therefore associated with the distance between crests (wavelength) of the wave. *The longer the wavelength, the lower the frequency – the shorter the wavelength, the higher the frequency.* This relationship between wavelength and frequency is more easily understood if you keep in mind that all waves travel at the same speed. So if you imagine them moving along and

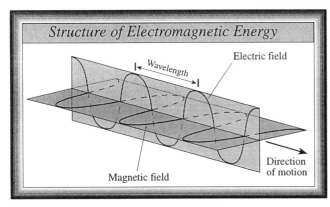

Figure 5-3 Electromagnetic energy, like sound energy, travels in the form of waves. Different types of EM energy have different wavelengths—the measured distance between the tops of the crests. It is a combination of electric and magnetic waves moving perpendicular to one another.

encountering a fixed, stationary detector, more crests of short wavelength radiation strike the detector in a given time interval than crests of long wavelength radiation.

It turns out that different types of radiation (i.e., different photons, or photons of different wavelengths or frequencies—however you want to think of it) are associated with different amounts of energy. Short wavelength radiation consists of more energetic photons than long wavelength radiation. For example, X-rays more easily penetrate through matter than ultraviolet radiation, visible radiation penetrates through matter more easily than infrared radiation, and so forth. The fact that ultraviolet radiation is responsible for sunburn whereas visible radiation does no harm to human skin, is evidence of the connection between energy content and wavelength. *The longer the wavelength, the less the energy content – the shorter the wavelength, the greater the energy content.*

Of course in these examples I am referring to sound, not radiation. Sound is the vibration in matter such as air, wood, or metal. But it travels at a much slower rate than radiation (only 760 miles per hour, compared to radiation's 186,000 miles per second). Radiation behaves like sound in that it has wavelength and frequency. Since a given wavelength is associated with a given frequency, just as in sound, knowing one allows for the determination of the other. Throughout this Book, I will consistently use the property of wavelength whenever it is important to talk about radiation. Just remember that frequency and wavelength are interchangeable. You are probably more familiar with the expression frequency than with wavelength. 94.5 on the FM dial, for example, refers to a frequency of 94.5 mHz (one **megaHertz** = one million cycles per second). This corresponds to a wavelength of about 10.5 feet.

Before exploring the fascinating subject of how astronomers use the properties of radiation to study objects in space, I will review the categories of radiation that scientists have agreed upon. These categories, of course, are based upon intervals of wavelengths of radiation.

Figure 5-4 This photograph of a portion of the sky was taken with film that recorded only the visible light coming from the stars. Since stars emit radiation at various wavelengths other than just visible, a photograph taken with film sensitive to those other wavelengths would reveal entirely different patterns than the one seen here. (Lick Observatory)

Visible Radiation I don't need to elaborate on visible light. You have seen numerous sources of visible light — candles, light bulbs, lightning, lasers, and so on. Obviously Sun and other stars give off visible radiation as well. The sky that fascinates us at night is adorned with those objects that reveal their presence by the visible radiation they emit (Figure 5-4).

There is more to visible light than meets the eye, however (no pun intended). You are aware of this fact whenever you see a rainbow in the sky, or play with one of those plastic diffraction disks that cause everything to have a rainbow around it. You are certainly familiar with the record album cover of *Dark Side of Moon*, by the popular Rock group *Pink Floyd*. It shows a beam of sunlight (or source of visible light) entering a **prism**, a triangular piece of glass that breaks the entering light up into separate wavelengths. This is the principle that operates to form rainbows in the sky (Figure 5-5). Tiny water droplets suspended in the atmosphere act like miniature *prisms* in breaking sunlight into its component wavelengths. Our eyes see (our brains record) the separate wavelengths as different colors. *A* **color** *is the language our brains use to tell*

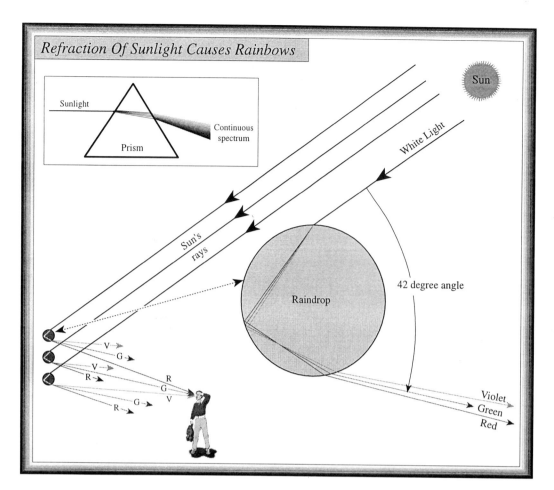

Figure 5–5 *Raindrops and prisms effectively do the same thing to sunlight as it enters one side and leaves the other. The wavelengths of visible radiation are separated so that the full range are seen as the colors of the "rainbow" when they emerge. In the case of raindrops, the 42-degree angle between entering and emerging rays of light is what gives the rainbow its angular radius of that same angle.*

us that our eyes have received (detected) particular *wavelengths of visible radiation*.

When you look around in sunlight, and see something that appears red, you might consider that it really isn't red. Well, certainly not in the sense that it is emitting red wavelengths. It is red because only the red wavelengths of visible light that weren't absorbed by the outer material of the object. All of the visible wavelengths coming from Sun are falling on the object, but the chemicals of which it consists selectively absorb out all of the wavelengths except for the red ones. Those are reflected away and into your eyes, interpreted by you as the object being red. This may seem like splitting hairs to you, or as being too technical to be of value, but most of our astronomical knowledge rests on this concept. So I am preparing your thinking for exciting discoveries ahead.

Infrared Radiation If you stand in sunlight, your skin detects the infrared radiation emitted by Sun as well. Other stars emit infrared radiation too, although not necessarily in the same amounts as Sun. Whereas hot stars emit more of their radiation at wavelengths shorter than visible radiation, cooler stars emit more radiation at wavelengths longer than visible radiation, which is infrared. Therefore, hot stars appear brighter in the sky if observed with instruments that detect short wavelengths, and cool stars appear brighter in the sky if observed with instruments that detect infrared radiation.

As I mentioned earlier, infrared energy is also called heat energy. This is not to say that objects that are hot emit infrared energy, nor that if you feel heat it is coming from something that is hot. Hot objects do emit infrared energy, but emit most of their energy at much shorter wavelengths. Do not associate heat energy with something that is hot. *Hot* is a subjective term. The term *heat* is not. It is a term that scientists use to denote a type of radiation. The amount of heat that an object emits can be measured. If you were able observe the nighttime sky with eyes sensitive to infrared radiation, you would see an entirely different sky than the one you see in visible light. You would have to form an entirely different set of constellations, and establish a new list of stars according to **apparent infrared magnitude**. The brightest stars you would see would be those that emit the greatest amount of infrared energy. The same would be true of all types of radiation. You may recall from Chapter 4 that this fact is one of the scientific objections to Astrology. The constellations of the zodiac — and by implication their interpretation — are based only on the visible light they emit. Astrologers are visible-light chauvinists!

Since cool objects emit more of their radiation at wavelengths longer than those of visible, astronomers use

94 SECTION 1 / Observing the Sky

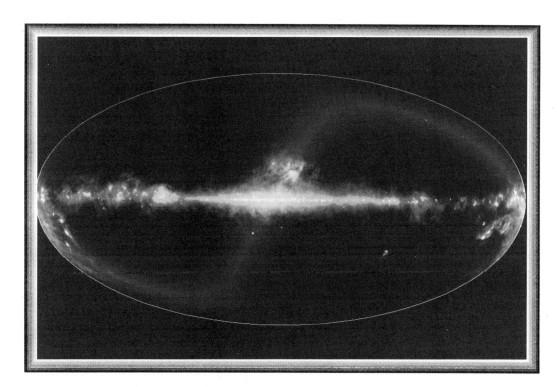

Figure 5-6 This infrared photograph of the entire sky was taken by NASA's Cosmic Background Explorer satellite (COBE). The plane of the Milky Way lies horizontally across the middle of the image with the center of the galaxy at the center of the image. The image is dominated by emission from interstellar dust in the galaxy. Images of such dust clouds are important to astronomers since it is within such clouds that star formation occurs. (NASA)

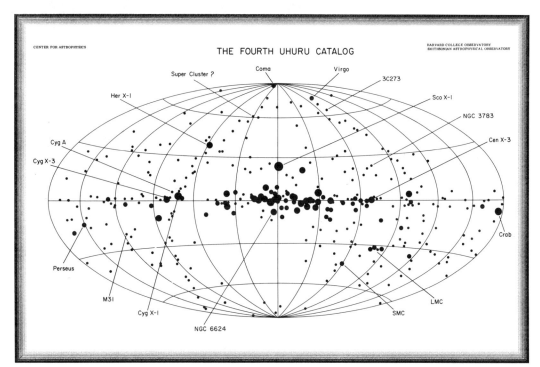

Figure 5-7 One of our early X-ray satellites (Uhuru) provided astronomers with information on locations of X-ray sources in the sky. The sizes of the dots of those sources plotted on this map represent their intensities. The one designated Cyg X-1 is believed to be a black hole swallowing a neighboring star. Compare this map with the infrared map above. (Harvard College Observatory/Smithsonian Astrophysical Observatory)

instruments sensitive to infrared in order to study clouds of gas and dust in **interstellar** space (between the stars). This is of special interest to astronomers because it is inside such clouds that stars form. Thus, infrared astronomy is the basis for our ability to study the birthplaces of stars, the OB clinics of star formation (Figure 5-6).

Ultraviolet Radiation You've also had experience with radiation whose wavelengths are shorter than those of visible light, but it hasn't always been a good experience. You feel the warmth of Sun as you lie on a beach — that is the infrared radiation coming from Sun. Later in the day you discover that your skin has turned red, a condition known as sunburn. While exposed to the light and heat coming from Sun, you were also receiving some of the **ultraviolet** radiation emitted by Sun. We are unable, with our bodies alone, to detect ultraviolet radiation — it is only after the fact that we learn of its earlier presence. Ultraviolet radiation has slightly shorter wavelengths than visible, so it has greater penetrating power.

Because ultraviolet radiation consists of shorter wavelengths than visible, its photons contain more energy, and these photons are therefore more capable of doing damage to delicate chemical arrangements such as human skin. This is rather obvious when we consider that ultraviolet lamps are frequently used in hospitals to sterilize instruments. In the field of astronomy, since ultraviolet radiation is associated with hot stars, astronomers use ultraviolet detectors to study the younger stars that are particularly hot. In order to generate photons that have high energy, high temperatures are needed.

X-Ray Radiation Stars also give off radiation at wavelengths even shorter than those of ultraviolet, and longer than those of infrared. The wavelengths that are shorter than the ultraviolet are called **X-rays**. You are familiar with X-rays, especially with their ability to penetrate. They are used in any number of practical applications, from diagnosing medical problems to detecting flaws in weld joints. Because they have great penetrating ability, X-rays are also dangerous to humans inasmuch as they are absorbed within the body. A human cell absorbing an X-ray can be damaged in the process. The value of detecting a serious problem within the body is greater in just about every case, however, than the chance that the X-rays will cause a new internal problem.

X-rays are associated with violent events in the universe, such as exploding stars, exploding galaxies, and the devouring of a star by a black hole. There are such incredible amounts of energy associated with explosive events that it is easy to understand why high-energy photons are generated in the process. Thus X-ray astronomy seeks to understand the processes of star death by searching for X-ray emissions in the sky (Figure 5-7).

Gamma Radiation There are many sources of harmful radiation in our environment. Intelligent exposure to them minimizes harmful effects. In general, radiation whose wavelength is shorter than that of visible light is capable of damage to the human body. The intensity of the source is important, of course, as well as the distance you are from the source. The very shortest wavelengths are **gamma rays**. You probably have very little contact with these highest energy photons, and hopefully never will. They are associated with atomic and nuclear processes, and radioactive decay. The most violent events in the universe, such as exploding galaxies, are detected by the use of gamma-ray telescopes.

Radio Radiation The longest wavelength — therefore lowest energy — photons are **radio waves**. Actually, the radio portion of the spectrum is a rather broad one. Since the inherently safe nature of radio waves makes them suitable for transmitting/receiving information, the **Federal Communications Commission (FCC)** is assigned the task of licensing the use of the wavelengths in this broad region. The FCC breaks the radio spectrum into usable intervals so that a given wavelength doesn't get so crowded that someone is unable to send a clear message to someone else. This is especially important in activities such as police and fire protection, aircraft control and even communications between Earth and astronauts in space.

It is easy to confuse radio waves with sound, and with what we hear on the radio. Sound, like radiation, consists of waves. But sound waves require a medium through which to travel, like air or water or wood. Radiation, on the other hand, does not. It requires only space. Sound energy containing information (like musical notes) can be encoded onto a radio wave and sent through space to be received and converted back into sound energy for someone else to listen to. So a radio receiver is simply a device for receiving radio energy, extracting the information contained thereon, and then attaching that information to sound waves. In order to detect the weak radio signals from space, astronomers use very large receivers (Figure 5-8).

Figure 5–8 The 1000–foot radio telescope located at Arecibo, Puerto Rico, has a larger collecting "dish" than all other radio telescopes combined. It has been responsible for notable discoveries, as well as having sent out into space *the only* intentional *message to any technical civilization capable of detecting it.* (National Astronomy and Ionosphere Center operated by Cornell University under contract with the National Science Foundation)

We will soon see how the *natural* information on radio waves (as well as other wavelengths of electromagnetic energy) tell astronomers much about the contents and behavior of celestial objects. One of the most important uses of radio energy in astronomy is in the detection of cool clouds of **hydrogen** gas that occupy interstellar space. Because this lightest-of-all gases makes up 90% of all the matter in the universe, its presence in space is fundamental to our understanding of the universe.

Because of the extensive use of non-wire communication systems throughout the world today, a small portion of the radio spectrum has received its own specific designation as the **microwave** region. You are familiar with this particular type of radiation because of its extensive use in the kitchen and the transmission of telephone conversations around the world. In the field of Astronomy, the study of microwave radiation from space has provided convincing evidence that a Big Bang was responsible for the origin of the universe.

Measuring Radiation at Earth's Surface

By now you should have gotten the idea that to fully understand the contents and behavior of objects in space, it is essential that we open not only our visible eyes onto the sky, but our ultraviolet, infrared, X-ray, and all the others as well. Only by doing so can we properly interpret what is going on out there. In principle, at least, we should take photographs of the sky at each and every wavelength of the spectrum, line them up in front of us, and examine each and every one of them prior to making any hypothesis about the nature of the objects in space. But there is an inherent obstacle to our being able to do that — our atmosphere.

Radiation and the Atmosphere As much as we must have air to breathe, from an astronomer's point of view, our atmosphere is a serious limitation to our study of space. Not only are there cloudy nights and turbulent winds that blur the images seen in telescopes, but the atmosphere selectively absorbs out most of the radiation coming from space even before it hits Earth's surface. Therefore, detectors sitting on Earth's surface tuned to receive those particular wavelengths "see" nothing.

Up until a few decades ago, astronomers on Earth's surface were limited to the study of only two types of radiation from space — visible and radio. That is because those are the only two intervals of wavelengths in the EM spectrum that get through the atmosphere largely undisturbed. The visible and radio **atmospheric windows** are clear because those types of radiation are not blocked by components of the atmosphere. The other windows of radiation are dirty. This is not an all-or-nothing mechanism, however — some small percentage of every type of radiation gets through the atmosphere. The fact that you feel Sun's heat and occasionally even get sunburn is evidence that some infrared and ultraviolet get through. The wavy lines in Figure 5-2 represent the different wavelengths of radiation that hit Earth's atmosphere from objects in space. The lower end of the wavy line represents the approximate height above Earth's surface at which that particular wavelength is mostly absorbed. Those that extend all the way to Earth's surface represent those wavelengths that are not absorbed by chemicals in our atmosphere.

Ozone Layer Those wavy lines that fail to reach Earth's surface represent those wavelengths that are more or less absorbed within the atmosphere. For example, you are aware of the concern and controversy over the depletion of the **ozone layer**, and the resultant "hole" in the *ozone layer* over the Antarctic. This layer of the chemical *ozone* is uniformly spread around Earth at a height of 15 to 30 miles above the surface. Ozone consists of three atoms of oxygen bound together as a molecule, and is responsible for absorbing some of the damaging ultraviolet wavelengths of radiation that are produced by Sun. In the process of absorbing ultraviolet radiation, ozone molecules are broken into oxygen atoms which in turn reform into normal oxygen molecules.

But it has been shown that certain products of modern technology that are released into the atmosphere perform the same breaking down of the ozone, which leaves the ultraviolet radiation from Sun free to penetrate through to Earth's surface. It is not particularly healthy for humans to be exposed to increased amounts of ultraviolet radiation. There is rather convincing evidence that an epidemic of skin cancer cases in the world is a result of what we are doing to Earth's atmosphere. For the point of view of the astronomer who wants to study objects in space that emit quantities of ultraviolet radiation, however, the ozone layer is an obstacle.

Artificial Satellites With the advent of artificial Earth satellites in the late 1950s, astronomers are now able to overcome the limitations of the atmosphere by sending telescopes into orbit around Earth above the absorbing layers of the atmosphere. From that vantage point, they record the receipt of wavelengths of radiation normally absorbed by the atmosphere, place the information in digital form on radio waves, and transmit the information to a receiving station on Earth's surface. Since the radio waves can penetrate through the atmosphere, astronomers extract from it the information originally contained within the radiation that would normally have been absorbed by the atmosphere. That information tells them about the objects responsible for emitting it in the first place. In other words, artificial satellites rescue information from space that would normally be lost in the atmosphere.

Consequently, within a very brief period of time (compared to the birth of modern science in the 1600s), incredible discoveries have been made and mind-boggling

★★★★★

events have been detected. It is as if for centuries you had been confined to a house having extremely dirty windows, allowing you only fuzzy images of what was outside the house. Then, magically, someone came along to wash your windows, revealing all the subtle details of trees and birds and flowers. And even some things you were completely unaware of, such as distant lakes and a mountain range.

You and I live during what is being called by the media the **Golden Age of Astronomy**. In fact, any astronomer today tell us that there is far more data provided by these artificial Earth satellites and new telescopes and instruments than there are astronomers to study it. In other words, there are probably profound discoveries hidden somewhere in the reams of computer printouts on some astronomer's desk, just waiting for someone to pick it up and recognize it for what it is. And we have only recently opened those windows of opportunity. Keep tuned.

Actually, there haven't been very many satellites sent up above the atmosphere for the purpose of studying those normally-absorbed wavelengths. Long lead times are necessary for making detectors sensitive to those wavelengths, and funds available for pure research compete with many other government programs. Ultimately, taxpayers absorb the cost of scientific research, most of which is not something many of us are anxious to vigorously support. Now that the **Space Shuttle** program has recovered from the *Challenger* disaster of 1986, satellite launches are again revealing vast new vistas for modern astronomers.

The **Hubble Space Telescope (HST**, for short), which initially received considerable bad press, has been an enormous success in spite of its shortcomings. In contrast to the early space program launches, which were one-shot affairs, the *Space Shuttle* vehicles are designed to be reusable cargo vehicles. In addition to performing in-orbit studies of both Earth and space, the Shuttle can carry and inject satellites into orbit either around Earth or into orbit to other planets or Sun. It then returns to Earth to pick up a new payload.

Measuring Radiation Most of the matter in the universe, it seems, exists in the form of stars. Stars are the basic building blocks of the universe. Stars do not give off just one type of radiation or another. As you know from experience, Sun gives off at the very least visible, infrared, and ultraviolet radiation just as other stars do. How, then, does one star differ from another? Because *different stars emit different wavelengths of radiation in different amounts.*

Generally, large, massive stars "burn" hot and emit shorter wavelengths of radiation (ultraviolet) — whereas smaller, less massive stars "burn" cool and emit longer wavelengths (infrared). We therefore refer to some stars as being "infrared" stars, some as being "ultraviolet" stars, and some (like Sun) as being "visible" stars. That is not to say that each star gives off only that particular type of radiation — only that most of the radiation it emits falls within the wavelength interval associated with that type of radiation.

Photometers measure the number of photons received from a star, or from any object for that matter. Astronomers tune the instrument to detect and record only those photons of a particular wavelength. They then scan the entire spectrum of wavelengths from ultraviolet to infrared emitted by a particular star, counting the number of photons received at each particular wavelength. With that information, they construct an **energy curve** for that star. The energy curve for Sun (Figure 5-9) is a smooth, uninterrupted curve running through three types of radiation — but it obviously "peaks" at visible wavelengths. Our star emits most of its energy at visible wavelengths.

Energy Curves Looking at the visible part of the spectrum in the same Figure 5-9, notice that the particular region within the visible spectrum where Sun's energy curve peaks is the yellow. Our Sun is a yellow star, in the sense that although it emits radiation at all wavelengths, it emits more (and consequently we receive more) yellow wavelengths than any other. Mathematically, the shaded area within that region of the spectrum under the curve is a measurement of the total energy emitted by Sun.

Now look at the energy curves for a hot star and a cool star. Notice a couple of interesting differences between those curves and that of our Sun. First of all, the area under the hot star's curve is greater than that of Sun, and the area under the cool star's curve is less than that of Sun. Are you surprised? Hot stars emit more total energy (photons) than Sun, and cool stars emit less total energy (photons).

And where does the hot star's curve peak? In the ultraviolet. Does that mean hot stars are invisible to us? No, they are still visible, because visible photons are still being emitted, just not in amounts as great as the ultraviolet. At what wavelengths in the visible portion of the spectrum does this hot star emit most of its radiation? At the shorter, or blue, wavelengths. So you shouldn't be surprised to learn that hot stars appear bluish!

Likewise, you shouldn't be surprised to learn that cool stars appear reddish! If only we adjusted our vision so that we could observe the sky at ultraviolet wavelengths, we would see hot stars appearing brighter than they do at visible wavelengths. And infrared vision would cause the cooler stars to appear brighter than they do at visible wavelengths. Isn't this exciting? It all makes so much sense. These are things you have undoubtedly experienced yourself, but never thought of explaining. We're doing science now. I hope you will take advantage of what you are learning, and locate some hot stars and cool stars in the nighttime sky!

Radiation and Sight In light of the above discussion, are you prepared to explain why — with all of the wavelengths of visible radiation that fall on Earth's surface — the human eye is most sensitive to those close to the green wavelengths of visible radiation? Why don't we see infrared or ultraviolet? Well, if that were our primary mechanism for detecting the environment, we would be marginally blind and wouldn't easily be able to negotiate our way around. There just aren't many of those wavelengths that penetrate through the atmosphere to reach Earth's surface to reflect off things so that they can be seen.

We could have developed much larger eyeballs to collect more of the ultraviolet or infrared photons so that we could see well enough to get around. Nature — according to the evolutionary theory — had to choose between making huge eyeballs to collect the feeble amounts of infrared and ultraviolet radiation, and making smaller eyeballs that use what is most readily available (visible radiation).

It "chose" the latter. Nature therefore "made" the smallest possible receptors (eyes) necessary for allowing us move around to find food and a mate — two fundamental requirements for survival of the species. Not all creatures on Earth share this strategy, of course. There is a wide range of uses of the electromagnetic spectrum by terrestrial life forms. Rattlesnakes, for example, have a small infrared detector that allows them to "see" warm prey in the dark.

What if an alien from a planet orbiting an ultraviolet star were to visit us? Or one from an infrared star? Presumedly, limited to their bodily organs alone, they would be blind. We would be blind on their planets as well. Hollywood doesn't seem to consider this in making science fiction movies. The aliens usually step out of their spaceship in broad sunlight to ask directions to our Leader. In a very real sense, we humans remain visible light chauvinists!

Telescopes

Galileo ushered in modern science through the use of the telescope in observing objects in the sky. It wasn't that he was the first to use the telescope. It was that he was the first to use it for research purposes — to observe objects in the heavens, record their features and behavior, and report his conclusions to a wide audience of other scientists. That is basically what modern astronomers do, except that the sophistication of modern telescopes is beyond the imagination. The design, placement, and operation of, and communication between modern observatories is an entire science in itself. You owe it to yourself to visit one sometime. Inasmuch as they are generally funded by universities or other nonprofit organizations, they usually have some type of public program that allows access to at least look at some of the instruments. You'll learn shortly why you won't in most cases be invited to look through any of the telescopes.

Placement of Observatories Given the fact that Earth's atmosphere is a hindrance to astronomical studies of objects in space, there are a few things that astronomers do to maximize success. First of all, major observatories are built atop mountain peaks in order to be placed above as much of Earth's atmosphere as possible. For example, the largest optical telescope in the world was just recently installed atop 13,600-foot Mauna Kea in Hawaii (see Figure 5-10). At this elevation, the telescope is above 60% of the

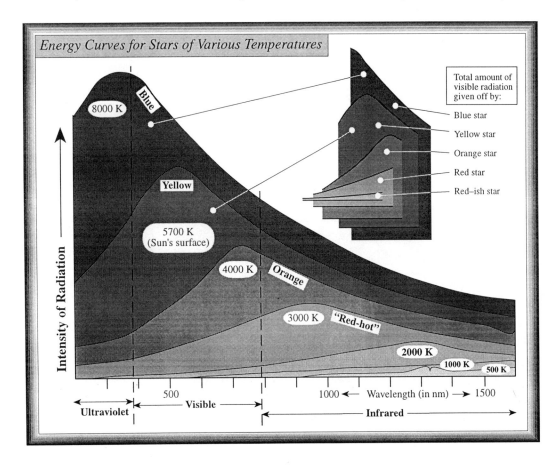

Figure 5–9 Astronomers use a photometer that can be "tuned" to different wavelengths to measure the amounts of radiation of each wavelength coming from a star. By plotting the amount of radiation at each wavelength on a graph, they obtain an energy curve for that star. By comparing the energy curves of different stars, one can understand the range of stars by type.

CHAPTER 5 / Telescopes and Radiation

atmosphere, and above 90% of the water vapor that is responsible for absorbing infrared radiation from celestial objects (notice the observatory on top of the mountain in Figure 5-2). The telescope is being used to explore those regions of our galaxy in which stars are thought to be forming (the infrared portion of the spectrum).

Seeing Conditions Astronomers (amateurs, as well as professionals) generally conduct their observing when the objects they want to observe are as high in the sky as possible (in other words, while crossing the meridian). So in Figure 5-11, the light from Object A travels through more of Earth's atmosphere than does the light from Object B. Object A will appear more blurred than Object B. This is why, incidentally, when you look carefully at the night sky, stars close to the horizon appear to "twinkle" more than those higher in the sky.

The astronomer regards the atmosphere as a dirty window, not just because its chemical makeup (either natural or human-made) filters out much of the radiation from space, but also because it is in constant motion. This atmospheric motion causes objectionable amounts of blurring on images taken with a telescope under these conditions. That is why most newly-built observatories are installed on high mountain peaks. There, they are not only above the radiation-absorbing layers of the atmosphere, but there is less turbulence to affect seeing conditions.

You are surely familiar with this distortion effect, since you have noticed that stars appear to "twinkle." You recall the children's song: "Twinkle, twinkle, little star, . . ." It is tempting — especially as a child — to think that the stars are actually twinkling, but that is not the case. Since the light from stars pass through the moving currents of air on the way to Earth's surface, it travels back and forth with those currents. By the time it reaches us, it does not appear to be coming from a single spot.

If a star is recorded on photographic film during a long time exposure, it is not recorded as a dot as is desired, but as a blurry circle. A larger object (such as a galaxy) appears fuzzy. Someone has likened the study of celestial objects from Earth's surface to the study of birds while sitting on the bottom of a lake! The quality of the sky conditions at any particular moment is called the **seeing** conditions. A **10** night means that seeing is superb, a **1** night means that you would probably not even want to set up your telescope.

The following is one of the ways in which an astronomer determines the *seeing* conditions before an observing session. Try it the next time you go out to observe the sky. The higher above the horizon you must look in order to see stars not twinkling, the worse are the seeing conditions. Usually, stars close to the *zenith* do not twinkle. Start by looking directly overhead and gradually lower your gaze toward the horizon. The further down toward the horizon you observe without seeing twinkling stars, the better the seeing conditions. This definition of seeing, of course, is above and beyond the obvious ones that limit any viewing session — weather conditions, scattered light from Moon, and the scattered light of metropolitan areas close to the viewing site. Are you surprised that viewing sessions are planned months in advance, especially at major astronomical observatories?

Figure 5-10 Having the largest optical telescope in the world on top of a 13,800-foot peak allows astronomers on Mauna Kea in Hawaii to obtain unsurpassed images of objects deep in space. Although the newly-installed telescope is presently only in its calibration stage, it has already made some important discoveries. It is housed in the dome on the far left. There are several other telescopes operated by astronomers from around the world on the peak as well. (Tom Bullock)

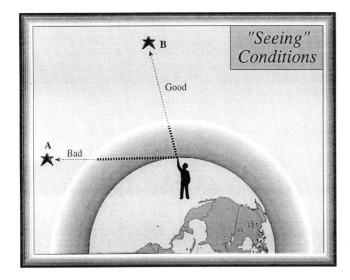

Figure 5-11 Astronomers attempt to study objects when they are as high in the sky as possible. Looking at radiation from an object close to the horizon results in considerable distortion because the radiation must travel through more of the moving layers of atmosphere. Looking at objects directly overhead allows the clearest view.

Airborne Telescopes You must have the impression by now that the higher above Earth's surface astronomers place their instruments, the clearer and brighter are the images of celestial objects. And also the wider the range of wavelengths that are detected and studied. And so you are correct. Hot air balloons lift delicate instruments to altitudes of 25 miles to study objects such as Sun. Rockets carry instruments up to similar altitudes to study the effects of Sun on the upper atmosphere of Earth (the Northern Lights).

The United States space agency — **NASA** — operates a large jet aircraft out of its headquarters at Moffet Naval Air Station in California solely for the purpose of astronomical observations. Inside the aircraft is mounted a 36-inch telescope. Called the **Kuiper Airborne Observatory**, it flies to altitudes of 50,000 feet while studying selected astronomical objects. This altitude places it above 99% of the water vapor responsible for the absorption of infrared radiation. It was during a routine investigation of the planet Uranus in 1975 that this airborne telescope discovered rings around that planet (Figure 5-12).

Telescope Features The easiest way to think of telescopes is that they are light funnels. *The goal of using a telescope is to collect as many photons as possible coming from an object, and focusing them to a single point at which a detector is located.*

Figure 5-12 NASA operates an airborne telescope in an aircraft that can get high above some of the radiation-absorbing layers of the atmosphere. Perhaps its most notable discovery were the rings of Uranus. Notice the opening for the telescope just in front of the wing. (NASA)

★★★★★

The collection and focusing of the photons is accomplished by either an *objective lens* or an *objective mirror* used in combination with an eyepiece. The detector can be the human eye, film in a camera, a photometer, a CCD, or any number of other instruments that analyze the characteristics of radiation.

The purpose of collecting photons is that the wavelengths of photons received, the number of each wavelength received, and their distribution over an area of the sky tell us something about the nature and behavior of the object(s) emitting those photons. In the next Chapter, I explain in more detail just what the connection is between radiation and matter that allows us to determine so much information about celestial objects. But what astronomers basically want to detect in their telescope is a *large, bright, clear* image of whatever they are studying. These three requirements are met in slightly different ways in different types of telescopes.

Refracting Telescopes I begin the discussion of the inner workings of telescopes with the very first telescope used, that of Galileo, the instrument that so revolutionized the way we obtain knowledge about the universe. Figure 5-15 is a cross sectional diagram of its operation. A close look at the shape of the objective lens tells you why the word "lens" was chosen for the glass installed in Galileo's newly-discovered telescope. The word lens comes from the Latin *"lentis"* or lentil. This type of telescope, in which radiation passes through a piece of glass (lens) on its way to the eyepiece is called a **refracting telescope**. The word *refracting* is used because **refraction** is the formal term used to specify the action of a lens in splitting up radiation into component wavelengths as it passes through.

The same general design is used in the second largest telescope of its type in the world: the 36-inch telescope at *Lick Observatory* near San Jose, California (Figure 5-13). When we refer to a telescope by its size (the 36-inch, in this case), we are expressing the diameter of the **objective lens**. That is the lentil-shaped piece of glass that collects the radiation and focuses it to a point. The objective lens of the Lick refracting telescope is 36 inches in diameter. Obviously, the larger the objective lens, the greater the number of photons collected, and the brighter the object will appear to the detector. *Brightness*, you recall, is simply a way of expressing *the number of photons received during a given interval of time.*

Figure 5-13 The 36-inch telescope on Mount Hamilton in San Jose, California, is the second largest of its kind in the world. Although it weighs several tons, it is so precisely balanced that it is easy to move with a single finger. (Lick Observatory)

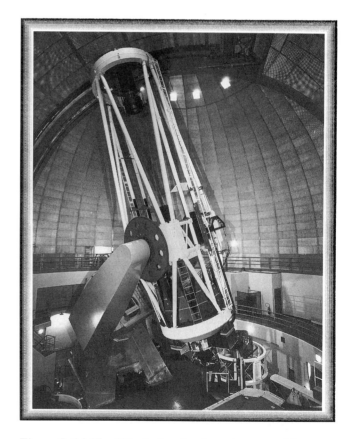

Figure 5-14 The 120-inch reflecting telesope on Mount Hamilton, although not close to being the largest of its kind, is notable because of the use of special electronics with which to image distant objects. The instruments are mounted at the bottom of the telescope where observers used to look through an eyepiece. Electronics have replaced the human eye in most cases of modern astronomy. (Lick Observatory)

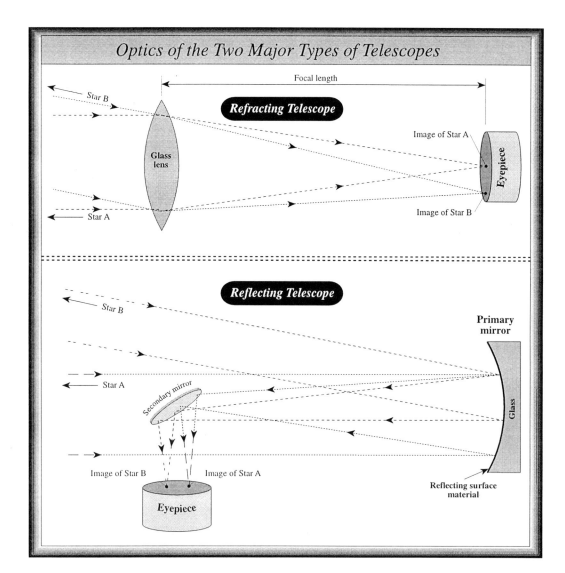

Figure 5–15 *A cutaway view of the two principle tpyes of telescopes used by astronomers reveals that the principal components of a telescope are designed to focus as much of the incoming radiation from an object to a point on a detector of some type.*

Collecting Power Astronomers don't just want to make objects brighter. They want to detect and study objects that are too dim to be seen with the human eye, either because they are too far away or are very cool objects. Recall that cool objects give off fewer photons than do hot ones. So the larger the diameter of a telescope, the more distant are the objects that it is capable of detecting. Mathematically speaking, the number of photons that a given lens collects depends upon the total surface area of the lens, which is proportional to the square of the diameter of the lens. We refer to the ability of a given telescope to collect photons as the **collecting power**.

This relationship between collecting power and diameter of lens tells us that the 36-inch Lick telescope collects 324 times more radiation than Galileo's 2-inch telescope (36 squared divided by 2 squared equals 324). It turns out that besides determining the brightness of an object, the diameter of the lens of a telescope also determines the degree of clarity of the image — the larger the diameter, the greater the clarity or what astronomers call **resolution**.

This emphasis on the diameter of a telescope as a measurement of its value may seem odd at first, because people most often associate *magnification* with the quality of a telescope. When you enter the telescope section of a store in a large shopping center, it is usually the magnifying power of the telescope that is emphasized in the displayed advertisement: "This telescope is capable of 300 power magnification!"

Magnification Of course, astronomers are interested in seeing the image of an object large enough so that it can be easily studied. But not if it is at the expense of it being so dim that fine detail cannot be seen. Let me pose the choice to you in another way. What would you prefer to see — a large, dark, blurry Saturn, or a smaller, brighter, clearer one? Since astronomers want to extract as much information as possible from the study of radiation, the brightness and clarity of an image are generally the most important considerations when using telescopes of any type. Therefore, the diameter of the objective lens is a much better

CHAPTER 5 / Telescopes and Radiation   **103**

indicator of the value of the telescope than that of magnification.

Because of the manner in which lens work — the field of study is called **optics** — the greater the distance between the objective lens and the detector, the greater the magnification of the object being studied. This measured length in the telescope is called the **focal length** of the telescope. So astronomers build telescopes with large diameters (for collecting power and resolution) and great lengths (for magnification power). The 36-inch refractor at Lick Observatory is 57 feet long, for example!

Eyepieces The magnifying power of a telescope is not fixed as soon as its length has been determined, however. Magnification can be changed by using different **eyepieces** in conjunction with the telescope. By using eyepieces of different focal lengths, the astronomer changes the apparent size of the object being studied. There is a cost involved, however. Keep in mind that the same number of photons are collected by the objective lens of the telescope, regardless of what eyepiece is on the telescope. So if the astronomer chooses an eyepiece that creates a large image size over an eyepiece that creates a small one, the same number of photons are spread out over the larger image size, with the result that the image is dimmer. It will also be fuzzier, because the resolution is affected by adding the more powerful eyepiece as well.

Resolution As I mentioned, astronomers refer to the ability of a telescope (or any optical device) to provide clear images of objects as the *resolving power*. The matter of obtaining fine detail in any telescopic image is so important that astronomers go to great lengths to improve it. Obtaining a bright enough image is pretty straight forward — just built a telescope with as large a diameter as possible. But brightness doesn't make much sense if there isn't enough detail in the image to allow for any conclusions about the object being imaged. Think of it this way. I hold up two fingers as I walk away from you observing my hand. At some point in my walking away, you are unable to tell if I have two fingers up any more. You lose the ability to resolve the difference between the two fingers (Figure 5-16).

So the resolution of an optical device is expressed as the minimum distance between two objects that the device can detect as two separate objects. In the example of my fingers, you could have picked up a pair of binoculars and seen both of my fingers as separate objects. And as I walked away further, even the binoculars would fail to resolve the two fingers as separate objects. Every optical device has a limit to its resolution. Resolution is also affected by the wavelength of radiation being detected. Longer wavelength radiation such as radio is inherently less capable of providing fine detail than is visible radiation. I will explain how radio astronomers deal with this problem shortly.

Every astronomer selects a combination of telescope-eyepiece design in order to study an image whose size and brightness and sharpness of detail allows him/her to obtain the maximum amount of information. Viewing objects with telescopes is an art. Each astronomer prides himself/herself in the choice of equipments in order to obtain a good image. If you have the opportunity to attend a sky-viewing session (usually called a **star party**) at which amateur astronomers offer the public an opportunity to look through their telescopes, you will be impressed with the manner in which they go about their art. I use the word "amateur" in the sense that they don't earn their living using the telescope (Figure 5-17).

Limitations of Refracting Telescopes Going back to the cutaway diagram of Galileo's telescope, and looking closely at the shape of the objective lens, you notice that its edge is shaped very much like that of a prism. A beam of visible light passing through a triangular piece of glass leaves the opposite face as all of the wavelengths of visible light in what we call the **continuous spectrum**. It is continuous because all of the wavelengths are present from the shorter, blue wavelengths to the longer, red wavelengths. You have seen the continuous spectrum in a rainbow, the result of visible light from Sun being broken up into different wavelengths by water droplets making up clouds in the atmosphere. Figure 5-5 illustrates how the rainbow is seen by an observer.

Since different wavelengths are spread out by the edges of a lens, not all of the photons are focused to a single point. Some converge in front of the eyepiece, some behind.

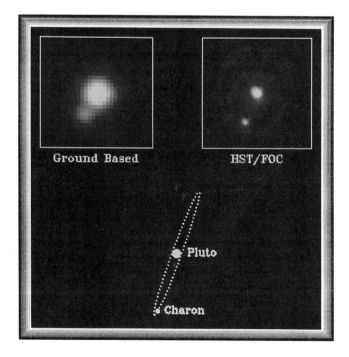

Figure 5-16 This HST image of Pluto and its satellite Charon clearly resolves the separation between them. For comparison, the best ground-based image ever taken to date is shown at the upper left. (NASA)

104 SECTION 1 / Observing the Sky

Someone looking into the eyepiece will be unable to see all of the photons that entered the telescope. And that, of course, limits the information desired. This effect is referred to as **chromatic aberration**, and is an inherent limitation of the refracting telescope. There are ways of minimizing the effect of chromatic aberration by using compound lens, but there are other disadvantages to this type that favor the use of another type of telescope, the *reflecting telescope*.

Reflecting Telescopes Faced with the limitations of the refracting telescope first used by Galileo, Sir Isaac Newton found that he could avoid chromatic aberration by allowing light to reflect off the surface of a highly reflective material to a focus point. In so doing, the light does not pass through any glass. Hence this type of telescope that uses an objective mirror is called a **reflecting telescope**. Figure 5-14 shows one of these, the 120-inch telescope at Lick Observatory. The functional characteristics are identical to those of refracting telescopes — the diameter of the objective mirror determines the collecting and resolving powers, the focal length determines magnifying power. The largest one in the world is located on Mauna Kea on the Big Island in Hawaii. Its collecting power is equivalent to that of a 33-foot mirror!

Reflecting vs Refracting You may be asking why reflecting telescopes are so much larger than refracting telescopes? Well, that is mostly a matter of engineering. Since light has to pass through the objective lens of the refracting telescope, the lens has to be rigidly supported along its edge, a difficult task in light of the tremendous weight of such a large piece of glass. Also, the weight of the glass has to be counterbalanced at the opposite end of the telescope in order to have the precise balance needed for rotating the instrument to keep up with the motions in the sky (Figure 5-13).

But with such heavy weights at either end, the tube sags, and the photons entering the telescope are no longer focused to the point where the detector is located. The result is a distorted image, and the amount of distortion changes as the angle of elevation of the telescope changes. If the telescope is pointed directly upward, there is no sag at all. But as it is pointed more and more toward the horizon, the sag is greater and greater. For that reason, the 40-inch refracting telescope located at Yerkes Observatory in Wisconsin is probably the largest that will ever be built, on Earth at least (Figure 5-18). On a smaller planet, or on Moon, the gravity is less, the sag is less, and therefore a larger refractor could be built.

So why doesn't gravity affect the reflecting telescopes? Looking at the cutaway diagram of Figure 5-15, notice that since light doesn't have to pass through the objective mirror, the mirror can be supported from underneath and thereby avoids the sagging problem. Well, at least to a point. Even the supporting mechanism eventually sags when it reaches a certain size. The largest single-mirrored telescope in the world is the 236-inch telescope in Russia. Some of the major telescopes of the world are:

- Mount Wilson   Hooker 100-inch
- Mount Palomar   200-inch
- Kitt Peak   McMath solar telescope
                                     Mayall 4-meter
- Mauna Kea   Keck Telescope 10-meter
                                     Keck II 10-meter

Figure 5–17 The very best way to be introduced to the wonders of the sky is to participate in a "star party." Amateur astronomers make their telescopes available to the public to look at planets, nebulae, and galaxies. Don't miss the opportunity. (Tom Bullock)

CHAPTER 5 / Telescopes and Radiation

- Russia — Bolshoi Teleskop 6-meter
- Chile — Cerro Tololo 4-meter reflector
  European Southern Obs 4-m
  New Technology Telesc 3.6-m
  Las Campanas 2.5-meter
- Australia — Siding Spring Mountain 2.3- and 3.9-meter
- Canary Islands — William Herschel Telescope 4.2-meter

There are some 63 telescopes equal to or greater than 60 inches (1.5-meters). There are about 70 telescopes in the 1.0- to 1.5-meter range, and more than 200 professional telescopes smaller than 1.0 meters.

Multi-Mirror Telescopes Despite their advantages over refracting telescopes, reflecting telescopes still suffer from having to operate within Earth's gravitational environment. The giant mirrors do sag somewhat, and their sheer masses prevent them from adjusting rapidly to changes in the surrounding air temperatures.

Since glass expands and contracts slightly with temperature change, and the shape of the mirror is critical in focusing radiation to a single point, astronomers must wait for the shape of the mirror to stabilize after sunset before making observations.

Thus the current fad of Earthbound telescope technology is the **multi-mirrored reflector telescope**, the aim being to use several small mirrors (each of which cools rapidly with temperature change) in place of a single large mirror. The first of its kind was the **Multiple Mirror Telescope (MMT)**, located in Arizona (Figure 5-19).

It had a set of six 72-inch round mirrors, all precisely controlled by computers so that the radiation falling on each was electronically focused to a single point. The telescope was equivalent to a single mirror 176 inches (15 feet) in diameter. There is probably no limit to the size of a telescope that can be built in this manner.

The **Keck Telescope**, a 36-mirror telescope designed by the University of California and the California Institute of Technology, was completed in April 1992, and installed on the 13,600-foot peak of Mauna Kea in Hawaii. It has the equivalent light-gathering power of a 33-foot mirror, making it the largest optical telescope in the world. This is almost twice the diameter of the world's third largest, the 200-inch (17 feet) at Mt. Palomar in California.

The telescope is used primarily as an infrared telescope, inasmuch as its location at 13,800 feet places it above 90% of the water vapor in Earth's atmosphere. It is water vapor that is responsible for filtering out most of the infrared radiation that attempts to enter through our atmosphere from celestial objects like Sun.

The Keck telescope is revolutionizing our knowledge about those objects and events in the universe that reveal themselves in the infrared portion of the spectrum. Examples of where progress is occurring — imaging planets in orbit around other stars, detailed study of star-forming regions, confirming the existence of a super massive black hole at the center of our Galaxy, and detailed studies of galaxies that improves our present theory of galaxy formation.

Just recently, for example, astronomers on Mauna Kea announced that they had detected galaxies colliding on a grand scale for the first time. Astronomers had previously observed many individual cases of two galaxies colliding, but an observation such as this should lead us to a better understanding of the early formation of galaxies.

New Developments in Telescope Design

The Keck Foundation, responsible for the funding behind the present telescope, has pledged enough money to build

Figure 5-18 The 40-inch refracting telescope in Wisconsin is the largest of its kind in the world. Because the weight of the mirror and supporting tube assembly cause a certain amount of sag in the tube, a larger one of this type will probably never be built on Earth. In a zero- or low-gravity location, like Moon, a larger one could be useful. (Yerkes Observatory)

Figure 5–19 The limitations of building larger and larger telescopes in a gravitational environment have led to inovative techniques. The Multi–Mirror Telescope in Arizona, the first of its kind, uses six small mirrors to collect the radiation from objects. Electronics control the mirrors so that they are all focused on the same object. This adaptive optics *technique was the first of its kind, and is now used throughout the world. (Smithsonian Institution and University of Arizona)*

a second 10-meter telescope, called **Keck II**, which is being sited 250 feet from the original and is scheduled for completion in 1996. Side by side, the twin telescopes will act as an infrared or optical **interferometer**, and provide an unparalleled combination of light-gathering power (equivalent to an 85-meter-aperture telescope) and 10 to 100 times better resolution.

Interferometry is a method by which two or more telescopes located some distance from one another, and simultaneously detecting radiation from the same object, can achieve greater resolution than a single telescope. As a matter of illustration, the expected resolution of the twin Keck telescopes is equivalent to separating a car's headlights from 16,000 miles away! Operating in this fashion, the combination would provide the ability to detect warm planets the size of Jupiter in orbit around the nearest 100 stars to Sun.

The special purpose **Hobby-Eberly Telescope** is being built by Penn State and University of Texas astronomers, and will have an 11-meter mirror made up of 91 computer-controlled segments. Used specifically for obtaining spectra of faint objects, the telescope, unlike conventional ones, will not move during exposures.

Rather, a primary focus module will track objects and focus light onto the mirrors once the telescope is pointed to the object being studied. It is being built at the McDonald Observatory in Texas by a consortium of three US and two German Universities, and is expected to be completed in late 1997.

The **Large Binocular Telescope** will consist of two 8.4-meter mirrors mounted side by side, and will see first light in 1997. A joint project of the University of Arizona and Italy's Arcetri Astrophysical Observatory, the two mirrors working in concert (as an interferometer) will provide the light-gathering power of a 12-meter telescope and the resolving power of a 23-meter telescope. It will be constructed on Mount Graham in Arizona.

The **National Science Foundation** is the US Government organization that advises the President and Congress on the progress of science in the United States, and distributes money from a budget authorized by Congress for science projects. Because of the high cost of building modern telescopes, it is participating with 5 other countries (United States-50%, United Kingdom-25%, Canada-15%, Chile-5%, Brazil-2.5%, Argentina-2.5%) to build two 8-meter telescopes as the **Gemini Project**.

The telescopes will operate as either visible-light or infrared instruments. One will be installed on Mauna Kea, the other on Cerro Pachón in Chile. The latter is six miles from Cerro Tololo, home to several large telescopes already. Present schedule call for first light for the Mauna Kea telescope in 1998 and for the Cerro Pachón telescope in 2000. The mirror for the Mauna Kea site has already been cast, and grinding is currently in progress.

Figure 5–20 Aerial view of the Cerro Tololo Inter-American Observatory near La Serena, Chile. The large dome house the 4-meter telescope, the largest in the southern hemisphere. (National Optical Astronomy Observatory)

★★★★★

Active Optics Imagine astronomers being able to analyze images of objects even as the radiation is being collected. If so, they might be able to adjust the telescope optics from moment to moment to correct for the effects of mirror distortion, temperature changes, and poor seeing!

Actually, some of these techniques, collectively called active optics, are already in place in the **New Technology Telescope (NTT)**, located at the European Southern Observatory in Chile (Figure 5-20). It contains a 141-inch (12-foot) mirror that is only 10 inches thick. Computers control the shape of the mirror as its temperature and orientation change, thereby maintaining the very best focus at all times. From its very first observing run, the NTT became the highest-resolution optical telescope in the world.

The very first multi-mirror telescope (the MMT, mentioned earlier) is currently being upgraded to take advantage of this new technology as well. Its six mirrors are being replaced by a single thin mirror 6.5 meters (256-inches) in diameter. When it sees "first light" in 1996, the new MMT will have twice the light-gathering power of the old instrument. The Keck twin telescopes will eventually use active optics as well, and ultimately surpass the resolution obtained by the NTT.

Joining the two Keck telescopes on Mauna Kea will be **Subaru**, an 8.2-meter visible-light and infrared telescope built by the National Astronomical Observatory of Japan. To maintain the shape of the 20-centimeter thick mirror, over 250 sensors and actuators will deform the mirror to maintain high-quality images. First light should occur in 1999.

Spun–Cast Mirrors A new technique for obtaining the exact shape of a telescopic mirror is **spin–casting**. An oven under the football stands at the University of Arizona turns like a merry-go-round, forcing molten glass outward in the mold to form a concave surface. Steward Observatory has cast the single 256-inch (12-foot) diameter mirror in this manner to replace the six 72-inch mirrors of the MMT telescope.

Ten tons of glass were melted down and poured into a mold, then spun at 7 times per minute to result in the proper shape. It took $1\text{-}1/2$ years to polish the mirrors. After cooling, the mirror will be ground to a final shape. Preparations for a second mirror have begun, and an 8-meter casting began in 1994. Eventually the oven will be able to handle mirrors up to 26 feet in diameter.

The second mirror to be cast at the University of Arizona will be installed in the Magellan Telescope, a project of the University of Arizona and the Carnegie Institution of Washington. It will reside at the Las Campanas Observatory in Chile. The mirror will be 6.5 meters in diameter, and will see first light in 1996.

One spun-cast mirror was just recently placed into operation in the 3.5-meter **WIYN Telescope** (for the University of Wisconsin, Indiana University, Yale University, and the National Optical Astronomical Observatories that funded it). It is located on Kitt Peak in Arizona, and initial results surpassed the expectations of astronomers.

The United States is not the only country interested in establishing a prominent position in this exciting field of bigger/more powerful telescopes. The **Very Large Telescope (VLT)** is being installed atop 8,700-foot Cerro Paranal in Chile. Current plans call for it to have four 8.2-meter mirrors that will in combination have the light-gathering power of a 52-foot (16.4-meter) mirror!

When operated as an interferometer, it will provide images with the resolution of a 120-meter telescope. Hundreds of computer-controlled supports will permit the use of mirrors that are only about 10 inches thick. Plans call for its completion in 1998.

What wonders will we see with these new telescopes? And what new discoveries will they make that challenge our currently-held ideas about the cosmos?

Operating a Telescope

Small refracting telescopes are operated in very much the same way as large refracting telescopes since the eyepiece (or other such detector) is located at the opposite end of the

Figure 5–21 An astronomer at Kitt Peak Observatory is shown in the prime–focus cage of the 4–meter telescope. Ideally, this is the best location for visual observations, since it is the first point at which radiation is focused from distant objects. (National Optical Astronomy Observatory)

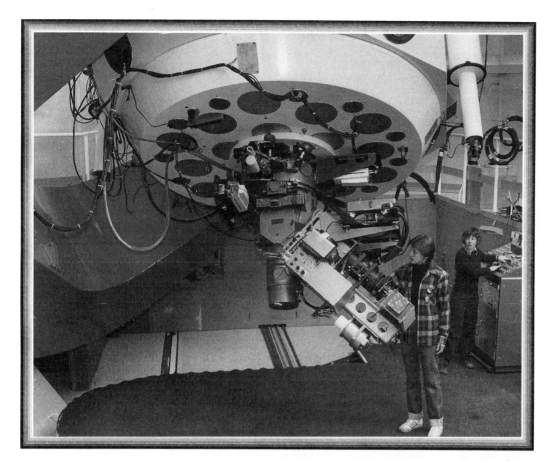

Figure 5-22 The 2.1-meter telescope at Kitt Peak is checked out by Drs. Catherine Pilachowski and Carol Christian in preparation for a night's viewing. Notice the vast array of electronics that is attached to the telescope where the human eye used to do the viewing. During the actual viewing session, the astronomers work in an adjacent room watching the display on various monitors while computers store the information collected by the telescope. (National Optical Astronomy Observatory)

tube from the objective lens (Figure 5-18). But large reflecting telescopes have a versatility that small ones lack. Refer back to the diagram of a reflecting telescope in Figure 5-15.

Recall that the major goal of a telescope is to focus as many photons onto the detector as possible. Common sense tells you that the more mirrors or lens that the incoming radiation must go through or be reflected off, the greater the chance that some of the photons entering the objective will eventually be scattered or absorbed and therefore not contribute to the brightness of the image.

So it is reasonable to conclude that the best location for any detector (including the eyeball) in a reflecting telescope is the point inside the tube at which the photons are focused. This point is called the **prime focus**. In a small reflecting telescope, you would have considerable difficulty placing your eye at that point, and even if you did, your head would be blocking the photons you are trying to detect. So a small diagonal mirror is placed at the prime focus in small telescopes. The mirror reflects the image out to the side of the tube where an eyepiece is located. This viewing position is referred to as the **Newtonian focus**, since Newton used it in his first reflecting telescope.

In a large reflecting telescope, however, like the one at Lick Observatory, a small observer's cage is suspended within the tube, blocking no more radiation than a diagonal mirror would anyway, and the astronomer can ride inside this cage while observing objects at the prime focus (Figure 5-21). This setup offers the finest view. Large or small telescope alike, the cage or secondary mirror must be secured to the telescope.

These supports are usually four in number, and necessarily interfere slightly with any radiation traveling toward the mirror. The result is the familiar light spikes associated with overexposed stars in astronomical photographs, such as that in Figure 1-4. Illustrations of this phenomena have a tendency to show up on Christmas cards as well, creating a "Star of Bethlehem" effect.

Modern Detection Devices Alas, technological developments in astronomy are removing the astronomer from the rather romantic-sounding chore of spending the night at the eyepiece of the telescope observing the wonders of the cosmos. Supersensitive devices are now installed where the eye used to be. The information (image) is then fed by wire into a warm room adjacent to the telescope, where it is stored in a computer and simultaneously displayed on a video screen. The TV camera is a stand-in for the human eyeball, and the wires running to the monitor and computer are acting as the optic nerve connecting eyeball to brain (Figure 5-22).

These supersensitive detection devices are called **charged-coupled devices** (**CCDs** for short). This is very device that is at the heart of every camrecorder, so popular

CHAPTER 5 / Telescopes and Radiation **109**

★★★★★

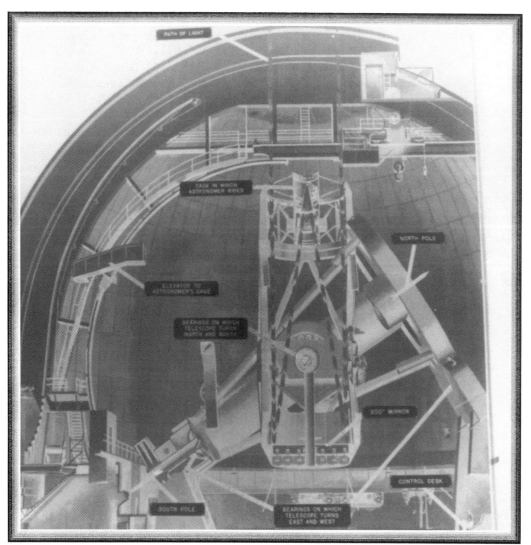

Figure 5–24 A cross-sectional diagram of the 200–inch reflecting telescope at Mount Palomar. The long axis of the mount is aligned to the NCP so that the telescope can track objects during long time exposures. (Tom Bullock)

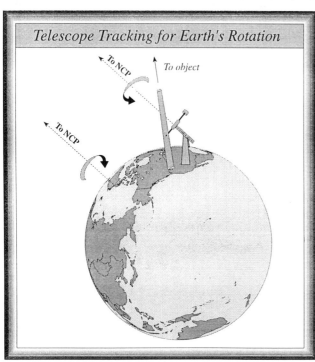

in photography today. It is the mechanism that detects the incoming light and sends a signal to the tape head to be recorded. CCDs free the astronomer from having to use the limited senses with which he/she is born, and instead allows him/her to use that which appears to lack limits — the brain. Recording the image or information is only the first step.

The astronomer then organizes the data in order to make sense of it and integrate it into the broader arena of astronomical knowledge. Or, the new data may even upset the field of astronomy by discovering something that contradicts the generally-held theory. This possibility of contradiction acts as a strong stimulus for astronomical research in the first place.

Tracking with Telescopes Earth rotates. So if you are looking through a telescope, the object you are looking at will appear to move through your field of view and then be lost

Figure 5–23 *In order to image dim objects, telescopes must make long exposures on film or electronics. Thus the mount is aligned to the NCP so that the telescope can track the object.*

110 SECTION 1 / Observing the Sky

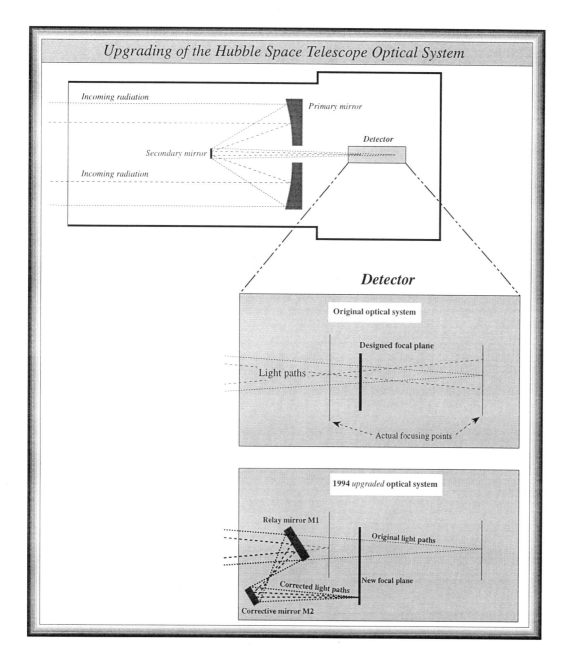

Figure 5–26 *A cutaway of the optical system of the HST, before and after the installation of corrective lenses during the spectacular Shuttle mission in late 1993.*

again. At the same time that a telescope is magnifying the size of an object, it is also magnifying the motion of either that object or the base on which the telescope is resting — i.e., Earth. It seems as if the object itself is moving, but rest assured that it is the result of Earth rotating.

To avoid having to constantly readjust the telescope to keep the object within view in the eyepiece, telescopes frequently have a small motor on the mount that turns the telescope at the same rate that Earth rotates. Hence the object

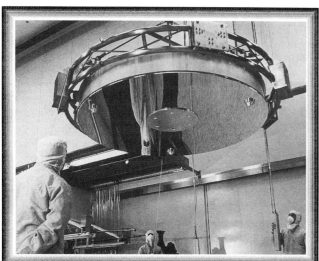

Figure 5–25 *The 96–inch mirror for the HST is carefully positioned during installation at Lockheed in Sunnyvale, California. It was only after the telescope was launched into Earth orbit that the slight flaw in the mirror was discovered.* (NASA)

CHAPTER 5 / Telescopes and Radiation

remains in the eyepiece as long as the motor is running. This is called **tracking**. The axis around which the telescope pivots must be pointed exactly toward the north celestial pole, since that is the point in the sky around which the sky appears to rotate (Figure 5-23).

Accurate tracking is very convenient for the casual telescope user, but it is absolutely essential for the professional astronomer. Most large telescopes, even with sensitive CCDs attached as detectors, require exposure times of hours in order to collect enough photons to obtain an image of a very dim object. If the tracking is off by even a slight amount, the resulting photograph will be blurry.

The photograph in Figure 1-3, for example, required a 2-hour exposure. Astronomers pride themselves in the skill needed to keep a large telescope properly tracking for the duration of an exposure. I hope you can appreciate the many considerations that go into the planning and conducting of an observing session. It isn't just a matter of removing the telescope from the back of the car, setting it upright, and looking through it.

Construction of Telescopes Above and beyond the extremely fine tolerances required in the construction and operation of a telescope, there are two other considerations that affect the ability of a telescope to provide detailed images of celestial objects. When astronomers speak of the cost of the building of a telescope, they are usually referring to the quality of the lens (or mirror), which, of course, does the lion's share of the work.

A telescope is only as good as the ability of the lens to collect photons and focus them to a point. To accomplish that task, the shape of the lens/mirror must be as close to being a perfect mathematical shape (usually that of a parabola) as possible. Grinding a lens/mirror to this exact shape is a time-consuming and delicate task. How close technicians get to exact is determined by how precise the grinding and testing equipment is.

The 95-inch mirror of the Hubble Space Telescope, launched by astronauts of the shuttle *Discovery* on April 24, 1990, was ground and polished to a slightly parabolic shape (Figure 5-25). If it were scaled up to the 3,000-mile size of the United States, there would be no hill or valley that deviates more than 4 inches from that parabolic shape. That's precision!

Unfortunately, the instrument used to check the mathematical shape of the mirror as it was being ground was itself defective, so that the photons collected at the outer edges of the mirror were not focused properly to the detector.

It was similar to the problem of chromatic aberration in refracting telescopes. The defect limited the amount of collecting surface of the objective mirror that astronomers could use, and that of course limited the collecting power of the telescope. Basically, astronomers were not able to carry out studies of the very faint and/or distant objects. But help soon arrived. A set of corrective lenses were attached to the HST in late 1993 by crew members of a Space Shuttle (Figure 5-26).

Radio Telescopes

Radio telescopes have been and continue to be extremely important in our attempt to understand the behavior of objects and matter in space. In addition to being able to detect that wavelength of radiation emitted naturally by clouds of cool hydrogen in space (at a wavelength of **21-centimeters**), they are also able to pinpoint the locations

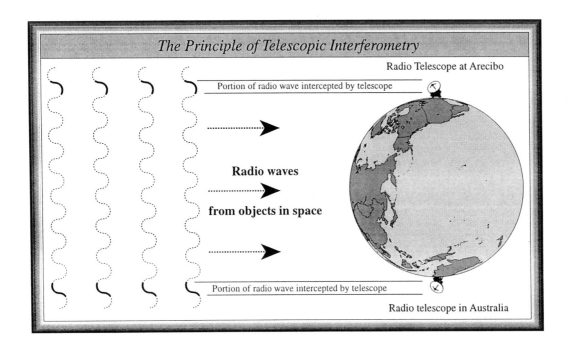

Figure 5-27 In order to compensate for the inherent low-energy and low resolution characteristics of radio waves, radio astronomers connect widely-separated telescopes electronically. By processing the signals received by both telescopes, astronomers can obtain detailed images of radio sources.

Figure 5-28 The Very Large Array is the world's most powerful radio telesope, and is located 52 miles west of Socorro, New Mexico. The 27 separate radio telescopes are linked electronically so that they operate as one huge telescope. This allows for superb resolution and the detection of extremely faint signals from space. (NRAO/AUI)

of some of the more violent events such as exploding galaxies and black holes swallowing neighboring stars. From a practical point of view, radio telescopes allow us to maintain communications with satellites in orbit around Earth or enroute to the planets, simply because of the fact that Earth's atmosphere is transparent to most radio waves.

Resolution Limitations Radio telescopes do, however, have serious limitations. You learned that the resolution of a telescope depends upon the diameter of the mirror or objective lens, but it also depends upon the wavelength of the radiation being detected. Basically, *the length of the antenna being used to receive a given type of radiation must be at least as long as the wavelength of that radiation.* Catching a thrown baseball requires only an open hand. Catching a thrown mattress requires outstretched arms. It is no accident that the radio antennas on your automobiles are all about the same length. The range of wavelengths of radio radiation within which the FCC allows radio stations to transmit can be absorbed on that standardized length.

Think of it this way. You would like to determine the outline of your teacher's body by throwing marbles at him/her standing in front of a blackboard. Would the outline obtained by using marbles be more or less clear if instead you used basketballs dipped in melted chocolate? The shorter the wavelength of radiation used to study an object, the more precise is the outline or boundary of that object. That's one reason why X-rays are used to study the fine detail of the body's interior. The other reason, of course, is due to the X-ray's great penetrating ability.

Largest Radio Telescope In order for radio telescopes to have reasonable resolution, they must have large diameters. A large telescope is also needed to catch enough radiation for a signal to be detected, since radio energy is the weakest of all types of radiation. The largest in the world, shown in Figure 5-8, is the 1,000-foot dish suspended in a valley in Arecibo, Puerto Rico. The radio telescope is known as the **Arecibo radio telescope**. Its collecting power is greater than that of all of the other radio telescopes in the world combined. Many of the amazing discoveries made by this telescope are subjects discussed in the remainder of this book. In addition, it has been used both as a transmitter to send a message to potential civilizations that may exist in space, and as a passive listening device for listening to possible signals from other civilizations.

Radio Interferometer Even with the enormous diameter, however, the Arecibo telescope cannot compete with optical telescopes as far as resolution is concerned. So astronomers have developed a technique to improve the resolution of radio telescopes by operating two or more of them in conjunction with one another even though located at great distances from one another. This is referred to as a **radio interferometer**. In fact, different countries of the world (the US and Australia, for example) coordinate their radio astronomy programs so as to effectively use Earth's diameter as the diameter of a radio telescope by having two telescopes on opposite sides of Earth observe the same object at the same moment in time (Figure 5-27).

It is as if each radio telescope detects a slightly different portion of the same radio wavelength, and a sophisticated computer program fills in the missing interval much as a graphic artist uses a French curve to draw shapes between points. A single interferometer is located in New Mexico, where there are 27 radio antennas, each of which are 82 feet in diameter. They are positioned in a Y-shaped pattern so as to simulate a radio telescope 25 miles in diameter (Figure 5-28). It is called the **Very Large Array (VLA)**, and has provided enormously important radio images of explosive events in very distant galaxies.

With the recent completion of the **Very-Long-Baseline Array (VLBA**, for short), practically the entire diameter of Earth is being used as a radio telescope. Ten integrated radio dishes are installed along a line running from Puerto Rico to Hawaii, forming a radio interferometer stretching along nearly one-fourth of Earth's circumference (Figure 5-35). It has a resolving power equivalent to the reading of a newspaper located 600 miles away! It is able to make radio maps that are 100 times more detailed then the best photographs taken from Earth.

Weak Energy of Radio Waves A further limitation of radio telescopes is the inherently low energy content of radio photons. In order to collect enough photons to focus on the antenna so that a signal can be generated in the receiver, a large "dish" is necessary. But then that makes for an additional problem, that of interference from human-made devices such as TV stations and even faulty ignition systems in automobiles driving near the site. So astronomers build their radio telescopes, as well as optical telescopes, as far from civilization as possible. The far side of Moon would be an excellent location for a radio telescope.

Source vs Object In the field of astronomy, it is important to distinguish between a **source** and an **object**. An *object* is something that can be seen with the eye or recorded on a detector sensitive to *visible* light (photographic film, CCD, etc.). Basically, it is something that you and I would say is visible. A *source*, on the other hand, is something that is detectable only by an instrument sensitive to a type of radiation *other than visible*. Most things that we study are detectable in visible as well as other types of radiation. As long as we have a visible image of the thing, we call it an object.

But there are important discoveries of things in space of which astronomers have no visible images – but of which they have radio energy images, or X-ray energy images, and so on. This should not surprise us, of course. There is no fundamental reason why everything has to give off visible radiation. Some things, especially those that are extremely far away from us and therefore inherently dim, are only detectable by the radiation that they emit in greatest quantities.

The reason that I mention this matter is because astronomers frequently ask us to believe that they have discovered something very important out there without being able to provide a photograph to prove that it is there. At least, not a *visible* photograph. So we are asked to accept as evidence a graphical plot of radiation received from another type of detector. Radio maps of the sky, for example, appear like the contour maps you might use in hiking in the mountains.

Because radio astronomers generally deal with low resolution telescopes, the plot of the strengths of radio signals from space show up as lines like those seen as contour lines on a topographical map. All points that lie along a given line are of the same strength. The "brightest" portion of the sky in radio energy is that point at the center of the concentric circles. Of course, optical astronomers are certainly going to be interested in searching that particular area for signs of visible light. If they eventually obtain an image with the optical telescope, they turn the source into an object.

Telescopes in Space

Judging by the results of the few successes so far, the greatest progress in astronomical knowledge in the foreseeable future will no doubt come from satellites orbiting Earth. Not only are most wavelengths of radiation absorbed by Earth's atmosphere, but 39% of the visible radiation from distant objects reflects off our atmosphere, never to be detected at the surface. Although more costly to launch and operate than ground-based telescopes, these orbiting telescopes nevertheless provide us with information not available at Earth's surface.

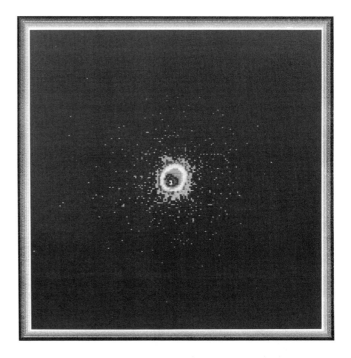

Figure 5-29 One of the first images taken by the ROSAT X-ray satellite was of Cygnus X-2. It is believed to be a neutron star orbiting a normal star about 3,000 light-years from Earth. The X-rays are emitted when material from the normal star is pulled into the gravity "well" of the neutron star. (NASA)

Optical Satellite The *Hubble Space Telescope (HST)* was mentioned earlier as an example of the high precision that goes into the grinding and polishing of mirrors for telescopes. Ironically, and unfortunately, after it was placed in orbit, investigators discovered that the 94.5-inch primary mirror had been ground incorrectly, and that light could not focus at the right point. Actually, the instrument that was used to test and therefore guide the grinding process was itself flawed.

While 85 percent of the light the telescope collected fell at the wrong place on the detector, the other 15 percent was right on target (Figure 5-26). Despite the bad press received, the original HST provided some remarkable data to astronomers. It took high-resolution photographs of Saturn's new Great White Spot, an impossible task with Earth-bound telescopes. It also revealed Pluto and its satellite Charon in much finer detail than the best data from Earth's surface. The image in the Photo far exceeds in resolution any taken from Earth.

But the telescope could not collect enough light from extremely faint objects to shed light on some of the fundamental questions astronomers had hoped to have answered. Questions about the size and age of the universe, mysteries about quasars, pulsars, and other extremely faint and distant objects had to wait. But not for long.

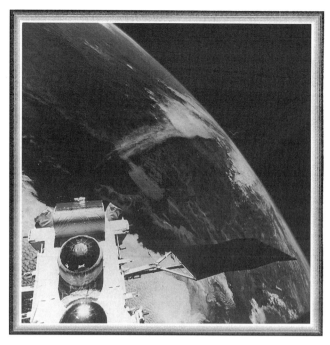

Figure 5-31 The Gamma Ray Observatory *is launched by the crew of the Space Shuttle* Atlantis *in 1991. A portion of southern California appears in the background. The photograph was taken by one of the astronauts from inside the cabin.* (NASA)

In December 1993, a space shuttle rendezvoused with the telescope, and astronauts replaced the camera with an advanced version that contains modified mirrors to compensate for the flawed mirror (Figure 5-27). In effect, the HST received some corrective lens!

The new images surpass all expectations, and are revolutionizing current thinking about the universe. You will encounter many of the new photographs during the rest of the Tour, and learn of the extent to which they are extending and changing our understanding of the universe.

Other opportunities to return to the telescope will occur throughout the 1990s, and second-generation instruments are being designed and built to correct for the primary mirror and to perform more advance astronomical experiments. After all, have you seen any improvements in VCR's during the last few years?

X-Ray Satellites In 1970, the United States launched a satellite (*Uhuru*, which means "freedom" in Swahili) to study sources of X-rays in space. As you know, X-rays cannot (fortunately for us!) penetrate through Earth's atmosphere. Among the 300 or so X-ray sources discovered, one (*Cygnus X-1*) behaves very much like what we think a black hole would be if it were swallowing a star or any other matter in space (Figure 5-7). I discuss this behavior in detail in Chapter 11. Suffice it to say that without an X-ray telescope above Earth's surface, we would be unable to observe such phenomena. Since black holes cannot be seen, their existence can only be inferred by their

Figure 5-30 U.S. and Dutch technicians prepare the Infrared Astronomical Satellite *for launch in 1983. It has provided astronomers with a wealth of information about the location of star-forming regions in our galaxy.* (NASA)

CHAPTER 5 / Telescopes and Radiation

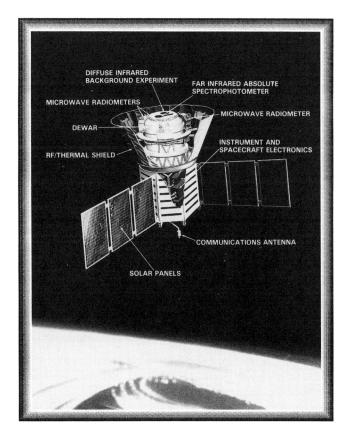

Figure 5-32 The COBE *satellite was launched in 1989 for the purpose of investigating events surrounding the origin of the universe. It has other instruments with which to study infrared regions of the sky. It was data provided by this satellite that early in 1992 was used to detect conditions in the early universe at which time the galaxies began to form.* (NASA)

Figure 5-33 The astronomical satellite Astro-1 *is prepared for operation by Space Shuttle astronauts. In the telescope are infrared, ultraviolet and X-ray detectors.* (NASA)

eating habits. So we have to look for the X-rays that they belch out as they swallow neighboring stars.

The **European Space Agency** (**ESA**) launched the *Exosat* X-ray telescope in 1983, and it proved valuable in detecting high-energy violence in the universe until it failed in April 1986. The US-German-British **Rosat** was launched in 1990. It is the largest and most precise X-ray telescope ever built. Now, almost three years after its launch, it has successfully completed the first all-sky survey, and has made more than 1,700 individual observations. The all-sky survey revealed 60,000 X-ray sources of which 25% were identified by comparing their positions with entries in astronomical catalogues that list visible objects in the sky. About 25% of those identified were normal stars and 50% were explosive galaxies and quasars (Figure 5-29).

Many of the X-ray sources detected by *Rosat* are supernova remnants — the debris expelled out into space by exploding stars. Thousands of years after the explosion, the hot gases forced away from the point of explosion continue to emit the X-rays that the detectors aboard *Rosat* collected. The first supernova detected by the satellite, G156.2 + 5.7, is estimated to be 26,000 years old and located between two spiral arms in our Milky Way galaxy.

The latest X-ray satellite to be launched was the *Japanese Advanced Satellite for Cosmology and Astrophysics* (*ASCA*, for short). Placed in orbit in February 1993, it is studying supernovae remnants, neutron stars, black holes, galactic cores, and clusters of galaxies.

Infrared Satellite The **Infrared Astronomical Satellite** (**IRAS**, for short), launched in 1983 and sensitive only to infrared wavelengths, mapped the location of warm dust clouds in our galaxy. It is at sites such as these that star-formation is believed to be taking place. More excitingly, however, was its discovery of clouds of material circling around a neighboring star, *Vega*. This material is thought to be the very stuff that will eventually form into planets circling *Vega*. It observed through most of 1983 until its helium coolant was exhausted (Figure 5-30).

Gamma Ray Satellites Likewise, several gamma-ray telescopes mounted in satellites have extended our understanding of celestial objects, but also have provided us with new mysteries to unravel. The **Gamma Ray Observatory** (**GRO** for short) was launched from a *Space Shuttle* in April 1991 (Figure 5-31). This 17-ton satellite has 10 to 50 times the sensitivity of any previous gamma-ray telescope. It will first map the entire sky and then study specific targets that appear interesting in the all-sky survey. The Soviet Union orbited two gamma-ray telescopes, the most recent being aboard COS B launched by the ESA in 1975.

Ultraviolet Satellites Since the ozone layer in Earth's upper atmosphere blocks our view of the ultraviolet radiation emitted by hot objects, several orbiting ultraviolet satellites have covered that region of the spectrum for us. The **International Ultraviolet Explorer** (**IUE**, for short) was launched in 1978, and carries an 18-inch telescope with attached spectrographs. It was placed in synchronous orbit so that it orbits at the same rate Earth spins. Thus it always remains over the NASA Goddard Space Flight Center in Maryland.

In June 1992, NASA's **Extreme Ultraviolet Explorer** (**EUVE**, for short) was launched into orbit. Its detectors are sensitive to shorter wavelengths of ultraviolet radiation than the IUE. So it studies very hot objects (from 100,000 to 2 million degrees F). This includes such objects as *white dwarf stars*, which were once like our Sun, but are now nearing the end of their stellar life cycles — cataclysmic variable stars, which are two-star systems in which one siphons material from the outer atmosphere of its companion — and cool stars with very hot outer atmospheres. Ultraviolet radiation is blocked by the ozone layer of Earth's atmosphere before it can reach the astronomical instruments on Earth's surface. The *EUVE* has already completed an all-sky survey, and is currently spending the first of several years on a detailed study of those objects that were detected in the first survey.

Since this particular range of wavelengths of the UV portion of the spectrum has never been studied before, no one knows what *EUVE* will see. It used to be believed that the cool hydrogen gas that permeates space between the

Type of Radiation/Origin	Telescope	Remarks
Gamma-rays:	Gamma Ray Observatory (*GRO*)	NASA, 1991–present
X-rays:	*Uhuru* Satellite	NASA, 1970
gas in clusters of galaxies	Einstein Observatory	NASA, 1978–1981
supernova remnants	*Skylab*	NASA, 1973–1974
solar corona	Exosat	ESA, 1983-1986
	Roentgen Satellite (*ROSAT*)	NASA–UK–German, 1990–present
	*Advanced X-ray Astrophysics Telescope Facility (AXAF)	NASA, 1996 (?)
Extreme ultraviolet:		
supernova remnants	Extreme Ultraviolet Explorer (*EUVE*)	NASA, 1992–present
	Far UV Space Telescope (*Faust*)	NASA, 1992, aboard Space Shuttle
Ultraviolet:		
very hot stars	International Ultraviolet Explorer (*IUE*)	NASA–ESA–UK, 1978–present
Visible:		
stars	Hubble Space Telescope (*HST*)	NASA–ESA, 1990–present
hot gas clouds		
Infrared:		
cool clouds of gas and dust	Hubble Space Telescope (*HST*)	NASA–ESA, 1990–present
planets and their satellites	Infrared Astronomical Satellite (*IRAS*)	NASA–Dutch–ESA, 1983–1984
	Infrared Space Observatory (*ISO*)	ESA, 1995-present
	*Space Infrared Telescope Facility SIRTF	NASA, 2000 (?)
Radio:		
electrons moving in magnetic fields	Cosmic Background Explorer (*COBE*)	NASA, 1990–1993
"glow" of Big Bang explosion		

Note: (*) Indicates that satellite is planned, but not yet launched.

Figure 5–34 A list of the satellites that have been launched (or are planned for launch) for the purpose of detecting radiation of different wavelengths before being absorbed by Earth's atmosphere should alone explain why we live during the Golden Age of Astronomy.

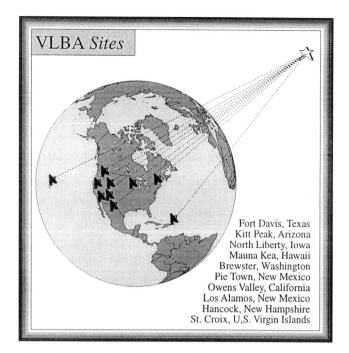

Figure 5-35 *The Very Long Baseline Array will be a series of radio telescopes that are electronically linked in such a way as to enable radio astronomers to use the entire diameter of Earth as one immense radio telescope.*

stars would block any UV radiation coming from space, but early experiments with *EUVE* reveal the presence of extremely hot, ionized gas that is transparent to extreme UV radiation. Astronomers believe that this area, called the "*Local Bubble*," is the result of a star that exploded some 100,000 years ago. So the *EUVE* is able to study the UV that passes through this hot gas.

Microwave Satellite Launched in November 1989 from Vandenberg Air Force Base in California, the **Cosmic Background Explorer** (**COBE**, for short) was specifically designed to study microwave radiation that is believed to be the remnant of the Big Bang explosion that "created" the universe 20 billion years ago. But it is also searching at various infrared wavelengths for radiation from the first stars and galaxies (Figure 5-32). It is also mapping the infrared distribution in the Milky Way and other galaxies, In that work, it is extending the observations of the *IRAS*.

Astronomers just recently announced to a startled scientific as well as public that *COBE* detected structure in the microwave background radiation that is the remnant of the Big Bang, providing the "missing link" between the origin of the universe and the formation of the galaxies that we see around us today. I say much more about this discovery while discussing the matter of the Big Bang theory in Chapter 15.

The Great Observatories Program The United States is in the process of launching the series of satellites designed to detect each of the types of radiation that fail to reach Earth's surface (Figure 5-34). Already in orbit, the *Hubble Space Telescope* is designed to cover the infrared, visible and ultraviolet portions of the spectrum. The *Gamma Ray Observatory* is currently exploring the gamma ray portion of the spectrum for astronomers.

In 1999, the **Advanced X-ray Astrophysics Facility** (**AXAF**, for short) will be launched to broaden the studies of the earlier satellites studying the X-ray portion of the spectrum. Housing a 47-inch telescope operating from 6 nm to 0.15 nm in wavelength, it will have about 100 times the sensitivity of the best X-ray telescopes yet launched. Because of budget concerns, however, Congress recently cut the funds for the AXAF by $60 million. NASA is having to scale back the complexity of the satellite.

The **Space Infrared Telescope Facility** (**SIRTF**, for short) does not yet have a firm launch date. As of mid-1993, all funding for this project — given highest priority among all astronomy programs for the 1990s by the National Research Council — is contingent on approval of a down-sized version of the original project. It is the fourth and last of the *Great Observatory Program* projects. The earlier *IRAS* satellite gave only broad coverage of the galaxy at infrared wavelengths. The *SIRTF* is designed to look more closely at those areas deemed to be interesting in the photographs provided by IRAS. If funded soon, it will be launched from a Space Shuttle about the year 2000, and will be capable of detecting planets a bit larger than Jupiter that orbit any of the nearby stars.

Improved Radio Telescope For more than 20 years, astronomers made important discoveries while using the 300-foot radio dish at Greenbank, West Virginia. But in November of 1988 a metal plate in the telescope failed and the entire instrument collapsed into a jumbled mass of twisted metal.

A new 320-foot telescope is presently being constructed as a replacement, and will, when completed, be the largest fully steerable radio telescope in the world. Because it is being built with state-of-the-art materials and computer/laser controls, the telescope will avoid the limitations of other large steerable radio telescopes around the world.

Satellite Astronomy Try to visualize the following possible development in astronomical observing. The astronomer will not even be present at the telescope during the observing session. A microwave link between the observatory and his/her office (or even home) allows for the receipt of the needed information without having to drive for miles to get to the observatory. Of course, technicians have to be on hand at the telescope to maintain the instruments. This approach to observing is already in place and operating. It is exactly the manner in which satellite astronomy is conducted, for example, with the Hubble Space Telescope. Astronomers control the telescope from their computer consoles on Earth, and receive data back from the telescope in a manner identical to that of a TV broadcast.

Summary/Conclusion

In order for astronomers to determine the contents of space and what those contents are doing, they build telescopes to which are attached various instruments that record the receipt of the various types of radiation in the electromagnetic spectrum. Different events in space emit different types of radiation. Different chemicals in the atmosphere are responsible for filtering out specific portions of that spectrum, preventing the astronomer from having access to all the information contained within that radiation that tells about the object that emitted it. Satellites with telescopes/instruments are sent above the atmosphere to detect the radiation before it is absorbed, which then relay the information contained within that radiation down to Earth's surface using a radiation that does penetrate the atmosphere.

The sole purpose of a telescope is to collect the photons that comprise radiation and focus it onto a detector so that information about the object that emitted it can be obtained. To that end, the major characteristics of collecting power, resolution, and magnification are always maximized. Since nothing is known to travel faster than radiation, and since radiation is such an efficient method of sending information from one place to another, astronomers assume that extraterrestrial civilizations would attempt to communicate with other civilizations using the electromagnetic spectrum. The SETI Program has been listening for signals from other civilizations using radio telescopes, since that is one of the two types of radiation not absorbed by the atmosphere.

Just as Galileo began a revolution in our thinking as he applied a new instrument to the study of the heavens, so we may be in the process of creating a new revolution as we launch new *eyes* into space. We cannot even begin to imagine what incredible discoveries will be made when they begin returning data to Earth. I will certainly have to rewrite portions of this book after the results are in, but that is progress. If our theories remain unchanged for long, we are doing our homework.

LEARNING OBJECTIVES *Now that you have studied the Chapter, you should be able to*

1. List the divisions of the electromagnetic spectrum into types of radiation, and the properties of radiation that allow for those divisions.
2. Explain the difference between a hot star and a cool star in terms of the amounts by wavelength of the radiation the star emits.
3. Describe each of the types of radiation, and the general features of objects that emit each type.
4. Describe the main functions of a telescope (light-gathering, resolving, and magnifying powers) and how each is achieved in the design of a telescope.
5. Compare and contrast the designs of reflecting and refracting telescopes, relating those designs to the potential of better telescopes in the future.
6. Explain the placement of telescopes for maximum performance, and the alignment necessary in order to obtain long-exposure photographs of dim objects.
7. Compare and contrast optical and radio telescopes, explaining the necessity that radio telescopes be large.
8. Explain how satellite astronomy has the potential for revolutionizing our knowledge of the universe, and list the proposed components of the Great Observatories Program.
9. Define and use in a complete sentence each of the following **NEW TERMS**:

★★★★★

Active optics 107
Advanced X-ray Astrophysics Facility (AXAF) 118
Apparent infrared magnitude 94
Arecibo radio telescope 113
Atmospheric window 97
Charged-couple device (CCD) 109
Chromatic aberration 105
Cosmic Background Explorer 118
Collecting power 103
Color 93
Continuous spectrum 104
Electromagnetic energy/spectrum 91
Energy curve 98
Extreme Ultraviolet Explorer (EUVE) 117
Eyepiece 104
Federal Communication Commission 96
Focal length 104
Frequency 92
Gamma rays 96
Gamma Ray Observatory 116
Golden Age of Astronomy 98
Great Observatories Program 118
Heat 91
Hubble Space Telescope 98
Hydrogen 97
Infrared radiation 91
Infrared Astronomical Satellite (IRAS) 116
Interferometer 106
International Ultraviolet Explorer 117
Interstellar 95
Keck Telescope 106
Kuiper Airborne Observatory 101

Microwaves 97
Multiple-Mirror Telescope 106
NASA 101
National Science Foundation 107
Newtonian focus 109
Object 114
Objective lens 102
Optics 104
Ozone layer 97
Prime focus 109
Prism 93
Radio radiation 96
Radio interferometer 113
Refracting telescope 102
Refraction 112
Reflecting telescope 105
Resolution 103
Rosat X-ray Telescope 116
Seeing 100
Source 114
Space Infrared Telescope Facility (SIRTF) 118
Space Shuttle 98
Spin-casting 108
Star party 104
Tracking 111
Ultraviolet radiation 95
Very Large Array (VLA) 113
Very Long Baseline Array (VLBA) 114
Visible light 91
Wavelength 92
X-rays 96
21-centimeter radiation 112

Section 2

Evolution of the Stars

Chapter 6
Atoms, Matter, and Radiation

Figure 6–1 This photograph of the Eagle Nebula (M16) in the constellation of Serpens illustrates the ability of modern telescopes to capture the beauty of space. The sky is riddled with stars and intricate patterns of gas and dust clouds. By analyzing the radiation coming from such areas, astronomers are able to deduce the events going on in them. (Lick Observatory)

CENTRAL THEME: How does radiation we receive from objects in space tell us about the atoms making up those objects, and what those objects are doing?

So by studying the radiation collected by telescopes, astronomers claim to know what is going on out in space. But exactly how does that take place? How can radiation in any of its forms carry information across the vast distances separating the stars and galaxies? Astronomers are making what appear to be fantastic claims today — the universe came into existence with a Big Bang 20 billion years ago, there are cannibalistic black holes in space that swallow other stars; Earth, Sun, and planets evolved from a cloud of gas and dust 5 billion years ago! How do we arrive at such theories? Remember, that is the focus of this book — what evidence supports our theories?

The very basis of just about every theory in astronomy is the study of radiation. Remember that with the exception of meteorites and moonrocks, we have no actual celestial material to study in our laboratories. To be somewhat philosophical about this matter, let me put the problem another way. You have never <u>seen</u> a star, Sun, or any other celestial object! You have merely <u>detected</u> the radiation coming from what you *presume* to be an actual object in space, and always at some moment in time much later than the time the radiation left the object.

How do we know that objects in space actually exist? Because we presume there is a relationship between objects and radiation — that an object either emits light or reflects it from another object that does emit it. But just what is <u>that</u> relationship? How do objects emit or reflect radiation? It is convenient, and easier to understand the answer to these questions, if we tune our thinking to a finer level than that of just objects. Think in terms of the very building blocks of matter, atoms and molecules.

Atoms and Light Light does not come from light bulbs — it comes from the atoms making up the filament inside the light bulb. The type of radiation emitted by a light bulb depends upon the type of filament used in that light bulb. Radiation does not come from Sun — it comes from the atoms making up the outer layers of Sun. And so it goes. I am not just trying to be clever with words. Astronomers need to know what kinds of atoms and molecules make up an object in order to fully understand what is going on inside of or in the vicinity of that object.

It is the interaction of light and atoms that allow us to collect information about the universe out there and then formulate theories as to how those atoms and molecules started doing what they are doing in the first place, and what they will be doing in the future. This all presumes, you are thinking, that the rules by which light and atoms interact here on Earth are valid way out there as well. I cannot prove that to be true. It is an article of faith that scientists have, faith that the universe out there operates according to the same laws of physics and chemistry as we have, by experimentation, found to be true here on Earth.

Assumptions If you are one of those who questions the validity of such faith, isn't it interesting that such faith was the basis for our having gotten men to the Moon and back again? And isn't it interesting that what we believe to be the process by which Sun emits radiation was duplicated here on Earth, and today accounts for all the stockpiles of hydrogen bombs in our nuclear arsenals? Galileo was one of the first to suggest that we should be able to assume the universality of laws of physics and chemistry. Kepler, and especially Sir Issac Newton, used that assumption to explain the motions of objects within the solar system.

So let me make amateur atomic physicists out of you, or at least give you a crash course in atomic physics so that you can appreciate how astronomers are able to extract a tremendous amount of information from the radiation received from a distant object. This treatment will be somewhat simplified. I prefer to spend more time discussing the exciting discoveries in astronomy than the details of atomic physics.

Atomic Structure

Let us consider the most simple of atoms, the hydrogen atom (Figure 6-2). It consists of a central **nucleus**, containing a particle with a positive charge, a **proton**, and a negatively-charged particle in orbit about the proton, an **electron**. It is the opposite charges of these two particles that holds the two together, the electron orbiting around the nucleus (which, for the hydrogen atom, is a single proton). The electron and proton are like two small magnets attracting one another. This attractive force is called the **electromagnetic force**. You will see why it is named that shortly.

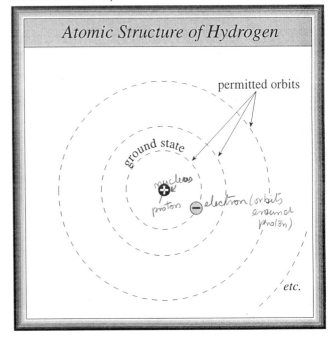

Figure 6–2 *The atomic structure of hydrogen, the simplest of all atoms, illustrates the attraction of oppositely-charged particles: protons and electrons.*

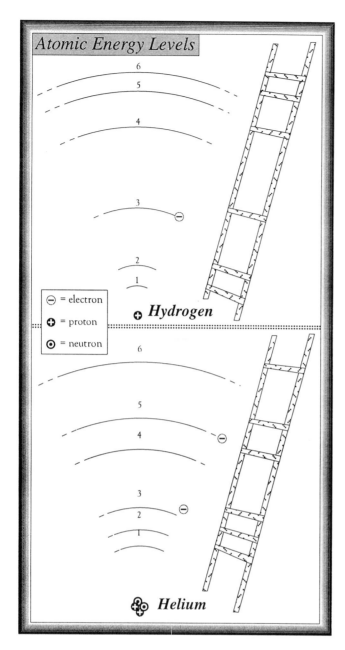

Figure 6–3 It is convenient to think of the energy levels of an atom as the rungs of a ladder. It requires energy to climb a ladder just as it requires energy to force an electron into higher energy levels. Each type of atom has a unique set of energy levels.

Electron Orbits It turns out that the electron's orbit is not a fixed orbit like that of Earth's. There are a large number of **permitted orbits** for the electron, orbits it can occupy if it possesses the proper amount of energy. But in order for it to leave the orbit closest to the proton, called the **ground state**, and jump to another orbit, it must first absorb energy. After all, the proton wants to hold it close, like Earth's gravity holds onto you. If you "climb away" from Earth, say up a ladder, you must burn some stored-up energy.

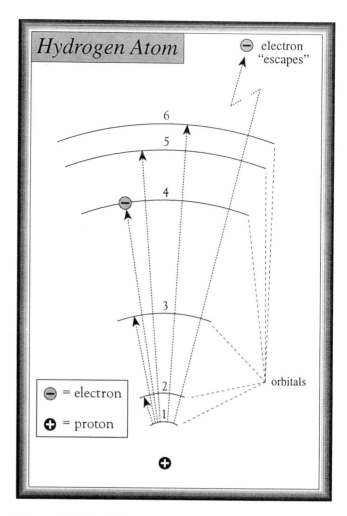

Figure 6–4 The hydrogen atom is the simplest of the atoms. The single electron may skip energy levels, or escape the influence of the proton entirely.

The permitted orbits of the hydrogen atom can be thought of as rungs of a ladder, on which the electron is located at any moment of time (Figure 6-3). While the electron occupies any orbit other than that of the ground state, we say that the atom is **excited** — excited in the sense that the electron does not like to remain in one of the higher-energy orbits. It prefers to drop back down to the ground state. Of course, in the act of dropping down, it must release the energy it absorbed in going up to the higher-energy orbit.

The electron may, if the necessary energy is available, jump from one orbit to another until it is entirely free of the attraction the proton has for it (broken line in Figure 6-4). At that point, the hydrogen atom consists of the single proton that was left behind. It is now an **ionized** atom, and it has a strong attraction for an oppositely-charged particle, another electron. If an electron wanders by, and gets close enough, it may be attracted and then captured by the proton. If so, it drops from one orbit level to another until it reaches the ground state, all the while giving up energy.

Just as you skipped steps as you anxiously ran upstairs

124   SECTION 2 / Evolution of the Stars

List of Chemical Elements and Atomic Numbers		
1 hydrogen	32 germanium	63 europium
2 helium	33 arsenic	64 gadolinium
3 lithium	34 selenium	65 terbium
4 beryllium	35 bromine	66 dysprosium
5 boron	36 krypton	67 holmium
6 carbon	37 rubidium	68 erbium
7 nitrogen	38 strontium	69 thulium
8 oxygen	39 yttrium	70 ytterbium
9 fluorine	40 zirconium	71 lutetium
10 neon	41 niobium	72 hafnium
11 sodium	42 molybdeum	73 tantalum
12 magnesium	43 technetium	74 tungsten
13 aluminum	44 ruthenium	75 rhenium
14 silicon	45 rhodium	76 osmium
15 phosphorus	46 palladium	77 iridium
16 sulfur	47 silver	78 platinum
17 chlorine	48 cadmium	79 gold
18 argon	49 indium	80 mercury
19 potassium	50 tin	81 thallium
20 calcium	51 antimony	82 lead
21 scandium	52 tellurium	83 bismuth
22 titanium	53 iodine	84 polonium
23 vanadium	54 xenon	85 astatine
24 chromium	55 cesuyn	86 radon
25 manganese	56 barium	87 francium
26 iron	57 lanthanum	88 radium
27 cobalt	58 cerium	89 actinium
28 nickel	59 praseodymium	90 thorium
29 copper	60 neodymium	91 protactinium
30 zinc	61 promethium	92 uranium
31 gallium	62 samarium	

Figure 6–5 There are 92 naturally–occurring chemical elements. Each has a specific set of energy levels, and is numbered by the number of protons grouped in the nucleus.

to go to bed as a child, so electrons can skip orbits either on the way up or on the way down. Of course, skipping more than one orbit requires more energy — the energy difference between orbits 2 and 4 is greater than that between orbits 3 and 4, for example.

Now here is a feature that makes atomic structure quite different from your experience with ladders and staircases, but which enables us to know so much about the interaction of light and matter. It turns out that the orbit levels, unlike the rungs of a ladder, are not equidistant from one another. The orbit levels are rather like a ladder whose rungs are at varying distances from one another, no two intervals being identical. The importance of this fact will become clear shortly.

Types of Atoms The universe is composed not only of hydrogen atoms, but combinations of 92 naturally occurring atoms. It is the different combinations of those 92 elements that make up all of the exciting objects we observe — banana slugs, stars, and redwood trees. We refer to a given type of atom by the number of protons in the nucleus — the **atomic number**. Hydrogen is atomic number 1, helium is number 2, lithium is number 3, uranium is number 92. All of the known naturally-occurring chemical elements are shown in Figure 6-5. I have been using the hydrogen atom as a model for any atom because it is the simplest in structure and because it is the most common element in the universe, making up 90% of all the atoms in the universe.

The reason for that is quite simple — mostly what is out there are stars, and stars are composed mostly of hydrogen. It is the fuel by which stars shine. Helium, the second lightest element, makes up 9% of all the atoms in the universe. Hydrogen and helium together control 99% of the stock. The things we see around us on Earth and with which you and I are most familiar are, therefore, not composed of the most common chemicals that make up the universe. The planets are leftover remains of the formation of Sun. Astronomers assume that same process occurs around other stars as well. We will examine this concept in more detail when we consider Earth's origin in Chapter 16.

Structure of Atoms Let us look at the structure of the helium atom and compare its features with that of the hydrogen atom. Figure 6-3 illustrates that the nucleus of the helium atom consists of two protons, as we would expect, and two newcomers called **neutrons** — particles lacking any charge at all. The importance of the neutrons is not important until we consider the manner in which stars die in Chapter 11. The two electrons of the helium atom can also occupy any of the permitted orbits, but notice that the spacing of the orbits is entirely different from that of the hydrogen atom. There is, therefore, an entirely different set of energies to be absorbed by the electrons of the helium atom in going from orbit to orbit on the way up, or given up on the way down. The helium atom can, like the hydrogen atom, be excited — it can be ionized once or twice, depending upon whether it loses one or two electrons.

The lithium atom differs from both the hydrogen and the helium atoms, not only in the number of protons in the nucleus, but in the pattern of electron energy levels. No two atoms are alike. Chemists have studied all 92 chemical elements in detail, and recorded the energy levels that each exhibits. They are not included here because such information is not necessary to your understanding of the subject at hand, but of course you could easily obtain them in a chemistry handbook.

Electron–Photon Connection Now reflect for a moment on my use of the word "energy" when referring to electrons jumping from one orbit to another, and substitute a word that you have encountered frequently in the book — the word is "*photon*." Previously, I described a photon as a bundle of energy, and specified each according to its wavelength, frequency, or energy content. Each wavelength photon has a particular frequency and a given amount of energy associated with it.

What is unique about an atom is that in order for an electron to move from one orbit to one that is further from the nucleus — a higher energy orbit — it must absorb only that photon whose energy content is exactly equal to the difference in energy between the two orbits. It can neither absorb one that has an excessive amount (shorter wavelength) nor one that is lacking (longer wavelength). If, in climbing a ladder, you do not expend enough energy to lift your foot high enough, you do not reach the next rung. If you expend too much energy, your foot lands between rungs. Only by releasing some energy (lowering your foot) can you place your foot correctly on the next rung.

Each chemical element has a set of photons (expressed in wavelengths) that its electron(s) absorb, and *no other element has an identical set of wavelengths that it absorbs*. It is the nature of each chemical element to have its own unique set of electron orbits. And just as a given chemical element absorbs only a given set of wavelengths, so also does it give up (emit) that same set when its electron(s) jump downward from one orbit to another. What is important to remember is that between any two orbits there is a specific amount of energy difference. Therefore, when an electron moves between them, that specific amount of energy (wavelength photon) is absorbed or emitted.

Just about every source of light you have ever seen — a light bulb, a candle, lightning — has been a variation on this theme. Some mechanism causes electrons in the atoms making up the source (the filament, wax, air molecules, etc.) to jump into higher energy levels, and when they return to lower levels, light photons are emitted. I said "just about" a moment ago because Sun and other stars do not emit light in this manner, but that is a story I get to soon enough.

For the moment, I am stressing the connection between the radiation we receive and the atoms from which it came:

By measuring the wavelengths of that radiation, scientists are able to determine (1) from what specific atoms – chemical elements – the radiation was emitted, and (2) between what orbits the electrons of those atoms jumped to have emitted the radiation.

Atoms and Sunlight

Let me put this theory of atomic structure into perspective by using the example of a familiar object, Sun (Figure 6-6). Deep in the interior of Sun, at the core, the pressure and temperature are high enough for atoms of hydrogen to fuse together to form an entirely new atom — helium. In the process, energy is released in the form of gamma-ray photons. These liberated photons, seeking lower pressure and lower temperature, attempt to escape out to Sun's surface 400,000 miles away. But due to the extremely high density of atoms within Sun, the gamma-ray photons

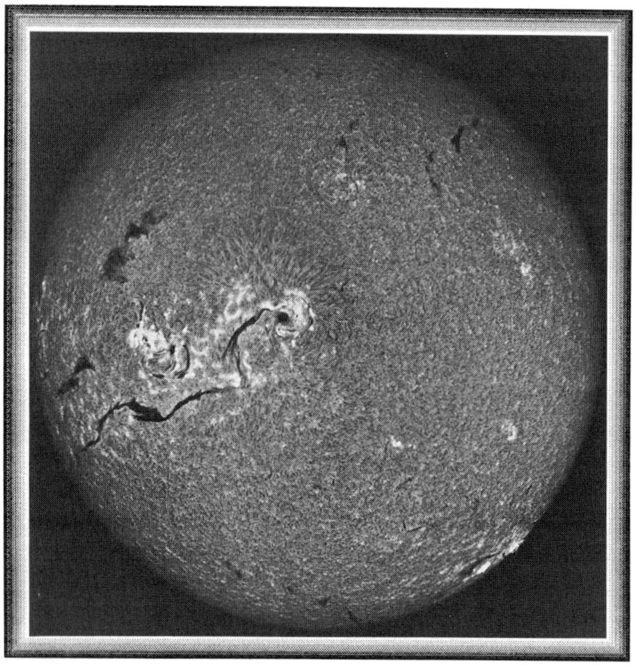

Figure 6-6 Sun is the source of the radiation that sustains life on Earth. But that radiation originates at the center of Sun, and eventually emerges at its surface to create the features that are readily observable with special filters attached to a telescope. (National Optical Astronomy Observatories)

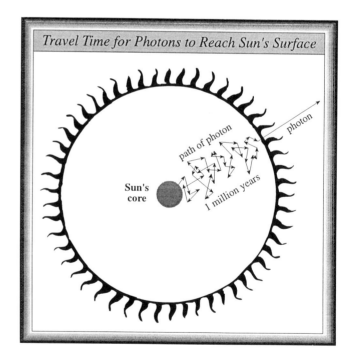

Figure 6-7 On the average, it requires one million years for a gamma ray photon to make the journey from Sun's core to its surface. But in the process, it loses energy and therefore changes wavelength. Overall, the gamma ray photons degrade into all of the types of photons. So coming from Sun's surface are all of the types of radiation in the EM spectrum.

Figure 6–8 Dr. Catherine Pilachowski, a staff astronomer at Kitt Peak, adjusts the spectrograph on one of the telescopes. The spectrograph is one of the most important tools available to the astronomer, inasmuch as it provides information about the basic characteristics of stars and other objects in space. (National Optical Astronomy Observatories)

travel only a short distance before being absorbed by an atom.

During the process of absorption, that atom speeds up, collides with an adjacent atom, and then re-emits a photon as it slows down due to impact. This photon is not necessarily re-emitted in the same direction that the first photon was travelling. This re-emitted photon is in turn absorbed, re-emitted, and so on and so forth. Because the density of atoms within Sun decreases with distance from the center, the overall movement of re-emitted photons is toward Sun's surface. The path that a given photon takes in getting to the surface is therefore a jagged one, and the average photon spends approximately 1 million years getting out to Sun's surface (Figure 6-7).

In the process of being alternately absorbed and re-emitted, what began the journey as a gamma-ray photon eventually reaches the surface as an X-ray, an ultraviolet, or even a radio photon. In other words, each photon loses some amount of energy in the process of getting to the surface. Some emerge having lost very little, and emerge as gamma-ray photons. Some photons go through so many alternate absorptions and re-emissions that they eventually emerge at Sun's surface as radio waves. Since there are countless numbers of gamma-ray photons being produced every second at Sun's core, every possible wavelength of the spectrum eventually emerges at Sun's surface, from the shortest gamma-rays to the longest radio waves.

Although all the produced photons eventually emerge at the surface, they are not present in equal numbers. We learned in Chapter 5 that for a star like our Sun, most of the photons emerging at the surface are those whose wavelengths lie within the visible portion of the spectrum, and at yellow wavelengths at that. For a hotter star, there are more ultraviolet photons present, for a cooler star more infrared photons. Nevertheless, if you were able to stand on the surface of Sun and measure the photons coming up from within Sun, all wavelengths would all be present.

Measuring Wavelengths One of the most important tools the astronomer has to extend the capabilities of the telescope is designed to measure the wavelengths of light that are received from a source of radiation. It is called a **spectroscope**. You are probably familiar with its close cousin, the common prism, which is nothing more than a highly-polished triangular piece of glass. As a beam of light travels through the prism (or spectroscope), the light is separated into its component wavelengths and spread out into a spectrum according to wavelength. If all wavelengths are present — as would be the case at Sun's surface and is the case with the common light bulb — the spread-out pattern of wavelengths is called a **continuous spectrum** (Figure 6-9).

Even if you have never seen an experiment with a prism performed, you surely have witnessed a rainbow in the sky. That is a good example of a continuous spectrum. The principle is the same for both, except that in the case of the

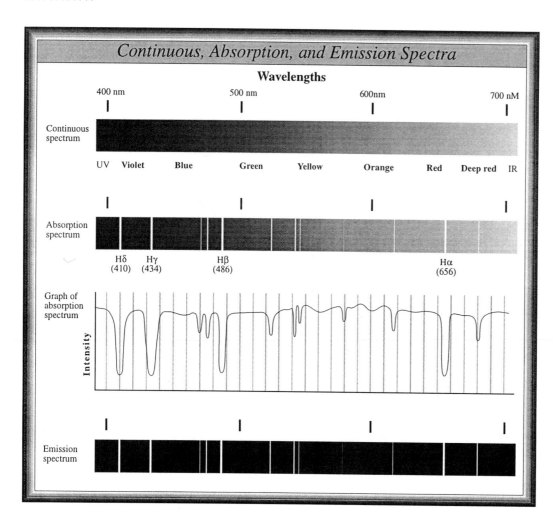

Figure 6–9 *Obtaining spectra from objects in space is key to understanding those objects. The three types of spectra, continuous, absorption and emission arise under different conditions. By obtaining a particular type of spectrum from an object, we know that those conditions exist in that object.*

rainbow, water droplets suspended in Earth's atmosphere act as miniature prisms in separating sunlight into its component colors (wavelengths). Color *is* wavelength. Perhaps it will help if you keep in mind that what you and I refer to as color is simply the brain's language for telling us that our eyes have detected certain wavelengths of visible light from something. Thus infrared radiation doesn't have a "color" associated with it because we cannot detect infrared with our eyes. However, we make reference to the fact that infrared radiation lies at the red end of the spectrum because its longer wavelengths are past the red end of the visible portion of the spectrum.

One might think that if all wavelengths of radiation are leaving Sun's surface, we should, at the surface of an airless Earth, receive all of the wavelengths in the form of a continuous spectrum. But that is not the case. Between the surfaces of Sun and Earth are two environments that absorb wavelengths of radiation — Sun's atmosphere (recall in Chapter 3 that Sun's atmosphere, the corona, is visible during total solar eclipses) and Earth's atmosphere. From Earth's surface, instead of observing a continuous spectrum from Sun, we see a continuous spectrum *with dark lines*. Each dark line represents a single wavelength photon absorbed by an atom present in either Sun's or Earth's atmosphere. We call such a spectrum an **absorption spectrum**, in view of the fact that something between us and the object emitting the light absorbed some photons (Figure 6-9).

At first, this might seem to be a great disadvantage to astronomers, for photons to be absorbed and the object caused to appear dimmer. True, the object appears slightly dimmer as a result of absorption taking place, but that is a small price to pay for the information gained about the object itself. Because the dark lines represent the very "signatures" of the chemical elements that make up the object. Associating the wavelengths of the dark lines with specific atoms is the very process by which we obtain so much information about stars and galaxies. In that sense, an absorption spectrum is very much like the *Universal Marketing Symbol* found on most products we purchase at the store. In order to visualize how absorption occurs, let's consider a single atom in Sun's atmosphere. I will use the hydrogen atom as my representative atom, both because it is the simplest of all atoms and because it outnumbers all of the other atoms in Sun by far.

Photon Absorption As photons of all wavelengths stream outward from Sun's surface at the speed of light, some pass

128   SECTION 2 / Evolution of the Stars

close to our chosen hydrogen atom in the corona. Out of all the photons available, the electron of that atom selects that particular photon that has just the necessary energy (expressible as a wavelength) to allow it to jump to a higher energy level. The remaining photons continue to stream past the atom until they reach Earth's surface, at which place a spectroscope can spread them out according to wavelength and recorded on film as a spectrum.

But the photon absorbed by the hydrogen atom in the corona is of course missing — which means that instead of a continuous spectrum at Earth's surface, an absorption spectrum is obtained. Dark lines appear in the spectrum because certain wavelengths have been absorbed by atoms that occupy the interval between the surface of Sun (where the photon was first emitted) and the surface of Earth (where the photons are collected). There are, for Sun as well as for other stars, many different types of atoms present in the corona, so many absorption lines are present. Since each chemical element has its own unique appetite for certain wavelength photons, it is possible to determine the chemical composition of Sun's (or star's) atmosphere. Astronomers must make note of the fact that some of the absorption lines are due to Earth's atmosphere, but since we know its chemical makeup quite well by now, it is easy to recognize which dark lines are due to Earth's atmosphere. The remaining dark lines must be due to the object under investigation (Sun, in this case).

Spectrum Analysis When I first introduced you to the electromagnetic spectrum using Figure 5-2, I chose to express wavelength in meters so as not to get too technical too soon. But since scientists began the study of radiation with those at visible wavelengths, which are very small, the unit of measurement adopted to express wavelength is the **Angstrom unit** (symbolized as Å). It is, by definition, equal to one ten-billionth of a meter (10^{-10} meter). You probably won't want to remember that figure, but as you encounter the use of the term in this Chapter, remember that it takes quite a few Angstroms to make a length that you can relate to in your everyday life. Wavelengths of visible radiation are extremely short — the antennae (rods and cones) in the human eye must be very small in order to detect their arrival.

The chemical elements are identified according to the particular wavelengths absorbed — by those that are missing in the spectra. How is it possible to know what chemical elements are associated with the different wavelength photons? Simple. When we heat up a chemical element in the laboratory until it glows, the light given off consists of the specific wavelengths that element emits *at that particular temperature*. A spectrograph splits up and records by wavelength the photons emitted, and the *photograph or plot of that set of wavelengths* is called that element's **spectrum** (Figure 6-9).

I distinguish between a photograph of a spectrum and a plot of a spectrum because of the different manners in which different types of radiation are recorded. Radio and X-ray telescopes, for example, record the receipt of energy in the form of plots, whereas optical telescope use photographic film. The important point to keep in mind here is that spectral lines can be measured along any portion of the EM spectrum. Astronomers are not confined to the use of optical telescopes in determining what chemical elements are in space.

Since the spectrum obtained by heating up the gas consists of those photons *emitted* when electrons dropped from higher to lower energy levels, we call such a spectrum an **emission spectrum**. Notice that an emission spectrum consists of *emitted* photons, whereas an absorption spectrum consists of all wavelengths *except those absorbed* by an intervening medium.

Spectrum of Hydrogen Not all hydrogen atoms absorb the same wavelength photons. Different atoms of hydrogen have electrons in different orbits, thereby absorbing different wavelengths as they move to higher orbits. Hydrogen, when heated up to a certain temperature, emits three visible prominent wavelengths of photons — through a spectroscope they appear as a red line (6563Å), a green-blue line (4861Å), and a violet line (4340Å). An illustration of what is seen in a spectroscope is in Figure 6-10. These are referred to as the **Balmer lines** of hydrogen.

Heating the hydrogen gas to a higher temperature causes a different set of wavelengths to be emitted, and since

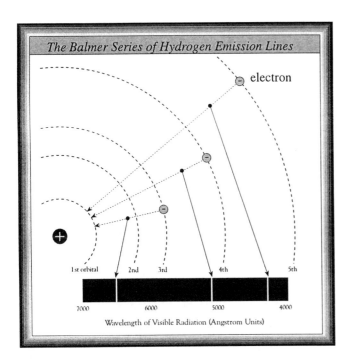

Figure 6–10 A pattern of spectral lines is emitted when the chemical element hydrogen is heated up enough to glow a magenta color. If this single color is viewed through a spectroscope, three wavelengths—the Balmer series—are seen.

they are higher-energy photons they are invisible to the human eye. But they are detected in the ultraviolet portion of the spectrum with wavelengths of 1216Å, 1026Å, and 973Å. Likewise, hydrogen gas heated to a lower temperature emits infrared photons at wavelengths of 18,751Å, 12,818Å, and 10,938Å.

Let's take the example of those lines emitted at visible wavelengths. We know from laboratory experiments that the Balmer lines are emitted when the hydrogen atom's single electron drops from the 3rd orbit to the 2nd orbit, from the 4th orbit to the 2nd orbit, and from the 5th orbit to the 2nd orbit. When we look at hydrogen's *absorption spectrum*, we see dark lines that represent the empty gaps left when hydrogen electrons absorbed the same wavelength photons while jumping up from the 2nd orbit to the 3rd, 4th, and 5th orbits.

In other words, *a given atom at a given temperature absorbs the same wavelength photons from an external light source as it emits when it is heated to that same temperature.* The emission and absorption spectra of a given atom are merely reflections of one another — one is a negative image of the other. They are rather like a photographic negative and photographic print. Therefore, the pattern of lines in the absorption spectrum of Sun tells us not only what chemicals are present in Sun's corona, but also the orbital levels between which the electrons of the various elements are spending most of their time jumping.

Spectra of Solids Up until now I have confined my discussion to the method of determining the chemical composition of gases that make up the atmospheres of stars like Sun. But what about solid objects like planets and asteroids? They don't "glow" like stars, and therefore don't emit radiation that can be measured. Nor is radiation absorbed in the process of passing through a solid object so that an absorption spectrum can be obtained and analyzed.

Some radiation is stopped when it hits a solid object like a planet, but some of it is reflected back out into space so that you and I can see it. But we can do more with that reflected radiation than just look at it. We can compare it to the radiation from Sun that strikes the planet directly, and whatever the difference is between the two must be due to the particular chemical elements in the surface of the planet.

Planets are illuminated by Sun. There is no other significant source of radiation in the solar system. There is really no such thing as moonlight. There is sunlight that reflects off Moon and onto Earth. But of course not all of the wavelengths of radiation that strike Moon's surface are reflected back out into space. Most of it (93% to be exact) is absorbed by the material that makes up Moon's surface. Only 7% of the visible light from Sun that hits Moon is reflected back to space.

Well, it is easy to see that if we determine what wavelengths from Sun fall directly on Moon (by looking directly at Sun), and what wavelengths come directly from Moon, the difference between the two sets must be due to whatever chemical elements are on Moon. And that is basically the technique that we have used not only for determining the compositions of the surfaces of planets, but asteroids and satellites (moons) of the solar system as well.

To take a simple example, observe the planet Mars in the sky some night. It appears red to us. Why red? It is not shining red — it is reflecting red light from Sun. But why just red light? Sun gives off all wavelengths of the visible spectrum. A rainbow is good proof of that statement. Of course — there is some chemical(s) in the surface material of Mars that absorbs most all of the blue and green and yellow wavelengths and allows the red ones to be re-emitted back out into space. If you travel around Earth a bit, you will find red soil in many locations. Very often, the redness is due to the presence of iron that has oxidized — what we call rust. Mars has rust on its surface!

Temperature

At this point in time, a couple of observations are in order. First of all, realize that what we are pursuing here is a method of determining the chemical composition of Sun's atmosphere, not the main body of Sun itself. This may seem like a serious limitation in our investigation of Sun and stars, but since neither Sun nor the stars are solid like Earth, there is considerable interchange of the gases in the bodies of stars and their atmospheres. So the chemical analysis of a star's corona is typically similar to that of the body of the star itself.

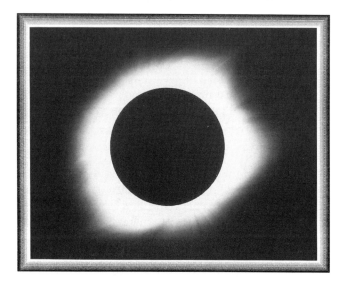

Figure 6–11 Sun's corona is visible during a total solar eclipse. The light we see is emitted when photons escaping at Sun's surface are alternately absorbed and then re-emitted by atoms making up the corona. (Lick Observatory)

Second of all, when we refer to the temperature of something, we usually think in terms of how hot or cool that thing is. We express a subjective judgement about something when we refer to its temperature as being cold or hot. But in science we require a more objective definition of the **temperature** of a object – *it is a measurement of the average speed of the atoms and molecules making up that object.*

So when we refer to a classroom being warm or at a temperature of say 70 degrees, we are in effect making a subjective statement about the sensation we have as a result of the motions of air molecules bombarding our skin. The word "warm" is subjective, whereas the expression "70° F" is specific because it can be translated into a numerical value for the average speed of the molecules of air. The **F** stands for **Fahrenheit**, a commonly-used scale for temperatures.

The definition of temperature I use here may seem unnecessarily technical, but it is important in astronomy, since it is impossible for us to measure temperatures of objects in space with thermometers. Instead, we determine the average velocity of the atoms making up an object through the study of the spectral lines in that object's spectrum.

This is particularly important when you realize that Sun's corona is such a rarefied gas that if I inserted a thermometer into it, there wouldn't be enough atoms bombarding the thermometer to cause it to register anything above zero. And yet astronomers claim that it is about 1 million degrees in temperature! In other words, I would freeze to death in Sun's corona (assuming I was shielded against the radiation coming from the surface of Sun itself)!

Here is another example to illustrate the importance of thinking of temperature in a different way as we explore outer space. Pick up any book discussing the planet Mars, and you'll find references to the temperatures on the Martian surface as being much colder than on Earth. Occasionally, the book mentions that near the Martian equator during the summer season, in the middle of the day, the temperature is as high as 68° F or so, comfortable enough for humans!

Well, unless you are thinking in terms of the speeds of the atoms making up the thin Martian atmosphere, you are probably going to be imagining yourself sunbathing on the Martian surface. I suggest you reconsider. The atmosphere of Mars is about 1% as thick as that of Earth's. So although the speed of the average carbon dioxide molecule making up the atmosphere is the same as that of an oxygen molecule in our atmosphere at the same temperature, there aren't enough CO_2 molecules bombarding your skin to cause a sensation of "warmth." You would freeze to death (not to speak of the deadly solar radiation able to penetrate through the thin atmosphere)! If your bare toes are sunk into the sands on the Martian surface, they will be warm since the sand is absorbing the infrared radiation from Sun and is able to transfer it to anything that is in contact with it. But your ankles would be frozen.

All is not lost, however. You would be perfectly comfortable in wearing a wet suit on Mars, during any season or time of the day-night cycle. You see, the thin atmosphere is not able to extract energy from your body as Earth's atmosphere is able to do here. The reason that you get cold outside during the winter is that the atoms of air surrounding you absorb the infrared radiation (heat) that your body radiates. Since the air is so thin on Mars, you need only prevent the infrared from escaping by covering your skin. Well, the point is that as you imagine yourself visiting the various environments that I am introducing you to in this book, put on a different thinking cap as you encounter my use of the word temperature.

Surface Temperatures of Stars The next question is obvious – "How is it possible to determine temperatures from analyzing spectral lines?" It turns out that when atoms collide with one another, their electrons may be forced up into higher energy orbits. Some of the energy of motion the atoms had prior to colliding is used to change the energy levels of the electrons during the collision. The speeds of the atoms after the collision must, therefore, be less.

The more rapidly the atoms are moving (i.e., the higher the temperature), the greater the number of collisions and the greater the number of atoms that are excited. We can make a loose comparison between atoms at various temperatures and automobiles on freeways – the greater the number and speeds of cars, the greater the number of collisions.

It turns out that electrons in higher energy orbits are more likely to absorb photons than those in the ground state. When nearest the nucleus, electrons are most strongly attracted to it because of their opposite charges. The wavelength absorbed by an electron in going from one orbit to another is different for each orbit change, so by determining the wavelength absorbed, we know which level the electron was in before and after the change. If the electron continues to absorb different wavelengths until it escapes from the nucleus entirely, then, of course, the electron can no longer absorb further wavelengths of radiation.

Absorption Graphs Because in any environment there are billions of atoms behaving this way, electrons in those billions of atoms are at a particular moment be in many different energy levels. Ultimately, we must use statistics to determine which energy levels the electrons occupy. We can't determine what each and every electron is doing. We can determine what most of the electrons are doing. The pattern of behavior is clear – *the higher the temperature, the greater the number of collisions amongst atoms, the greater the number of electrons forced into higher energy levels, the greater the number of photons absorbed by electrons, the greater the number of electrons in higher energy levels, the darker the absorption line in the spectrum* (Figure 6-12). A measurement of the degree of darkness of a spectral line is therefore a measurement of the degree of absorption that took place at that particular wavelength.

★★★★★

But as the temperature continues to climb higher and higher, more and more of the atoms are ionized (lose one or more electrons), fewer photons are absorbed, and the lighter the absorption line in the spectrum. All of this technical jargon can be eloquently expressed by a single mathematical graph for a particular chemical element showing the degree of absorption of photons at different temperatures. This is an **absorption graph**.

As the temperature of the chemical element increases, more and more absorption occurs, resulting in darker and darker absorption lines. But eventually, at some point during temperature rise, atoms begin ionizing, and a slow drop-off in the darkness of the lines results as more and more atoms ionize. *Once an atom is completely ionized, there are no longer any electrons to absorb photons that result in dark lines.* The top of the peak in an absorption graph thus represents the temperature at which the greatest amount of absorption occurs (i.e., the darkest absorption lines occur). The greatest absorption of hydrogen atoms, for example, occurs at a temperature of about 20,000°F (Figure 6-12).

Each chemical element has its own specific pattern of behavior regarding absorption of photons, because each

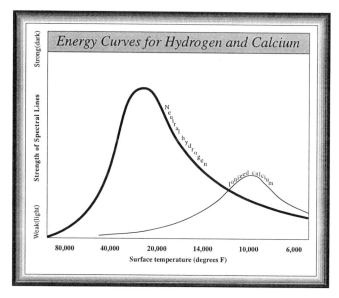

Figure 6–12 The energy curves for hydrogen and calcium, two chemicals commonly found in stars. The strengths of the spectral lines change as the temperature of the chemicals increases, but at different rates for different chemicals.

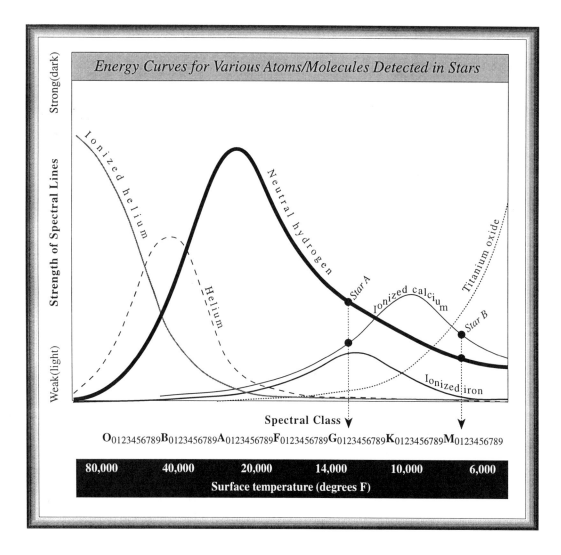

Figure 6–13 Plotting the spectral line energy curves for chemicals commonly found in stars allows astronomers to determine the surface temperature of a star whose spectrum has been obtained. Instead of using numbers to express that temperature, astronomers prefer to use a sequence of letters called the spectral classification.

132   SECTION 2 / Evolution of the Stars

element has a different number of electrons and different pattern of energy levels. So an absorption graph showing the degree of absorption of photons at different temperatures for a chemical element like calcium is different from that of hydrogen (Figure 6-12).

Since all absorption curves use a common temperature scale, we can place them on one single graph (Figure 6-13). Astronomers use this combined absorption graph to determine the temperature of a star's atmosphere (corona). Assume that an astronomer familiar with absorption spectra studies the spectrum of *Star A*, recognizes the absorption lines of <u>hydrogen</u>, measures their degree of darkness to be <u>moderately</u> <u>strong</u>. He/she finds those of the chemical <u>calcium</u> to be <u>moderately</u> <u>strong</u> as well. A glance at the graph (Figure 6-13) shows that in order to meet both of those conditions, the star's atmosphere must have a temperature of about 12,000°F. Both gases in the star's atmosphere must be at the same temperature.

Likewise, the spectrum of a *Star B* might have <u>weak</u> (light-ish) lines of <u>hydrogen</u> and <u>moderately</u> <u>strong</u> (dark) lines of <u>calcium</u>, indicating (by the graph) that its atmosphere has a temperature of about 7,000°F. Astronomers possess a powerful method of determining a vital characteristic of every star (or indeed any object that emits radiation) whose spectrum we can obtain, in spite of the fact that the object may appear as a single dot in the sky!

Specifying Stars by Temperature Astronomers prefer to refer to a star's temperature not by a number such as 7,000°F, but by a letter-code system called the **spectral sequence**. The very hottest stars fall on the sequence as O-type stars, the very coolest as M-type stars, and those between according to the letter sequence shown in Figure 6-13. Inasmuch as determining the vital statistics of stars is fundamental to understanding any celestial object, scientists who specialize in this aspect of astronomy are called **spectroscopists**.

They are like experts in fingerprinting. They measure the strengths (degree of darkness) of absorption lines very accurately, then classify a star's temperature according to a letter and subclass number between any two letters (from zero as the hottest subtype, to 9 being the coolest subtype). So a star whose temperature is halfway between that of an O-type and a B-type is referred to as an O5 type. Our Sun, for example, is a G2-type star.

The **spectral type** designation of a star is a fundamental language in referring to stars. If you overhear astronomers discussing stars, you will undoubtedly hear the frequent use of spectral type. Different stars are referred to according to their spectral type, which expresses their surface temperatures. Later you will learn that spectral type is a fundamental variable in the evolutionary history of a star. By knowing the spectral type and energy output of a star, the astronomer can describe the manner in which the star was born, how long it will live, and the manner in which it will die. As an example of a star's spectrum, Figure 6-14 shows that of Sun. Obviously, it requires some evaluation to determine what lines are responsible for what chemicals in Sun.

If you tell me the surface temperature (using spectral analysis) of a star, I can tell you the complete history of that star — from birth to life to death. And so generations of students have learned the sequence of stars by temperature (**O B A F G K E**) by remembering the phrase — **O**h, **B**e **A** **F**ine **G**irl(**G**uy), **K**iss **M**e! Or perhaps you can make up a catchy phrase yourself. But don't forget what the spectral type of a star expresses about that star — its <u>surface temperature</u>. The spectra of some well-known stars in each of these major temperature groups are shown in Figure 6-15.

Emission Spectrum of Sun's Corona In an earlier section, while discussing absorption spectra, I stated that photons leaving Sun's surface and passing through Sun's corona are occasionally absorbed by atoms making up the corona. Just what happened to that absorbed photon as it was moving outward through the corona? The electron that absorbed it jumped to a higher orbit, and may even have absorbed additional photons to make additional jumps. But it may just as likely have dropped to a lower orbit and released a photon in the process. Recall that electrons tend to migrate to the lowest possible energy state — the ground state.

The higher the temperature, the less likely the electron of a given atom will reach the ground state. But if the density of atoms is slight, as it is in the case of a star's atmosphere, most electrons do drop to lower energy levels. That, of course, means that a photon will be re-emitted. Now doesn't that simply cancel out the photon that was absorbed in the first place? Doesn't the re-emitted photon fill in the dark absorption line so that a continuous spectrum is created? No, because when the photon is re-emitted, the chances of it being re-emitted in the original direction of travel is extremely unlikely. True, photons re-emitted by <u>some</u> atoms go off in that direction to make the absorption line less dark. But most re-emitted photons go out in random directions, not necessarily toward Earth.

Now imagine yourself an observer on Earth studying Sun's corona at the time of a total solar eclipse. You receive that re-emitted photon off to the side of the face of Sun. Do you see? These are the photons you receive when you see Sun's corona during a total solar eclipse! Sun is the source of photons that causes the atoms in its atmosphere to get excited, then when the electrons drop back down they give off the photons that enable us to see the location of those atoms. Thus we receive the re-emitted photons of the atoms making up the corona at the time of a total solar eclipse. The spectroscope separates the specific wavelengths into an emission spectrum, enabling us to determine the chemical composition of the corona in a slightly different manner.

This emission spectrum tells us more specifically the

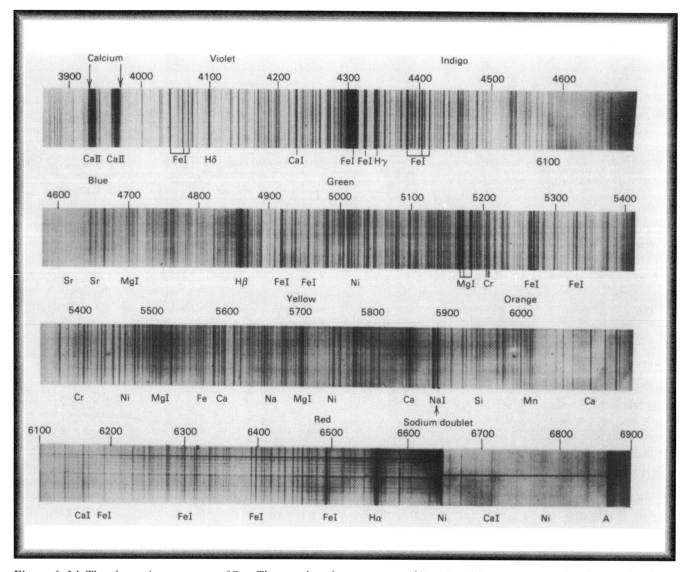

Figure 6–14 The absorption spectrum of Sun. The wavelengths are expressed in units of Angstroms, with the shorter (blue) ones in the upper left, and the longer (red) ones in the lower right. (Lick Observatory)

temperature, chemical composition, and relative frequency of the types of atoms in different parts of Sun's corona. The analysis of the emission lines according to wavelength and brightness provides the same information as an absorption spectrum, but emission spectra can be obtained from various parts of the corona under the conditions of a total solar eclipse, whereas only an average reading of the corona can be obtained when we study Sun's face directly in obtaining an absorption spectrum. Notice that an emission spectrum is the result of hot gases heated by a source of energy. In the case of the emission spectrum obtained during a total solar eclipse, the source of energy is the photons generated at Sun's core and which eventually emerge at Sun's surface to heat the corona.

Emission Nebulae The same situation occurs throughout the Milky Way galaxy within clouds of gas such as the one shown in Figure 6-16. In this case, the stars embedded in the cloud emit the photons that are being absorbed by the atoms of gas making up the cloud. Since we have a photograph of the gas cloud, photons are obviously being emitted by the atoms making up the cloud. And there had to be a source of energy to cause the electrons to jump up in the first place. So the stars are the likely candidates for providing the photons. Because of their telescopic appearances as fuzzy patches of light in the sky, these clouds of gas and dust are called *nebulae* (singular = *nebula*).

If you look closely at the photograph in Figure 6-16, you see embedded in the cloud the bright stars that are responsible for the light being emitted by the atoms making up the cloud. A spectrum of the gas cloud itself is an emission spectrum, whereas the spectrum of one of the stars embedded in the cloud is an absorption spectrum. Recall that most objects in the universe are so distant that they appear as fuzzy little dots of light on a photographic plate. So if we obtain a spectrum of a newly-discovered dot,

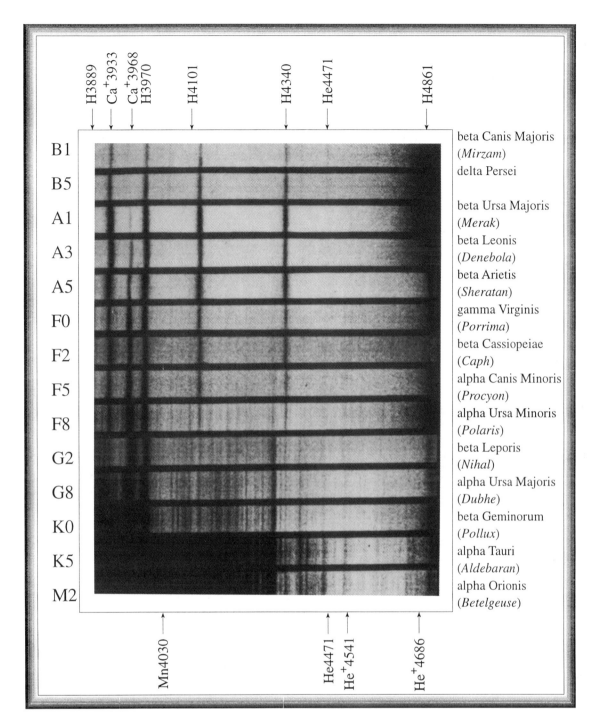

Figure 6–15
These are photographs of the absorption spectra of stars typical of the classes shown in the vertical column. Compare these with the graph of Figure 6–12. If you obtained or were given a spectrum of a star, you could determine that star's surface temperature simply by comparing it with those on this chart. The star whose photograph that most closely matches that of the unknown star has the same temperature as the unknown star whose spectrum you have. (Lick Observatory)

and we find that it is an emission spectrum, we are confident that it is associated with hot gases.

If we obtain an absorption spectrum, we conclude that there is an absorbing cloud of gas between the source of radiation and ourselves, even if we cannot directly see that absorbing layer of gas. If there is reason to believe that the dot is a star, then of course the layer of gas that is absorbing the wavelengths is presumed to be the atmosphere of the star. In any case, it makes our task of understanding what is out there in space easier if we keep these distinctions between absorption and emission spectra in mind.

Conclusions Drawn from Obtaining Spectra

Chemistry of the Universe Our method of determining temperatures and chemical compositions of objects in space is rather technical, but it is one example of the "*How do we know?*" theme I stress often in this Book. There are so many mind-boggling and controversial conclusions drawn from these studies that to accept or reject them out of hand without knowing how the conclusions were drawn

★★★★★

Figure 6–16 An emission nebula emits radiation when hot stars embedded in the nebula heat up the gases of which the nebula is composed to the point that they glow. In color photographs, the color of the glow reveals what chemicals are in the cloud. (Lick Observatory)

from observation would be like saying "Don't bother me with facts." So what is an example of one of these mind-boggling "grand conclusions" drawn from spectral analysis?

I touched briefly on one important conclusion earlier while discussing radio telescopes — hydrogen atoms comprise 90% of all atoms in the universe, while helium atoms comprise 9% of all atoms in the universe. That means, if my arithmetic is correct, that the chemicals you and I are not only most familiar with but also <u>made of</u> together make up a mere 1% of all atoms in the universe — iron, gold, carbon, aluminum, oxygen, and so forth.

Those chemical elements around which our lives revolve are rarities in the cosmos (Figure 6-17). Yes, there is hydrogen in the water we drink and in the body fluids upon which we depend for survival, so hydrogen is also important to us. But certainly judging the rest of the contents of the universe based upon what we find immediately around us is misleading and limiting. The reason for this unequally weighted (unbalanced) universe is simply that most matter (as far as we know) is in stars. Mostly what is happening in the universe is that stars are shining.

I am not suggesting that is the most important thing happening, because I would like to think that my ability to think and feel and cry and laugh are equally as important. But isn't it interesting that without stars (or at least one, our Sun), I might not be here to think or feel in the first place! As long as there is a connection between life and body, the existence of stars is absolutely essential for thought and feeling.

Source of Energy for Stars And there is another connection between the universe out there and you and I, that follows from our determination of the chemical composition of the universe. Stars shine. Stars are made up mostly of hydrogen and helium. Isn't it therefore reasonable to assume that Sun's shining has something to do with its chemical composition? Yes, of course. And so it was that a group of scientists proposed in 1938 that in the act of converting hydrogen into helium at its core, Sun gives off light and heat so that we might live. Ah ha, you are asking, "If that is true, what happens when Sun runs out of hydrogen, and all of it has turned into helium?"

Is it all too obvious? Sun must surely die, as all stars

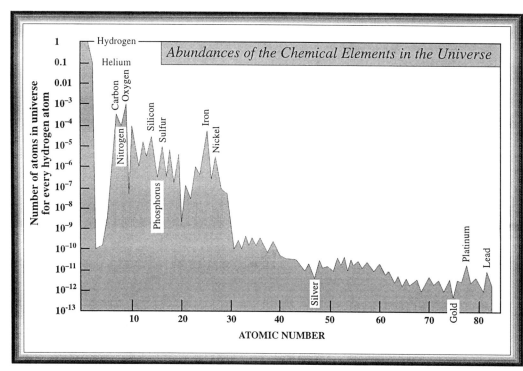

Figure 6–17 The analysis of spectra from stars and galaxies allows astronomers to determine the percentage breakdown of the chemical elements in the universe. This graph reveals the general rule that the heavier the chemical element, the less common it is. In Chapters 9, 10 and 11 we will learn why this is so.

must, at one time or another. We live off the death of Sun. Because Sun converts hydrogen into helium and thereby emits radiation, we are able to live on Earth's surface. Sun dies that we might live. This is an example of what I have referred to as the "*What does it mean?*" Actually, there are many features of Sun that affect us, above and beyond the life-sustaining radiation that we receive. Chapter 7 explains.

Motions of Objects in Space

Knowing the motions of objects in space is obviously important toward knowing where those objects are coming from and knowing where they are going. Knowing past, present, and future allows us to theorize about the evolution of things. You may be familiar with (hopefully not had the personal experience of) the use of radar in tracking speeding motorists (Figure 6-18).

The device used is referred to as a *Doppler gun*, named for the principle upon which the gun operates — the **Doppler effect**. Radar is a portion of the electromagnetic spectrum, a subdivision of the radio region. The Doppler gun sends out radar photons of a fixed wavelength in the direction of vehicles on a highway. Some of the photons reflect off the body of a vehicle and back into the gun's detector.

The gun then simply compares the wavelength of the outgoing signal with that of the incoming signal. If the vehicle is stationary with respect to the gun, the outgoing and incoming wavelengths are identical and the dial registering the speed of the vehicle reads zero. If, on the other hand, the vehicle is moving toward the gun, the reflected (incoming) radiation has a shorter wavelength than that of the emitted (outgoing) radiation, and the difference between the two wavelengths is converted into the speed at which the vehicle is moving relative to the gun.

The vehicle, in effect, compresses the waves in the direction of its motion — i.e., toward the gun. If the vehicle moves away from the gun, the reflected wave stretches out, and the difference between the outgoing and incoming wavelengths is expressed in miles per hour. Actually, this same principle is used throughout industry and scientific fields to determine motions of objects (e.g., tracking aircraft).

Since both radiation — traveling 186,000 miles per second — and sound — traveling 760 miles per hour at sea level — exhibit the property of being wavelike, they both exhibit the Doppler effect. It can be directly experienced when listening to an approaching or receding source of sound, such as a race car or airplane in flight. We use the term **pitch** to express the wavelength or frequency of sound. So the pitch of an approaching car sounds higher than that of the car after it has passed the listener. To the driver of the car, the pitch of the whirling engine would remain the same.

Determining Motions of Objects In astronomy, the Doppler effect is used extensively in determining the motions of objects in space, whether they be planets, stars, nebulae, or galaxies (Figure 6-19). Obviously we do not send radar signals out to reflect off distant objects in the same way that police officers track automobiles. We can use radar for objects within the inner part of the solar system, such as the Moon and nearby planets, but the travel time for the signals to reach even the nearest star prevents their

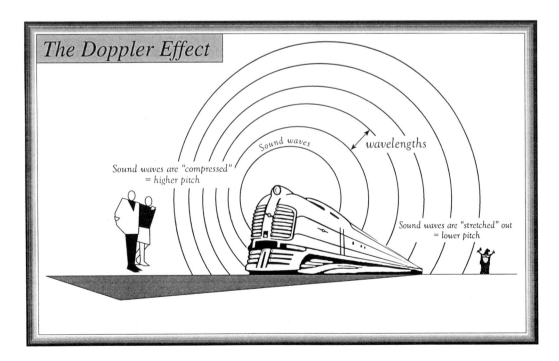

Figure 6-18 The Doppler effect causes the spectral lines in an object's spectrum to be shifted toward shorter or longer wavelengths, providing astronomers with knowledge of the rate of the object's motion with respect to Earth. This is similar to the method of detecting speeding motorists.

CHAPTER 6 / Atoms, Matter, and Radiation

★★★★★

Figure 6–19 The study of the spectra of stars near Sun tells us their motions by the Doppler effect. Our conclusion from this study is that Sun is located on the outer edge of one of the spiral arms of the galaxy. In this photograph of another galaxy, you can imagine Sun as one of the dots of light two-thirds of the way out from the center of the galaxy. (Lick Observatory)

use beyond the solar system. Rather, we accurately measure and then compare the wavelengths of radiation received from deep-space objects with the known wavelengths of chemicals we find in the spectra of those distant objects.

Earlier, we explored the concept that the atoms of different chemical elements absorb specific wavelength photons. If, when we measure the wavelengths of absorption lines in a star's spectrum, we find that all of them have been stretched out (lengthened) by a specific amount, we assume that it is due to that star moving away from us. By comparing the wavelength of that stretched-out photon with that of the wavelength produced by the same chemical in the laboratory, we determine how fast it is receding (moving away) from us.

Conversely, if all of the wavelengths are compressed (shortened), then we measure the rate at which that star is approaching us. Anything giving off or reflecting light can be measured in this manner. Deep space objects are too far away to actually see approaching or receding — we deduce their movements by measuring the shift of spectral lines.

Sun's Position in the Galaxy In Chapter 12, we apply the Doppler effect to the problem of positioning ourselves accurately within the galaxy. By observing the motions of stars in the vicinity of Sun, and by knowing the behavior of objects acting under the influence of gravity, we conclude that we are not located in the dense center of the Milky Way galaxy, but rather out in the suburbs of the spiral arm. Our star, Sun, does not appear to occupy a particularly important spot within the galaxy. We are not strategically placed. But an even more profound conclusion is reached in studying the spectra of more distant galaxies.

Expansion of the Universe Because galaxies consist by and large of billions of stars, the absorption spectra obtained from them represent the average chemical compositions of all the stars in them. In measuring the wavelengths of the lines in the spectrum of an average galaxy, we find that all of them have been lengthened to a lesser or greater extent. Each and every galaxy's spectrum reveals the same stretching out of wavelengths.

According to the Doppler effect, such lengthening of spectral lines can only be explained by concluding that each galaxy is moving away from us. This is not to say that we are located at the center of the universe. Observers in any of those distant galaxies measure the same for our galaxy. So the universe appears to be expanding, as if our galaxy is but a single raisin in an expanding loaf of raisin bread.

Now I am sure that a lot of questions come to mind at this point, like what is the universe expanding into and from what point is it that we are expanding. But we are not yet prepared to discuss this conclusion in detail. I must save that until Chapter 15, after discussing the origin and nature of the raisins (galaxies) themselves.

I will let you ponder this thought, however — if the universe is expanding and causing galaxies to rush away from each other today, then it seems reasonable to conclude that the galaxies were closer together last year, even closer a century ago, and so on and so forth. Knowing the rate at which they are presently moving away from one another, we calculate how long ago it was that all the galaxies, the entire contents of the universe, were in one spot at one time. That, according to the calculations, happened some 20 billion years ago, at which time the galaxies had their common birth together in what is called the **Big Bang theory** for the origin of the universe.

Although many astronomers find such a theory just too simple and bizarre, considerable evidence has been accumulated during the past few decades to support it. As much as we might like to reject it out of hand, the evidence is impossible to avoid. After all, the universe is already here. Whatever manner in which it came into existence already took place. The universe does not really care how we personally feel about how it came into existence. We cannot let our own personal wishes interfere with what we observe, as long as we admit to the reliability of our senses and intellect in understanding what is going on out there.

Summary/Conclusion

Building on the characteristics of radiation that we learned in Chapter 5, and assuming that the same laws of physics and chemistry with which we are familiar here on Earth also operate in space, we now have a method of determining many of the characteristics of the objects in space. Spectral analysis tells us the chemistry, temperature, and motions of objects in space. Basically, it tells us what the atoms in the universe are doing.

The conclusions so far are profound — most of the universe consists of hydrogen and helium, stars generate their energies using these simplest of atoms, the universe of galaxies is expanding, and life-forms consist of the rarer chemical elements in the universe. So we are slowly piecing together a clearer understanding of the conditions in the universe within which intelligent life has arisen. So far, we can appreciate the importance of spectral analysis in concluding that Sun is not a unique star in the universe. This may be the strongest argument so far that intelligent life might be out there as well.

LEARNING OBJECTIVES: *Now that you have studied this Chapter, you should be able to:*

1. Describe the construction of an atom, and how the electron energy levels determine the emission and absorption of radiation.
2. Describe the difference in appearance and origin of the three types of spectra (continuous, emission, and absorption) as seen through a spectroscope.
3. Explain how an energy curve is obtained for each chemical commonly found in stars.
4. Use the energy-level diagram of a hydrogen atom to explain how the Balmer series of wavelengths are produced as both emission and absorption lines.
5. Explain in simple terms how we obtain a absorption spectrum from Sun or a star, and an emission from Sun's corona during a total solar eclipse.
6. Determine the spectral type of a star, given the strengths of the absorption lines of at least two chemicals found in the spectrum of that star.
7. Explain how a spectrum allows astronomers to determine the chemical composition, surface temperature, and motion of a star.
8. Explain how the Doppler effect allows astronomers to determine the rate of motion of objects in space.
9. Describe three important conclusions about the universe obtained through the analysis of the spectra of stars and galaxies.
10. Define and use in a complete sentence each of the following **NEW TERMS**:

Absorption graph 132
Absorption spectrum 128
Angstrom unit 129
Atomic number 125
Balmer lines 129
Big Bang Theory 138
Continuous Spectrum 127
Doppler effect 137
Electromagnetic force 123
Electron 123
Emission spectrum 129
Excited 124
Fahrenheit 131
Ground state 124

Ionized 124
Neutrons 125
Nucleus 123
OBAFGKM 133
Permitted orbits 124
Pitch 137
Proton 123
Spectral sequence 133
Spectral type 133
Spectroscope 127
Spectroscopist 133
Spectrum 129
Temperature 131

Chapter 7

Sun, Our Nearest Star

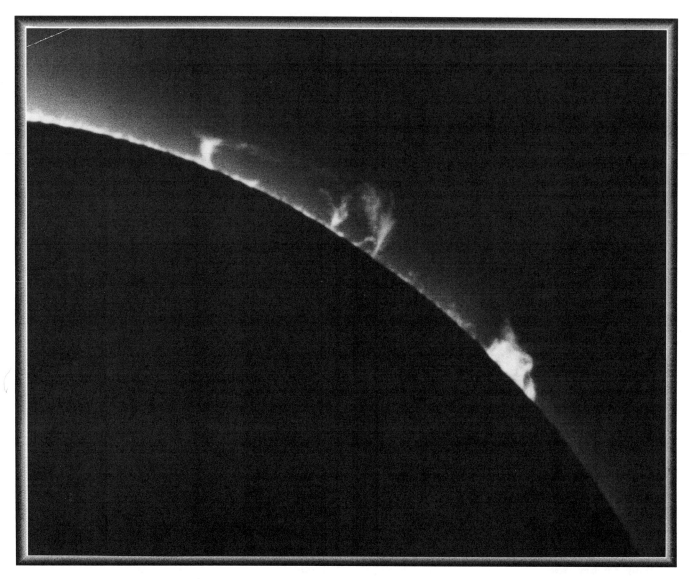

Figure 7–1 A close–up view of the limb of Sun during a total solar eclipse reveals some of the violent activity that routinely occurs on our nearest star. (Lick Observatory)

CENTRAL THEME: What features can we observe on the sun, and how can we explain them in terms of the manner in which it produces radiation and causes it to flow to the surface?

Our star, Sun, is an average star. In terms of its size, shape, temperature, and chemical composition, we find nothing that sets it apart from the multitude of stars visible to us in the sky. This has not always been the case, since its closeness to us is deceptive in that its scattered light drowns out the rest of the stars during the daytime. And at night it is easy to forget about the nearest star while we gaze at those dots of light, knowing well that Sun will eventually rise to give us the much needed light and heat required for our survival.

Most likely there were ancient thinkers who entertained the notion that the stars were simply more distant suns. But their intuitive awareness of the necessity of Sun, especially with the primitive food-gathering methods used, prevented serious consideration of that notion. Besides, common sense told them that all things were orbiting Earth, created for their benefit, and therefore additional suns were unnecessary. Stars were placed up there simply as a reminder of the sky's permanence (and our own impermanence).

Well, of course, there is still much to learn from and about our star, Sun. We certainly haven't explained all of the phenomena that we observe on or around Sun. Being as close as it is, however, it provides us with an opportunity to model stars in general. By accumulating enough understanding about our star, we can more readily understand other stars — their likenesses and their differences. It becomes the basis for our being able to formulate a comprehensive theory of stellar evolution — the birth, life, and death of stars. Not all stars were born like ours. Not all are living like ours. And not all will die like ours. The pattern is quite clear, thanks to our knowledge of Sun.

But I would be amiss if I were not also to include in my discussion of Sun its effects on our inhabited planet and on other planets as well. So where applicable, I attempt to arouse your appreciation of Sun's total effect on Earth and its inhabitants. It seems almost obscene to dissect Sun, peeling off layer after layer until we get to the core. It is such a familiar and dependable object. I have a feeling almost bordering on fear that I will discover something about Sun that will cause me to think of it in human terms, that it too is capable of illness, or error, or even death.

It is almost like moving onto a piece of property in the mountains, sight unseen, and then hesitating to get a geological report on the property for fear that it is in an earthquake zone, or an avalanche area, or an area that is regularly buzzed by flying saucers. But then I recall what I learned long ago — Sun has been around for about 5 billion years, and is expected to be around for another 6 or 8 billion years. Phew! My fears subside.

Observing Sun

It is surprising how few people have had or taken the opportunity to look at Sun through a telescope (Figure 7-1). Many have seen Moon or Saturn or Jupiter or one of the other marvels of the sky. But Sun? Is there anything to see besides a round bright ball anyway? Well, I hope you take any opportunity that comes your way to look at some of the finer detail on Sun's surface — it really is a beautiful sight!

Sunspots Even a smallish telescope allows you to observe events on Sun's surface, although a special filter must first be installed on the telescope to prevent blindness. What you can observe just about anytime you choose to look are dark spots — or at least areas on Sun that appear darker than the rest of Sun's surface. We call them **sunspots**. They are darker because they are cooler areas, at a temperature of only 8,000°F rather than that of 12,000°F for the rest of the photosphere of Sun (Figure 7-2). They typically measure about 6,000 miles across, about the size of Earth. They often occur in groups. At any particular moment, there may be hundreds of sunspots distributed on Sun's surface, or there may be none at all. Now let us assume that you want to take the observations a little more seriously. Let's assume that you are going to spend a few minutes every day observing Sun, with notebook in hand. What might you see over various intervals of time? If you observe Sun each day for a period of several weeks, you begin to notice patterns in the <u>number</u>, <u>location</u>, and <u>behavior</u> of sunspots *after only a few days*.

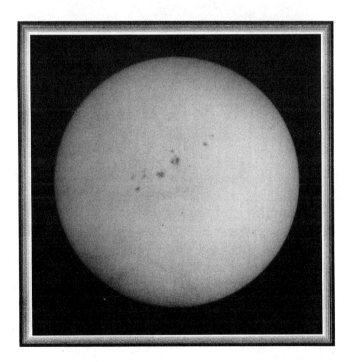

Figure 7-2 Sunspots can easily be seen through a small telescope with a special filter attached. They are cooler areas of Sun seen against the hotter regions. (Tom Bullock)

Rotation of Sun First of all, you notice that sunspots move as a group toward one edge of Sun and eventually disappear, while new sunspots appear from the opposite edge,

★★★★★

Figure 7–3 The McMath solar telescope is 153 meters long, and uses a 60-inch mirror to reflect Sun's image down the angled tube to observing rooms 300 feet beneath Earth's surface. The steady temperature at that location allows for exceptionally crisp images of Sun's surface. (National Optical Astronomy Observatory/National Solar Observatory)

move across the face of Sun, disappear, and so on. We conclude that Sun rotates, taking sunspots with it. Now you might think that if you identify a particular sunspot, you could wait until it went all the way around once and thereby establish Sun's rotational period. But alas, a single sunspot doesn't normally last long enough to survive one complete rotation of Sun. Sunspots come and go, lasting anywhere from 1 to 100 days. In addition, the periods of rotation you determine for different sunspots differ in value. It is possible, however, to measure the angular distance that a given sunspot moves across Sun in a given period of time, and a ratio of angular change to 360° times that period of time provides the *rotational rate of Sun at the latitude of that particular sunspot.*

I emphasize this latter point because Sun does not rotate uniformly. If you perform the same experiment with several different sunspots, all located at different latitudes on Sun's surface, you discover that Sun rotates faster along its equator than at higher latitudes. The measured amounts are about 27 days at the equator, 29 days near the poles. This behavior is called **differential rotation**. Actually, you encountered this concept in the very first Chapter when you learned that the Milky Way galaxy rotates in a similar manner, although in the case of the Galaxy it is the stars making up the galaxy that orbit at different rates around the galaxy's center.

Therefore, it is more appropriate to refer to our Galaxy as exhibiting differential revolution. That Sun rotates at all should not surprise you, and in the next Chapter I will explain more thoroughly why we believe that all stars rotate as well. Why Sun rotates in a differential manner is not well understood, but astronomers are convinced that this particular fact determines much of Sun's behavior.

Figure 7–4 Two solar astronomers work with the solar image at the McMath solar telescope at Kitt Peak, Arizona. One of their tasks is to count the number of sunspots on a daily basis. (National Optical Astronomy Observatories)

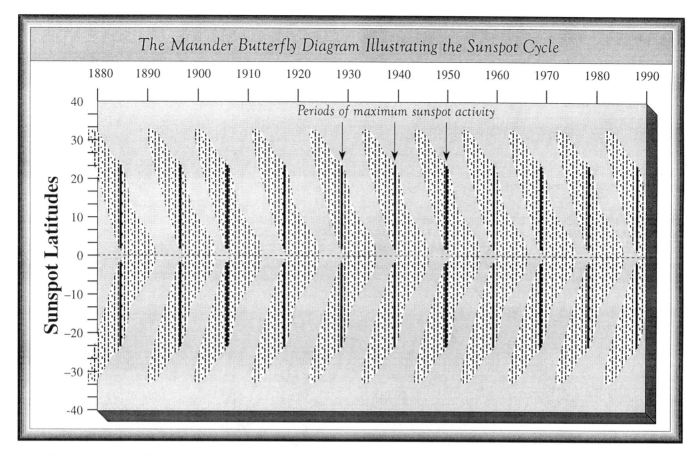

Figure 7-5 The Maunder "butterfly" diagram is a plot of the number of sunspots on Sun's surface at a given latitude over a period of time. The plot reveals a repetitive pattern that allows astronomers to better understand Sun's dynamics.

Sunspot Cycle You have been busy recording the number and location of sunspots in your notebook. A pattern you notice in the behavior of sunspots *during several weeks' viewing* is that they are generally lined up along two lines parallel to Sun's equator. Also after the end of several weeks' viewing, you notice a pattern in the number of sunspots appearing on Sun — they have been either increasing or decreasing. And is there a pattern of behavior if you carry out observations of Sun from several weeks out to several years? Yes. Imagine plotting on a Sun globe (as opposed to an Earth globe) the positions of all sunspots on a particular day. Do that each day for several years, a globe for each day. Line the globes up so that you can walk along and look for patterns of sunspot behavior.

Yes, astronomers do that sort of thing. The premiere facility for monitoring solar activity is the McMath solar telescope located outside Tucson, Arizona (Figure 7-3). It is designed to track Sun from sunrise to sunset every sunny day of the year, enabling astronomers to keep a moment to moment vigil on Sun's activity. In fact, astronomers have been doing that for several decades. That is what the activity is about in the photograph of Figure 7-4. The obtaining of sunspot counts by location on Sun and by date is a very serious matter amongst astronomers. The results are displayed in the form of a table called the **Maunder "butter-fly" diagram**, named after E. Walter Maunder, who was the first to recognized a pattern in the behavior of sunspots (Figure 7-5).

The term "butterfly" is used simply because the plot of the positions of sunspots relative to Sun's equator over periods of years suggests the wings of a butterfly. Spend a few minutes studying the diagram. It basically summarizes the conclusions I suggested you would make if you observed Sun each day over a period of several years. Each dot on the diagram represents a certain number of sunspots counted at the latitude and time (month and year) noted at the side of the diagram. What should appear obvious in the "butterfly" diagram is that:

- At any given time, an equal number of sunspots appear symmetrically at approximately the same northern and southern latitudes;

- At the beginning of each cycle (say in 1965), Sunspots first appear at great distances from the solar equator (usually within 40 degrees);

- As old sunspots disappear after a few days, the newly-formed sunspots appear closer to the solar equator;

CHAPTER 7 / Sun, Our Nearest Star

★★★★★

- The greatest number of sunspots occurs sometime between the beginning of one cycle and the beginning of the next cycle (for example, in 1969); and

- The time interval between one cycle and the next averages 11 years.

The sunspot cycle is measured from minimum sunspot number to minimum again, or from maximum sunspot number to maximum again. In either case, it's about the same time interval. The length of the interval varies from cycle to cycle, as you can deduce from Figure 7-21 if you want to look ahead a bit, but averages 11.2 years. I'll offer an explanation for this inconsistent behavior later in the Chapter. A detailed summary of the most recent sunspot activity is shown in Figure 7-6, suggesting that we are currently experiencing low sunspot activity.

Solar Activity along the Edge Some amateur astronomers can afford to purchase an additional gadget to attach to their telescopes that allows for observations of the more violent events that occur on Sun. It is called a **hydrogen-alpha filter**, so called because it allows a single wavelength (6,563Å) of visible radiation to pass through the filter into the eyepiece. All other wavelengths coming from Sun are filtered out. That particular wavelength is one of the three prominent wavelengths emitted when hydrogen gas is heated up (see Figure 6-10). Specifically, it is the photon emitted when the electron in the hydrogen atom drops from the third energy level to the second energy level. Now it turns out that violent events on Sun's surface always involve hot hydrogen gas. That is not surprising, since Sun is 90% hydrogen.

So the hydrogen-alpha filter allows us to observe those violent events, filtering out the radiation due to less interesting events. It is just another way of increasing the contrast between what we want to see and what we don't want to see. Figure 7-1 is a photograph taken of what we see when such a filter is installed — flame-like structures along the edge of Sun that we call **prominences**. Why prominences? Because they are prominent, of course. What do I mean by prominent? Recall that Sun's diameter is about 100 times that of Earth. So the prominences are typically 10 to 20 times taller than Earth's diameter. It is during a total solar eclipse, by the way, that is the very best time for observing outbursts on Sun's surface. That is one reason that so many amateur astronomers look forward to going to a total solar eclipse.

Look closely at the photograph of Figure 7-7. Also taken with a hydrogen-alpha filter in place, it reveals dark wavy lines on Sun's surface. We call them **filaments**. Since they are darker than the rest of Sun's surface, they must be cooler. They are actually prominences seen face-on. When we see prominences along the edge, we note their appearances as arch-like features. But from above, they appear as lines. It is rather like seeing Golden Arches from street level as opposed to seeing them from a high-flying airplane. So they are, whatever their origin, cooler than Sun's surface.

These are about the only observations you can make of Sun unless you have some rather fancy equipment to attach to a telescope, or you have access to an observatory that specializes in solar studies. There is one such telescope located within the United States that is accessible to the public at least for the purpose of touring the facility and admiring the equipment used to give us our best views of Sun so far.

Detailed Studies of Sun

The solar telescope at Kitt Peak near Tucson, Arizona, is the largest of its kind in the world (Figure 7-3). It monitors Sun daily throughout the year, photographing, obtaining spectra, and watching for unusual events that may occur. The reason it is located in Arizona is to take advantage of the large number of cloud-free days in that particular region. A large moveable mirror at the top receives sunlight and reflects it down the sloped tube 300 feet beneath Earth's surface. There, it can be reflected by a second mirror off into any one of several viewing rooms, in each of which a separate experiment or observation can be conducted.

Sunspots and Magnetic Fields When a spectrum is taken of the region around a sunspot, a curious result is obtained — the spectral lines are each split up into two or more lines. This is a phenomena with which we are familiar on Earth. It is called the **Zeeman effect**, and results from the presence of strong magnetic fields in the region being

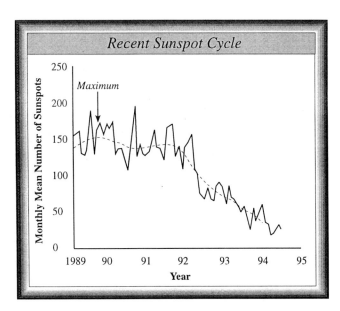

Figure 7-6 A detailed examination of the most recent plot of sunspots reveals that the last maximum occurred in late 1989. There was an unusual secondary buildup in 1991.

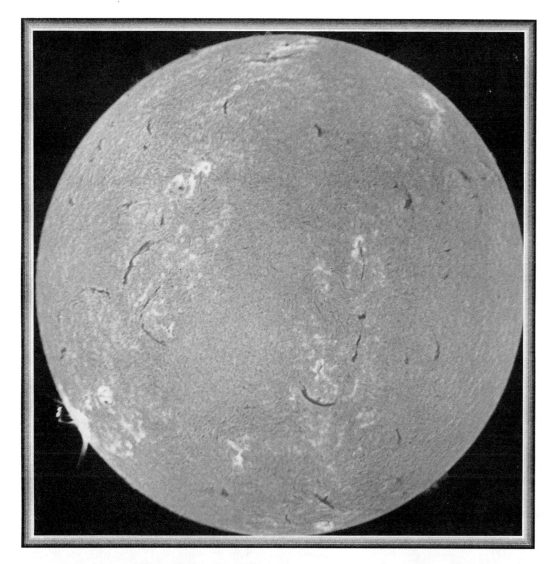

Figure 7–7 This spectacular image of the solar surface taken with the largest solar telescope in the world reveals the fine detail of activity normally present. Aside from the general texture caused by hot gases emerging from Sun's interior, the dark wavy lines are the loops of gases called prominences seen face–on. (National Optical Astronomy Observatory/National Solar Observatory))

Figure 7–8 Comparing the pattern of iron filings in the vicinity of a magnetic with a close–up photograph of sunspots strongly suggests that powerful magnetic fields are associated with sunspots. (Tom Bullock)

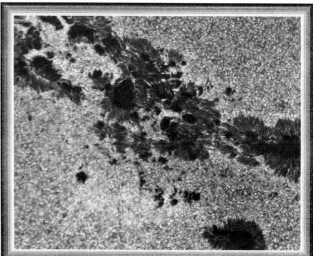

Figure 7–9 An exceptionally clear image of a complex sunspot group and granulation of Sun. Compare the patterns of the hot gases surrounding the sunspots with the pattern of iron filings in Figure 7–7. (National Optical Astronomy Observatory/National Solar Observatory)

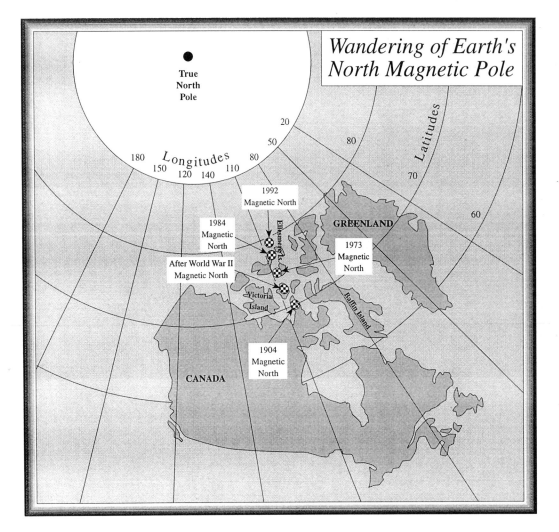

Figure 7–10 Earth's magnetic axis is not lined up with its rotational axis, so a compass—which lines up along Earth's magnetic field lines—does not point to true North. In addition, the magnetic axis wanders as molten material within Earth shifts position.

studied. Sunspots, in other words, are cooler regions on Sun's surface at which intense magnetic fields exist. Since Sun's magnetic fields are basic to an understanding of Sun's observable phenomenon, we should take a short diversion to review what you learned about magnets when you were younger.

Properties of Magnets The magnetic field surrounding a magnet cannot be seen directly, but its presence can be observed by sprinkling iron filings around it. Figure 7-8 illustrates the concept, although in actuality it is a three-dimensional field rather than the two-dimensional one shown. Visualize a series of magnetic force lines running at intervals from one pole of the magnetic to the other, three-dimensionally (Figure 7-9).

Earth is such a magnet. There are **magnetic force lines** running from the north magnetic pole to the south magnetic pole. If you drop some iron filings in the force field surrounding Earth, they too line up along the force lines. In fact, that is exactly what happens when you use a magnetic compass. The needle of the compass is behaving as if it is an iron filing, lining up along Earth's field lines. That is a way of locating the magnetic poles. But don't confuse the magnetic poles with the rotational poles, what we refer to as the geographical poles upon which the coordinates of latitude and longitude are based.

A compass doesn't point in the direction of the rotational poles (true north and south), but to magnetic north and south. The difference between "true" and "magnetic" north is rather slight to most people (up to 12 degrees), but is terribly significant to those who require precise navigational information. The north magnetic pole is presently located amongst the Parry Islands in northern Canada (Latitude = 79° N, Longitude = 100° W).

There is good evidence that it slowly wanders along Earth's surface (Figure 7-10). And you have probably heard that the poles have reversed in the past, the north becoming the south, and vice versa. Well, obviously, that refers to the magnetic poles, not the geographical poles. Later you will find that different planets in the solar system have different angles between their rotational axes and magnetic axes. Uranus is the strangest of all, there being a 60° angle between them. We can't yet explain that observation.

Sunspot Pairs Just as a magnet has a "north" end and a

Figure 7–11 A field of spicules on Sun's surface photographed in red hydrogen–alpha light. A spicule is a short-lived (minutes), narrow jet of gas spouting out of the chromosphere. (National Optical Astronomy Observatories)

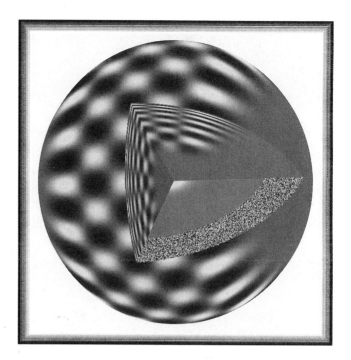

Figure 7–12 A computer representation of one of nearly ten million modes of soundwave oscillations of Sun. By measuring the frequencies of many such modes, and using theoretical models, solar astronomers can inter much about the internal structure and dynamics of Sun. (National Optical Astronomy Observatories)

"south" end, or positive and negative **polarity**, respectively, so also does a given sunspot have a positive and negative *polarity*. This is an important feature of a sunspot, because it gives us a strong clue as to the origin and behavior of sunspots. When astronomers determine the polarities of sunspots appearing daily on Sun's surface, they find that sunspots frequently appear as a pair in the northern hemisphere at the same time that another pair appears in the southern latitude at approximately the same latitude angle. But the two pairs have reversed polarities! That is, if the easternmost sunspot in the northern pair has a positive polarity, the other member of the pair is negative. And the pair in the southern hemisphere is the reverse of that. Figure 7-22 illustrates the observation frequently made — not always, but frequently enough to suggest that something important is going on. I'll deal with this concept in greater detail later in the Chapter.

The presence of magnetic fields in the environments of both Earth and Sun is the result of the same basic process, astronomers believe. When an object spins around a molten interior, friction between the atoms and molecules strips away electrons (ionization) so that charged particles flow along with the rest of the material. But flowing ions create magnetic fields. Electricity is simply a fluid of electrons, a flow of charged particles. There are magnetic fields around all of the wires in your home when electrons are flowing through them.

Sun's interior, being very hot, is a fluid of ions, or what we call a **plasma**. That is pretty easy to understand as far as Sun's magnetic field is concerned. But what about Earth? Well, it turns out that Earth is partially molten inside, allowing for a slow, fluid-type movement to occur, resulting in a magnetic field. I discuss this process in more detail in Chapter 17. In the meanwhile, entertain the possibility of a connection between the semi-molten interior with the occurrences of volcanoes and Earthquakes.

Detailed Surface Features Figure 7-7 is a dramatic photograph of Sun's visible surface, the photosphere. The general appearance reminds us of salt and pepper sprinkled on a table top. Each of the dark and light globs is about 500 miles in size. Again, we would correctly guess that the dark globs are cooler regions than the light globs. "Glob," by the way, is not a technical term. Astronomers call this general grainy pattern of the photosphere **granulation**, and each glob is called a **granule**.

Obtaining spectra of these granules with the Kitt Peak solar instrument reveals them to be the tops of convection currents, the light ones being the tops of ascending currents and the dark ones being the tops of descending currents. We know this by applying the Doppler principle to

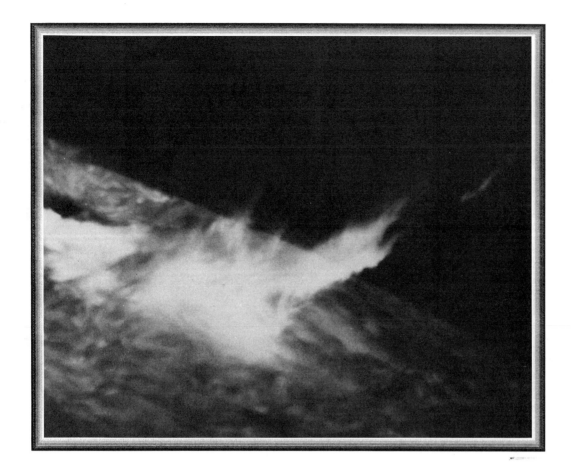

Figure 7–13 The most violent event in the solar system is a solar flare. This is a close-up photograph of one in progress. They only last for an hour or so, which means that solar astronomers must be at the ready in order to image them. (National Optical Astronomy Observatories)

radiation received from the granules. The light from the light-colored granules reveals a Doppler shift toward the shorter wavelengths (approaching us), whereas the dark granules reveal movement away from us.

So the currents appear to be acting like elevators bringing up to Sun's surface the radiation produced in the core, releasing the energy in the form of sunlight. They then return as a cooler current to the interior to pick up more photon passengers. This is, in principle at least, identical to a pot of boiling liquid on a conventional stove. The liquid is hottest where it is in contact with the heating element. That heat turns some of the water to steam which forces currents of liquid to move upward toward the surface, there to release heat to the air above. Then the cooled liquid returns to the bottom of the pot to be reheated.

As the hot gases from the interior break through Sun's surface, they create short-lived flame-like structures called **spicules** (Figure 7-11). Recent studies of these suggest that they may be largely responsible for the high temperature of the corona.

With the development of sophisticated telescope and computer techniques, an entirely new field of astronomical research has recently opened up the study of the large-scale motions of the currents within Sun. **Helioseismology** uses the Doppler effect to measure the general motions of the gases at Sun's surface, and reveals patterns of up and down pulsations that are not yet clearly understood (Figure 7-12).

Then there are the bright white regions found in the vicinity of sunspots. They are obviously hot regions — at least hotter than the surrounding areas. These are called **plages**, which in French means *"beach."*

Solar Violence In terms of violent events, nothing in the solar system, not even the worst act of humans, comes close to **solar flares**. When I use the word *violent*, of course, I am using it in the technical sense of *energy released*. Usually occurring in the vicinity of sunspots (and therefore the plages), these appear to consist of hot gases emerging from Sun's interior, releasing as much energy in several minutes as would 100 million hydrogen bombs with a yield of 100 megatons each. You are probably asking yourself — "If flares release that much energy, aren't they going to be visible to an observer, or at least to an amateur astronomer with a simple telescope?"

Yes, if the observer happens to be observing Sun — with appropriate filters in place — at the time of the outburst. The problem is, the flares don't last too long — one or two hours, at the most. Unless you just happen to be close to a telescope at the time, or you carry a telescope around with you for that very purpose, you'll miss it. Since such outbursts are extremely important to both solar astronomers and segments of industry (I will explain later), a worldwide communication system is on 24-hour alert to notify interested parties whenever a flare erupts.

Solar flares follow the sunspot cycle. That is, they are most common when there are lots of sunspots (sunspot maxima), and even appear to erupt in the immediate vicinity of sunspots. Through a telescope, they appear as rapidly-growing white (hot) regions. Seen against the solar limb, they appear as rapidly-rising "flames" of white gases (Figure 7-13).

There is no doubt that flares are extremely energetic. But astronomers are not exactly sure what causes them. The fact that they occur in the vicinity of sunspots provides us with a reasonable theory, however. They seem to result from the sudden releasing of energy stored in the magnetic field lines above Sunspots. Visualize those lines as twisted rubber bands used to power the propeller of a child's toy airplane. There is enough energy stored up in the twisted bands to lift the airplane off the ground. In Sun, there seems to be a limit to the amount of twisting that the magnetic lines can withstand before they break and suddenly release the tremendous energy stored therein.

In addition to great amounts of visible, ultraviolet, and X-ray radiation emitted at the moment of a flare, <u>electrons</u> and <u>protons</u> are accelerated up to velocities of 50,000 miles per second, about one-half the speed of light. As these charged particles move outward away from Sunspots, some are absorbed by atoms in the chromosphere or corona, thereby raising the temperature of the corona. Some stream outward into the solar system and beyond. This "breeze" of charged particles from Sun is called the **solar wind**.

Solar Wind During the few days following a flare outburst, solar wind particles may get entrapped in the magnetic lines of force surrounding Earth, and spiral along the lines toward the magnetic poles. When they encounter the atoms and molecules that make up Earth's atmosphere, their *energy of motion is transformed into energy of radiation*, and they excite those atoms and molecules of air. When the electrons drop down into lower energy levels, visible light is given off (refer back to Figure 6-10). These displays of light in the sky are seen in the form of the **aurora Borealis** and **aurora Australius**, commonly called the **Northern** and **Southern Lights** (Figure 7-14). During these **geomagnetic storms**, radio blackouts and severe electrical power surges are common. Solar flares are one of the few truly astronomical events that directly affect the immediate environment of Earth.

Solar wind particles are also common during low sunspot activity, so there must be other causes for them besides the flares. In a sense, we on Earth live in the outer extremity of the solar corona. The high temperature of the corona itself is probably responsible for an almost continuous supply of solar wind particles. Recall from Chapter 6 that photons emerging from Sun's surface (photosphere) can be absorbed by the atoms making up the corona. Some are absorbed by the electrons orbiting the nuclei, exciting the atom, resulting in the release of photons that allow us to see the corona during a total solar eclipse. But other photons are absorbed by the nuclei of the atoms themselves, imparting to them greater velocity (higher temperature). If the energy absorbed is great enough, the atom achieves escape velocity and escapes from Sun for the remote parts of the solar system, perhaps even to be captured by the magnetic field of Earth (Figure 7-15).

These solar wind particles do not move directly outward from Sun, however. The magnetic field lines spiral outward like water from a rotating sprinkler. Since the ejected charged particles follow the lines of force, they flow

Figure 7-14 A spectacular photograph of an aurora in the Southern Hemisphere taken by astronauts aboard one of the Shuttle flights. Notice how the pattern within the aurora suggests the presence of magnetic fields. (NASA)

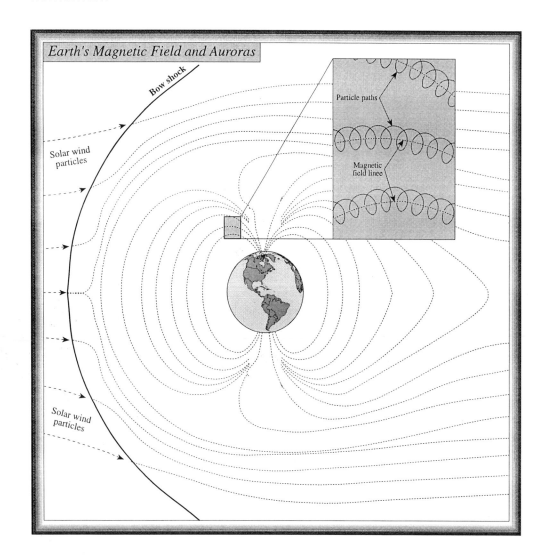

Figure 7–15 *The auroras occur when charged particles from Sun (solar wind) are caught within Earth's magnetic field lines. They gradually "spiral" in toward Earth until they meet charged particles in Earth's atmosphere. The interaction between oppositely-charged particles results in the light being emitted.*

outward in that same pattern. It is estimated that on the average, Sun loses one-half million tons of matter each second in this manner. That seems like an enormous rate of weight reduction. But assuming that this process has been going on for 5 billion years, and will go on for another 7 billion years until Sun dies, Sun will have lost only one-thousandth of 1% of its original mass.

So our interest in the solar wind has nothing to do with its adverse affect on Sun, but on its effects within the solar system, especially on Earth. The solar wind contributes to the behavior of the tails of comets. Why the tails of comets always point away from Sun was not understood until the nature of the solar wind was discovered. Studies of the solar wind are important for the matter of the International Space Station *Alpha*, as well. *SkyLab* fell to Earth prematurely in 1979 because of the fact that the solar wind heated up Earth's atmosphere to the point that it expanded outward to encroach on *SkyLab*'s orbit around Earth. Actually, the solar wind affects just about every object in the solar system, more on some, less on others. Therefore, astronomers are particularly interested in knowing its origin and behavior in order to explain conditions and phenomena observed on planets and satellites.

Satellites Study Sun In the mid-1970s, two satellites named *Helios* were sent into the solar wind within 26 million miles of Sun so that we might study its behavior shortly after leaving Sun. Much was learned. But they could not explore the regions above or below Sun's poles. In 1990, NASA launched the satellite **Ulysses** on a four-year journey to explore Sun's poles, something impossible to do directly from Earth (Figure 7-16).

The rocket used to launch the satellite wasn't powerful enough to eject it away from Earth's orbital plane. So when *Ulysses* encountered the immense gravity of Jupiter in 1992, the "slingshot effect" threw it out of the plane so that it could eventually look "up" at Sun's south pole (June – October 1994) and then around and across the plane to look "down" at Sun's north pole (June – September 1995) (Figure 7-17).

This was a long mission, but *Ulysses* was busy throughout. Some data was obtained at Jupiter, since the probe, unlike the previous satellites *Pioneers* and *Voyagers*, passed over the polar areas. Even while far from celestial objects,

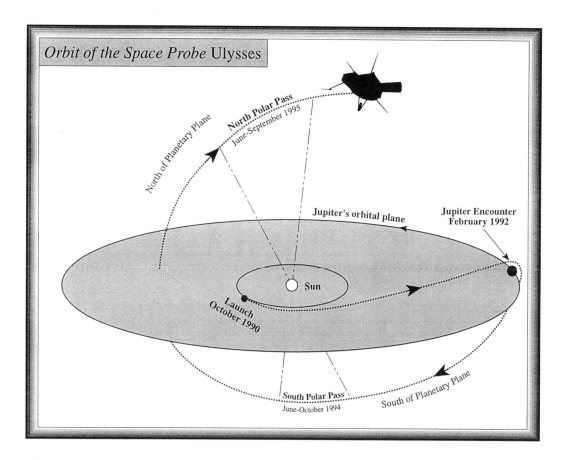

Figure 7–16 The Ulysses *mission to study Sun away from the ecliptic plane has added to our understanding of Sun's behavior, as well as raise new questions.*

its nine instruments measured magnetic fields, cosmic rays and dust, gamma and X-rays, and many other properties of interplanetary space. Scientists on the Ulysses mission found some surprising new discoveries about the polar regions of Sun. Among them, the spacecraft revealed two clearly separate and distinct solar wind regimes with fast wind emerging from the solar poles. Scientists were surprised by observations of how cosmic rays make their way into the solar system from galaxies beyond the Milky Way.

The magnetic field of the Sun over its poles turned out to be very different from previous expectations using ground-based observations. In addition, *Ulysses* found a beam of particles from interstellar space that was penetrating the solar system at a velocity of close to 50,000 miles per hour. On September 29, 1995, Ulysses completed its first solar orbit and began its voyage back out to the orbit of Jupiter, where it will loop around and return to the vicinity of Sun in September 2000. At that time Sun will be in a very active phase of its 11-year solar cycle and Ulysses will find itself battling through the atmosphere of a star that is no longer docile. Stay tuned.

Detailed Look at Prominences Prominences, as I mentioned earlier, are rather easy to observe. They are seen as arch-like features along the edge of Sun, and as dark wavy lines on the face of Sun (Figure 7-18). Observing one at a particular moment will not, however, reveal its true nature. Because of its great size, it is tempting to associate it with violent activity much as we do with the flares. But that is not usually the case. What one needs to do is take a series of photographs over a period of hours or days. This series, when projected one after the other as if it were a movie strip,

Figure 7–17 The Ulysses *satellite is seen above Sun's north pole in 1995. Data returned to astronomers on Earth is still being analyzed to determine how Sun behaves in those regions that cannot be studied directly from Earth.* (NASA)

is called **time-lapse** photography.

Although prominences superficially appear to consist of material ejected from the surface of Sun upward, time-lapse photography reveals that the material is most often moving downward in graceful arcs, obviously along lines of magnetic force. Since the loops of the prominences most often arch between two sunspot groups, it is easy to imagine that whatever disturbance that is responsible for Sunspots, plages, and flares is responsible for prominences as well. Perhaps the disturbance moves up along the looped magnetic field, causing the gas atoms in the vicinity to radiate energy as they de-excite. Obviously, they too are related to Sunspot cycle.

Coronal Activity Inasmuch as Sun is not a solid object, one can think of Sun's atmosphere, the corona, as simply an extension of the photosphere. There isn't a clear boundary that separates the two, except for the scientist's need to classify nature in accordance with sometimes arbitrary criteria. So there is constant interchange and interplay between the two.

The corona is most easily seen during the time of a total solar eclipse. I say "most easily seen" because astronomers developed a technique to create an artificial eclipse in order to study the corona any time they want. The device is a small circular disk that is placed in the telescope where the image of Sun appears (the focus point). The disk is of the size to exactly block out Sun so that the corona can be observed and studied. The disk is simply acting like an artificial moon. This device is called a **coronagraph**.

If you have the opportunity to observe more than one eclipse, you will notice that the corona is not always shaped the same. When Sun is experiencing sunspot maximum, it is nearly circular, with streamers radiating outward in all directions (refer back to Figure 6-11). The streamers are the locations of the hot gases as they absorb and re-emit the photons coming from the photosphere. During sunspot minimum, however, the corona extends farther out away from the equatorial regions than from the polar regions.

These observations tell us of the importance of magnetic fields in structuring the corona. When sunspots are at maximum, magnetic fields are at maximum. Therefore, the atoms of gas line up along the symmetrically-oriented lines and cause the corona to look circular. During sunspot minimum, the magnetic fields are most intense along the equator, thereby causing the gas atoms to be more prominently displayed above the equator.

With that as background information, we are now able to offer an explanation for the high temperature (1 million degrees or so) of the corona. If magnetic fields can be likened to the release of energy by twisted rubber bands, then what happens to that stored energy if it isn't released suddenly in the form of flares? That's right. It gradually leaks out to heat the corona.

Figure 7-19 The Solar Max mission to study Sun added immeasurably to our understanding of solar dynamics, in addition to taking some 30,000 images from above Earth's atmosphere. (NASA)

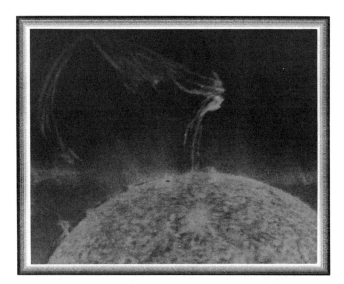

Figure 7-18 Solar prominences such as this are not as violent as flares, but they last for much longer and are therefore much more easily observed. On just about any day, prominences can be seen along Sun's edge if a hydrogen-alpha filter is in place on the telescope. (NASA)

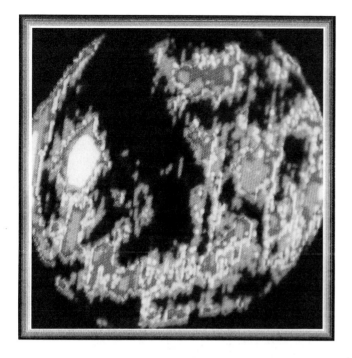

Figure 7–20 X–ray studies of Sun by Skylab astronauts in 1973 revealed for the first time changes in the X–ray emitting areas of the corona, now called coronal holes. They are clearly related to sunspots, and may be the regions from which solar wind particles can escape from Sun's surface to eventually interact with the planets. (NASA)

In 1980, one of the most sophisticated satellites ever built was launched for the purpose of studying Sun during sunspot maximum (Figure 7-19). But the **Solar Maximum Mission (Solar Max,** for short) satellite went haywire and was useless until the crew of the Space Shuttle *Challenger* pulled it into the cargo bay, repaired it, and then relaunched it. During the first nine months of operation, it returned to Earth some 30,000 photographs of the corona taken in X-ray and ultraviolet wavelengths.

What these photographs revealed for the first time was the existence of **coronal holes**, regions of the corona in which the magnetic lines of force open out into interplanetary space rather than looping back onto Sun's surface. In X-ray photographs, these coronal holes appear as darker regions, indicating they are cooler areas (Figure 7-20). Since the lines lead outward away from Sun, these are the convenient routes of escape for the solar wind particles. So it is now believed that coronal holes are the source of the high-energy subatomic particles constituting the solar wind. The cause of the holes themselves is not well understood.

Studies of Sun's corona by use of a coronagraph are not quite as good as studies conducted during a natural eclipse, but they do provide useful information, especially during sunspot maxima. Why then? Because the greater the number of sunspots, the greater the number of the violent events on Sun. Sunspots in themselves are not violent events, but, for reasons to be explained later in the Chapter, give rise to violent events. At the same time, I will pull all of these observations together into one concise theory that attempts to explain Sun's somewhat erratic behavior.

Sun–Earth Relationships

In 1893, a British astronomer, E. Walter Maunder, studied European historical records of sunspot activity. Naked-eye sunspot sighting reports exist for periods going back as early as 28 BC. To his surprise, very few sunspots were observed during the period from 1645 AD to 1715 AD, now referred to as the **Maunder Minimum**. Since Maunder's original discovery, astronomers and historians have concluded that during that same period very few aurora were observed in Europe, and that the corona seen during eclipses of that period was absent or very weak. The graph in Figure 7-21 illustrates, based upon historical records, the variations in solar activity during the past.

What is intriguing about this is that there appears to be

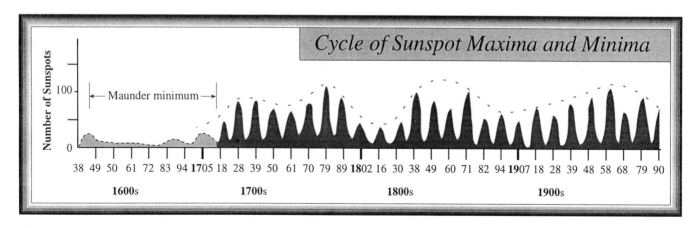

Figure 7–21 Long–term variations in the sunspot cycle reveal an even larger pattern of changes. Some scientists believe that these long–term variations are related to ice ages that may be linked to Sun's erratic behavior.

CHAPTER 7 / Sun, Our Nearest Star **153**

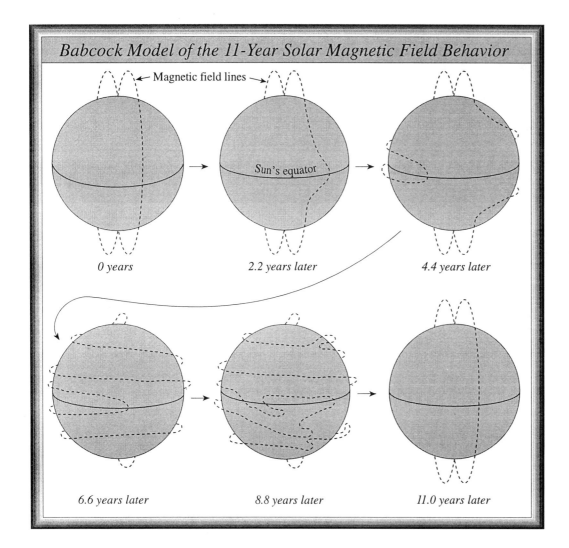

Figure 7–22 Incorporating all of Sun's features, the Babcock model proposes that Sun's cycle progresses as a result of its magnetic field lines getting stretched and twisted by the differential rotation of Sun. Once the lines are twisted enough to break, the magnetic field breaks down and a new one forms. This takes an 11–year period of time on the average.

a relationship between solar activity and climatic variations on Earth. Tree-ring studies and radiocarbon dating methods show that climatic conditions were measurably altered during the Maunder minimum period, aside from the severely cold winters reported about that time. Likewise, tree-ring studies reveal a 22-year drought cycle in the southwestern United States.

The entire relationship between solar events and biology and climate is attracting new research. Changes in solar activity may be due to changes in the pattern of the convection currents which bring up to the photosphere the energy produced at the core, or changes in the energy output at the core itself. One of the goals of the *Ulysses* mission is to gain a better understanding of the relationship between Earth's magnetic field and solar events. This undertaking is a small portion of an *International Solar-Terrestrial Physics* program.

Solar Behavior: the Babcock Model

The method of science is to go from observation to theory to model. So now that we have observed the various phenomena on Sun, let's attempt to piece them all together into a model that includes the prediction of future behavior. Fortunately, someone has done the work for us already. It is called the **Babcock model** after the astronomer who proposed it in the 1950s.

Magnetic Lines of Force Sun is a giant magnet. Visualize the magnetic lines of force running between Sun's two magnetic poles, one of which has a positive (+) charge, and the other a negative (−) charge. Visualize that they run not only outward from the surface of Sun, but also just under the photosphere, rather like an elaborate system of sewer pipes. At the very beginning of a particular sunspot cycle, the lines run directly between the poles as straight lines.

The beginning of a sunspot cycle, in other words, coincides with the creation of a "fresh" or newly-formed set of lines of force. These lines, embedded within the photosphere, rotate along with Sun. But the photosphere rotates differentially — that is, the equatorial regions rotate faster than those closer to the poles. So as time goes on, the lines begin to stretch out as they are pulled ahead by the faster-moving plasma along the equator (Figure 7–22).

This stretching out of the magnetic lines in Sun is

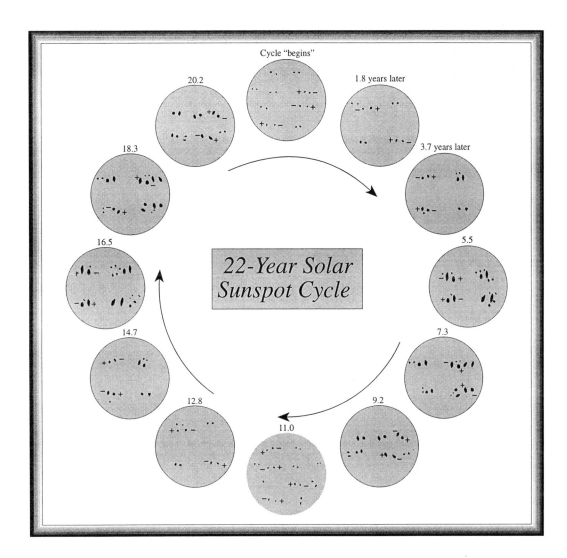

Figure 7-23 *This diagram represents the changing pattern of sunspots according to size, number, location and polarity. Note especially the change in polarity between "leaders" and "followers" sunspots at the end of each 11-year cycle.*

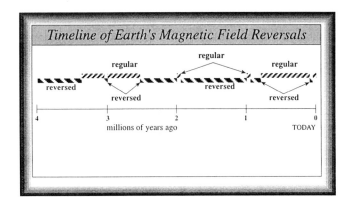

Figure 7-24 *The reversals of Earth's magnetic field are not well understood, but the mechanism that is responsible is certainly not unique in the solar system. Sun's magnetic field reverses at the conclusion of each 11-year sunspot cycle. This chart represents Earth's magnetic reversal history.*

likened to throwing a handful of pine needles in a mountain stream — the current is faster in the middle, and therefore carries the needles caught there downstream faster than those along the edges of the stream. After many solar rotations, the field lines are stretched into a powerful east-west field, still just under the surface of the photosphere. The field lines gradually curl into tubes, and if a kink in a tube occurs, it is pushed up as an arch through the photosphere. The tube penetrates through the surface at two locations. So the magnetic nature of the tube is exposed, just as if movement along an Earthquake fault line pushes sewer pipes up through the pavement.

For reasons not clearly understood, the intense magnetic fields associated with the two "openings" in the photosphere repel the hotter gases, preventing the convection currents from bringing hot gases out at those openings. That means, of course, that the gases in the vicinity of the two openings are cooler. Sunspots, get it! And usually in pairs! And associated with strong magnetic fields as well! It all fits! The polarity of the spots is obvious. The spot associated with the tube running to the north pole is positive, the other negative. And if the same tube becomes kinked in the opposite hemisphere, two sunspots occur there with the opposite polarities. The symmetry between the two hemispheres is simply due to the symmetry with which the two hemispheres rotate and therefore stretch out the lines of force.

Sunspot Cycle Since the Babcock model is doing so well in explaining the observed phenomena, let's now tackle the pattern of sunspot occurrence at gradually decreasing latitude during the 11-year sunspot cycle. At the beginning of the cycle, the greatest stretching and kinking of the lines of force occurs at the higher latitudes. As differential rotation proceeds and the lines closer to the equator get wrapped around Sun and therefore twisted and distorted, kinks occur at locations closer and closer to the equator, and in greater numbers (Figure 7-23).

Eventually, as lines of force break up and decay, the number of kinks decreases, a new magnetic field forms, and a new cycle begins. So as the last sunspots of one cycle fade away close to the equator, sunspots that are a part of the next cycle begin to form at the higher latitudes. During that short interval between the two cycles, there are sunspots close to the equator and at higher latitudes at the same time.

One other thing. As the next cycle begins, sunspot pairs have the opposite polarities. That is, if the easternmost spot was positive during the old cycle, then it is negative during the new cycle. In effect, the magnetic field of Sun reverses itself with each new cycle, the north pole becoming the south pole and vice versa. The period of complete reversal of sunspot polarity is therefore 22 years, a period which is more consistent than the 11-year period between cycle (Figure 7-23).

Sun's energy output is not constant, a fact that has been known for several years now. The changes in brightness accompany sunspot cycle. It is brightest when Sunspots are at a maximum, a fact which at first glance seems contradictory. But the bright eruptions called **faculae** that appear at sunspot maximum more than make up for the darkening caused by sunspots. Based on records of past cycles, the current cycle that we are in was supposed to have peaked in 1990 or 1991. But the curve started to turn downward in June 1989. This is the earliest one on record (Figure 7-6).

Journey into Sun

The exterior of Sun is somewhat easy to observe. The features that appear on the surface and in the atmosphere can be photographed and analyzed with batteries of sophisticated instruments. But that is certainly not the case with Sun's interior. I have over and over again referred to the fact that Sun shines by the process of converting hydrogen into helium. But now it is time to substantiate that claim and to examine the conditions under which the process is thought to take place.

Nature of Solar Material So what is going on at the core of Sun, anyway? In Chapter 6, I merely mentioned that photons in the form of gamma rays are produced there, eventually escaping at the surface as photons of all wavelengths (Figure 7-25). But how are they produced, and why there? First, let us talk a bit about the nature of the material within Sun. It is tempting to think of the material within Sun in the same way that we think of material on Earth, but that is not appropriate.

To groups of young children, I refer to Sun as a "big ball of burning gases." A big ball it is, but "gases" implies that Sun is made up of a collection of atoms that move about freely, not rigidly held in a given configuration. The difference between a gas and a liquid is one of temperature – water (a liquid) heated up turns into steam (a gas). And a solid? Ice (a solid) heated up turns into water (Figure 7-26).

The temperatures within Sun are high, varying from 15 million degrees F at the core to 12,000° F at the photosphere. At those temperatures, atoms cease to exist as we know them on Earth. The electrons are totally stripped away from the nuclei (complete ionization). Under those conditions, the nuclei and electrons move about as charged particles in what one would call a fluid if we could be present. A fluid of charged particles is called *plasma*. In physics, plasma is referred to as the "Fourth State of Matter" – after solid, liquid, and gas. Inasmuch as the universe consists mostly of stars, we can say that most of matter in the universe is in the form of plasma.

The electricity that comes out of the sockets in your home is nothing more than a stream of electrons, so electricity is an example of plasma. We can create plasma on Earth. Perhaps, then, you can imagine what it would be like to be within the interior of Sun. It would rather be like

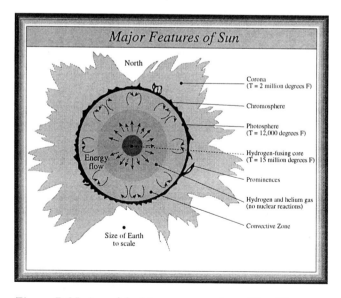

Figure 7-25 A model of the cross-section of Sun illustrates some of the features observed on the surface as well as the mechanisms by which energy produced in the core gets to the surface.

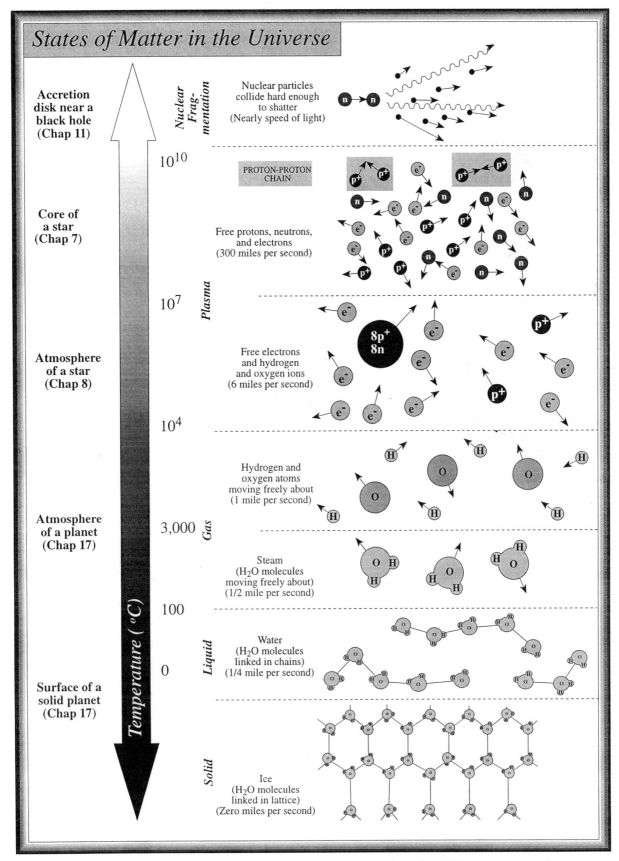

Figure 7–26 The "state" of a particular type of matter in the universe depends upon its temperature. As we explore different parts of the universe, we learn that plasma is the most common "state," even though it is not commonly found on Earth's surface. Most stars are made of plasma.

swimming in a pool of liquid electricity! The density of Sun's plasma must be greatest at the core, at which place it should appear as a solid. It would be least dense in the photosphere, at which place an airplane could fly through the upper layer. The greater the pressure due to gravity, the greater the density. That is how we can model the densities of plasma within Sun.

Material at Sun's Core Since Sun's chemical composition consists mostly of hydrogen, the plasma can be visualized as a fluid of protons and electrons, the basic ingredients for the reaction responsible for sunlight. It is only at Sun's core that the pressures are great enough to force the protons close enough together so that a reaction can occur. You see, protons don't like to join together. Since they have like charges, they repulse one another and resist close contact. But if the pressure is great enough, if the protons are moving rapidly enough, they can collide. Again, think of a crowded freeway.

Jupiter, the most massive planet in our solar system, is almost identical to Sun chemically. But it does not shine. That is because there isn't enough pressure at the core to force the protons close enough together for a reaction to occur. At Sun's core, because of the great pressure, two

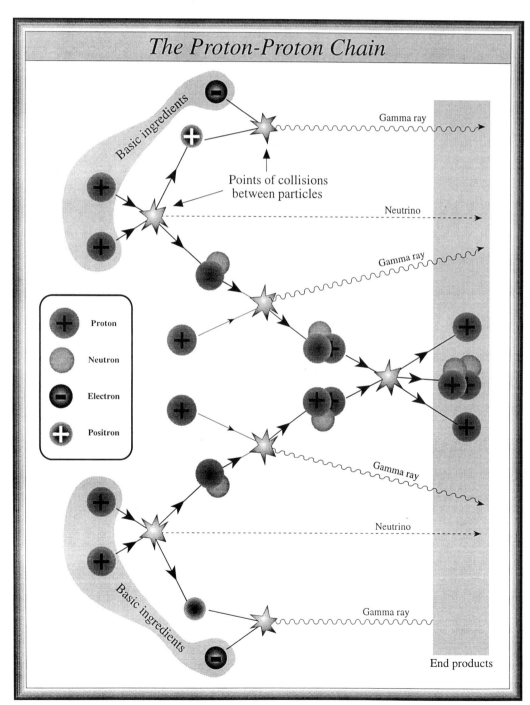

Figure 7–27 The proton–proton chain is the most likely model of energy generation within Sun. The process has been duplicated on Earth in the form of hydrogen bombs, and various portions of the reaction can be studied in atomic accelerators.

Figure 7–28 The Stanford Linear Accelerator Center (SLAC) *in California is one of the world's leading facilities for investigating the nature of sub–atomic particles and the forces that bind them together to make matter. Much of the research conducted there provides astronomers with an understanding of the nature and behavior of objects in space such as Sun.* (Stanford Linear Accelerator Center)

protons are forced close enough together (and move rapidly enough) to result in a collision (Figure 7-27). When that collision occurs, the two protons combine to form a heavy hydrogen nucleus, called **deuterium**. It consists of a proton and a newly-created particle, the neutron.

In the process of the two protons colliding to form a deuterium nucleus, there are two by-products. These are a *neutrino* and a *positron*. I return to these shortly. Another available proton then collides with the heavy hydrogen nucleus (deuterium) to form a <u>lightweight</u> helium nucleus ("lightweight" because it has only one neutron). At the time of this collision, a *gamma-ray* photon is produced that begins its journey toward the photosphere, eventually becoming sunlight that nourishes us. The lightweight helium nucleus in turn collides with another lightweight helium nucleus that formed in identical fashion, and this final fusion results in the formation of a <u>normal</u> helium nucleus and *two protons* ("normal" because an atom normally has as many neutrons as protons in its nucleus).

The helium atom remains in Sun's core as a kind of "ash," and the two protons are free to enter into additional collisions as just described. Looking at the entire process, then, the overall reaction is the conversion of protons (hydrogen nuclei) into helium nuclei plus radiation in the form of gamma rays plus positrons and neutrinos. We call this process the **proton-proton chain** – "proton-proton" because protons are the basic ingredient, and "chain" because the protons that emerge out of the final collision between two lightweight helium nuclei enter into further collisions to sustain the process. The process can go on and on as long as there are protons and the necessary conditions of pressure and temperature.

Anti-Matter At the beginning of the proton-proton chain, there is the production of a strange particle called a **positron**. It is what we might call an "antiparticle." That is, it is identical in nature (mass, size, etc.) to an electron except that it has the opposite charge – positive instead of negative. Within **particle accelerators** like **Stanford Linear Accelerator Center**, located in Palo Alto, California, we discovered that for every particle making up matter, we can create for a <u>brief</u> moment its opposite: an **antiparticle**. This is

★★★★★

accomplished by accelerating protons, electrons, or neutrons to speeds close to the speed of light, and studying the by-products (Figure 7-28). Scientists have detected antiprotons (negative instead of positive), antineutrons (which spin in the opposite direction as do neutrons), and a whole host of other odd characters that make up matter.

But I should emphasize my use of the word "brief." The interesting feature about antiparticles is that as soon as they are created in the accelerator, they immediately (within one-billionth of a second or so) find their opposite and the two join together in wedded bliss, at which time they disappear and leave behind energy that can be detected. The entire Earth is made up of matter containing normal particles, so the antiparticles don't have a difficult time finding an opposite partner to marry.

Incidentally, this particle/antiparticle annihilation process is thought to be the ultimate form of energy production possible in our universe. There are no by-products of the reaction, no ash, nothing to clean up. Just pure energy (radiation). No wonder, then, that the *Enterprise* of "Star Trek" fame uses antimatter as its fuel while searching out the Klingons to destroy! If positrons could exist long enough to establish orbit around antiprotons, there could exist antimatter. Alas, that doesn't appear possible on Earth. Some speculate, however, that half the universe out there consists of anti-stars and anti-galaxies, and even anti-students!

So you now know what happens to that positron generated at Sun's core. Yes, it finds an electron to which it weds, leaving behind a photon to escape to Sun's surface and eventually to enter (perhaps) your eye as sunlight. Theoretically, then, half the sunlight striking Earth is a product of particle/antiparticle annihilation. So it is not just a feature of science-fiction stories.

Mass Loss in Sun Before I explain the remaining participant in the proton-proton chain, the *neutrino*, I would like to stimulate your appreciation of one of the most important discoveries of this century — the interchangeability of mass and energy. Matter can be converted into energy, and energy can be converted into matter. The former is the basic process by which stars shine. The latter has been accomplished only within the confines of atomic accelerators like that at Stanford. If we weigh the initial ingredients in the proton-proton chain and compare that sum with the weight of the final products, the helium, we find that there is a slight difference. The exact numbers are as follows:

4 H atoms = 6.693×10^{-27} kg (kilogram)
1 He atom = 6.645×10^{-27} kg

Difference in mass = 0.048×10^{-27} kg

Of course, that difference in mass appears extremely slight. But when one considers the enormous number of atoms that participate in this fusion process in order to produce the energy observed at Sun's surface, it turns out to be about 5 million tons per second! Sun is (has been, will be) reducing its mass by 5 million tons per second. Where does it go? Einstein tells us with his famous equation

$E = mc^2$, which says that the amount of energy (**E**) that can be obtained from a given amount of mass (**m**) is equal to that amount of mass times the speed of light (**c**) squared.

Since the speed of light is 186,000 miles per second, it should be obvious that a tremendous amount of energy can be obtained from a small amount of mass.

By the way, this equation is not a one-way street. It also says that energy can be converted into mass. Now don't think that you can go out and manufacture some gold out of sunlight. You will need a particle accelerator to get even a hint of this opposite reaction occurring. The principle continues to be evoked extensively in science fiction, as in the movie *The Fly* and the transporter system of "Star Trek."

How Do We Know? Now you are asking — "How do we know what is going on at the center of a star that is one million times larger than Earth where the temperature is 15 million degrees?" To which I respond — "Because our study of atomic physics and Sun's chemical composition leads us to that conclusion." So far, that is the only mechanism that can possibly explain what we observe. We may well be wrong. We could be entirely off base. Isn't it interesting, though, that at the same time we developed the proton-proton model of the basic process by which the sky at night shines with points of light, we tested out the model on the very surface of Earth, and succeeded!

I am speaking, of course, not of atomic power plants, but of hydrogen bombs. Atomic bombs, built earlier using the same knowledge of atomic physics, operate on the principle of **fission**, the breaking apart of large atomic nuclei such as uranium and plutonium. But the process of protons joining together to form a larger helium nucleus is a **fusion** process. In order to create the 15 million degree temperature necessary for the fusion process to proceed inside of the H-bomb, an A-bomb is wrapped around the hydrogen fuel and detonated. The A-bomb is a trigger for the H-bomb. So in a certain manner of speaking, Sun (and every other star) is a continuously exploding H-bomb.

The Neutrino I have saved the *neutrino* for the finale because it is that which relates to and ties in with the possible variation of solar activity with time — which in turn influences Earth's climate. You may recall from Chapter 7 that the average gamma-ray photon at Sun's core takes a million years to get to the surface. A measurement taken today of the number of photons escaping from the photosphere is therefore directly related to the rate at which the proton–proton chain took place one million years ago. That rate, in turn, is related to the temperature in Sun's core at the time. The higher the temperature, the faster moving are the protons and electrons and other particles, the greater the number of collisions and fusions occurring, the greater the number of photons created, the greater the number of photons eventually leaving Sun.

Now, it is a matter of simple mathematics (inverse square law) to determine the total energy output of Sun, a number called the **solar constant**. Knowing Earth–Sun distance and the amount of sunlight falling per unit time on a given surface area of Earth, we can calculate that constant. In order for Sun to have that value today, the temperature of Sun's core one million years ago must have been about *15 million degrees*. Keep this number in mind for a few minutes, because I am about to compare it with another measurement taken of the temperature at Sun's core.

Neutrinos also tell us about the rate at which the proton–proton chain takes place at Sun's core. Neutrinos, unlike photons, are not influenced by overlying layers of plasma within Sun. They come out directly from the core

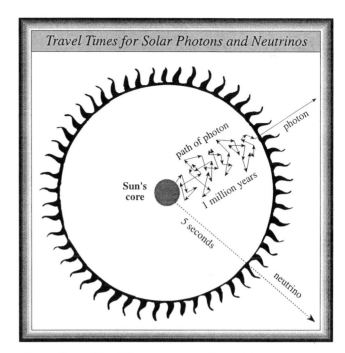

Figure 7–30 The difference in travel times for gamma ray photons and neutrinos leaving Sun's core provides astronomers to compare the variable rate at which Sun produces energy. Sun may be a long–term variable star.

without being absorbed, blocked, or reflected elsewhere. To the eyes of a neutrino, the universe is empty. A neutrino would (statistically speaking) have to go through 6 trillion miles of lead before it would hit an atomic nucleus or electron to be absorbed (and therefore detected). Knowing that, you can understand why it is named after the Italian for *"little elusive one."*

Neutrinos appear to be massless little particles that have no electric charge and that travel close to the speed of light. They should, therefore, be thought of as bundles of energy (like photons) rather than as particles of mass (like neutrons). Now if we count the number of neutrinos that leave the surface of Sun at any particular moment, just as we do for photons, we can calculate the temperature at Sun's core not as it was a million years ago, but what it is today (or 8 minutes plus 2 seconds ago, to be precise). And that enables us to determine if the rate at which Sun's energy is created varies with time. In other words, we wonder if the solar constant is truly constant. After all, there is good evidence that climatic conditions on Earth in the past were different from what they are today, such as the Ice Ages. Are such changes due to changes in Sun, or on Earth, or both?

Figure 7–29 It was in this tank one mile underground in the Homestake Gold Mine in Lead, South Dakota, that the number of neutrinos released from Sun was much lower then theory predicts. An international collaboration continues to collect date to solve this puzzle at a new laboratory in Italy. (Brookhaven National Laboratory)

Solar Neutrino Problem But if neutrinos don't run into anything, how can we detect them? After all, detecting something implies something touching a detector — photons being absorbed by your eye, for example. Well, there is a solution — liquid cleaning fluid. A chemist determined that chlorine atoms are particularly sensitive to the presence

★★★★★

of neutrinos. If a neutrino passes close enough to a *chlorine* atom, the atom is transformed into an atom of *argon* — another gas. Since the probability of a close encounter of the *chlorine* kind is extremely low, chemists have to use a huge number of chlorine atoms in order to detect such an event within a human life span. So a huge tank of chlorine atoms is the necessary detector.

But there are other types of radiation coming from events in space (X-rays, gamma rays, cosmic rays, etc.) that would swamp any such experiment conducted on Earth's surface, and mask any reactions due to neutrinos passing by. So a 100,000-gallon tank of chlorine atoms (actually a cleaning fluid called perchloroethylene) is installed deep underground in the Homestake Gold Mine in Lead, South Dakota (Figure 7-29). Forty experimental runs were conducted in the 1970s, chemists having to flush out the tank after each run, separating the argon atoms from the chlorine atoms.

The number of argon atoms present are directly related, of course, to the number of neutrinos passing through the tank during each run. The **solar-neutrino experiment** reveals that the temperature at Sun's core today in order to have generated the number of neutrinos passing through the tank is about *12 million degrees*. Compare that with the figure I asked you to keep in mind a few minutes ago (*15 million degrees*). Conclusion: Sun's core is about 3 million degrees cooler today than it was one million years ago (Figure 7-30)!

Three experiments confirm the lack of neutrinos coming from Sun. *Gallex* is a 30-ton gallium detector located under a 10,000-foot mountain peak in northeastern Italy. It has detected only 63 percent of the neutrinos predicted by standard models of Sun's interior. The *Homestake* experiment sees only 27 percent. And the *Kamiokande II* is a water-based detector located in a mine 200 miles from Tokyo which has detected only 47 percent of neutrinos expected. The answer to the problem may lie in the realm of particle physics. We may have to revise our solar models to incorporate these new observations.

If experiments do confirm that Sun's core temperature has lowered during the past one million years, then we might want to consider the consequences. We have not yet, of course, begun to feel the effects of a lowering of Sun's energy output. We are still basking in Sunlight generated long ago. But sometime within the next million years, assuming the neutrino experiment is not flawed, conditions on Earth will change. We don't know when that will be. The temperature at the core could have lowered only a few thousand years ago, in which case we have hundreds of thousands of years before Sun's output will be diminished. On the other hand, if it lowered close to a million years ago, watch out! Any time now we will begin to feel the effects.

Now I am just making this dramatic to excite your interest. Scientists doubt that the problem is something that has to be solved immediately. There could be an experimental error, there could be something about neutrinos we don't understand, and, of course, we may not be aware of all of the processes going on within Sun. In any case, we stand to gain from the experiment and its consequences. As we find out more about neutrinos, we find out more about Earth-Sun relationship, and who can predict what else we may stumble upon?

LEARNING OBJECTIVES: *Now that you have studied this Chapter, you should be able to:*

1. Describe the features visible on Sun, including their behavior with time.
2. Explain how Sun is able to generate a magnetic field, and how this magnetic field influences the behavior of Sun.
3. Explain how the differential rotation of Sun acting on the magnetic field of Sun causes the observed features on Sun and their behavior with time.
4. Describe the process by which Sun generates energy at the core via the proton-proton chain, and how this energy escapes from the surface as radiation.
5. Explain how the study of neutrinos allows us to determine the present temperature at Sun's core, and how this differs from the temperature obtained by measuring the radiation coming off the surface.
6. Explain the nature of the Northern and Southern Lights, and their connection with Sun's behavior.
7. Trace the flow of energy from Sun's core to Earth, and describe how the features of Sun relate to this flow.
8. Explain the importance of anti-matter in the production of Sun's energy.
9. Define and use in a complete sentence each of the following **NEW TERMS**:

Anti-particle 159
Aurora Australius 149
Aurora Borealis 149
Babcock model 153
Coronagraph 152
Coronal holes 152
Deuterium 156
Differential rotation 142
Faculae 156
Filaments 144
Fission 160
Fusion 160
Geomagnetic storms 149
Granule 147
Granulation 147
Helioseismology 148
Hydrogen-alpha filter 144
Magnetic force lines 146
Maunder butterfly diagram 143
Maunder minimum period 153

Neutrino 160
Northern (Southern) Lights 149
Particle accelerator 159
Plage 148
Plasma 147
Polarity 146
Positron 159
Prominence 144
Proton-proton chain 158
Solar constant 160
Solar flares 148
Solar Max 152
Solar-neutrino experiment 161
Solar wind 149
Spicules 148
Stanford Linear Accelerator Center 159
Sunspot 141
Time-lapse photography 151
Ulysees 150
Zeeman effect 144

Chapter 8

Characteristics and Properties of Stars

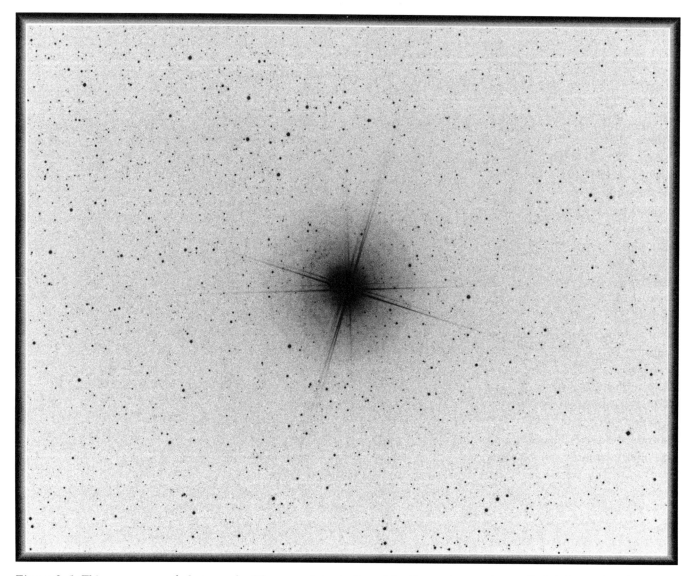

Figure 8-1 This over-exposed photograph of the nearest star to Sun, alpha Centauri, *reveals very little information about the star. It is only by using a spectrograph that astronomers are able to deduce the vital statistics of stars.* (National Optical Astronomical Observatories)

CENTRAL THEME: How are astronomers able to determine the physical properties of stars in order to classify them into types?

So at long last we arrive at the subject of stars, to what appear to be the basic building blocks of the universe. Up to this point, I have pretty much related to stars as mere dots in the nighttime sky. With this Chapter, I would like to begin to attach personalities to those stars. Not each and every star. But representative stars of the categories that astronomers have assigned them to. I would like you to be able to associate various properties with a given star that you see in the sky. I would like you to be able to identify a particular star and then associate it with a certain size, temperature, color, and life history. If you are able to do that by the end of the Chapter, I guarantee that your excursions outside at night will be enriched.

With the previous Chapter's survey of the star we know best — Sun — you will find it easier to visualize all of the stars in the sky as suns as well. But to fully appreciate stars as suns requires some mental activity — some reasoning and assumption-making for which scientists are famous. This is true simply because through even the largest telescope, the nearest star to us (beyond Sun) appears as a simple dot of light. The stars beyond Sun are just too far away to reveal their sizes and therefore surface details. To arrive at knowledge of the stars' personality traits requires some understanding of basic principles of physics and chemistry.

You might have been surprised in Chapter 1 to learn that some stars are larger than our Sun, some smaller; some are hotter, some cooler; some are more massive, some less; some are older, and some younger. In general, there doesn't appear to be anything unusual or unique about our Sun. It is an average, garden variety star — or so it seems. I wonder, then, if there are any consciousnesses out there, living around other stars, wondering about or studying our star as we are studying theirs?

One of the side benefits of learning and participating in astronomy is that the stars in the sky begin to take on personalities, rather than just being impersonal dots in the sky. It is comforting when you can go out on a starry night and look up, and be able to recognize some familiar friends. But in addition to knowing the name and the constellation a star is in — which was the emphasis in Chapter 2 — there is another language that astronomers use to describe a star's inherent properties. It is the language used in atlases of stars or in articles in popular astronomy magazines, when reference is made to a particular type of star.

The language has much to do with the lifestyle of the star — how it formed the way it did, how it emits the radiation it does, and even what its anticipated death will be like. Thus astronomers make a connection between the inherent properties of a star and its total life span. People are not so easily categorized. But stars are basically simple in their structures and behaviors. So when I use the designation F3V to refer to a star at the end of this Chapter, I expect that you can visualize it according to *size, color, temperature, age, energy output, mass,* and *stage of life* that it is in. All of that information is encoded in that simple designation F3V.

Determining Inherent Brightnesses of Stars

Let us briefly review a few of the vital statistics we have already learned about stars. In Chapter 2, in order to express the brightness of a star as it appears to us on Earth, we assigned a number to that star, the *apparent visual magnitude* (m_v). In practice, we arrive at that number by attaching a photometer to a telescope focused on a star, much as we obtain a light-level reading in a simple camera with a light meter. That number, of course, does not express anything about the *inherent* qualities of the star being studied. A star might appear bright in the sky because it is close to us, even though it may not give off large amounts of radiation. Or it may appear bright in the sky because it gives off enormous amounts of radiation, even though it is located at a very great distance from us.

It is rather like a teacher looking out into a classroom of students and trying to determine who the tallest member of the class is. Those in the front rows appear larger simply because they are closest. Those in the back rows appear to be the smallest. Ideally, I should like all students to rise and stand next to one another at a fixed distance away from me, say along the back wall of the classroom. At that stage of the experiment, I'm able to compare all students with one another, and it is easy to select the tallest student from the group.

In a sense, we do the same thing with stars — not physically, of course, but mathematically. We move them mathematically from their actual distances to a standard distance of 32.6 light-years, and then ask ourselves — "How bright would each star appear to us — as measured by a photometer — if it were located 32.6 light-years away from us rather than its actual distance? The brightness calculated for that new standard distance is used to express the inherent quality of brightness of the star. The star's apparent brightness at 32.6 light-years can be used to express how much energy the star is giving off. Here's how astronomers actually do it.

Trigonometric Parallax Before the star's apparent visual magnitude at 32.6 light-years can be calculated, we first need to know how far away the star is. We need to know how far away it is from that standard 32.6 light-year distance. How can a star's distance be determined? Well, recall from Chapter 4 (Figure 4-12) that the concept of parallax arose as a part of the controversy between Galileo and the Church. Since the heliocentric theory placed the planets orbiting Sun, nearby stars should appear to jump back and forth at six month intervals. Neither Galileo nor the Church realized how far away even the nearest star is, so parallax was not discovered until a powerful enough telescopes was built in 1838.

It is now possible to calculate the distances of nearby stars by using that very concept you learned in high school mathematics. Because the technique involves the use of trigonometry, we call it **trigonometric parallax.** We use the word "trigonometric" in order to distinguish it from another type of parallax that astronomers use, and to which I introduce you later in the Chapter.

Like many of the principles of physics and chemistry you encounter in this Book, you use the principle of parallax frequently without knowing there is a name associated with it. Astronomers take two photographs of an identical region of the sky separated by a six-month interval of time. Any stars close to us appear to have moved during that interval of time — they appear to be at slightly different locations on the two photographs with respect to the more distant stars. Knowing the magnification used in taking the two photographs, astronomers determine the angular separation between the two dots on the photographs.

In Figure 4-12, the geometry of determining the distance to a star by trigonometric parallax is portrayed from the perspective of an observer outside the solar system. The parallax angle is the small angle (p) formed by the two lines between the observer and the nearby star at 6-month intervals of time. Since the distance to the star is enormously large compared to Earth's orbit around Sun, the base of the triangle formed by the angle (p) — twice the Earth's average distance from Sun — is also very small compared to the star's distance. But that distance is known (approximately 186 million miles. There is a principle in geometry that says: *"If an angle of a right triangle and the opposite side are known, any other dimension of that triangle can be determined."* Since those two quantities are known for this situation, the distance from Earth to the star can be calculated. The actual equation is quite simple:

$$d = 1/p,$$ where

p = measured angle in seconds of arc
(one second = 1/3600th of a degree of arc)

d = distance in *parsecs*
(parsec is a unit of measurement = 3.26 light-years)

So we are able to measure the distance of a star from Sun. You may recall another characteristic of parallax — the farther away a star is, the less parallax it reveals (recall moving your finger further and further away from your eye?). For stars further away than about 300 light-years, even the most powerful telescope on Earth is unable to detect parallax angles. So the distances to objects beyond that distance must be determined by other methods. Even though parallax is a valuable tool for determining distances to stars, it is limited only to those closest to Sun. Remember that the diameter of our galaxy is 100,000 light-years, so parallax allows us to calculate the three-dimensional organization of a very small portion of the entire galaxy. But it is a start.

Another Unit of Distance Being naturally curious, you've probably been wondering where the standard distance of 32.6 light-years came from, right? It is not all that important to explain how astronomers came up with that particular unit as a standard measurement, but as long as you asked, I will explain. As I just explained, the equation we use to determine distances to stars once we have their

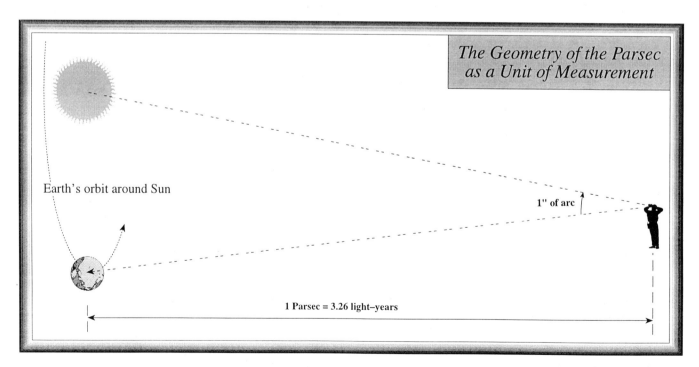

Figure 8–2 *The definition of the unit of measurement called the parsec is illustrated in this diagram. It is the distance one would have to go away from the solar system in order to measure Earth's angular distance from Sun as one second of arc.*

parallax angles is $d = 1/p$. If the angle p is expressed in units of *seconds of arc*, the answer d is in units of distance that we call **parsecs**. It is equal to 3.26 light-years. Therefore, since the arc angle is measured directly off the photographic plates that are taken at 6-month intervals, it makes the task of calculating the distance very easy by working in units of parsecs instead of light-years.

The reason that a parsec is equal to 3.26 light-years is because that is the distance one must move away from Earth-Sun system in order to measure the angular separation between the two as one second of arc (Figure 8-2). (This is accomplished as a calculation, of course, not by sending scientists out into space!) Astronomers adopted ten parsecs as the standard distance since there are no stars within a distance of one parsec from Sun (the nearest star is located about 4.2 light-years, or 1.3 parsecs). Because the calculation of parallax is so integral a part of the investigation of space, the use of *parsec* as an expression of distance is widespread. You will see it used extensively in popular astronomy magazines, for example. Are you glad you asked?

Decease of Brightness with Distance So we calculate a star's distance using parallax and determine its apparent visual magnitude using a photometer. But we want to know how bright that star would be at another distance from us — the *standard distance of 32.6 light-years*. How can we do that? Well, the calculation is based on a principle with which you are so familiar that you are not even aware of its significance — the **inverse-square law**.

You are certainly aware that the farther away a source of light moves from you, the dimmer it appears. That much is obvious. What you perhaps didn't know is that a mathematical relationship exists between how much farther away it gets and how much less bright it appears (Figure 8-3). The equation expressing this relationship looks very much like the equation that expresses the force of attraction between any two objects — *Newton's Universal Law of Gravity* — since they are both inverse-square law relationships:

$$B \sim 1/d^2$$

This equation says that the *apparent brightness* (**B**) *of a source of light is proportional to* (~) *the inverse of the distance squared* (d^2)

Notice in the equation that as **d** gets greater and greater, **B** gets smaller and smaller. If **d** is doubled, **B** is one-fourth as great — if **d** is tripled, **B** is one-ninth as great, and so on. That is why the relationship is called the inverse-square law — the brightness is inversely proportional to the square of the distance. Knowing the apparent visual magnitude of a star at its actual distance (calculated by parallax), we calculate its apparent visual magnitude at the standard distance of 32.6 light-years using the inverse-square law.

Absolute Visual Magnitude That *new* apparent visual magnitude calculated for the star's location at the standard distance is called the **absolute visual magnitude** — abbreviated M_v. We therefore define *absolute visual magnitude* as *the apparent visual magnitude an object has when it is relocated to a distance of 32.6 light-years from Earth*. It is referred to as absolute inasmuch as it is a true expression of the star's output of energy (radiation), since it is calculated on the basis of all stars being located at a common distance of 32.6 light-years from us. In actual practice, there is an equation that expresses the relationship between the three variables m_v, distance (d), and M_v is as follows:

$$M_v = m_v + 5 - 5\log_{10}(d)$$

Notice that if two of the variables in the equation are known, the third can be determined. The procedure is illustrated in Figure 8-4. Five stars whose distances are known are moved to the standard distance — mathematically, of course. The absolute visual magnitudes determined from the inverse square law are placed next to their new locations. Some interesting conclusions about these stars can now be drawn from observing the diagram. The star *Sirius*, of course, appears brightest to us in the sky because its apparent visual magnitude is (-1.5). But the star *Deneb* gives off the greatest amount of energy, since it would be the brightest star of the five if they were all located at the same distance from us.

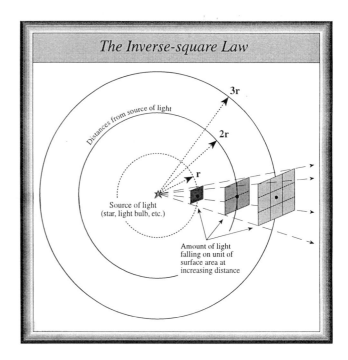

Figure 8–3 The inverse–square law is diagrammed to illustrate why the decrease of radiation with distance is a mathematical concept. The inverse–square law applies to gravity and sound as well as radiation.

CHAPTER 8 / Characteristics and Properties of Stars

★★★★★★

To know that a star has an absolute visual magnitude of such and such, however, doesn't yet allow us to know the difference between that star and Sun. So let's now move Sun out to that standard distance and calculate its absolute visual magnitude. We move it from its present 93 million miles to 32.6 light-years (about 200 trillion miles) away. At that new distance, Sun's apparent visual magnitude is +5. That figure is Sun's absolute visual magnitude. Using the scale for apparent visual magnitudes in Chapter 2, we see that at that distance Sun is hardly noticeable in the sky to an observer on Earth. In other words, *if Sun were moved to the standard distance of 32.6 light-years from Earth, it would be only slightly brighter than the dimmest star in the visible sky!*

Comparing Luminosities of Stars Even though calculating the M_v of a star provides a number with which to compare that star's energy output with those of other stars, that comparison is not yet quantitative. We can say that a star whose M_v is +7.5 gives off less energy than one whose M_v is +4.6, but how much less? How much more energy does *Deneb* emit than our Sun – for example? Recall from Chapter 2 that the apparent visual magnitude scale is based on a scale in which there is a 2.5 energy difference between any two magnitudes. That is, we receive 2.5 times less radiation from a +4 star than from a +3 star. The same difference of 2.5 times more or less energy applies to the scale of absolute visual magnitudes, as well. For example, a star whose absolute visual magnitude is +3 *gives off* 2.5 times more energy than does a +4 star. Rather than saying that one object "gives off such and such more or less energy" than another object, astronomers prefer to say that it is more or less **luminous** by a given amount.

So now we can *quantitatively* compare Sun (in terms of luminosity) with the other stars in Figure 8-4. Let us take the extreme case – the star (*Deneb*) whose absolute visual magnitude is -7.1. Since there are approximately 12 magnitude differences between a +5 star (Sun) and a -7 star (*Deneb*), the difference in energy emitted (*luminosity*) is 2.5 raised to the 12th power. That number calculates out to be about 60,000 (Figure 8-5).

Size and Temperature *Deneb* is about 60,000 more luminous than our Sun! How is that possible? Your intuition may tell you that its temperature must be 60,000 times hotter than that of our Sun. Or it may be that *Deneb* is 60,000 times bigger than our Sun. But it is more likely a combination of both factors – *size* (surface area) *and temperature*.

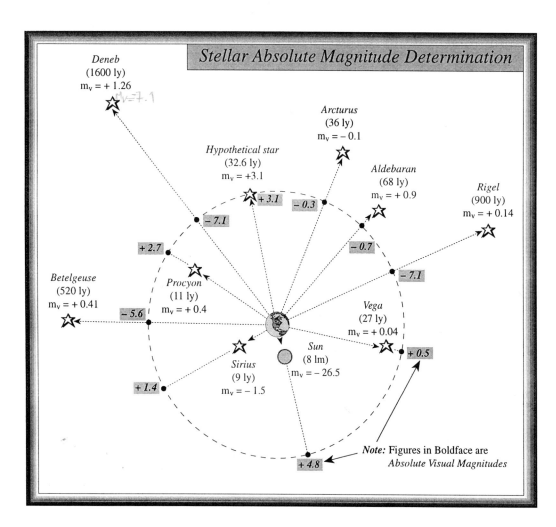

Figure 8–4 The method of determining the absolute visual magnitude of a star once we know its distance and apparent visual magnitude is illustrated in this diagram. Astronomers mathematically move the star from one location to a known distance to obtain its brightness at that new distance.

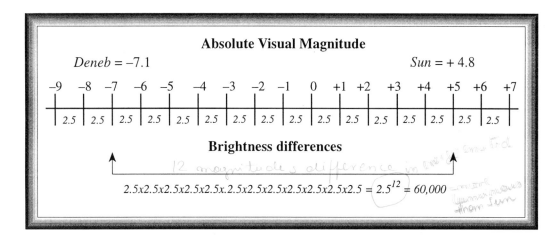

Figure 8–5 The luminosities of two stars can be compared once we know their absolute visual magnitudes. This is possible because there is a difference of 2.5 energy units between any two magnitude levels. A star like Deneb emits about 60,000 times more energy than Sun.

Consider the burners on your kitchen stove. There are big burners and small burners. When turned to the same power level, they are at the same temperature. And yet the large burners give off more energy than do the small burners, simply because they have larger surface areas from which to emit radiation (mostly infrared). It is possible, with some experimentation, to adjust the burners so that the same amount of energy is emitted by both a large burner and a small burner (if, for example, the small burner is set on *high* while the large burner is adjusted to the *medium-low* setting).

Likewise, *Deneb* could be giving off 60,000 times more energy than Sun either because it is much *hotter* or because it is much *larger* or a combination of *both*. But how are we to separate out these two factors, so that we know for a particular star the contribution of each factor in making the

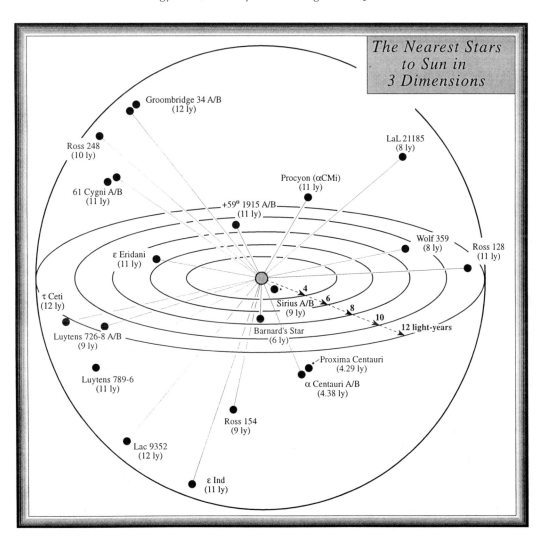

Figure 8–6 Here are our nearest neighboring stars in space, shown according to their distances from us in light–years. As you will learn later in the Chapter, just because they are close to us doesn't mean they are easily seen in the sky. Most are very dim stars.

CHAPTER 8 / Characteristics and Properties of Stars

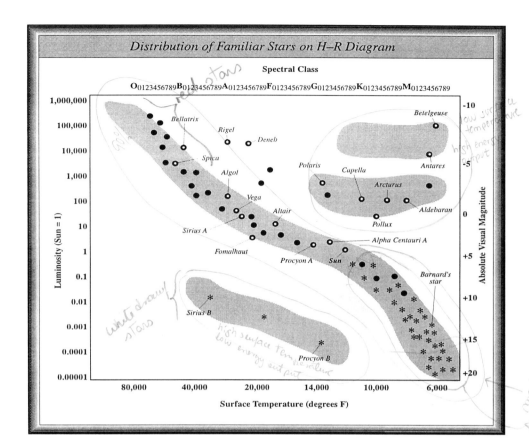

Figure 8–7 *Once we have obtained the surface temperature of a star by spectral analysis, and its distance by trigonometric parallax, we can plot stars according to those variables. What is striking about that plot, called the H–R diagram, is the clustering of stars in particular regions of the chart. This implies something important about stars.*

energy output of the star what it is? To our advantage, there is an interesting *relationship between two characteristics* of all stars that leads us not only to the solution of this problem, but also to understanding the behavior of stars as they evolve from birth, life, to death.

That relationship is expressed in the *H–R Diagram*, the subject of the following section. The two characteristics of all stars that are compared in the H–R Diagram are those of *surface temperature* and *absolute visual magnitude*. The surface temperature of a star — you recall — is determined as we analyze the spectrum of that star (spectral analysis). The absolute visual magnitude is calculated once a star's distance is determined by the method of trigonometric parallax. Remember that M_v is an expression of the total energy emitted by a star — an expression of the star's luminosity.

Separating Temperature from Luminosity

In Chapter 6, I elaborated on the method by which astronomers calculate a star's surface temperature — by analyzing its spectral lines. Astronomers usually express that temperature not as a certain number of degrees, but by the spectral type of the star. Every star has a surface temperature and a total energy output.

Assume that astronomers have established a long list of stars, and that each star's absolute visual magnitude and surface temperature have been determined and recorded. Taking a random sampling of those stars and positioning them on a diagram whose variables are those of surface temperature (spectral type) and absolute visual magnitude (luminosity), we obtain a plot like that in Figure 8-7. This is referred to as the **H–R diagram**, named after two astronomers, *Hertzsprung* and *Russell*, who discovered this important relationship about stars in 1910. Notice the patterns of the plotted stars.

Plot of Stars by Type Obviously something important is expressed in a diagram relating two variables when the plot ends up in a pattern like Figure 8-7. At least 90% of stars are plotted in a broad band running from upper left to lower right. That pattern is interpreted to mean that most stars obey a simple yet important rule: *the hotter a star is, the more energy it gives off.* And since they constitute the largest population of stars, astronomers refer to them as **main sequence stars**. The line drawn through the main sequence stars represents that particular group of stars on most H–R diagrams so that the stars don't have to be drawn in all of the time. You will encounter the H–R diagram frequently throughout the remainder of the Book, since it is *the astronomer's tool for expressing the life history of stars*.

Most stars seen through a telescope are main sequence stars. Look again at the plot of just the main sequence stars in Figure 8-7. There is also an important pattern in the plot of the main sequence stars themselves: they are not distrib-

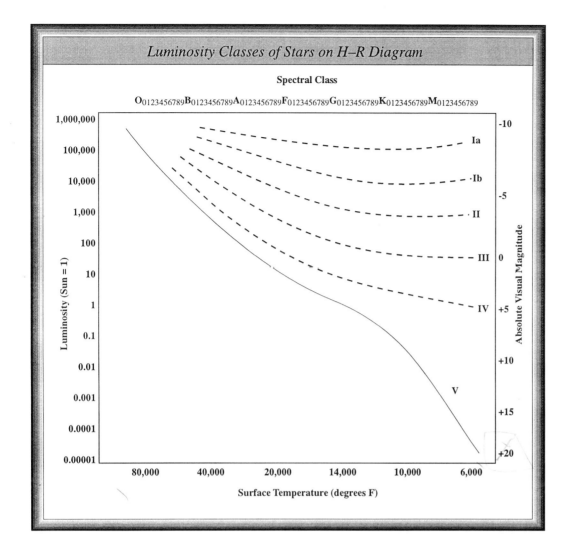

Figure 8–8 For convenience, astronomers note particular groupings of stars off the main sequence as luminosity classes. Two stars can have the same surface temperature, but emit different amounts of energy (expressed as absolute visual magnitude). This grouping of stars by luminosity allows astronomers another category in which to place stars in an attempt to explain their origin and behavior.

uted evenly along the main sequence line running from high temperature, high luminosity to low temperature, low luminosity — most are concentrated at the lower-right corner (low surface temperature, low luminosity). Approximately 90% of main sequence stars are located in that portion of the H-R diagram, which allows us to conclude that approximately 81% of all stars are low surface temperature, low luminosity stars (90% times 90% equals 81%). This rather startling discovery has important consequences throughout the field of astronomy, not the least of which is deciding which stars are most likely to have intelligent civilizations around them.

The rule followed by main sequence stars (the hotter they are, the more luminous they are) isn't really startling to you, I'm sure — it is something you would have guessed without reading an astronomy textbook. It seems obvious that hot objects should give off more energy than cool objects. Have we come all this way only to arrive at the obvious? Well, notice in Figure 8-7 that not all stars obey the rule however. The remaining 10% of plotted stars are scattered throughout the remainder of the H-R diagram. Of those 10%, however, most are in the lower-left corner (*high surface temperature, low energy output*), and the upper-right corner (*low surface temperature, high energy output*). So although it is easy to understand the behavior of the vast majority of stars (main sequence stars), explaining the characteristics of the minority requires a little insight. Let's reflect on those two groups of minority stars for a moment.

Red Giants and White Dwarfs We might ask ourselves, how can a star have a low surface temperature, and yet emit enormous amounts of radiation? Because it is huge — it has a large surface area from which energy is radiated. Likewise, a star can have a very high surface temperature while emitting a small amount of radiation simply by being small. Since cool, huge stars emit most of their visible radiation at longer wavelengths, they appear reddish in color. Therefore, we call stars located in the upper-right corner of the H-R diagram **red giants** (Figure 8-9). The small stars located on the lower-left of the diagram are extremely hot, in fact "white hot," so we refer to them as **white dwarf stars**. These hot, small stars are not numerous on the diagram, but they are very common in the universe.

★★★★★★

Figure 8–9 The first ever photograph of a star showing an actual surface. This is the red giant star, Betelgeuse, measured to be larger in diameter than Jupiter's orbit around Sun. If Sun were replaced by Betelgeuse, Earth would be inside the star. (STScI/NASA)

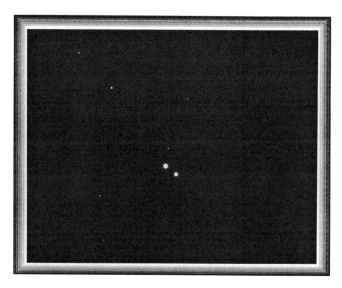

Figure 8–10 The star 61 Cygni is one of the nearest stars to Sun, as you saw in Figure 8–5. It is an excellent example of a binary star system—two stars in orbit around one another. To the naked eye, you see only a single dot of light. (Lick Observatory)

Classifying Stars by Luminosity It is convenient to place stars into subcategories according to surface temperature and size, and in the process establish another characteristic of stars — **luminosity class**. Notice in Figure 8-8 that four lines run parallel to and above the main sequence line at various intervals. Each is assigned a Roman numeral from I to IV. Each line represents a luminosity class — a star on or close to a line is assigned the luminosity class of that line. So a star has a luminosity class in addition to its spectral type. The spectral type expresses the surface temperature of a star, whereas the luminosity expresses the total energy output of the star and therefore tells us something about the size of the star.

Two stars can have the same surface temperature but the larger one will give off more energy and therefore be plotted higher up on the H-R diagram. The larger of the two stars, in other words, has the lower-numbered luminosity class. Because of the importance of knowing this particular characteristic of a star, astronomers choose to include the luminosity class of a star (expressed as a Roman numeral) along with the spectral type designation of that star — for example G2IV. Main sequence stars are assigned to luminosity class V. Since 90% of all stars are main sequence stars, the V designation for these stars is often times implied. So the designation of stars on star charts often include only the spectral class of stars — since most are main sequence stars.

Main Sequence Stars As I noted above, there are far more low-temperature main sequence stars than there are high-temperature main sequence stars. In fact, what we observe in plotting a random sampling of stars on the diagram is that *The higher the surface temperature of a main sequence star,* *the less common is that star.* The universe favors the evolution of low-temperature stars. But why? Why are stars distributed on this diagram in such a specific way? Why the groupings into white dwarfs and red giants?

Answers to these questions tell us quite a bit about the life-cycle of stars, and are therefore important topics in modern astronomy inasmuch as stars are the basic ingredients of the universe. The answers to those questions lie in a property of stars that we haven't yet considered. So we must first take a diversion to another subject to see what part it plays in the drama of stellar evolution — that subject is the *mass* of a star. But keep in mind that our *ultimate goal in determining the masses of stars is to return to the diagram of Figure 8-7 in order to see what adding mass to the diagram tells us about stellar evolution.*

Binary Stars: Determining their Masses

When we refer to the amount of material making up an object — the number and type of atoms and molecules — we are referring to the *mass* of the object. It is tempting to use the word "weight" interchangeably with mass, but to do so, strictly speaking, is incorrect. *Mass under the influence of another mass has weight.* That "influence" for large objects like stars is of course gravity. That "influence" for atomic particles like electrons and protons is the electromagnetic force. The mass of your body has weight while you remain on Earth. But as you venture away from Earth, you gradually lose your weight even though you retain your mass. We cannot bring stars to Earth in order to weigh them, so we instead refer to their masses. Unfortunately, it is not easy to determine the mass of a star.

172 SECTION 2 / Evolution of the Stars

Consider this situation — a star is all alone in space, with nothing close enough to be influenced by its gravity. How can we determine the amount of material (mass) in this star from our distant location? We can determine its size, but that does not tell us its mass. A balloon can be large without being massive. So in order to determine its mass, we must be able to measure its influence on another object.

There is an example within our solar system. The masses of the planets are fairly accurately known, since their influences on one another are easily determined by measuring the changes in their orbits as they pass one another. But Pluto's mass was a mystery for a long time, since we didn't see any evidence of anything close to it. But in 1978 an object was discovered close to Pluto on a photographic plate taken of Pluto itself. By taking consecutive photographs of the position of the object relative to Pluto from week to week, it became obvious that it was a moon in orbit around Pluto. At the same time, its plotted orbit around the planet allowed astronomers to determine Pluto's influence on Moon's orbit and therefore Pluto's mass.

Visual Binary Stars Similar techniques can be used with stars. We find stars in the presence of other stars, close enough to one another that they influence one another. Stars found in close company to one another are what astronomers refer to as **binary star systems**. If there are three or more stars orbiting one another, we refer to it as a **multiple star system**. By measuring the stars' movements with respect to one another, we determine their masses. I am aware that earlier I emphasized the vast distances that exist between stars, and that I made a big deal out of the seemingly emptiness of space. And it may seem that I am now contradicting those earlier claims that stars are indeed close to one another after all.

Well, I am just fine-tuning those early claims. What I am now saying is that binary and multiple star systems are indeed very widely scattered in space, but that stars within such a system can be close indeed. It turns out that at least 50% of the stars we observe in the sky at night are members of such binary or multiple star systems, although it is anything but obvious to the casual observer. So how do astronomers locate these systems of stars? Let's take a familiar example.

Optical Binary Stars On your next outdoor adventure to a dark location to observe the nighttime sky, locate the group of stars called the *Big Dipper*. It consists of seven magnitude +2 stars contained within the constellation of *Ursa Major*, the Great Bear (Figure 8-11). Next to the middle star in the handle of the Big Dipper — a star called *Mizar* — there is a faint, barely-visible magnitude +6 star called *Alcor*. To the naked eye, *Mizar* and *Alcor* appear so close together that we might at first want to believe that they are also close to one another in space.

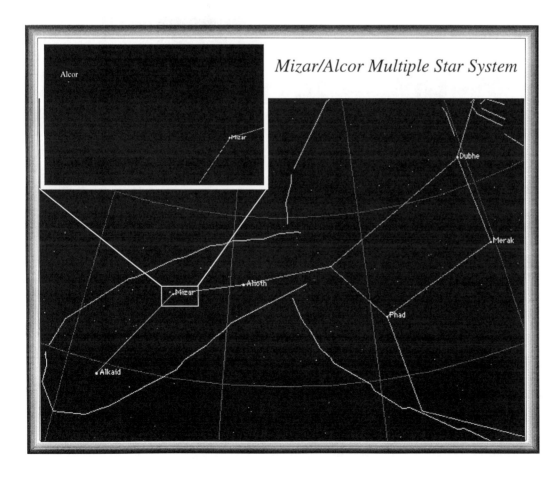

Figure 8-11 In the familiar pattern of the Big Dipper, there is a good example of a multiple star system. The two stars Mizar and Alcor, visible to the naked eye, together form an example of a visual binary system. But each star is itself a binary star system. (Carina Software)

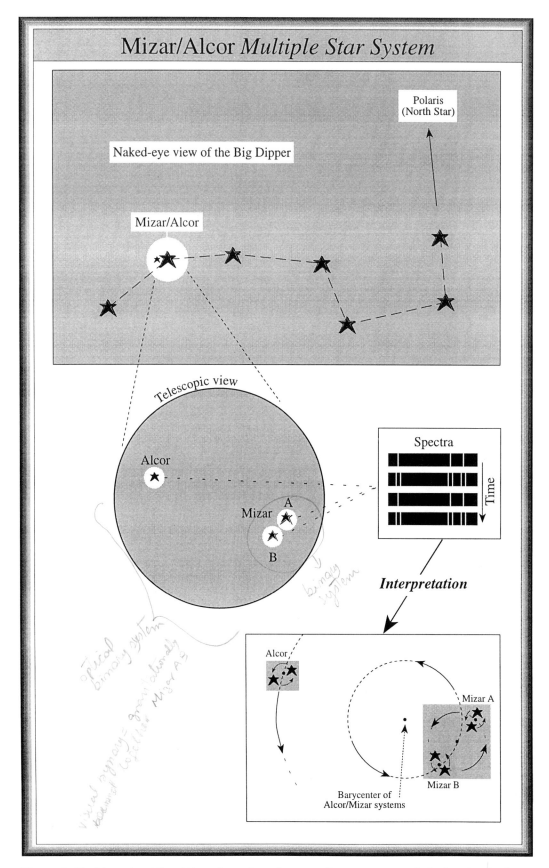

Figure 8–12 Observing Mizar/Alcor through a telescope allows one to conclude that the single dot we see as Mizar is itself a visual binary system. Another technique allows us to conclude that those two dots are in turn double star systems.

But that is not necessarily the case. In fact, they are not. They are actually at different distances from us — there is simply a chance near-alignment. They only appear close together, much as telephone poles along a residential street appear close together to someone looking down the length of the street. Two stars that only appear to be close together are referred to as an **optical binary** system, although it is *not a true binary* in the sense that they orbit one another. Mizar and Alcor, however, are a true binary system, but observations of the stars over a period of time are necessary to arrive at that conclusion.

Let's look a little closer at *Mizar* and *Alcor* — say through a modest-sized telescope. What is now obvious is that there is a third star appearing even closer to *Mizar* than *Alcor*. We are led to the conclusion that when we look at *Mizar* with our naked eyes, we are actually looking at the combined light of two stars, *Mizar* and its faint companion. The resolution of our unaided eyes is not sufficient to enable us to see them as separate stars. It is rather like someone holding up two flashlights and walking away from you until the two lights appear as one. Telescopes, because of their superior resolution, allow us to see many of the stars in the sky as binary or "double" stars. In fact, astronomers use this particular technique to determine the resolving power of a telescope. The better the resolution of a telescope, the better its ability to detect binary pairs. Astronomers call it "splitting a double."

Astronomers photographed *Mizar* and its faint companion at intervals of several months over several years of time. From looking closely at the behavior of the two stars over that long period of time, it was obvious that the two stars are orbiting one another. Thus *Mizar* and the faint companion star are a true, binary system, called a **visual binary system**. We *see* them as separate stars, and their movements over long periods of time tell us that they are gravitationally bound together (Figure 8-12). That's why we call them a *visual binary*.

Center of Revolution What do we mean by "orbiting" one another? Do Earth and Moon orbit one another? Well, strictly speaking, they don't. *Moon and Earth orbit (revolve) around a common center of gravity*. Let's consider a simple analogy. Figure 8-13 shows two objects of unequal masses. Their orbits cause them to revolve around the point that is the mathematical balance between the two. We call this point the **barycenter** of the system.

This situation is similar to that of a child's seesaw. In order to balance the board properly, the more massive child (greater weight) sits closer to the balance point. Likewise with Earth-Moon system. What moves around Sun in an elliptical orbit (Kepler's first law) is not the center of Earth, but the barycenter of Earth-Moon system. But doesn't that suggest that as Moon and Earth orbit that common center, they wobble and trace wavy paths around Sun?

Yes, but that is much more true for Moon than for Earth. Earth is so much more massive than Moon that the barycenter is located a mere 3,000 miles from Earth's center. From an external observer's point of view, Moon would nevertheless appear to orbit Earth. But if one wanted to be extremely technical (not a good way to make friends, incidentally) one could say that Moon orbits Earth-Moon barycenter — not Earth itself. This is true of

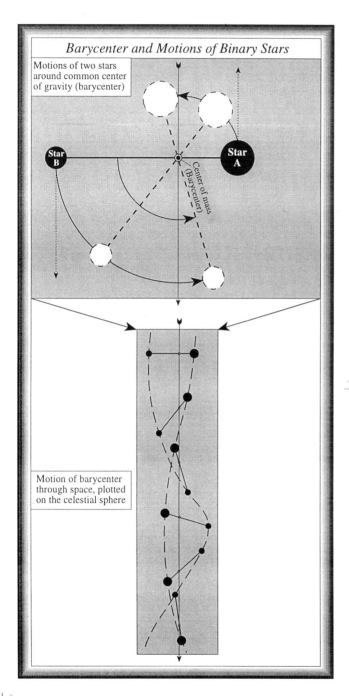

Figure 8-13 Since two unequal masses orbiting about a common center of gravity (barycenter) and moving through space at the same time reveal a wobbly motion to an outside observer, astronomers can determine their masses by recording and plotting the extent of the wobble.

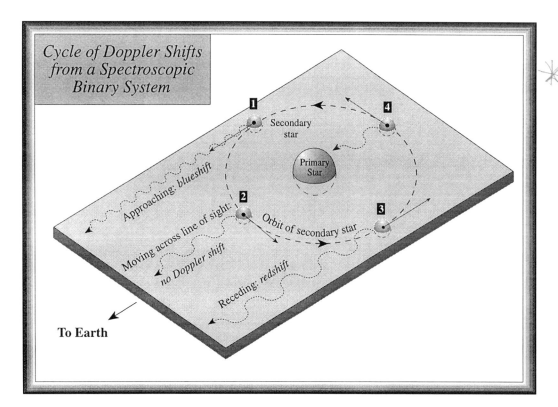

Figure 8–14 The largest number of binary stars are detected not directly, but through analysis of their spectra. Two stars orbiting their barycenter alternately approach and recede from Earth. The Doppler effect shows up in the spectrum of the single dot that is being studied.

all orbiting objects, even the planets orbiting Sun. Technically-speaking, Earth orbits the Earth–Sun barycenter. This technique, by the way, is one of the techniques being used to detect planets around nearby stars.

Back to *Mizar* and its faint companion. Since we photographed these two stars over a long time interval, it is easy to determine how long it would take them to orbit around their barycenter once — their orbital period. If we determine their common distance away from us by parallax, we can then calculate their **separation distance** — how far the stars are from one another. We use *Newton's universal law of gravity* to make the calculation.

The law relates the masses and separation distance of two objects to the force of attraction between them. The greater their masses, the greater the force of attraction, the faster they move in their orbits, and the shorter their orbital period. The greater their separation distance, the lesser the force of attraction between them, the slower they move in their orbits, and the longer their orbital periods.

Think of it this way. Imagine yourself on a seesaw at a known distance from the balance point, and someone walks up and places a large package for you at the other end of the seesaw at a fixed distance from the balance. And assume that the package exactly balances your weight (which you know). Can you determine the mass of the package according to where it is placed? If it is closer to the balance point than you are, its mass must be greater than your mass. And vice versa.

This is identical to how astronomers use binary stars to determine their masses. Except that they use the relationship between distance from the balance (barycenter) and the orbital period. This is how astronomers measure the masses of two stars found in a visual binary system. Since both stars are visible, their spectral classes are easily obtained and astronomers can thereby relate a given spectral-type star with a given mass.

Spectroscopic Binary Stars Whenever we decide to study an object in the sky, the very first thing we do is obtain a spectrum of that object, for obvious reasons. A spectrum provides us with the raw data upon which our understanding of that object's composition and behavior is based. And so we have done that with both *Mizar* and its faint companion. In both cases, we encounter an interesting observation that leads us to the conclusion that each of them is itself a binary system. This is a case of a binary system orbiting another binary system!

In Chapter 6 you learned that a star normally reveals an absorption spectrum, a set of missing photons measured by wavelength that indicate the composition of and conditions in the star's atmosphere. If the star is moving away from us, the Doppler effect causes all of the wavelengths of missing photons to be slightly longer. If the star is moving towards us, they are compressed into shorter wavelengths (Figure 8-14). In the case of both *Mizar* and its companion star, we obtain a set of single lines at one time, and a set of double lines at another time (see Figure 8-15). Astronomers obtain spectra from these stars on a regular basis, and are thus able to calculate a precise time interval between occasions of double lines.

Our interpretation of what is happening is this — remember that we are measuring the light from a single dot.

This single dot must in fact be two stars so close together and/or so far away from us that we are unable to see them as two separate stars with even the largest telescopes. But the alternate double, then single, then double, then single lines suggest that the two orbiting stars alternate in approaching us, then receding from us, then approaching us, etc. When they are both along our line of sight, of course, they are neither approaching nor receding, and the absorption lines are simply the sum of those produced by both stars. We see their individual lines only when the Doppler shift causes them to separate into separate lines.

It is unlikely that both stars have identical compositions and temperatures, so seldom do all lines appear to split to become double. Only those lines common to both appear as double. The point is, though, that we can obtain the spectra of two different stars even though they appear as a single dot of light in the sky. Furthermore, since we record the time interval between successive double-line spectra — or between single-line spectra, which is the same time interval — we know their orbital period. And the amount of separation of the lines during the double-line spectra is a measurement of the speed at which the two stars travel in their orbits.

Again, as in the case of the visual binary system, these bits of information allow us to calculate the masses of the two stars. Such two-star systems are called **spectroscopic binaries**, inasmuch as we obtain the information necessary to confirm their existence as two separate stars form the spectrum of a single dot. Thus *Mizar* and its faint companion are a "double-double" system — as a visual binary system they are orbiting one another, but each is in itself a spectroscopic binary system. *Alcor* is also a spectroscopic binary star, although it is too distant from *Mizar* to be a visual binary, as I noted previously.

Light Curves of Stars Although not very commonly found, there is a third type of binary system that in addition to providing the masses of stars, tells us a characteristic of stars that is unobtainable from either visual or spectroscopic systems. But let's first deal with the matter of determining the masses of stars with this technique. What if I were to tell you that there is a star in the sky whose apparent visual magnitude changes from +2.1 to +3.4 every couple of days, and that a reward of $200,000 awaits anyone who locates the star. Assuming that you are not allowed to use star maps or astronomical instruments, are you willing to invest the time necessary to find the star?

Before you get stirred up for action, consider the task at hand. You must first record the positions and the apparent brightnesses of all of the stars in the sky. Then you must return to each star on a schedule to find the one that

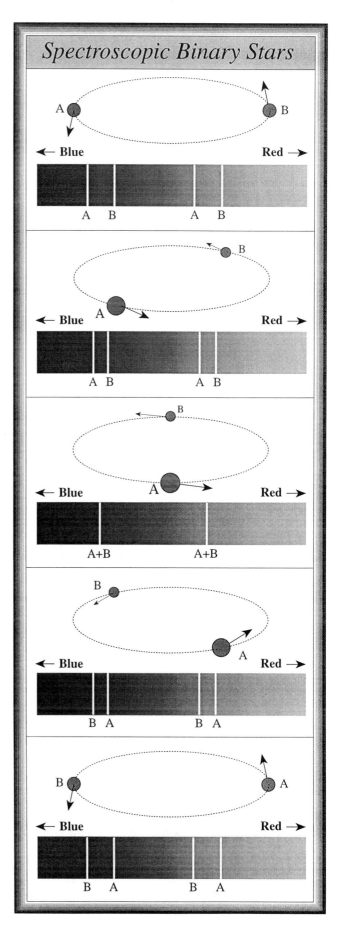

Figure 8–15 This diagram shows how spectroscopic binary stars are detected through their spectra. Spectral lines alternately split and recombine as the stars change in their positions with respect to Earth.

CHAPTER 8 / Characteristics and Properties of Stars

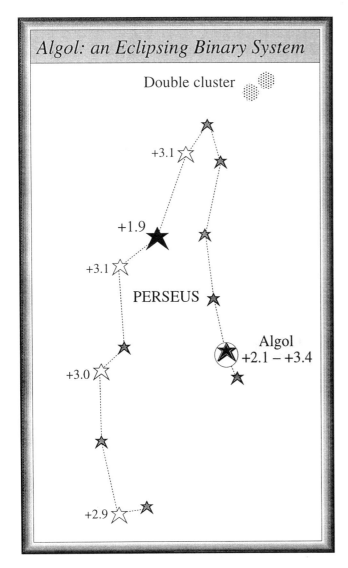

Figure 8–16 The star Algol *is the best example of an eclipsing binary star. Try this experiment. After locating the group of stars called Perseus in the sky and identifying the star Algol, use the apparent visual magnitudes of surrounding stars to judge the brightness of* Algol. *Do that every night for several nights and record your findings. You should find that its brightness varies with time.*

has changed in brightness. To do that, you must become familiar with each star in comparison to its neighboring stars in order to recognize the mystery star when its brightness has changed. And of course, the star isn't necessarily going to be visible on any particular night. Different stars appear in the sky at different times of the year (seasonal stars). It appears that the task will take you at least several months. No, I think that I'll pass on that one.

Yet there is such a star — it is called **Algol**, which in Arabic means "the demon's head." (Figure 8-16) The prefix "*Al*" in Arabic means "*the*". The English word "*ghoul*", commonly used during our celebration of Halloween, derives from the Arabic. Although officially first detected in 1669, there is reason to believe that the Arabs long before had noticed its variation in brightness, and therefore assigned a name appropriate for its behavior. After all, early cultures considered the sky to be permanent and unchanging and eternal. How would the Church have reacted had Galileo known of *Algol's* changes in brightness and had confronted the Church with that observation? Unfortunately, the Arab discovery of *Algol's* behavior did not fall into Galileo's hands.

Having discovered such a star, astronomers first obtain a spectrum of the star in order to determine its basic characteristics. In the case of *Algol*, the spectrum reveals all of the characteristics of being a spectroscopic binary star — single lines, double lines, etc. But why the light variations? That becomes obvious when we use a photometer to determine the apparent visual magnitude of the star over a long period of time.

Eclipsing Binary Stars Plotting the observations made with the photometer on graph paper gives us what we refer to as a **light curve**, a plot of light received (m_v) against time. The data reveals a rather interesting pattern in the light variation. Our interpretation is that in addition to the two stars alternately moving away from and toward us, their orbital plane just happens to lie along our line of sight. Thus the two stars alternately eclipse one another. This is called an **eclipsing binary system**. The probability of this chance alignment is quite low, so most spectroscopic binary systems are not like *Algol*. *All eclipsing binary systems are spectroscopic systems, but very few spectroscopic systems are eclipsing systems.*

Let's look a little closer at the manner in which we conclude that *Algol* actually consists of two stars. Figure 8-17 illustrates how the pattern of apparent visual magnitude plots is explained. Astronomers interpret the light curve of *Algol* by suggesting that two stars of different temperatures are orbiting one another. When the two stars are not aligned, we receive the combined light emitted by both.

When an eclipse occurs, the light from one is absorbed by the backside of the nearest star, and the amount of light we get is less. When the hotter of the two stars passes in front of the cooler star, there is a decrease of light, but not as much as when the cooler star passes in front of the hotter star. That is how we explain the two unequal dips in the light curve. Being a spectroscopic binary system at the same time, astronomers can determine the masses of the two stars as explained previously for visual and spectroscopic binary systems.

Determining Diameters of Stars In addition to determining the orbital period of the two stars by knowing the time interval between successive dips, we can also calculate the diameters of the two stars in an eclipsing binary system. By knowing the orbital velocities of the stars (from measuring the Doppler effect in the spectral lines), and the times required for each star to pass in front of each other (the

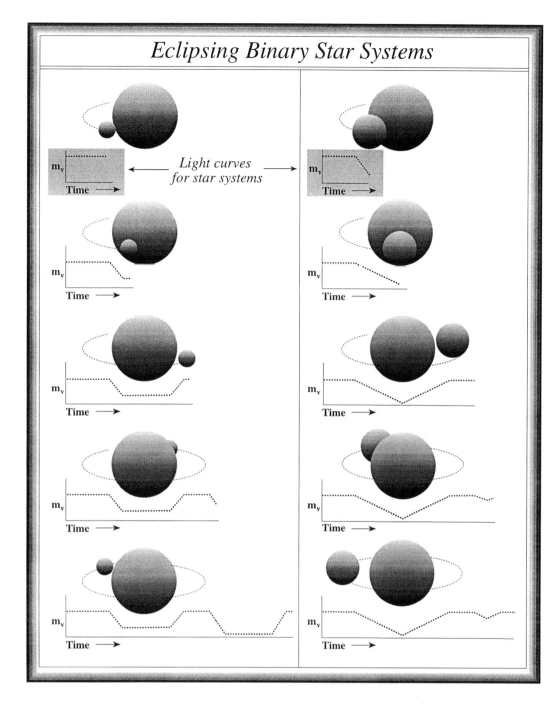

Figure 8–17 This diagram illustrates the connection between the geometry of two stars orbiting a common center of gravity, and their variation in brightness as recorded by astronomers using a photometer attached to a telescope.

width of each dip in the light curve), the diameters are easily calculated (Figure 8-17).

Aren't you a little impressed with the information that astronomers can obtain from the study of a single little dot of light in the nighttime sky? Gradually we are collecting the data about stars which allows us to visualize them as objects with features and behaviors, not just as tiny dots in the sky. Most importantly, in this subject of binary stars, we have seen how they allow us to determine stellar masses so that we can add that characteristic to those of surface temperature and energy output on the H-R diagram.

Interpretation of the H-R Diagram

Having determined the masses of stars in binary star systems, and knowing the spectral types of those stars by spectral analysis, we are in a position to return to the H-R diagram of Figure 8-7 and add the additional characteristics of mass and size to the stars plotted thereon. Figures 8-18 and 8-19 are the result. We now discover another pattern – *for main sequence stars only, the greater the surface temperature, the greater the mass*. Or what amounts to the same thing – *the greater the energy output of a star (luminos-*

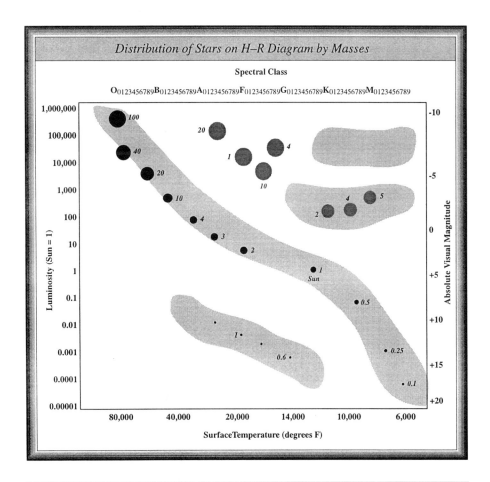

Figure 8–18 When the additional characteristic of mass is added to the H–R diagram developed earlier in the Chapter, another pattern emerges—for main sequence stars, the higher the mass the greater the mass. This discovery allows astronomers to explain why stars differ in the amounts of radiation they emit. Masses of stars indicated are expressed in terms of Sun's mass.

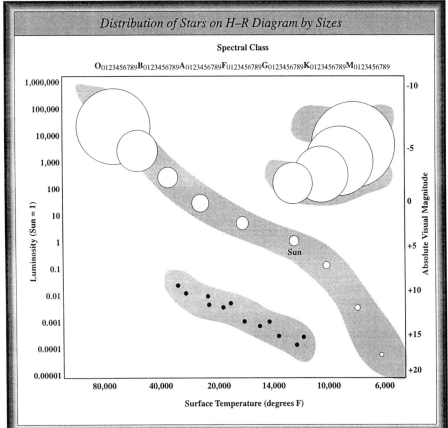

Figure 8–19 Plotting light curves for eclipsing binary stars also provides astronomers with a technique for determining the diameters of stars. Kepler's laws of motion are used to determine the speed of stars around the barycenter, and combined with the length of the eclipse, provides the diameter.

ity), the greater is its mass. This conclusion may not initially seem extraordinary, but it now allows us to offer an explanation as to why it is that stars are organized on the H-R diagram as they are.

Explaining Main Sequence Stars We learned in Chapter 7 that at the center of Sun (and presumedly centers of other stars, as well) the fusion of hydrogen into helium takes place because of the high temperatures and pressures that exist there. In the process, there is a release of gamma-ray photons which attempt to escape to the surface to become "sunlight" (or "starlight" in the case of stars). As they attempt to escape, they are alternately absorbed and re-emitted by the particles within the star, all the while changing wavelength as they eventually emerge at the surface mostly as visible photons (at least for our Sun). The photons, in effect, exert pressure outward as they are absorbed, since they act as tiny bundles of energy bumping into particles of matter. We call this **radiation pressure**.

As far as these photons are concerned, they would like nothing better than to blow the star apart from the inside out. But a star is made up of a huge amount of matter, and this matter presses toward the center by the force of gravity. It is this very pressure that at the center forces protons close together to fuse into helium nuclei, a by-product of which are the photons that are attempting to blow the star apart.

So there is a delicate balance between the pressure exerted outward by the escaping photons, and the pressure exerted inward by the force of gravity. This is rather like saying that the inside wants to get out, and the outside wants to get in! This balance is referred to as **hydrostatic equilibrium** — "hydro" because the star's plasma can be considered a fluid; "static" because the size of the star remains constant; and "equilibrium" because there is a balance between the two opposing forces. All main sequence stars are in the state of hydrostatic equilibrium. That explains why 90% of all stars are main sequence stars.

What do stars do? They shine. But in order to shine for long, they must be in a state of hydrostatic equilibrium. As long as a star has fuel to burn (specifically, protons to fuse together to become helium nuclei), it liberates photons so we are able to see it. Let's elaborate on this concept a bit, since it allows us to visualize the process by which a star first "turns on," and the changes that it goes through as it dies.

Hydrostatic Equilibrium Let us imagine a normal (main sequence) healthy star like our Sun in a state of hydrostatic equilibrium. Assume that for some reason the proton-proton chain at Sun's core speeds up, causing an excess number of photons to head for Sun's surface? Well, the photons would push Sun outward and blow it apart if Sun didn't respond in some way to the increased internal pressure. So it expands. Actually, the increased internal pressure pushes the outer layers of Sun outward, which increases Sun's surface area, and which thereby allows for release of the extra photons. If the increased energy output at Sun's core is only temporary, Sun eventually contracts back to its original size.

Now assume the opposite situation — some additional matter is added to the outer layers of Sun. Say that as Sun circles the galaxy and encounters a massive cloud of gas and dust, it collects some of it. The additional mass wants to crush Sun, to cause it to collapse. But the additional pressure also forces protons at the core closer together, speeding up the rate at which they fuse together. This increased pressure increases the rate of the proton-proton chain and the production of more escaping photons. With the increased pressure of these additional escaping photons, Sun resists the collapse that the additional mass initially threatened to do.

In both cases, of course, Sun's energy output increases. But the point is that stars for most of their lives shine steadily due to this delicate balance being maintained. Even though it seems intuitively obvious anyway, we are now able to explain why it is that the more massive stars give off more energy and have higher temperatures – *The greater the mass, the greater the pressure, the faster the proton-proton reaction, the more photons that escape, the more photons absorbed by atoms in the star's interior, the faster moving are the atoms absorbing the photons, the higher the temperature.* Whew! Isn't nature ingenious!

Determining Distance with a Spectrum With the use of the H-R diagram, we now have an additional method of calculating distances to stars. Again, to know the distance to a star is fundamental to knowing the inherent properties of the star. Assume that a star is located beyond the 300 light-year limit for the use of the trigonometric parallax method. The star, in other words, does not change position with respect to the background of stars at 6-month intervals. The spectral type of the star is easily determined by reading its spectrum (Figure 6-15). Having only the spectral type, all we can say is that its location on the H-R diagram is somewhere along a vertical line at the surface temperature determined from its spectrum. *We don't know its absolute visual magnitude because we don't know its distance.*

But isn't it reasonable to assume that a given star is a main sequence star, in light of our conclusion that 90% of all stars whose distances we do know are main sequence stars? So we place it on the main sequence at the spectral type determined for it, and we read its absolute visual magnitude off the vertical scale of the H-R diagram. In other words, all main sequence stars of a given temperature have the same absolute visual magnitude (M_v). But M_v is based on distance and apparent visual magnitude (m_v).

Since m_v is always available for a star, we can calculate the distance to the star. In Figure 8-20, the example I use is that of a spectrum revealing a star to be an **F5V**-type star. Its M_v is therefore be about +2.7. Knowing its m_v and its M_v allow us to calculate its distance. This method of determining the distance to a star by its placement on the H-R

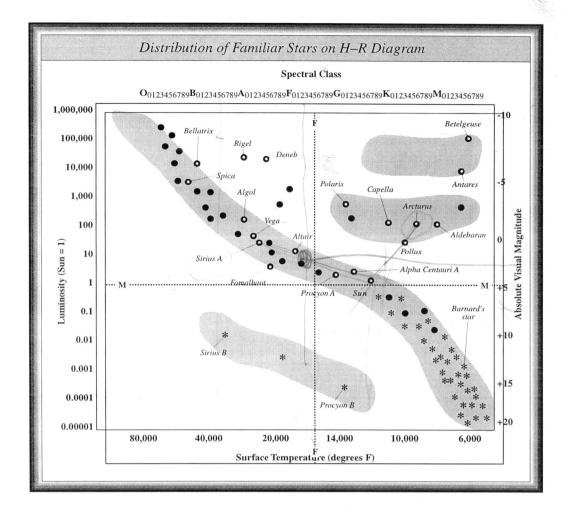

Figure 8–20 The H–R diagram can be used to determine the distance to a star that is too far away for trigonometric parallax to be used. Determining its spectral type and assuming its luminosity class from the spectral lines allows for the determination of its absolute visual magnitude. Its distance follows directly from knowing that and its apparent visual magnitude.

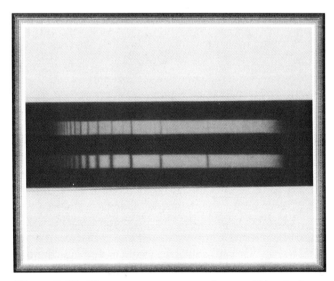

Figure 8–21 These two spectra are of stars of identical spectral type but of different sizes. The giant star, whose atmosphere is thin, does not scatter photons that pass through. The lines are thin and sharp (top). The smaller main sequence star has a dense atmosphere, so its spectral lines are broad (lower). The thickness of spectral lines tells us the luminosity class of a star. (Lick Observatory)

diagram is called **spectroscopic parallax**. Too big of an assumption, you say? Perhaps — but there is another factor to consider that makes the technique quite convincing. Astronomers can determine the luminosity class of a star by carefully examining the spectral lines of the star.

It turns out that the spectral lines of a main sequence star are much broader than those of a giant star — which are thin and distinct (Figure 8-21). The atmosphere of a giant star is much less dense than that of a main sequence star, and therefore there is less scattering of the photons trying to get through. So the luminosity class of a star can be determined from the widths of the spectral lines of that star, and its approximate placement on the H-R diagram therefore established. This is an additional piece of data that a star's spectrum provides the astronomer. Determining the distance to an object in space is an extremely important task for astronomers, because without it, we lack knowledge of the luminosity of the object. And lacking the luminosity of the object, we lack knowing what might be going on in the object that is responsible for whatever energy that it gives off.

Twenty–Four Nearest Stars

Star	Apparent Visual Magnitude (m_v)	Distance (in ly)	Absolute Visual Magnitude (M_v)	Spectral Type
Sun	−26.8		4.83	G2V
α Centaurus A	0.1	4.3	4.38	G2V
B	1.5	4.3	5.76	K0V
Bernard's Star	9.5	5.9	13.21	M5V
Wolf 359	13.5	7.6	16.80	M8V
Lalande 21185	7.5	8.1	10.42	M2V
Sirius A	−1.5	8.6	1.41	A1V
B	7.2	8.6	11.54	white dwarf
Luyten 726-8A	12.5	8.9	15.27	M5V
B (UV Cetus)	13.0	8.9	15.8	M5V
Ross 154	10.6	9.4	13.3	M5V
Ross 248	12.2	10.3	14.8	M6V
ε Eridanus	3.7	10.7	6.13	K2V
Luyten 789-6	12.2	10.8	14.6	M6V
Ross 128	11.1	10.8	13.5	M5V
61 Cygnus A	5.2	11.2	7.58	K5V
B	6.0	11.2	8.39	K7V
ε Indus	4.7	11.2	7.0	K5V
Procyon A	0.3	11.4	2.64	F5V
B	10.8	11.4	13.1	white dwarf
BD +59⁰ 1915 A	8.9	11.5	11.15	M4V
B	9.7	11.5	11.94	M5V
Groombridge 34 A	8.1	11.6	10.32	M2V
B	11.0	11.6	13.29	M6V
Lacaille 9352	7.4	11.7	9.59	M2V
τ Ceti	3.5	11.9	5.72	G8V
BD + 5⁰ 1668	9.8	12.2	11.98	M5V
LUYTEN 725-32	11.5	12.4	15.27	M5V
Lacaille 8760	6.7	12.5	8.75	M0V
Kapteyn's Star	8.8	12.7	10.85	M0V
Kruger 60 A	9.7	12.8	11.87	M3V
B	11.2	12.8	13.3	M4V

Note: *shading indicates binary system*

Figure 8–22 A list of the 24 nearest stars to Sun. By comparing their characteristics with those of Sun, and assuming that they are representative of stars throughout the galaxy, we get a sense of how Sun stacks up amongst the stars.

Comparison of Stars

We have spent a lot of time being theoretical and technical. Let's take a short breather and relate the concepts just developed with something close to your experience — the nighttime sky. Try to think of each of those dots of light in the sky having the properties we have been discussing — surface temperature, luminosity, size, and so forth.

Nearest Stars to Sun Carefully study Figure 8-22, which contains a list of the 24 stars closest to Earth. Their personality traits (properties) are included that you can become familiar with them. You should be able to make a few interesting generalizations about the stars that are our nearest neighbors:

- Most (11 out of 20) cannot be seen with the naked eye (i. e., $m_v > 6$),

- Most (15 out of 20) are the cooler K- and M-type stars, and

- Most (16 out of 20) emit less energy than our Sun (i. e., $M_v > 4.8$).

Conclusion: If our local region of space is typical of the universe in general, then we must conclude that the universe tends toward smallness (small mass) when it comes to making stars. We arrived at the same conclusion earlier when we plotted a random selection of stars on the H-R diagram. Can this fact be explained by anything we've already learned about stars? Well, not exactly — but it does make sense if we consider a familiar example. A large, gas-guzzling American automobile typically has a 20-gallon gas tank. But because it only gets perhaps 10 miles to the gallon, it only travels 200 miles on a full tank. A small, gas-efficient automobile having only a 10-gallon gas tank, but getting 30 miles to the gallon, can travel 300 miles on a full tank.

Likewise with stars. Large-mass stars contain substantial quantities of fuel in the form of hydrogen, but they

Characteristics of Main Sequence Stars							
Spectral Type	Surface Temp (^0K)	M_v	Luminosity (Sun = 1)	Radius (Sun = 1)	Mass (Sun = 1)	Average Density Water = 1	Years on the Main Sequence (Approximately)
O5	40,000	−5.8	405,000	18	40	0.01	1×10^6
B0	28,000	−4.1	13,000	7.4	15	0.1	11×10^6
A0	9,900	+0.7	80	2.5	3.5	0.3	440×10^6
F0	7,400	+2.6	6.4	1.4	1.7	1.0	3×10^9
G0	6,000	+4.4	1.4	1.0	1.1	1.4	8×10^9
K0	4,900	+5.9	0.46	0.8	0.8	1.8	17×10^9
M0	3,500	+9.0	0.08	0.6	0.5	2.5	56×10^9

Figure 8–23 Applying basic laws of physics and chemistry to the characteristics of stars allows astronomers to calculate how long a particular type will "survive" with the fuel it has. This list of life spans for the various classes of stars is referred back to several times during the course of the Book.

"burn" the fuel quite rapidly and consequently use it up much more quickly than do low-mass stars. This process of "fuel" consumption in stars can be quantified. We have extensive knowledge of the behavior of gases in high-temperature environments. We also have computer programs that handle enormous amounts of data in simulating the conditions within a star.

Basically, we tell the computer what the distribution of matter is within a star that has a given mass, and tell it what the conditions are that affect the matter (laws of physics and chemistry). We then ask the computer to tell us what happens to that matter over a long period of time. From such experiments, we determine the average life spans of main sequence stars of a given spectral type (and therefore of a certain mass). A list of the lifespans is in Figure 8-23.

Low-Mass Stars Is it now obvious why low-mass stars dominate any survey taken of the stars? They simply live longer. Only if a star is still around can it be included in a survey. Is it reasonable to conclude, therefore, that the O- and B-type stars on our diagram are young in age? Yes — as long as you think of several million years as being the criteria for youth amongst stars. O- and B-type stars eventually die, but even at death they have been in existence for only a short period of time.

Our Sun, a G-type star, has a total life span of some 10 billion years. Since it has — according to our best estimates — been around and shining for 5 billion years, we think of it as being a middle-aged star. We have another 5 billion years during which to plan for an alternative to our dying with Sun. And what about those very populous, low-mass, M-type stars?

Well, each one experienced youth after forming, and some are no doubt in that stage today, but the greater possibility is that any given M-type star we observe is quite old. Most stars in the universe are, therefore, much older than our Sun. In fact, they are probably <u>much</u> older than our Sun. In Figure 8-24 is a photograph of a cluster of stars found in the Milky Way galaxy. It contains approximately 500,000 M-type main sequence stars, the average age of which is 15 billion years. That suggests that the Milky Way galaxy has been around for at least 15 billion years.

Brightest Stars in the Sky So *most of the nearby stars are low-mass, low-temperature, low-energy stars invisible to the naked eye.* Why, then, is the sky so magnificently studded with bright stars — especially during the wintertime? To answer that question, let us look at a list of the 20 brightest stars seen in the sky, together with their characteristics (Figure 8-25). Again, we can make a few generalizations about the stars on the list:

Figure 8–24 A globular cluster such as this, M13, contains exclusively old, cool, red dwarf stars. The ages of such slow-burning stars are on the order of 15 billion years or so, three times older than Sun. (Lick Observatory)

- Most (11 out of 20) are very luminous B- and A-type stars,

- The cooler K- and M-type stars (5 out of 20) are giant stars and are therefore quite luminous, and

- Most are located at great distances from us.

Conclusion: *The stars visible to us in the nighttime sky are mostly not main sequence stars, and therefore are not typical of the stars making up our galaxy.* They are magnificent to look at, but are visible to the naked eye only by virtue of being very large and luminous. Therefore, we want to avoid using them as a basis for studying the nature of stars in general. Are people who use star patterns to forecast their futures making that very mistake!

Personalities of the Stars Do you recall my having mentioned a F3V-type star at the very beginning of this Chapter? Are you now able to visualize some of its characteristics? Mass, surface temperature, size, color, luminosity, and what stage of life it is in? I hope so. Because in the next few Chapters I use that nomenclature in describing the birth, life, and death of stars, and the locations of the different types of stars in the Milky Way galaxy. As an exercise in testing your learning, close your eyes and try visualizing a B7II-type star, a M4I-type star, and a G2V-type star.

Perhaps a nagging question has occurred to you throughout this discussion of main sequence stars — "What about the remaining 10% of stars on the H-R diagram?" "How do red giants and white dwarfs fit into the picture of stellar evolution?" It turns out that the red giants and white dwarfs

Twenty Brightest Stars

Star	Apparent Visual Magnitude (m_v)	Distance (in ly)	Absolute Visual Magnitude (M_v)	Spectral Type
Sirius	−1.47	8.7	+1.4	A1V
Canopus	−0.72	98	−3.1	F0II
Rigil Kentaurus	−0.01	4.3	+4.4	G2V
Arcturus	−0.06	36	−0.3	K2III
Vega	0.04	26.5	+0.5	A0V
Capella	0.05	45	−0.6	G8III
Rigel	0.14	900	−7.1	B8I
Procyon	0.37	11.3	+2.7	F5IV
Betelgeuse	0.41	520	−5.6	M2I
Achernar	0.51	118	−2.3	B5V
Hadar	0.63	490	−5.2	B1III
Altair	0.77	16.5	+2.2	A7IV
Aldebaran	0.86	68	−0.7	K5III
Acrux	0.90	260	−3.5	B1IV
Spica	0.91	220	−3.3	B1V
Antares	0.92	520	−5.1	M1II
Fomalhaut	1.15	22.6	+2.0	A3V
Pollux	1.16	35	+1.0	K0III
Deneb	1.26	1600	−7.1	A2I
Beta Crucis	1.28	490	−4.6	B0.5IV

Distance determined by spectroscopic parallax

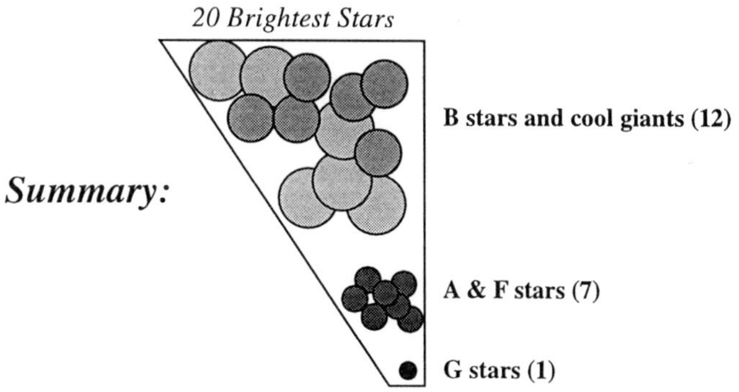

Summary:

20 Brightest Stars

B stars and cool giants (12)

A & F stars (7)

G stars (1)

Figure 8–25 A comparison of these stars that are brightest in the sky to those that are nearest to us (Figure 8–22) is very revealing. The splendor of the nighttime sky is not due to the nearby, common-type stars. Rather, the brightest stars are the short-lived, high-luminosity stars that are quite distant.

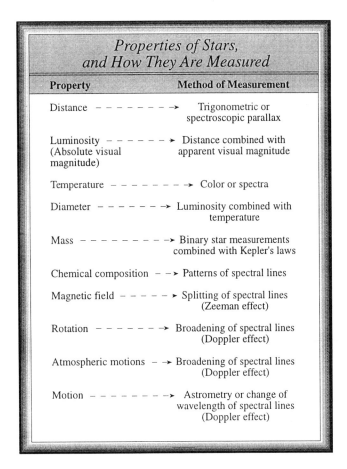

Figure 8-26 The basic properties of stars and how astronomers measure them.

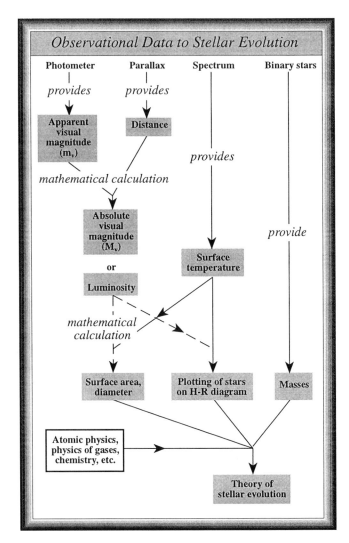

Figure 8-27 A summary of how various characteristics of stars are weaved together into a comprehensive theory to explain how they are born, live out their lives, and eventually die.

are stages in a star's death. But I'm going to delay discussing stellar death until we first learn how stars form, and how they become main sequence stars in the first place. Suffice it to say that as stars die, they leave the main sequence as they go through drastic changes in surface temperature and/or luminosity. At any point between life as a main sequence star and death as a stellar corpse, it will be found somewhere on the H-R diagram. Each star plotted on the H-R diagram that is not on the main sequence, in other words, is in the act of dying.

Summary of Stellar Characteristics In the course of developing a theory for the birth and death of stars in the next few Chapters, I refer back to many of the principles of stellar properties discussed in this Chapter. The diagrams in Figures 8-26 and 8-27 will serve as a review of those principles and how astronomers determine the basic properties of stars from spectra and the like.

Extraterrestrials?

Do you feel like speculating a bit? Let us assume that our estimate of 20 billion years for the age of the universe is correct. Let us further assume that the scientific theory of the origin of life on Earth is correct, and that our estimate of 5 billion years for the age of the solar system is also correct. Assume that the conditions necessary for the evolution of life are not unique to our solar system. Assume further that on the average any star requires 5 billion years to produce life that develops technology to our present level. Then it seems logical to conclude that if we ever do contact another civilization in space, it will be far more advanced than we are (Figure 8-28).

Most stars, in other words, are older than ours, and therefore any civilization on a planet orbiting one of them is likely to be older as well. The oldest stars in our galaxy are those low-mass M-type stars, 15 billion years old. Since they are 10 billion years older than our Sun, any civilization there could conceivably be 10 billion years more advanced than we are! What would such a civilization be

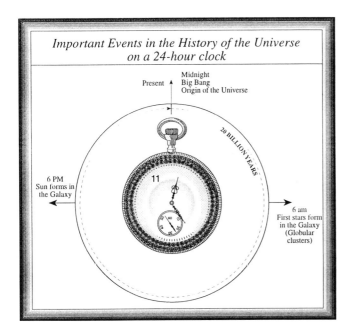

Figure 8-28 Imagine the presumed 20 billion year history of the universe being compressed to a 12-hour clock. The Big Bang occurs at 12 noon, the Milky Way galaxy forms at 3 PM, Sun forms at 9 PM, and "modern" civilization is a fraction of a second before midnight. Given the enormous expanse of time within which life has presumedly evolved on Earth, any extraterrestrial civilization with whom we establish contact will most likely be considerably more advanced than us.

like? Well, of course, we can't say. But if we imagine what our own civilization might be like far into the future, we might be willing to conclude one of two things — either we will have destroyed ourselves, or we will be very, very advanced — perhaps to the point of not even existing in our bodies!

Are We Average? So what does that mean in terms of the guys out there? Either we are one of the most advanced civilizations in the universe, just about to add ourselves to the list of civilizations that eventually became advanced enough to destroy themselves, or the guys out there are so far advanced that they either don't care about us or are waiting for us to demonstrate some sign of maturity before inviting us to join the galactic community.

In either case, it is certainly not reasonable to believe that they will visit us in some type of large, metallic-looking, frying pan-shaped object that comes out of the sky, sets down on Earth, opens its doors, from which a gangplank lowers, and out of which walk some cute little creatures who hum together with a well-rehearsed phrase like "Take me to your leader." Certainly the universe is more imaginative than that. Or is it?

Contacting Distant Civilizations As long as I am off on a tangent, consider this. As far as we know, there has been a single, serious attempt to send a signal out into space for the purpose of contacting another civilization. Assuming there is *someone* there to receive the message encoded therein (the content of the message will be discussed in detail in Chapter 20). It was sent by the United States in 1974 from the world's largest radio telescope in Puerto Rico and beamed toward the star cluster shown in Figure 8-24. That particular group of stars was chosen to be the target of the message both because of the ages of the stars therein, and because by the time the radio beam arrives, it will have spread out in a cone so as to be detectable by any civilization living around any one of the 500,000 stars contained within the cluster.

Now lest you be frightened enough to want to purchase weapons to protect yourself against any invading aliens, since the cluster is located 25,000 light-years from Earth, the message requires another 24,980 years or so to arrive there. Hopefully, the message hasn't been intercepted by anyone closer. If so, they probably know of our existence anyway from the amount of radio and TV-radiation leaking out of our atmosphere. I will expand upon these notions of contacting distant civilizations in Chapter 20. I haven't yet presented a strong enough case for their having a place to live in the first place, since it is doubtful that life can evolve on a star itself.

Summary/Conclusion

Now that we have been able to compile a list of characteristics of stars, including their masses, energy output (luminosity), size, and lifespans, we are in a better position to evaluate the types of stars that would be suitable candidates for our SETI program. Using our experience with the origin of life on Earth, which required some 5 billion years to reach our present level of development, it appears that most stars in the universe live at least that long. In fact, a survey of the stars right around us suggest that most stars live much longer than that, so that we might be led to believe that if we ever do meet up with some guys out there, they will be incredibly more advanced than us.

★★★★★★★

LEARNING OBJECTIVES: *Now that you have studied this Chapter, you should be able to:*

1. Outline the methods astronomers use to find the following physical properties of stars: (a) apparent visual magnitude, (b) distance, (c) surface temperature, (d) chemical composition, (e) amount of energy emitted (absolute visual magnitude), and (f) mass.
2. Show by drawing a simple diagram how the distance to a nearby star can be calculated by the technique of parallax.
3. Explain how the inverse-square law governing radiation allows astronomers to determine the energy output of a star from knowing its apparent magnitude and distance.
4. Describe the method of calculating the energy output of a star using Sun as a standard candle once that star's absolute magnitude is known.
5. Sketch a H-R diagram for stars, indicating the position of the main sequence, the sun, white dwarfs, and giant stars of different luminosities.
6. Use the H-R diagram to describe stars according to their relative luminosities, surface temperatures, and sizes.
7. Describe the methods astronomers use to calculate the masses of stars (visual, spectroscopic, and eclipsing binaries).
8. Compare and contrast the nearest 20 stars to Sun with the 20 brightest stars in the sky, and offer generalizations as to the most common stars to be found in the universe and why.
9. Explain how the addition of the masses of stars on the H-R diagram allows astronomers to infer the life cycle of stars.
10. Compare the life-spans of stars according to type, and speculate as to the possibility of contacting civilizations around each of those types.
11. Define and use in a complete sentence each of the following **NEW TERMS**:

Absolute visual magnitude 167
Algol 178
Barycenter 175
Binary star system 173
Eclipsing binary system 178
H-R diagram 170
Hydrostatic equilibrium 181
Inverse-square relationship 167
Light curve 178
Luminosity 168
Luminosity Class 172
Main sequence stars 170

Multiple star system 173
Optical binary 175
Parsec 167
Radiation pressure 181
Red giant 171
Spectroscopic binary system 177
Spectroscopic parallax 181
Separation distance 176
Trigonometric parallax 165
Visual binary system 175
White dwarfs 171

Chapter 9

The Birth of Stars

Figure 9–1 The Horsehead Nebula in Orion consists of a denser–than–average dust cloud silhouetted against a glowing gas cloud. It is within such regions as this that stars are in the process of forming. (NASA)

CENTRAL THEME: What is the evidence astronomers have to explain how stars are born out of the material between the stars?

Star-life is obvious. There they are in the nighttime sky, shining (emitting radiation). It is relatively easy to photograph them, name them, measure them, catalogue them, classify them, and plot them. But to develop a hypothesis or theory about how they came to be in the first place — assuming they have a beginning — is more difficult. The task requires that we collect and integrate an assortment of observations of chemists and atomic physicists as well as astronomers, develop computer programs that apply laws of physics to mathematically-modeled stars, and that we finally step back and look at the general pattern of evidence. Fortunately, recent HST photographs have provided dramatic evidence for star-birth.

We've already accomplished much of the work in developing the H-R diagram that organizes stars according to their characteristics of mass, temperature, and luminosity. We have also proposed the method by which stars shine — the proton-proton chain. The question now before us is — "How did that reaction begin in the first place? Where did the hydrogen come from, and what made it begin to fuse together to form helium?"

Locating Young Stars

Get out a road map of the city in which you live. Imagine plotting on it the location of all newly born babies who are less than eight hours old. Do you think a plot of their locations would reveal a uniform distribution throughout the map? Or would they tend to be clustered in some kind of pattern? You got it. Most locations will be plotted in the immediate vicinity of the hospital delivery rooms in which they are born. Some will be distributed randomly around the map, highlighting those who intentionally or unintentionally are delivered at home. A few babies may even be born on the way home from work, or on the way to the hospital. A rare exception might be one born during a flight to another country.

Galactic Star Clusters Let us do the same with stars. In the previous Chapter we found that O- and B-type stars must be young stars because they burn up their fuel rapidly. To be still around and shining, they must be young. You are probably familiar with a group of these young-ish stars already. In the late fall or winter evening sky there is an easily-seen small group of stars commonly mistaken for the Little Dipper. It is called the **Pleiades**, or Seven Sisters (Figure 9-2) — located in the constellation of Taurus the Bull (designated M45 in Figure 9-3). It is typical of the groups of stars found close to the Milky Way band (the galactic plane), for which reason we call them **galactic clusters**. The fact that the stars within these clusters are gravitationally bound together — all being located relatively close together in space — is significant. That fact suggests they had a common birth in time.

An important clue supporting the conclusion that they had a common birth in time is apparent as you look carefully at the brightest stars in the Seven Sisters in the photograph of Figure 9-2. The stars appear to be surrounded by hazy, cloudy material. This cloudy material is not visible to the naked eye, although its light contributes to the overall brightness of the stars that are imbedded in it and that are visible to the naked eye. In color photographs, the fuzzy clouds appear *bluish*. To astronomers, that immediately identifies the cloud material as consisting of tiny dust particles. This dust is interpreted as being the remains of the material from which the stars within the cluster just recently formed. It is the debris leftover from star formation. The fact that it is found within the vicinity of what appear to be young stars is evidence of this hypothesis.

Clouds of Dust Around Stars Astronomers call blue dust clouds **reflection nebulae**, inasmuch as the dust particles reflect light from the bright stars located in the vicinity (Figure 9-4). It turns out that the dust particles in space are quite small in size — about that of the shorter wavelengths of visible light. That means blue photons arriving from the source are more easily absorbed and re-emitted by the grains of dust. The longer red photons pass through the

Figure 9–2 The Pleiades, *or Seven Sisters, is the best visible example of an open or galactic cluster.* (Lick Observatory)

★★★★★★★★

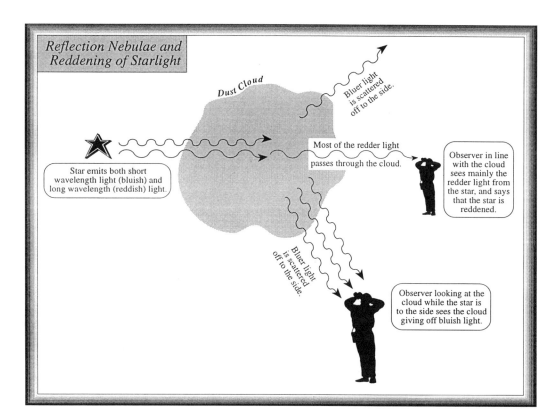

Figure 9–4 *This diagram illustrates why clouds of dust surrrounding hot, young stars appear blue on color photographic plates.*

Figure 9–3 *The location of the beautiful stellar nursery called the Orion Nebula (Shown as M42 in the diagram). It is visible to the naked eye, although it appears as a somewhat dim dot. It is one of the most beautiful objects to observe through a telescope. The group of young stars called the Pleiades is shown circled. (Carina Software)*

Figure 9-5 A full view of the Orion nebula, a site in which star formation is taking place. This cloud can be seen with the naked eye as a fuzzy little dot in the constellation of Orion. Notice the diffraction "spikes" caused by light passing by the mirror supports in the telescope. (Lick Observatory)

dust mostly unaffected.

The grains of dust are acting, in effect, as tiny antennae. You recall from Chapter 5 that the length of an antenna must be at least as long as the wavelength of radiation it is designed to collect or detect or intercept. Since we see only blue wavelengths reflected off the dust particles, we can determine their maximum size. Knowing the distance to the *Pleiades* and therefore to the dust clouds surrounding them, astronomers can determine the density of the dust according to how much blue light is reflected. The calculation results in the estimate that there is one grain of dust in a cube 200 yards on a side.

It is difficult to imagine such a thin cloud of dust reflecting so much light, but then again the clouds are quite huge (several light-years in diameter) and we are observing them from quite a distance away (approximately 400 light-years). It is dust like this, incidentally, that pervades the spaces between stars and inhibits our view of the very center of our galaxy. For although it is found concentrated in clouds in the vicinity of young stars such as the *Pleiades*, there are also scattered clouds and concentrations of clouds of dust all along the band of the Milky Way. In fact, much of this dust is seen not directly, but deduced from the spectra of stars lying within or behind such clouds.

Since blue photons are more effectively scattered by interstellar dust grains than red photons are, more red than blue photons coming from stars within or behind the dust clouds arrive in our telescopes. This effectively makes the stars appear redder than they would appear were the dust not present (Figure 9-4). The absorption lines still allow us to determine all of the stellar characteristics provided by spectra, but there is less contrast between the absorption lines and the fewer blue photons.

Apparent visual magnitudes for stars must be corrected for this scattering effect — called **interstellar reddening**. We deduce from the study of this effect on numerous clouds that the dust grains are probably combinations of graphite, iron, silicon, carbide, silicates, and ices. The important point here, however, is that the reddening of starlight tells us of the presence of dust along the plane of our galaxy. So *interstellar reddening* provides us with the connection between young stars and clouds of dust.

Emission Nebulae Almost without exception, clouds of dust are found in close proximity to or intermingling with clouds of gas. The manner in which gas clouds reveal their presence is entirely different than that of the dust. As with the case of the *Pleiades* and its associated reflection nebula, there is a good example of a gas cloud that can be seen with the naked eye — although it is not obvious that it is a cloud. It is located not too far from the *Seven Sisters*. It is located in the well-known constellation of *Orion, the Hunter* (Figure 9-3).

You may be familiar with Orion's belt, which consists

CHAPTER 9 / The Birth of Stars

of three bright stars in a fairly straight line. The cloud of gas under consideration lies below the belt, amongst a group of stars that form what appears to be a sword hanging from Orion's belt. To the naked eye, it appears like it is just another star. But seen through a telescope, it is one of the most beautiful objects in the entire sky – the **Orion nebula** (Figure 9-5). Atoms of gas are not large enough to reflect photons. But if the atoms are close (dense) enough to collide or if they are excited by photons emitted by nearby hot stars, then they emit photons.

In interstellar space, gas atoms are not very densely packed – only about ten atoms per cubic inch. But in the Orion Nebula there are many hot, young stars whose short-wavelength radiation (mostly UV) excites the gas atoms, causing them to emit radiation when electrons drop to lower energy levels. We refer to such clouds as **emission nebulae**. The amount of UV radiation may be so intense that gas close to the source of the radiation may be ionized. Such ionized gas regions are called **HII regions**. Regions more distant from the hot young stars may contain gas atoms that have only been excited by the radiation. These are called **HI regions** (Figure 9-6). There are many, many examples of beautiful gaseous nebulae in the sky, and your participation at a star party during the summer months allows you to see many of them. Don't expect to see in the eyepiece of the telescope the colors that are so dramatic in photographs. Remember that the color-sensitive cones in our eyes do not work under darkened conditions. But the subtle shapes and textures of light and dark within the cloud make it exciting to observe. Try it.

Emission nebulae appear red in color-sensitive film. That fact tells us straight away that the dominant chemical in such clouds is *hydrogen gas*. Recall that in Chapter 6 we learned that when hydrogen gas is heated to a sufficiently high temperature, it emits visible radiation of a magenta color. Passing that emitted light through a spectroscope reveals that the magenta color is actually a combination of three wavelengths, one of which is a red color. That particular wavelength is designated the hydrogen alpha (Hα) wavelength. The color of the hydrogen alpha line is exactly the same as that of emission nebulae. That is easily verified, of course. Spectral analysis of the light coming from the nebulae confirms this to be the case.

As with dust clouds, we detect concentrations of gas in

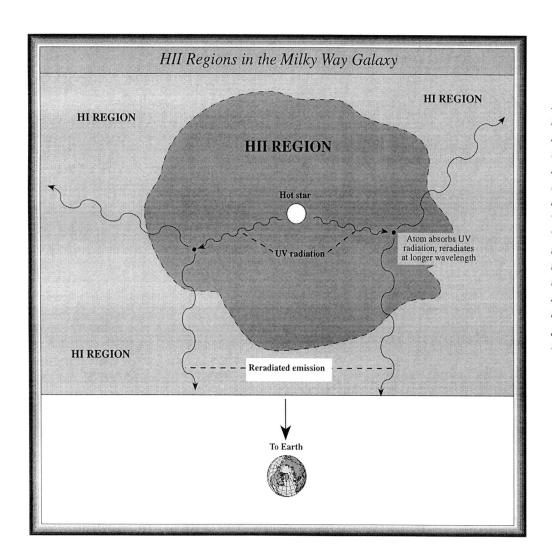

Figure 9-6 Emission nebulae are caused by hot stars whose ultraviolet radiation is absorbed by surrounding gas. The radiation excites the hydrogen atoms to glow with the Balmer series of wavelengths. The radiation can be strong enough to ionize regions of gas immediately around the star (HII regions), or merely excite regions of gas farther away (HI regions).

space without necessarily observing them in the vicinity of hot stars. We learned earlier that hydrogen gas also reveals itself by emitting the 21-centimeter wavelength of radio energy. Radio telescopes confirm the fact that hydrogen gas is most commonly found along the galactic plane — in the very regions of the galaxy that dust clouds are found. And there is another method of detecting the presence of gas clouds in space without observing them directly.

Occasionally, while obtaining the spectrum of a star, astronomers find absorption lines that as a group have Doppler shifts different than those of the star. Our interpretation of this pattern of lines is that there is a concentration of gas between us and the star being studied. Since the gas cloud is moving at a different rate than the star, the Doppler shift of the lines due to the star will be different than the Doppler shift of those due to the cloud — one set of lines will reveal a greater shift than the other. Clouds of gas detected in this manner are referred to as **absorption nebulae** (Figure 9-7).

Interstellar Medium Astronomers determine the chemistry of gas clouds by identifying lines in the spectra. The patterns of lines reveal the presence of such chemicals as hydrogen, helium, carbon, nitrogen, oxygen, iron, sodium, calcium, and other lesser-known chemicals. These are, as you might expect, the most commonly found elements in stars as well. So scattered amongst the stars are varying concentrations of gas and dust — what is generally called the **interstellar medium**.

If the gas/dust clouds are concentrated enough and

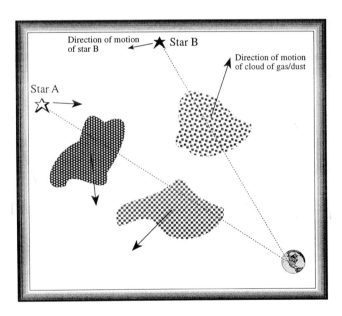

Figure 9-7 Clouds of gas/dust are detected even when they are not visible. If spectral lines in a star's spectrum reveal a different Doppler shift than those of the star itself, astronomers deduce the existence of a cloud of gas/dust between the star and Earth.

are close enough to hot stars to cause them to shine so that we can observe them visually, we call them **interstellar clouds**. Sometimes *interstellar clouds* are seen when they reflect light (reflection nebulae), sometimes when they are heated up enough to emit light (emission nebulae), and sometimes when they are positioned between us and a star (absorption nebulae).

Once you understand the extent to which clouds of the gas and dust are distributed around the galaxy, you will never look at the nighttime sky in the same way. Looking along the band of the Milky Way at night is like scanning those clouds. Look at it carefully sometime. Notice that it is not a smooth band of white running across the sky — rather it consists of spurs and branches. That is a result of the intermingling of stars, gas, and dust. Where the gas clouds dominate or are closer to us — the band is bright with light. Where the dust clouds dominate or are closer to us — it is dark. Some regions are so dense with dust that they are silhouetted against the stars or luminous gas clouds behind them. Figures 9-1 and 9-8 show examples of two of the regions of dust that are called **dark nebulae**.

Clouds of Complex Chemicals Evidence of the presence of gas and dust in space does not in itself convince us that stars form from them. We would like to detect regions of high gas/dust density within which the necessary conditions exist for the collapse of the cloud to become stars. The most likely place for this to happen is in what are called **molecular clouds**, so named because of the discovery of an assortment of atoms and molecules (combinations of atoms) therein (Figure 9-9).

The heaviest atom detected to date is tin, discovered by the HST. It is believed that this and other heavy atoms are formed by stellar winds. Newly-developed techniques with radio telescopes allow astronomers to detect the pattern of radio wavelengths that uniquely defines each of the molecules present in the clouds. Remember that clouds of gas and dust are normally quite thin, with considerable distance between the atoms and/or molecules that make up that gas or dust. So how do the atoms get close enough together for chemical bonds to connect them into the arrangements we detect as molecules?

Simple. By being closer together than the average. In other words, the molecular clouds must be in the act of collapsing — allowing the atoms to get close enough together for complex molecules to form. We don't observe the density of the clouds directly — we observe the molecules and their behavior and deduce the density of the cloud from that information. Astronomers suspect that the molecules form on the surfaces of the dust grains — the dust grains act as a collecting area for the gas atoms to come together and form larger arrangements (like molecules). As astronomers study these concentrations of gas/dust containing complex molecules, such as in the interior of the Orion nebula (Figure 9-11), they gather more and more convincing evidence that stars are in the process of forming there.

CHAPTER 9 / The Birth of Stars   **195**

★★★★★★★★

Figure 9–8 A dark cloud of dust obscures the stars that are more distant. This is a dark nebula. (Lick Observatory)

When we consider that life is an arrangement of complex chemicals, and that we try to understand how it first got started on Earth, we are impressed by the discovery that the universe is capable of manufacturing complex arrangements of atoms in the chill vacuum of space. Perhaps there is something to be learned from studies of molecular clouds about how brain cells can develop under even better conditions (like on Earth's surface)!

Star Formation and the Interstellar Medium The reason that I am making such a big deal about the interstellar medium is that it is believed to be the raw material out of which stars form. Think about it. Young, newly-born stars are found embedded in and surrounded by clouds of gas and dust. *The chemical composition of the clouds closely approximates the chemical composition of the stars themselves* (Figure 9-10). That simple observation remains the strongest argument for the origin of stars within the interstellar medium.

Just where those clouds came from in the first place is a subject for a later Chapter. Suffice it to say that much of the gas and dust in the interstellar medium originated as the matter was ejected into space *by newly-forming stars* and *by old, dying stars*. But most of it is no doubt left over from the very event that created the universe – the Big Bang. This is a new topic that I will explain later in this Chapter and in following Chapters. For the moment, I am attempting to establish the connection between star birth and the interstellar medium, and to explain the process of star formation itself. How does a huge, extended cloud of gas and dust collapse to become a dense, brightly shining star? Why have some clouds already formed into stars, whereas other clouds are still out there in their extended forms?

Composition of the Interstellar Medium in Our Region of the Galaxy				
Element	Atomic #	Total (atoms, molecules, dust grains), % by mass	Gas Atoms % by mass	Concentration (total ÷ gas)
H	1	78.3	78	1
He	2	19.8	20	1
O	8	0.8	0.2	4
C	6	0.3	0.02	15
N	7	0.2	0.03	7
Ne	10	0.2	0.2	1
Ni	28	0.2	0.1	2
Si	14	0.06	0.03	2
S	16	0.04	0.03	1.3
Fe	26	0.04	0.001	40
Mg	12	0.015	0.0007	21
Ca	20	0.009	0.000002	4500
Al	13	0.006	0.00003	200
Ar	18	0.006	0.001	6
Na	11	0.003	0.0006	5

Figure 9–9 The chemical composition of the interstellar medium reveals a pattern observed in stars—the lighter elements are more common than the heavier ones. Notice especially the ratio between the amounts of gas and dust for each chemical element.

Molecules Detected in Interstellar Clouds

| \multicolumn{10}{c|}{Number of Atoms} |
2	3	4	5	6	7	8	9	11	13
H_2	H_2O	NH_3	C_4H	CH_3OH	CH_3CCH	CH_3COOH	CH_4OCH_3	HC_9N	$HC_{11}N$
CO^+	O_3	H_2C_2	CH_2NH	CH_3SH	CH_3CHO	CH_3C_3N	CH_3CH_2OH		
CO	N_2H	H_2CO	CH_2CN	CH_3CN	CH_3NH_2		CH_3CH_2CN		
CH	HCO^+	H_2CS	$HCOOH$	NH_2CHO	CH_2CHCN		CH_3C_4H		
CH^+	HCS^+	$HNCO$	$HNCO$		HC_5N		HC_7N		
CN	HCN	$HNCS$	CH_2CO		CH_3C_2H				
CS	C_2H	C_3H	HC_3N						
C_2	SO_2	C_3O	H_2C_2O						
CH^+	H_2S	C_3N							
OH	HCO	$HOCO^+$							
NO	HCO^+								
NS	OCS								
SO	HNO								
SiO	HOC^+								
SiS	$NaOH$								

Figure 9–10 One of the significant discoveries in astronomy in recent years is that the environment of space contains rather complex chemical substances. This finding supports the theory that the origin of life in the universe is not an unusual process, but is a consequence of the fundamental laws of chemistry and physics operating on the basic ingredients in interstellar space. This is a list of the molecules found to date using radio telescopes.

Interiors of Birthing Stars

Let us apply some basic laws of physics as we consider what is going on inside one of the interstellar clouds. It consists of atoms, molecules, and grains of dust moving in random patterns. At any particular moment in time, the motion of a single particle is just as likely toward the outside edge of the cloud as inward toward the center. The particles are far enough away from one another that there is no collective force of attraction between them. So the billions of particles moving in random directions and at random speeds do not act as a group, but rather as individuals. Therefore the cloud remains extended and does not collapse.

But if some mechanism were to force particles in one part of the cloud closer together, the collective attraction of that more densely-packed group of particles would begin to

Figure 9–11 This is an enlargement of the very center of the Orion nebula, revealed to be the site of intense star formation. Compare this photograph with the brighter section in the center of the larger view of the cloud of gas/dust in Figure 9–5. (Lick Observatory)

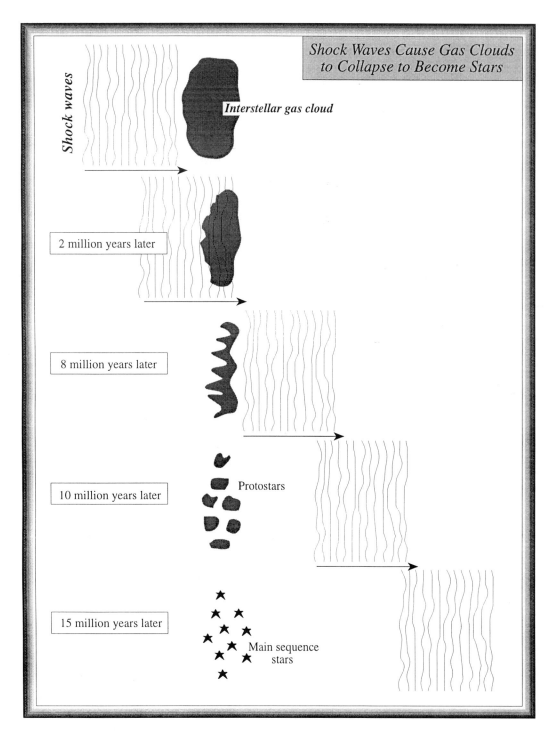

Figure 9–12 Clouds of gas/dust do not collapse on their own. But there are processes in the universe, such as violent outbursts of radiation from exploding stars, that create shock waves that are responsible for the collapse. As the shock wave passes through the cloud, it "pushes" the atoms and molecules close enough together to allow gravity to take over in making the cloud collapse.

affect the entire cloud. As a portion of the cloud is "pushed" or falls together, its mass is concentrated in a growing glob, which in turn attracts other particles in the cloud, and so on. In other words, for a cloud of gas and dust of a given mass, there is a *minimum density required before gravity takes over and pulls the cloud together into a smaller, more concentrated form called a "star."*

Initiating the Collapse To emphasize the point I just made, I will state it another way. In order for a cloud to collapse to become a star, the random *motions* of the particles comprising the cloud must be *overcome by density* so that gravity can do its thing. So what forces the cloud to that critical density? There are several mechanisms possible, but the two that astronomers believe are the most common are *spiral density waves* (to be discussed in Chapter 12) and *supernovae outbursts* (to be discussed in the next Chapter).

In either case, a traveling compression wave "hits" the cloud of gas and dust, and starts it on its inward journey to become a collection of stars (Figure 9-12). These compression waves are analogous to the sonic boom heard when a rifle is fired or a jet plane flies overhead. In space, of course,

198   SECTION 2 / Evolution of the Stars

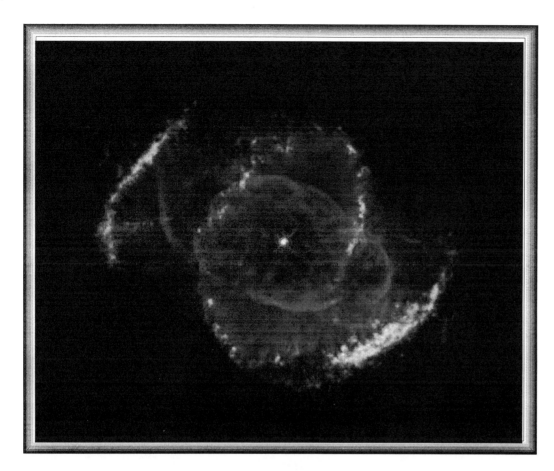

Figure 9–13 The HST took this stunning photograph of the Cat's Eye Nebula, a fossil of a star that ejected its layers of gas out into space. Notice the "bubble–like" structure of the clouds of ejected matter. It is waves such as this that collide with pre–existing clouds to initiate collapse of the cloud to form stars. (NASA/STScI)

there is no medium through which sound can travel. But matter ejected violently from an exploding star forms a radiation shock wave that compresses neighboring clouds of gas and dust (Figure 9-13). S*piral density waves* are also thought to be disturbances that cause the spiral structure of our galaxy as well as other spiral galaxies (Figure 9-14).

Evidence of Pre–Star Stage I have established a relationship between clouds of gas and dust and newly-formed stars, and offered a theory by which one evolves into the other. But what about the missing link? Can we observe objects whose characteristics place them at the stage between cloud and star — what astronomers call **protostar** stage? Yes, we can. And, in fact, some of the observations of the *protostar* stage are very recent and therefore hot off the press.

Look carefully at the photograph in Figure 9-16, for example. This is an enlargement of a small portion of the emission nebula shown in Figure 9-15 (can you locate the portion in this photograph that is in shown in Figure 9-16?). The small, dark globs outlined by the luminous gas cloud behind are called **Bok globules** in memory of astronomer Bart Bok (1906-1983), who first studied them. Although they are a only a few light-years across, they contain several solar masses worth of material — which is thousands of times more dense than the average within the interstellar medium.

Many of these *Bok globules* are found to contain — or at least be in the vicinity of — young stars. This implies, of course, that they are associated with the process of star formation. The famous *Horsehead Nebula* in the constellation of Orion (Figure 9-1) is probably a Bok globule as well. The dark "head" of the "horse" (appearing as if its head is sticking out of the stable door), which consists of dense dust, stands out in bold contrast against a large emission nebula located more distant and behind the dust cloud.

As matter in these collapsing clouds falls inward, energy is released in the form of infrared radiation. Remember, anything going downhill (collapsing) releases energy. Initially, the infrared radiation escapes from the collapsing cloud. But as the cloud gets denser and denser, more and more of the radiation is absorbed by the atoms to make them ionized (lose their electrons).

The radiation released by collapse is therefore no longer available to resist the infall of material, and further collapse occurs quite rapidly (on the order of hundreds or a few thousands of years). By now, all of the atoms are ionized, and the protostar is composed of plasma. The temperature at the center is high enough for the proton-proton chain to begin, and radiation now pushes outward to resist further collapse. The protostar has reached the state of hydrostatic equilibrium. A star is born!

In addition to radiation pushing its way out to the star's surface to announce the star's presence in the universe, the

★★★★★★★★

Figure 9–14 *A distant galaxy observed face–on reveals concentrations of stars in a spiral structure that suggests a process that initiates the collapse of clouds to become the stars.* (Lick Observatory)

the *Pleiades* (Figure 9-2) is an excellent example of this situation, as is what is referred to as the "Bubble" nebula (NGC 7635) shown in Figure 9-17. In this latter case, the gas and dust that is being propelled back out into interstellar space by the birthing star is clearly visible.

Another example of the stellar birthing process, called a **Herbig-Haro** object, appears in Figure 9-18 in a sequence of photographs taken over a period of a few years. The time required for a cloud of gas and dust to collapse to become a new star is on the order of tens of thousands of years. Yet these photographs taken at intervals of several years show striking evidence of change. In such *Herbig-Haro* objects astronomers see convincing evidence of clouds of gas and dust being "pushed" about by the escaping radiation from the surrounding young stars (Figure 9-19). Apparently, such objects appear where jets of gas escape from young stars and collide with clouds of interstellar matter. The atoms in the cloud become excited, then release radiation as electrons drop to lower energy levels. As the jets of gas fluctuate, the clouds change in brightness and shape.

star clears away any material around it that failed to join in the initial collapse. In other words, newly-formed stars are embedded in clouds of leftover debris. Not all of the gas and dust in the original cloud is going to end up in the protostar. The escaping radiation of the newly-formed star provides the outward pressure — forcing gas atoms and dust grains back into the interstellar medium, there to await another opportunity to become a star when a compressional wave comes along. The earlier-mentioned group of stars called

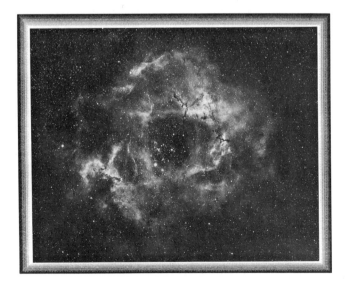

Figure 9–15 *The Rosette nebula (NGC 2237) is a fine example of an emission nebula within which star formation is taking place. The dark "splotches" are the Bok globules shown in enlargement in Figure 9–16. Compare the two photographs so that you can locate the globules in both.* (Lick Observatory)

Figure 9–16 *A close–up photograph of the dark "sploches" in the Rosette nebula shown in Figure 9–15 reveals them to contain stars in the process of forming. Presently, the clouds of dust prevent us from seeing the individual stars.* (Lick Observatory)

Figure 9-17 The "Bubble" nebula is believed to be the result of a collapsing birthing star emitting radiation that "pushes" the remaining portion of the gas/dust cloud back out into interstellar space. (Lick Observatory)

Infrared Radiation and Birthing Stars From the foregoing description of star birth, it is obvious that in order to locate birthing stars astronomers should look for sources of infrared radiation in the sky. But infrared radiation — by and large — is absorbed by our atmosphere, so satellites in orbit around Earth must perform the task for us. In 1983 the **InfraR**ed **A**stronomy **S**atellite (*IRAS*, for short) was launched for the purpose of photographing such objects. The photograph in Figure 9-21 was taken by the IRAS. A new star (arrow) is seen emerging from the cloud of gas and dust from which it was born. The satellite found that this cloud, called *Barnard 5*, contains as many as five protostars.

Interestingly, there is a group of stars whose characteristics are exactly what we expect of protostars — they are just about to "turn on" as main sequence stars. In Figure 9-20, the stars are plotted on an H-R diagram so that their characteristics as a group are easily seen. The stars plot just above the main sequence line, which is exactly where we expect protostars to be located. Consider the two characteristics *surface temperature* and *luminosity* a protostar reveals to an observer who is patient enough to make the necessary observations (and calculations of their numerical values) over the tens- or hundreds-of-thousands of years during which the protostar evolves to main sequence star.

Figure 9-18 In this dramatic sequence of photographs, A Herbig–Haro *object is shown. This is thought to be the result of clouds of gas and dust surrounding birthing stars being "pushed" about by the escaping radiation from the forming stars.* (Lick Observatory)

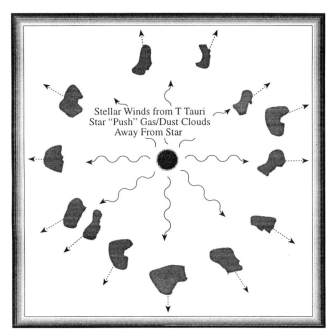

Figure 9–19 The observed detail and behavior of gas and dust clouds surrounding Herbig–Haro *objects convinces astronomers that as a star is "born," and begins to generate energy by the proton–proton chain, stellar winds drive outward away from the star any residual gas and dust in the vicinity. Thus stars end up with but a small fraction of the total amount of material in the cloud from which it formed.*

Imagine plotting those changing characteristics on an H-R diagram. Astronomers refer to such a plot of a star's evolution as an **evolutionary track**. When the cloud is fully extended, both values are small — a plot on the H-R diagram would be far off the lower-right corner. As the cloud heats up during collapse, the luminosity also increases — successive plots on the diagram are therefore further up and to the left. At the point during collapse when the emitted radiation is absorbed within the cloud rather than released out to space, the surface temperature continues to rise, but luminosity decreases. The plot on the diagram of this point in time is shown as the downward loop of the *evolutionary track*.

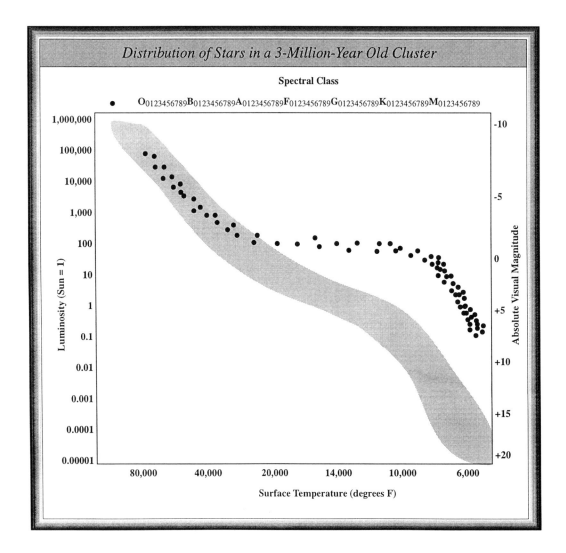

Figure 9–20 The plotting of the stars in a star cluster on a H–R diagram according to their characteristics of surface temperature and luminosity reveals the "age" of the cluster. It is presumed that since the stars are bound by gravity to the cluster, they must have had a common birth. Most of the stars in this particular cluster have not yet even arrived at the main sequence, suggesting that they are just about to "turn on" as stars. The reason that they are distributed up and down the main sequence as they are is because more massive stars evolve faster than less massive stars.

Figure 9–21 The IRAS satellite took this photograph of a birthing star. Since such stars are inherently cool, not yet having initiated the proton–proton chain, they emit most of their radiation in the infrared. Since Earth's atmosphere absorbs such radiation, satellites are quite useful in detecting birthing stars. (NASA)

The horizontal lines in Figure 9-22 represent the evolutionary tracks of a few protostars as they collapse from cool, extended clouds (low surface temperatures, low luminosities) to main sequence stars. The three stars differ in the mass of the original cloud from which they collapsed. The plots just above the main sequence are of a group of stars called **T Tauri** stars—further evidence that stars experience a beginning. T Tauri stars are low-mass stars that have just recently formed, but are still collapsing and have not yet reached the main-sequence. They are probably just about to begin the proton-proton chain in their cores. They are usually found amongst O- and B-type stars in galactic clusters like the *Pleiades*.

Recent HST Discoveries Forgive me please – I've been teasing you about evidence for the birthing of stars. The HST has recently provided dramatic evidence of the stellar birthing process in remarkably detailed photographs of nebulae. In Figure 9-24 a small portion of the Orion Nebula reveals two young stars. They are surrounded by gas and dust trapped as the stars formed, but were left in orbit about the star. These are possibly protoplanetary disks, or

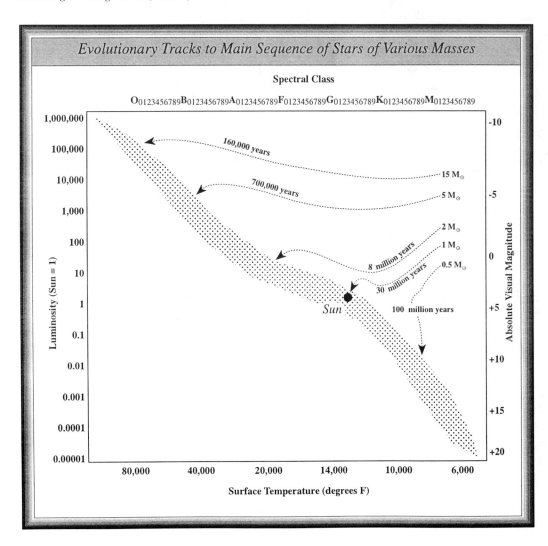

Figure 9–22 The plotting of the positions of protostars of different masses as they evolve on the H–R diagram assists astronomers in understanding the overall process of star formation. The fact that we observe individual stars at various locations along these "evolutionary tracks" supports the conclusion that our theory of star formation is correct.

CHAPTER 9 / The Birth of Stars

★★★★★★★★

Figure 9–23 The HST photograph of detail within the Eagle nebula (M16), also shown in Figure 6–1, reveals many of the processes associated with birthing stars. As stars form within the dark clouds of dust and gas, radiation from nearby stars "boils" away the edges of the cloud, revealing the birthing stars within. These dense globules are called EEGs. (NASA/STScI)

"**proplyds**," that might eventually evolve into planets. The *proplyds* that are closest to the hottest stars of the parent star cluster are seen as bright objects, while the object farthest from the hottest stars is seen as a dark object. The field of view in these photographs is only 0.14 light-years across.

Look carefully at the photograph in Figure 9-23. Undersea corral? Enchanted castles? Space serpents? These eerie, dark pillar–like structures are actually columns of cool interstellar hydrogen gas and dust that serve as incubators for new stars. The pillars protrude from the interior wall of a dark molecular cloud like stalagmites from the floor of a cavern. They are part of the "Eagle Nebula" (also called M16), a nearby star-forming region 7,000 light-years away in the constellation *Serpens*. The pillars are in some ways akin to buttes in the desert, where dense rock have protected a region from erosion, while the surrounding landscape has been worn away over millennia.

In this celestial case, it is especially dense clouds of molecular hydrogen gas (two atoms of hydrogen in each molecule) and dust that have survived longer than their surroundings in the face of a flood of ultraviolet light from hot, massive newborn stars (off the top edge of the picture). This process is called **photo evaporation**. The ultraviolet light is also responsible for illuminating the convoluted surfaces of the columns and the ghostly streamers of gas boiling away from their surfaces, producing the dramatic visual effects that highlight the three-dimensional nature of the clouds. The tallest pillar (left) is about a light-year long from base to tip. As the pillars themselves are slowly eroded away by the ultraviolet light, small globules of even denser gas buried within the pillars are uncovered.

These dense globules have been dubbed "**EGGs**," an acronym for "Evaporating Gaseous Globules." But it is also a word that describes what these objects are. Forming inside at least some of the EGGs are embryonic stars — stars that abruptly stop growing when the EGGs are uncovered and they are separated from the larger reservoir of gas from which they were drawing mass. Eventually, the stars themselves emerge from the EGGs as the EGGs themselves succumb to photo evaporation.

Resembling interstellar Frisbees, the photographs in Figures 9-24 and 9-25 are of disks of dust around

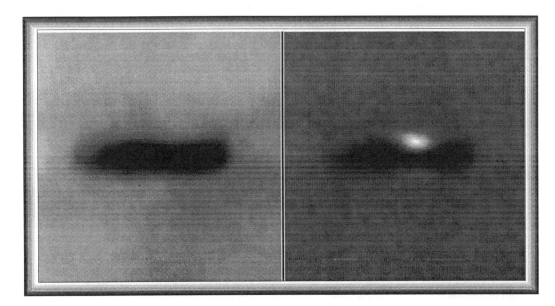

Figure 9–24 These images, taken by the HST of regions within the Orion Nebula, reveal what are interpreted to be circumstellar disks of dust and gas, seen edgewise. Not only are these the sites of newly-formed stars, but it is suspected that the disks of dust and gas around the stars could be in the process of evolving into planets. (NASA/STScI)

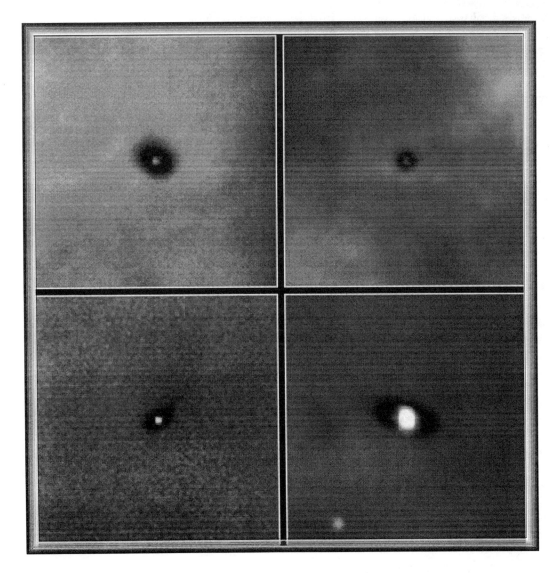

Figure 9–25 The HST images of the disks of dust around young stars within the Orion Nebula suggest that at least half of the stars in the nebula have similar disks. This makes the prospects of locating planets around other stars quite promising. (NASA/STScI)

newborn stars in the Orion nebula, located 1,500 light-years away. Because the disks are viewed edge-on, the stars are largely hidden inside. The disks may be embryonic planetary systems in the making. At 17 times the diameter of our own solar system, these disks are the largest of several recently discovered in the Orion nebula (Figure 9-25). Clearly visible in these images are nebulosities above and below the plane of the disk; these betray the presence of the otherwise invisible central stars, which cannot be seen directly due to dust in the edge-on disks.

Brown dwarfs By now you should be able to visualize the process of gravitational collapse that results in the birth of a star. What eventually stops the collapse, of course, is the radiation that is generated as the proton-proton chain begins (when the pressure and temperature are great enough). But what happens if the mass of the collapsing cloud is too small to create the high pressures and temperatures needed to initiate the reaction at the core of the cloud (the calculate limit is 0.08 solar mass)? Well — look under your feet. Earth presumedly formed along with Sun — from the same cloud of gas and dust. But Earth doesn't shine. Planets are at the other extreme — a collapsing cloud with sufficient mass evolves into a star. A collapsing cloud with insufficient mass evolves into a planet.

Jupiter is a good example of this hypothesis. Chemically speaking, Jupiter is almost identical to Sun — mostly hydrogen and helium. Jupiter has the necessary chemical ingredients to be a star. But the pressures and temperatures in Jupiter's core are insufficient to "ignite" the hydrogen. And yet it is collapsing — at a very slow rate. So there are probably many, many examples of collapsed clouds in the Milky Way galaxy that are caught somewhere *between being very feeble stars and being big gaseous planets like Jupiter*. Might we refer to them as planet-like/starlike objects?

Astronomers have been searching for objects that meet these criteria for some time now, but they prefer to call them **brown dwarf** stars. Their inherent low luminosities require the use of very large telescopes, and therefore it was only with the recent completion of the Keck telescope on Mauna Kea that the first brown dwarf star was detected. Designated **PPL 15**, it is located in the young star cluster *Pleiades* which is some 400 ly from Earth. Based upon its apparent and absolute visual magnitudes, astronomers calculate that its diameter is 25,000 miles and its mass about 8% that of Sun.

Proof of its being a brown dwarf star is derived from the detection of lithium in its spectrum. Computer models of the internal behavior of plasma within stars of different masses (and therefore temperatures) predict that the chemical element lithium is broken down into lighter elements by the fusion process and mixing in normal stars with hot interiors (Figure 9-26). So the measuring of the amount of lithium in star *PPL 15* by astronomers allows them to conclude that it is not a full-blown star. Its luminosity disqualifies it from being in the category of a planet like

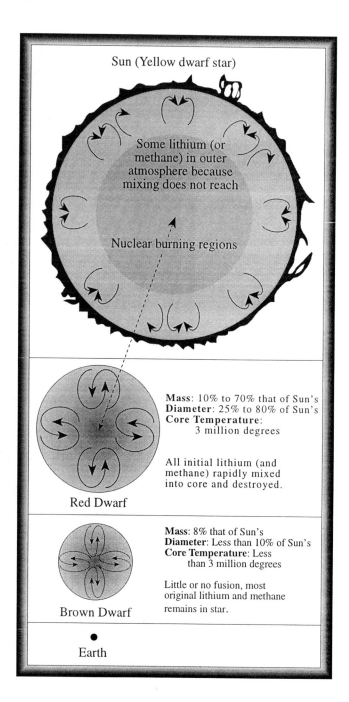

Figure 9-26 A comparision of the theoretical interiors of three types of stars: a normal star like Sun, a cooler, less-massive red dwarf, and a brown dwarf—reveals how the amount of lithium or methane in a star's atmosphere is dependent upon the amount of mixing that occurs within the star's interior. Since cooler stars have a greater amount of mixing, the mass of a star can be determined from the amount of lithium (or methane) that is present in the star's atmosphere. This is the basis for detecting brown dwarf stars.

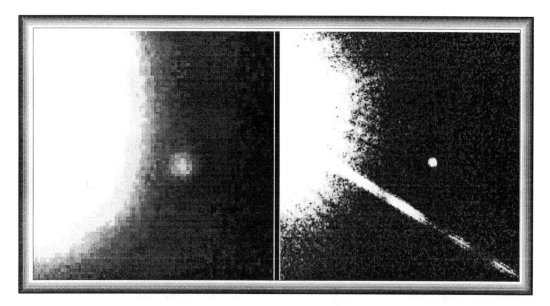

Figure 9-27 After a 30-year searach, astronomers have made the first unambiguous detection and image of an elusive type of object known as a brown dwarf. The evidence consists of an image from the 60-inch observatory on Mt. Palomar (left) and a confirmatory image from the HST (right). The brown dwarf, called Gliese 229B, is a small companion to the cool red star Gliese 229, located 19 light-years from Earth in the constellation Lepus. Estimated to be 20 to 50 times the mass of Jupiter, GL229B is too massive and hot to be classified as a planet as we know it, but too small and cool to shine like a star. (NASA/STScI)

Jupiter. Still, the 0.08 solar-mass of PPL 15 is too close to the critical limit to be entirely convincing. The star **Teide 1** (also located in the Pleiades), with a 0.02 solar-mass, is more convincing.

But an even more certain candidate for a brown dwarf was discovered just recently around a star closer to home: about 20 light-years from Sun. A red dwarf star, designated *Gliese 229*, appears to have a binary companion (called **Gliese 229B**) located some 44 AU's away — comparable to Pluto's distance from Sun (Figure 9-27). Its surface temperature is warmer than Jupiter but cooler than the coolest star. This perfectly fits the brown dwarf model. But the clincher is the presence of methane in its spectrum. Like lithium in the star *PPL 15*, methane molecules are broken apart by the high temperatures in normal stars. *Gliese 229* itself appears to be about 5 billion years old.

The spectrum of *Gliese 229B* is astonishingly like that of a gas giant planet like Jupiter — an abundance of methane. Methane is not seen in ordinary stars, but it is present in Jupiter and other giant gaseous planets in our solar system. The Hubble data obtained and analyzed so far already show the object is far dimmer, cooler (no more than 1,300 degrees Fahrenheit) and less massive than previously reported brown dwarf candidates, which are all near the theoretical limit (eight percent the mass of Sun) where a star has enough mass to sustain nuclear fusion.

Brown dwarfs are a mysterious class of long-sought objects that form the same way stars do, that is, by condensing out of a cloud of hydrogen gas. However, they do not accumulate enough mass to generate the high temperatures needed to sustain nuclear fusion at their cores, which is the mechanism that makes stars shine. Instead brown dwarfs shine the same way that gas giant planets like Jupiter radiate energy, that is, through gravitational contraction. In fact, the chemical composition of *Gliese 229B's* (GL229B, for short) atmosphere looks remarkably like that of Jupiter.

Planets in the Solar System are believed to have formed out of a primeval disk of dust around the newborn Sun because all the planets' orbits are nearly circular and lie almost in the same plane. Brown dwarfs, like full-fledged stars, would have fragmented and gravitationally collapsed out of a large cloud of hydrogen but were not massive enough to sustain fusion reactions at their cores. The orbit of *GL229B* could eventually provide clues to its origin. If the orbit is nearly circular then it may have formed out of a dust disk, where viscous forces in the dense disk would keep objects at about the same distance from their parent star. If the dwarf formed as a binary companion, its orbit probably would be far more elliptical, as seen on most binary stars.

The initial Hubble observations will begin providing valuable data for eventually calculating the brown dwarf's orbit. However, the orbital motion is so slow, it will take many decades of telescopic observations before a true orbit can be calculated. GL229B is at least four billion miles from its companion star, which is roughly the separation between the planet Pluto and our Sun. Astronomers have

★★★★★★★★

been trying to detect brown dwarfs for three decades. Their lack of success is partly due to the fact that as brown dwarfs age they become cooler, fainter, and more difficult to see.

You would have to take a journey to the center of such a brown dwarf to determine if the proton-proton chain is actually in progress there. Since brown dwarfs are probably slowly collapsing as Jupiter is, they should emit infrared radiation (as Jupiter does). The *IRAS* satellite has detected a few suspects for this category of star, but they have not yet been confirmed as brown dwarf stars.

So the evidence is complete. Even though we have never observed a star go through all stages from extended cloud to main sequence, we have observed and studied different objects in the various stages along the way. The evidence fits together nicely. We are now prepared for the violence of stellar death.

Summary/Conclusion

The strongest evidence for the birth of stars from clouds of gas and dust is that we find, by various techniques, that young stars are always found inside or in the vicinity of such clouds. Furthermore, recent HST photographs have provided astronomers with striking evidence of the various processes that take place in the regions surrounding stellar birth. Theory and observation go hand and hand toward painting a picture of the entire process from cloud of gas and dust to brightly shining new star. In addition, it appears that dust clouds especially offer an environment in which simple chemical elements in space marry together to form more complex ones — perhaps even those that are the basic building blocks of life. So if extraterrestrial civilizations evolve within the immediate environments of stars, and stars routinely form throughout the widely-distributed nebulae, then there must be plenty of sites for us to explore while searching for Them.

LEARNING OBJECTIVES: *Now that you have studied this Chapter, you should be able to:*

1. Describe the observational evidence for the presence of gas and dust between the stars.
2. Describe the effects of interstellar dust and gas on starlight.
3. Describe the mechanism of gravitational collapse, and how it may be initiated.
4. Describe the conditions within star-forming regions of space, and offer evidence for the presence of birthing stars there.
5. Explain the importance of molecular clouds in terms of the location of star-forming regions.
6. Define and use in a complete sentence each of the following **NEW TERMS**:

Absorption nebulae 194
Bok globules 199
Brown dwarf 206
Dark nebula 195
Emission nebulae 194
Evaporating Gaseous Globules (EGGs) 204
Evolutionary track 202
Galactic clusters 191
Gliese 229B 207
HI Region 194
HII Region 194
Herbig-Haro object 200

Interstellar clouds 194
Interstellar medium 194
Interstellar reddening 193
Molecular cloud 195
Orion nebula 194
Photo-evaporation 204
Pleiades 191
PPL 15 206
Proplyds 204
Protostar 199
Reflection nebula 191
Teide 1 207
T Tauri variable star 203

Figure 9–28 Hubble's variable nebula (NGC 2261), as the name implies, shows evidence of variability in the amount of radiation emitted, suggesting that the cloud of gas and dust may be in the process of collapsing toward star formation. (Lick Observatory)

Chapter 10

The Death of Stars

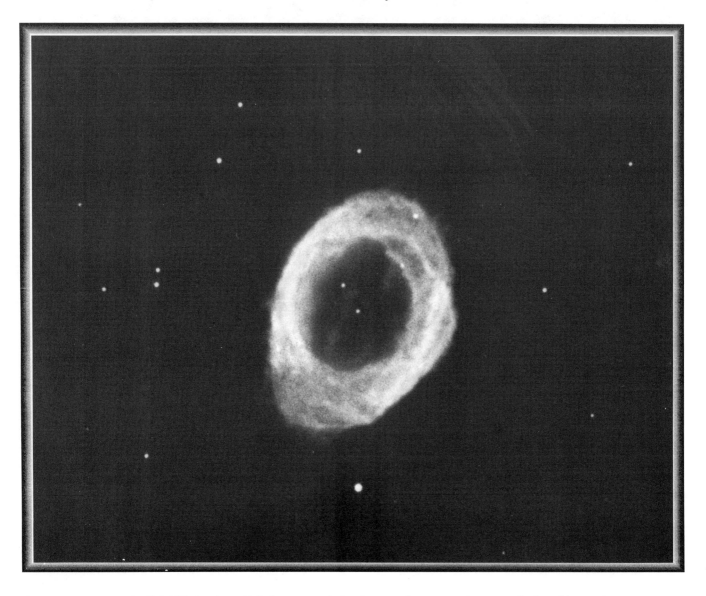

Figure 10–1 The "Ring" (M57) *is a beautiful planetary nebula, the stage between red giant and white dwarf. The white dwarf is seen at the very center, and is the remnant of the core of what used to be a normal shining star like Sun.* (Lick Observatory)

CENTRAL THEME: **What stages does a star go through as it dies, and how do astronomers detect the remains?**

Whereas the birth of stars is not very obvious, requiring circumstantial evidence available in the vicinity of clouds of gas and dust, the death of stars is directly observed — not very often, but nevertheless obvious to the astronomer. That is because a very small percentage of stars explode in an event called a **supernova** (plural of supernova is supernov**ae**). At the time of the explosion, there is such an immense amount of radiation emitted that the event is visible even if it occurs in a distant galaxy. The photographs in Figure 10-2 reveal how a typical supernova appears in another galaxy before and during the explosion.

In comparing the two photographs, you notice that the only difference between them is the existence of a bright dot near the arrow in the upper photograph. Within a matter of a few days, what must have been a star on its last leg exploded and emitted as much light as the combined light of all the remaining stars in the galaxy. This is how we directly detect the sudden death of a star. But you are probably wondering why it is that we don't observe such events closer to us — within our own galaxy. The answer is that although we keep our eyes open for them to occur, there has not been one since 1604, at which time one was reported and described by both Kepler and Galileo.

Actually, astronomers would like nothing better than to have the opportunity to study a supernova erupting close to us. If it happened to a star whose characteristics we had determined prior to the explosion, we would be better able to understand at what stage of a star's life it occurs — to what type(s) of star(s) it occurs — and perhaps the mechanism responsible for it. We are unable to do that with supernovae in distant galaxies — we are unable to resolve enough individual stars in distant galaxies to compile lists of stars along with their individual characteristics. We have done that for some stars in a few of the nearest galaxies, but the chance of any of those particular stars exploding is not very good. The stars in the distant galaxies appear to be mashed together, making individual study impossible.

On the other hand, we don't really want a nearby star to explode anyway — even if it were to provide a wonderful opportunity for astronomers. If it occurred too close to us, it might upset the delicate balance maintained in Sun and cause — oh, I don't even want to consider the possibility. In any case, supernovae are rare events, but they do remind us of the violence of which the universe is capable.

It is probably unfair of me to introduce you to the death of stars by discussing supernovae, because the great majority of stars do not die as supernovae — they die quite uneventfully (Figure 10-1). I simply wanted to catch your attention. We will return to supernovae in the next chapter, because the remnants of the explosions are so important to modern astronomers that the subject deserves a chapter of its own. So we will first follow the course of death of the average star — like Sun. It seems to me that this has some relevance to us. After all, we depend on the good health of Sun — even the slightest change in the energy output of Sun would have drastic consequences for life on Earth. Evolution toward death for stars like Sun is an extremely slow process, and is therefore not something the limited life span of humans allows to be easy observed. Astronomers detect only small, subtle changes in a star's behavior that they interpret as signs of the aging process. But there are additional tools available to us that allow for us to confirm that stars do indeed eventually die – *atomic physics* and the *placement of observed stars on the H–R diagram.*

And so that is how we will approach the subject of dying stars. Our knowledge of atomic physics tells us how matter and energy behave under given circumstances, and our study of actual stars in the sky allows us to look for those characteristics and behaviors that atomic physics describes or predicts. Here again is that powerful method of the scientific method — the marriage of theory and observation leading to knowledge.

In spite of the fact that we cannot observe the complete life span of a star from birth to death, there are innumerable examples in the sky of stars at the different stages along the way. The principles of atomic physics allows us to place them along a pattern. Of course you and I do this in our

Figure 10–2 Supernova explosions are rare in our own galaxy, but fortunately there are billions of galaxies to monitor for what is one of the most violent events in the universe. These before and during photographs of a supernova in galaxy NGC 7331 reveal the tremendous amount of energy emitted during the explosion. (Lick Observatory)

lives. I have never seen a single person go through all stages of life — from birth to child to teenager to young adult to senior citizen to death. I theorize the progression through those stages by observing different people at the various stages.

My approach, then, is to alternate discussion between the *sub-atomic events* happening inside of stars and the *observed characteristics of stars* in the sky. Since the latter are portrayed and neatly expressed on the H–R diagram, you will find it helpful to refer to it whenever it is mentioned in the text.

Interiors of Dying Stars

While alive and well on the main sequence, a star converts hydrogen into helium in its core — and gives off radiation in the process. For most of the star's life, hydrogen throughout the interior can migrate to the core to replenish the hydrogen that is consumed during the production of helium and energy by the proton-proton chain.

In other words, there is sufficient mixing of the plasma in a star's interior to enable the "fuel" to get to the central core where the reaction occurs (where the pressures and temperatures are great enough). Because the helium nucleus — two protons and two neutrons — is larger than the hydrogen nucleus (one proton), the helium plasma is not so easily mixed. It tends to collect around the core as a kind of "ash," thereby presenting an obstacle to hydrogen nuclei trying to get to the core.

So a star begins to die not when it runs out of "fuel," but when the remaining "fuel" can't get to the core where it can participate in the proton-proton chain to make helium. With less and less hydrogen available to "burn," the fusion reaction slows down and the helium core begins to collapse. During the normal life of the star, the outer layers of the core have been kept extended because escaping radiation "pushes" the plasma outward. But as the reaction slows down and the amount of emitted radiation declines, there is less pressure to keep the plasma away from the center. As the core collapses, it releases energy to the hydrogen plasma that surrounds it.

Collapsing Helium Core Anything that collapses releases energy. Water flowing over the face of a dam is, in a sense, an example of Earth collapsing. We tap that released energy and use it to run turbines which in turn run generators which in turn produce electricity. To make an object return to its original expanded size requires energy. Sunlight fall on the oceans, evaporates some of the water to form clouds,

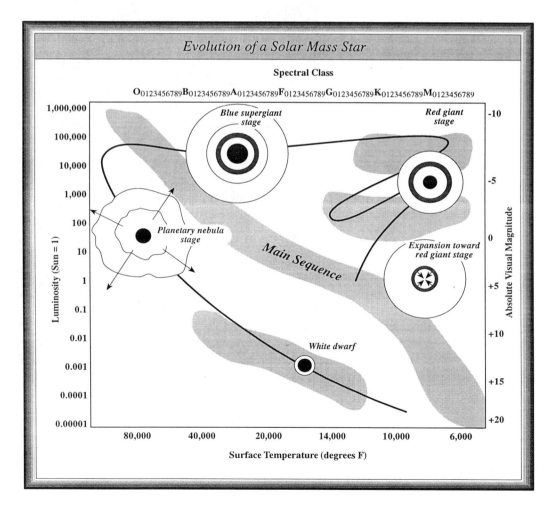

Figure 10–3 A timeline on the H–R diagram summaries what is believed to be the sequence of events that accompanies the death of a solar mass star. Stars less or more massive than Sun have slightly different timelines. You may want to refer back to this diagram as the specific steps are discussed in the Chapter.

which drop water in the form of rain or snow behind the dam, water flows over the face of the dam, ... etc.

The plasma surrounding the collapsing helium core eventually absorbs enough energy released by collapse to be at very high temperatures — high enough to allow the proton-proton chain to begin outside the collapsing core. The reaction fuses hydrogen into helium, just as it did in the core. This additional flood of energy pushes the outer layers of the star outward. The star expands, becomes cooler, and emits more energy. A star in this extended state is what we immediately recognize as a red giant. In Figure 10-3, those stars plotted in the upper-right corner (low temperature, high luminosity) have the characteristics of red giants predicted by atomic physics.

Also notice on the H-R diagram in addition to stars whose characteristics make them red giants, there are stars located between the main sequence and red giant stage. Well, of course, those must be examples of stars that are presently expanding as the core collapses. Observation matches theory. But why — you ask — do stars stop expanding when they arrive at red giant stage? Why doesn't it continue to expand forever? Recall from above that the mechanism responsible for the expansion is the collapsing core and the subsequent energy released to re-ignite the hydrogen. But the core cannot collapse forever.

Nature of Collapsed Core To understand why this is so, let us consider the atomic makeup of the collapsing core. Present in the core are helium nuclei (Figure 10-4) and electrons bound close together by the great pressures exerted inward by the star's gravity. Unlike gravity, which is always an attractive force, the electromagnetic force can be either attractive or repulsive — like charges repulse, unlike charges attract. So as the core collapses, and electrons are forced closer and closer together, the force of repulsion between electrons increases to the point that they can get no closer.

The point at which the repulsive force between electrons prevents further collapse is called the **Coulomb barrier**. According to calculations, the *Coulomb barrier* kicks in when an average star's core collapses to about the size of Earth. At that point, the helium nuclei and electrons are compressed so close together that the amount of matter contained within the volume of a golf ball weighs several tons! The matter in the core is extremely dense, primarily because electrons are not bound in orbit around atomic nuclei, and therefore the electrons and helium nuclei are packed very closely together.

We don't have matter of this type on Earth. We cannot make this type of material in our laboratories. The pressures needed to strip electrons out of their orbits and force them close to the nuclei around which they normally orbit are just not available here. You see, the matter with which you and I are familiar on Earth is just not typical in the universe. That is one reason why the subject of astronomy can be at the same time both extremely exciting and confusing. We are so used to Earth's environment and the forces operating here that it is initially shocking to our common sense to be asked to consider quite different environments and states of matter.

Highly Dense Matter Just consider for a moment the nature of matter found on Earth. A scale model of the typical atom (or molecule) is portrayed by a sesame seed — representing the nucleus of the atom — surrounded by dust specks — representing electrons — located 500 yards away from the seed. In other words, *matter* of which Earth (and its inhabitants) are made *consists mostly of space* — the space between the nuclei and electrons of atoms. In our model, there is a vast 500-yard emptiness between the seed and the dust particles. In the matter making up Earth, the emptiness is between the nuclei and orbiting electrons.

The key to understanding the characteristics and behavior of the corpses of dead stars is to understand what happens when this "emptiness" is invaded by neighboring particles of matter (Figure 10-4). You see, under normal circumstances, neighboring atoms can get only so close to one another before their orbiting electrons begin repulsing one another. The orbiting electrons are analogous to sentries on duty along the castle walls, keeping a look out for approaching enemy soldiers. If an attack takes place, the sentries attempt to repulse the soldiers and prevent them from invading the castle's interior to steal the princess (or prince) in the tower. The castle's interior is safe as long as the repulsive force of the sentries is greater than the attacking force of the enemy soldiers.

The outer layer of your hand consists of the outer electrons of the atoms making up "hand stuff." The outer layer of this page of paper you are reading is a like layer of electrons. As you try to "touch" the page with your finger, the repulsive force between the outer layer of electrons in your finger and the outer layer of electrons in the page prevent the two layers of electrons from actually "touching" one another. As long as you are limited to the mere strength of your arm muscles, you are unable to get any of the electrons to touch one another. In fact, you have never really "touched" anything in your life — in the sense that your outer electrons came into physical contact with the outer electrons in any other object!

But you are saying to yourself — "Why it certainly feels as if I am touching things all the time!" Because electrons are charged particles, they "sense" the presence of nearby electrons, and transfer that "sense" along a nerve path that ends in the brain. The brain, in turn, interprets the message as "touch." In terms of physics, we say that the electron, being a moving charged particle, carries around with it a magnetic field.

An electron's magnetic field is affected by ("senses") an invading electron in the same way that if you hold two magnets in your hands and attempt to bring together their like-charged poles, you will feel the repulsion between them. If you pressed hard enough on this page — in

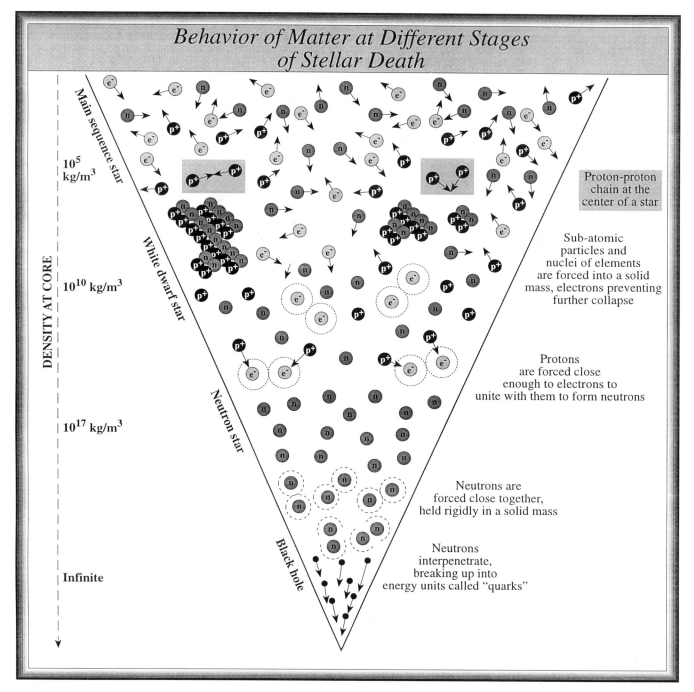

Figure 10–4 A good understanding of events preceding, during, and following the death of a star can best be understood by considering the behavior of the very atoms that make up the star. Very strange kinds of arrangements are possible for objects as big as stars. This diagram too will come in handy during the Chapter. How subatomic particles interact inside of a star ultimately determines the type of death it will have.

principle at least — you could overcome the repulsive forces between electrons, strip them out of their orbits around the nuclei, and force them closer together. Unfortunately, the amount of force necessary to do that is not easy to come by on Earth. But it happens in space all the time as stars die.

In terms of the model of the atom previously introduced, if the forces are great enough, the dust grains of neighboring atoms are forced to intermingle with one another, and even forced closer to the sesame seed itself. That entire region of "emptiness" can — in effect — be filled with dust grains from neighboring atoms. In fact, even neighboring sesame seeds can occupy that previously empty space. As more and more dust specks and sesame seeds are packed close together, the density (mass per unit volume) of the substance increases.

That is exactly why the density of the collapsed core of

a star is so great. The electrons and positively charged nuclei are forced by gravity to rub shoulders with one another — until the repulsive forces between electrons allow no further contraction. It is ultimately the repulsive force between electrons that controls the spacing between particles in the core.

Red Giant Stars So what happens to the star after the red giant stage? Although the core collapses to become extremely dense, the outer layers of the star, having been pushed outward to great distances, are quite thin — much thinner than the air you are presently breathing. So although red giant stars are immense in terms of size, they are not particularly substantial in their outer portions as far as densities go (Figure 10-5).

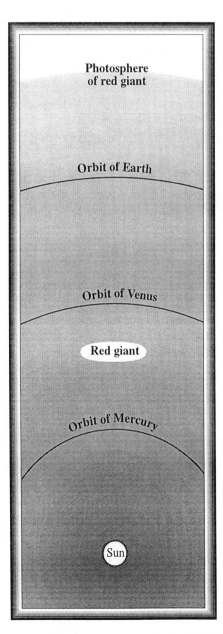

Figure 10–5 If Sun were to be replace by a red giant star such as Betelgeuse in the constellation of Orion, Earth would be inside the star. When a star expands to red giant stage, it is immense in size, but not substantial in density. Nevertheless, the increased luminosity would destroy life on any planets circling the star.

There are some fine examples of red giant stars to be easily seen with the naked eye. They are delectably red, although of course their huge sizes cannot be directly observed. *Antares*, in the constellation of Scorpius, is seen toward the south during the summer months. *Betelgeuse*, in the constellation of Orion, is seen toward the south during the winter months (Refer to the Star Charts in the Appendix).

To give you a feeling for what red giant stars are typically like, consider replacing our Sun with *Betelgeuse*. Its surface (outer layers) would extend out past the planet Mars! Earth would be inside *Betelgeuse*! *Antares* is similar to *Betelgeuse*. But although such giant stars are inconceivably dense in their cores, they are quite unsubstantial in their outer layers. Properly insulated from the heat, a properly-designed spaceship could fly right through the outer layers of a red giant star. Just stay away from the core. And it is because the outer layers of a red giant star consist of such thin gases that they do not stop expanding outward even when the core stops contracting. But let's get back to the core before we consider where those expanding layers go.

White Dwarf Stars The contracting core has, by now, heated to extremely high temperatures. That heat flows outward, gradually pushing (by radiation pressure) the outer layers of the red giant to the point at which they become gravitational separated from the core itself. The core is finally revealed for what it is — an extremely dense, white-hot star, about the size of Earth. This is the *white dwarf* star, mentioned earlier as one type of dead star (Figure 10-6). In essence, *white dwarf stars are nothing more than the collapsed cores of main sequence stars*, having formed inside what appeared to an outside observer as a red giant star. *Extremely dense objects* such as this are called **compact objects**. We encounter two other types of *compact objects* in the next Chapter.

You would think that extremely hot white dwarf stars would be easy to locate in the sky. But they are not. In fact, there is not a single one to be seen with the naked eye. And the first to be observed with a telescope was only found in the year 1862. It is a companion star to the star *Sirius*, the brightest star in the constellation of *Canis Major* as well as the brightest star in the sky. In other words, when you look at *Sirius* in the sky, you are looking at the combined light of two stars — a hot blue star (*Sirius* itself) and a white dwarf star. The star *Procyon*, the brightest star in Orion's other dog, the constellation of *Canis Minor*, also has a white dwarf companion.

An analysis of its spectrum and its behavior with respect to *Sirius* allows astronomers to create a profile of the companion of Sirius. It has a mass of about 5 percent greater than Sun, but its diameter is only about twice that of Earth. A simple calculation allows us to determine the density of the star — 1.6×10^5 times that of water. A teaspoon of the star's material weighs about 50 tons! That's dense!

★★★★★★★★★

Separation of Red Giant from White Dwarf The next questions are obvious — "What eventually happens to the outer layers of the red giant as they are pushed away from the white dwarf star?" And "What eventually happens to the white dwarf?" The most interesting way to get an answer to the last question is to ask an amateur astronomer to focus her/his telescope on a planetary nebula. These are beautiful examples of the expanding layers of gases surrounding white dwarf stars.

When they were observed telescopically by early astronomers, they appeared as fuzzy round objects — much as some of the planets in our solar system appear through small telescopes. So the name **planetary nebula** was assigned to them, even though they have nothing to do with planets. The easiest planetary nebula to locate in the sky, and certainly one of the finest objects to observe in the entire sky, is called the *Ring Nebula* in the constellation of *Lyra*, the Harp. (Figure 10-1).

The Ring — as it is affectionately known by astronomers — appears in the eyepiece of a telescope more like a smoke ring than a spherical bubble. That is because we look through the thin layers of the bubble when we look near the center (position B in Figure 10-7), and through the thicker layers away from the center (position A in Figure 10-7). The white dwarf star located at the very center of the bubble is quite dim (m_v = +10 or so), and can therefore be observed only with very large telescopes.

As you would guess, its total energy output is very low. The photograph in Figure 10-1 was taken with a long time exposure in order for the white dwarf star to be recorded on the film. Planetary nebula are quite commonly found. And in each case, a white dwarf star is found at the center —

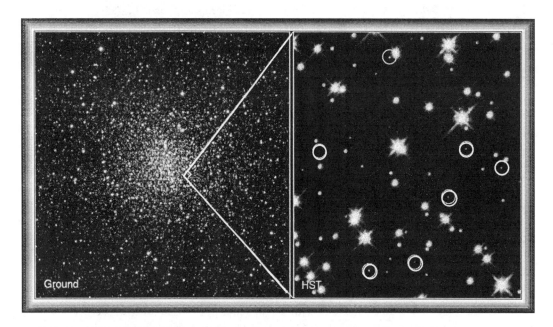

Figure 10–6 Since white dwarf stars are not very luminous, they are difficult to detect at great distances. But the capabilities of the HST enable astronomers to image several white dwarf stars in an old cluster of stars. Compare the ability of a ground-based telescope view of the cluster (left) with that of the HST of a small portion of the cluster (right). (NASA/STScI)

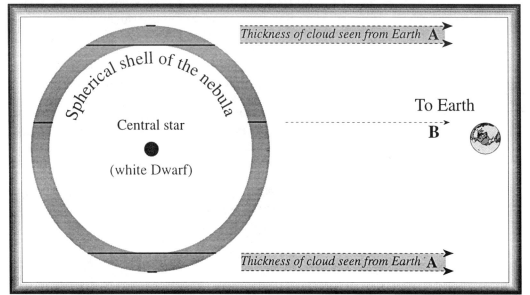

Figure 10–7 Although planetary nebulae like the Ring look like smoke rings, they are more like expanding soap bubbles. What we see are the thicker edges of the gas/dust cloud. The face-on portions are not dense enough to be visible.

verifying the theory that they are born inside of red giant stars. Red giants give birth to white dwarfs (Figure 10-7). Or perhaps it would be better to say that white dwarfs and red giants are respectively the collapsed center and expanded outer portions of a dying star.

Given additional time, the expanding shell continues to dissipate and to thin out — getting less and less dense, getting dimmer and dimmer, and gradually mixing with the general debris (mostly hydrogen) of space. As you will shortly learn, some of this ejected debris mixes with contracting proto-stars to contribute to the matter from which new stars form. Stellar death is not necessarily a completely destructive event, even if inhabited planets orbiting the star are severely charred as the star's outer layers expand past the planets. Some of the debris from the death of one star contributes to the conditions necessary for the evolution of life around newly-forming stars.

If this recycling process seems unlikely, you might be interested in knowing that astronomers have reason to believe our Sun is a third-generation star. That is, the material of which our entire solar system is presently made originated from debris ejected by two previous generations of stars and which in the process of expanding outward into space mixed with clouds of hydrogen gas left over from the Big Bang. We encounter further evidence of this birth-life-death-birth-life-death cycle as we proceed further into the subject of stellar evolution.

Black Dwarfs? But let's not leave the white dwarf star out in the cold. What happens to it? Since it is left behind in the cold — the general coldness of space — the white dwarf star can only gradually cool off. It is certainly very hot, but it has no further nuclear reactions upon which to call for additional energy. It gradually cools — according to theory — to become a **black dwarf**. A couple of clarifying comments should be made at this point. Space is mostly empty. The average distance between stars in a galaxy like ours is 5 light-years. Scaled down to human dimensions, stars are represented by golf balls scattered at 800-mile intervals!

Only in the immediate vicinity of a star is there enough radiation to cause high temperatures. At Pluto's distance from Sun, for example, the temperature is about minus 370° Fahrenheit! So any hot object like a white dwarf star located in the cold of interstellar space eventually cools off to the temperature of the surrounding medium. A cup of hot coffee, if left alone on the breakfast table, cools off to the temperature of the surrounding air in the room. The temperature of the environment that surrounds an object is called the **ambient temperature**.

Another point of clarification involves my use of the word "gradually" when I referred to white dwarf stars cooling off to the *ambient temperature* of space. The mathematical equations that express the principles of atomic physics tell us that because of the extreme densities of white dwarf stars, extremely great time intervals are required to cool off to black dwarf stage. Calculations tell us some 30 to 40 billion years are necessary. So are you surprised to learn that we have *never found a black dwarf star*? I hope not. Because according to our calculations, which we perform in Chapter 15, the universe itself is only 20 billion years old.

The universe hasn't been around long enough for a single white dwarf to have cooled off sufficiently to become a black dwarf. This observation (should I say lack of observation?) in itself *suggests a finite age for our universe*. In other words, if the universe has always existed — if the universe is infinitely old — and our theory of stellar evolution is correct, then we expect to find infinitely old stars. We expect there to be some black dwarf stars out there to observe. But we haven't found a single one, although we routinely explore the sky with our giant telescopes night after night. We aren't conducting searches specifically for black dwarf stars — we'd just expect to encounter them occasionally as we study the general contents of space.

I know — you are saying to yourself — "Well, of course we haven't! Since they are black, they can't be seen!" And of course you are correct. They can't be "seen." But being "seen," as we have so often found, is not the sole criteria for determining the presence or existence of something. So

Figure 10–8 The proper motion of Barnard's star is clearly revealed in these two photographs taken at a year's interval. Can you detect the amount of movement of the star in question? Barnard's star has the greatest proper motion of any star, mostly due to the fact that it is a nearby star. (Lick Observatory)

how do we detect something that doesn't emit radiation of any type? Answer: by its influence upon something else. We learned in Chapter 8 that a fairly good percentage (maybe 60%) of stars exist as members of gravitationally-bound pairs — binary stars.

Tracking Motions of Stars The stars we observe in the sky, although appearing stationary on the celestial sphere from night to night, move at different rates around the center of our galaxy — and therefore show evidence of changing position on the celestial sphere over long periods of time (Figure 10-8). *The change of position of an object with respect to the backdrop of distant stars is called the object's* **proper motion**. Consider how a lone star's *proper motion* should appear on successive photographs taken over a long time period (like several years). Plots of its position over time will appear in a straight line.

This is not a surprising discovery, since Newton's first law says that objects move in straight lines unless acted upon by an outside force. Assume now that we take successive photographs of a star whose binary partner is a black dwarf — and therefore invisible to us? A plot of its changing position relative to the background of more distant stars is no longer a straight line, but rather a wavy line. The waviness is due to the two objects alternately revolving from one side of the barycenter to the other. A binary star system detected in this manner is called an **astrometric binary system** (see Figure 10-9).

By accurately measuring the extent of the waviness of the plotted line, and by knowing the spectral class of the visible star (and therefore its mass), we can determine the mass of the companion dark object. The greater its mass, the greater is its effect on the visible star. So here is a method of detecting the existence of black dwarf stars.

In the process of investigating hundreds of astrometric binary systems, we haven't yet found one that meets the characteristics of a black dwarf. Nevertheless, "Absence of

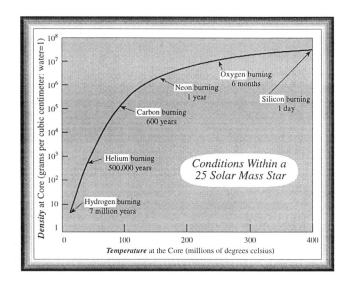

Figure 10-10 As the temperature in the core of a dying star increases, heavier elements form as lighter ones "burn" and fuse together. This process causes the density of the core to increase. This diagram is a summary of the major chemicals that form inside of dying stars, and the length of time of the "burning" that creates them.

evidence is not evidence of absence." So astronomers are not willing to use this as a strong argument for a finite age for the universe. It only suggests that conclusion as a temporary argument. In a later chapter we encounter much more convincing evidence that the universe is only 20 billion years old.

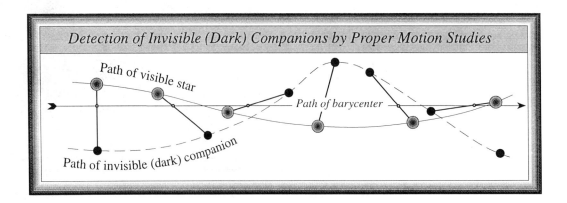

Figure 10-9 Two stars in orbit around a common center of gravity and moving through space together will reveal wavy motions against the backdrop of more distant stars. If a black dwarf star were in orbit around a normal star, we would see the wavy motion of the normal star and deduce the presence of the black dwarf star. This is called an astrometic binary system.

Red Giant to White Dwarf

I have occasionally maintained that astronomers theorize that stars are responsible for manufacturing — and distributing via planetary nebulae, at least — the chemical elements that are most abundant in and around us. And that those are the very chemical elements that participated in the origin and evolution of life on Earth.

This theory has the appearance of some type of "cosmic theology," so it would seem important that we not adopt it without considering the evidence to support it. Actually, the theory has many far-reaching consequences and implications besides that of the chemical environment of Earth and how life fits into that environment. So the following discussion is not a comprehensive coverage of *nucleosynthesis* — we will encounter it in future Chapters as well.

Making of Heavier Elements To understand how chemical elements are manufactured inside of stars, let us return to the core of a star as it expands to red giant stage. The helium core is contracting and releasing energy, which in turn ignites the surrounding layers of hydrogen, which in turn pushes the outer layers of the star outward into space. As the contracting helium core gets hotter and hotter, it eventually reaches the temperature at which helium nuclei can no longer resist one another, and they suddenly fuse together to form a heavier atom — carbon. In effect, what the star does is heat what was initially the "ash" to a high enough temperature so that it becomes another "fuel."

The reaction involves the combining together of three helium nuclei to form one carbon nucleus. The three sets of 2 protons and 2 neutrons in each helium nucleus combine together to form of carbon nucleus of 6 protons and 6 neutrons. But carbon is a very different substance than helium, with very different properties. Since the formation of the carbon took place at the level of nuclei as they are added (synthesized) together, we refer to the process as **nucleosynthesis** (Figure 10-10). The proton-proton chain itself involves nucleosynthesis, since hydrogen nuclei (protons) fuse together to form helium nuclei (two protons bound to two neutrons).

The reaction in which helium forms into carbon, like that of the proton-proton chain, generates enormous amounts of energy. But it requires a much higher temperature than that necessary for synthesizing hydrogen into helium. It is only when the core of a star collapses that the necessary temperature can be reached. Actually, the fusion of helium into carbon is quite a sudden event — it is more like a detonation. Astronomers refer to this stage in the death of a star as the **helium flash**.

Since the red giant star is in an extended state by the time the *helium flash* occurs, however, it does not blow apart. The energy suddenly released by the reaction is absorbed within the extended layers of the star. This increased energy must go somewhere, and it goes to making the star hotter. If we trace the star's behavior on the H-R diagram at this particular time of its life, we find it moving from right to left — from red giant star toward hotter temperature giant star. In other words, it increases in temperature while emitting about the same total amount of energy (luminosity remains the same). Notice how stars plotted in that region of the H-R diagram fall along the predicted path of an object that gets hotter and hotter while retaining the same luminosity.

Carbon Detonation As the helium in the core synthesizes into carbon, releasing energy in the process, the core expands. But once the *available* helium is used up, the newly-formed carbon core begins collapsing (remember the concept of hydrostatic equilibrium?). "Collapse" means to get hotter and hotter, and this increased temperature of the core does two things: it (1) ignites any helium surrounding the core, and it (2) causes the carbon to fuse with residual helium to form oxygen.

This stage of stellar death during which carbon and helium synthesize into oxygen is called **carbon detonation**. Again, as with the helium flash, energy is released. And so the pattern continues — the core builds up successive layers of the heavier and heavier atoms as higher and higher temperatures are reached (Figure 10-11). The star takes on the form of a layered onion — each layer consisting of the chemicals built up by successive reactions.

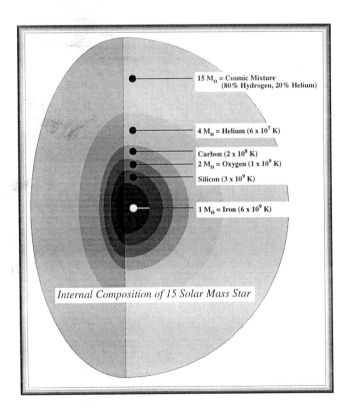

Figure 10–11 For a 15 solar mass star, the calculated layers of built–up chemical elements inside as it dies is shown in this cross–section. Notice that the heaviest of the chemicals—the iron—is located at the center of the star.

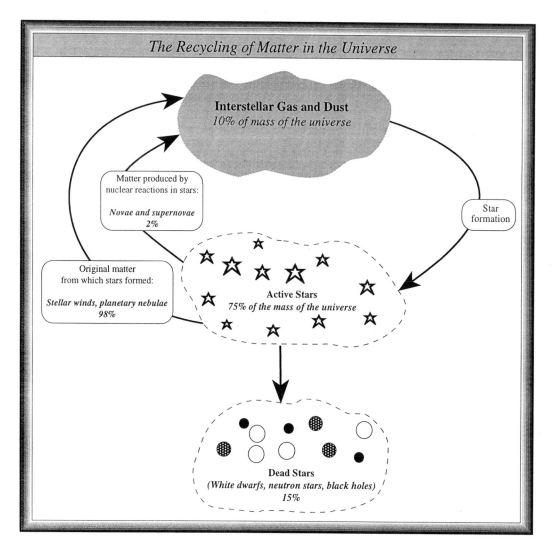

Figure 10–12 Dying stars eject a percentage of their expanding outer layers back out into space, where it can participate in the formation of a new generation of stars. Gradually, however, matter is removed from this recycling process by the formation of dead stellar remnants such as white dwarfs, neutron stars, and black holes.

If you are wondering why it is that higher and higher temperatures are needed to make heavier and heavier atoms, remember that like charges repulse one another. The greater the number of like charges, the greater the repulsion. Heavy nuclei require greater speeds (temperatures) than light nuclei in order to fuse together to release energy. And that is why — determined from spectral analysis of stars and galaxies — the chemical composition of the universe is heavily skewed toward lighter elements. Higher temperatures are less commonly available in the universe than lower temperatures. A list of the chemical elements in the universe by commonness shows how obvious that is (refer back to Figure 6-17).

So how far can a given star go in synthesizing successively heavier and heavier chemical elements in its core? The limit — not surprisingly — depends on the *mass* of the star. Less massive stars do not have the necessary amounts of matter to compress the cores to high enough temperatures to synthesize the heavier nuclei. Even a star only 40 percent more massive than our Sun cannot reach the temperatures necessary to synthesize elements heavier than iron from the lighter nuclei. Nature as provided a limit — a barrier — in the form of iron.

Nature has divided the 92 chemical elements into two groups — synthesizing elements 2 to 26 (helium to iron) from lighter elements *releases energy*, whereas synthesizing elements 27 to 92 (cobalt to uranium) *requires energy*. Within stars whose masses are less than 40 percent more massive than our Sun, the process of nucleosynthesis manufactures only those elements up to and including the element iron. You must wait until the next chapter to find out where and how the elements heavier than iron are manufactured and distributed throughout the universe.

Cosmic Recycling of Elements Keep in mind that the nuclei heavier than helium are synthesized only during star death — after a star leaves the main sequence. Main sequence stars convert hydrogen into helium. That is all our Sun is doing at the moment, and has been doing for the past 5 billion years or so. Our Sun will manufacture the heavier elements only after it swells to red giant stage in another 6 or 8 billion years. But if that is so, from whence came the calcium in our bones? The iron in our blood? The gold in our crowns? Not from Sun. Sun is only making

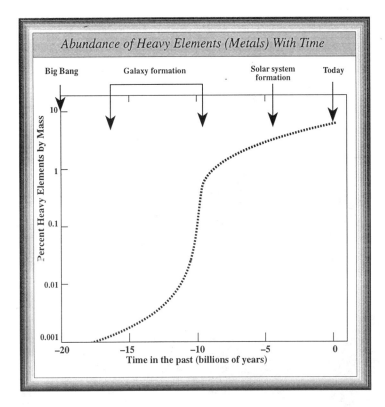

Figure 10–13 This diagram is a plot of what is believed to be the percentage of the heavy elements that have been collectively built up by dying stars over the history of the universe. Stars that are forming today, therefore, have a higher percentage of the heavier elements to begin with than stars that formed in the early universe.

Figure 10–14 This HST image shows a small portion of a nebula called the "Cygnus Loop." It is the expanding blastwave from a stellar explosion that occurred about 15,000 years ago. The wave is moving from left to right across the field of view, and has recently hit a cloud of denser than average interstellar gas. This collision drives shock waves into the cloud that heats interstellar gas, causing it to glow. Heavy elements created at the time of the explosion are detected in the spectra of the glowing cloud. (NASA/STScI)

helium at this stage in its life. Our Sun and system of planets and satellites are mere youngsters in the universe.

According to the scenario developed by astronomers, there was a 15-billion-year interval of time between the beginning of the universe and the collapse of the cloud of gas and dust that became our solar system. During that 15-billion-year stretch, stars formed, lived, and died. At death, as we have seen, stars went through planetary nebula stage and "puffed" material out into space. That material contained not only some of the helium formed during the main sequence lifetime of the star, but also some of the heavier elements formed in the interior after it became a red giant (Figure 10-12).

Spectral analysis confirms that to be the case. Spectra of planetary nebulae reveals the presence of a high percentage of the heavier nuclei. And so the debris from the death of the Milky Way galaxy's first generation of stars mixed with clouds of hydrogen left over from the Big Bang to form second generation stars, which in turn built up even greater abundances of heavier elements, and then ejected some of that material back into space to mix with other clouds of hydrogen to form third generation stars, and so on. Our Sun, it is believed, is such a third generation star.

If that scenario is correct, then every atom in your body that is not a hydrogen atom came from a dying star that went through a cycle of birth, life, and death long before Sun was a twinkle in the eye of a cloud of gas and dust (Figure 10-13). And it is entirely possible that the dying star, in addition to manufacturing and distributing the chemical elements of which we are made, was a supernova. It might well have generated the shock wave that caused the presolar nebula to begin to collapse in the first place (Figure 10-14).

Chemicals and Life Shall we reflect on this theory for a moment? To some, the recycling of debris from dying stars to the raw ingredients upon which life is based seems almost revolting. But reflecting on it for a moment allows you to see it as being very much like the cycle of life and death on Earth's surface anyway.

At birth, you were tiny. Now you are large. How did you get that way? You ate. Your body chemistry broke down the food you ate — the corn flakes, the bananas, the Big Mac's — into usable components.

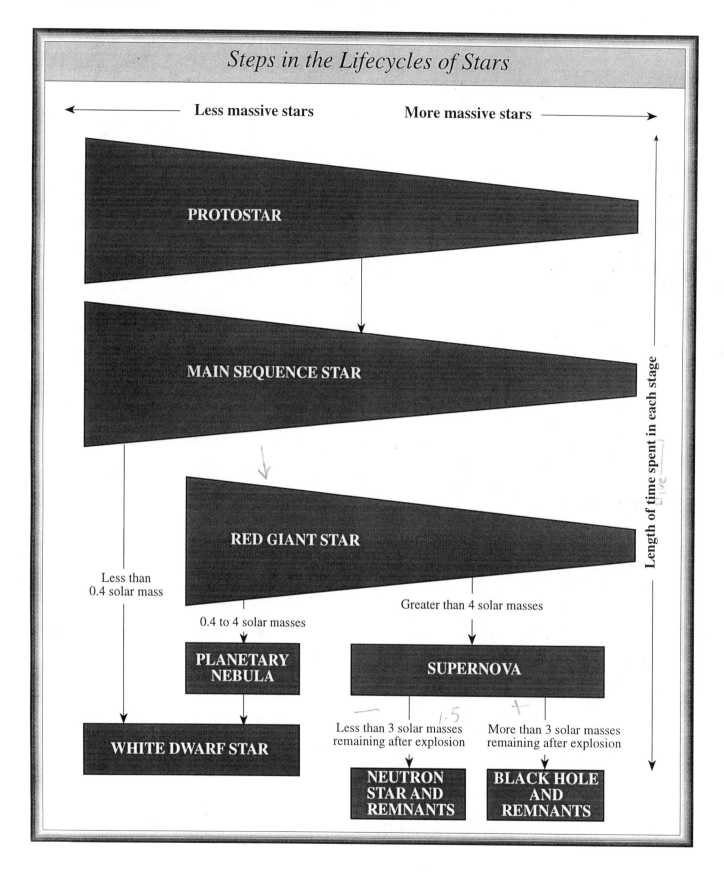

Figure 10–15 The type of death that a given star has depends entirely on its mass. This diagram summaries the stages that different mass stars go through from birth to death. The vertical axis illustrates the length of time that the given star spends in each stage. Massive stars, for example, spend relatively little time as a protostar, main sequence star, or red giant star.

From those components, your body made not only new cells to replace old and damaged ones — like those of a skinned knee — but those necessary to enlarge your body so that you could enter the competitive and reproductive stages necessary for the survival of the species.

And from whence came the food you ate? From Earth and Sun. It came not just from seed or egg, but also from nourishments provided by the environment — air, water, soil, decomposed organic material, and inorganic chemicals. We grow from Earth compost. Other things die that we might live — just as it is in outer space. Doesn't that make you feel a closer kinship with Earth and universe out there? Truly we come from dust and return to dust, but on the cosmic scale, it is star dust from which we come and to which we return.

Summary/Conclusion

So the fate of a normal star is to collapse to extreme density and then turn dark and cold, expelling some gas and dust in the process. But that can't be all there is to it, since I introduced this chapter by referring to stars that explode. Like people, stars have different kinds of deaths. The factor determining which type of death is in store for a particular star is mass — which determines the pressure, temperature, and rate of reactions at the core (Figure 10-15).

The quiet type of death just discussed is that predicted for the least massive stars — those up to 40% more massive than our Sun. That includes all stars from F- to M-type spectral classes. Most stars — perhaps 98 percent — fall within this range of types. So by far the greatest number of stars die quietly. Nevertheless, it is that small percentage of stars that do explode that create the conditions necessary for the eventual evolution of life in the universe — according to the current scientific theory of the origin of life. We see how that happens in the next Chapter.

LEARNING OBJECTIVES: *Now that you have studied this Chapter, you should be able to:*

1. Explain how the H-R diagram provides clues about the evolution of a star from life to death.
2. State the basic concept upon which the death of stars must occur.
3. Describe the theoretical behavior of matter within the three types of stars as they run out of fuel and approach death.
4. Describe the theoretical behavior of dying stars as seen from a distance, and how that behavior is reflected in the plot of stars on the H-R diagram.
5. Describe the physical changes that the sun is expected to go through (internal and external) from its present state to that of a black dwarf.
6. Describe a method of detecting an unseen object around a visible star through its gravitational influence on that star.
7. Define and use in a complete sentence each of the following **NEW TERMS**:

Ambient temperature 217
Astrometric binary system 218
Black dwarf 217
Carbon detonation 219
Compact object 215
Coulomb barrier 213

Helium flash 219
Nucleosynthesis 219
Planetary nebulae 216
Proper motion 218
Supernova 211

Chapter 11

Pulsars, Neutron Stars, and Black Holes

Figure 11–1 A supernova was detected in the galaxy NGC 4303 in 1961. By determining its apparent visual magnitude and the distance to the galaxy, astronomers were able to calculate the amount of energy released during the star's explosive death. (Lick Observatory)

CENTRAL THEME: What happens when stars explode, and what evidence do we have that neutron stars and black holes exist as remnants?

Although some stars recycle some of their chemical elements back into space in the process of evolving through red giant, planetary nebula, and white dwarf stages, they do not in themselves account for all of the chemical elements found on Earth. The synthesis of chemical elements heavier than that of iron requires temperatures and conditions that are not available inside red giants, planetary nebulae, and white dwarfs. We look elsewhere for a mechanism that allows for sufficiently high temperatures to cause heavy nuclei to stick together to form even heavier nuclei without falling apart again.

It might be easier to explain the difference between synthesizing lighter elements in less-massive stars and synthesizing heavier elements in the more massive stars if I relate it to the difference between atomic and nuclear weapons. Atomic bombs (and atomic power plants) operate by virtue of the energy released when heavy chemical elements (like uranium, number 92) are split apart (fission). It requires energy to form uranium from lighter elements, but that same energy can be released if it is broken into component elements again.

The hydrogen bomb, on the other hand, uses the fusion process similar to that occurring in stars to release energy as helium is formed from the synthesis of hydrogen. So a normal star, in the process of dying, builds up and ejects into space in the form of a planetary nebula only those elements of iron and lighter elements. That is why in Figure 6-16 we observe through spectral analysis an unusually large percentage of iron in the universe. There is a buildup of that particular element because it is the last one to be made with a release of energy.

So what is the origin of chemicals such as gold, silver, lead, and even uranium from which we make weapons? These are obviously heavier than iron. And where are the extremely high temperatures necessary for their manufacture available? Supernovae.

Supernovae

Supernovae are not common — fortunately for life in the universe. If the nearest star to our Sun, Alpha Centauri, were to explode as a supernova, the radiation emitted could well spell doom for us on Earth. At maximum, the absolute visual magnitude of a supernova can be -20. At a distance of 32.6 light-years such a supernova would approach the brightness of Sun. Imagine how bright it would appear from a distance of only 4.3 light-years, the distance to Alpha Centauri!

From the distance of even distant galaxies, supernovae are easily seen (Figure 11-1). In fact, at the time of such an outburst, the amount of radiation emitted by the exploding star may well exceed the entire energy output of the galaxy in which it is contained — a family of 100 billion stars! This often allows astronomers to calculate the distance to the galaxy, since the energy output of supernovae (absolute visual magnitude) are generally similar in amount.

To understand why stars explode in the first place, we must resort to atomic physics to understand behavior in the core of the star itself. In the previous Chapter I led you to the conclusion that the common not-so-massive stars build up shells of heavier elements within their cores, but cannot compress iron and lighter elements together to form anything heavier. It is at this point of a normal star's life that its core is a white dwarf star and the outer layers are being pushed outward into space as a planetary nebula.

So we might then ask ourselves — "What happens to the more massive stars (40% more massive than our Sun and greater) after they build up their cores to the point that they consist mostly of iron?" Unfortunately, we can provide answers based only on theory — not on observation. We have had only a single opportunity to study a particular star prior to its becoming a supernova, and even that did not provide us with a comprehensive understanding of why the star exploded in the first place. I will explain the discovery of this supernova shortly.

Not that there is any serious contender to the theory that supernovae are exploding stars, but simply to remind you that science proceeds on the basis of model-building and hypothesis-forming. Supernovae could just as well be the result of warring, super-advanced civilizations occasionally missing their targets and hitting instead their sun — upsetting hydrostatic equilibrium and causing the star to explode. I guess we could still call that hypothesis a variation of an exploding star theory, except that it is due to an unnatural process — intelligence, or lack thereof. Any-

Star	Date	Max Brightness (m_v)
Centaurus supernova	A.D. 185	-7?
Cassiopeia supernova	369	-2?
Lupus supernova	1006*	-5
Crab Nebula supernova	1054*	-2 to -6?
Tycho's supernova	1572*	-4?
Kepler's supernova	1604*	-2?
Cassiopeia A supernova	1680*	+2 to +5?
Eta Carinae nova	1843	-0.8
T Corona Borealis nova	1866	+1.9
GK Persei nova	1901	+0.2
DQ Herculis	1934	+1.3
Nova Cygni	1975	+1.9
o Ceti (Mira) variable star		+2.0 to +10
Large Magellanic Cloud	1987	+2.4

Figure 11–2 Stars that have suddenly burst forth in brightness are shown in this diagram. Only those of the years 1006, 1054, 1572, 1604, and 1680 are believed to have been the result of stars in our galaxy going supernova. Stars that flare up but do not explode (novae) are also shown, as well as one variable star that varies between visibility and invisibility.

Figure 11-3 *The Cygnus loop nebula (NGC 6992) is a concentration of gas and dust expelled by a supernova explosion millions of years ago. It is quite dispersed in space by now, but continues to emit the X-rays that identify it as a supernova remnant.* (Lick Observatory)

Figure 11-4 *The famous* Crab *nebula is the remains of a star that exploded in the year 1054 AD. In stark contrast to the Cygnus loop nebula in Figure 11-3, the gases have not yet dispersed very far out into interstellar space.* (Lick Observatory)

way, what is the evidence that connects supernovae with stellar death?

Observational Evidence of Supernovae Supernovae are observed in other galaxies. We use them as distance indicators. They are, fortunately for us, rare in any particular galaxy. Historical records suggest supernovae in our own galaxy in the years 1006, 1054, 1572, 1604, and 1680 (Figure 11-2). Prior to the 1960s, however, the only evidence we had of the remains of such explosions were clouds of gas and dust. In fact, there are a number of such **supernovae remnants** easily observed in the sky with a modest-sized telescope. The photographs in Figures 11-3 and 11-4 are supernova remnants. But other than their appearances — which suggest expansion outward from some central point of explosion — how can we be sure that these clouds are the leftover remains of exploding stars?

One method is to obtain spectra of different portions of the cloud, and examine them for Doppler-shifted spectral lines. If it is indeed the remains of an exploded star, we should see evidence of the cloud expanding outward from the point of explosion. Taking the spectrum of an object is usually one of the first tasks of an astronomer, so this has been done for all of these clouds. And our suspicions are confirmed — they are indeed expanding outward.

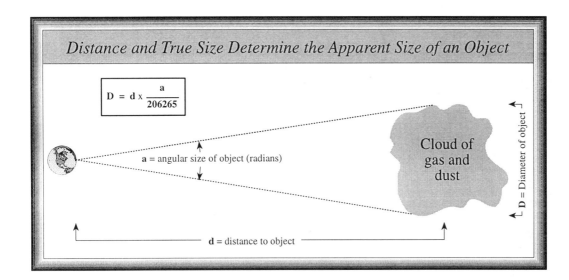

Figure 11-5 *The true diameter of an object like a nebula can be determined with the use of geometry once we know its distance and measure its angular diameter in the telescope. This diagram illustrates the method.*

In the case of the **Crab Nebula** (Figure 11-4), the rate of expansion is calculated to be about 1,000 miles per second! This knowledge, in turn, allows us to calculate how long ago the star exploded so as to produce the nebula of gas and dust at its present size. The procedure is rather straightforward — if we know the size of the nebula from the center (at which place the explosion presumedly occurred) out to the edge, and we know how fast that edge is moving outward, simple mathematics tells us the length of time that edge of material took to get to its present position. The size of the cloud, which is what we want to know, cannot be determined without first knowing its distance from us. In other words, the apparent size of something in the sky tells us nothing about the size of the thing itself — unless we know how far away it is. The angular diameter (a) of an object in space, its linear diameter (D), and its distance (d) from us are related to one another according to the equation and geometry shown in Figure 11-5.

Distance to the Crab Nebula The angular diameter of the Crab nebula is easily measured by physically measuring it on a photograph taken at a known magnification. Calculating the distance to the cloud, however, is a little more demanding. The subject of measuring distances arises in the next chapter in a very important way, so for the moment assume we know the distance to the Crab nebula. Knowing the angular diameter and distance allows us to calculate its true diameter (Figure 11-5).

Dividing the diameter of the nebula by its rate of expansion, we obtain an age of about 1,000 years for the Crab Nebula. Now that is rather interesting, you see, because modern astronomer had uncovered an ancient Chinese manuscript showing a map of that region of the nighttime sky known in western civilization as the constellation of Taurus, the Bull. There on the map, at the very location of the Crab Nebula, was a reference to a "guest star" that appeared in the year 1054 AD. The Chinese used the term "guest star" to refer to appearances of stars that suddenly appeared in the sky where no star previously had been seen.

This particular one was only one of many that the Chinese had recorded even before the birth of Christ. What made it unusual, however, was the fact that it was visible to the naked eye *during the daytime* for 23 days. It was visible at night for nearly 2 years! The energy emitted at the moment of the explosion must have been tremendous! So the connection between the Chinese report and the presence of the Crab nebula is obvious — isn't it? The age of the nebula — as determined from its rate of expansion — is very close to the time since the Chinese reported the appearance of the guest star in 1054 AD — and in the same position in the sky. It seems pretty convincing.

The story isn't over, however. In 1967, the astronomical community was astonished by the discoveries of a young graduate student at Cambridge University in England, Jocelyn Bell. She detected several pulsating radio sources in the sky that were definitely coming from deep space. Further observations and research have led to the conclusion that these are the remains — the corpses — of supernovae. They are called *neutron stars*. Their low-light emissions prevent astronomers from seeing them directly. Their unusual characteristics — as calculated by what radiation we can measure — challenge the best of astronomers. So evidence that they do indeed exist in the form that astronomers say they do has come about rather like a good mystery novel.

Theoretical Basis for Neutron Stars

The early decades of this century saw the discovery of the basic constituents of all matter — electrons, protons, and neutrons. Prior to that time, it was not known that the diversity of substances on Earth and in space are related at the deepest level. That is, at the level of the atom, all substances are made of those three building blocks.

★★★★★★★★★★

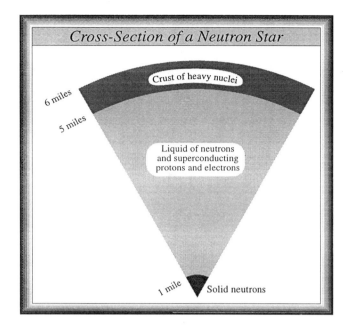

Figure 11–6 The theoretical cross–section of a neutron star. It contains the mass of the star from which it originally collapsed, and yet is only 6 or so miles in diameter. This makes for highly-dense matter.

Like children playing with new toys, astronomers began experimenting with possible combinations those building blocks can make. To their astonishment, they calculated that if the *Coulomb barrier* could be breached, a new form of matter would appear. In order for this to happen, however, there must be great force exerted on those particles — enough force to overcome the repulsive force exerted by the like-charge electrons, which are trying desperately to keep one another at bay. Only inside of stars, so it seemed, were there sufficient pressures for this condition to occur.

Calculations further suggested that the minimum mass of a star needed for the gravitational pressure to be great enough for this event to happen is *1.4 solar masses*. In other words, any star whose mass is greater than 1.4 solar masses cannot die as a white dwarf — the weight of the outer layers of such a massive star prevents electrons from keeping a safe distance. Instead, they are forced close enough to protons that are interspersed with the electrons that fusion of the two particles takes place.

You recall from the previous chapter that the reason a normal star stops at the red dwarf stage in the act of dying is that the Coulomb barrier resists further contraction. The red dwarf was collapsing because no new reactions were occurring in the core, and therefore no radiation was being produced to prevent the outer layers from falling inward. Eventually the red dwarf reaches a smallest size as a white dwarf.

At that point, it can do only one thing — cool off to the temperature of surrounding space. So the 1.4-solar-mass limit essentially separates all stars into two different groups of stars, a profoundly significant discovery about the behavior of the universe. It should not surprise you to learn, therefore, that the person who made the discovery not only received the Nobel prize in Physics, but was honored by having the limit named after him — the **Chandrasekhar limit**.

Neutron Stars The fusion of electrons and protons together in these massive stars creates a particle that already exists in abundance in the universe — neutrons. The particle was known to exist prior to the mathematical discovery, but it was not known that it could form by combining an electron and a proton. It certainly seems logical — the electron is negative and the proton is positive.

Adding a plus and a minus together yields a zero, or neutral charge — hence a neutron. What the calculations suggested was that a white dwarf star compressed past a certain critical point turns into a star made of neutrons. It would be smaller than the white dwarf, since neutrons occupy less volume than electrons and protons together (refer back to Figure 10-4).

The neutron star should measure about 6 miles in diameter or so, and yet contain the same amount of mass

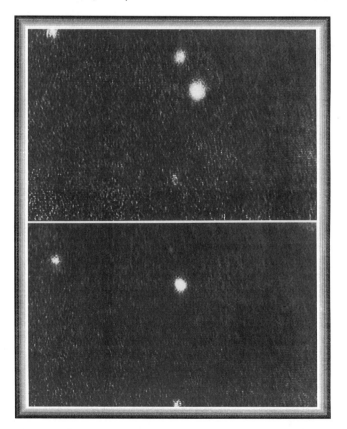

Figure 11-7 The discovery of a pulsar at the center of the Crab nebula was one of the most exciting discoveries in modern astronomy. With it, astronomers were able to match theory to observation. These two photographs were taken when the beam of radiation was lined up with Earth (upper) and again when it wasn't (lower). (Lick Observatory)

it contained as a white dwarf (Figure 11-6). The density, therefore, would be quite great — approximately 10 billion tons per cubic inch! This was — I emphasize — only a theoretical possibility based upon calculations using the known laws of atomic physics. There was no evidence that such objects existed anywhere, or even that it was possible for them to exist, in a practical sense. Just because the laws of physics (or chemistry, or biology) allow for something to happen doesn't mean that it will. So from the 1930s to the 1960s the subject of **neutron stars** was forgotten — put on the back burner to await new discoveries.

Discovery of Pulsars Radio astronomy was born out of the development of radar and radio communications during World War II. By the 1960s, the techniques had been refined to the point at which very faint radio emissions from stars could be detected and precisely plotted against time — much like using a photometer to plot light variations in an eclipsing binary system. In 1967, Jocelyn Bell, using a radio telescope, detected a source of radio energy that pulsated off and on every 1.337301173 seconds. No object could be seen with optical telescopes at the point from which the pulses of radio energy were coming, so at first it was impossible to explain the discovery. Because the radio energy arrived in pulses, it was nicknamed a **pulsar** (Figure 11-7).

There were attempts to explain the signals — some of which were quite bizarre. Because many scientists think

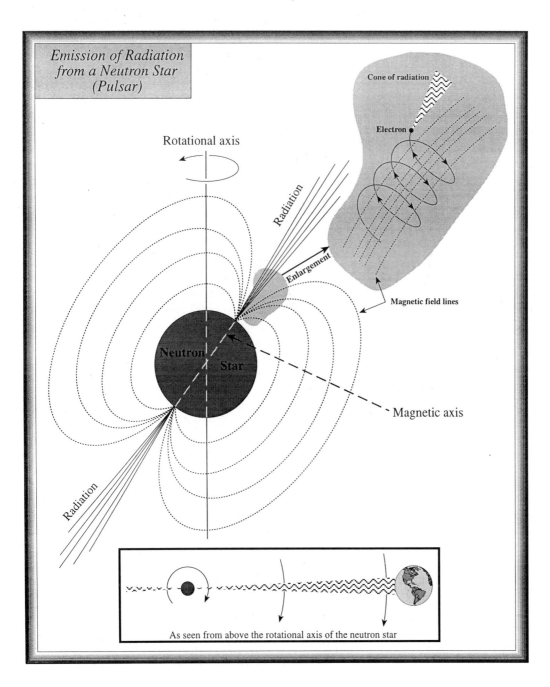

Figure 11-8 The model of a pulsar's geometry, illustrating how the spinning neutron star can be aligned such that the beam of radiation sweeps past Earth once with each rotation of the neutron star.

★★★★★★★★★

that intelligent civilizations in the universe might communicate with one another by using coded messages such as the Morse code we use, Ms Bell and her colleagues looked very closely at the pattern of pulses to see if there was any semblance of intelligent content. In other words, they looked for patterns in the pulses that would indicate an intelligent origin rather than a natural origin. If, for example, the pattern of pulses went something like: - -- --- ---- ----- etc., it might suggest an intelligent origin. If, on the other hand, the pattern of pulses were like this: -- -- -- -- -- etc., it could be the result of some natural process. We will discuss this idea in greater detail in chapter 20 when the subject of communication with distant civilizations arises.

While the pulses were being examined for intelligent content, the pulsar was also called an *LGM*, short for **L**ittle **G**reen **M**en. It would probably be more exciting to you if I told you that the pulses were distress signals from a dying race of intelligent beings — a cosmic 911 call. But in a matter of weeks, additional pulsars were detected in other regions of the sky, and the suspicion that they were of intelligent origin gradually disappeared. So what are they, then?

As I develop the scientific model to explain pulsars, try to visualize a lighthouse at the seashore on a foggy night. Each time the light points in your direction, you see a pulse of light. Passengers in an airplane flying overhead will not see the pulses, since they are not along the line of sight of the beam of light. This is the easiest way of visualizing pulsars. But just try to imagine the object believed to be responsible for the pulses received by Jocelyn Bell — an object 10 miles in diameter, density of 10 billion tons per cubic inch, rotating once every 1.337301173 seconds! Objects with those characteristics had been predicted long before the discovery of pulsars, but it was not known they would reveal themselves in that fashion. But how do the neutron stars attain those almost unbelievable properties in the first place?

Increasing Rotation Rate of a Collapsing Object A principle of physics to which I make frequent reference throughout the rest of the Book is that of the principle of the *conservation of angular momentum*. We first encountered this principle of physics while dealing with the subject of Earth ocean tides, and how they result in the slowing down of Earth's rotation and the subsequent recession of Moon away from Earth. You are aware of its effect on human activity as well. Visualize an ice skater spinning on a skate. As he/she draws his/her arms inward — closer to his/her body — his/her body spins faster and faster.

As the mass of a rotating object moves closer and closer to the center of rotation, the object rotates faster and faster. Theoretically, at least, as you lift your arms from your sides and place them over your head, Earth's rotation decreases very slightly — and speeds up again as you lower them. The principle is also being demonstrated while you watch divers spinning at different rates as they go off the boards at swimming/diving meets. Or while you watch water spin faster and faster as it goes down the bathtub drain.

So if a star's core is spinning slowly as it begins to collapse toward white dwarf stage, it will certainly be spinning more rapidly as it arrives at that collapsed stage — when it is about the size of Earth. And if the white dwarf continues to collapse even further (a subject about to be discussed), it will spin even faster. Now a rapidly spinning object like a white dwarf generates a powerful magnetic field — much as we encountered in the case of Sun, except much more powerful. Certainly white dwarfs are not the same as Sun in terms of their size and chemical composition, but they do have the common property of being made of plasma — a fluid of charged particles

Magnetic Field of a Neutron Star Yes, the plasma in a white dwarf is to a great extent in a solid mass, but even so there will be a flow of charged particles within the star. And a flow of charged particles is what is responsible for the generation of magnetic fields (the dynamo effect). And just as Earth's magnetic axis does not line up with its rotational axis, there is very likely an angle of tilt between a collapsed star's magnetic axis and its rotational axis (Figure 11-8). The magnetic axis wobbles as the star spins around the rotational axis. The radiation emitted along the magnetic axis therefore goes out into space in a rather complicated pattern. But what kind of radiation do we expect to receive from such a bizarre object as this?

Here's an entirely new concept for you. It happens that neutrons are stable as long as they exist within the nucleus of an atom or under the extreme pressures within a neutron star. A free neutron survives for a mere 10 minutes before it spontaneously breaks into its component parts — an electron and a proton. This process of decay probably occurs on the surface of a neutron star where the pressures are not great enough to keep the neutron intact.

Theory tells us that these charged particles — protons and electrons — are hurled outward from rapidly spinning, high-density objects like neutron stars. It is much like marshmallows placed on the outer edge of a rotating phonograph record. Except that particles on the neutron star's surface will move outward from the ends of the magnetic axis and along the magnetic lines of force — not from everywhere on the surface. Being charged particles, protons and electrons move along magnetic field lines, just as iron filings lined up along magnetic lines in our magnet experiment of Figure 7-8.

Synchrotron Radiation Emission As electrons and protons hurl outward along the magnetic lines, they emit continuous streams of photons of many different wavelengths, including those in the visible portion of the spectrum. We are quite familiar with this particular process by which radiation is emitted by high-speed charged particles caught in magnetic fields. There is a special name for such radiation — **synchrotron**.

Optical Pulsars

Location	Identification	Period (Sec)	Age	In Supernova Remnant?
Crab	PSR 0531 + 21	0.033	900 years	Yes
Vela	PSR 0833 - 45	0.089	11,000 years	Yes
LMC	PSR 0540 - 69.3	0.050	Unknown	Yes
Vulpecula	PSR 1937 + 21	0.0016	Old	No

Figure 11–9 A list of some well–known pulsars and their pulsation rates. Notice that there is a relationship between the age of a pulsar and its pulsation rate. This suggests that neutron stars gradually slow down over time.

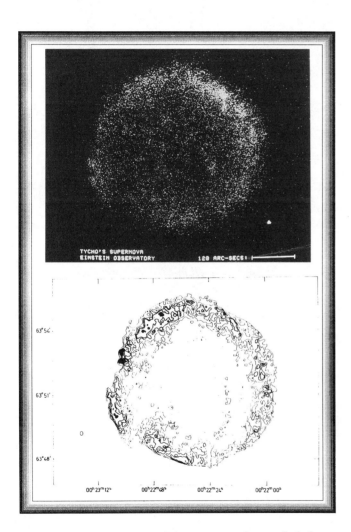

Figure 11–10 Two views of the remnant of an exploded star known as "Tycho's Supernova." The upper view is an X-ray image made by the Einstein satellite. The lower view is a map of radio emission from the same object. Both views show the expanding "bubble" of hot, gaseous material ejected into space by the stellar explosion. (Smithsonian Astrophysical Observatory)

Scientists learned about synchrotron radiation in the process of building and operating *synchrotron accelerators*. These are huge devices that allow scientists to study the nature and composition of atoms by accelerating electrons to high speeds inside a huge magnet, and then crashing them into a target of atoms that split them apart into component parts.

So streaming outward from the ends of the magnetic axis of a spinning neutron star are two beams of radiation — going outward in opposite directions. If the orientation of the star's magnetic axis is such that Earth falls along the extension of this axis in one direction or another, we on Earth observe a pulse with each rotation of the neutron star. That is, the interval between pulses is simply the time required for the neutron star to rotate once. This is called the **lighthouse effect**.

But only if we are so correctly aligned will we be able to observe the pulses of energy. So a pulsar is the manifestation of the presence of a neutron star. We don't see neutron stars directly. We see the pulses of synchrotron radiation that (we presume) are emitted by neutron stars. The probability of our being correctly aligned so that we receive the pulses from a given neutron star are, of course, very small. So the fact that we have discovered over 300 already suggest there must be quite a large number of them in our galaxy.

Are you convinced that neutron stars exist? It is tempting, but a single piece of observational evidence — pulsars — to support a theoretical model is far from convincing to the scientific community. Especially when it is being asked to believe in these incredibly dense collapsed stars whose properties defy our commonly held beliefs about the universe. To believe in extraordinary events requires that extraordinary evidence be presented. Well, there is additional evidence available making the existence of neutron stars almost a certainty. It is the observational evidence connecting exploding stars to the pulsars.

★★★★★★★★★

Supernova/Pulsar Connection After discovering pulsars in various regions of the sky, and believing their behavior indicated the existence of neutron stars predicted back in the 1930s, astronomers began observing suspected remnants of supernovae. The Crab nebula was one of the first to be investigated, and sure enough, a pulsar was found at almost the precise center! The connection between supernovae, neutron stars, pulsars, and supernovae remnants now became quite evident. The connection between supernovae and their remnants is not strong for all observed supernovae, but that is to be expected. For most pulsars, there has been so much elapsed time since the explosions that created them, the remnants have spread out and dissipated into space to the point that they are no longer visible.

Some 100 supernova remnants have been found in our galaxy, but they are extremely difficult to detect optically since the gases are so thin and diffuse. It turns out, however, that they are very strong sources of X-rays. These X-rays are no doubt generated by the electrons ejected outward by the original explosion and which are still spiraling around the intense magnetic fields within the gaseous cloud.

You learned in Chapter 5 that Earth's atmosphere absorbs X-ray radiation quite efficiently. Consequently, to study X-ray sources such as supernova remnants requires the use of X-ray detectors lifted above the atmosphere. Three X-ray satellites were launched in the late 1970s specifically to study objects such as the Crab nebula and other supernova remnants. The upper photograph in Figure 11-10 was taken by one of those satellites — the **Einstein X-Ray Observatory**. Launched in 1978, it detected thousands of X-ray sources in the sky, many if which are believed to be supernovae remnants. The lower photograph in the same figure is of the same object, but is a plot of the radio energy coming from the supernova remnant.

Spinning Rates of Pulsars Not all supernovae remnants reveal the presence of pulsars at their centers. But that is also expected, since Earth must line up along the magnetic axis of the neutron star in order to allow us to detect the pulses of radiation. Fortunately for astronomers, the neutron star within the Crab Nebula is correctly aligned, making it one of the most thoroughly studied objects in all the sky. If you find yourself at an evening observing session with amateur astronomers, you will undoubtedly hear one of them make reference to "The Crab".

The pulses from The Crab are extremely precise in their timing. We have found, in fact, that the interval between pulses is accurate to nine decimals — 1.337301173 seconds to be exact. That surpasses even a Rolex watch! It is even used for calibrating the accuracy of atomic clocks (see Figure 11-8). But the intervals between pulses has been found to be lengthening by about 40 billionths of a second per day. This fact suggests that the neutron star's rotation is slowing down.

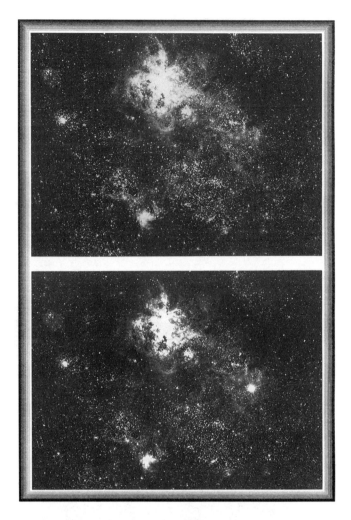

Figure 11–11 Before and after photographs of Supernova 1987A in the Large Magellanic Cloud were made at the Cerro Tololo Inter–American Observatory in Chile. It was not visible from most of the Northern Hemisphere. (National Optical Astronomy Observatories)

Evolution of Pulsars This discovery is actually not too surprising. The energy emitted by the neutron star in the form of radiation, visible, X-ray, as well as other types, must be at the expense of energy of rotation. Radiation is, of course, energy. And energy and mass are interchangeable. So loss of energy in the form of radiation is equivalent to a loss of mass. Imagine the situation of the spinning ice skater again. Can you visualize what happens if he/she loses a heavy belt buckle while spinning — the buckle flying off the skater and into the spectator stand? Do you visualize the skater slowing down? So astronomers theorize that the shorter the interval between the pulses of a pulsar, the younger that pulsar must be.

A change in the pulsation rate of a pulsar, called a **glitch**, has been observed in over 90 pulsars. But only 4 of the 553 known pulsars reveal themselves optically — the others are detected by radio and X-ray telescopes (Figure 11-9). So perhaps visible pulsars represent only a tempo-

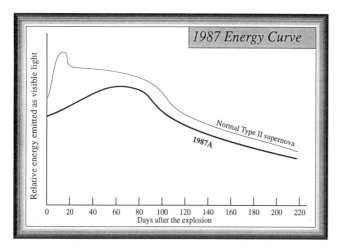

Figure 11–12 A light curve plotted for Supernova 1987A is compared to a curve for a "normal" supernova (those studied in other galaxies). What is apparent is that 1987A had some peculiarities of its own.

rary stage in the lifetime of a neutron star. Only while in youth (as the Crab pulsar) do neutron stars emit pulses at visible wavelengths. As they grow older and slow down, they emit energy at decreasing wavelengths.

So it was astounding in 1982 when astronomers discovered a pulsar spinning 20 times faster – 642 times per second – than the Crab pulsar. Even for a neutron star, this rate of speed is close to being enough to cause it to fly apart. Many of these "millisecond" pulsars have been discovered. Radio signals from the millisecond pulsar PSR 1957 + 20, the second fastest known, disappear every 9 hours for a period of 50 minutes.

Astronomers interpret this behavior as being caused by a companion star eclipsing the pulsar, or else material is being pulled off the companion onto the pulsar. Calculations suggest that the companion star is being evaporated by the pulsar, and will disappear in a billion years or so. Hence the pulsar is referred to as the "Black Widow pulsar" because of its similarity to the female Black Widow spider that eats its mate.

There is much yet to be learned about neutron stars and their behavior as pulsars, and especially about the manner in which they are created within the supernovae themselves. The connection between supernovae and neutron stars appears obvious, but just how neutron stars form is far from clear.

Detection of Supernovae

A supernova should not be thought of as an explosion in the normal sense of the word. It suddenly appears in the sky because its brightness just happens to fall within the limits of the instrument used to observe it. In the case of the naked eye, that limit is +6 on the apparent visual magnitude (m_v) scale. The supernova builds up to a maximum brightness and then slowly decays. Fortunately, many supernovae have been detected prior to reaching maximum brightness, allowing astronomers to obtain light curves for them. Recall that we used light curves for detecting eclipsing binary stars. Astronomers have found that light curves of supernovae reveal a definite pattern. This suggests that whatever mechanism causes stars to explode is common to all.

Nearby Supernova SN 1987A On February 24, 1987, astronomers were treated to the brightest supernova to occur in 383 years. Since it was the first one detected during the year 1987, it is designated **SN 1987A**. It occurred not

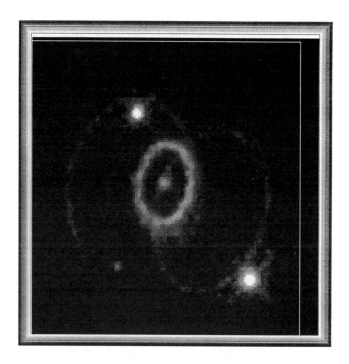

Figure 11–13 This striking HST picture shows three rings of glowing gas encircling the site of supernova 1987A, a star which exploded in February 1987. Though all of the rings appear inclined to our view (so that they appear to intersect) they are probably in three different planes. The small bright ring lies in a plane containing the supernova, the two larger rings lie in front and behind it. The rings are a surprise because astronomers expected to see, instead, an hourglass–shaped bubble of gas being blown into space by the supernova's progenitor star (based on previous HST observations, and images at lower resolution taken at ground-based observatories). One possibility is that the two rings might be "painted" on the invisible hourglass by a high–energy beam of radiation that is sweeping across the gas, like a searchlight sweeping across clouds. The source of the radiation might be a previously unknown stellar remnant that is a binary companion to the star that exploded in 1987. (Dr. Christopher Burrows, ESA/STScI and NASA)

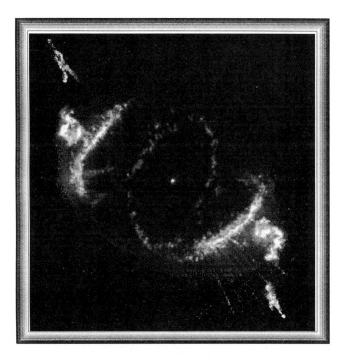

Figure 11-14 This Hubble Space Telescope *image is of one of the most bizarre objects studied by astronomers. The nebula (NGC 6543) is the result of a star having expelled much of itself during an outburst.* (NASA/STScI)

in our galaxy, but in the **L**arge **M**agellanic **C**loud (**LMC**, for short) — one of the nearest galaxies to the Milky Way. Because the star is positioned only about 20 degrees from the South Celestial Pole, it could not be observed and studied from major observatories in the northern hemisphere. But astronomers located in the southern hemisphere had a field day.

It was first seen by Ian Shelton working at the Las Campanas observatory located high in the Andes mountains in Chile. At maximum brightness, it reached an m_v of +2.4 (Figure 11-11). Since the LMC is known to be at a distance of 160,000 light-years from Earth, the M_v could be calculated — it was -14 at brightest. As weeks went by, it became obvious that there was something odd about this supernovae. The light curve, rather than rising immediately to peak brightness, showed a leveling off when its m_v got to about magnitude +4.5. Then a few weeks later increased to the maximum of +2.4 (Figure 11-12). This puzzling delay in its rise to fame was understood when the cause of the collapse of the star was discovered.

Astronomers routinely take photographs of different sections of the sky in hopes of finding something new, and then place them in photographic libraries that are accessible to other astronomers. After SN 1987A was discovered, prints of that region of the sky were studied carefully, and a star was found at the exact location of the supernova. Cataloged as *Sanduleak -69°202*, it turned out to be a very hot, blue supergiant star. Now this is strange — our model for supernovae is that red supergiants collapse and rebound as the explosion, not blue supergiants.

So astronomers have had to revise their model. It is theorized now that Sanduleak -69°202 was a normal red giant whose strong solar winds had blown away its outer cooler layers to expose the deeper hot and blue layers visible in the earlier photographs. So the delay in the light curve reaching a peak is thus explained by this less-than-normal red giant star collapsing and exploding — much of the energy was used to blow the smaller and denser star apart, and less was emitted as light.

That it was a supernova is not in doubt. Five pulses of neutrinos over a 12-second interval were detected just 18 hours before SN 1987A was first observed — evidence of the supernova's violent death pangs. Theory tells us that a certain pattern of neutrinos should accompany the collapse of the core of a star just prior to its supernova outburst. But we had never been in a position to confirm that prior to SN 1987A, because supernovae have always occurred too far away — off in another galaxy. The pattern of neutrinos was found in two detectors — one in Ohio and one in Japan — after scientists heard of the supernova outburst. And that pattern confirmed our theory and also told us that a neutron star should have formed in the process. At the moment, however, the debris from the explosion is blocking our view of whatever lies within.

The HST has detected a ring of ionized gas surrounding the spot where SN 1987A occurred (Figure 11-13). Ejected at the time of the explosion, it is being heated up and ionized by ultraviolet radiation emitted by events associated with the explosion. Eventually, this gas ring will be destroyed as the blast wave that is expanding outward from the point of explosion catches up with it. The blast wave will cause the gas ring to thin out even faster, and of course the gas will cool off in the process. This will probably occur within 100 years or so.

Eventually, astronomers should be able to peek into the interior of the cloud and observe the corpse of Sanduleak -69°202. Will we detect the predicted neutron star, or will we have to revise our model for neutron star formation? Stay tuned. I can predict that if astronomers do confirm the presence of a neutron star where SN 1987A occurred, there will be a lot of celebrating and the news will appear everywhere. This will be the observational evidence needed to connect neutron stars and supernovae.

Meanwhile, astronomers have been able to verify another theory that has been drifting around for years without solid footing. Analysis of the spectra of the region around SN 1987A reveals the presence of some of the heavy chemical elements that theory suggested should form within such a supernova. So SN 1987A has contributed to the universe some more of the chemicals upon which life is dependent.

HST and Supernovae The HST is extending our understanding of supernovae through the detailed examination of some of the remnants of exploded stars. Because of its superior light-gathering ability above the absorbing layers of Earth's atmosphere, it is capable of resolving the fine detail in the very inner portions of the clouds of gas and dust ejected outward at the time of the explosion. The HST recently captured an image of one of the most bizarre objects in the sky (Figure 11-14). Eta Carinae is a nebula formed by a stellar outburst in the year 1843. It was assumed, up until a few years ago, that when the explosion occurred, the entire star was blown into gas and dust. But infrared studies found a star located near the center. The HST image reveals a rather well-defined edge to the expanding gas/dust cloud that is presently about 2 light-years in diameter.

Rather than being entirely filled with matter, the cloud appears to be a thin shell of matter — rather like a planetary nebula. Shock waves created at the time of the explosion cause the knots and filaments within the nebula. Individual clumps of gas/dust about the size of our solar system are visible in this high-resolution image. Features never before seen elsewhere appear within the nebula as jets and ladder-like "rungs." Astronomers are not yet able to explain the mechanism for these unusual structures.

Cause of Supernovae

In spite of numerous observations related to supernovae and their remnants, it is still far from clear just what the connection is between supernovae and neutron stars. Nevertheless, let us at least brainstorm a bit — using basic laws of physics as our guide. Essentially, we are attempting to explain how the repulsive force between electrons in a star's core can be overcome so that electrons can combine with protons to form neutrons. Obviously, a great force is necessary.

Theory tells us that if the mass of a collapsing core is great enough, iron atoms — the heaviest that can be made in the core of a star — will collide with one another and break up into protons, neutrons, and neutrinos. These, in turn, create the pressure that is suddenly released — violently blowing the outer parts of the stars into space.

Meanwhile — back at the core — the surviving protons and electrons are forced together (action = reaction) to form the neutron star. So although some percentage of the star's original mass is ejected out into space as the supernova remnant, much of it remains behind in the form of a neutron star. It is the large number of protons and neutrons ejected outward and that interact with the atoms in the outer layers of the star that are responsible for the making of chemical elements heavier than iron during the supernova explosion. Their speeds are so great that the chances of collision with and absorption by lighter nuclei are quite good. And it is, after all, only a matter of combining protons and neutrons together with other protons and neutrons that makes for the buildup of the elements.

Figure 11-15 Nova Cygni occurred in 1975. These two photographs show during (upper) and after (lower) images of the outburst of radiation. Using a photometer, astronomers measured it changing from m_v +2 to m_v +15 between photographs. (Lick Observatory)

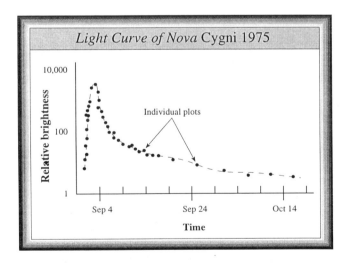

Figure 11-16 The light curve for Nova Cygni. Notice how there is a sudden brightening of the star, followed by a gradual decline. This information must be incorporated into models attempting to explain how such events occur.

★★★★★★★★★

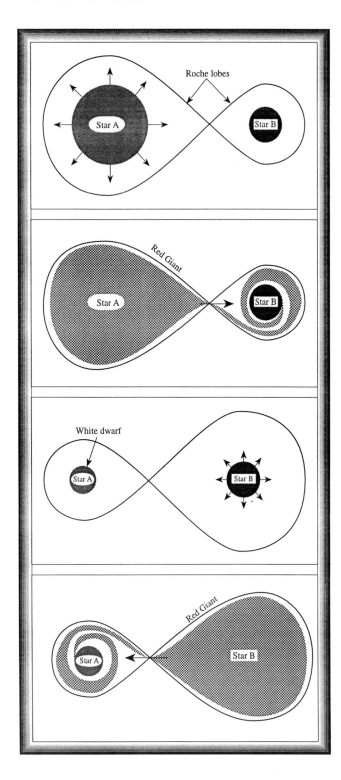

Figure 11–17 *The geometry of the Roche lobe is illustrated in this diagram. This allows astronomers to explain the outbursts associated with novae.*

So the heaviest elements are made during the very act of the star blowing outward into space. As far as we are concerned, this feature of the death of stars is quite convenient because presumably heavy elements are needed elsewhere in space if they are going to participate in the arising of intelligent civilizations. If the heavy elements were to remain on the neutron star, there would be no mechanism for distributing them outward into space so they can mix with clouds of gas and dust which eventually become new stars (and perhaps planetary systems).

Novae

The appearance of a "new" star in the sky in the year 1572 prompted Tycho Brahe to question the current belief in the permanence and perfection of the heavens. He wrote a small book – *De Stella Nova* (The New Star) — the year following the discovery in which he explained how the lack of parallax proved that it was more distant than Moon, and that it must therefore be a new star. We now believe that such new appearances are the result of behavior involving old stars rather than new, young stars.

But we retain the use of the word **novae** to refer to these observed phenomena, unless we find that it is indeed a supernova. The appearance of a novae and is identical to that of a supernovae. One would know the difference only if

- the distance to the object is known, allowing the absolute visual magnitude to be calculated, if

- the light curve is obtained, or if

- the spectrum of the object is obtained.

The M_v of a supernova is significantly greater than that of a novae. That is why "super" is in the word *super*nova. The supernova isn't necessarily brighter, because brightness depends upon distance as well as M_v. The light curve of a novae is characteristically different than that of a supernovae, as well as the spectrum. The star of 1572 was most likely a supernova (instruments for obtaining light curves and spectra were not available at the time). But there are many good examples of novae that have been imaged by modern telescopes. Nova Cygni was detected and imaged in 1975 (Figure 11-15). Comparing its light curve (Figure 11-16) with that of Supernova 1987A (refer back to Figure 11-12), it is apparent that the manner in which radiation is emitted in the two circumstances is different.

Cause of Novae As in the case of supernovae, novae are not well understood, but there is growing evidence that most novae are associated with old stars located in *close binary systems*. This evidence is the result of our recent ability to observe space at X-ray wavelengths by using

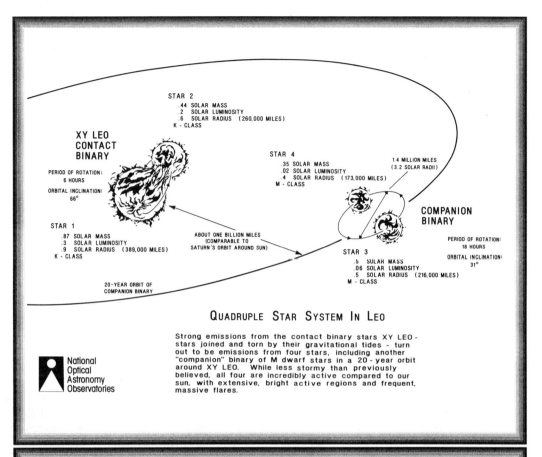

Figure 11–18 *This diagram illustrates a contact binary system, in which stars orbiting close to one another apparently exchange material with one another and emit radiation in the process. (National Optical Astronomy Observatories)*

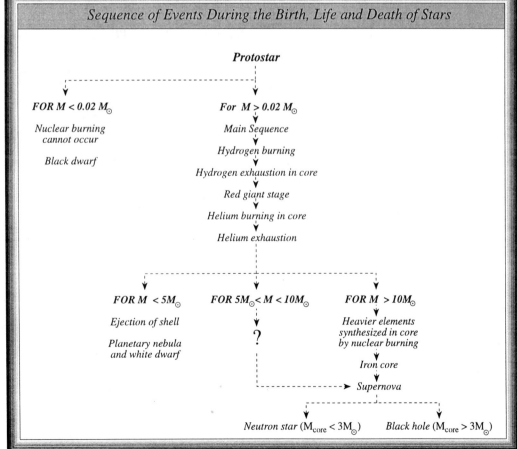

Figure 11–19 *This diagram is a summary of the three types of deaths that a star can experience, depending upon its mass. Since most stars are the not–so–massive M–type stars, the violent deaths for stars we are dealing with in this Chapter are uncommon. Yet you are about to learn that the debris they eject out into space has been instrumental in the arising of life in the universe.*

CHAPTER 11 / Pulsars, Neutron Stars, and Black Holes 237

satellites such as the *Einstein X-ray Observatory*. X-rays, you recall, are emitted by very hot objects or violent events. Compact objects such as white dwarfs, neutron stars, and black holes freely emit X-ray radiation under certain conditions.

In Chapter 8, we learned that binary stars are common and useful: they are used to determine the masses and diameters of stars. Obtaining that information is critical in our goal of explaining the placement of stars on the H-R diagram and thereby understanding the life cycle of stars. Although a good 60% of all stars are members of multiple-star systems, the ones of interest to us in explaining novae are those in which two stars are close enough to one another for mass in the form of gas to flow from one to another.

The two stars are essentially transferring gas back and forth between one another. In order for this to occur, the average distance between them has to be about the distance between Earth and Moon — about 240,000 miles. The region around a star within which its gravity dominates everything contained within is called the **Roche Lobe** (Figure 11-17). So if two stars are close enough so that their *Roche Lobes* touch one another, matter is pulled off the expanding shell of one star and transferred to the surface of the other star.

Consider the following situation. An interstellar cloud collapses and breaks up into two components to form a close binary system. One star is more massive than the other and is therefore the first to die. In the act of dying, it goes through a series of stages before arriving at white dwarf stage. Eventually, the second, less-massive star runs out of hydrogen fuel and swells to red giant stage. But as the outer layers of the red giant encroach upon the immense gravitational field created by the highly dense white dwarf star, gases are pulled away from the outer cool layers and onto the white dwarf.

Origin of Radiation from Novae The principle of *conservation of angular momentum* says that the matter pulled from the red giant will not fall directly onto the surface of the white dwarf, but will first spiral into a whirlpool around the star prior to hitting the surface. The same principle explains the behavior of water as it tries to go down the drain of your bathtub—it doesn't go directly down the drain when you pull the plug, but rather forms a whirlpool around the drain prior to going down. When astronomers refer to the *whirlpool of gases* around a white dwarf, they use the term **accretion disk** since it is accreting (being added to) gases from the neighboring red giant.

The impact of the gases falling onto the disk heats the gases already in the disk to very high temperatures, at which time nuclear reactions occur. This fusion process is much like that going on in the very cores of main sequence stars. The disk is not a solid as such, but the concentration of gases making up the disk is dense enough to act like one when the gases are pulled onto it by the gravity of the neighboring star. The resulting pressure from the "collision" is equivalent to that at the center of a normal star, allowing for the proton-proton reaction to occur in the accretion disk.

According to theoretical predictions, the radiation emitted in this reaction is not in a smooth, continuous flow, but in a rapidly flickering manner. This is exactly what astronomers observe. A nova brightens within a few days or weeks, drastically fades within months, and then gradually fades for years later. Even years after the outburst of radiation, the shell of gas blown away by the reaction in the disk becomes detectable through optical telescopes.

A similar situation can happen to a neutron star in a close binary system. As in the case above, matter from a close red giant is pulled toward the magnetic poles of the neutron star by its strong magnetic field. When the gases hit the surface of this incredibly dense object, they are heated to such high temperatures that thermonuclear reactions occur — and pulsating X-ray energy is emitted. Several of these systems have been detected by orbiting X-ray telescopes.

Contact Binaries Less violent binary star systems exist as well. If two not-so-massive stars orbit one another close enough, they may transfer matter back and forth between one another without the violent outburst associated with novae. Figure 11-18 illustrates the geometry of one of the better studied examples of **contact binary star** systems.

Summary of Star Death

Before we explore the third and most bizarre stage in stellar death, it is appropriate to summarize the types of deaths that we have already considered (Figure 10-19). Not-so-massive stars like our Sun die slowly as they exhaust their fuel, expanding to red giants on the outside, collapsing to white dwarfs on the inside. The outer shell continues to expand outward as a spherical cloud of gas and dust (planetary nebula), while the collapsing core builds up heavier and heavier atoms until the temperature prevents any heavier atoms being formed. The exhausted white dwarf slowly cools to become a black dwarf.

More massive stars are not so fortunate. After reaching the red giant stage, they too build up heavier elements in the core, but only up to the element iron. At that point, rapid collapse of the iron core causes a sudden explosion that blows much of the star's matter out into space and compresses the remaining material at the core into a neutron star. The neutron star loses energy of rotation as it emits electromagnetic radiations detected as the pulsar. Eventually, the spinning neutron star slows to a stop — it is no longer detectable by distant observers. The universe appears to be designed so that dying stars eventually become invisible — as black dwarfs or non-spinning neutron stars. This design seems to apply to another type of stellar death as well — the *black hole*.

Black Holes – Theoretical Basis

We approach the subject of black holes much as we did white dwarfs and neutron stars – theoretically and observationally. We look at the theory first. Imagine the collapse of the core of a massive star, the explosion of the outer layers as a supernova remnant, and the compression of the central core into the form of a neutron star.

We might ask ourselves – "Why does the core of the neutron star stop collapsing after the protons and electrons have combined together to form the more compact form of matter, the neutron?" Well, try visualizing the neutrons as small golf balls – tiny little balls of matter. The limit of collapse of such a bunch of balls is reached when all of the balls are touching one another, and there are no hollow spaces available for adding a single ball.

In this example, we are imagining neutrons as having edges or boundaries that limit just how small a volume a given number of balls could occupy. And we are correct as far as we have gone in our thinking. But now, let's change our analogy for the neutron from that of a golf ball to that of a ping-pong ball. How can you stuff additional ping-pong balls into a full bucket if it is important to do so? Step on the balls and crush them into the bucket, of course!

So you are wondering – can we do that with neutrons as well? We can't, but nature – stars in the act of exploding – apparently can. You see, neutrons in a compact configuration act much like electrons in a compact configuration, except that it is not a situation of like charges repulsing one another – it is a simple matter of several objects (neutrons) not being able to occupy the same volume at the same time.

Structure of Neutrons What we are visualizing is a mass of neutrons rubbing "shells" with one another. If the explosion of the star (supernova) is powerful enough, the core is forced inward to collapse past the point at which the "shells" resist one another, and the core shrinks even further. This explosion "inward" might seem strange at first – doesn't the *ex* in explosion mean outward? Yes, and in fact the word that we should use to describe the inward force is **implosion**.

This inward force is simply Newton's third law in practice – action equals reaction. So how much smaller does the implosion force the compressed neutrons to get, you say? If you are still thinking in terms of ping-pong balls in a bucket, you're probably thinking that the limit to the number of crushed balls that can be put into the bucket depends only on the thicknesses of the shells of the balls.

But alas – it is here that our analogy between neutrons and ping-pong balls breaks down and fails us. Because when the "shell" of the neutron is crushed (or penetrated or whatever), the neutron ceases to exist as a particle of matter, and reveals itself for what it is at the most basic level of matter presently known – a collection of small energy units named **quarks**. In other words, a neutron as analyzed in the chemistry laboratory reveals properties of mass, size,

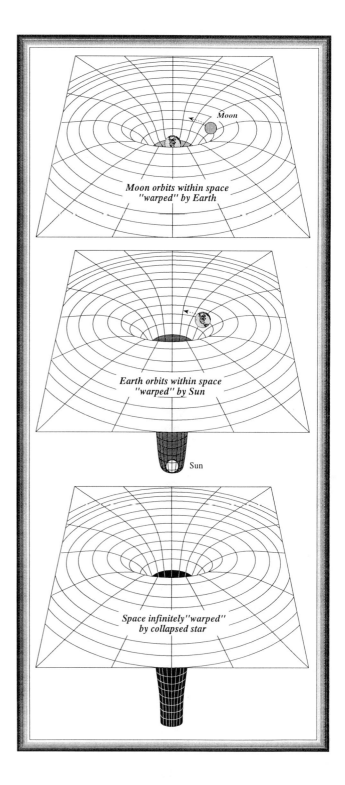

Figure 11–20 Representation of the various degrees of space "warped–ness." Moon orbits within Earth's slight warping of space (top), Earth orbits within Sun's moderate warping of space (middle), and infinitely–warped space within region of a collapse star (black hole).

speed, and so on. But as soon as it is broken apart in high-speed particle accelerators, it no longer reveals those properties, but instead in its place reveals a collection of energy bundles.

This has been one of the most fascinating discoveries of modern science, and is still under investigation at the present time. The implications of the discovery of quarks not only allows us to investigate and understand black holes, but the very nature of the origin of the universe from which even black holes came from. That is an exciting topic that we will tackle in the next section of the Book.

At first it may seem strange to you that matter when broken down into finer and finer pieces eventually becomes energy bundles, but reflect on the equation $E = mc^2$ that we encounter so often in astronomy. That equation states the *equivalence of mass and energy* — a certain amount of mass has the potential of being converted into a certain amount of energy, and vice versa. The classic movie, *The Fly*, is founded in this concept — the mass of a person can be turned into radiation (energy), beamed somewhere else, and then turned back into mass again — hopefully into the original form of the person! The *transporter* of "Star Trek" fame operates on the same principle.

Existence of Quarks If my attempt to spark your curiosity has been successful, at this point you should be asking yourself a stream of questions. Such as — "How do we know this to be true? How do we know that neutrons are composed of quarks? Has anyone ever seen one?" These questions are, of course, examples of the importance of taking a science class in the first place — what is the evidence that is available that we can use to base a belief on?

I realize that I've stressed the importance of requiring evidence for stated claims throughout the book, but in the limited space of the present book, I must ask that you take the evidence for quarks on faith. Hopefully you will want to explore the subject on your own. It is, without the slightest exaggeration, one of the most important scientific topics today. The interest in and popularity of Steven Hawking's book *A Brief History of Time* throughout the world is but a minor indication that the average citizen is interested in the affairs of modern science.

There is a field of study devoted exclusively to investigating the inner structure of neutrons, protons, and other subatomic particles — how they behave toward one another, and how they relate to the world that we see with our everyday eyes. That field of study is called **quantum mechanics**. It represents the biting edge of modern science, with implications that could rival those of the discovery of non-Earth-centered-ness by Galileo (or that we at least attribute to him).

We will encounter the principles of quantum mechanics again in Chapter 15 when we attempt to explain the origin of the biggest black hole of all — the universe. I am accepting (on faith) what scientists tell me about experiments that lead them to believe that quarks exist. It could

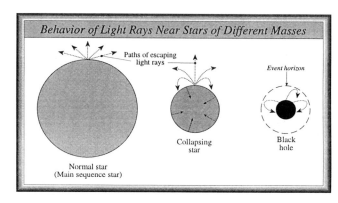

Figure 11-21 According to Einstein's concept of gravity, the "escape velocity" of an object is the speed at which something must travel in order to "climb" out of the "warped" space created by a larger object. A black hole is a place at which space is infinitely warped, and therefore even radiation cannot escape.

all be a giant hoax — but ultimately every theory stands the test of observation. And so far what astronomers have observed in space in the form of black holes works in neatly with what is predicted by the existence of quarks.

Collapse to a Singularity Let us return to the core of a supernova. It is there that neutrons are compressed to the point at which they lose their identities as particles of mass and take on new identities as bundles of energy — quarks. What happens to the star, you ask? Well, since it is no longer a "thing" in the sense of having mass, it disappears. *But its presence is still detectable.* You say I'm playing with your mind? Well, just hang in there, because it gets even more bizarre.

Listen — the star no longer exists in the form of matter, but its equivalence in the form of energy ($E = mc^2$) is still present and can be detected. The core of the star shrinks down to an infinitesimally small speck, creating what in mathematics is called a **singularity**. You encounter this concept whenever you consider dividing any number by zero — you obtain infinity (∞) for the answer. The reason that the concept of a *singularity* arises in the case of the black hole is because certain properties of the black hole have the numerical value of infinity. So singularity implies infinity — and vice versa.

Warping of Space-Time As applied to the subject of black holes, the concept of singularity means that space is infinitely warped in the vicinity of an infinitely collapsed object. That is the definition of a **black hole** – *infinitely warped space*. Singularity is the mathematical treatment of infinitely warped space as well as other topics that include infinity. Let me provide an analogy that may help you to visualize the concept of a black hole. Imagine a trampoline, the surface of which is a tightly stretched, elastic rubber sheet (Figure 11-20). You place a golf ball on the sheet and observe how it causes a dent where it sits. You place a baseball somewhere else on the sheet, and observe a deeper dent. You place a bowling ball on the sheet and observe an even deeper dent — and so on. Imagine that — prior to your placing balls on the rubber sheet — there was an ant that had lived its entire life on the tightly-stretched sheet. It was not aware of up or down. It knew only the surface of the flat sheet — and it knew it well.

So one fine day, while out for a walk, the ant is suddenly aware that it is uncontrollably sliding along the sheet faster (accelerating) toward a round object — the golf ball you placed on the sheet — just ahead. It slides right into the golf ball. Now being an inquisitive and scientific-thinking chap, the ant wants to know what caused it to run into the ball. Its immediate conclusion, I would guess, is that there is a mysterious force of attraction between it and that round object toward which it slid. Meanwhile, you and I — observing from above the flat plane of the sheet — know that it was for an entirely different reason — the gravity of Earth pulling the ant "down." But the ant cannot escape its flat world of the sheet to see that other explanation.

Neither can you nor I "escape" space in order to observe that gravity is not an "attraction" particles of mass have for one another. But it is rather the "warping" of space in the vicinity of things (like planets and stars and galaxies) that have mass (or its equivalent). Earth, having a large mass, bends or "warps" space around it. You and I live in space — I don't mean "outer" space, but space in a broader sense — and it is a property of warped space that matter caught therein goes toward that part which is most warped.

The surface of Earth, for example, is that part of space most warped by Earth. So you stay here not because you want to stay here, not because Earth holds you here, but because the properties of space force you to constantly move to the place of greatest warped-ness — Earth's surface. The greater the mass of the object, the deeper the warp, and the stronger the force with which something is held to the object's surface.

Degree of "Warped-ness" There is a convenient method of stating just how deep warped space is around an object. It is the **escape velocity** of the object — the velocity to which something must be accelerated in order for it to climb out of the warped space created by the object. For example, in order for us to send a rocketship to Moon, we must accelerate it to a velocity of 7 miles per second. *Seven miles per second is the escape velocity of Earth.*

So the rocketship "climbs out of" Earth's space-warp and "falls into" Moon's space-warp. And — assuming the rocketship properly decelerates upon arrival — then it settles softly down onto Moon's surface. In order to return to Earth, the rocketship must "climb out of" Moon's space-warp by accelerating to Moon's escape velocity — about 1 mile per second — by blasting off Moon's surface. Moon's escape velocity is less than that of Earth because of its smaller mass.

And what if we wished to send astronauts clear out of the solar system? Well, we can launch them off Earth's surface at the escape velocity of 7 miles per second, but that leaves them within Sun's space-warp. They become an

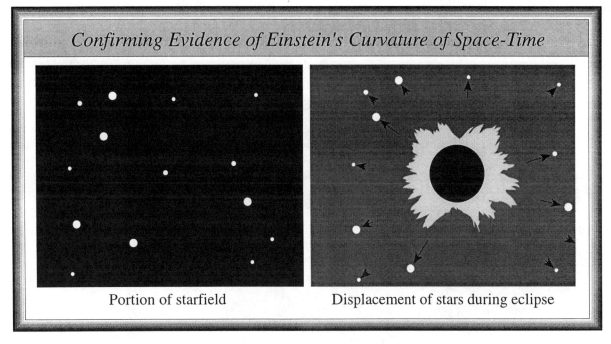

Figure 11–22 These diagrams illustrate the "proof" of Einstein's theory of gravity. The star field is the same for both right and left diagrams. But at the moment of totality during a total solar eclipse, the measured positions of stars in the diagram to the right are displaced slightly away from Sun. The interpretation of the observation is illustrated in Figure 11–23.

CHAPTER 11 / Pulsars, Neutron Stars, and Black Holes   241

artificial planet circling Sun — orbiting within Sun's space-warp. In Figure 11-20, you see how the concept of space curvature explains Earth's orbit about Sun.

Earth's warp is caught within Sun's warp, so that if it were not for Earth's revolution around Sun, Earth would fall into Sun like a ball rolling down the side of a large bowl. So after the astronauts climb out of Earth's space-warp, they must accelerate their spaceship to a velocity of 18 miles per second in order to escape the solar system and visit another star. The escape velocity of the solar system, in other words, is about 18 miles per second.

Let's return to the trampoline analogy by imagining placing on it an infinitely heavy object so that the rubber membrane stretches out to an infinitely deep "warp" — like an infinitely deep funnel. This is an analogy for the black hole. Because the warp consists of infinitely stretched space, the escape velocity — the velocity needed to climb out of the warped space — is infinite. Therefore, nothing can climb out of or "escape from" the black hole — including light itself (Figure 11-21). To an outside observer, therefore, infinitely-warped space appears black because light cannot come from within. Thus the expression "black hole."

Let's back up for a moment and consider some finer details of these strange objects (?). Actually, the ideas about warped space and black holes aren't really new. Einstein first proposed the idea of warped space in the year 1916, and supporting evidence for it was found in the year 1919.

Einstein's Theory of Gravity The concept of warped space and black holes is an excellent example of how science works at the finest level. It is an excellent example of how new discoveries in science cause us to think of the universe in new ways. Sir Issac Newton was the first to formulate the concept of gravity in a quantitative manner with the equation we used back in Chapter 3 to explain Earth tides: $F = G m m / d^2$. As a matter of review, you recall that what the equation says is that two objects with masses of (m_1) and (m_2) attract one another with a force (F). The strength of that attraction is dependent on how far the two masses are from one another — the term (d).

Einstein's concept of gravity as the bending or warping of space does not replace this equation. We just <u>think</u> differently about what the terms mean. Instead of F representing an attractive force, it is thought of as a "well" in the rubber membrane — the bending of space in the vicinity of the masses m_1 and m_2. Recall too that matter and energy are equivalent according to $E = mc^2$.

So in the equation expressing the amount of force associated with gravity ($F = G m m / d^2$), the *m* term values can be replaced by the equivalent *E* term values to show how light is also affected by the presence of mass. In other words, the equation is used to express the attraction between two bodies of mass, the attraction between a body of mass and a beam of energy (radiation), or even the attraction between two beams of energy.

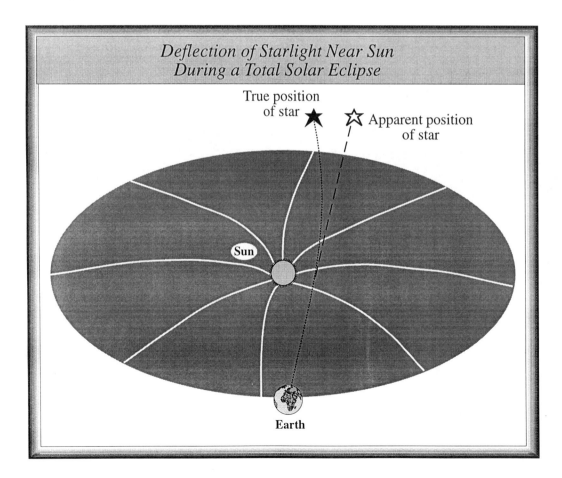

Figure 11–23 An interpretation of Einstein's prediction—and ultimate observation—that stars close to Sun at the time of a total solar eclipse appear displaced away from Sun due to the warped–ness of space created by Sun.

Now you may be saying to yourself — "So what's the big deal?" "Why does it matter whether we call it a force of attraction or the bending of space?" Well, it is very important because Einstein's new concept allows scientists to explain observations that otherwise wouldn't be possible. Black holes are one of those observations. The goal of science is to explain the phenomena of nature that we observe.

Einstein hadn't observed black holes, however. He was trying to explain another observation (the constant speed of light) that didn't fit in with the current scientific model for the behavior of space. During the process of working out the mathematical solution to the problem, he recognized that the equations predict that light passing by a massive object is distorted by the warped space around that object. In order to verify this prediction and at the same time to offer an explanation why the speed of light is constant to any observer anywhere in the universe, he did what all good scientists do — he proposed that astronomers perform an experiment.

The testing of a prediction is the strength of the scientific method. The laws of nature favor no one — they behave consistently in all identical situations. That is why the scientific community is skeptical about claims that people bend spoons without touching them (*psychokinesis*), that people project their minds out of their bodies to visit distant locations (*astral projection*), that people perform intricate surgical operations on people without opening the body (*psychic healing*), that people climb out onto limbs — and so on.

A newly-discovered law of nature predicts consequences of that law being true. If the predicted action cannot be observed or if the observed action doesn't conform to the predicted action, the law's validity is questioned. In the examples of claims just mentioned, there are no proposed laws involved — there are only random and inconsistent events. They are claims that something happened, not that something will happen. And most claims that something will happen are not specific enough to match the observed action.

Of course, one can claim that there is a natural law involved — it's just that scientists haven't discovered it yet. And certainly that is a possibility. Meanwhile, however, the field is wide open for anyone to come along and claim that he/she has a special gift to remedy any problem you might have — at a discount price, of course. Newspapers and television programs cover wonderful examples of fraud and dishonesty in the search for new natural laws.

Science demands that all claims must be publicly displayed in scientific journals and at scientific conferences for the scrutiny of the entire world. Why don't we demand rigorous evidence from anyone claiming to have stepped aboard a flying saucer? Why don't we demand rigorous evidence from someone claiming they sell a toothpaste that "all dentists" recommend? Again scientists say — "Extraordinary claims require extraordinary evidence."

Proof of Einstein's Theory Einstein was not asked to propose an experiment to support his theory of warped space near a massive object. He knew that he must if his theory was to be accepted by the scientific community. He was — at the time of this discovery — a nobody. The theory, he knew, would turn the scientific community on its ear — so radical and far-reaching were the consequences if it were true.

So the proposed test of the theory was simple — take a photograph of Sun during a total solar eclipse. It was published in a letter in Germany during World War I. A single copy of the letter was smuggled through neutral Holland to British astronomers. The next eclipse was on May 29, 1919. The British astronomers organized two expeditions to carry out the observation — one to northern Brazil and one to an island off the coast of West Africa.

Since the sky is dark during the period of totality, stars in the field of view of the camera appear on the photographic plate. Einstein suggested that astronomers compare the positions of the stars relative to one another on the photograph with the positions of those same stars on the accurate star charts available to all astronomers. He predicted that those stars close to Sun at the time of the eclipse

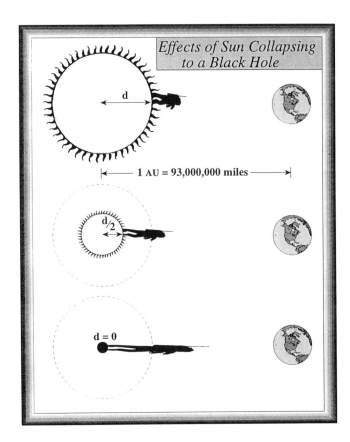

Figure 11–24 Were Sun to collapse to a black hole, Earth would continue to orbit around the singularity. The difference is that someone can now get closer to the center of Sun than when it was extended. But as a consequence, the "force" of gravity would be greater on the person.

CHAPTER 11 / Pulsars, Neutron Stars, and Black Holes

would be displaced a very slight amount (1.75 arc-second) away from Sun, compared to their positions at other times.

The British astronomers obtained photographs at both expedition sites. The measured positions for the stars at the time of the eclipse were indeed displaced as predicted, and measured to be within 6% of Einstein's predicted position (Figure 11-22). Einstein was taken from total obscurity to world celebrity with a single success almost overnight. He remained a celebrity until his death in 1955 (he was offered the presidency of the newly-formed State of Israel, but politely refused).

How exactly does this experiment prove Einstein's theory of warped space, you are asking? Let's look at the situation from a distance — as if we are able to see a beam of light from the side. We observe and measure the positions of stars against the backdrop of more distant stars in the positions shown in Figure 11-22.

Now place the eclipsed Sun in the position amongst those stars as shown. If space is indeed warped by the presence of Sun, then rays of starlight that would normally arrive on Earth would miss Earth entirely. But other rays of starlight that would normally miss Earth now hit it. From the point of view of observers on Earth, the new rays of starlight <u>appear</u> to come from a point farther away from Sun than the star is actually located.

In Figure 11-23, it appears that the ray of starlight is bent as it goes around Sun. That is one way to visualize it. But we must visualize it three-dimensionally — the ray of starlight travels in space, and if space is bent, then the ray of starlight must travel in that bent space. You and I live and work and play in that bent space, so we don't *see* the extent to which it is bent. We would have to leave space to do that. But how do we leave space? For the moment, at least, we must content ourselves with being able to <u>feel</u> the presence of warped space. That is, after all, what you feel when you climb mountains or stand up to go to the refrigerator.

Schwarzschild Radii for Black Holes of Various Masses	
Mass of Black Hole	**Schwarzschild Radius (radius of event horizon)**
1 Earth mass	1/3 inch
1 Jupiter mass	9 feet
1 Solar mass	2 miles
2 Solar masses	4 miles
3 Solar masses	6 miles
5 Solar masses	9 miles
10 Solar masses	18 miles
50 Solar masses	92 miles
100 Solar masses	184 miles
1,000 Solar masses	1,840 miles
One million Solar masses	10 light-seconds
One billion Solar masses	3 light-hours
One trillion Solar masses	117 light-days
The universe	3×10^{13} light-years

Figure 11-25 *The calculated sizes of various objects to which their masses would have to shrink in order for them to become black holes.*

Figure 11-26 *If Earth were to shrink to the size of a child's marble, it would "stretch" space enough to create a black hole.* (Tom Bullock)

The importance of the eclipse in this experiment is that it is only under that condition that stars can be seen close to a nearby massive object. Moon is not massive enough to deflect the rays of starlight sufficiently to be detectable. This experiment has been conducted on numerous occasions since that first one of 1919, and the results are always the same. In fact, more precise measurements are possible today because of the increased sophistication of astronomical techniques.

Radio astronomers have used radio interferometers, for example, to measure the precise positions of radio sources near Sun during total solar eclipses. The results agree with the predictions of Einstein's theory to within 1 percent. We encounter some other consequences of Einstein's theory when we discuss the possible "shapes" of curved space. That is a very important topic in astronomy, since it is tied in with theories about the birth and ultimate fate of the entire universe.

Event Horizon How would a black hole appear up close? Let us imagine Sun suddenly collapsing to become a black hole — not by an explosion, but by a sudden turning off of the proton-proton reaction at the core. What would happen to Earth if it no longer has Sun to orbit around? Nothing. Well sure — it would be dark and cold. But it would continue to orbit that which took the place of Sun — the black hole. Everything in the solar system — other planets, asteroids, and comets — would continue to orbit the black hole. You were probably thinking that Earth would be sucked into it, right? That image was probably

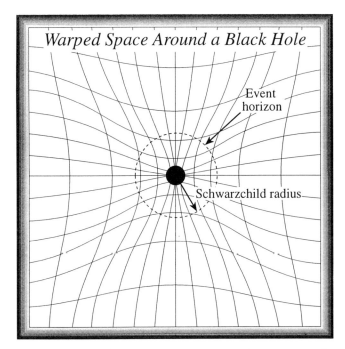

Figure 11-27 Looking "down" from outside of space itself at warped space around a black hole allows us to visualize the concept of the event horizon (the dashed circle). From within that boundary—the distance from the singularity of which is the Schwarzchild radius—nothing can escape.

planted in your mind by Hollywood — it certainly adds more drama to the situation.

But if you look at the variables in the equation expressing Sun's space-warp (or gravity), $F = G\,m_1 m_2/d^2$, you notice that none of them changes just because Sun collapses to a point. Neither Sun's mass (m_1) nor Earth's mass (m_2) nor the distance between them (d) changes — so F (the amount of force) remains the same. The difference is that since Sun has collapsed, an object (for example, a comet) can get closer to Sun's center of gravity than it could prior to Sun's collapse. As it does so, the value of d in the equation gets smaller and smaller, and F gets larger and larger. As the comet approaches ever closer, it eventually invades the boundary of no return — the point at which the escape velocity exceeds the speed of light (Figure 11-21).

Compare Sun now with its hypothetical collapsed state as a black hole in Figure 11-24 again. The closest an object can get to Sun now is the surface. That is where the force of gravity is greatest (greatest degree of space warped-ness). If something were able to go deep within Sun, the force of gravity would get less and less. At the very center of Sun, you would be weightless! Think about it as you stand there, surrounded by an equal amount of Sun matter in every direction you look. So you are being "pulled" with the same strength of gravitational force no matter what direction you look. Therefore you are weightless.

But if the entire mass of Sun were to shrink past the size of a 4-mile diameter ball, then it would become a black hole. The mass of Sun is still there — it is just in the form of energy. But now an object approaching Sun is no longer limited to staying Sun's surface distance away. Now it can invade what used to be the space occupied by the interior volume of Sun, and the gravitational force is correspondingly stronger (because the term d is smaller).

If anything gets closer than that 4-mile diameter for Sun, it is lost to the black hole and no longer in a common universe with us. We cannot receive messages from it. Only spaceships traveling at "warp speeds" (a science-fiction term for speeds in excess of the speed of light) can get out of the black hole. It is this boundary between what can and what cannot escape that is usually referred to when the expression "black hole" is used — it is from within that boundary that no radiation (and certainly no mass) escapes.

Formally, this boundary is called the **event horizon**, since events occurring inside are beyond our ability to observe. Its size depends upon the original mass of the object that collapsed to form the black hole, and is expressed in terms of the *radius* of the event horizon. It is called the **Schwarzschild radius** after the astronomer who worked out the mathematical equations needed to determine the size for a given mass (Figures 11-25 and 27).

According to calculations, a star of three solar masses collapsing to 10 miles in diameter would become a black hole. Its *Schwarzschild radius* is therefore 5 miles. The sizes to which various object must collapse to become black holes is shown in Figure 11-25. Earth, were it to shrink down to the size of a child's marble (Figure 11-26), would be a black hole. This is not to say that Earth can become a black hole. These are theoretical calculations simply to illustrate the concept of event horizon.

Appearance of Black Holes Are you wondering how an event horizon appears if you approach it in space? It appears as a black sphere against the black background of space studded with stars. In other words, you recognize it by its lack of light — like determining the shape of your hand held out in front of you against the dark sky at night. The shape is determined by the pattern of stars that you see around your hand. From whatever direction you approach the event horizon, you will see it as a black circle. In 3-dimension, it is a black sphere. You are not allowed to see inside the infinitely deep tunnel — the space warp. It is <u>inside</u> the black sphere.

To visualize the connection between the event horizon and the trampoline analogy, imagine looking down into the warped membrane from above the trampoline. Imagine drawing a circle around the entrance to the warp. Call that circle the event horizon (Figure 11-27). Anything inside of that circle is beyond the reach of the outside observer. What theory predicts for the interior of the black hole is infinite space — another universe! But that thought is difficult to entertain, so used are we to thinking that the volume of a sphere is a value determined after knowing the diameter of the sphere. But that mathematical expression is valid only

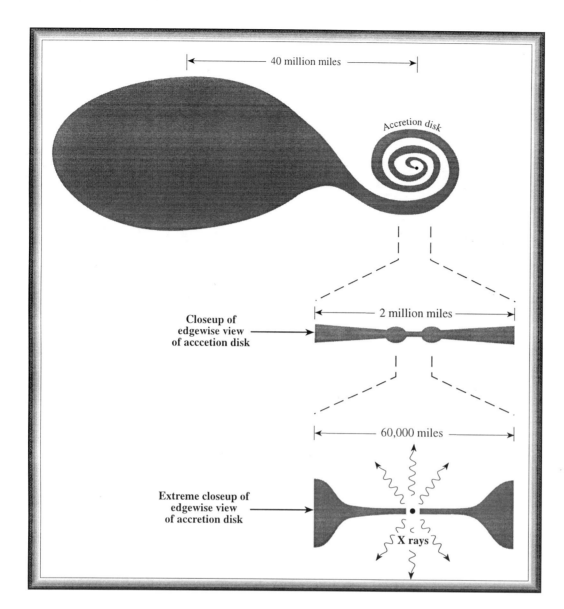

Figure 11–28 The X-rays coming from the regions of black holes are believed to be produced when gases are pulled from the outer layers of a neighboring star and onto the accretion disk surrounding the black hole. This heats the gases to very high temperatures, which in turn generates X-rays.

for three-dimensional space. That is why a college geometry course is usually referred to as a course in Euclidean (3-dimensional) geometry.

Space is stretched out inside the event horizon. So there is no limit to the amount of stuff that can go into it — rocks, planets, even stars. Of course, if the entrance (the event horizon) is only 10 miles across, a star can't be swallowed all at once. Tidal forces pull the near side of the star away from the far side, so the object is stretched out into a long, threadlike stream of atoms as it goes into the black hole, never to return. Even the atoms, according to theory, cannot survive passage beyond the event horizon. Matter is reduced to its very quark nature, just like the matter of the collapsing star that created the black hole in the first place. Just think, there are things in the universe that aren't really there!

Black Holes – Observational Evidence

Okay, the theory sounds fine. And it is certainly fun to speculate about how strange the universe can be. But how do we know that such bizarre things actually exist? And if they are black, how can we detect them to confirm the theory? Since stars are incredibly far away from us, there is no way we are presently able to see a black hole blocking out stars behind it.

Only if a black hole were close to us, and/or extremely large, would it be detectable in that manner. No, we must be more imaginative than that. We must consider the effect such an object has on a neighboring object, assuming there is one close enough to be affected. Perhaps we can detect the consequences of a black hole swallowing a neighboring star, for example! In view of the vast distances between stars, can that ever occur?

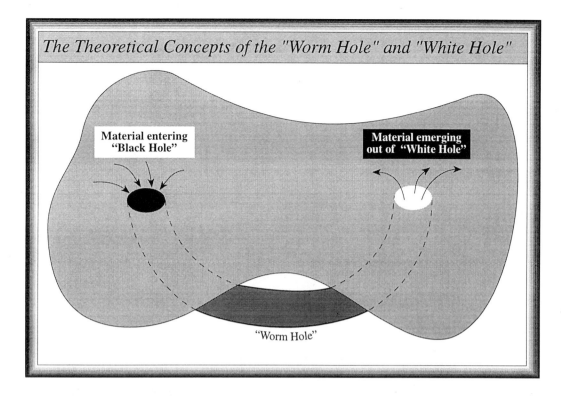

Figure 11–30 An illustration of the concept of the "worm hole." According to this idea, matter going into a black hole travels through a "worm hole" and emerges out through a "white hole" in someone else's universe, or our universe sometime in the past, present, or future.

Recall our discussion of binary stars. Evidence suggests that perhaps 60% of all stars within our galaxy are members of multiple star systems. So let's imagine a binary star system consisting of two stars of unequal masses — one that is very massive and one that is not very massive. According to our model of stellar evolution, the massive star evolves and dies to become a black hole.

So now there is a low-mass star and a black hole orbiting a common center of gravity. Eventually the low-mass star begins to die, and in the process expands outward to red giant stage. Now if the original stars were close enough together, the outer layers of the expanding low-mass star will eventually encroach on the warped space around the black hole. The outer layers will be pulled into the black hole — but not without a stretch (refer back to Figure 11-24).

Accretion Disks of Black Holes Watch the water going down the drain in your bathtub again. It doesn't go straight down, remember? It swirls around and around before it goes down the drain. It's that same principle again — the *conservation of angular momentum*. The same occurs in the vicinity of a black hole if matter from a neighboring star or anywhere

Black Hole Candidates Within the Milky Way Galaxy

Object	Location	Companion Star	Orbital Period	Possible Mass of Compact Object
Cygnus X-1	Cygnus	O-type supergiant	5.6 days	7-16 Solar masses
LMC X-3	Dorado	B3 main-sequence	1.70 days	10 Solar masses
A0620-00	Monoceros	K-type main-sequence	7.75234 hours	More than 3.18 Solar masses
V404 Cygni	Cygnus	G- or K-type main sequence	6.47 days	**More than 6.26 Solar masses**
Nova Muscae 1991	Musca	K-type main-sequence	10.4 hours	More than 3.1 Solar masses
GS 2000+25	Vulpecula	Red dwarf	8.3 hours	**6-14 Solar masses**

Figure 11–29 The three best candidates for black holes in our galaxy (in **bold** type). The others are less certain because of the possible tilts of their orbital places with respect to our view of them. Their calculated masses—based upon their influences on their visible neighbors—conforms to the theory of stellar evolution developed earlier.

CHAPTER 11 / Pulsars, Neutron Stars, and Black Holes

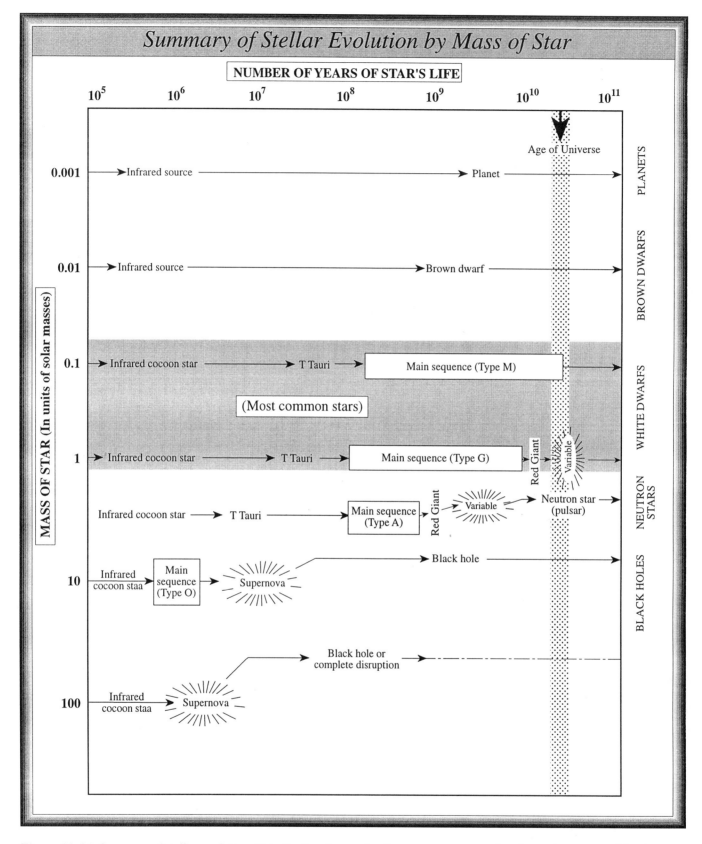

Figure 11–31 Summary of stellar evolution. Note the time frames for the events accompanying the various types of death.

else comes close enough. Gases pulled from the outer layers of the star accumulate just outside of the event horizon in a doughnut-shaped ring — the *accretion disk* again. As the inner parts of the ring are pulled into the black hole, the outer part is supplied by fresh material stripped away from the star.

Laws of physics and gas dynamics predict that gases falling onto the accretion disk heat up to temperatures on the order of hundreds of millions of degrees! At these temperatures, nuclear reactions similar to those at the core of a main sequence star take place, with bursts of X-rays sent out into space in all directions (Figure 11-28).

Since the accretion disk is outside of the event horizon, X-rays headed anywhere other than the black hole can safely escape, to be detected by eager astronomers on Earth. The pattern and intensity of the emitted X-rays provide the necessary clues as to what is happening in the situation. You recall that this very process is also thought to occur in conjunction with novae, in which case there is either a white dwarf or neutron star involved.

Searching for Black Holes So we should be looking for sources of X-rays in close proximity to giant and supergiant stars. But Earth's atmosphere blocks out X-ray radiation, requiring us to send X-ray instruments above it in balloons, rockets, or satellites. The *Einstein Orbiting Observatory*, named in honor of the very person who indirectly predicted the existence of black holes in the first place, was the first to detect a candidate for black hole category. Of the 3,000 or so X-ray sources mapped by *Einstein*, one is particularly interesting because it is close to an O9I-type blue supergiant star. The behavior of the X-rays is what we expect to receive from an accretion disk like the one just described. There is not universal consensus in the astronomical community that it is a black hole. It is called **Cygnus X-1** because it is located in the constellation of *Cygnus*, the Swan, and it is the **1**st **X**-ray source found in that constellation (Figure 11-29).

To be rigidly scientific about it, however, in addition to detecting a pattern of X-ray radiation suggestive of the presence of a black hole, astronomers attempt to determine (or at least estimate) the mass of the black hole — the mass of the object accompanying the companion star that is visible (e.g., the O9I star in *Cygnus X-1*). If the dark companion is 3.3 solar masses or greater, then we are much more convinced that it must be a black hole, since 3.2 solar masses is the theoretical limit for a neutron star. The greater the mass, the more likely it is a black hole and not a neutron star.

To determine the mass of the dark companion requires that we obtain spectra of the visible star, calculate the pattern of Doppler shifts within, and from that (using Kepler's laws of motion) calculate the mass of the dark companion that must be alternately speeding up and slowing down the motion of the visible star as they mutually revolve around a common barycenter. Unfortunately, this calculation reveals only a lower limit to the mass of the black hole, since the orbital plane of the binary star system may be tilted with respect to Earth. There is presently no satisfactory solution for determining what that tilt may be.

Therefore, we have established a list of black hole candidates based upon their minimum and estimated masses (Figure 11-29). Those whose minimum masses are greater than 3.2 are the very best candidates (**V404 Cygni** and **GS 2000+25**), followed by four whose masses are around 3, and approximately 6 with known minimum masses less than 3 but possible masses greater than three (e.g., *Cygnus X-1*).

The Interior of Black Holes Almost everyone has heard incredible *stories* about what happens inside of black holes, and that is just the category in which they should remain for the time being — stories. You have certainly heard a few yourself. Here are a couple I've heard. One says that if I go into a black hole, I emerge into another universe — a parallel universe — to meet my parallel Tom Bullock. Another says that if I go into a black hole I travel to another part of our universe, and arrive in the future or even the past — a time-warped subway system (Figure 11-30). Astronomers have even named these time/space warps — they call them **worm holes**.

These stories assume, of course, that I and my rocket-ship are capable of withstanding the tremendous tidal forces near the black hole. The story that a black hole is a time tunnel — à la Kurt Vonnegut — affords you the opportunity to meet your parents before they choose to conceive you. And to stop Adam and Eve from taking a bite on that fateful apple-eating day. These stories make for decent drama for the uninformed layperson and the beginning astronomy student, and it may be that such scenarios are eventually confirmed as being theoretically possible. That is, we may eventually collect evidence that convinces us that such stories are possible — if not likely. I won't be surprised.

At the moment, however, there is a major difficulty in speculating about life in a black hole. Laws of physics are similar to the laws we establish to govern people — they tell us the conditions under which they are applicable and are operating. Our laws of physics tell us that they do not necessarily work within the environment of black holes (infinitely warped space) — they may, but they may not. So if we work on the assumption that they <u>do</u> operate once the event horizon is trespassed, then the possibilities predicted for what we could experience can get pretty bizarre.

Having discovered one black hole, astronomers are of the opinion that there must be large numbers of them within our galaxy alone. They are certainly not the rule — remember that only the very massive stars die as black holes. But even a small percentage of the large number of stars in the Milky Way galaxy still amounts to a sizeable number of black holes. One percent of one hundred billion is one billion!

★★★★★★★★★

Aside from the fact that they complete our understanding of the ultimate collapse of matter and our theory of stellar evolution, black holes are important to us in another way. Now that there is every reason to believe that they exist as the collapsed remains of massive stars, there is also good reason to believe that galaxy-sized clouds can collapse to form black holes as well! These would be *massive black holes*. This, as you are about to see, may be the very mechanism responsible for the formation of galaxies in the first place.

Gravity Waves There is an interesting anecdote to attempts to detect black holes in space. Theory says that the presence and behavior of black holes can be detected by another method, one that has been used in ongoing search but that has yet to confirm them. What happens if you wave electrons up and down? They emit radio waves — this is the process responsible for radio stations broadcasting your favorite rock-and-roll music over the air. Radio waves are examples of electromagnetic waves, generated when electrically-charged particles are moved around.

But besides electrical charge, objects also have mass. So what happens when you wave masses around? You generate **gravity waves** — predicted by Einstein. Unfortunately, gravity waves are extremely weak, and would only be strong enough to detect on Earth if a black hole were acting on another object in space. One day we may have the technology to detect such direct evidence of the existence of black holes.

Summary/Conclusion

What is rather amazing about the life and death cycles of stars is that they show how clearly the arising of life in the universe is so intricately interwoven with the evolution of stars. Most stars are the not-so-massive, long-lived stars. These provide the long time-periods of the pouring out of steady amounts of radiation that life-forms require as they evolve into more complex forms. In the process of passing through red giant stage, however, stars build up chemical elements heavier than helium, and eject some percentage of them back out into space. They then mix with clouds of gas and dust to provide some of the additional raw materials from which life is thought to evolve.

Although less common, the short-lived, massive stars serve an important function in that when they die as supernovae, they create the chemical elements heavier than iron and eject those too into space. It appears that life in the universe is dependent not only on the good graces of stars to provide life with the necessary radiation at the cost of its own life. But in the act of dying it creates and redistributes the raw materials upon which live subsists. So our search for life elsewhere is encouraged by our discoveries that the life cycle of stars appears to be interwoven with the life cycle of conscious life (Figure 11-31). Perhaps we are too young as a civilization to understand what role black holes play in the scheme of things. That is something I would like to talk to an extraterrestrial about!

LEARNING OBJECTIVES: *Now that you have studied this Chapter, you should be able to:*

1. Describe the theoretical basis for exploding stars, and offer observational evidence for them having been detected.
2. Explain how the recent supernova 1987A offered additional evidence to support the model for exploding stars.
3. Explain the difference between a supernova and a nova.
4. Describe the theoretical basis for the existence of black holes, relating them to Einstein's concept of gravity.
5. Describe the experiment used to confirm Einstein's theory that gravity is the bending of space.
6. Describe the conditions thought to exist around and within a black hole, and how they may allow us to detect their presence in space.
7. Offer evidence for the existence of black holes in space.
8. Define and use in a complete sentence each of the following **NEW TERMS**:

Accretion Disk 238	**Implosion** 239	**SN1987A** 233
Black Hole 240	**Lighthouse effect** 231	**V404 Cygni** 249
Chandrasekhar limit 228	**Neutron star** 229	**Worm hole** 249
Contact binary system 238	**Novae** 236	
Crab nebula 227	**Pulsar** 229	
Cygnus X-1 249	**Quantum mechanics** 240	
Einstein Orbiting X-ray Observatory 232	**Quarks** 239	
Escape velocity 241	**Roche lobe** 238	
Event horizon 245	**Schwarzschild radius** 245	
Glitch 232	**Singularity** 240	
Gravity wave 250	**Supernova remnant** 226	
GS 2000+25 249	**Synchrotron radiation** 230	

Section 3

Evolution of the Galaxies

Chapter 12

The Milky Way, Our Home in Space

Figure 12–1 The spiral galaxy M33 in the constellation of Triangulum is seen face–on. It resembles the appearance of the Milky Way galaxy as seen from a similar distance. It is difficult to see this galaxy with a small telescope because the feeble amount of light is spread out over the size of the full Moon in the sky. (Lick Observatory)

CENTRAL THEME: What are the components of and the organization of the Milky Way galaxy, and what do they tell us about its origin and evolution?

We have arrived at a watershed in my presentation of astronomy. Up to this point, my goal has been to present to you the methods of science and the evidence collected by scientists to support the theory that stars are born, live out their lives, and eventually die. Keep in mind that whatever our personal or collective conclusion, we cannot prove anything. We present evidence based upon observation, mathematical calculation, and experimentation — but we are still limited to the use of our inaccurate senses, our sometimes faulty reasoning skills, and our imprecise instruments.

To be sure, if we could magically have access to an astronomy textbook written in the year 2268 AD, we would not even recognize much of its table of contents. If reflections on the history of human knowledge are any judge of what we can expect for the future, much of the contents of this book will be the subject of jokes and laughter in the year 2268 — just as you and I chuckle about the theory of Earth-centered-ness that dominated human thought less than 400 years ago.

Now don't slam the book shut and refuse to read further — it is a stepping stone to future understanding, a survey of what we know so far. And upon it will be built the new theories of the next century, and so on and so forth. In light of this conclusion — as I have said all along — the importance of keeping up with current knowledge of astronomy or any other subject is not just to know what is believed, but to know *how those beliefs affect the way we choose to engage with life.*

Those beliefs must, consciously or unconsciously, affect our personal definition of the meaning of life. At the same time, what may be equally as important, is that we are aware of the process by which we claim to obtain knowledge about nature and the universe. Thus we may apply similar processes to our attempts to improve our relationships with other people — personally and internationally.

If we can assume that our broad outline of stellar evolution is accurate, then of course there are some important considerations we might want to integrate into our long-term strategies for the survival of our planet and civilization. What might be of equal importance is to reevaluate our responsibilities for our stewardship of this tiny speck of a planet upon which we evolved — with or without Divine intervention. What I am saying is that if in fact we are components of the process of stellar evolution, then seemingly the goals and values of our civilization should reflect that belief. Nothing appears to be permanent — you, I, wealth, beauty, stars, nor the universe itself.

Evolution of the Universe

This concept — that the universe itself may have had a beginning and may eventually come to an end — is the terrain found on the other side of the watershed we are crossing. The first Chapters dealt with the life cycle of stars. Now our attention is focused on the life cycle of the universe — its birth (past), life (present), and death (future).

As in the case of stars, my objective is not to brainwash you into believing any particular theory. After all, I am not running for office. And the universe is going to exist regardless of what theory I use to explain how it got here in the first place. Rather, my objective is to tell you about the evidence that has accumulated, propose some theories that attempt to explain that evidence, and then let you conclude what you will.

I have probably said enough about the Big Bang theory already that you suspect me of believing in it. Actually, I don't care for it. It seems too simplistic and too naive. I would prefer to believe that the universe is timeless — that it never had a beginning and that it will never have an end. It always has been and always will be. But the evidence so far is strongly against what I want to believe. And so perhaps I can't have my own way about it after all. I could be stubborn and ignore the evidence and hope that eventually evidence to support my own notions will be discovered. But somehow the ostrich philosophy doesn't appeal to me — I like to keep up with things that are happening.

Just what should our approach be in tackling such a huge subject like the origin of the universe? It would seem reasonable enough to study the contents of the universe and look for indications that an origin occurred. I have done that to a minor extent already. In Chapter 11, we learned that the origin of chemical elements can best be explained by the chemical processes going on inside of stars, living and dying. The conclusion we reached was that since the basic reaction by which stars shine is the conversion of hydrogen into helium, there must have been a time at which hydrogen was the only chemical in existence.

Another tidbit of evidence for a beginning was briefly mentioned in conjunction with the Doppler effect. The observation that all galaxies are moving away from one another suggests that there was a time at which they were all together and at a point from which they retreated. So it should be obvious that galaxies must be our next topic of discussion, inasmuch as they are the things retreating from the point of origin.

Stellar evolution occurs within the galaxies as they separate further and further from one another. The question before us therefore is this: "Assuming that galaxies did begin as clouds of hydrogen expelled from the Big Bang, how did these clouds evolve into the diverse assortment of galaxies we observe today? We begin our quest for understanding close to home by examining how our own galaxy is shaped, what its contents are, and how it behaves. Can we explain all of those properties in terms of a theory that says it was once a cloud of hydrogen? Well, not all. There are certainly a lot of missing pieces, just as the theory of evolution of life on Earth. It is a combination of the patterns observed in the universe and the laws of chemistry and physics that explains the patterns that count.

★★★★★★★★★

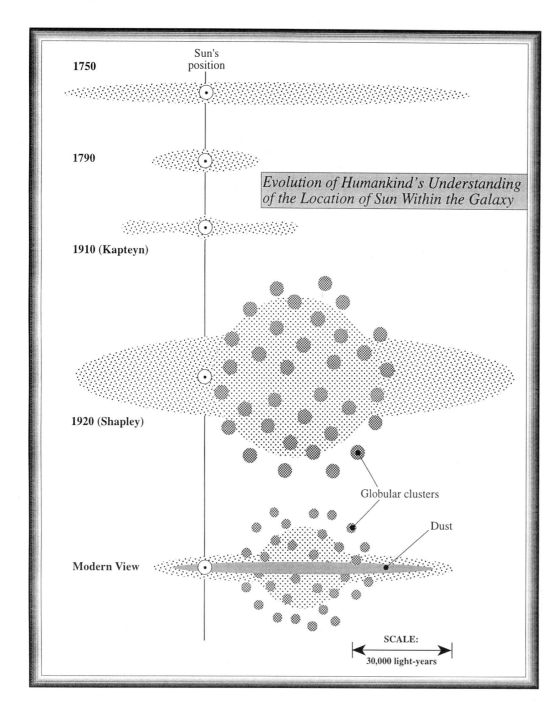

Figure 12–2 This diagram illustrates the evolution of humankind's knowledge about the extent of the universe of stars around Earth. Prior to the 1920s, the universe and the galaxy were one and the same. It is only recently that we have been able to reach deep into space and realize how utterly vast it is.

Discovery of our Galaxy

It comes as quite a surprise to many people to learn that up until the early 1930s, astronomy textbooks did not contain a chapter on our galaxy — or *any* galaxy, as far as that is concerned. Up until the 1920s, it was generally believed that galaxy and universe were synonymous — they were one and the same. It is only within the last 60 years that the existence of other galaxies was even known! True, the observed stars and nebulae were thought to form a large structure, but that structure <u>was</u> the universe.

Galaxy and Universe as One The generally-accepted size for the universe (galaxy) was about 50,000 light-years in diameter and about 6,000 light-years thick. The exact numbers were debated to and fro, but there wasn't much data to go on. Some fuzzy patches of light that we today know as other galaxies or systems of stars were visible through telescopes, but there was as yet no way to know or determine their distances. Without that vital piece of information, astronomers were unable to model a 3-dimensional view of the universe. Everyone agreed that the universe was flattened. The fact that there is a greater concentration of stars along the band of the Milky Way was enough evidence for that conclusion to be drawn (Figure 12-2).

Thus it was believed (for the sake of convenience) that the fuzzy patches were simply members of the one large system known as the universe. Some scientists speculated that the patches of light were other, separate systems of stars. But their opponents pointed out that if that were the case, then they should be evenly distributed around us. But they are not — they are concentrated along the Milky Way band. Furthermore, it was generally believed that Sun was pretty much located close to the center of the universe. This conclusion was not based on the belief that the universe revolved around Sun, but on the uniform manner in which stars are observed to be distributed in the sky.

The basic problem was that no one was in a position to determine just how far away the distant stars and fuzzy objects were. They were in the telescopes for everyone to see, but there were no techniques available to know their distances — so it was easiest to assume that they were all part of a big system that was conveniently called the universe. Try it out for yourself. Go outside and stare at the sky and ask yourself — "How far away do those dots and fuzzy patches appear to be located? So finding an accurate method of determining distances to things in space became the basis for another revolution in human thinking.

Observing the Milky Way Check out something else while you are out under a dark sky. Although the stars appear to be randomly placed in the sky, they appear to be more densely populated as you look toward the band of white we call the Milky Way (see Figure 2-3). We now know that the band consists of millions of stars, a fact first recognized by Galileo using his first telescope. Now look along the Milky Way band itself, and attempt to locate a direction in which the band appears particularly bright (populated) with stars. The band runs completely around Earth, so you will have to perform this experiment again later in the year when the rest of it is not drowned out by Sun.

At the conclusion of your experiment, you arrive at the same conclusion that generations of sky-watchers concluded — the Milky Way band is rather uniformly bright in all directions. Therefore Sun must be located fairly close to the center of this vast system of stars. If we are located close to the edge of the system, the density of stars will appear greater as we look toward the center (Figure 12-3). You will observe something else about the sky — the density of stars is less in the directions away from the Milky Way band. Therefore, we must live in the center of a flattened system

Figure 12-3 This photograph is typical of those taken in or near the middle of the Milky Way band. If you like challenges, try counting the number of stars in this one small region of the sky. There is also a cloud of gas called the North American nebula in the photograph. Can you find the pattern for which it is named? (Lick Observatory)

Figure 12-4 The study of a special type of star in the globular clusters such as this one—M92—allowed astronomers to conclude that Earth was not located even close to the center of the galaxy. (Lick Observatory)

(universe) of stars. This was the state of the knowledge in the scientific community in the early 1920s.

Observing Globular Clusters Meanwhile, the American astronomer Harlow Shapley (1885-1972) was hard at work studying a particular type of star cluster – the **globular cluster**. We encountered these groups of very old stars (K- and M-type) in chapter 8 when we were searching for examples of the oldest stars in the universe (Figure 12-4). Shapley didn't know anything about their ages – he was only interested in their locations in the sky. What he discovered about their distribution in the sky revolutionized our thinking about the size and extent of the universe.

Here we encounter one of the most profound problems in all of astronomy – determining the distance to an object. We will stumble through this problem throughout the rest of the book. Since our theories about the origin and eventual fate of the universe are based in a fundamental way upon our knowledge about how far away objects are, we should be particularly alert as we look at our methods. If those methods are based upon faulty assumptions or faulty reasoning, the entire edifice of the Big Bang theory could come tumbling down.

Determining Distances to Objects

We have already learned how to use two methods of distance determination – *trigonometric parallax* and *spectroscopic parallax*. Trigonometric parallax is based solely on the assumption that objects positioned in space can be treated as if they are dots on a piece of graph paper – and on the assumption that rules of trigonometry can be used to determine the angles and distances between them. Spectroscopic parallax is based upon the assumption that stars whose spectra are exactly alike occupy the same position on the H-R diagram – and further that radiation leaving their surfaces obeys the inverse square law. None of these assumptions are controversial, so astronomers consider the distances determined by these two methods to be reasonably accurate.

Cepheids as Distance Indicators In 1912, one of the most famous of women astronomers, Henrietta Leavitt, discovered a particular type of star that varied in apparent visual magnitude (m_v) in quite a regular fashion. Because the first such star she found (δ Cephei) is located within the boundaries of the constellation *Cepheus*, all stars imitating its behavior are called **Cepheid variable** stars (Figure 12-5). Ms. Leavitt plotted the stars' m_v's against time. Some Cepheids go from maximum m_v to maximum m_v in a single day – some take as long as 50 days to do the same. At the time, she did not know why they varied. She assumed they were going through an unstable phase in their life cycle.

Importantly, she recorded a large number of Cepheids embedded in and assumed to be associated with one of the

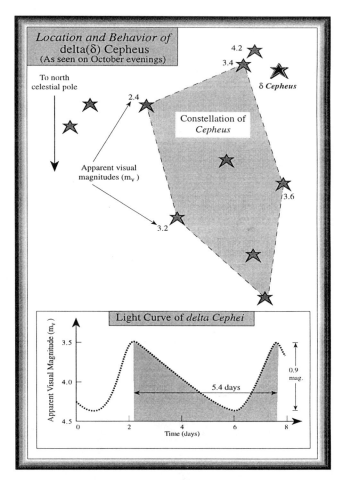

Figure 12-5 By plotting the apparent visual magnitude of stars against time, astronomers frequently identify a particular class of stars called Cepheids. They can be used for determining the distances to star clusters and nearby galaxies because there is a relationship between their pulsation periods and their luminosities.

fuzzy patches of light then called a *nebulae*. What she discovered in this particular case was that the brighter Cepheids had longer periods of variability than the dimmer ones. In fact, she found that there is a direct relationship between the average m_v of a Cepheid and the period of time it takes to go from maximum back to maximum again.

Big deal, you say? Well, don't you see – if the Cepheids are indeed associated with the nebula, then they are all approximately the same distance from us. Therefore, there must be the same direct relationship between the period of variability and the *absolute* visual magnitude as between the period of variability and the *apparent* visual magnitude. But there is still a problem – we have values for the m_v's of the Cepheids (use of the photometer), but we don't know the M_v's without knowing the distance to the nebula and therefore the Cepheids in the first place!

And that is where Shapley entered the picture. In effect, he determined the distances to some of the Cepheids that are close to Sun using the method of spectroscopic parallax. Having done that, he was then in a position to

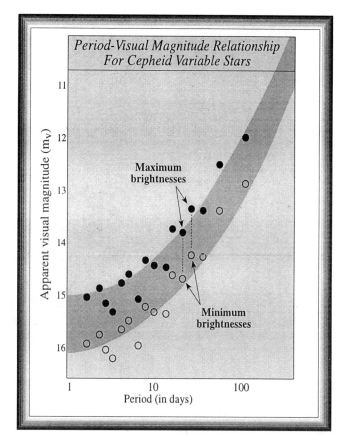
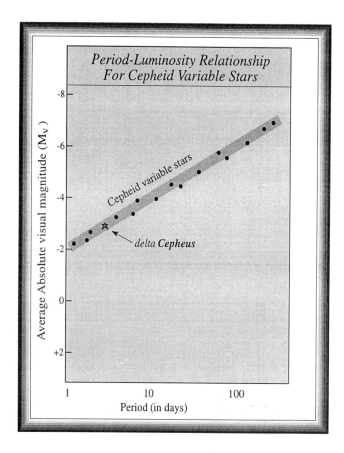

Figure 12–6 Shapley's original plot of Cepheid variable stars reveals the connection between period of pulsation and apparent visual magnitude. Whenever patterns such as this occur in nature, there is some kind of physical law operating.

Figure 12–7 Once the distances to a few Cepheid variable stars were found, it was possible to establish a relationship between period of pulsation and absolute visual magnitude. This diagram illustrates that relationship, and hence can be used to determine distants to objects in space containing these types of stars.

associate a particular Cepheid's pulsation period with that star's M_v — the assumption being that all Cepheids with the same pulsation period has the same average M_v, or average luminosity. When the observed M_v's for a number of Cepheids are plotted against time, we obtain what is known as the **period–luminosity** relationship. An obvious conclusion from looking at such a graph is that for Cepheid variable stars, the *longer the period of pulsation, the greater the luminosity* (Figure 12–6).

Given this discovery, the technique for finding the distance to a given globular cluster was straight forward:

- (1) focus the telescope on a Cepheid variable star located within one such cluster,

- (2) record the apparent visual magnitude each night for a period of several days or weeks,

- (3) determine the average value of that magnitude and the time interval between maximum and maximum,

- (4) look up on the period–luminosity diagram the absolute visual magnitude of a Cepheid with the period so determined, and

- (5) use that M_v and the average m_v and the inverse square law to calculate the distance to that Cepheid (and presumedly the cluster in which it appears) (Figure 12–7).

You may be wondering — isn't it possible that the Cepheid variable stars we use to determine the distance to a cluster just might be located somewhere between us and the cluster, and not be associated with the cluster at all? Of course it could. But we don't base our determination of distance on a single star. We use as many as we can observe, and average out the distances calculated for each. If any of them are significantly more or less than the average, then we ignore those values. In actual practice, although stars appear to be packed incredibly close together in photographs of the sky, the chances of alignment of a Cepheid in front of a globular cluster is extremely small.

CHAPTER 12 / The Milky Way, Our Home in Space

★★★★★★★★★★

The use of Cepheid variable stars for determining distances can be used not only with stars clusters within our own galaxy, but other galaxies as well. As long as we can identify a star in another galaxy as having a light curve pattern similar to those in our own, we can determine the distance to the star and therefore to the galaxy to which it is gravitationally attached. In the next Chapter, you will learn that this information is vital to our understanding of the rate of expansion of the universe, and therefore its age.

The galaxy M100 (Figure 12-8) is a spiral galaxy in the Virgo cluster of galaxies. Just recently, the Hubble Space Telescope, operating above the turbulence layers of Earth's atmosphere, was able to image several Cepheid variable stars in the outer regions of the spiral arms (Figure 12-9). By photographing the stars at various intervals of time, it enabled astronomers to determine the most accurate distance measurement to the galaxy (and therefore the cluster of which it is a member) ever. It is 56 million light-years from our Galaxy.

Locations of Globular Clusters Having catalogued the globular clusters according to their positions in the sky and their distances from us, Shapley then plotted their locations relative to Sun in a 3-dimensional model. What was obvious to him was that they were not distributed uniformly around Sun — they appeared to be concentrated in one particular direction in the sky (refer back to Figure 12-2 to see how this is explainable).

Figure 12–8 The spiral galaxy M100 is the most distant galaxy in which Cepheid variable stars have been measured accurately. This allows for a more accurate determination of distance to the galaxy, which is about 56 million light-years away. (NASA/STScI)

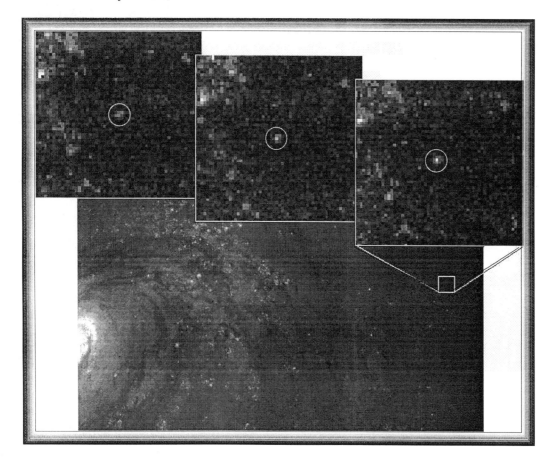

Figure 12–9 A Cepheid variable star is clearly visible in these three photographs of the outer region of the spiral arms of the galaxy M100. Notice the change of brightness of the star during the intervals between exposures. The pattern of this change allows astronomers to identify it as a Cepheid variable star. (NASA/STScI)

SECTION 3 / Evolution of the Galaxies

Proper motion studies of stars had already shown astronomers that stars move relative to one another — a conclusion in keeping with Newton's laws of motion. Objects in space move around a center of gravity. So if there is a geometrical center of the system of stars comprising the known universe, the globular clusters should be revolving around that center. At any particular moment, they should be distributed in a halo surrounding the center of the universe (Figure 12-10).

If indeed Sun is located at or close to that center, globular clusters should be equally distributed around Sun. But as I just pointed out, Shapley found that they were not — so Sun must not be located close to the center. The center must be located at the center of the swarm of globular clusters which was — according to Shapley's model — in the direction of Sagittarius. The calculated distance from Sun to that center was about 30,000 light-years! We are located in the suburbs after all.

Discovery of Island Universes As if that were not enough to bash our claim of importance in the universe, another dramatic discovery was unfolding about the same time. In the mid 1920s, another American astronomer — Edwin *Hubble* (1889-1953) — was using Cepheid variable stars for determining distances to some of the nebulae long seen in the sky. His reason for undertaking this task was based not only on the discovery of the period–luminosity relationship of Cepheids, but on the recently built 100-inch reflecting telescope at Mount Palomar in Southern California.

Being the most advanced telescope in the world at the time, its great resolving power enabled Hubble to isolate and identify individual Cepheids in nebulae which previously appeared as tiny, fuzzy patches through even the best of telescopes. Now he was able to resolve individual stars within the fuzzy patches, and some of those stars revealed the familiar pattern of Cepheid behavior. When he calculated the distances to the stars — and therefore the patches themselves — he found that the values exceeded those currently accepted for the size of the universe itself. They were located, in other words, outside of the known universe — or at least outside the system of stars and patches that were thought to make up the known universe!

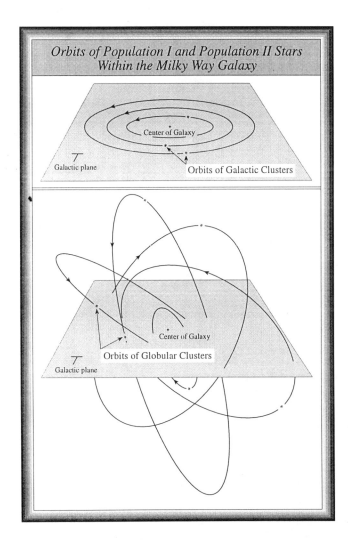

Figure 12–10 Different types of star clusters orbit our galaxy in different patterns. Clusters of young stars orbit within the plane of the galaxy, while the older star clusters orbit within a spherical shell around the center.

Figure 12–11 The Andromeda galaxy (M31) is one of the nearest galaxies to our Milky Way, and is visible to the naked eye from a dark location. It is shaped very much like our galaxy, and has similar contents as far as stars are concerned. (Lick Observatory)

★★★★★★★★★★

Well, that just wasn't possible. If we use the word "universe" to mean all that we observe in the sky, then Hubble was simply discovering that (1) the universe was much larger than currently believed, and (2) the universe consisted of clumps or families of stars beyond our own family of stars. We belong to one family, but there are many other families out there as well.

Since all of the stars seen with the naked eye are members of our system, and since the greatest concentration of those stars are seen along the Milky Way band, astronomers decided to use the expression "Milky Way" for the name of our family of stars. Of course, we now call the separate families of stars "the galaxies." The Milky Way is our galaxy. The Andromeda nebula, visible to the naked eyes of the earliest people on Earth, is another galaxy. Curiously enough, since the Andromeda galaxy (Figure 12-11) had been such an interesting target for telescopic viewing — because it was a patch of light instead of a single dot — you will see its designation on star charts even today as the "Great Andromeda Nebula." Even though we know that it is some 2 million light years from the Milky Way galaxy, its early description as a nebula has stuck in our vocabulary.

Not all of the patches or nebulae were found to be exterior galaxies. Some were found to be located within the accepted dimensions of the Milky Way. But almost overnight, the known universe went from a mere 50,000 light-years in diameter to millions and millions. And who knew how many additional little patches were out there yet to be discovered! So improvements in instruments (telescopes, in this case) did it again! What Galileo started with his puny little telescope is being furthered at this very moment, as astronomers design even larger and more sophisticated telescopes not to look for anything in particular, but knowing — or at least having faith — there are some incredible things out there yet to be discovered.

As you realize the importance of what Hubble was just about to discover with the 100-inch telescope, you'll be able to appreciate the excitement astronomers feel when they begin using a newly-constructed telescope for the first time. It must be like Christmas morning! By the end of the 1920s, discoveries with the Palomar telescope not only increased our humility in locating Sun out in the boondocks of the Milky way, but also revealed that the Milky Way is but one of billions of galaxies.

An Expanding Universe of Galaxies Are you wondering about this newly-discovered arrangement of galaxies? You may be asking — "If Newton's laws of motion are applicable to the universe as a whole, then around what point are these newly-discovered galaxies revolving?" They are, after all, considered to be objects in space just as stars are considered to be objects in a galaxy. If stars orbit the gravitational center of a galaxy, what do the galaxies orbit around? The answer to that question gets us a little ahead of the flow of the story — I will deal with it in the next Chapter. Suffice it to say that Hubble's discovery of a method for determining distances to the nebulae-turned-galaxies could be married together with another feature of the nebulae-turned-galaxies that had been recorded previously.

One of the very first studies performed on a newly-discovered object is that of obtaining a spectrum of its radiation. Spectral analysis is fundamental to knowing the chemistry and behavior of any object in space. With the larger telescope, it was possible to obtain much better spectra for these nebulae-turned-galaxies than were previously available, and what was discovered was that the lines in all of the spectra of the galaxies were shifted toward the longer wavelength end of the spectrum (red end). If we assume that the shifts in wavelength are the result of the Doppler Effect, then all of the galaxies are moving away from one another! The universe is expanding!

What the galaxies are "moving around" is the event that caused them to be expanding in the first place — what we have been referring to as the Big Bang. But let us leave it at that — there are many more discoveries to introduce you to before the other details of the Big Bang can be fully appreciated. But at least you can understand why Hubble's name is so revered in the astronomical world — why even an orbiting satellite was named for him. He made three important discoveries about the large-scale structure of universe that still have us working hard to understand. But we need to return to the interior of our own galaxy to figure out what is going on in the more local regions of space.

Properties of Star Clusters

As the 100-inch telescope was aimed at the nebulae located inside the Milky Way galaxy, astronomers found them to be concentrations of bright stars, gas, and dust. These nebulae, you recall, are found primarily along the Milky Way band in the sky. So discoveries in the 1920s were leading astronomers to the conclusion that stars are very often contained within groups of stars within the galaxy.

Astronomers identified two general types of groupings, or clusters – *globular* and *galactic* (or *open*). The **galactic clusters** are located along the galactic plane (hence the name given them). The globular clusters, on the other hand, are found away from the galactic plane — in a halo around the flattened shape of the galaxy (Figure 12-10). It was these that Shapley used to determine Sun's position in the galaxy. Astronomers believe that the locations of the two types of star clusters should provide important clues about the galaxy's origin since they differ profoundly with respect to the types of stars contained therein.

Globular Clusters Spectral analysis of the stars in the globular clusters reveals them to be primarily lower main-sequence stars — the G-, K-, and M-type stars. Those stars that stand out most brilliantly are red giants. Plotting a large number of stars of a globular cluster on an H-R diagram

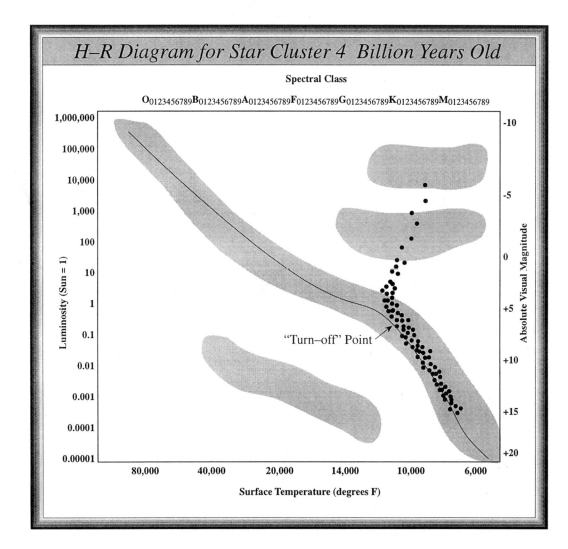

Figure 12–12 The stars of a globular cluster are plotted on an H–R diagram. What is interesting is the pattern, which, according to stellar evolution models, indicate that the stars—and therefore the cluster itself—is quite old. Many of the massive stars are headed toward red giant stage.

allows us to arrive at some important conclusions about this type of cluster (Figure 12-12). We assume that since all of the stars in a cluster are bound together by their mutual gravitational attraction, they must have a common birth in time — they are all the same age, in other words. If that is so, then their distribution throughout the H-R diagram must be due to the fact that they have different masses and therefore have had different rates of evolution. Recall that massive stars do not live as long as less-massive stars. "Live" means time on the main sequence.

The reason there are no stars on the upper part of the main sequence (high surface temperature and luminosity) is that the massive stars that were once there are in the process of dying — they are the red giants plotted on the upper-right portion of the diagram. The dashed tracks connecting those two regions represent the changes those stars would have taken while evolving from main sequence stars into red giant stars. A diagram of plotted stars like this can be used for obtaining an approximate age for the entire cluster of stars — again assuming that all stars had a common birth and therefore have a common age.

The point at which the stars of a cluster appear to leave the main sequence — called the **turnoff point** — can be associated with a star of a particular mass. Obviously, stars more massive than that particular star have already left the main sequence toward death. Those less massive have not yet left. This plot of stars in a cluster allows astronomers to estimate the age of the cluster. Presumedly, the stars in a cluster had a common origin from the same cloud of gas and dust. They all have the same age. The most massive formed rapidly, and have left the main sequence toward death. The least massive formed slowly, and are still main sequence stars.

Look at the plot of stars of the cluster in Figure 12-12. The star that is just about to leave the main sequence and head for red-giant stage is of spectral type K0. According to the chart in Figure 8-23, this type star lives on the main sequence for about 17 billion years before consuming its hydrogen fuel and swelling into red giant stage. Therefore, it must be at least 17 billion years old — it arrived on the main sequence 17 billion years ago. Now the age of that star must be the approximate age of the cluster itself. If it were any younger, some of the stars more massive than it would still be on the main sequence. If it were any older, then that star and others less massive would already have left the main sequence.

★★★★★★★★★★

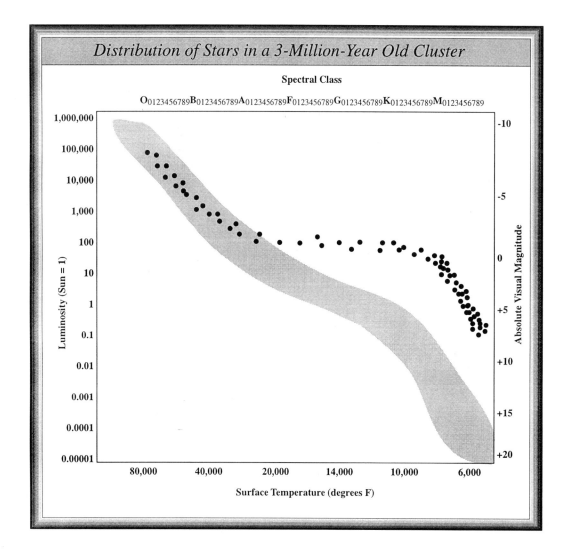

Figure 12-13 The stars of an open cluster are plotted on an H-R diagram. What is interesting is the pattern which, according to stellar evolution models, indicates that the stars—and therefore the cluster itself—is quite young. Many of the less-massive stars have not yet collapsed to the point that they are main sequence stars. They are proto-stars. Compare this plot with that of the globular cluster stars in Figure 12-12.

Galactic Clusters Let's approach the galactic clusters in the opposite manner. Rather than me telling you what spectral analysis tells us about the types of stars in them, let's look at the plot of stars in a typical galactic cluster on an H–R diagram, and then ask ourselves what kind of stars they are. Figure 12-13 is such a plot for the cluster known as the *Seven Sisters* – the Pleiades. They are easily seen during the fall and winter in the constellation of *Taurus*. They are frequently mistakened for the *Little Dipper* constellation.

In this plot, the distribution of stars along the main sequence is significantly different than that of globular clusters. Is it obvious that the Pleiades is a younger group of stars than the globular cluster? None of the hot stars have evolved very far away from the main sequence. Since we know that such massive stars burn up their fuel rapidly and hence do not remain on the main sequence very long, this cluster must be young.

As we did in the case of the globular cluster, we identify the spectral type of the star located at the *turnoff point*. It appears to be a B0 star. Using Figure 8-23 again, we find that a star of that type remains a main sequence star for approximately 11 *million* years. Therefore, the open cluster called the *Pleiades* is about 11 million years old. The bright stars of this cluster that appear so brilliantly in the sky are young O- and B-type stars. Spectral analysis tells us that directly, but I want you to appreciate how two different ways of studying star clusters can lead to the same conclusion.

Actually, there is a more direct way of concluding that the stars in the *Pleiades* are young stars – observe them through a telescope. What is easily seen is similar to what is revealed in the photograph of the *Seven Sisters* in Figure 9-2. Looking carefully around the bright stars, you notice some nebulosity – a telltale sign that there are clouds of gas and/or dust near the stars. But the presence of gas and/or dust suggests that star birth must be occurring in the vicinity, and of course young stars are going to be found close-by. Globular clusters, on the other hand, contain very little visible gas or dust, and therefore do not have the necessary ingredients for star formation. They consist of old stars – which we concluded from looking at their star plots on an H-R diagram.

None of this should surprise you too greatly. We learned earlier that unlike the open clusters that are located

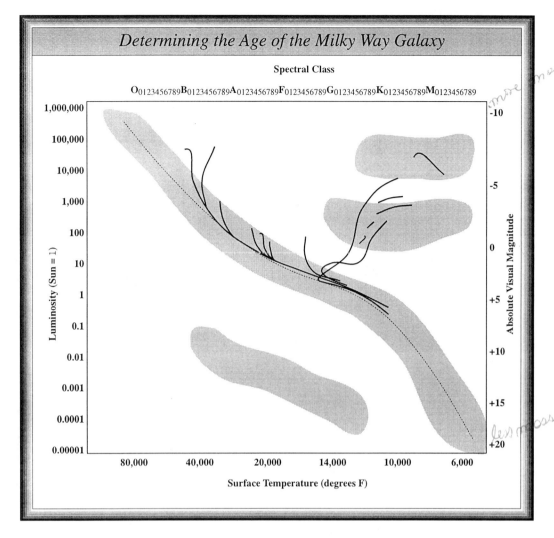

Figure 12–14 *When all globular clusters in the Milky Way galaxy are plotted on a single H–R diagram, astronomers are in a position to determine an approximate age for the galaxy. The age of that star at the lowest turn-off point must be the minimum age of the galaxy.*

almost exclusively along the gas-and dust-rich Milky Way band, the globular clusters are mostly located away from that band in a halo around the center of the galaxy. There are no great concentrations of gas or dust in the halo — it was used up in the making of the stars forming the globular clusters.

Age of Milky Way Let us carry this method of determining the age of a star cluster one step further. The further down on the main sequence a star cluster's turnoff point is, the older that star cluster is. If we plot all star clusters belonging to the Milky Way galaxy on a single H–R diagram, and calculate the age of that cluster whose turnoff point is lowest on the main sequence, we can approximate the age of the entire galaxy (Figure 12-14).

The galaxy is at least as old as its oldest star cluster. When we do that, we arrive at a figure of 15 billion years for the age of our galaxy, the Milky Way. It may be older than that, but it can't be any younger. That conclusion puts Sun in better perspective. Since its estimated age is 5 billion years, the Milky Way was already 10 billion years old by the time a cloud of gas and dust collapsed to form Sun and planets.

Chemistry of Star Clusters Aside from the locations of the different star clusters with respect to the Milky Way's shape, and the types of stars populating those clusters, there is another characteristic of stars that provides important clues about the Milky Way's origin — chemical composition. Since 99% of the universe consists of atoms of hydrogen and helium, it is convenient to lump the remaining 90 elements into the category of **metals**. This is not what comes to mind when you think of the word "metal," but scientists occasionally enlarge the definition of an existing word rather than invent a new one.

The percentage of *metals* contained in a star — not surprisingly — is calculated from analyzing the spectrum of the star. The darker the absorption lines associated with a particular chemical, the higher the percentage of that chemical that is present in the star. So we have another classification system into which stars can be divided — one that is based upon metal content. This classification is referred to as **Population type**. The four Population types found in our galaxy and the percentage of metals in each is shown in Figure 12-16. Study it carefully.

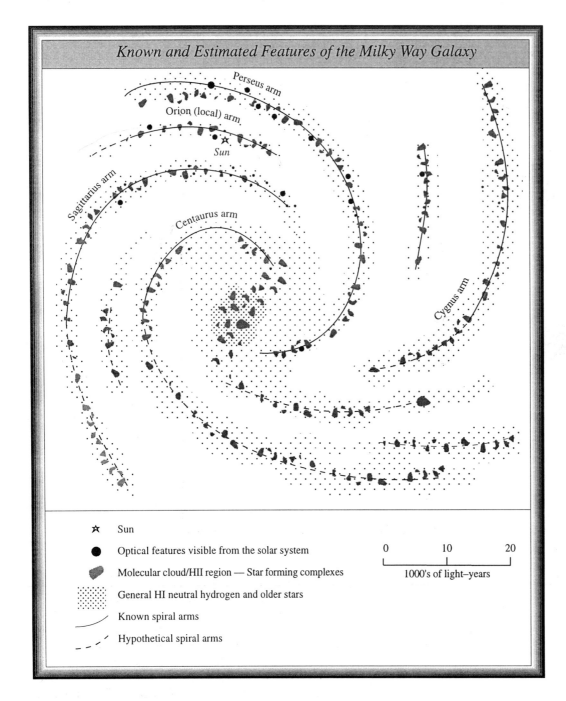

Figure 12–15 A diagram of the major features of the Milky Way galaxy, and Sun's location in it.

Population Types What is rather interesting about this classification system is that it is closely associated with the ages of stars in the galaxy and their placements within the galaxy. That is, you notice that **Population II Extreme** stars (old and metal-poor) are located in the globular clusters, **Population II Intermediate** stars (oldish and metal-poor-ish) are located in the nucleus of the galaxy, **Population I Intermediate** stars (youngish and metal-rich-ish) are located in the disk of the galaxy, and the **Population I Extreme** (young and metal-rich) are located within the spiral arms of the galaxy (Figure 12-16).

Population Type and SETI One of the interesting applications of categorizing stars according to their Population types is in the *SETI Program*. Astronomers do not assume that a given star in the galaxy is just as likely to harbor intelligent life as any other star. We just found that stars with a given Population type share a specific age, location within the galaxy, and chemical composition. Using those characteristics and a couple of reasonable assumptions, we can assign a priority to each star as to the probability that it has intelligent life in its vicinity.

Assuming that (1) Earth is typical of planets in orbit around stars in the universe, that (2) life here required 5 billion years to evolve to intelligence, and that (3) the evolution of intelligent life requires the presence of an

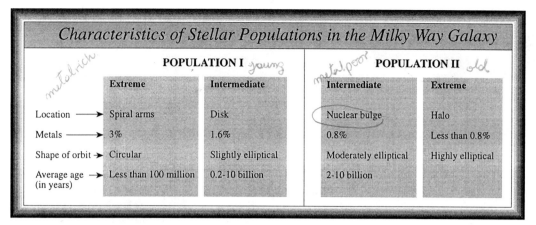

Figure 12–16 A chart of the characteristics of stars in the two Populations of stars shows drastic differences between them, suggesting that their evolutions have been different. This ultimately leads astronomers to a model for the formation of the galaxy.

abundance of heavy elements to form the building blocks of life, some stars are more probable to have intelligent life around them than others. Stars that live less than 5 billion years (are on the main sequence for less than 5 billion years) may experience an attempt by life on their planets to evolve toward the complexities of intelligence, but there will be insufficient time for it to achieve the level required to send or receive messages across interstellar distances.

Life may get a start around numerous stars in the universe, but if there is insufficient time for that life to achieve intelligence (in the sense that it can communication with us), it doesn't do us much good to try to establish contact with it-either by sending messages or listening for them. You noticed in Figure 8-23 that stars hotter than spectral type F0 live for less than 3 billion years. Calculations suggest that stars hotter than F5 stars live for less than 5 billion years as main sequence stars. All Population I Extreme and some Population I Intermediate stars are ruled out as probable locations for intelligent life.

Likewise, those stars that have a low percentage of metals in their chemical compositions are unlikely to possess planets on which there are the necessary chemicals for forming the chemical building blocks of life (whatever they may be). It is difficult to imagine that intelligent life can evolve out of the very lightest of gaseous elements like hydrogen and helium. The older stars (Population II Extreme and some Population II Intermediate) constitute another group of stars that astronomers rule out as having intelligent life. Calculations suggest that K3 and cooler stars are in this category. That leaves a small range of stars (by spectral class) of F5 to K3 that are the most likely stars around which to find intelligent life. This short analysis certainly doesn't guarantee success, but it certainly makes the task of the astronomer easier. After all, there are approximately 100,000,000,000 stars in our galaxy alone. We would like to at least have a starting point for a search, rather than wasting time on unsuitable stars. We can always go back and search for signals from what we think are unsuitable stars later on.

Shape of the Milky Way Galaxy

Spiral arms of the galaxy? Where did they come from? Just as we have improved our ability to determine the shapes and contents of other galaxies since the 1920s, we have also discovered that many galaxies reveal spiral shapes reminiscent of giant pinwheels. The photograph in Figure 12-1, for example, reveals some of the marvelous detail of one of these spiral galaxies. So too we have developed methods of determining the shape of our own galaxy, overcoming the inherent limitations we face by virtue of living inside of it and not having the means to view it from outside. We can more easily determine the shape and contents of a galaxy like that of Andromeda than our own.

Do you recall why it was that astronomers prior to the 1920s had assumed that our Sun was located close to the center of the universe (Milky Way)? Because the Milky Way band appeared equally bright in all directions. But if we are indeed located out in the suburbs — as Shapley proposed — how then do we explain the observation that the band of the Milky Way is uniformly bright in all directions? If there is a greater concentration of stars between us and the center of the galaxy than there is between us and the outer edge because we are located in an off-centered position, why isn't the sky in the direction of the center of the galaxy brighter with stars?

If it were just a matter of there being stars along the flattened plane of the galaxy, we would expect the direction toward the center to be brighter with stars. But what we didn't know before the 1920s was that between those stars are vast clouds of gas and dust, all of which block from our view the light of the billions of stars behind them. Our view of the 100 billion or so stars in the galaxy is inhibited by the presence of these mostly unseen clouds of gas and dust. In effect, from our position in the outer edge of the galaxy, we visually see objects only a short distance (percentage wise) into the galaxy in all directions.

To illustrate this point, using a model of our galaxy that is 12 inches in diameter, we can observe and study and catalogue visually all of those stars that are within a quarter's worth of distance around us. And yet there it is — a complete

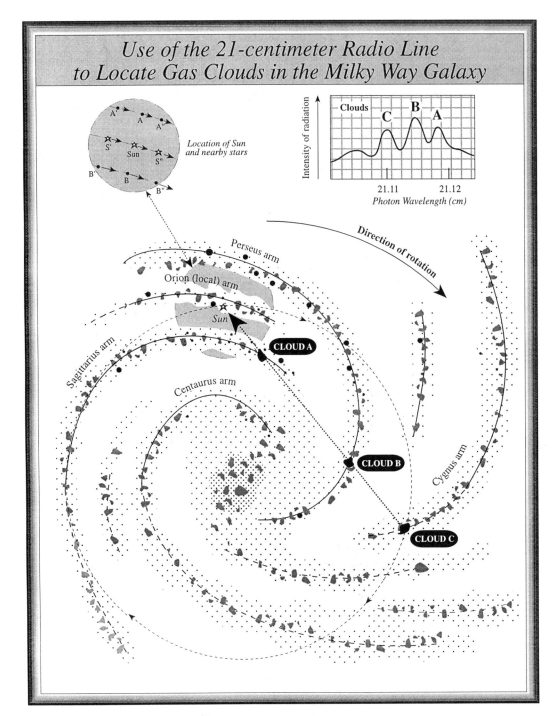

Figure 12–17 Since Sun is surrounded by clouds of gas and dust that inhibit our view of our galaxy's larger structure, radio astronomers use radio telescopes to detect and locate clouds of hydrogen. By mapping those locations, and knowing that hot, young stars are located in such clouds, the spiral arms of our galaxy have been mapped.

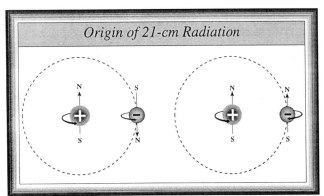

Figure 12–18 An illustration of the origin of the 21–cm wavelength of radiation from hydrogen in interstellar space. When atoms collide, the electron "flips" and emits a photon in the process.

Figure 12–19 A photograph of a spiral galaxy (M81) seen face–on. (Lick Observatory)

Figure 12–20 A spiral galaxy (NGC 5866) seen edgewise. Notice the dust along the galactic plane. (Lick Observatory)

model of the Milky Way galaxy. If we can see a mere quarter's worth, how then do we know what the contents of the rest of the galaxy are?

Location of Spiral Arms First of all, we learned in Chapter 9 that hot, young, luminous stars are almost always found in the vicinity of clouds of gas and dust. Figure 12-15 shows a diagram of the location of hot stars and gas and dust clouds that are visible from Earth and Earth-orbiting satellites (black dots in the diagram). The map certainly suggests that in our particular vicinity of the galaxy there is a pattern of spiral-ness in the distribution of hot stars and gas and dust clouds.

But how safe are we in assuming that such a pattern is extended throughout the rest of the galaxy? Well, let us assume that where we locate clouds of gas, we can expect to find hot stars and dust as well. In other words, if it is reliable to associate gas clouds with star formation — as we found we could in the regions close enough to us for study — then the assumption is a good one. So that is what we do. We map the locations of gas clouds within the Milky Way galaxy, using radio telescopes tuned to the 21-centimeter wavelength, and then fill in those areas with patterns of stars similar to those found in the regions right around us.

But how do we map the locations of these clouds of gas? Because 90% of all atoms in the universe are hydrogen atoms, and hydrogen atoms reveal their presence in the form of the 21-centimeter wavelengths of radiation — detectable with the use of radio telescopes. Figure 12-17 illustrates the use of this technique. Astronomers program the radio telescope to slowly scan the sky (even during the daytime), and chart recorders plot the strengths of the signals coming from each direction to which the telescope is pointed. Then a map of the distribution of hydrogen clouds is made from this data. This is possible only because radio radiation is unaffected by stars and dust clouds within the galaxy — quite unlike the visible radiation that is easily absorbed or scattered.

The origin of the 21-centimeter wavelength, by the way, is due to the behavior of neutral hydrogen atoms in space. When two atoms collide, one or both of the orbiting electrons will flip in their direction of spin (relative to the spinning proton), and the slight difference in energy level is released as a photon (Figure 12-18). The "blips" of 21-centimeter radio energy coming from hydrogen gas clouds don't tell us directly how far away the clouds are — they tell us only the strength of the signal and the direction from which it came. We need additional information to determine distance. To obtain that, we invoke the by-now familiar Kepler's Third Law — objects close to the center of the galaxy move at different rates than those further out toward the edge of the galaxy.

CHAPTER 12 / The Milky Way, Our Home in Space   **267**

★★★★★★★★

From our position two-thirds of the way out from the center, along a given line of sight the radio telescope receives 21-centimeter radiation from any number of gas clouds. But since each is located at a different distance from the center, each is moving at a different rate both relative to the center of the galaxy and to us. So the "blips" of energy coming from separate clouds will be Doppler-shifted by different amounts. The pattern of "blips" can – in conjunction with Kepler's Third Law – therefore be used to determine how far from us each cloud is located.

Other Spiral Galaxies So far so good. But perhaps we should test the assumption that where there is hydrogen, there are hot, young stars. So, too, we should test the applicability of Kepler's Third Law to this problem of determining the distances of the gas clouds from us. Simple. We simply observe nearby spiral galaxies – determine the distribution of hydrogen gas using radio telescopes, and the distribution of hot, young stars using spectral analysis. It checks out perfectly.

Measuring the Doppler effect evident in the spectral lines of gas clouds in the spiral arms of the galaxies also confirms our forgoing conclusion. If you look carefully at the photograph of the spiral galaxy in Figure 12-19, you see present in the spiral arms all of the very features that we associate with the regions of space close to Sun – hot, young stars, emission nebulae, and dust clouds. They are all there. So it appears that our assumption of commonness is a valid one.

Oftentimes, due to the randomness with which the galaxies are oriented in space, we observe spiral galaxies not face-on as in the example of Figure 12-19, but rather edge-on – as in Figure 12-20. The darkish band running straight through its entire diameter is the result of dust clouds within the spiral arms blocking from view (absorbing) the light from the vast number of stars located behind those clouds. In addition, we see some of the globular clusters that are located in the halo of the galaxy.

You know, it is rather like our being forever confined to a single house in a large housing tract. We are unable to walk out the door to see how our house appears to our neighbors. But by observing their houses through our windows and recognizing features in them that are similar to those of our own, we get a general idea of what ours must look like. That is what we do with other galaxies. We assume that *features we observe in other galaxies from the outside that appear similar to features we observe in our galaxy from the inside must be similar or identical.* If nothing else, we at least assume that the laws of physics and chemistry are identical. Comparing other galaxies to our own has been possible only within the past 60 years. Imagine what discoveries will be made during the next 60 years!

Commonness of Types of Stars I should make another important point before we use all of these discoveries to formulate a model of how our galaxy got to be in the first place. That point revolves around a conclusion arrived at in Chapter 8, when we compared the characteristics of the nearest stars to Sun to those of the brightest stars in the nighttime sky. The conclusion we reached was that luminous stars – such as O- and B-type stars – are easily seen over great distances, but that the less luminous K- and M-type stars are more numerous. Using the data observed and recorded for *stars in our immediate neighborhood*, we conclude that 90% are main sequence F through M stars, 9% are white dwarf stars, 0.5% are red giant stars, and 0.5% are everything else.

All of the stars more luminous than red dwarfs and white dwarfs together account for only 24% of all stars. Yet they account for 99% of the luminosity emitted by our entire galaxy. *76% of the stars provide only 1% of its radiation!* So when we observe another galaxy, we must keep in mind that what stands out and appears obvious is not typical of what is actually there. Mostly what we see are the hot, young stars. Mostly what is there are the less luminous red dwarf stars and white dwarf stars. Knowing that, we soon see, helps us to understand some additional features of spiral galaxies.

Evolution of the Milky Way

Although the evidence for the origin and evolution of the Milky Way galaxy is both scant and circumstantial, the nature of the scientific method is to form a model based upon whatever data is available – then make revisions as new information and observations are made available by evolving technology. The following is a sketch of the current model. Don't treat it as the final product. There will be much more to come in the future.

Flattening to a Disk Considering the distribution of stars by age, astronomers conclude that the stars in the halo – globular cluster stars – were the first to form. Those are the oldest stars yet found in the Milky Way galaxy. Stars in the nucleus formed shortly afterwards, followed by those between the spiral arms. Finally, those within the spiral arms formed and continue to form even today. To explain how the sequence of star formation could have proceeded in this manner, let's assume the existence of a huge, spherically-shaped cloud of hydrogen gas. It was expelled by whatever event caused the Big Bang, and still retains some degree of spinning even as gravity acts to make it collapse.

At the outer edges of this collapsing cloud the density of the gas gets great enough for separate, self-gravitating masses to form into the globular clusters. Turbulence within the larger inner cloud prevents star formation from occurring until it flattens into a disk, while at the same time it spins faster and faster. Why should it flatten into a thin disk? Because the principle of the *conservation of angular momentum* says that the cloud rotates faster and faster as it collapses. The principle of *centrifugal force* throws outward

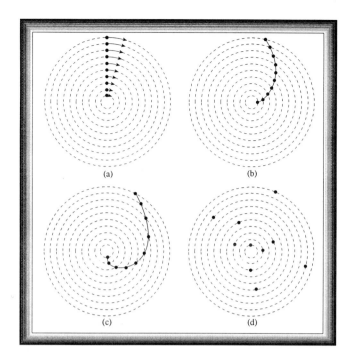

Figure 12–21 An illustration to explain why the spiral arms of a galaxy cannot be due to the simple winding up of arms that originally stuck straight out from the galactic nucleus. Using Kepler's laws of motion, the stars over a sequence of time intervals would end up being randomly distributed throughout the disk.

away from the axis of rotation those gas atoms furthest from the axis of rotation, but allow those close to the axis of rotation to collapse toward the center.

As conditions settle down in the center of the cloud, the clouds of hydrogen collapse to form the stars in the nucleus. Only when an equilibrium state is reached between centrifugal force (throwing material outward) and the gravitational force of attraction (attracting material inward) could the hydrogen clouds along the disk collapse into stars. Obviously not all of the original hydrogen has yet formed into stars. Our galaxy is still evolving.

Aside from the ages of the stars in the various regions of the galaxy, there are a few additional bits of evidence to support this model of the galaxy's formation. The globular clusters, containing the first stars to form at the perimeter of the original cloud, were left behind as the remaining gas collapsed and flattened and spun faster and faster. If this model is accurate, then comparison studies of the Doppler-shifted lines in the spectra of globular clusters should indicate a faster rate of motion for them in their orbits around the center of the galaxy than for those stars in the disk.

The Doppler shift, of course, measures relative motion between observer and object. If the hydrogen clouds from which the disk stars eventually formed spun faster and faster as they migrated toward the disk after the globular clusters formed and were left behind, then we (being a disk star ourselves) should detect a higher velocity for the orbits of globular clusters than for the disk stars that accompany our own movement around the center.

Motions of Globular Clusters And so it is. Figure 12-10 illustrates the motions of the globular clusters relative to the plane of the galaxy. Notice that their elliptical orbits carry them through the galactic plane, around the gravitational center of the galaxy, and back out again into the halo. In view of the fact that globular clusters, according to this model, pass through the galactic plane twice with each orbit, can you explain why it is that they are found mostly in the halo, and not along the galactic plane?

One reason is the presence of gas and dust along the galactic plane which blocks our view. But the major reason is explained by Kepler's Second Law — if we treat the globular clusters as behaving like planets orbiting the galaxy's center. They move fastest when they are closest to the center of gravity (galactic center), and slowest when they are furthest away in the halo. And, of course, that explains why there is a higher likelihood of our observing them when they move slower away from the galactic plane (in the halo). In other words, we observe most of the globular clusters away from the galactic plane — that is why they give the impression of forming a halo around the flattened portion of the galaxy.

Formation of Spiral Arms I'm sure that it has occurred to you to ask how spiral arms develop in a galaxy. It wouldn't seem in the model I just described that arms should develop in the disk. And yet when you look at a galaxy like that in Figure 12-19, the spiral arms look natural — they *look* like they have wound up from what might have been arms that were originally sticking straight out. Well, appearance may be deceiving. Kepler's laws of motion tell us that if the arms did begin straight out from the nucleus, differential motion over a long period of time would cause the stars to be randomly spread around in the disk (Figure 12-21).

And why is it, at the same time, that there are two subgroups of Population I stars — those in the disk (*Intermediate*) and those in the arms (*Extreme*)? Astronomers explain these two observations with a single theory — referred to as the **density wave theory**. What, another theory? Would you prefer to have some facts instead? Sorry, the subject of astronomy doesn't allow for much of those — at least not those that attempt to solve the big problems in astronomy.

Let's briefly review the contents of the disk portion of a spiral galaxy. The spiral arms stand out in photographs because of the presence of hot, young, luminous O- and B-type stars — in addition to luminous gas clouds. But a study of our local region in the galaxy reveals that the cooler, older, less luminous K- and M-type stars are much more numerous and are spread rather uniformly throughout the disk.

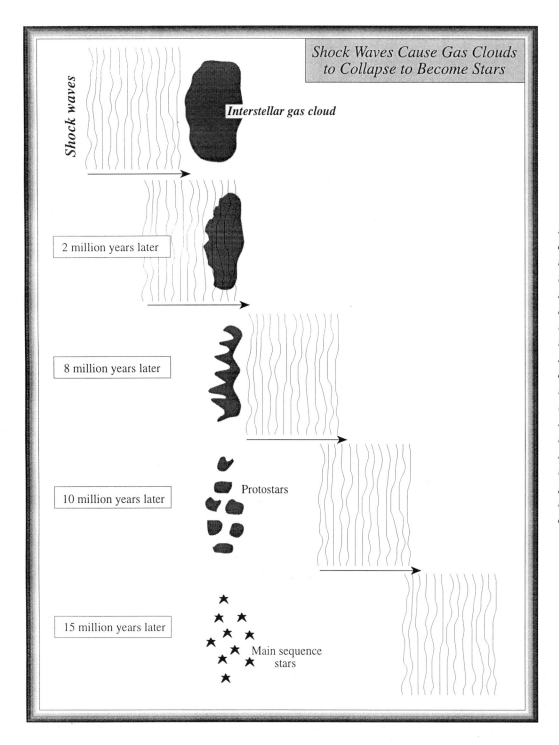

Figure 12–22 *This diagram illustrates the mechanism that is theorized to account for spiral arms in a galaxy. Spiral density waves run into nebulae which in turn collapse to form stars. Short–lived stars die before the waves get too far away, whereas less–massive stars survive while the waves move far away. Thus the spiral arms are simply those regions of a galaxy at which hot, bright, short–lived stars are located.*

In other words, photographs of spiral galaxies like that of Figure 12-19 would have us believe that stars in the disk of a spiral galaxy are concentrated along spiral arms. But what we learn from studying the types of stars within the spiral arms themselves and also in those regions between the spiral arms (which appear almost empty of stars), is quite different. *The spiral arms only appear to have more stars because of the type of stars located there.* In actual fact, there are as many stars distributed along the disk between the arms, but they are the less luminous stars that are invisible from afar.

But again — why is that? Let us imagine waves of compression moving through the disk of a galaxy — these waves being the result of gravitational variations within the disk. I can supply an analogy that perhaps will clarify how such a wave of compression can act, although the actual mechanism for causing the waves themselves is not so easily supplied. I commute to work along a long stretch of road uninterrupted by stop lights. But it is very often bumper to bumper traffic at 55 m.p.h..

Now if a car unexpectedly slows down or pulls off the road to repair a flat tire — for example — a ripple effect is generated in the flow of traffic behind that car. Cars slow down to avoid hitting the derelict car, and that causes cars to back up sometimes for miles. Even for hours after the incident, cars are still slowing down at the same spot, even though the flat is fixed and the car is happily on its way to work too.

Interacting Waves and Clouds Let's apply this analogy to the behavior of clouds of gas and dust, moving around the center of the galaxy, as they encounter the waves of compression. These waves are theorized to spiral outward from the center of the galaxy, moving at a rate that is slightly slower than the movement of stars and clouds of gas and dust around the center of the galaxy. Occasionally, the stars and clouds of gas and dust encounter and pass through these waves of compression. While located between successive waves, the clouds are not dense enough to collapse by their own gravity. But when they encounter the wave itself, the clouds are profoundly affected.

In the act of encountering a wave, the leading edge of a gas/dust cloud is compressed enough to initiate the conditions favorable for gravitational collapse — and fragments of varying masses eventually collapse to become stars. Stars that are already circling the galaxy pass unaffected through these waves, since stars are too massive to be

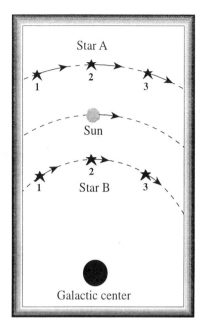

Figure 12–23 By measuring the rates of motions of stars near Sun using the Doppler effect, astronomers determine the speed at which Sun revolves around the galactic nucleus. The arrows represent the speeds at which the stars are moving. Star A at location 1 reveals a blue shift in its spectral lines, and a red shift if it is at location 3. Can you tell the types of shifts for star B at locations 1, 2 and 3?

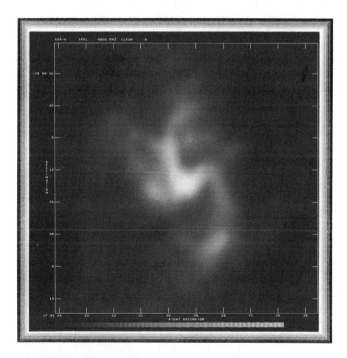

Figure 12–24 The VLA provided this plot of radio energy coming from the region of the center of our galaxy in the direction of Sagittarius. It reveals the presence of considerable amounts of hot, ionized gases constrained by magnetic fields into filamentary arches. This suggests some type of violent event ejecting material outward. (NRAO/AUI)

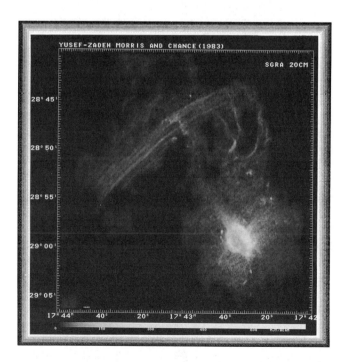

Figure 12–25 A detailed map of the radio energy coming from the center of the region shown in Figure 12–23 reveals the inner edge of a disk of hot, ionized gas and dust with the gas flowing into a central object. (NRAO/AUI)

affected by such weak waves. Only diffuse clouds of gas and dust are influenced by density waves (Figure 12-22).

The newly-formed stars continue to move slowly under the gravitational attraction of the galaxy itself. But since the most massive stars are so luminous and live for such a short period of time (cosmically speaking, of course), they are located close to the density wave responsible for their formation. The more long-lived the star (the less massive it is), the farther away from the wave it is found. Since the long-lived stars are not very luminous, their presence in the spaces between waves — where they are most likely to be found — are difficult to observe.

The spiral arms of a galaxy show up brilliantly in photographs not because of the greater number of stars there, but because they represent the locations of the density waves near which the very luminous O- and B-type stars are concentrated. The spaces between the arms appear empty not because of fewer stars there, but because of the low luminosities of the stars that are there.

For example, our Sun is estimated to be 5 billion years old. At its distance from the center of the galaxy, it requires 250 million years to circle the galaxy once. Therefore, Sun has circled the galaxy 20 times in its short lifespan of 5 billion years, having passed through many density waves in its lifetime. At the moment, however, it is located between two waves. If we were close to one, the nearest stars would be the luminous, hot, young stars.

Careful examination of other spiral galaxies provides us with supporting evidence for the density wave theory. Frequently we see vast dust clouds along the inner edges of their spiral arms. These would appear to be regions in which the gas/dust clouds are experiencing the first stages of collapse, the greater concentrations of material appearing dark to us. We conclude, then, that the spiral arms of a galaxy are not permanent features. Rather, they are constantly renewing themselves as massive stars die close to density waves, and new stars form as gas and dust clouds encounter the waves.

Rotation of Galaxies Do you wonder how astronomers came up with that 250-million-year period for the time Sun takes to circle the galaxy? In fact, how do we know that our galaxy is rotating like a giant pinwheel? Galaxies certainly <u>look</u> as if they should be spinning — at least as seen from the outside. Figure 12-19 can't look too unlike the Milky Way galaxy, and its appearance suggests it should be rotating clockwise as we look at it.

So you recommend that we take successive photographs of it over a period of time, and look for any changes in the positions of stars or gas clouds? That would certainly be reasonable if galaxies rotated fast enough, or our lifetimes were great enough to allow us to take photographs at intervals of a million years or so — but neither is the case. There is a solution, however — the Doppler effect.

We find a galaxy that appears edge-on to us, and take spectra of the two outer edges of the spiral arms. If it is spinning, one arm's spectral lines are shifted toward the red end, and the other arm's spectral lines toward the blue end. That immediately tells us the direction and speed at which it rotates. If the galaxy is exactly face-on to us, of course, rotation cannot be determined since there are no portions of the galaxy are either receding from or approaching us.

Rotation of the Milky Way That is fine for other galaxies, but what about our own? From inside the Milky Way galaxy, we cannot use the same method because we cannot see very much of the contents of the galaxy. The clouds of gas and dust inhibit our view. But the Doppler effect itself is useful, inasmuch as it allows to measure the velocities of objects right around us. In this case particular case, we measure the Doppler shifts of the stars surrounding Sun in our immediate neighborhood of the galaxy (Figure 12-23).

The numerical value of the gravitational force pulling on a star within our galaxy is a function of the total number of stars contained within that star's orbit about the center. In effect, we treat all of those stars as if they are equivalent to one single huge mass located at the center of the galaxy. The larger the number of stars, the faster a star moves around the galaxy in order to escape the fate of being pulled into the center. In other words, those *stars more distant from the galaxy's center move faster than those closer to the center because there is a greater mass of stars within their orbits.*

So if we measure the motions of the stars in our immediate vicinity — both radial and proper motions — we can determine the differences in the rates of motion of those stars relative to Sun. Knowing the differences in the relative velocities of a number of stars around us, we can determine our own rate of motion around the center of the galaxy. Knowing the size of our orbit by knowing our distance from the galactic center (thanks to Shapley and mathematics), and knowing how fast we are moving, we can determine the period of time it takes to complete one orbit about the center. That is how we arrive at the 250-million-year figure.

Metal Content of Stars An important detail has been left unexplained as I discussed the shape of our galaxy — that has to do with the metal content of stars. To briefly summarize our theory of the origin or metals as it relates to stellar evolution, stars for most of their lives (main sequence lifetime) do only one thing — turn hydrogen into helium. Only while evolving off the main sequence and passing through red giant stage do stars fuse helium into the heavier elements — up to and including the element iron.

As stars evolve past red giant stage, they eject (via the planetary nebula) much of those newly-formed elements back out into space to mix with hydrogen clouds. Those clouds in turn collapse to become new stars. Massive stars not only fuse helium into elements as heavy as iron, but — while exploding as supernovae — fuse iron into the very heaviest of metals and enrich hydrogen clouds with those as well (refer back to Figure 10-10).

We should find — if this scenario is correct — that stars forming today in the clouds in the spiral arms are rich in the metals formed inside of and dispersed out into space by previous generations of stars. The longer this process goes on, the greater should be the percentage of metals that are found within a gas and dust cloud (refer back to Figure 10-12). This theory predicts that the metal content of stars forming within nebulae today is greater than stars that formed long ago. In other words, the metal content of the galaxy is increasing. Stars change the lighter elements into the heavier ones.

The distribution of stellar Population types in the galaxy supports this theory, especially in the observation that Population I Extreme stars reveal the largest percentage of metals. The Population II Extreme stars contain the smallest percentage of metals. When the galaxy was young, there had not yet been sufficient time for stars to have manufactured and ejected out into space large amounts of the metals. So the oldest stars are metal poor, having formed from those earlier clouds that were poor in ejected elements from dying stars. Remember that red dwarf stars — the oldest stars around — are still alive and well, turning hydrogen into helium. That's all they've been doing for 15 billion years or so.

The Galactic Nucleus Something strange is happening at the very center of our galaxy. In addition to containing the older stars — the Population II Intermediate stars — the nucleus shows signs of behavior that may provide important clues to the manner in which our galaxy first formed. Fortunately, the clouds of gas and dust between Sun and the nucleus of the galaxy do not strongly affect radio, infrared, X-ray, or ultraviolet radiation, because otherwise we would be unaware of events in this intriguing region of our home in space. Figure 12-24 is a **radiograph** of the radio emission coming from the inner 10 light-years region of the center of our galaxy. A radiograph is just like a photograph, except that it is a plot is the locations of radio energy rather than a plot of the visible radiation as revealed in a photograph. This particular radiograph was collected by the VLA in New Mexico. It is produced by hot, ionized gas. The emission is strongest where the gas is most dense. The bright region in the middle is the very compact radio source precisely at the center of the galaxy.

The radiograph in Figure 12-25 reveals that near the outer portion of the galaxy's nucleus there is a ring of emission nebulae, and clouds of hot, flowing gases. Inside of this ring appear to be two clouds of neutral hydrogen revolving around the center but expanding outward — as if ejected outward from the very center.

Coming from a direction thought to be that of the galactic center, radio maps reveal the strongest source of radio energy in the galaxy. It is called **Sagittarius A** — named for the constellation associated with the direction to the center of the galaxy. Outward from *Sagittarius A*, the same map reveals the presence of hot interstellar gas clouds similar to those found in the spiral arms of our galaxy. Perhaps there is star formation going on in the downtown portion of our galaxy after all, the nearby gas clouds detected as emission nebulae being heated up by the newly-born hot, young stars.

Coinciding with the radio location of Sagittarius A is a source of infrared, ultraviolet, and X-ray radiation. The patterns of these radiations suggest that the source is only a few light-years in size and a mass estimated to be several million solar masses. At the moment, such phenomena are not easily explained with our present laws of physics — unless we invoke the use of something we have already encountered — the black hole. In this case, it would have to be a **massive black hole** to explain all of the phenomena we observe.

Astronomers model the source by assuming that a considerable amount of matter in the form of gas and dust and stars is being swallowed by a massive black hole located at the center of the galaxy. The radiation comes not from the black hole itself, but from the accretion disk where infalling matter collides, heats up to very high temperatures, and in the process emits the various types of radiation we detect. This is a rather tantalizing idea, and certainly worthy of further discussion. But it will be easier to understand after we investigate the contents of other galaxies, and search for similar behavior in their own downtown regions.

Recent Developments It might appear as if astronomers have a neatly worked-out theory for our galaxy's formation. It is a nice theory, but new observations and discoveries abound, and it is quite likely that profound changes to the theory will be forthcoming — perhaps even while this book is being printed for distribution. A good example of this is the recent discovery of one astronomer that the stars in the nucleus of our galaxy are actually older than those in the globular clusters that are located in the halo of the galaxy.

If this is confirmed, then it seriously contradicts the notion that the galaxy formed from a single collapsing and rapidly-spinning cloud of gas and dust. Rather, it would suggest that the galaxy indeed evolved from the merger of many smaller galaxies. Instead of the standard picture of the galaxy forming from the outside in, it may have been a matter of a formation from the inside out. Computer simulations of such mergers confirm that galaxies, by shape and content, may indeed form in this manner. Stay tuned.

★★★★★★★★★★

Summary/Conclusion

It certainly has been humbling to discover that there is more to the universe than just the Milky Way. The contents of our galaxy offer a range of star types by temperature, size, lifespan, and chemical content to keep us wondering about life out there for quite some time. But to realize that our galaxy is typical amongst the billions of galaxies is to move the question of life out there into the realm of probabilities. From a purely practical viewpoint, we now have an understanding of the placement of stars in our galaxy according to the characteristics that we associate with life.

LEARNING OBJECTIVES: *Now that you have studied this Chapter, you should be able to:*

1. Describe the properties of the Milky Way galaxy that prevented astronomers from knowing its structure and Sun's location in it for such a long period of time.
2. Describe the method used by Shapley to determine Sun's position within the Galaxy and the existence of galaxies beyond the Galaxy.
3. Explain the basis for the discovery that the universe is expanding.
4. Describe the properties of stars contained within the two types of star clusters.
5. Describe the four general types of stars found within the Galaxy, and how they offer an explanation for the manner in which the Galaxy formed.
6. Describe the manner in which astronomers determine the spiral structure of the Galaxy, and offer an explanation as to how that structure is maintained with the density-wave theory.
7. Describe the evidence for violent events at the center of the Galaxy, and offer possible explanations for that violence, including the existence of a supermassive black hole.
8. Make rough sketches of the Galaxy from the top and a side view, labeling its major components.
9. Define and use in a complete sentence each of the following **NEW TERMS**:

Cepheid variable star 256
Density wave theory 269
Galactic (open) cluster 260
Globular cluster 256
Massive black hole 273
Metals 263
Period–luminosity relationship 257
Population types 263

Population I extreme star 264
Population I intermediate star 264
Population II extreme star 264
Population II intermediate star 264
Radiograph 273
Sagittarius A 273
Turn-off point 261

Chapter 13
Galaxies Beyond the Milky Way

Figure 13–1 This HST photograph of about 2,000 galaxies provides us with a snapshot of the universe when it was less than 1 billion years old. It is the most distant view of the universe astronomers have to date. The field of view of the photograph covers the width of a dime 75 feet away. The HST had to stare at the tiny spot in the sky for 10 consecutive days, over 150 Earth orbits. These 10 straight days of observing time were broken into 342 exposures, lasting 15 to 40 minutes each. (NASA/STScI)

CENTRAL THEME: How are the other galaxies categorized according to structure and content, how are they distributed in space, and what evidence do we have of their evolution?

★★★★★★★★★★★★

When you have the opportunity to look through sizeable telescopes set up by amateur astronomers at remote locations (star parties), your excitement initially arises because of the opportunity to see for the first time, close-up, craters of the Moon, Saturn and its rings, Jupiter's bands of gases and four visible satellites, crescent Venus, or the polar icecap of Mars. Then you get excited when you see detail in the gas and dust clouds — the nebulae — often named for their appearances: the Lagoon nebula, the Swan nebula, the Ring nebula, the Crab nebula. Then there are the different types of star clusters — the Pleiades (a galactic cluster) and the globular cluster in Hercules. These are all local objects, requiring only modest-sized telescopes to see easily.

It is when you look beyond the Milky Way to see other galaxies, scattered without limit throughout the sky, that a feeling of humility begins to set in. Unfortunately, the view of a galaxy in the eyepiece of a telescope can be disappointing if one is used to looking at the beautifully clear and bright photographs of galaxies displayed in magazines and on television. Those have been taken with telescope-camera combinations which, for reasons explained in Chapter 5, allows for gradual buildup of light on the photographic film. The human eye doesn't collect light in that fashion, so the image of a galaxy in a telescope is usually small, dim, and rather fuzzy. My experience is that if you look at a photo of a particular galaxy for a few seconds just prior to looking at it through a telescope, your brain will fill in some of the missing parts for you, and it will be more obvious what you are looking at in the eyepiece. Try it.

Observing Galaxies

Naked-Eye Galaxies If you know exactly where to look and you are located away from the lights of a metropolitan area, there are actually two galaxies visible to the naked eye. Not surprisingly, the **Andromeda galaxy** is located in the constellation of *Andromeda*, and appears as a faint smudge of light in the nighttime sky. A good-sized amateur telescope reveals it to have all of the features of the Milky Way — spiral arms, a bright nucleus, and dark dust and bright gas clouds in the spiral arms. Figure 12-11 is a beautiful photograph of that galaxy.

The Andromeda galaxy is a spiral galaxy with about the same size and shape and contents as our own galaxy. But

Figure 13–2 When you next visit a country in the southern hemisphere, take the opportunity to see the nearest pair of galaxies to the Milky Way with your naked eyes. Since the Magellanic Clouds are close to the south celestial pole (SCP), you must travel to the equator or farther south before you can see them. They are the two patches of grey to the right of the SCP. (Carina Software)

★★★★★★★★★★★★

Figure 13–3 The Large Magellanic Cloud *(LMC)*, one of the nearest galaxies to the Milky Way. It is located about 200,000 light–years from Earth. (Lick Observatory)

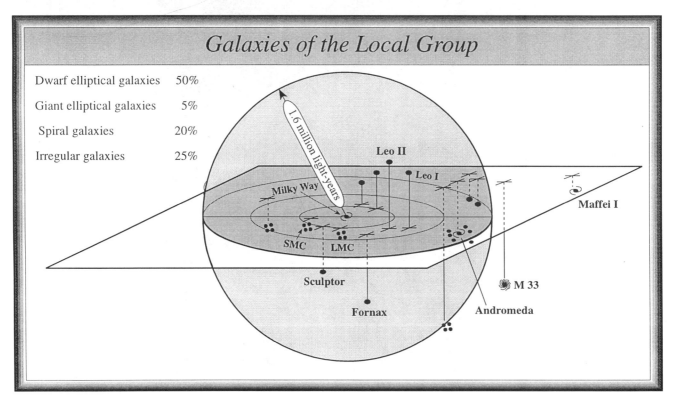

Figure 13–4 The Milky Way galaxy is member of a small group of 21 galaxies called the Local Group. They are distributed in space as shown in this diagram. The Milky Way is not located at the gravitational center of the Group. It is shown centered here for the purpose of illustrating their directions from our galaxy.

278   SECTION 3 / Evolution of the Galaxies

it is not the nearest one to us. That distinction goes to the **Magellanic Clouds** — actually two tiny galaxies that are visible to observers only if they are located below 15° North latitude. They are easily seen while visiting the Galapagos Islands (Chile) in the South Pacific, for example. They are called clouds because they appear as if they are clouds within our atmosphere — except they do not move as you would expect clouds to move (Figure 13-2).

Not knowing what they were, Magellan — the first European to record their presence in the sky as he made his famous journey into the southern hemisphere to circumnavigate the globe — called them clouds. And so the name has stuck. Through a telescope they appear to lack the symmetry and spiral pattern shapes of the Milky Way and Andromeda galaxies. Nor do they appear to be as densely populated with stars. Our estimates are that they contain only 100 million stars or so — compared to 100 billion stars in our galaxy. Because of their small sizes and irregular shapes, we call them **irregular galaxies** (Figure 13-3).

The Local Group But single galaxies are rare in the universe. Most are members of larger groupings of a few to a few thousand galaxies. The Milky Way, the Andromeda galaxy, and the Magellanic Clouds are three members contained within a small family of galaxies called the **Local Group**. By "family" I mean, of course, that gravity holds them together such that they revolve around a common center of gravity. In other words, galaxies move with respect to the celestial sphere.

But their tremendous distances result in such small changes over the 60 years or so that we have been studying other galaxies, that no proper motion has yet been measured. However, we can detect motion in the radial direction through the measurement of shifts in the spectral lines in their spectra. Those studies reveal the motions of galaxies within such groups such as that of which we are a member. Figure 13-3 shows the relative positions of the 21 galaxies in the Local Group.

Notice in Figure 13-4 that the Milky Way is not situated at the center of the Local Group, and that most of the galaxies are smaller globs of stars than the Milky Way and Andromeda galaxies. Most are **dwarf elliptical galaxies**, so-called because of their small sizes and shapes. If the content of the Local Group of galaxies is typical of what we might find anywhere else in space as far as the types of galaxies is concerned, then dwarf elliptical galaxies must be the rule in the universe. Indeed, estimates suggest that 70% of all galaxies in the universe are at least shaped in a similar manner. The percentages of galaxies in the Local Group according to type are as follows:

So that you have a general idea of the features of galaxies that astronomers spend their time determining, spend a moment examining Figure 13-5. It summarizes the major characteristics of our nearest neighbor galaxies in space.

Clusters and Clusters of Galaxies The Local Group is contained within a larger grouping of galaxies, the **Coma-Sculptor Cloud**. It contains a dozen or so small groups of

Characteristic	Ellipticals	Spirals			Irregulars
		Sa	Sb	Sc	
Dust	None	●———→ Some ←——●			Some
Percent hydrogen	0	1	3	9	20
Prominent populations	II	II (halo, center); I (spiral arms)			I
Rotation periods (million years)	?	60	140	200	300
Dominent color	Red	Red (halo, center); blue (spiral arms)			Blue
Spectrum of central region	K	K-G	G	F	F-A
Luminosity (Sun = 1)	Giants = 10^{11} or less Dwarfs = 10^5 or more	●———→ 10^8 to 10^{10} ←——●			$10^7 - 10^9$
Diameter (1,000 ly)	Giants = 200 or less Dwarfs = 1 or more	●———→ 5 to 50 ←——●			1 - 10
Mass (Sun = 1)	Giants = 10^9 to 10^{12} Dwarfs = 10^6 to 10^9	5×10^{11}	3×10^{11}	1×10^{11}	1×10^9

General Characteristics of Galaxies

Figure 13-5 This diagram summaries the general characteristics of the types of galaxies that are located close enough to the Milky Way that the features can be measured.

★★★★★★★★★★★★

Figure 13–7 (Right) This is a two–dimensional plot of the Virgo Supercluster of galaxies showing the location of the Local Group (of which the Milky Way is a member) and some of the larger clusters of galaxies.

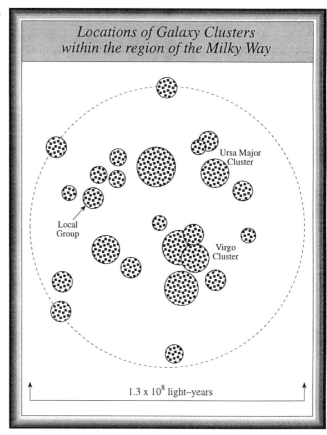

galaxies similar to the Local Group, and is shaped rather like a cylinder. That is, the small groups are distributed along a line that is about 50 million light-years long and 5 million light-years across. The ends of the cylinder of galaxies appear in the constellations of *Coma Berenices* and *Sculptor* — hence the name of the Cloud. The Local Group is located roughly in the center of the Cloud. Beyond the boundaries of the Coma-Sculptor Cloud are other galaxy clouds and large *clusters of galaxies*. The nearest large cluster of galaxies, the **Coma Cluster**, contains somewhere in the neighborhood of a thousand galaxies, and lies about 60 million light-years from the Milky Way (Figure 13-6).

Surveys of deep space reveal the existence of over 2,700 galaxies within 4 billion light-years of us. *Clusters of galaxies*, in turn, appear to form *superclusters of galaxies*. The Virgo Cluster, for example, sits at the center of a larger structure called the **Virgo Supercluster** (Figure 13-7). Galaxies appear to reside in groups, which in turn reside in clusters, which in turn reside in superclusters, which in turn reside in super-superclusters — and so on. There is a large scale structure to the universe, the exact form of which we are at the moment only beginning to identify but not yet

Figure 13-6 Examine this photograph carefully. Almost all of the white "smudges" are distant galaxies that are members of the Coma *cluster. (Lick Observatory)*

280 SECTION 3 / Evolution of the Galaxies

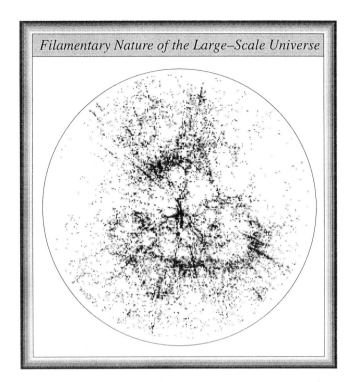

Figure 13-8 (Left) This is the latest "pie" slice of the nearby universe made by extremely careful observations. Earth is at center, and over 14,000 galaxies out to about 500 million light-years are plotted according to direction and distance. (Margaret Geller and John Huchra, Harvard-Smithsonian Center for Astrophysics) recent that 13.22 picture.

understand (Figure 13-8).

In order to get a sense of the distribution of galaxies in space, consider something with which you are already familiar — your own home. Think of your home as a galaxy in a subdivision (Local Group of galaxies) within a small town (Coma-Sculptor Cloud) that lies astride a major highway that leads into a populated valley (Virgo Supercluster) in which there are two large cities (Virgo and Ursa Major Clusters). Seven years ago, a group of astronomers discovered that just over the hill from this valley was another major population center. It is a vast collection of galaxies that is about 300 million light-years away from us and located in the direction of the constellations of Hydra and Centaurus. It is called the "Great Attractor" for reasons to be explained shortly.

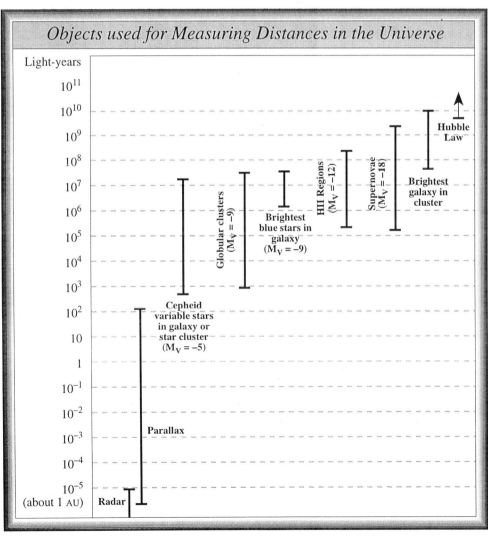

Figure 13-9 The determination of distance is extremely important, but gets increasingly difficult with greater distance. The objects used to establish distances to objects are shown in this diagram according to how far away that particular object is useful.

CHAPTER 13 / Galaxies Beyond the Milky Way **281**

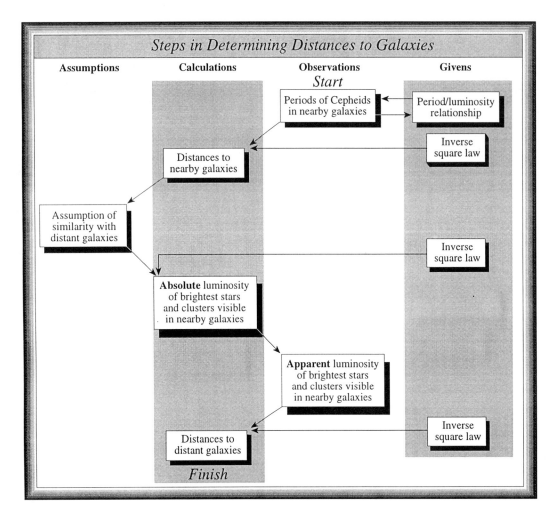

Figure 13–10 *Since measurements of the inherent characteristics of objects in the universe are dependent on accurate measurements of distances to those objects, it is important to keep our assumptions clear. This flow diagram illustrates the method of distance determination and the assumptions used in the process.*

In space, gravity is the dominant force. So evidently during the early history of the universe, matter was not distributed in a random manner — the hydrogen gas was lumpy. The result was that as gravity pulled clouds of hydrogen together to form the present-day galaxies, there existed a general pattern to their eventual distribution. This overall pattern is revealed in the plot of the locations of over 14,000 galaxies shown in Figure 13-8).

The lumpiness of the early universe may have resulted from the randomness of the Big Bang itself, much as the debris from a firecracker is scattered rather randomly around the point of the explosion. You will learn in Chapter 15 that recent observations by a satellite sent aloft specifically to measure the amount of lumpiness in the early universe confirmed that it was lumpy enough after the Big Bang to account for the formation of galaxies.

So far, this Chapter has been an excursion into the universe of galaxies, into the far reaches of space outside of the Milky Way galaxy. I casually included the distances to the nearby groupings of galaxies and clusters of galaxies so that you might have a sense of the organization of the universe, but those distances are difficult to obtain and are estimates at best. We can't really claim to know much about the galaxies if we don't know how far away they are.

Galaxies, like stars, have characteristics such as mass, luminosity, and velocity whose values are dependent upon our knowing their distances.

Unlike the case of stars, however, the distances to the galaxies supply a basic ingredient for establishing the history of the entire universe and the origin of the universe itself. Without that knowledge, we would be at a loss to explain how the galaxies formed in the first place. So it is time to tackle the topic of distances to galaxies. Since major conclusions about the origin of the universe are based upon the accuracy of those distances, you might like to know how much confidence to put into them.

Distances to Galaxies

As with stars, we determine the intrinsic properties of galaxies only if we know their distances from Earth. But how do we determine just how far away a given galaxy is? Most look small in even large telescopes, and if you think of them as being huge objects in their own right, you naturally think of them as being far away. But that is not a reliable method. That method certainly doesn't give us a numerical value. In the last Chapter, we learned that the discovery of galaxies outside the Milky Way was originally determined by studying *Cepheid variable stars*. An object that allows us to determine the distance to the parent object in which it is located is called a **distance indicator** (Figure 13-9).

Identical Objects Have Identical M_v's Besides Cepheid variable stars, objects commonly used for this purpose are bright stars, globular clusters, gas clouds, novae, supernovae, and the brightest galaxy in a cluster. The basic technique used for all distance indicators is the same — we assume that the *absolute visual magnitude* of an object in another galaxy is identical to the *absolute visual magnitude* of an identical object located whose distance within our own galaxy is known.

If, for example, we measure the *apparent* visual magnitude of a globular cluster in a distant galaxy, we can determine its distance since we can assume that its *absolute* magnitude is identical to that of globular clusters in our own galaxy. We assume, as a starting point, that identical objects emit identical amounts of radiation - *the M_v's of identical objects are identical*.

Certainly there is some degree of error in this method, but until we can send a space vehicle out there with a tape measure in tow behind it, we have no other method. Naturally we are not going to base our entire estimate on a single object. We calculate the distances to several different globular clusters in the same galaxy — or better yet, to several separate distance indicators — then an average of all of those distances cancels out the over-errors and under-errors in the method. Thus we obtain an rather accurate measurement of the distance to the entire galaxy in which the objects are located. The more distance indicators located in the galaxy that we use for determining distance to the galaxy, the more accurate our estimate.

As we study galaxies at increasingly greater and greater distances, however, the more difficult it is to identify distance indicators in them. Globular clusters are eventually so fuzzy that they are indistinguishable from gas clouds. They appear as tiny smudges in the photographs, so we aren't sure which type of distance indicator it is. And therefore we don't know what value of M_v to assign to it. If we are fortunate, we may eventually observe a supernova go off in the galaxy — it is obvious what they are, since they suddenly appear. One would think that by the time we peer far enough into space that we observe only clusters of galaxies — that appear as clusters of smudges in the telescope — we would lose any ability at all to determine distances to the galaxies.

There is a solution, however. As you recall, galaxies cluster together in space. The more distant we look, the more obvious the clustering becomes. Now it just happens that a giant elliptical galaxy is very often located close to the center of such clusters. If we assume that all giant ellipticals are approximately the same size — and therefore have the same M_v — then we can use that giant elliptical as a distance indicator. By doing so, we determine the distance to all of

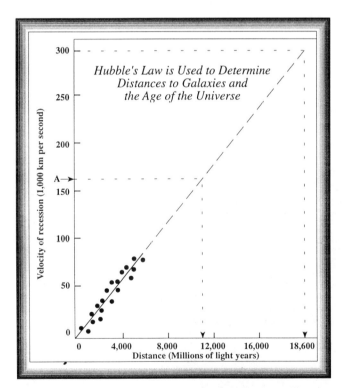

Figure 13–11 Hubble's law is used to determine distances to very distant galaxies that are too small to reveal the presence of a distant indicator. Astronomers measure the recessional velocity of the galaxy by the Doppler effect, and then measure the distance from this graph.

Cluster NAME	DISTANCE (millions of lys)	DIAMETER (millions of lys)	NUMBER of galaxies counted	DENSITY (# of galaxies per cubic million lys)	Recession VELOCITY (miles per second)
Local Group	2	3	27	2,000	0
Virgo	60	13	2500	20,000	+700
Pegasus I	200	3	100	40,000	+2,200
Cancer	250	13	150	20,000	+3,000
Perseus	280	23	500	10,000	+3,500
Coma	300	25	800	1,500	+4,300
Hercules	400	1	300	700,000	+6,400
Ursa Major I	600	10	300	7,000	+10,000
Leo	800	10	300	7,000	+13,000
Gemini	900	10	200	4,000	+15,000
Boötes	1400	10	150	4,000	+25,000
Ursa Major II	1500	10	200	15,000	+26,000
Hydra	2100	?	?	?	+35,000
3C 123 and cluster	5000	?	?	?	+80,000

Figure 13–12 A list of the clusters of galaxies that are closest to us shows some of their characteristics. The distances to the more distant ones were found by using Hubble's law.

the galaxies in that cluster, since the distances between the galaxies within the cluster are small compared to the distance between us and the entire cluster. Figure 13-10 summaries the manner in which astronomers collect their data in order to determine distances to galaxies. But what can we do if the galaxy under study is a loner — lacking membership in a cluster?

Motions of galaxies Galaxies within a cluster orbit around a common center of gravity, just as stars in a galaxy orbit the gravitational center of the galaxy. These motions are quite complicated, and of course astronomers cannot observe the motions directly. But the Doppler shift in the spectra of a galaxy can tell us of that galaxy's motion relative to nearby galaxies, and gradually we get a sense of how they are moving in relationship to one another. There is another even more interesting motion of galaxies aside from that around a center of gravity of its parent cluster.

When I first introduced you to the Doppler effect, I tried to excite your interest in modern topics in astronomy by casually mentioning the fundamental discovery made in the late 1920s that all galaxies reveal redshifts in their spectral lines. This fact is interpreted as being the result of the Doppler effect — which means that all galaxies are moving away from us. What I didn't tell you at the time, however, was that there is a distinct relationship between the rate at which a galaxy is moving away from us (amount of the redshift) and its distance from us.

When Edwin Hubble measured each galaxy's redshift and distance, and then plotted all galaxies on a graph according to those two characteristics, he discovered that the more distant galaxies are moving away from us at a faster rate than are the nearer ones (Figure 13-11). *The further away a galaxy is from us, the faster it is moving away from us.*

This discovery is so important and fundamental to modern astronomical thought that it is called **Hubble's law**.

Using this fundamental law — which presumably all galaxies in the universe <u>must</u> obey, for reasons soon to be explained — we can determine the distance to a solitary galaxy in space. Even if we are unable to identify any distance indicators within it, we can nevertheless obtain a spectrum and calculate the amount of the redshift for that galaxy. We then place it on Hubble's diagram for that measured redshift, and read on the horizontal distance scale the value for its distance from us. Known galaxies (those whose distances and redshifts are known) are used to establish the numerical relationship between the distance and redshift, and then that relationship is applied to any galaxy whose distance we want to determine.

Determining Distances From Redshifts What is easier than carrying a copy of the graph around, however, is just remembering the numerical value of the *slope* of the line running through the plots of galaxies on Hubble's diagram. You are familiar with the word *slope* in reference to the steepness of a hillside, for example. What amounts to the same thing, but treated mathematically, is the "steepness" or *slope* of a line on a graph. The **slope** of a line is defined as the *length of a horizontal line divided by the vertical distance between the two ends of the horizontal line.* Thus the slope of the line on Hubble's diagram of galaxies is expressed as the recessional rate of galaxies (horizontal scale) divided by distance to the galaxies (vertical scale). The value of this slope is called the **Hubble constant**. Once a galaxy's recessional rate is determined from measuring the shift in its spectral lines, its distance can be determined by dividing the recessional rate by the *Hubble constant*. Figure 13-12 shows some of the clusters of galaxies whose distances were

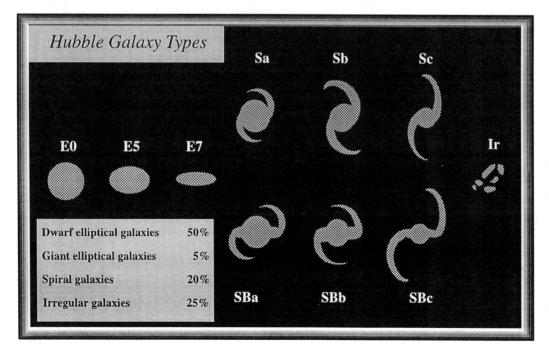

Figure 13–13 Hubble's original classification of galaxies by shape was thought to represent galaxies according to their evolutionary development from one type to another. There is now reason to believe that other factors influence the shape of a galaxy, not just the aging process.

determined in this manner.

The accuracy of the Hubble constant (the value of the slope of the line) is only as good as the measurements of distances for the galaxies plotted on the diagram. Some astronomers, using a particular method of calculating distances, calculate the constant to be *10 miles per second per million light-years.* Other astronomers obtain a figure twice that amount. It is anything but constant, but that fact is due to the inherent limitations of our measuring techniques.

To summarize, if we know a galaxy's redshift, we know its distance. The reason this method works is because Hubble's law is an expression of the rate at which the universe is expanding—meaning that all of the galaxies are moving away from one another. Hubble's law is used not only to calculate distances to extremely distant galaxies, but to provide evidence for an actual beginning of the universe. No wonder Hubble has a law named after him!

Age of the Universe I've used Hubble's law — without bothering to mention it — whenever I referred to the universe being 20 billion years old. And now you are wondering how I arrived at that particular number, right? Here's how. The fastest anything can travel (as far as we know) is the speed of light — 186,000 miles per second. If we adopt the *10 miles per second per million light-years* value for the Hubble constant, then we calculate (using Hubble's law) how far away a thing would be if it were going away at the speed of light (because it is very distant and sharing in the expansion of the universe). The answer is 18.6 billion light-years (that is close enough to 20 billion). In other words, if we divide the speed at which light travels by the Hubble constant, we obtain the distance to the origin of the universe (Figure 13-11).

Can you see the consequences of adopting a value for the Hubble constant other than the 10 figure I just used? How old is the universe if we adopt the 20 figure? And can you now see the importance of determining distances in astronomy? Can you also see why it is that we will never get to observe the moment of the Big Bang? It is going away from us <u>at</u> the speed of light! I am confident that this observation of the expanding universe and the Hubble constant causes a lot of questions to arise in your mind, but I hope to settle them when we get to Chapter 15 when we tackle the origin of the universe. There are other aspects of galaxies that require our attention before we get into matters of the creation of the universe.

Classifying Galaxies

Astronomers have done (and continue to do) with galaxies what they did (and do) with stars — classified them according to whatever patterns they observe in studying them. The reason we do that, of course, is that patterns tell us something about the processes that operate in nature — in this case the origin and evolution of galaxies. For the Milky Way galaxy, at least, the pattern of star location by age, metal content, and spectral class allowed us to explain its evolution from a spinning and collapsing cloud of gas and dust. But other galaxies have different characteristics, and may therefore have different origins and/or evolutions. What we would like is a comprehensive theory of galactic evolution to account for all of the different types of galaxies we observe. To do that, we group the galaxies into categories according to size, shape, and contents.

Hubble's System The diagram in Figure 13-13 shows Hubble's original classification system. Recall that he was the astronomer distinguished for the discovery that other galaxies besides the Milky Way existed. As you study his classification system, do not be misled into thinking that it is a pattern that expresses the evolution of galaxies. That is, there is no evidence to suggest that irregular galaxies evolve into spiral galaxies and then into elliptical galaxies, or vice versa. As we will learn shortly, the contents of the different types of galaxies do not support that conclusion. Nevertheless, the diagram assists us in developing a comprehensive model for galaxy formation.

Elliptical Galaxies The general characteristics of the three broad types of galaxies are summarized in Figure 13-5. As

Figure 13–14 This photograph of an elliptical galaxy (NGC 97) is typical of its type. Notice the globular clusters in a halo around the galaxy. (Lick Observatory)

I discuss each type of galaxy, you may find it useful to refer to this chart and also the Hubble classification of Figure 13-13. The elliptical galaxies appear roundish or elliptical in shape. Their varied degrees of flatness are expressed by the designations E0, E1, E2, E3, E4, E5, E6, and E7. E0 galaxies are spherical in shape, while E7 is quite flat. Although it would seem that the flattened elliptical galaxies (e.g., E5, E6, and E7) are simply E0 galaxies flattened out like saucers, they can also be shaped like watermelons or even hot dog buns — longer than they are wide, and wider than they are thick. Two-dimensional viewing has its limitations.

Elliptical galaxies can be quite small, or they can be the very largest of galaxies. Elliptical galaxies, of all the galaxies, have the greatest range in terms of size and numbers of stars. What all elliptical galaxies have in common — revealed by spectral analysis — is a common type of star — the old, Population II-type star found in the halo and nucleus of our Milky Way galaxy. In color photographs, elliptical galaxies therefore appear reddish, because old, M-type stars are reddish in color. As you would expect, there is practically no gas and dust in the elliptical galaxies. If there were, young stars would be present as well.

These conclusions about star-types in elliptical galaxies are obvious when you look carefully at a photograph of one of these old-timer galaxies (Figure 13-14). There are no dark splotches — that would mark the locations of dust clouds — evident at all. And radio telescopes searching for evidence of hydrogen gas (through 21-centimeter studies) have found them to be practically void of the raw material for star formation. Therefore, either elliptical galaxies (1) were the *earliest galaxies to form*, and have gone through many generations of star formation — turning the debris of dying stars into the coolish, less-massive stars, or they (2) *formed more rapidly* than other types of galaxies, and went through several generations of stars more rapidly than the others. We require more evidence before attempting to decide which is correct.

Irregular Galaxies The irregular galaxies, in contrast to elliptical galaxies, are so loaded with gas and dust that star formation proceeds rapidly. They contain old stars, but the vast majority are young, hot stars. The presence of luminous gas clouds out of which stars form verifies that active star formation is most certainly going on. Color photographs reveal, as you would expect, an overall blue color. Red splotches indicate the locations of hydrogen clouds. They are unlike elliptical galaxies in another feature — they have no symmetry. They have more of a chaotic appearance, perhaps because their small sizes make them vulnerable to the gravitational disturbances of nearby galaxies.

Figure 13–15 The "Whirlpool" galaxy (M51), classified an Sc galaxy, reveals a smaller irregular galaxy gravitationally bound to it. (Lick Observatory)

Figure 13–16 An example of a spiral galaxy (NGC 253) seen not face–on as in Figure 13–15, but at more of an angle. (Lick Observatory)

The reasoning behind that theory is based on our studies of the two irregular galaxies we know best — those two that are the nearest galaxies to the Milky Way. There is good evidence that the *Magellanic Clouds* are "attached" to the Milky Way in the sense that there is a "bridge" of stars connecting the three galaxies. This feature is similar to that apparent in the *Whirlpool* galaxy, which is located near the end of the handle of the Big Dipper (Figure 13-15). At the end of one of the spiral arms is what appears to be an irregular galaxy. Such objects as this and the Magellanic Clouds may occur by near collisions between galaxies.

Because they are small and contain fewer stars, irregular galaxies are difficult to study if they are located far away, and impossible to detect at great distances. In fact, only the four irregular galaxies in the Local Group are close enough to be studied well. That is also true of the dwarf elliptical galaxies. Recall that in our Local Group, dwarf elliptical galaxies predominate. The important consequence of this is that what we observe at the greatest distances are the largest galaxies, the giant elliptical galaxies and the **spiral galaxies**. These latter two types may not be the most common galaxies in the universe, but, because they are large and contain lots of stars, they are the easiest to observe and photograph. Keep this fact in mind, because a survey of the most distant regions of space may not reveal its true contents — there may be lots of dwarf galaxies out there, but they won't show up on our photographs. This is referred to as a **selection effect** in taking population samples of stars or galaxies.

Spiral Galaxies When the average layperson thinks of a galaxy, he/she usually visualizes a spiral galaxy. That is probably due to the fact that our Milky Way is a spiral galaxy, and the media prefers to display images of galaxies shaped like it. Because the Milky Way is a spiral, we understand the type quite well. From the perspective of types of stars involved, a spiral galaxy is a hybrid galaxy consisting of an irregular galaxy (the spiral arms) attached to an elliptical galaxy (nucleus). The spiral arms are regions of active star formation, whereas the nucleus is fairly void of gas and dust.

Therefore, the nuclei of spiral galaxies consist of old, Population II stars, whereas the spiral arms consist of the young, hot Population I stars. Because of the striking appearances and sizes of the spiral galaxies, they appear to dominate the photographs taken of galactic clusters located far away (refer back to Figure 13-1) — along with the giant elliptical galaxies. Presumedly, as I mentioned, there are dwarf elliptical and irregular galaxies located amongst the visible galaxies in that cluster.

While Hubble was busy photographing and classifying galaxies, he noticed that spirals were not uniformly the same in shape, so he subdivided them into two general types — **normal** and **barred**. The *normal spiral galaxy* (**S**-type) and the *barred spiral galaxy* (**SB**-type) differ from one another in that the barred has spiral arms that are "attached" not directly to the elliptically-shaped nucleus, but to bar-like extensions on either side of the nucleus (Figure 13-17). The origin of these bars is not well understood, but probably has to do with irregular rotations within the nucleus of the galaxy itself. A third type of spiral galaxy, not commonly found, is the **SO** galaxy, which has no spiral arms, and therefore very little gas and dust. But since it has an obvious disk component with a prominent nucleus, it is considered to be intermediate — in shape and contents — between an elliptical and a spiral galaxy. There is very little visual difference between a SO and an EO galaxy — one must look quite carefully for indications of a disk component.

Do keep in mind that galaxies are not large solid objects, but consist of millions or billions of individual stars — each of which has its own orbit around the gravitational center of the galaxy. So even stars within the nucleus of a galaxy are in orbit around the center. With respect to the types of stars contained therein, normal and barred spirals are identical. But each type does show another subdivision based upon the *size of the nucleus* and the *extent to which the arms are wound around the nucleus*. The letters **a**, **b**, and **c** are used to indicate these variables amongst spiral galaxies.

Sa and **SBa** galaxies have tightly-wound spiral arms, large nuclei, and less gas and dust. The **Sc** and **SBc** have loosely-wound arms, small nuclei, and more gas and dust.

Figure 13-17 An example of a barred spiral galaxy, NGC 7479 is classified as a SBb galaxy. Notice the "bars" of stars running out from the nucleus. (Lick Observatory)

The **Sb** and **SBb** are intermediate between the two. The Andromeda galaxy and Milky Way galaxies are both *Sb* type galaxies. These letter designations are merely part of a language that astronomers use when referring to galaxies. They are not precise — nature has not divided the galaxies up into neat little packages like chemical elements. They are human inventions for the purpose of understanding them better. You will hear the language used at an observing session when it comes time to look at galaxies through the telescopes.

Origin of Galaxy Shapes Several questions arise at this point. Amongst the elliptical galaxies, why are there degrees of flattening — and amongst the spirals, why are there different degrees of the winding up of the spiral arms? One would initially think that it has something to do with the degree of rotation — the greater the rate of rotation, the more flattened the galaxy and/or the tighter the arms. Certainly when you look at a sequence of photos of spiral galaxies, it appears that way — as if an Sc galaxy speeds up to become an Sb and then further to become an Sa galaxy. That would also seem reasonable in light of our having learned earlier that the shape and contents of the Milky Way galaxy can best be explained in terms of rotation and the conservation of angular momentum.

The weight of available evidence today leads us to conclude that spiral galaxies at least are still dynamic and evolving, simply because star formation and the recycling of chemical elements back into interstellar space are going on. Elliptical galaxies, on the other hand, seem to have reached some kind of equilibrium state, in which very little stellar activity takes place. Because all three types of galaxies contain <u>some</u> old stars, we must conclude that there is no significant age difference between them — they are all about the same age (presently estimated at about 15 billion years old). Obviously, then, in order to account for the differences in the relative amounts of young and old stars, we must look for mechanisms that influenced the rate of star formation in the galaxies over their 15 billion year lifespans.

Unfortunately, rotation does not in itself explain all of the differences between galaxy types. Rotation is certainly important, but there must be other forces at work in the shaping of galaxies. For example, elliptical galaxies do rotate. It is just that they have not necessarily formed disks (like the E0, E1, and E2 galaxies). The difference, then, is between the galaxy consuming the gas in making stars prior to forming a disk, and forming a disk prior to the gas and dust being used up by early and rapid star formation.

The reason that the rate of star formation is so critical here is that stars, once formed, settle down to orbit the center of the galaxy outside of the disk. Clouds of gas, in contrast, act as fluids and settle into a disk with time. The essential question, therefore, is whether there is sufficient *viscosity* within the material to allow for the dissipation of energy so that collapse to the plane of the disk can occur. Stars orbiting the galaxy do not collide frequently enough to dissipate energy — gas particles do.

So we naturally ask ourselves — why did star formation occur at a faster rate in some galaxies than in others? The most likely answer is that it depends on the density of material in the cloud of gas and dust that is to become a galaxy. In those regions of highest density, collapse and subsequent star formation occur at a faster rate than in the less dense regions. The initial rotation of the cloud would

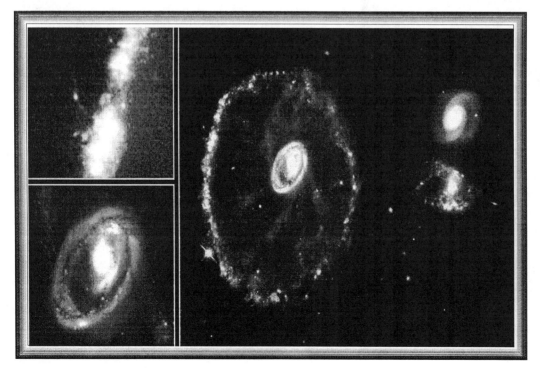

Figure 13–18 A rare and spectacular head–on collision between two galaxies appears in this HST image of the Cartwheel galaxy, located 500 million light–years from Earth. (NASA/STScI)

certainly influence the rate of star formation as well, since turbulence of the cloud material is influenced by rotation, and rate of star formation is influenced by turbulence.

New Theory of Galaxy Formation Within the last few years, an entirely new explanation for the differences between galaxy types has been gaining ground. This theory basically says that spiral galaxies formed first – during the initial era of galaxy formation shortly after the Big Bang. The elliptical galaxies subsequently formed through the merger, collision, or interaction between two or more spiral galaxies. This idea should not completely surprise you – we learned that the average distance between galaxies is about 20 galaxy diameters. Compare that to the average distance between stars as a function of star diameters (remember the golf balls?). It is about 10^7 star diameters! Large telescopes reveal hundreds of galaxies that appear to be colliding. One of the most famous examples is known as "The Mice" because of their appearance. Other interacting galaxies create the impression that they are colliding by what appears to be material shared by both galaxies – rather like a bridge of material between the two.

Galactic Collisions When galaxies collide, there is little chance that stars actually come into physical contact with one another. It is more like the galaxies pass through one another. That fact is easily understood when the distances between stars is considered. The gas clouds in either or both can be influenced by the encounter, to be sure. But it is the gravitational fields of the galaxies that have the most visible effects – distorted arms, peculiar tails, and bridges of material. These effects are not seen in real time, of course – the collisions take place over millions of years of time.

The evidence supporting the theory that mutual gravitational disturbances are responsible for the odd features we see is the simulation of such collisions in super computers. In a large modern computer, only 5 minutes or so are required to reproduce the conditions of the encounter and run them through millions of years of time. If galactic collisions are common, we should expect to observe an occasional head-on collision. Computer models of such events suggest that the damage should look something like a "smoke ring" in space. That is just what the HST imaged in 1994. The Cartwheel galaxy, located some 500 million light-years away, appears to have formed from just such a head-on collision. The intruder galaxy might well have been one of the two objects seen to the right of the ring in Figure 13-18.

Like a stone tossed into a pond, the collision sent a ripple of energy into space, plowing gas and dust in front of it. Expanding at 200,000 miles per hour, the ripple leaves in its wake regions of intense star formation. The HST resolves bright blue knots that are gigantic clusters of newborn stars and immense bubbles of gas and dust blown out into space by supernovae going off like a string of firecrackers. Images such as this provide astronomers with the opportunity to study how extremely massive stars are born in large fragmented gas clouds.

The Cartwheel galaxy presumedly was a normal galaxy like the Milky Way prior to the collision. This spiral structure is beginning to reemerge, as seen in the faint arms or spokes between the outer ring and the bulls-eye shaped nucleus. The ring contains at least several billion new stars that would not normally have been created in such a short time span. The ring also reveals the effects of thousands of supernovae on the ring structure. One flurry of explosions blew a hole in the ring and formed a giant bubble of hot gas. Secondary star formation on the edge of this bubble appears as an arc extending beyond the ring.

More direct evidence of galaxy collisions is being provided by the HST. An image of the galaxy NGC 7252, for example, taken in 1993, reveals a pair of long tails that astronomers interpret as evidence of gravity's effects during the merger of two galaxies about one billion years ago. But what makes this discovery particularly important is the detection of about 40 extremely bright clusters of young stars near the heart of the merged galaxy. This would imply that star formation was the direct result of the merger of the galaxies. This fits in well with existing models of star formation that require some kind of initial "push" to force gas and dust clouds to collapse to become stars.

To Merge or To Pass By But the encounters I have discussed so far are ones in which the collision is followed by separation. This no doubt occurs because the relative velocity of approach is so great that despite the slowing down of the galaxies as they gravitational distort one another, their forward momenta carry them further and further away after the encounter.

But we are looking for evidence that elliptical galaxies are the result of the merging of galaxies – in which two or more galaxies collide and turn into a single galaxy. If the relative velocity of the colliding galaxies is small at the offset, on the other hand, then they may not have sufficient momenta to escape one another's gravitational attraction – they don't possess *escape velocity*. They will pass one another, slow down, fall back together, and collide again and again until they eventually merge into a single galaxy.

If the mass of one of the galaxies is significantly greater than that of the other, the effect will be dramatic. The larger of the two will tear the outer portions of the other away, and then begin to eat away at the denser nucleus. This dense core, being more compact, is more difficult to tear apart, and so it "falls" to the center of the larger galaxy's nucleus and is "digested" by it. This is referred to as **galactic cannibalism**. As bizarre as this scenario seems, there is every reason to believe that it occurs frequently amongst the billions of galaxies. Again, we use computer simulations in the attempt to match prediction with observation. There are many interacting galaxies that have the very visible features that are simulated by computer. There are certain giant elliptical galaxies that even have several nuclei, suggesting that they devoured more than one galaxy during their lifetime. Those galaxies consumed by larger neighbors are called **missionary galaxies**.

The best example of this process just happens to be a nearby neighbor of the Milky Way: the Andromeda galaxy. Astronomers working with recent HST images found two bright spots at the heart of M31 (Figure 13-19). The dimmer of the two appears to mark the exact gravitational center of the galaxy, whereas the brighter is at least 5 light-years from the true center. This brighter spot is thought to be the remnant of a smaller galaxy that fell into the galaxy a billion or so years ago. The smaller galaxy's core is the only surviving relic of the collision. A problem with this scenario is that the remnant core should be torn apart by the massive black hole theorized to dwell at the exact center of M31. Further investigation is necessary to resolve the conflict.

And it shouldn't surprise you to learn that the Milky Way reveals tantalizing clues that it has cannibalized galaxies in the past, and therefore may be in the process of doing so at this very moment to the Magellanic Clouds. Some astronomers have discovered a group of young stars (Population I) outside of the disk of the galaxy where Population II stars are expected to be found.

One possible interpretation is that our Galaxy cannibalized a smaller galaxy and the gravitational compression triggered star formation in the victim's gas while still outside the disk portion of our Galaxy. If further observations support this initial discovery, then it will be easy to entertain the theory that the same fate awaits the two nearest galaxies to us. The Milky Way galaxy may eventually "swallow" the Magellanic Clouds. And to think that we used to call ourselves quiet and passive!

HST and Distant Galaxies A principle goal of the Hubble Space Telescope (HST) is to trace galaxy evolution through direct observation. Because it is collecting radiation from above Earth's turbulent atmosphere, it is detecting objects that are too faint to be seen from Earth's surface. If they are galaxies, they are faint either because they are very distant or because they are not very luminous. Recently, astronomers reported that HST imaged some suspected ancestors of today's galaxies.

The images reveal that star-forming galaxies were far more prevalent in the clusters of the younger universe than in the clusters we see near us today. That is, the more distant clusters are longer ago, existing when the universe was younger than it is today. The images show a full range of galaxy types in the universe of 4 billion years ago — elliptical, spiral, irregular, and distorted forms. Some images reveal galaxies in collision, tearing material from one another. Others appear to be merging into single systems. These galaxy interactions appear in the form of "tails" that distort the shapes of some of the galaxies. The tails are probably caused by tidal effects where the gravitational pull between closely passing galaxies stretch and disrupt their stellar distributions.

The result is that many ancient spirals might have merged to form giant elliptical galaxies or simply been torn apart and dispersed by violence that accompanies such encounters. That theory works toward explaining why there are so many elliptical galaxies around today, and why there are so many more (per unit volume of space) long ago. A third possibility to explain why it is that we see a shortfall of spirals in the nearby clusters is that when star formation subsided in many spiral galaxies, the galaxies may have faded to the point that they cannot be seen even in the nearby clusters.

Expansion of the Universe and Galaxies It might have occurred to you that there is a contradiction between Hubble's law and the theory proposed to explain the shapes of galaxies. If the galaxies are moving away from one another (Hubble's law), then how do we explain the observation that galaxies collide? We do by refining our thinking to consider two motions of galaxies in space — the motion of each galaxy within its parent cluster of galaxies, and the motion of the cluster itself.

To be more precise, Hubble's law applies to clusters of galaxies, not to every individual galaxy we observe and study. That is, we measure blueshifts for some galaxies within larger clusters, but overall the cluster is receding from us. So it would be more accurate to say that what astronomers have determined through spectral analysis is that *clusters of galaxies are all moving away from one another.* And that is what we mean by the expansion of the universe.

Figure 13-19 This HST image of the nucleus of the Andromeda galaxy (M31) reveals the presence of a double nucleus. This may be due to M31 having "cannibalized" a smaller galaxy, although the massive black hole believed to be at the center would have torn such a galaxy apart by now. Another possibility is that dust might dim the core to create the illusion of a pair of separate star clusters. (NASA/STScI)

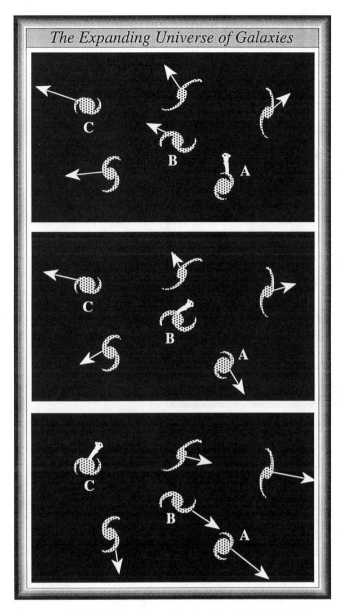

Figure 13–20 The fact that all galaxies are moving away from us does not place us at the center. The motion is relative. Astronomers in those other galaxies detect the Milky Way receding from them.

It is common for people to misunderstand what astronomers mean by the expansion of the universe of galaxies. It is tempting to ask just what it is that the universe is expanding "from," and "toward" what it is expanding. The answers to these questions will be made clear in the next Chapter, but for the moment, realize that the motions of the galaxies we detect from Earth are <u>relative</u> motions: observers in those distant galaxies will detect the Milky Way galaxy moving away them (Figure 13-20 Earth cannot be considered to be stationary and located at the center of the universe away from which the galaxies are receding. All astronomers in the universe will detect the expansion of the universe.

Galactic Rotation

As we looked at the various shapes of galaxies earlier, I'm sure it occurred to you to ask just how astronomers measure a galaxy's rotation in the first place. Certainly the time required for one to spin enough to reveal even the slightest change in two consecutive photographs exceeds several human lifetimes.

Doppler effect. The Doppler effect is our tool. While observing a nearby galaxy, at least, we can focus in on and obtain spectra of its two outer edges. The direction and rate of rotation is then determined by the amount and direction of the shift in spectral lines in the two spectra. For the Andromeda galaxy in Figure 12-11, for example, the spectrum of the left edge reveals redshifted lines, whereas the spectrum of the right edge reveals blueshifted lines. The right edge is approaching us and the left edge is receding from us. It must be rotating!

The amount of the shift in spectral lines tells us that the galaxy *on the average* rotates about 250 miles per second. I say on the average because of course the Andromeda is made up of at least 100 billion stars, each of which moves at its own particular rate around the gravitational center of the galaxy, but which also contributes to the overall rotation of the galaxy we measure.

Of course, if the galaxy is not seen exactly edgewise, we must compensate for the angle of tilt as we determine its rotational rate. If it is seen exactly face-on, there is no way to determine its rotational rate by the Doppler effect, since there is no relative movement between the opposite edges. We could, of course, take photographs of such face-on galaxies and store them in a sealed vault, and leave instructions to astronomers thousands of years from now to make comparison studies with their own photographs of the same galaxies.

Rotation Curves While looking for a connection between galactic shape and rotation rate, astronomers found something peculiar. Do you recall <u>Kepler's second law</u>? It says that <u>objects orbit the center of gravity of another object</u> (or group of objects), and that the rate is directly proportional to the masses of <u>the object or objects and inversely proportional to its distance from the center of gravity of that object</u>(s). When we plot the orbital velocities of stars within a galaxy with respect to their distances from the center of that galaxy, we obtain **rotation curves** such as that in Figure 13-21.

What these curves reveal is that the velocities of stars close to the nucleus of the galaxy is less than that of stars farther away from the center. That is what we expect to find, since stars further from the center have more mass (in the form of stars) around which to orbit. The nucleus of a galaxy contains a higher concentration of stars than do the spiral arms. And we also expect to find that stars quite distant from the nucleus will have lower velocities — simply because

★★★★★★★★★★★

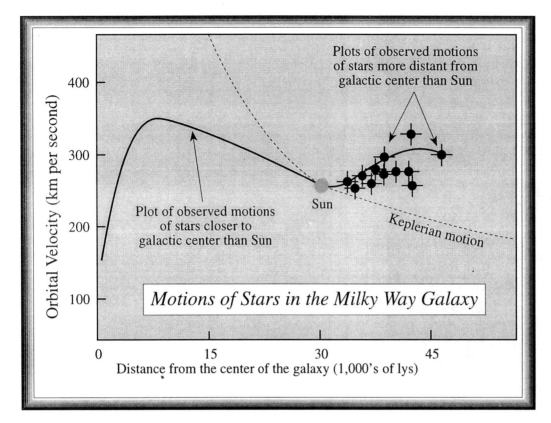

Figure 13–21 *A rotation curve for the Milky Way galaxy reveals that stars in the outer reaches orbit faster than Kepler's laws say they should. This implies there is considerably more matter in the galaxy than has been detected.*

of their great distances from the center. On the contrary, the rotation curve suggests that approximately the same velocity is maintained for stars that are progressively farther and farther away from the nucleus!

Our conclusion is — there must be a significant amount of non-luminous (invisible) matter within galaxies that causes stars farther out from the nuclei to orbit faster than they should, based on the amount of matter we are able to observe. In fact, astronomers estimate that to account for the observed velocities at the outer edges of some galaxies — including our own — there must be ten times more matter associated with them than can be detected. Astronomers are therefore led to theorize the existence of significant amounts of **dark matter** in the halos of spiral and elliptical galaxies — matter that produces the gravitational pull to which the orbiting stars respond. During the study of the even more distant clusters of galaxies, astronomers obtain similar results — the galaxies appear to orbit the centers of their parent clusters at rates greater than can be explained solely by the number of galaxies we observe directly.

The situation is analogous to a person observing a mountain stream of snowmelt that carries on its surface fallen tree leafs. Assume that you are unable to observe the water directly, and do not even know of its existence. But you are able to observe the motions of the leafs as they float by your position on the bank on the stream. As you ponder what might be causing the leafs to move, you observe their patterns of movement relative to surrounding objects (the bank of the stream, for example). You notice that the leafs change their direction of motion at those locations where the bank takes a turn. So you theorize that whatever carries the leafs must somehow be influenced by the bank itself. Gradually you could piece together some basic properties of the medium responsible for the movement of the leafs. This is similar the situation confronting astronomers as they study the behavior of galaxies.

For example, you recall earlier my mention of the *Great Attractor*, a large supercluster of galaxies that is located beyond the nearest supercluster to us, the Virgo supercluster. Now I am prepared to explain why it is so named. By measuring the motions of the galaxies in our neighborhood using the Doppler Effect, astronomers discovered that the local galaxies share a streaming motion toward this great concentration of previously unknown galaxies. That finding, in itself, is not too surprising. What is remarkable is that the 7,500 or so galaxies that make up the *Great Attractor* are spread out over a very large volume of space, and cannot themselves account for the gravitational pull that results in the high speed at which we (in the Milky Way galaxy) are moving toward it (200 miles per second). Astronomers calculate that 50,000 Milky Way–sized galaxies are needed to pull us at that rate of speed. This additional mass could be in the form of dark matter. We just don't know (Figure 13-22).

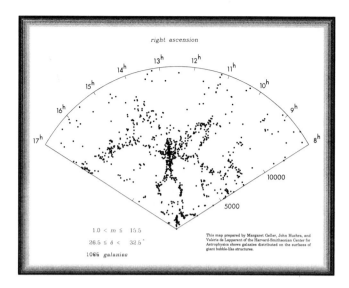

✹*Figure 13–22* A plot of 1,065 galaxies located near the Milky Way galaxy reveals that galaxies are not randomly distributed in space, but are in patterns that suggest a larger mechanism determined where they formed. Dark matter most certainly plays a role in this mechanism. (Margaret Geller, John Huchra, and Valérie de Lapparent of the Harvard–Smithsonian Center for Astrophysics)

Nature of Dark matter Exactly what can this dark matter be, you are asking? Well, that question is on the cutting edge of current astronomical research. Some astronomers believe that it consists of low-mass stars or burnt-out stellar corpses like white dwarfs, neutron stars, and black holes. Since all of these objects have very low luminosities, it is easy to imagine that there are a far greater number of them beyond our ability to detect. Besides, since they are the normal end products of stellar evolution, their presence in galaxies would not be surprising. Some astronomers, on the other hand, believe that the dark matter is of a very different nature — perhaps elementary particles (protons, electrons, neutrons, etc.) left over from the Big Bang that did not participate in the original collapse of the galaxy, but was instead left behind in the halo of the galaxy.

Think of it this way. Imagine an large number of Jupiter-sized lumps of stuff scattered around our Galaxy's halo. They would be too small and therefore too dim to be observed from Earth, and yet they could add up to more mass than is contained within stars. Astronomers call these hypothetical objects **machos** (for **ma**ssive **c**ompact **h**alo **o**bjects). Some astronomers are searching for these objects by looking for the deflection of starlight that passes close to them. Estimates are that if the halo of a galaxy is made of *macho* stuff, about one star in five million should have its light deflected at any particular moment. So far, the search has failed to find evidence of the existence of any in the Large Magellanic Cloud.

Another type of search for dark matter is being conducted on the assumption that it is not in the form of ordinary matter, but something that falls within a class of objects called **w**eakly **i**nteracting **m**assive **p**articles, or **wimps** for short. More massive than the protons and neutrons that make up ordinary matter, *wimps* are theorized to interact only occasionally with the matter of which you and I and the planet are made. So millions of them could be passing through our bodies (and Earth) at this very moment. In that respect, they could be like the neutrinos that are by-products of Sun's core reactions. A group of astronomers at Stanford University is preparing to detect that occasional wimp that does interact with ordinary matter. A large crystal of germanium will be cooled to within a few thousandths of a degree of absolute zero, so that its atoms are hardly moving. If a wimp strikes the crystal, the struck nucleus will recoil and raise the temperature of the crystal ever so slightly. Scientists are prepared to measure that slight rise in temperature.

Thus there could be an enormous supply of the raw materials necessary for star formation remaining within present-day galaxies — elliptical as well as spiral. Elliptical galaxies only look old and senile — there is plenty of life left. The trouble is, according to this theory, the dark matter is cold (low velocity). Therefore, the chance of these particles collecting together into dense clouds to collapse by gravity is quite unlikely. The search for *dark matter* has extremely important consequences for the Big Bang theory, so we will also deal with the theory of dark matter in Chapter 15 when we examine various theories for the origin of the universe.

Evolution of Galaxies

Our current attempts to develop an adequate theory for galaxy formation are frustrated by the very nature of the subject. Galaxies are huge objects, and contain many different types of objects. Stars, on the other hand, are quite simple in their makeup. Modeling the interior of a star by computer is therefore quite simple because the behavior of gases inside a star is comparable to the experiments that we can conduct in the laboratory on Earth. But a galaxy is not so easily modeled. Nevertheless, with the development of supercomputers, more and more simulations of galactic behavior are being studied. Be patient.

Dark Matter to Galaxies While speculative at this point in time, astronomers are gradually piecing together a theory that attempts to wrap all of the observations of galaxies into a neat package. The **dark matter particle theory** explains many of the features we observe in galaxies, including the observation that they do not rotate as expected. Generally, the theory goes like this. Soon after the Big Bang, occasional disturbances within the matter generated by the explosion acted as seeds for the gravity-induced collapse of clouds of dark matter. Although this matter was distributed fairly uniformly during the early stages of the universe, gravity ultimately caused it to develop structure.

★★★★★★★★★★★

Many of the collapsing clouds were the building blocks for the formation of larger clouds from which the giant elliptical and spiral galaxies formed. As those large clouds further fragmented to form individual stars within the cloud, many of the cold dark particles were left behind in the voids between the stars, where it is theorized they still exist today.

According to this theory, then, the difference between the formation of a giant elliptical and a spiral is simply a matter of mass. The more dense concentrations of dark matter — having greater gravitational collapse — formed the giant ellipticals. The less dense clouds collapsed more slowly, having a chance to develop spiral structure as they rotated faster and faster by the principle of conservation of angular momentum.

Dwarf Elliptical Galaxies But, you say, what about the most common type of galaxy — the dwarf elliptical? How did they form within the framework of the cold dark matter theory? Evidently, the dwarf elliptical galaxies — lacking the mass of their sibling giants — represent the average disturbance within the clouds of dark matter after the Big Bang. The unusually high-density disturbances that *were not rotating rapidly collapsed rapidly* to become the old star-rich giant ellipticals. Those that *were rotating rapidly collapsed more slowly* to become the spirals — with star formation still going on today. The density wave would cause the spiral structure to persist indefinitely.

Different density wave patterns would result in the different kinds of spirals — Sa, Sb, Sc. This is, of course, consistent with the observations we made about our own spiral galaxy in the previous Chapter. But the greater number of disturbances within the clouds of dark matter were of a magnitude that caused the formation of the not-so-massive galaxies — dwarf elliptical, for example.

Just as in the case of stars, the universe tends toward smallness. Computer studies show that if the average disturbances were rotating faster than a critical amount, rather than forming a dwarf spiral or elliptical galaxy on their own — as you might expect — they slowly dissipate and perhaps be absorbed within another disturbance to become a larger galaxy. In other words, there may be a critical mass and limit to the degree of rotation necessary in order for a cloud of dark matter to collapse by virtue of its own gravity. This theory is supported by the observation that ellipticals are found in the densest parts of galactic clusters, whereas spirals are found in low-density regions and outside the clusters where higher rates of rotation of disturbed clouds of dark matter are common.

Formation of Galaxies

Both the *merger-collision-interaction* and *dark particle matter* theories attempt to explain how the galaxies got the variety of shapes we observe today. They deal with the evolution of galaxies, but not their origin. That particular subject spills over into the subject of the origin of the universe itself, so we will encounter the subject of galaxies once again when we attempt to tackle the question of the creation event. The origin of galaxies is not well understood, but to take a peek at the starting point for an explanation is to look at the largest-scale structure of the universe possible at the moment.

Large-Scale Structure of Universe Now look carefully at Figure 13-22, a plot of a sample of some 1,065 distant galaxies as they are distributed in space. This is about as close as modern astronomers can get to showing us a photograph of the universe as a entity unto itself! There is something odd about their arrangement in space — something yet to be explained about the manner in which the galaxies formed. The locations of the galaxies appear to take the form of **filaments** or strings, between which are great voids of space in which there are fewer galaxies.

Seen 3-dimensionally, this plot of galaxies would appear similar to that of a bowl of soap bubbles or a chunk of Swiss cheese. The galaxies appear along what would be the regions along which soap bubbles touch (filaments), but not in the *voids* inside the bubbles themselves. According to the Big Bang theory, the universe began in an extremely dense state some 20 billion years ago, and then began to expand uniformly outward in all directions. The matter contained within the expanding space eventually became the galaxies of stars we observe. But we find it difficult to explain how it was that galaxies were able to form out of expanding gas.

The filaments and voids represent the largest known structures in the universe. But we are at a loss to explain how they originate and how they relate to the formation of galaxies in the first place. Did the filaments form first and the galaxies form from the gas that collected along the filaments? Or did the galaxies form first in clusters and then some mechanism created voids between the clusters as they expanded away from one another? We don't yet know. Some calculations performed on supercomputers predict that galaxies will form along long filaments or in pancake-shaped walls that are separated by large voids of space.

Cosmic Strings A recent theory has been proposed to explain how it was that the galaxies formed in the pattern of voids and filaments we see in Figure 13-22. During the first few microseconds after the Big Bang, the then-tiny universe separated into regions in which the laws of physics evolved at different rates. It is akin to the manner in which water turns to steam when it boils, and turns to ice when it freezes. These are called **phase changes**, and are accompanied by obvious differences in the nature of the matter involved in the change. Look carefully at the crystal patterns in an ice cube — these represent water molecules aligned in different directions. Likewise, in the early universe, at the boundaries between different domains, "**cosmic strings**" formed.

These strings would have the capability of locking up immense mass and energy in long filaments of matter. Think of this analogy. You blow up a balloon, but some of the rubber is folded over and "stuck" on itself — so there are concentrations of rubber along the creases where the rubber is stuck together. It is along these creases, or *cosmic strings*, that galaxy formation took place in the early universe. No one has ever seen or detected one of these *cosmic strings*, but there are some researchers who are currently building the apparatus with which they could be detected. Stay tuned.

Summary/Conclusion

It seems apparent that the evolution of stars has taken place within the framework of larger units called galaxies. Their different sizes and shapes attest to the varying conditions within the clouds of gas and dust from which they formed. Since looking out at increasing distances is equivalent to looking back in time, it appears that the more distant galaxies are younger, and more violent, than those right around us in time. So perhaps the universe, when it was young, was not favorable to the evolution of life. For the moment, at least, we will probably want to confine our search for others to our own galaxy. But knowing that there are plenty of other families out there to search gives us encouragement of eventual success.

LEARNING OBJECTIVES: *Now that you have studied this Chapter, you should be able to:*

1. Describe the physical characteristics of spiral, elliptical and irregular galaxies, including how they compare in size, shape, mass, color, types of stars, and amount of gas and dust.
2. Sketch the general distribution of galaxies in the Local Group, including the Milky Way galaxy.
3. Describe the general distribution of galaxies in space.
4. Explain how the rotation of a galaxy is determined, and how that value implies the presence of a great deal of unseen matter within galaxies.
5. Describe the general process by which galaxies formed after the Big Bang, and how that is supported by the appearances and behavior of observed galaxies.
6. Explain how the distances to the distant galaxies are determined, and the inherent limitations of the methods used.
7. Explain how Hubble's law is derived from knowing galactic redshifts and distances, and how the Hubble Constant can be used to determine the distances to galaxies and the age of the universe.
8. Explain by using the raisin bread analogy how Hubble's law implies that the universe is expanding.
9. Compare and contrast the normal and active galaxies in terms of appearance, energy output, and distance.
10. Define and use in a complete sentence each of the following **NEW TERMS**:

Andromeda galaxy 277
Barred spiral galaxy 287
Coma cluster of galaxies 280
Coma-Sculptor cloud 279
Cosmic strings 295
Dark matter 292
Dark matter particle theory 293
Distance indicator 283
Dwarf elliptical galaxy 279
Filaments (voids) 295
Galactic cannibalism 289
Hubble constant 284
Hubble's law 284

Irregular galaxy 279
Local Group 279
machos 293
Magellanic Clouds 279
Missionary galaxy 290
Normal spiral galaxy 287
Phase change 295
Rotation curve 291
Selection effect 287
Slope of curve 284
Spiral galaxy 287
Virgo supercluster of galaxies 280
wimps 293

Chapter 14

Active Core Galaxies, Quasars, and the Origin of Galaxies

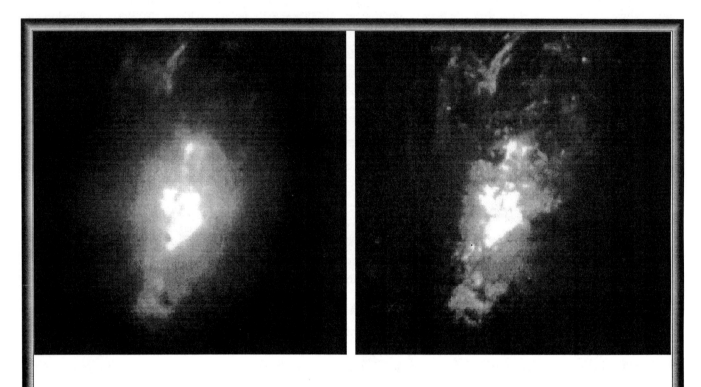

Figure 14–1 The HST has provided these outstanding images of the nuclear region of the galaxy NGC 1068, located at a distance of approximately 60 million light–years. For the purpose of comparison, the image on the left was taken prior to the upgrading of the telescope, and that on the right taken after the installation of the corrective lenses. This galaxy is the prototype of a class of galaxies, known as a Seyfert. In active galaxies such as this, typically the core shines with the brightness of a billion solar luminosities, and the brightness of the core fluctuates over the period of a few days. This implies that the energy is being released from a region only a few light–days in extent. The most likely source for this enormous amount of energy is a "super massive" black-hole with a total mass of 100 million stars like Sun. (NASA/STScI)

CENTRAL THEME: What are quasars, the most violent objects in the universe, and what do they tell us about the origin and evolution of galaxies since the Big Bang?

Fortunately for astronomers, not all galaxies fit into the neat and tidy packages of elliptical, spiral, and irregular. I use the word fortunately because it is — as with people — the peculiar cases that often times provide us with the greatest insight as to origin and behavior. Those galaxies that do not fit into Hubble's classification system (Figure 13-13) are referred to as **peculiar galaxies**. Percentage-wise they are rare. But a small percentage of a very large number — the number of galaxies in the universe is estimated to be about 100 billion — is still a large number.

Problems in Explaining the Origin of Galaxies In the last Chapter, I painted only a brief sketch of the current theory of how galaxies came to be. One would expect that as we observe deeper and deeper into space we should find younger and younger galaxies, since looking out into space is equivalent to looking back into time. But we don't: we observe only **quasars** and quasar-like objects, objects that are extremely energetic outbursts that occur in the centers of some galaxies (Figure 14-1). We are therefore left with two problems: how did the galaxies come to be, and how did the peculiar galaxies (including quasars) come to be?

There is a growing body of evidence that these two problems are related to one another. The general picture that is emerging is that the formation of galaxies and their active nuclei occurred nearly at the same time. The details of the connection are not clear, but observational data is arriving in vast amounts as a new generation of telescopes comes on line and the HST continues to perform in spectacular fashion.

Peculiar Galaxies

To date, astronomers have catalogued over 10,000 peculiar galaxies. Without my having mentioned it, I mentioned a few of them in the last Chapter while explaining the evolution of the shapes of galaxies. I theorized that when the universe was younger and more crowded, galaxies frequently collided, merged, or interacted so as to form the variety of galaxy shapes observed today. As evidence, I offered photographs of colliding galaxies — galaxies that clearly do not fit the Hubble scheme. It is not just their distorted shapes, however, that place them in the classification of *peculiar*. The energies emitted by peculiar galaxies are a telltale sign of violence unmatched by galaxies such as the Milky Way.

Active Galaxies A subgroup of peculiar galaxies includes those that have very *bright nuclei and/or emit unusually great amounts of energy from their nuclei* — quite unlike the normal galaxies in Hubble's classification system. Many of these active galaxies look rather quiet and mundane in visible-light photographs, but reveal their violent natures when studied with X-ray satellites, infrared satellites, and/or radio telescopes.

Seyfert galaxy An excellent example of an active galaxy is the **Seyfert galaxy**, named after Carl Seyfert (1911-1960), who first discovered their peculiar properties. Over 100 of them have been catalogued. They appear in visible-light photographs as normal spiral galaxies. But an extremely

Object	Radio*	Infrared*	Optical*	X-ray*	Gamma Ray*	Luminosity**
Supernova	10^{27}	10^{28}	10^{31}	10^{30}	$10^{26} - 10^{27}$	10^{7}
Normal galaxy	10^{31}	$10^{33} - 10^{35}$	$10^{35} - 10^{37}$	$10^{31} - 10^{32}$	$10^{31} - 10^{32}$	10^{10}
Seyfert galaxy	$10^{33} - 10^{35}$	$10^{36} - 10^{39}$	$10^{35} - 10^{38}$	$10^{34} - 10^{36}$	10^{41}	10^{13}
Radio galaxy	$10^{35} - 10^{38}$	$<10^{36}$	10^{37}	$10^{34} - 10^{38}$	10^{37}	10^{12}
Quasar	$10^{37} - 10^{33}$	$10^{39} - 10^{40}$	$10^{38} - 10^{40}$	$10^{37} - 10^{40}$	10^{39}	10^{14}

Luminosities of Active Galaxies

***Rate of radiation in watts**   ****Luminosity is in units of Sun = 1**

Figure 14-2 A chart showing the amounts of energy at different wavelengths emitted by different objects in the universe reveals that the most distant objects are also the most violent.

short exposure still reveals a bright, luminous starlike nucleus. The spectrum of the nucleus reveals not absorption lines, as we would expect if there are just stars present — such is the case with normal galaxies — but emission lines with strong Doppler shifts.

The emission lines indicate the presence of hot gases moving at high velocities (thousands of miles per second) within the nucleus and/or out of the nucleus. Intense X-rays, infrared, and radio radiation coming from the nucleus verifies this conclusion, and suggests that some type of explosive event occurred (or continues to occur) there.

The amount of violence (energy) associated with peculiar galaxies is more fully appreciated when we compare the energy output of different celestial objects in the various regions of the electromagnetic spectrum (Figure 14-2). They are arranged according to the amount of energy emitted. This is very significant because the more luminous objects are also the more distant objects. Not only are the more distant galaxies moving away from us at faster rates, but they also emit more radiation. The types of radiation they emit, however, varies from object to object. This is not surprising, since different processes and events are responsible for the emission of different types of radiation.

Radio Galaxies Radio energy, as we learned while learning about pulsars and neutron stars, is often associated with violent events. It is emitted when electrons are accelerated to velocities close to the speed of light and then trapped around lines of force in a strong magnetic field. Well, there is a category of active galaxies (10,000 or so have been recorded) that are so "bright" in their output of radio energy that we refer to them as **radio galaxies**.

Be careful not to place these different types of *peculiar* galaxies in narrow, confining categories. We create different categories for galaxies (e.g., normal, Seyfert, radio, quasar, etc.), for our own convenience. There is a great deal of overlapping of features amongst them. For example, some Seyfert galaxies emit as much radio energy as do some radio galaxies — and some radio galaxies emit as much visible radiation as some normal galaxies. The categories simply allow astronomers to better understand the origin of galaxies and the processes responsible for their behaviors.

Types of Radio Galaxies Our interest is in the patterns that arise when we compare variables such as shape, distance, radio-energy output, visible-energy output, etc. Radio galaxies are subdivided into two types — those emitting their energies from within the nucleus (**compact sources**), and those emitting their energies from regions located on opposite sides of what are frequently normal-looking elliptical galaxies (**extended sources**).

A good example of a *compact radio source* is the giant elliptical galaxy M87, located in the constellation of *Virgo* (Figure 14-3). We refer to it as *Virgo A* because it was the first (A) radio source detected in the constellation of *Virgo*. A short-exposure photograph of the galaxy reveals a small,

Figure 14-3 In this straightforward photograph of the elliptical galaxy M87, everything look fairly normal. (Lick Observatory)

Figure 14-4 But in this short exposure photograph, it is clear that something violent is going on in the nucleus. (Lick Observatory)

bright core that reminds us of a Seyfert galaxy — and a luminous jet extending away from the core (Figure 14-4). An ultraviolet image made by the Hubble Space Telescope of the optical jet shows the structure of the jet, which appears as a series of knots extending out from the nucleus of M87 (Figure 14-5). The hot plasma within the jet is channeled by magnetic fields. A large, bright knot midway along the jet shows where the jet is disrupted. The bend near the end of the jet indicates where it rams into a wall of gas.

Cygnus A A good example of an *extended radio source* is the very first one discovered: *Cygnus A*. Optically, it looks like two blobs running into one another (Figure 14-6). A

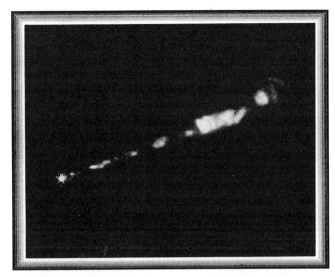

Figure 14–5 And in this remarkable image taken by the HST, *it is clear that the "jet" in M87 is discontinuous, as if the violence is periodic.* (NASA)

Figure 14–6 Cygnus A, *imaged optically, looks like two "blobs" running into one another. They are at the very center of the photograph.* (Lick Observatory)

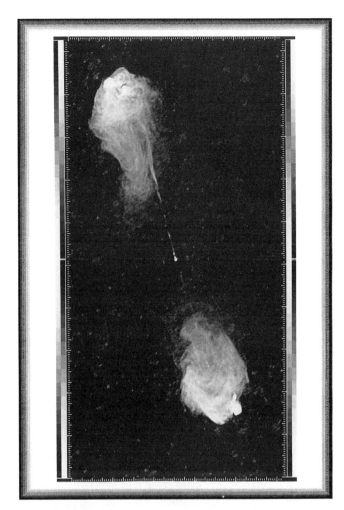

Figure 14–7 A radiograph of Cygnus A *reveals the two radio lobes. The dot between the lobes is the location of the two "blobs" in Figure 14–6.* (NRAO/AUI)

radio contour map (Figure 14-7) reveals that the radio energy comes not from the core of the blobs, but from regions to either side of the blobs — what astronomers call **radio lobes**. Naturally, astronomers are tempted to associate the radio emission with whatever is happening in the lobes themselves. If the radio energy results from high-speed electrons caught in magnetic fields, then obviously a violent event(s) occurred close by to give the electrons their high energies. Electrons do not just go off at high speeds on their own.

Actually, if you look carefully at the radio contours, you see evidence that suggests the high-speed electrons come from the colliding (?) lobes themselves. It appears that the radio-emitting electrons are more numerous further away from *Cygnus A* than close to it — as if they are piling up as they run into something. Our interpretation is that the electrons, ejected outward from the center of *Cygnus A*, encounter intergalactic matter in the form of gas and dust which slows them down and causes them to pile up. This answers the question of the distribution of radio energy

CHAPTER 14 / Active Core Galaxies and Quasars

coming from the vicinity of the object, but it certainly doesn't explain how the electrons got accelerated to such high speeds in the first place.

So our conclusion for both of these radio galaxies (as well as the other 9,998) is that explosive events are common in the cores of galaxies. But isn't that exactly what we found to be happening in the very core of our own galaxy? True — but there is an important distinction. The extent of the violence in nearby galaxies like our own is minor compared to that occurring in Seyfert galaxies and radio galaxies.

The HST and Peculiar Galaxies The HST's view of the core of NGC 7457, a typical galaxy 40 million light-years away, reveals that the stellar density of the nucleus exceeds earlier estimates from ground-based observations by a factor of 400. The HST also discovered a unexpected chain of luminous knots in the core of the most distant known galaxy, called 4C41.17. Located some 12 billion light-years from Earth, it appears that the inner region of this primeval — we see it as it was 12 billion years ago — galaxy is highly disturbed. 4C41.17 is also a well-studied radio galaxy, and reveals in radio radiation the presence of twin jets of high-speed particles moving away from the nucleus.

There are two models to explain the combined radio and visible observations. High-speed jets of particles causing the radio emission are also compressing gas and dust clouds they encounter along the way, triggering new star formation. The luminous knots seen by the HST would then be giant star clusters, each containing 10 billion stars or so. The other model supposes that the optical emissions come from light scattering off clouds of gas and dust. The light itself could be coming from an accretion disk surrounding a massive black hole at the galaxy's center. Perhaps when the HST receives its upgrading later this year, we'll get a clearer picture of what is going on there.

Pattern of Violence in Universe Allow me to briefly summarize the pattern I have been trying to establish in your thinking about galaxies. The local galaxies show evidence of outbursts of energy in their nuclei. The more distant galaxies show not only greater discharges of energy and electrons from their nuclei, but their appearances often times show signs of distortion and disruption. The more distant galaxies are observed further back in time, suggesting that the *galaxies that existed long ago were more violent than those existing today.*

But of course we came from them — they represent what we used to be. So the pattern seems obvious. The universe of galaxies has evolved. Early in their history, galaxies in the process of forming were violent — birthing pains, you might say. Gradually they settled down, and whatever mechanism was responsible for the violence in the nuclei slowly diminished. That is where our galaxy is today.

The Quasar Controversy

One of the most exciting frontiers of modern astronomy involves the discovery of quasars in 1961. Even though we have gone as far as naming microwave ovens, TV sets, and other household appliances after these strange-behaving objects, we don't yet fully understand them. Yes, we certainly have accumulated an enormous amount of information in the three decades since their discovery. To date, over 5,000 have been detected and catalogued. But they are nevertheless quite mysterious to us. You'll see why shortly.

Early Attempt to Explain Quasars In fact, some scientists insist that the only way to explain them is to give up some of our most cherished assumptions about the nature of the universe. This, in turn, might create a revolution in science that is as far-reaching in its implications as the Copernican Revolution was in its day. To date, however, quasars simply require that we stretch the laws of physics that have served us so well for so long. We must always be prepared for the breaking point. But since their characteristics and behaviors fill some of the gaps in our knowledge about the origin of the universe, they are at the moment — at least — a valuable discovery for astronomers.

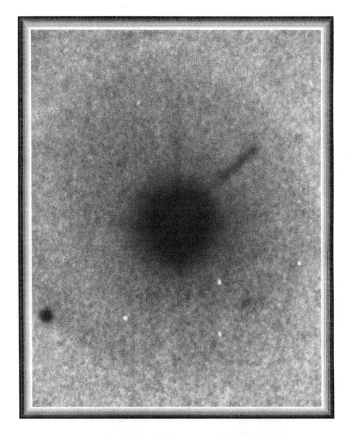

Figure 14–8 Quasar 3C 273 was the first such object discovered. Later images revealed that it too has a jet of luminous material moving at great speed out of the nucleus. (University of Hawaii, Institute for Astronomy)

One of the popularized theories for explaining the existence of quasars suggests that they are the other ends of black holes — the points in space out of which emerges the matter and/or radiation that went into a black hole somewhere else in space. So the idea is that matter pulled into a black hole located somewhere else — either inside or outside our universe — comes (back) into space with a burst of fireworks that we observe as the quasar (Figure 14-8). But I'm getting ahead of myself. I haven't even explained the nature of the fireworks yet. We will get back to this bizarre theory once we have found out what we are dealing with.

Discovery As radio astronomy was making its debut in the 1950s as a result of rapid advances in electronics during World War II, maps were gradually drawn of sources of radio energy coming from space. Since long wavelength radiation is inherently limited in providing good resolution, astronomers were unable to locate these sources exactly on a map of the sky. All they knew was that the radiation was coming from a rather large portion of the sky. They were therefore unable to associate the sources of strong radio energy with any particular visible objects on photographs taken of the same region of the sky — at least for most sources. Sun is easily identifiable as a radio source because it is so large. Likewise, the center of the Milky Way was easily identified because it covers a rather large area of the Milky Way band.

Even without being able to explain these radio sources, astronomers compiled lists of them in the *Cambridge catalogue of radio sources*. In the early 1960s, astronomers calculated that Moon would, on a given date, move in front of one of the strongest radio sources in the sky, and thereby provide a unique opportunity to pinpoint the exact spot from which the radio energy was coming. This source is identified as *3C 273* (the **273**rd object recorded in the **3**rd **Cambridge catalogue**). **Occultations** as this occur when one object moves in front of and covers up or hides (*occults*) another object.

We encountered this situation earlier when we learned that a particular type of binary star system — eclipsing binary — involves one star blocking out some of its companion's starlight as it moves in front of it. Occultations and eclipses are words used to describe the same kind of event. Since Moon's orbit is known with great precision, it was easy to determine the exact position of Moon's edge at that instant when the source of radio energy (3C 273) was occulted — and again when the opposite edge uncovered the source. This technique is very commonly used in astronomy. In Australia, a large radio telescope was used to observe the passage of Moon across the position in the sky of the bright radio source, and a precise location for it was determined.

Optical astronomers then went to work looking for an object at that exact location on photographs taken previously of that region of the sky. There was a tiny dot at that location. They went to work to obtain and then analyze the spectrum of the object at that position on the visible photograph. Actually, examined under high magnification with the 200-inch Palomar telescope, it was more than a dot — it appeared to have a luminous jet-like feature extending out to one side. It was temporarily called a **quasi-stellar radio source** (**quasar** for short) because it had the appearance of a star — but unlike a star was a strong source of radio energy. The spectrum of the quasar baffled astronomers, since it consisted of emission lines and absorption lines that did not match those of any of the chemical elements with which they were familiar (Figure 14-10).

Characteristics of Quasars

Interpreting Spectra You may recall that emission lines are associated with hot gases that emit photons as a result of a source of energy like nearby hot stars. If the dot portion of the quasar image was a star, we would expect to obtain an absorption spectrum from it. Astronomers recognized both emission and absorption lines in the spectrum of 3C 273 (now referred to as a quasar). And the pattern of lines? If the pattern of lines was not due to any known chemical substance, then what could they be due to? The answer came in 1963 when Maarten Schmidt of Mt. Palomar Observatory recognized the lines as those of hydrogen.

What had confused previous astronomers was that most of the lines of hydrogen normally seen in the visible part of the spectrum had been redshifted into the infrared portion of the spectrum. But since the initial spectrum obtained of the quasar was taken in the visible portion of the spectrum, the few lines that were visible were not recognizable. It was rather like trying to identify the year and make of an automobile by observing the tip of the rear bumper sticking out of the garage.

But why such behavior of the lines? Why were they located mostly in the infrared? Well, the most reasonable explanation was that the Doppler effect was causing the original wavelengths to be lengthened to the point that most were in the longer wavelength, infrared portion of the spectrum. Since there is a direct relationship between the *amount of the lengthening* of the wave and the *speed* at which the object emitting the wave is moving, astronomers were able to calculate that this first quasar (whatever it was) was going away from us at the incredible speed of 30,000 miles per second (16% the speed of light)!

Now really, is that actually possible? Yes it is — if you assume the quasar is moving in accordance with Hubble's law. What does that law say? Hubble found that there was a significant relationship between the amount of shift in the spectral lines of a galaxy and its distance — the more distant a galaxy is, the faster it is going away from us. This, you'll recall, is interpreted by astronomers as being the result of the universe expanding.

But remember that our interpretation of redshifts is not so much that the galaxies are moving away rapidly, but that

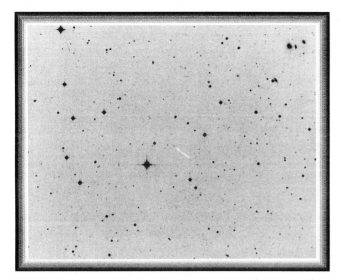

Figure 14–9 Quasar Q0051–279, discovered in 1987, was, at the time, the most distant object yet discovered. (Royal Observatory, Edinborough, U.K. Schmidt Telescope Unit, plate 8715204)

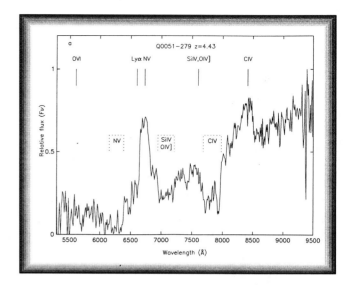

Figure 14–10 The spectrum of quasar Q0051–279 reveals spectral lines so far shifted toward the red end of the spectrum as to place it near the assumed beginning of the universe. (National Optical Astronomy Observatories)

space is stretching out and carrying the galaxies along with it. The galaxies are embedded in space. So the great redshift of the spectral lines of the quasar was simply a valuable tool for determining its distance from us. If we choose to use the Hubble constant of *10 miles per second per million light-years* for determining the distance to 3C 273, we find that it is 1.8 billion light-years away from Earth.

Distances There was not (and still is not) any way of determining the distance to a quasar by observing a known feature in the quasar itself, since there are no observable features identifiable as a distance indicators. Most quasars show up on photographic plates as mere dots. But if we assume that quasars are acting like the observed galaxies, obeying Hubble's law, then we can determine their distances (Figure 13-11), as I just did for quasar 3C 273.

Of the 5,000 already discovered, studied, and catalogued, the quasar that presently has the greatest redshift is receding at a velocity of 170,000 miles per second, corresponding to 90% the speed of light! And every year sees another astronomer announcing the discovery of an even farther quasar with an even greater redshift. The present record-holder is quasar PC 1247+3406, located 14 billion light-years from Earth.

Energy Outputs Determining both the distance to a quasar (using Hubble's law) and its apparent visual magnitude (using a photometer) allows us to calculate its absolute visual magnitude — the apparent magnitude it would have if it were located 32.6 light-years from us. The values for the full range of quasars we have catalogued varies from –24 to –28. In other words, if a typical quasar (say M_v = –26) were located 32.6 light-years from us, it would appear as bright as Sun (m_v = –26) does in the sky. And we can also use that absolute magnitude number to arrive at another incredible conclusion about quasars.

The absolute visual magnitude, as we learned earlier, represents the energy output of an object. So if we determine the number of magnitude intervals between the typical quasar and Sun, we can use the concepts of Chapter 8 to compare its energy output with that of Sun. Again using the typical quasar (M_v = –26) for illustration, since Sun's absolute magnitude is +4.8, the difference is approximately 31 intervals (–26 to +5 is 31 intervals). We then calculate the amount of energy this represents in units of the energy output of Sun:

$$(2.5)^{31} = 2 \times 10^{12} = 2{,}000{,}000{,}000{,}000 \text{ Suns of energy}$$

The typical quasar gives off 2 trillion times more energy than Sun, or the equivalent of 2 trillion Suns!

I am not suggesting that a quasar is a collection of stars which are identical to Sun. I'm only comparing it with Sun in order to get a value for its energy output. Whatever it is, the typical quasar acts as if it were a collection of 2 trillion stars each of which emits an amount of energy comparable to Sun. You learned in Chapter 12 that stars live in families called galaxies. If quasars are galaxies too, appearing tiny in photographs because they are located very far away, then the average quasar contains 20 times more stars than a galaxy like ours — the Milky Way. In terms of energy output, *the average quasar is equivalent to 20 Milky Ways of stars!* And that is only considering the visible energy output. Now that we have orbiting satellites that can detect the full range of radiation types that objects emit, we are finding that quasars give off much of their energy at non-visible wavelengths (Figure 14-11).

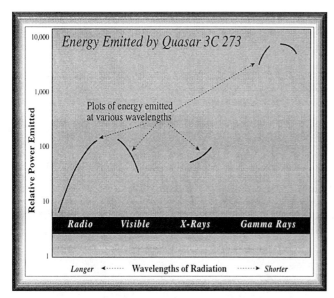

Figure 14–11 A plot of the energy emitted by 3C 273 across the electromagnetic spectrum reveals the violence involved in whatever powers the process. Radio and gamma ray radiation is frequently associated with violence.

Sizes of Quasars Of course you may be saying — "Okay, I can live with the idea that there are giant galaxies out there emitting enormous amounts of energy. They appear faint to us because they are very distant." But if we take that stance, we immediately encounter a serious problem. When astronomers record the apparent visual magnitudes of many of the quasars taken over long periods of time, and then look carefully at the values, they find that those quasars do not shine at a steady rate — they vary or pulsate in brightness.

What this observation indicates to us is that the source of energy for the quasar is confined to a very small region of space — about the volume of our solar system. What that means is that in order to explain the origin of the energy emitted by some quasars, we must imagine the equivalent of 20 galaxies worth of stars being confined to the volume of our solar system!

How is it that variations in energy output tell us the size of an object, you ask? Assume for the moment that a typical quasar is an extremely large galaxy like the one shown in Figure 14-12. We on Earth receive the combined light emitted by all of its stars, and, conforming to the behavior indicated on the light curve above, observe that the combined light varies with time. In order for that to happen, the light output of each star must vary with time — something that is difficult to imagine even at the offset.

But even if that were possible, how is it that the varying light from a star on the opposite side of this enormous galaxy is so coordinated with the varying light of a star on the near side that by the time the two are combined together they are in phase with one another? And for that to be true of all 2 trillion stars making up the quasar? Since it takes time for light to travel from one place to another, the combined light variations of all the stars would cancel out the highs and lows so that by the time the light arrives on Earth, it would be a steady amount — the average of all the ups and downs of all the stars.

This situation is comparable to listening to the musical instruments of a marching band in a very large stadium. The more spread out the band members are, the more out of tune the instruments sound. When they are globed together, they sound fine. Sound, too, has a finite speed. So it sounds as if the musicians are playing out of tune, in fact it is because the arriving sound waves are not in sync.

If the object responsible for emitting the energy of quasars is small, the time required for the light to travel from the far side to the near side is less than the time interval between highs and lows in the light curves, and we receive the light as varying. This is the case for many quasars, so we conclude that we have established an upper limit as to how large they can be. That upper limit is about a light-day in diameter — about the diameter of our solar system.

Problems in Explaining Quasars

When astronomers first encountered the problem of explaining how 20 galaxies worth of energy could come from a volume of space equivalent to our solar system, they looked very carefully at the assumptions used in arriving at that conclusion. The laws of physics as then understood just did not seem to allow for such behavior in our universe. The obvious place to begin the questioning was at the

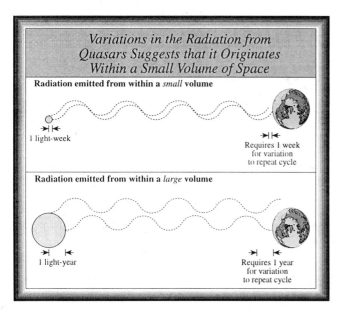

Figure 14–12 The fact that the light from quasars does not arrive in a steady stream, but instead in a wavy pattern, indicates that the region from which the light originates must be quite small. This diagram illustrates why the light-emitting region could be large and result in a wavy pattern at the same time.

★★★★★★★★★★★★

Figure 14–13 This HST image shows, to the surprise of researchers, that the environment surrounding quasars is far more violent and complex than expected, with evidence for galactic collisions and mergers. The most widely accepted model is that a quasar is powered by a supermassive black hole in the core of a more or less normal galaxy. However, confirming this model has been difficult because a quasar is so bright it drowns out the light from the stars in the suspected host galaxy. HST images like this one reveal that these apparently naked quasars have distinct companion galaxies that are so close that they will merge with the quasars in no more than ten million years. This photograph also reveals a galaxy that has been distorted by the gravitational pull of the quasar. This is clear evidence for interactions between the quasar and its nearby companion galaxy. This would mean that the quasar seen with a host galaxy has been caught in the act of merging with its companion. (STScI/NASA)

redshift level. It is, of course, the interpretation of the redshifts that tells us that the quasars are moving away at such high velocities in the first place.

Isn't it possible that something else may be responsible for the lengthening of the spectral lines other than the Doppler effect? Is it possible that quasars are not moving away as rapidly as we think they are, and that another mechanism is causes the spectral lines to lengthen as the radiation travels to Earth?

We know that the Doppler effect causes shifts in wavelengths of radiation. We use it invaluably in tracking aircraft and satellites and other moving objects — even in keeping our highways free of speeding motorists. Well, it certainly is possible that *something else* influences the radiation coming from objects (quasars and galaxies alike) in space that is in no way related to the movements of those objects or even the objects themselves. Perhaps neither the quasars nor the galaxies are not moving away from us at all! Or perhaps it is just quasars that do not obey Hubble's law, and thus are not at the great distances we suspect they are.

If they are not far away, then they do not give off the great amounts of energy we calculate they do! It is *only because we think quasars are far away that we calculate such great energy outputs for them.* But what then is the mechanism responsible for the shifting of their spectral lines? We do not know. We do not have a better model or explanation for the redshifts of quasars (or galaxies). We are left only with the Doppler effect to explain the redshifts — for better or for worse.

Distance Problem For the moment, at least, we are left with the hypothesis that quasars move away from us at great velocities. But why must we automatically assume that they are located at great distances just because they are moving away at great velocities? Well, we don't have to. We do so because we commonly use Hubble's law to determine distances to normal galaxies. So it seems reasonable that we should be able to apply it to quasars as well. But perhaps the quasars — whatever they are — are not like galaxies at all, and therefore do not obey Hubble's law like galaxies do (Figure 14-13). The "law" part of Hubble's law doesn't mean that everything in the universe has to obey it.

An object obeys Hubble's law only if it is a part of the large-scale structure of the universe and therefore participates in the expansion of the universe. Yes, Sun and Earth are also participants in the expansion of the universe, and theoretically — at least — are expanding away from one another. But they are so close to one another that the rate of recession cannot be measured. Or let us say that their movements within the solar system and Galaxy are so much more dominant as to overwhelm the very slight motion of expansion of the universe itself. It is only if we measure across vast distances in space that the amount of the "stretching" of space adds up to the point of being measurable.

Quasars as Local Events? But certainly something can be moving away from us rapidly without being located at a great distance from us! Let's ponder that possibility. Let's assume that quasars are actually located quite close to us — say just outside the disk of the Milky Way galaxy (say 40,000 light-years or so). Let us further assume that they are stars moving rapidly away from the galactic nucleus — some at greater velocities than others — having been fired out of the nucleus by some gigantic explosion long ago. This explains most of the characteristics we observe about a typical quasar — point source of light, great redshift, small, etc. This is what we call the **local hypothesis**, since it attempts to explain quasars in terms of them being close to us.

Does that explanation of quasars seem possible to you? We don't have to explain the energy output of quasars —

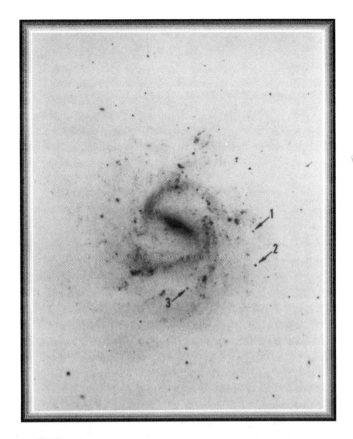

Figure 14–14 Close to this barred spiral galaxy are three quasars whose redshifts are significantly greater than that of the galaxy. And yet it appears as if the quasars are somehow "linked" to the galaxy. If they are at the same distance from us, but have different redshifts, then Hubble's law is violated. (Lick Observatory)

local galaxies, they should reveal blueshifted spectral lines. But *we have never observed a quasar with shortened (blueshifted) spectral lines*. All of the quasars appear to be going away from us — which implies that they cannot be associated with other galaxies in any way. Of course we could avoid this contradiction entirely by assuming that our galaxy is unique — the only galaxy that had an explosive event capable of ejecting stars.

You must know by now what scientists (especially astronomers) think about assumptions of uniqueness. There is implied faith in the scientific method that what happens here can happen elsewhere. The universe is assumed to be uniform in space and time — to have similar events occurring everywhere. We seem to end up with a black eye whenever we assume that there is something unique about *where* we are and *how* we are. The Copernican Revolution was a profound lesson for us.

As if the foregoing objections weren't enough to destroy your acceptance of the *local hypothesis*, there is one final objection — astronomers have failed to detect proper motion for a single one of the hundreds of quasars discovered. If they are close to us and their velocities away from us are indeed tens of thousands of miles per second, then surely we should observe slight changes of position on the celestial sphere. But we haven't.

Alignment Problem Actually, there are a few other observations of quasars that make the mystery even more intriguing. Some photographs of quasars show them to be located quite close to normal-looking galaxies — galaxies whose redshifts are significantly less than those of the neighboring quasars (Figure 14-14). Their closeness would suggest that perhaps both galaxy and quasar are located at the same distance from us, perhaps even gravitationally bound to one another — and not that it is just a matter of chance alignment. If they are at the same distance, then one would be unable to explain the difference between redshifts — unless there is another mechanism causing the Doppler shifts in the quasars.

Most astronomers argue that these few cases where normal galaxies and quasars appear close together are chance alignments — statistically such occurrences are as common as we find them to be. Others, a minority at the moment, argue otherwise. They insist that galaxy and quasar are next-door neighbors in space, and the fact that they have such different redshifts suggests that the entire matter of redshifts — for quasar and galaxy alike — needs to be reexamined. So it boils down to a matter of statistics. As someone once remarked, "Statistics lie, and liars use statistics." In other words, it is easy to prove the probability or improbability of the closeness of galaxy and quasar by using different methods of statistics.

Some photographs even show what appears to be luminous material connecting the galaxy with the quasar, arguing again for a common distance to both. But there is considerable debate as to whether these smudges are

they are dim not because they are far away but because they are merely very distant stars. We even learned in Chapter 12 that some kind of explosive event is presently occurring in the nucleus portion of our galaxy. The variations in light output may have something to do with the types of stars involved, especially since they must have been subjected to tremendous forces to have caused them to be moving at speeds up to 170,000 miles per second.

But here's the rub — How can a star possibly hold together while experiencing such an explosion, especially since it consists of plasma — not solid material? You would think that stars in the vicinity of such an explosion would be vaporized — turned back into clouds of gas and dust from which they came. Do you think your automobile could survive a forceful push in getting it to go 170,000 miles per second? And aside from that objection — how about other galaxies? Do they also eject stars outward by explosive events in their nuclei? And if so, are some of the quasars we observe those coming from nearby galaxies? And if that is so, why do we obtain only redshifted spectral lines from quasars — never blueshifted?

If some quasars are coming toward us, "fired" out of

actually material in space or an artifact created by the process used in obtaining the photographs. These kinds of effects are common when working at the very limits of the telescopes and the instruments used to image faint objects.

Proof of Great Distances The Local Hypothesis was Recently, strong evidence has been offered for the great distances of quasars. Back in the 1920s, when Einstein proposed that gravity is the warping or bending of space in the vicinity of objects having mass, he predicted a rather peculiar consequence of that notion. This predicted behavior of warped space was actually a variation of the experiment used to test his hypothesis when the apparent displacement of stars near the eclipsed Sun was measured (refer back to Figure 11-22). In this case, however, the prediction involved observing very distant objects — far beyond the local stars used in the eclipse test. The telescopes of Einstein's time were not powerful enough to detect the predicted behavior of distant objects. In fact, at the time (early 1920s), astronomers did not even know that very distant objects existed!

Let me explain this predicted behavior. Looking at Figure 14-16, assume we on Earth are observing a very distant quasar. Streams of photons are emitted outward in all directions by the quasar, although we on Earth receive only those sent directly toward Earth. Let imagine a massive galaxy located exactly between the quasar and Earth, so that the stream of photons we would normally detect now are absorbed by the far side of the intervening galaxy. If Einstein is correct in his theory about gravity warping space around a massive object, then streams of photons that would normally miss Earth enter the warped space around the massive galaxy and curve around and hit Earth.

To an observer located on Earth, the quasar therefore appears twice on the photographic plate. Seen from the perspective of Figure 14-16, the quasar appears to be

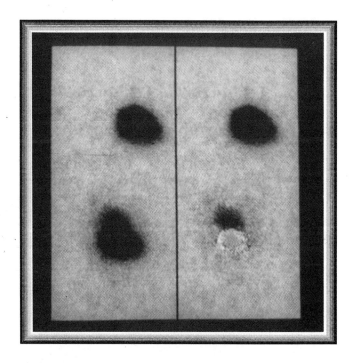

Figure 14–15 *The gravitational lens effect is revealed in these computer–enhanced images of quasar 0957 + 561. The two images of the quasar appear as black blobs in both photographs. In the right photograph, the lower blob has been subtracted from the upper blob to reveal the remainder—the giant elliptical galaxy whose mass warps the space that causes the lens effect.* (University of Hawaii, Institute for Astronomy)

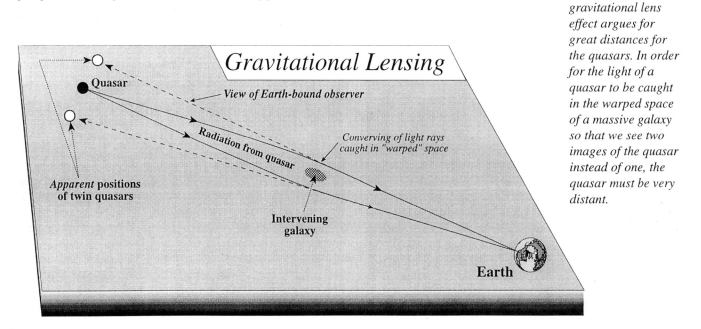

Figure 14–16 *The gravitational lens effect argues for great distances for the quasars. In order for the light of a quasar to be caught in the warped space of a massive galaxy so that we see two images of the quasar instead of one, the quasar must be very distant.*

Figure 14-17 *The gravitational lens effect is very noticeable in this spectacular HST photograph of a cluster of galaxies that are collectively warping the local space through which the light from more distant galaxies is caught.* (NASA/STScI)

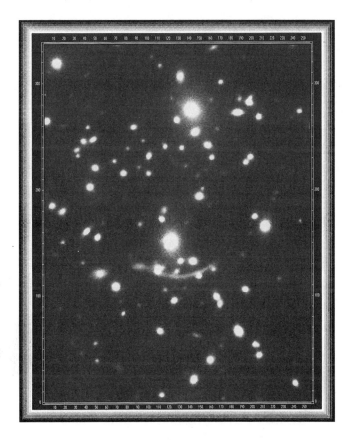

Figure 14-18 *A view of galaxy cluster* Abell 370, *showing a large "luminous arc" formation. Close examination of the light spread out along the arc reveals that the spectral lines are the same along its full length. This, along with other arguments, convince astronomers that a single quasar's radiation is being spread out into the arc by the warped space around a massive spherical galaxy.* (National Optical Astronomy Observatories)

located on both sides of the massive galaxy. So astronomers observe two images of the same quasar in the same photograph! Bizarre, eh! The warped space in the vicinity of the massive galaxy acts like the lens of a telescope, focusing rays of light to a single point. So this predicted behavior of space is called the **gravitational lens effect**.

Now, the importance of the *gravitational lens effect* is that it only occurs if there are great distances between Earth and the massive galaxy and between the massive galaxy and the quasar. So we can observe the effect only if quasars are indeed located at the great distances implied by their redshifts. Well, guess what! We found our first such double-quasar image in 1979, verifying Einstein's prediction! The actual photograph of *0957 + 561A and B* is shown on the left in Figure 14-15. In the photograph on the right, the top quasar image has been subtracted from the bottom quasar image to reveal just the intervening object that is responsible for the bending of space. In this particular case, that object appears to be a giant elliptical galaxy whose redshift is about one-fourth that of the quasar.

When you think about it, there is no reason why the light from a distant quasar should be "bent" into just two images. After all, the "warping" of space around an object is a three-dimensional process. So why just two images? It is most likely because the intervening galaxy is flattened, so that the "warping" is greatest where the galaxy's mass is most concentrated — i.e., the most flattened portion. And what would the result be if the intervening galaxy were spherical in shape? Yes, it would be in the shape of an arc or ring — there would be multiple images of the more distant quasar, perhaps even smeared out into the form of an arc or ring of light.

So guess what! Look at the photographs in Figures 14-17 and 14-18! Arcs! Just as Einstein predicted! To date, tens of "multiple" quasars have been discovered. From the

CHAPTER 14 / Active Core Galaxies and Quasars **307**

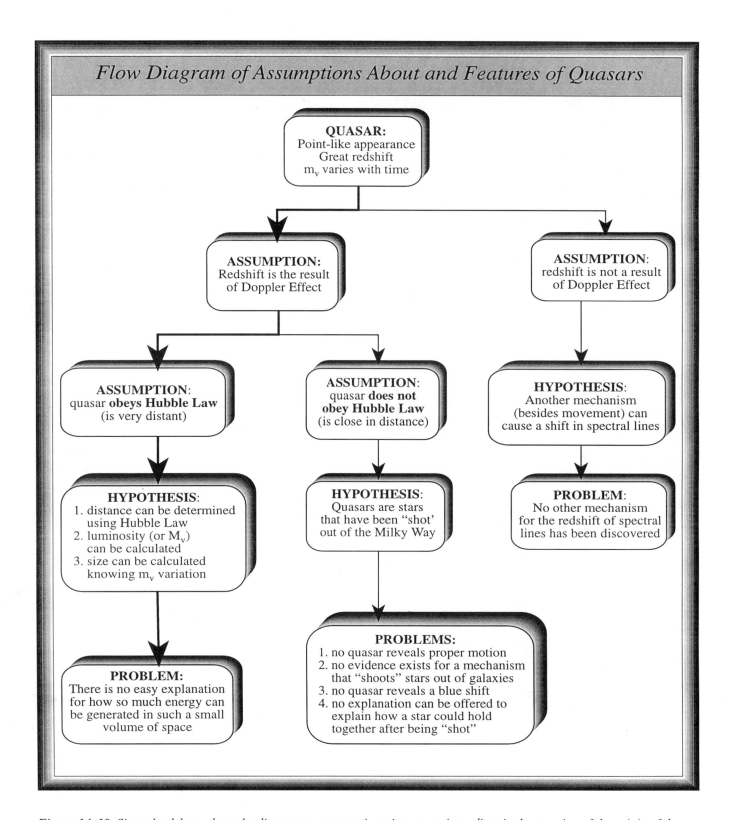

Figure 14–19 Since the debate about the distances to quasars is an important ingredient in the question of the origin of the universe, this flow chart of the assumptions and conclusions that are made during the debate will make the issue clearer. The currently–held preference is indicated by the bold line.

numbers found so far using available techniques, astronomers estimate that at least a thousand of them should be eventually found. So the evidence is gradually surfacing to support the hypothesis that quasars are the most distant and therefore youngest objects in the universe.

Best-Guess Model The flow chart in Figure 14-19 illustrates the assumption-testing we just performed on quasars. No matter what route we take, we eventually confront conclusions about quasars that are difficult to explain. So astronomers go for the easiest — for the one that works best within the framework of existing theories about the behavior of matter and energy. So it is that today, the "best-guess" model for quasars is that they are far away — the most distant objects in the universe — and that they emit tremendous amounts of energies from within small volumes of space.

So the obvious question now is — "What exactly is responsible for the great energy output of quasars?" In Chapter 13 we found that there are galaxies whose initial appearances remind us of normal spiral galaxies, yet whose nuclei are excessively bright — the Seyfert galaxies. It may be, then, that quasars are merely extreme examples of the Seyfert galaxies — that is, quasars may be the brilliant nuclei of galaxies whose distances are so great that we are unable to observe the fainter, outer portions of the galaxies.

That doesn't explain the source of energy for the brilliant nuclei, of course — it merely allows us to establish a connection between galaxies with which we are familiar and objects about which we still a little confused. It is comforting to work on the assumption that galaxies — normal as well as Seyfert — and quasars are in the same ball game, obeying the same rules.

Since the spectrum of an object gives us such an enormous amount of useful information about that object, we should go back to examine the spectra of quasars in more detail. Their visible spectra reveal emission lines of familiar ionized chemical elements such as carbon, magnesium, oxygen, neon, silicon, helium, and, of course, hydrogen. They are quite similar to those we obtain from planetary nebulae, novae, and Sun's corona, suggesting that they are due to hot gases surrounding a source of energy.

In addition, most quasars have narrow absorption lines in their spectra — usually with redshifts less than or the same as those of the emission lines. We can interpret these absorption lines as being due to cool clouds of gas moving outward from the central source of energy. Since those cool clouds of gas are coming toward us relative to the hot gases responsible for the emission lines, the absorption redshifts are different than the emission redshifts.

All observational evidence considered, our current model is that of a central source of energy generating all wavelengths of the visible spectrum (continuous spectrum), simultaneously ejecting outward clouds of hot gases (emission spectrum) — some of which have already moved far

Figure 14–20 In a normal photograph, the "Whirlpool" galaxy M51 doesn't appear to be experiencing any violence. (Lick Observatory)

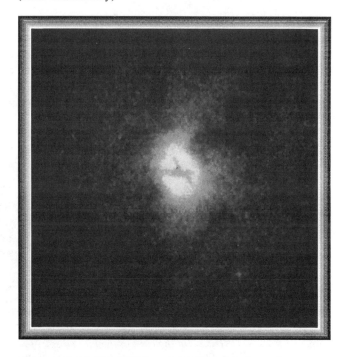

Figure 14–21 The HST imaged M51 from above the atmosphere, and in this photograph an "X" marks the location of the center of that galaxy. The dark material forming the "X" is believed to be associated with a massive black hole. (STScI/NASA)

CHAPTER 14 / Active Core Galaxies and Quasars

★★★★★★★★★★★★★

Figure 14–22 This HST image (right) reveals the faint host galaxy that a bright quasar dwells within. The wealth of new detail in this picture helps solve a three-decade old mystery about the true nature of quasars, the most distant and energetic objects in the universe. The HST image shows clearly that the quasar, called 1229+204, lies in the core of a galaxy that has a common shape consisting of two spiral arms of stars connected by a bar-like feature. The host galaxy is in a spectacular collision with a dwarf galaxy. The collision apparently fuels the quasar "engine" at the galaxy center - presumably a massive black hole—and also triggers many sites of new star-formation. Though a previous ground–based observation (left) first identified the barred spiral galaxy in 1229+204, Hubble shows clearly the galaxy's structure and reveals details of the collision. (STScI/NASA)

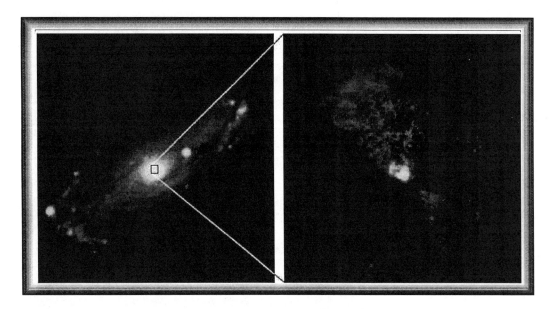

Figure 14–23 A HST view of the core of the barred spiral Seyfert galaxy NGC 5728 reveals a spectacular bi–conical beam of radiation that is ionizing the gas in the central region of the galaxy. Because this is an active galaxy, the core might contain a super massive black hole surrounded by a disk of gas. However, a dense ring of gas blocks HST's view of the black hole and glowing accretion disk. The radiation escapes along the open ends of the gas "donut" to several thousands of light–years from the nucleus. (STScI/NASA)

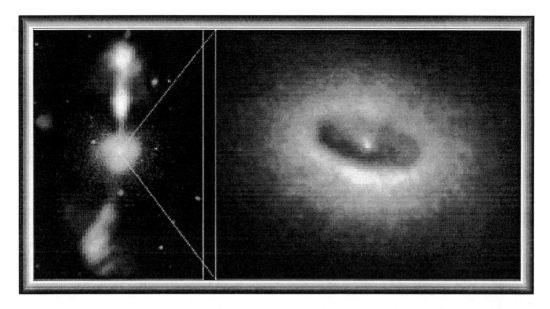

Figure 14–24 The giant elliptical galaxy NGC 4261 is one of the twelve brightest galaxies in the Virgo cluster, located 45 million light-years away. Photographed in visible light (center of left image) the galaxy appears as a fuzzy disk of hundreds of billions of stars. A radio image (left image) shows a pair of opposed jets emanating from the nucleus and spanning a distance of 88,000 light-years. The HST image (right) reveals a giant disk of cold gas and dust fueling a possible black hole at the core of the galaxy. Estimated to be 300 light-years across, the disk is tipped enough (about 60 degrees) to provide astronomers with a clear view of the bright hub, which presumably harbors the black hole. The dark, dusty disk represents a cold outer region which extends inwards to an ultra-hot accretion disk which is a few hundred million miles from the suspected black hole. This disk feeds matter into the black hole, where gravity compresses and heats the material. Hot gas rushing from the vicinity of the black hole creates the radio jets. The jets are aligned perpendicular to the disk, like an axel through a wheel. This provides strong circumstantial evidence for the existence of black hole "central engine" in NGC 4261. (NASA/STScI)

enough away from the energy source to have cooled off to the point that they are now absorbing photons (absorption spectrum). The spectra of some quasars, in fact, reveal several sets of emission and/or absorption lines, suggesting that successive shells of gases are expelled from the region of the energy source — cooling off as they move farther and farther away.

Energy Source for Quasars? Have you noticed that we still haven't explained the central energy source itself? Well, you'll understand why when you realize that something like 10 average-sized galaxies would have to be converted into energy by the proton-proton chain just to equal the radio energy being emitted by some quasars. Mechanisms capable of such energy output are at the forefront of modern astronomical theory, and the conclusions are far from final. The candidates being considered for explaining the central energy source — separately or in combination — are such physical processes as intense gravitational fields, powerful magnetic fields, immense explosions, or rapid rotations.

The most intriguing possibility, however, the one that receives the highest rating amongst astronomers, is that of a rotating black hole whose mass is equivalent to one billion solar masses. Matter in the form of gas or entire stars could be falling inward toward the black hole's event horizon, so that energy is emitted in voluminous quantities as nuclear reactions occur within the accretion disk (Figures 14-20 and 14-21).

Quasar-Black Hole Connection Several observations support this connection between quasars and massive black holes. First of all, models of stellar evolution include the almost certainty that black holes exist in abundance in the universe. In connection with that prediction, there are sufficient candidates for black holes to convince the larger body of astronomers that black holes actually exist. If a star can collapse to become a black hole, then it seems reasonable that under conditions that existed in the early universe, massive clouds of hydrogen gas could have collapsed under their own weight, creating massive black holes. It is these that we could now be detecting as quasars (Figure 14-23).

Initially, these massive black holes would have "swallowed up" much of the material that wandered too close. But gradually as matter settled in orbit around these gravitational centers, the violence and intense radiation subsided. The gas then formed into the stars that we observe in the galaxies immediately around us. What I am suggesting, of course, is that quasars are the ancestors of the present-day galaxies. Our galaxy was once a quasar. The massive black holes, then, are simply the gravitational

✯✯✯✯✯✯✯✯✯✯✯✯

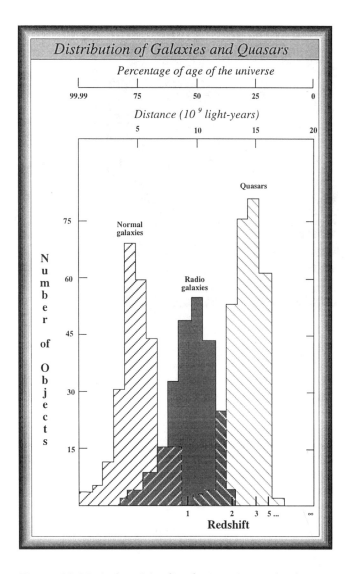

Figure 14–25 A plot of the distribution of normal galaxies, radio galaxies, and quasars according to their distances from Earth reveals a clear pattern of evolution: quasars were an integral part of the birth of galaxies.

centers around which the hydrogen gas necessary for star formation could accumulate.

This model for the formation of some galaxies, at least, is supported by the kinds of objects we observe at increasingly greater and greater distances. The local galaxies are not very violent in terms of emitted energy. But the Seyfert galaxies and radio galaxies discussed in the last Chapter are intermediate in distance (and therefore in time) between the nearby galaxies and the quasars. They may represent, therefore, the intermediate stage of the gradual evolution of quasars into normal galaxies.

Perhaps, then, the Milky Way galaxy has passed through all of those stages while evolving from quasar to normal spiral galaxy. The fact that we have reason to believe that a massive black hole exists at the nucleus of our galaxy is further reason to accept this hypothesis. If so, then we may want to pay close attention to the behavior of the nuclei of other galaxies close to us, since according to this theory they too may contain massive black holes. For example, the *Hubble Space Telescope* recently found evidence that a black hole, equivalent to 2.6 billions solar masses, exists at the center of M87 – a giant elliptical galaxy. The images from the HST show that stars are strongly concentrated toward the center of the galaxy, as if drawn into the center and held there by the gravitational field of a massive black hole. Even more convincing are the photographs such as Figure 14-24 in which just about every feature associated with a massive black hole is visible except the black hole itself.

Direct Detection of Massive Black Holes Other than modeling the behavior of massive black holes when they "swallow" stars, gas, and dust, and then looking for evidence in the form of conventional radiation-detecting schemes, there is a more direct method. Do you recall my discussion of *gravity waves* in Chapter 11 in conjunction with stellar black holes? Well, just as we may detect a stellar black hole in space as it creates gravity waves, so also we may detect gravity waves created by massive black holes in the centers of galaxies. Astronomers are continuously looking for patterns of change in wavelength of radio signals transmitted between Earth and the various satellites sent out into space (e.g., *Ulysses, Galileo*).

Hubble Diagram-Quasar Connection The Hubble diagram, in addition to portraying the expansion of the universe, can also be seen as an evolutionary diagram. As we observe more and more distant galaxies, we see them in a universe that was younger and younger. Quasars no longer exist today, since we do not find any close to us in distance. Seyfert and radio galaxies do not exist today either. All presumedly have become the normal galaxies that we see around us (Figure 14-25). We'll be unable to verify this for millions of years in the future, during which time we should observe the gradual changes in the present-day galaxies. From those observed changes, we will perhaps be able to extrapolate backward in time to model the evolution of galaxies from their primordial beginnings. Meanwhile, we must be content with looking for patterns in the types of objects at increasing distances from us.

The Birth of Galaxies

There are two steps in the formation of galaxies. First, material must assemble into galaxy-size clumps. Then, the material must be converted into stars. The first process is probably dominated by dark matter, which, according to calculations, comprises 90 percent of the matter in the universe. The second process, the formation of stars from primordial gas, is observed directly in such locations as the Orion Nebula. What makes this possible are the large amounts of energy released during the early stages of star formation. When astronomers use the energies emitted by birthing stars within our own galaxy to calculate the amount of energy that an entire galaxy of birthing stars would emit, the results are interesting.

Assuming that the initial birthing period lasts between 100 million and one billion years, the calculated energy is about 10^{61} ergs — enough so that we should be able to see the galaxy across the entire observable universe. But as I mentioned earlier, we have been unable to detect such young galaxies — at least directly. The HST has recently been working at the limits of its ability to image faint objects (+30 magnitude), and has successfully detected gravitational lensing presumedly caused by clumps of young galaxies. But the most intriguing possibility is that we are observing a population of protogalaxies in the form of the highly redshifted quasars! In other words, it may be that nearly every young galaxy developed an active nucleus right away, and underwent a quasar phase at the very offset — at the same time early star formation took place (Figure 14-22). Whatever the "engine" is that powers the quasar, it overwhelms the star formation so that we are unable to witness that particular aspect of the birthing process.

Summary/Conclusion

The pattern of the more distant galaxies being more violent than the nearby galaxies ties in neatly with our discovery of quasars (Figure 14-25). Although considerable debate still rages over the precise "engine" that drives the energy output of quasars, the evidence is clearly pointing to some profound new realizations about the universe in which we live. There doesn't seem to be any connection between quasars and the search for life elsewhere, except that they provide evidence for the birth of the universe. In other words, it appears as if we must live with evidence that the galaxies evolved from the earliest of conditions in the universe. So why should life be so unique as not to evolve as well?

LEARNING OBJECTIVES: *Now that you have studied this Chapter, you should be able to:*

1. Describe the observational features of quasars, comparing them with other objects such as stars, normal galaxies, and active galaxies.
2. Explain how the evidence convinces us that they are quite small in size, yet emit more energy than anything else in the universe.
3. Outline the method used to determine the distances to quasars, and discuss its uncertainties.
4. Explain how the evidence for their distances, and therefore energy outputs, is the subject of great debate.
5. Explain why astronomers currently accept quasars as being the most distant objects in the universe, and therefore the first things to have formed after the Big Bang.
6. Explain the apparent connection between normal galaxies, active galaxies, and quasars.
7. Sketch a possible model that explains the observed features of a quasar.
8. Define and use in a complete sentence each of the following **NEW TERMS**:

Active galaxy 297
Compact sources 298
Extended sources 298
Gravitational lens effect 307
Local hypothesis 305
Occultation 301

Peculiar galaxy 297
Quasi-stellar radio source (quasar) 297
Radio galaxy 298
Radio lobes 299
Seyfert galaxy 297

Chapter 15

The Origin of the Universe

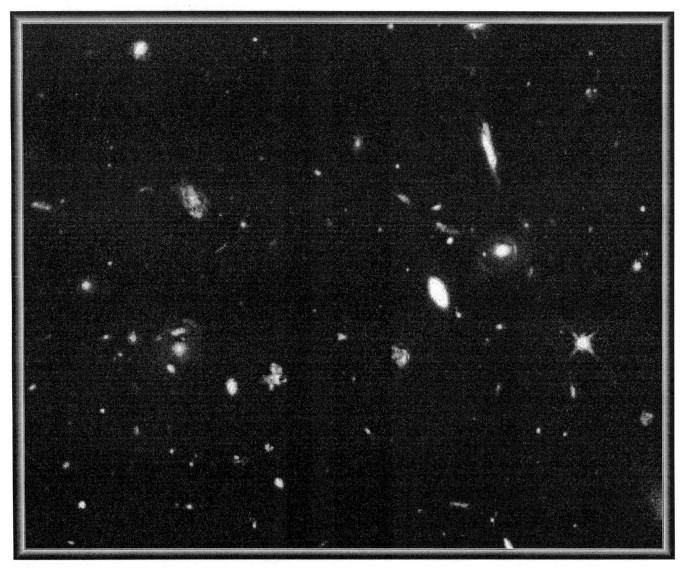

Figure 15–1 Astronomers using NASA's Hubble Space Telescope have solved a 20-year-old mystery by showing that a class of galaxy once thought to be rare is actually the most common type of galaxy in the universe. Small irregular objects called "blue dwarfs" were more numerous several billion years ago, outnumbering the spiral galaxies like our Milky Way, and giant elliptical galaxies as well. This means the blue dwarfs are a more important constituent of the universe and figure more prominently in the evolution of galaxies than previously thought. Most of these faint objects are extremely blue in color, a strong indication that they are undergoing a brief, rapid burst of star formation. (NASA/STScI)

CENTRAL THEME: What evidence do we have that the universe had an origin, and what assumptions do we use in arriving at that conclusion?

Finally we arrive at the biggest question of all: Where did the universe itself come from? By "universe," we mean all space, matter, energy, and time. The answer to this question is inseparable from all of the answers to all of the questions asked up to this point. In order to explain the evolution of the entire universe, we have had to survey the contents of the universe — the stars and galaxies. Any theory that attempts to explain the very origin of matter and the space it occupies must necessarily account for what we observe. This study and description of the origin and large-scale structure of the universe is called **cosmology**. It is not a new topic. Cultures throughout the centuries have proposed cosmologies in order to explain the chaos of their experiences — earthquakes, hurricanes, volcanoes, and the like.

Modern astronomers are no different, except that they have a wider range of instruments with which to tackle the question. Developments in observing techniques have been growing at such a rapid pace that we must — as always — be prepared to change our theories as new observations are made. But what usually happens in science is not the complete reversal of a theory when new discoveries are made, but a refinement and enlargement of that theory.

The question before us, however, is like no other. In order to approach it adequately, we must obtain information from a variety of sources — from the farthest reaches of space to the very smallest of subatomic particles. So we may have to confess at the very offset our inherent limitation in tackling such a big question at all, and prepare ourselves for frequent reversals in our beliefs and theories.

While examining the life histories of stars and galaxies, we encountered features of the universe that strongly hinted that the universe itself is evolving from a point of origin (Figure 15-1). As a matter of summary:

- Stars, in the act of shining, convert hydrogen to helium. When they die, they create and distribute out into space the heavier elements. Those newly-created chemicals mix with clouds of hydrogen, which eventually collapse to form new generations of stars. Heavier elements are being manufactured at the expense of lighter ones. By implication, there was a time when only hydrogen gas was present in the universe.

- Spectra taken of galaxies reveal a uniform redshift of spectral lines, suggesting that all galaxies are receding from one another. By implication, the galaxies were at one time all crammed together in a single blob of matter. Hubble's law allows us to calculate that the blob first appeared about 18.6 billion years ago.

- Observing galaxies that are progressively farther and farther away is equivalent to observing them as they were longer and longer ago — we observe cross-sections of the past history of the universe. Those most distant objects — quasars — are profoundly different from the local galaxies. By implication, the present universe of galaxies evolved from quasars — quasars are the early forms of the galaxies. By implication, the quasars evolved from something that, together with the above two observations, constitutes an origin of matter in the universe (Figure 15-2).

Size and Extent of the Universe

I recall as a young boy laying in an open field at night, looking up at the star-studded sky, and wondering — how far does space go out there? Does it go on forever, or does it eventually come to an end? Those same questions are relevant now, because they bring to light a fundamental property of the universe that leads to the modern cosmological model of the Big Bang — the curvature of space.

Traveling in an Open Universe In order to determine what choices we have as to the nature and extent of our universe, let us perform a simple thought experiment. We

Figure 15–2 If you look carefully at this photograph, you will notice that most of the "smudges" are galaxies. This is the Hercules cluster, containing about 300 galaxies. The field of study called cosmology attempts to explain how these clusters came to be in the first place, and what mechanism is responsible for the manner in which they are distributed. (Lick Observatory)

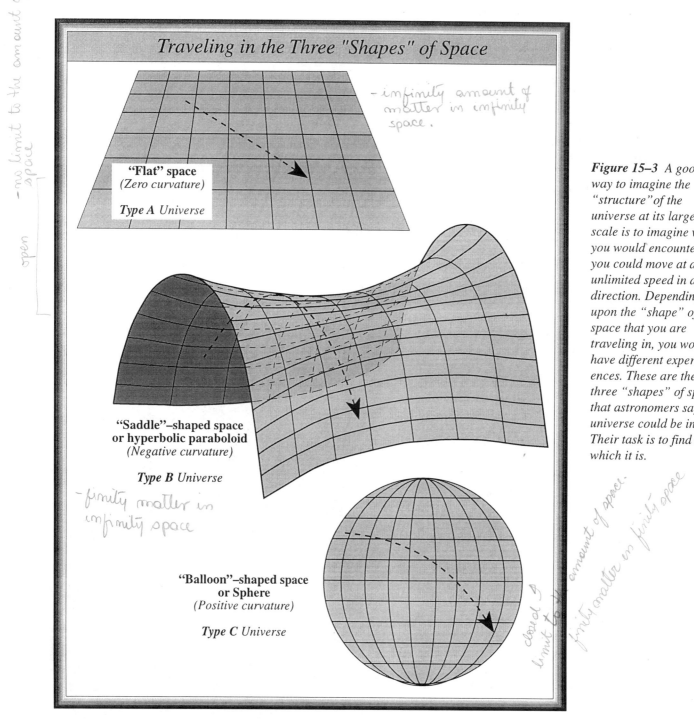

Figure 15–3 A good way to imagine the "structure" of the universe at its largest scale is to imagine what you would encounter if you could move at an unlimited speed in any direction. Depending upon the "shape" of space that you are traveling in, you would have different experiences. These are the three "shapes" of space that astronomers say the universe could be in. Their task is to find out which it is.

are passengers aboard a spaceship that is not limited to traveling at speeds less than that of light. So we can easily zip quickly out of our solar system, out of the Milky Way galaxy, and cruise through intergalactic space. Our assignment — let us assume — is to survey and record the contents of the universe. We photograph each of the galaxies that we encounter, record a few brief notes about its appearance and contents, and list it in our notebook.

We travel on and on and on. And no matter how long we travel, we still encounter new galaxies to photograph and record. There is no end to the galaxies, and there is *no end* to the space in which the galaxies exist. In such a universe, there is an *infinite amount of matter in infinite space*. Let's call that a **Type A** universe.

Another obvious possibility is that at some time during our journey, the galaxies appear to thin out and get farther and farther apart — until eventually there is not a single remaining galaxy to photograph and record. We have reached the limit of matter in space, but we are still traveling in space. And no matter how long we continue to travel, we encounter no additional galaxies. If we do, then we are back in a *Type A* universe again.

316   SECTION 3 / Evolution of the Galaxies

The case that I am presently describing is one in which there is *a limit to the number of galaxies but not to the amount of space* — there is *finite matter in infinite space*. Let that be a **Type B** universe (Figure 15-3). Both Type A and Type B universes are what we refer to as **open** universes, in the sense that there is no limit to the amount of space in the universe. There may be a limited amount of matter in the universe, but there is an unlimited amount of space within which that matter can move.

Closed Universe But there is a third scenario. Out we go, traveling — as before — without consideration for speed or time elapsed. We are traveling along — still recording galaxies — when a member of our crew observes that the galaxy we just encountered and recorded looks terribly familiar. We look back at page 1 of our logbook, and lo and behold — there is an entry for the very same galaxy we just passed, recorded when we first started our journey! Continuing along, we repeat our photographs and listings of the galaxies we observe. *We have been through this region of space before.*

It wasn't a matter of us losing our way. We travelled in a straight line — or so we thought. What happened, in fact, was that we made a complete circuit around the universe and began over again. Space is not "flat," so light and matter cannot travel in a straight line because it moves within the fabric of curved space! Does that seem strange to you? Oh, good — I thought I am the only one being astounded by the possibilities the universe has to offer.

The universe of space — in other words — might be folded back on itself in any direction we travel, so that if we go out far enough in one direction, we return to the same starting point. This is in spite of our thinking that we are traveling in a straight line. This **Type C** universe, therefore, is a **closed** universe, because there is a limit to the amount of space in the universe. Admittedly, this scenario does not fit well with our common sense. But for people a few hundred years ago, common sense also dictated that Earth could not possibly be round.

If Earth were round, travelers venturing too far would eventually go around to the opposite side and slide off. They couldn't imagine traveling around Earth because they had no concept of gravity as a force — that a person could travel to the opposite side of a sphere and remain upside down while standing on its surface. The surface of Earth, in other words, is analogous to the Type C universe. It goes on and on forever, but eventually repeats itself. It is limitless but *bounded*.

Before you give into temptation and reject a *Type C* universe simply because you have never encountered the idea before, give me a chance to explain. After all, the universe is going to be what it is, regardless of what you want it to be. It is already here. We're merely looking for clues that allow us to explain its origin and structure. The *Type C* universe is *finite matter in finite space*. There is only a certain number of galaxies in the universe because there is only a certain amount of space in which those galaxies can exist.

In discussing the largest scale of the universe, we must keep clear in our mind what we mean when we use the term "end" — matter or space. Space can exist without matter — most of the space between stars is essentially empty of matter. But matter doesn't exist without space. If there is matter outside of space, we don't have access to it because we don't know how to leave space in order to get to wherever *no space* is.

In fact, we can think of the entire universe as an enormous black hole — we are inside the warped space, surrounded by galaxies and the like. Perhaps the most distant objects (quasars) are just now entering through the event horizon. They behave quite differently than the other galaxies, since they just left the turbulent accretion disk of another universe we can never know.

The Type B universe, although somewhat contrary to common sense, is also considered to be a good model for the universe in which we live, especially if the universe is expanding. The galaxies are headed outward away from the event horizon from which they emerged, to eventually occupy as yet unoccupied space.

Curvature of Space

When we encountered the curvature of space around massive objects like black holes, we didn't have to consider the "shape" of that curvature. It was convenient for me to represent it as a funnel in elastic space. For the universe as a whole, however, the "shape" of space is very important. Obviously, if in the Type C universe space curves back onto itself, it must have some type of "shape." As you are about to learn, the "shape" of the universe is strongly connected its origin and ultimate fate. So if we strive for an explanation for the origin of the universe, we necessarily must pin down a "shape" for the universe.

Mathematicians tell us that the universe is one of three shapes, or *curvatures* — positive, negative, and zero. We don't presently know which of the three the universe has. That is the subject of an enormous amount of research in astronomy. The three shapes are predicted in the equations derived by Einstein while he developed his famous relativity theories in the early 1900s.

Cause of Space Curvature The large-scale universe contains mass in the form of galaxies. So just imagine that collectively, all of those galaxies bend space into a particular shape. So far, we have encountered two independent observations as evidence for the bending of space — the gravitational lens effect (quasars), and the displacement of stars close to Sun during total solar eclipses. Therefore, we can immediately rule out zero curvature for the shape of our universe. Zero curvature means "flat" space — flat in the sense that light would travel in a straight line in such a universe. But if the space in which light travels is bent, then the light can't travel in a straight line. That leaves us with a universe that has either positive or negative curvature.

Positive Curvature Mathematically, space that has positive curvature can be represented by a sphere, like a balloon — not the inside of the balloon, but the surface itself. Imagine yourself confined to the surface of a balloon — moving along the surface represents your movement through space. Moving along what you think is a straight line results in your completely circling the balloon and beginning over again.

You are immediately asking yourself — "But what does the outside of the balloon represent?" "And the inside?" Well, if I have convinced you that the universe is expanding, that the galaxies are getting farther and farther away from one another, then consider those questions as you blow up the balloon. Why not take it a step further and draw dots to represent galaxies on the surface of the balloon.

As the balloon gets larger and larger, the dots get further and further from one another. If you were traveling along its surface — from one dot to another, say — the distance between them would increase as the balloon inflates. At the same time, the dots move away from the "space" inside the balloon to "space" outside the balloon. In other words, the dots are moving farther away from the center of the balloon.

Therefore, the "space" outside of the balloon represents future time, the "space" inside represents past time. The surface represents present time. When we look out into space, we look back into time — time at which the galaxies were closer together. So when we look at a distant object, we are seeing it as it existed in our universe when the universe was smaller in volume than it is now. And what are the galaxies expanding into? What is outside the balloon? As the galaxies continue to recede from one another, the volume of space in the universe increases — the surface of the balloon enlarges to encompass "new" time, the future. The present moves into the future.

In that sense, time can be thought of as the expansion of the universe. If the universe stops expanding, time stops. If time stops, the universe stops expanding. The expansion began when time began, or time began when the universe began to expand. Chicken or egg — that is our choice.

Is it possible that all events that will ever happen to you are already fixed and determined in that vast reservoir of future time outside the balloon, waiting for the universe of matter and space to expand outward to encompass them so that they "happen?" If the universe were not expanding, there would be no time. If the expansion of the universe began at a single point (of mass <u>and</u> space) at the center of the balloon, then time must have begun when the universe was "created" at that point. Later, we can speculate as to where that point came from.

Negative Curvature Now let's consider negatively-curved space. The mathematical model of space that has negative curvature is the **hyperbolic paraboloid** (Figure 15-4). Since it looks similar to a saddle used on horses, it is often referred to as the "saddle." In this model, time is not represented by what is outside of or inside of the shape. Time is instead represented by the dispersion of the galaxies on the surface of the shape itself — or the stretching out of the shape that causes galaxies to move away from one another.

Keep in mind that the shape shown in Figure 15-4 is only a small section of an infinitely large shape, since space with negative curvature is unlimited in extent. In your imagination, simply extend the shape outward in all directions. Another thing that might not be so obvious — the mathematical characteristics of a hyperbolic paraboloid do not allow for any part to bend around to fold back upon itself. In other words, you are prohibited in this kind of universe from traveling out into space and returning to your original position along what seems like a straight line. In whatever direction you choose to go, you go to infinity.

The model illustrates why this shape is referred to as a hyperbolic paraboloid. The ribs of the model are parabolas of different sizes, so constructed that when put together, they form the two shapes known collectively as a hyperbola. Both parabolas and hyperbolas are infinite shapes — they do not close back on themselves. The model consists of only a section of the infinite shape. Your imagination can

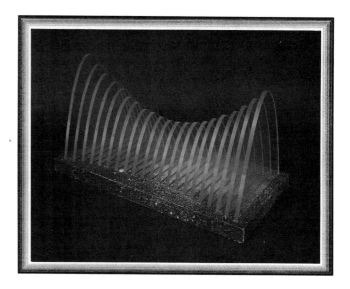

Figure 15–4 A universe in which there is infinite space can be represented by a mathematical shape called a hyperbolic paraboloid. This is a photograph of a model of such a shape. "Space" is represented by the surfaces of the plastic ribs, and the regions inside and outside of the ribs is not defined. In order to visualize it being infinite, imagine the ribs extending outward in all directions. (Tom Bullock)

fill in the extensions of the model outward in all directions.

Recognizing the connection between the two possible shapes for the universe (sphere and saddle) and the three possible space journeys (Types A, B, and C) that we took earlier should now be easy. Since the sphere (with finite surface area) represents a universe with a finite volume of space, then the sphere must be the shape of the Type C universe. The hyperbolic paraboloid (with infinite surface area), on the other hand, represents a universe of infinite volume of space.

But in that infinite space, there can be either a finite number of galaxies or an infinite number of galaxies. So the saddle represents the shapes of both the Type A and B universes, since both contain infinite space. The choice between a Type A universe and a Type B universe is a choice between infinite matter (Type A) in infinite space or finite matter (Type B) in infinite space. Either the infinite surface of the saddle is covered by an infinite number of dots, or there is only a finite number of dots.

Origin of the Universe

With the shape of space as background information, we are now in a position to discuss the origin of the universe. Any theory for the origin of the universe must include the shape of space and the amount of matter in that space. There is, in other words, a relationship between the origin of the universe and the large-scale structure of the universe.

Although I have obviously provided more evidence for the Big Bang theory than any other toward explaining the origin of the universe, let us rid ourselves of any prior prejudice and give all competitors an equal chance. It doesn't really matter, does it? The universe is already here. We can't change the way in which it got here. Obviously our fascination with the subject (as is true of all cultures, no doubt) comes from our need to embrace a *meaning of life* — a component of which necessarily comes from observing and explaining the physical world around us.

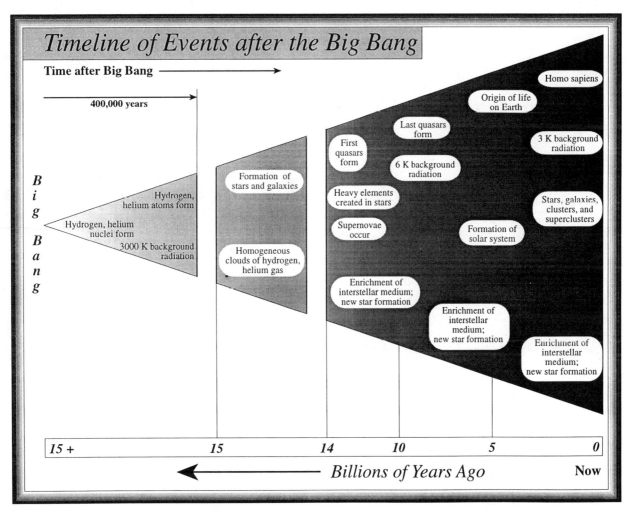

Figure 15–5 According to the Big Bang theory, time and energy came into existence 15 to 20 billion years ago. During the interval since that moment, the energy turned into the matter that formed into the objects that we see around us today. The major events of the building up of matter into larger objects are illustrated in this diagram.

★★★★★★★★★★★★

Big Bang Theory Both Type B and C universes, since they account for finite matter in the universe, are Big Bang universes. The Big Bang theory says that all matter and space were confined to a single dot about the size of an atom which "exploded" about 20 billion years ago. At the moment of the explosion, space began to expand as the matter in the original dot dispersed with the expansion.

The shape of the expanding space was determined by the speed at which the initial "explosion" caused space to begin expanding. If that speed was great enough, the universe will continue to expand infinitely into the future. But if the speed was somewhat less, then expanding space will eventually slow down, stop, and begin to collapse — perhaps to set off another Big Bang when all matter and space collapse to the size of an atom again (this event is referred to as the **Big Crunch**). Although we are able (with Hubble's law) to determine the rate of expansion, we don't know yet if it is great enough to allow for infinite expansion because we don't know the total amount of matter in the universe resisting that expansion. Let me explain.

Expand Forever or Big Crunch Imagine yourself throwing an apple up into the air away from Earth's surface. After reaching a maximum height, it falls back down to Earth. How might you get it to go higher? By either throwing it harder (adding more energy of motion to it), or by decreasing the mass of either the apple or Earth. Applying the same force to a walnut, for example, will cause it to go higher. Throwing the apple with the same force from Moon's surface would cause it to go higher. Likewise, at the moment of the Big Bang, space and its contents were thrown "outward."

In an expanding universe, there are two competing forces. Gravity attempts to slow down the movement of galaxies away from the "point" of the Big Bang," while the expansion itself spreads the galaxies farther and farther apart. Although space today (20 billion years after the Big Bang) is still stretching outward — taking galaxies along with it — galaxies continue to attract one another and thereby resist further expansion. All objects in space attract one another. Collectively, all galaxies are pulling on one another, as they have been doing since the the Big Bang. Earth's gravity pulls against the apple even as it goes upward away from Earth. If you were capable of giving an apple an initial velocity of 7 miles per second, it would escape Earth entirely. 7 miles per second is Earth's escape velocity.

In the same manner, the universe could also be slowed down enough by the attraction of galaxies for one another so that it "falls" back to the "point" from which it got its original "push." Indirectly, we determine the escape velocity of the universe by determining the collective pull of all galaxies on one another — which depends upon the total number of galaxies in the universe. We know how fast the universe is expanding, so by knowing the mass of the universe, we know whether or not the galaxies have "escape velocity." In essence, we can *predict the ultimate fate of the universe if we accurately know the mass of the universe and the rate at which the galaxies are receding from one another.* When we calculate the total mass of the universe, we come up short by a factor of 90%. That is, we have only observed/detected only 10% of the mass the universe must have in order to eventually fall back onto itself to form another superdense particle like that from which the present universe sprang. This is the *Big Crunch. (Figure* 15-6).

Assuming that the Big Bang theory is correct, then, our failure to detect sufficient mass in the universe suggests that the universe will expand forever. Accordingly, the shape of the universe must be that of a saddle. If we eventually detect

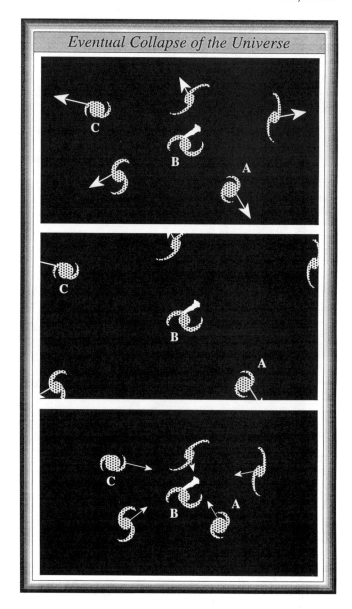

Figure 15–6 If there is enough mass in the universe that is collectively resisiting the expansion of the universe, the galaxies in the distant future may eventually stop their outward expansion, and begin to "fall" back together. Astronomers on Earth will then detect a "blue–shift" in the spectra of galaxies.

the additional 90% of matter needed to close the universe, then the shape of space must be that of a sphere, and we can look forward to the *Big Crunch*. So you can readily see why it is that the current search for dark matter (Chapter 13) is such a hot topic in astronomy. The predicted future of the universe is dependent upon the outcome of the search for machos, wimps, and other exotic particles that may inhabit the halos of galaxies.

So you are asking yourself — "Why don't we just measure the shape of space?" Then we would know what the fate of the universe is. If we find the shape to be that of a saddle, we know the universe will expand forever. If we find the shape to be that of a sphere, we know that we're in for the Big Crunch when the universe falls back onto itself! Ah, that it could be as easily done as said. If only we could get outside the universe to look back and see the shape of space!

Actually, there have been proposed methods of determining the shape of space from within, but they require accurate measurements of the distances to the most remote galaxies in the universe, and we just aren't able to do that yet. Nor do we know how much stuff is out there that can't be seen or detected. The Hubble Space Telescope (HST) is providing more data to astronomers daily, but to date there is no consensus as to what the data suggests about the shape of the universe.

Steady State Theory I have neglected the Type A universe — a universe which contains infinite matter in infinite space and is shaped like a saddle. Where would it have come from? Nowhere. It has always been here, and will always be here. It had no beginning and will have no end — either in matter, space, or time. This is called the **Steady State theory**. The "steady" in Steady State does not mean that nothing is happening. Obviously you and I are evolving toward death as are all other material objects in the universe like stars. The "steady" part has another meaning.

According to the Steady State-ers, stellar evolution proceeds as theorized by Big Bang-ers — planets evolve, and even life evolves. But there has never been a time when things weren't happening. Well, you say, if that is so, then how can Steady State-ers explain the expansion of the universe? How can it be steady if it has evolved from a dot? In order for a Steady State universe to exist, its density must be constant with time — its density at the largest scale, that is. This is referred to as the **Perfect Cosmological Principle** (**PCP**, for short) — a large volume of space at any particular moment in time has the same density and average contents as a similar volume of space at any other moment in time (Figure 15-7).

In other words, we should see — as we look farther and farther into space (therefore farther and farther back into time) — the same kinds of things distributed at the same densities as those we see close to us in distance (and time). But the universe is expanding. How can its density remain the same if the galaxies are rushing away from one another?

The Steady Staters respond: "New matter is being born into the universe at a rate just necessary to maintain a constant density." "As galaxies rush away from one another, new material in the form of hydrogen atoms spontaneously appears in space to fill up the voids left behind by receding galaxies." Born into the universe — you say? How absurd. From where? The Steady Staters reply — "From the same place you get your entire universe in your Big Bang theory!"

In other words, the Steady Staters are not bothered by the observation that the universe is expanding. That is an accepted fact upon which any cosmological model today

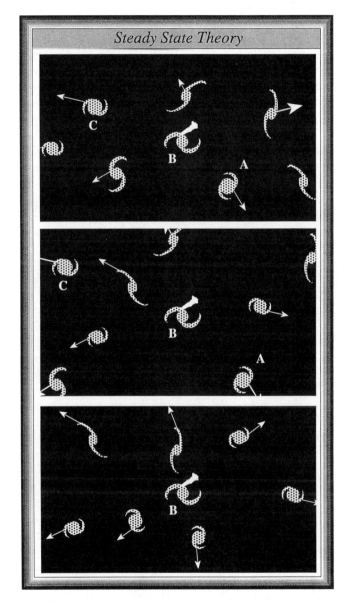

Figure 15–7 The Steady State theory accepts an expanding universe, but to account for the assumption that by "steady state" the density of matter in the universe must remain the same over time, it proposes that new matter is constantly "born" into the universe. This is not in the form of galaxies as shown here, but in the form of hydrogen atoms.

must be built — at least until someone provides evidence that the redshifts of galaxies is due to something other than the Doppler effect. The amount of matter needed to be added to the universe in order for a constant density to be maintained is not great enough to be detected. It amounts to one hydrogen atom per year suddenly appearing or being "born" into each volume of space equivalent to a cube 60 miles on a side! Conveniently for the Steady Staters, an experiment to test this prediction may be forever beyond our capability to conduct.

Mass of the Universe But what about the amount of matter in the universe? How does that fit in with the choice between the Big Bang and Steady State theories? To date, astronomers have found only a finite amount of matter in the universe — it amounts to about 10% of the matter necessary for a closed, Big Bang universe. That is just our current knowledge. But it is almost a foregone conclusion that astronomy is only in its infancy in terms of studying the realm of space beyond our own galaxy.

After all, it has been a mere 60 years or so since we even knew that other galaxies existed! So astronomers are quite confident that the remaining 90% of matter will eventually be found. In fact, astronomers have given a name to that 90% of matter that has not yet been found — the **missing mass**. It is not missing in the sense that we had it and lost it — we're just so confident that it's out there somewhere that we gave it a name. The discrepancy between the prediction that it should be out there and we aren't able to find it leads astronomers to name the subject the *missing mass problem*. It is one of the most sought-after answers in modern astronomy.

But if we admit to a basic ignorance about the amount of matter in the universe isn't it also possible that we may eventually find evidence for an infinite amount of matter in the universe — thereby finding evidence for the Steady State theory? Yes, of course. So our present-day estimates for the amount of matter in the universe is not a strong argument in favor of the Big Bang — finite matter universe — theory. So what is the convincing evidence, if there is any at all?

Test of Cosmological Theories

Choosing a theory for the origin of the universe requires that we consider the full range of topics that we have already considered — and then some. Essentially, we treat the matter rather like a court of law — lining up the evidence for each theory, critically examining each in turn, and drawing the most reasonable conclusion from the available evidence. The "true" explanation may not even be in court. But we can deal only with what we know and work within the constraints that the scientific method demands. "Your Honor, the evidence."

Stellar Evolution and Old Stars Our current model of stellar evolution predicts that the ultimate fate of a normal star that runs out of fuel and collapses to a cold stellar corpse should be detected as a black dwarf. Not only have we not found any such remnants of old stars, but the oldest stars we have detected are about 15 billion years old. Either

- (1) our model of stellar evolution is faulty,
- (2) methods of detecting black dwarfs are deficient, or
- (3) the universe is not infinitely old (as the Steady State theory claims).

This is not proof for the Big Bang theory, of course. This just impeaches witnesses for the Steady State theory.

Quasars Earlier, we found that the least complicated model for quasars locates them at great distances from us — they are the most distant objects in the universe. If so, then the universe was quite different long ago than it is today. There are no quasars close to us in distance. And that conclusion contradicts the PCP of the Steady State theory. The PCP says that matter should be distributed uniformly through space and time — the universe long ago was essentially the same as it is today.

But the great distances of quasars implies that the universe long ago was quite different from what it is like today. There are no longer any violent quasars around. They have apparently settled down to become galaxies like ours. So astronomers offer this as Strike One against the Steady State-ers — *if we assume the quasars are as far away as their redshifts indicate* (refer back to Figure 14-19).

Radio Galaxies By a similar argument, radio galaxies can also be used as evidence that the universe long ago was significantly different than the universe is today. Assuming that their redshifts are good indicators of their distances by using Hubble's law, radio galaxies appear to be more densely populated at great distances than at small distances from us. In the previous Chapter, we speculated that this observation suggests that radio galaxies simply represent a temporary evolutionary stage between quasar and normal galaxy. In any case, the distribution of radio galaxies in space constitutes Strike Two against the Steady State theory.

Echo of the Big Bang Strike Three is a ninth-inning finale — an observation that removes the Steady State theory from the championship playoffs — at least for the time being. You can observe the evidence anytime you want by changing channels on your television set until you settle on one of the unused channels. You observe static that appears in the form of what looks like "snow." Scientists estimate that about 10% of that random "snow" or radiation hitting the CRT of the TV set comes from the explosion of the Big Bang that is just now arriving at Earth. It is called the *cosmic background radiation*. In order to understand how astronomers are led to this conclusion, however, we must again explore the universe of the very small. Strike Three requires some fancy pitching.

Conditions after the Big Bang

To appreciate the importance of this strike-out evidence, we need to examine and understand the events and conditions that existed just after the Big Bang itself. It is ironic and certainly a twist of human reasoning that astronomers are finding answers to questions about the largest aspects of the universe by exploring the universe at its smallest scale — the subatomic world of the atom. Actually, this isn't really quite that surprising when you think about it. If the universe at the moment of the Big Bang was in fact a mere dot containing all of the matter, energy, and time of the universe that was about to be, then it was rather like an atom.

Role of Particle Accelerators in Cosmology The conditions thought to exist just after the Big Bang are created and investigated within special facilities like that at Stanford University in Palo Alto, California. It is called the *Stanford Linear Accelerator Center* (SLAC, for short) because it uses a linear track to accelerate subatomic particles (protons, electrons, neutrons) close to the speed of light and then to cause them to collide against a target (refer back to Figure 7-28).

Other particle accelerators, like those at the **Fermi Lab** in Illinois and the *European Center for Nuclear Research* (formerly known as CERN) on the border between France and Switzerland use circular tracks. But they fundamentally differ only in the energies they are capable of adding to the particles participating in the collision. At the moment of impact between subatomic particles and atomic nuclei in the target material, constituents of the subatomic particles (should we call them sub-sub-atomic particles?) fly off in different directions to be detected and measured for speed, mass, charge, spin rate, and so on.

Now it may initially seem strange that in order to investigate the largest object of all — the entire universe — we must investigate the tiniest bits of matter that exist. But it is simply a matter of studying matter at the temperatures that must have existed at and soon after the moment of the Big Bang. The faster the particles are moving at the moment of collision, the closer they are to simulating the conditions at the very origin of the universe. Keep in mind that temperature is a measurement of speed.

Since our model of the Big Bang requires that all of the matter in the universe be packed within a volume less than the size of an atom, the temperature at that moment had to be incredibly high. That is, whatever was present at the moment of the Big Bang was moving very fast. At SLAC, for example, the subatomic particles are moving at 99.99999997% the speed of light at the moment of collision.

And yet the scientific community felt so strongly that a more powerful accelerator would provide valuable information about conditions closer to the moment of creation that they convinced Congress to fund the *Superconducting Supercollider* (SSC, for short) particle accelerator. It was under construction outside Waxahachie, Texas, when the congressional budget axe hit, cancelling the project completely. The 54-mile-long track would have made it the most powerful accelerator on Earth.

Matter–Energy Connection The laws of physics governing subatomic particles tell us that at the moment of the Big Bang, the universe was filled with radiation in the form of gamma rays. Evidently the superdense dot of matter that preceded the Big Bang "explosion" became unstable and converted into its equivalent — energy — according to the now familiar equation $E = mc^2$. This conversion of matter into energy was the explosive event that initiated the expansion of space.

Thus the earliest seconds after the explosion saw the universe as a small volume of extremely dense gamma-ray photons. It reminds one of the first portion of the Christian version of Creation in the Holy Bible – *"Then God said, 'Let there be light,' and there was light."* The gamma-ray photons possessed the potential for turning back into matter, but the temperature was too high (1500 billion degrees) to allow subatomic particles to exist for long. If they did attempt to condense out of the high concentration of gamma-ray photons to become particles of matter, they were immediately bombarded by gamma ray photons and returned back to their gamma ray state.

The universe cooled as it continued to expand, and subatomic particles condensing out of the fog of gamma ray photons could survive for increasingly longer periods of time before being returned to their energy state. After about 1 million years of expansion, according to our estimates, the temperature in the universe decreased to the point where the subatomic particles survived without being returned to their energy state. We call this time at which gamma ray photons and subatomic particles could survive indefinitely side by side the **time of decoupling**. By "subatomic particles" I mean — of course — protons, neutrons, and electrons. Those are the basic ingredients of the matter which is presently spread about the universe.

You had best not be asking what eventually happened to those subatomic particles — that's what Chapters 1–14 have been about. But as a matter of review, the subatomic particles formed the hydrogen that went into the stars participating in the forming of the quasars that evolved into the radio galaxies that evolved into the normal galaxies — in one of which we live. But not all of the gamma-ray photons turned into subatomic particles.

Many of the gamma ray photons created at the moment of the Big Bang survived the time of decoupling to travel through space between the stars and galaxies. Some of them are traveling even as you read, having never encountered anything that absorbed them (remember that space is mostly empty). They are, in a sense, like tiny galaxies that we see at great distances — they comprise the light from the most distant object of all, the Big Bang itself (refer back to Figure 15-5)!

So according to Hubble's law, that light should be strongly redshifted, since the Big Bang is going away from us at the speed of light. Another way of visualizing this leftover radiation is to think of the universe as gradually cooling off as it expands. So the original high-energy photons (gamma rays) slowly lost energy (lengthened wavelengths) with time, eventually became X-rays, then ultraviolet, then visible, then infrared, and then after 20 billion years of expansion, microwaves.

This scenario of events was based upon the calculations of astronomers in 1948, at which time accumulated evidence indicated that the universe had evolved from a superdense state. Knowing the rate at which the universe seemed to be expanding, and knowing how large the universe was, astronomers estimated just how long it had been expanding from a single dot. They then calculated the wavelength at which the original gamma-ray photons would now be detectable — about one centimeter. Since this wavelength is within the microwave portion of the electromagnetic spectrum, the predicted radiation left over from the Big Bang was called the **microwave** or **cosmic background radiation**.

Discovery of the Big Bang Echo In the early 1960s, a research team at Princeton University began building a telescope with which to search for this specific wavelength of background radiation in hopes of confirming the Big Bang theory. Before they could even test the instrument, two research scientists working for Bell Laboratories, Arno Penzias and Robert Wilson, accidentally discovered the radiation. They were working with a newly-built radio telescope designed to conduct experiments with the first Telstar communications satellite. But when they first turned it on, they received only a puzzling static microwave wavelengths (Figure 15-8).

Thinking they had a defective part in the sophisticated system, they dismantled it, checked each component, and reassembled it. But the static remained. As they talked around the scientific community, they learned that the Princeton researchers were just about to search for that very noise! What convinced them (and even us to this day) that the static is indeed the remnant radiation from the Big Bang is that it comes from every direction in the sky.

It is uniformly distributed throughout the sky — rather like a background haze in the universe. If it were coming from a single object, a group of objects, or if it were even some scattered radiation from a supersecret Pentagon communication facility, it would not be evenly distributed in space.

Figure 15–8 Drs. Robert Wilson (left) and Arno Penzias (right) are standing in front of the Bell Telephone Laboratories horn reflector antenna with which they made the historic discovery of the 3 degree microwave background radiation in 1965. What makes the discovery fascinating is that they were not looking for it, but accidently picked up what had only been theoretically predicted but never searched for. (Courtesy of AT&T Archives)

Current Studies of Background Radiation A satellite was launched in 1989 specifically for the purpose of exploring the universe at infrared and microwave wavelengths. The goal of the **Cosmic Background Explorer** (**COBE**, for short) was to measure the strength and degree of "lumpiness" of the background radiation from above the distorting layers of Earth's atmosphere. The degree of "lumpiness" is related to the current problem in astronomy of not being able to easily explain the origin of the galaxies in an expanding universe.

If the universe began expanding from the very beginning — material rushing outward and thinning out as it went — then what eventually caused some of that material to come together to form galaxies? If the background radiation is smoothly distributed throughout the sky, and it originated when the universe was quite young, why don't we see within that microwave radiation evidence of the precursors to the galaxies? Shouldn't the radiation reflect the early tendency of radiation-becoming-matter to form into lumps that eventually became the galaxies? This has been a rough area for astronomers. They'd like to know just when those "lumps" began to form, and how.

On April 24, 1992, the scientific world and world at large was startled by the announcement by the leader of a research team of the Lawrence Berkeley Laboratory that the COBE had detected lumpiness in the background radiation (Figure 15-9). At the time, it was hailed as one of the major discoveries of science — ever! The research team worked on the project since COBE began taking measurements, and the announcement of the discovery even withheld for over a year while the team checked for errors in the hundreds of millions of measurements upon which the discovery is based. Prior to being turned off in December 1993, COBE confirmed the prior findings.

Now that we know about the background radiation's existence, astronomers have been able to account for some of that "static" on our television screens when we have it set for a channel that is off the air. Knowing the strength of the background radiation at Earth's surface, we estimate that about 10 percent of those little white spots that sparkle on the screen are caused by the background radiation. So if you want "proof" that the universe began with a Big Bang, turn on your TV set!

The fact that the discovery of the microwave background radiation is currently offered as the best evidence for a Big Bang because it is uniformly scattered throughout the sky may require a short explanation.

Mapping the Past History of the Universe

Go outside and look outward in any direction you wish – *you are looking back into time in the direction of the Big Bang.* The Big Bang occurred everywhere, not just in a single direction. If someone asks you to point in the direction that the Big Bang occurred, just point anywhere. You are correct

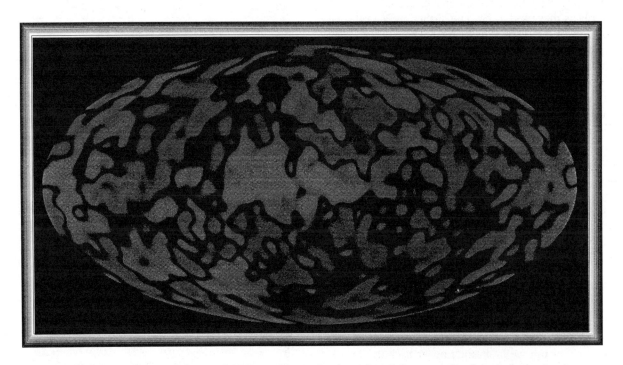

Figure 15–9 This image taken by the COBE satellite is simply a plot of the strengths of the 3 degree background radiation in all areas of the sky. The long axis of the oblong shape lies along the celestial equator, and the oblong shape represents the entire sky. What is significant is that the background radiation is not uniform, but rather lumpy. This indicates that the energy/matter in the early expanding universe began collapsing to become the eventual galaxies we observe today. (NASA)

regardless of where you choose to point. All directions are back in time.

If that is so, you may ask yourself — if we could build a telescope powerful enough to look outward 20 billion light-years, wouldn't we be seeing things that existed 20 billion years ago? And if the universe began 20 billion years ago, wouldn't we be looking at the Big Bang itself? In principle, yes we would. But in practice, we can't. There are two reasons.

Recall from Chapter 13 that Hubble's law tells us that an object going away from us at the speed of light is located 20 billion light-years away from us. That, in fact, is how we calculated the age of the universe. So the Big Bang event is receding from us at the speed of light. But that means we are unable to detect it directly, since it is going away from us at the same speed at which it is sending information (radiation) back to us.

The second reason we can't see the Big Bang with a super powerful telescope is more subtle. We cannot observe it because the atoms (or their energy equivalent) that make up our eyeballs were there! You cannot use your eyeballs to look back into time to see the birth of the very atoms which eventually were to make up your eyeballs! Figure 15-11 illustrates this concept. Assume space to be shaped like a balloon as in a Type C universe. Galaxies are distributed randomly in space, and are shown as dots on the cross section of the balloon's surface.

This is the universe that presumedly exists today — in our time. But we cannot see other galaxies as they exist today, only as they used to be. The farther away a galaxy is from us, the longer ago we are seeing it. The balloon, as we are looking at it, is a God's-eye view of the universe — an instantaneous view of the contents of space at a single moment of time. We mortals are not so fortunate — we must see the contents only as they used to be, as they were when the universe was smaller.

The concentric circles within the largest circle represent the universe as it was at various stages of its evolution. A line from the center (time of the Big Bang) out to each galaxy is a trace of that galaxy's evolution with time. So, for example, points **A**, **B**, and **C** represent the Milky Way 5 billion years ago, 10 billion years ago, and 15 billion years ago, respectively. Likewise, points **a**, **b**, and **c** represent a galaxy presently located 5 billion light-years away from us during those same time intervals. But since we observe that galaxy only as it used to be, we are seeing it as it was 5 billion years ago. The curved line between **a** and the Milky Way, which is our present position, represents the path of the radiation that left the galaxy 5 billion years ago and which has just now arrived in our telescopes.

The further out we look, the further back in time we look. So the curved line running from the Big Bang to us represents the path of the hypothetical light rays that left the Big Bang 20 billion years ago. But since the universe is constantly expanding, this curved line is stretching out. The dots along this line representing galaxies can be thought of as raisins in the loaf of bread we used in Chapter 13 to illustrate the expansion of the universe.

While you contemplate this diagram, you should keep in mind that this is only a two-dimensional drawing that attempts to illustrate something that is inherently four-dimensional. You perhaps are wondering how this diagram can illustrate the universe if it reverses its expansion and begins to collapse back upon itself. If time is expressed as the expansion of the universe, will we have to relive in reverse all of our life experiences as the universe collapses? Must we repeat our college education in reverse?

No. Time would reverse itself only in a mathematical sense. By the way, I assume you have not gotten the impression that you might awake one morning to news that the universe reversed itself while you were asleep! The slowing down and reversal would occur over an extremely long time period. Speaking of reversal, is it possible for astronomers to determine whether or not the universe is slowing down, and whether it will eventually collapse?

Change in Rate of Expansion Think about it — the galaxies right around us are moving away from one another at a certain rate. The very distant galaxies are also moving away from one another at a certain rate. What if we observe the expansion rate of the very distant galaxies to be significantly greater than that of the nearby galaxies?

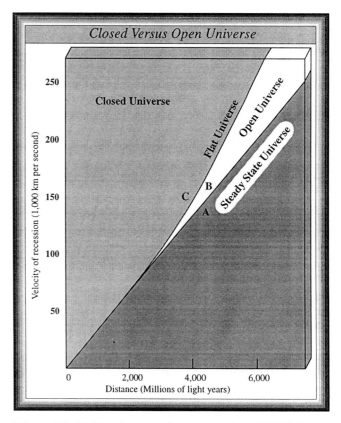

Figure 15–10 Plotting galaxies accurately on Hubble's diagram would allow us to determine the fate of the universe.

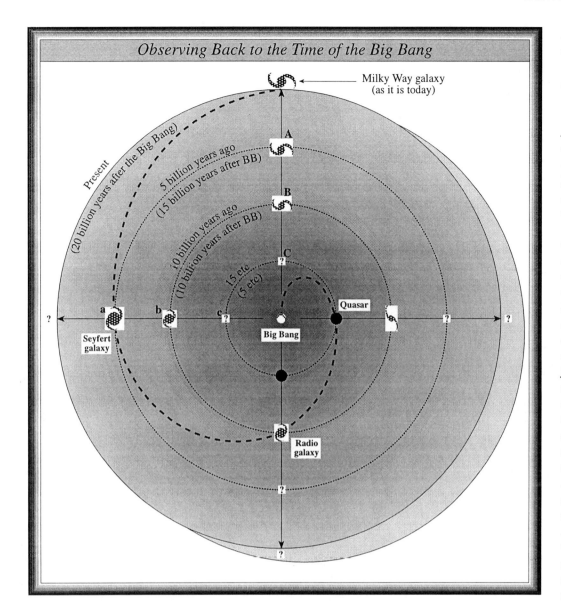

Figure 15–11 *This diagram illustrates the cross–section of a spherically–shaped expanding universe. The center depicts the point from which the expansion began—the Big Bang. 20 or so billion years later, the evolving galaxies have expanded out to the outer circle. When astronomers look out in space, they look back into time. This is equivalent to looking deeper and deeper into the interior of the sphere. The dashed spiral line represents a hypothetical ray of light from the Big Bang as it makes its way through the expanding space until it reaches us 20 billion years later. We, detecting the light, say that we are seeing the Big Bang. As the hypothetical ray of light travels, it is Doppler–shifted, however. So what began the journey as gamma radiation now arrives at Earth as microwave radiation.*

Doesn't that suggest that the universe is slowing down in its expansion? Yes, of course. The balloon long ago was expanding at a faster rate than the balloon of today. So the balloon is slowing down in its expansion, and will eventually reach a maximum size. But how can we measure a difference in expansion rates?

Take another look at Hubble's law. It expresses the radial velocity of galaxies moving away from us at given distances from us. The fact that the more distant galaxies are moving away from us at greater velocities than the closer ones is interpreted by the uniform expansion of the universe (the loaf of raisin bread is uniformly expanding).

But if the more distant galaxies are expanding *away from one another* at a much greater rate than the nearby galaxies are expanding *away from one another*, then the distant galaxies will reveal much greater radial velocities than if the universe were expanding uniformly (the loaf is slowing down in its expansion). Let me illustrate this on a diagram. If the universe is slowing down in its expansion, we would expect the plot of the most distant galaxies to appear along line **C** in Figure 15-10 rather than along the line **A** that is for a uniformly — expanding universe.

Actually, the universe should be slowing down slightly even if it isn't going to slow down enough reverse itself. It is possible for something to slow down without ever stopping. This is because the mutual gravitational attraction between all the galaxies resists the expansion. But that slight slowing down might never reach zero.

It can slow down more and more and more, but never get to the point at which expansion stops. In mathematical language, we say that the rate of expansion "approaches zero asymptotically." It's like cutting a pie in half, cutting the half in half, and so on. There is always a bit of pie remaining to cut in half. So line **B** represents the expected plot of galaxies in a universe that expands forever (Type B universe). Line **A**, representing the plot of galaxies in a

★★★★★★★★★★★★★★

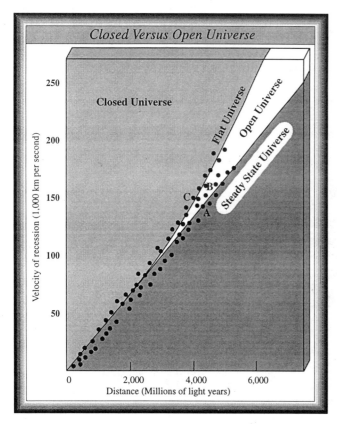

Figure 15-12 *The diagram in Figure 15-10 is repeated here, but with plots of increasingly distant galaxies plotted as accurately as possible. The plotted points do not allow us to choose between the three possible "fates" for the universe because the plots for the very distant galaxies are so scattered around the diagram. This is because the accuracy with which we can measure the distances to distant galaxies is very limited.*

uniformly expanding universe, is associated with the Steady State theory.

So you see, we have stumbled upon a method of finding out whether our universe is A, B, or C. We plot galaxies that we measure on a diagram like this, and then look to see if the line along which the plots lie conform to that of A, B, or C. But alas, it isn't as easy as it might appear. Do you recall the difficulty of determining the distances to the most distant galaxies? The quasars are not useful at all, since their distances are measured from Hubble's law in the first place.

Figure 15-12 is a plot of the most accurate data for distant galaxies available today. Is it obvious that any one of the lines A, B, or C could be drawn through the plotted points? Would you argue strongly for one of the three lines? In order to use the technique to determine the future of the universe, we will have to wait until more accurate methods of determining distances to distant galaxies have been found. Meanwhile, it is comforting at least to know that we have a method available. When the accurate distance measurements arrive, we'll be there waiting.

Big Bang vs Steady State Does it seem that I have given equal treatment to the Steady State theory? Well, I tried. But the evidence available to us today just doesn't provide us much evidence to support it. The accidental discovery of the microwave background radiation in 1965 was the final nail hammered into the coffin of the Steady State theory.

If that is so, you might yet wonder why we even bother to spend time discussing it. No one likes a one-party system of government, especially when the politics of that one party are difficult to believe. Likewise, with a topic as comprehensive as cosmology, which attempts to explain the existence of the entire universe, we like to have a point of comparison.

Since observations needed to test theories of cosmology are made at the very limits of the instruments presently available to astronomers, and since much of the reasoning is based upon untestable assumptions, we like to keep the Steady State theory around just in case.

As new data and observations become available, we attempt to fit them both into the Big Bang and into the Steady State theories. If it fits only the Big Bang, then we feel that we must be on the right track. In other words, astronomers do not believe that the currently available data is so convincing for the Big Bang that the game is over. We may be just about to begin a new inning. In light of that situation, let's consider research that might just clarify the game's conclusion.

Current Research in Cosmology

Some of the most exciting research in cosmology today falls within the scientific discipline called *quantum mechanics*. The raw data for this field of study comes from the particle accelerators mentioned previously. But at a fundamental level, quantum mechanics deals with the very most basic laws by which matter and energy behave. Since the very laws that govern the universe are presumed to have been "born" into the universe at the moment of the Big Bang, understanding the laws are paramount to understanding the origin of the universe.

Forces in the Universe There are four known forces that govern the behavior of all objects in the universe (Figure 15-13). *Gravity* is the weakest, although its effects are the most obvious to us. It is the force that holds Earth in orbit around Sun, Moon around Earth, and you and I on Earth. *Electromagnetic forces* are stronger still, holding together the atoms and molecules within our bodies. If gravity were stronger than electromagnetic forces, then jumping up into the air would cause you to stretch out from your feet to your head, with your feet returning first to Earth's surface followed by the rest of your body—like a stack of dominoes.

While exploring the nature and behavior of stars and galaxies, we have been able to confine our interest to those of gravity and electromagnetism. But in attempting to recreate the conditions that existed just after the Big Bang, astronomers deal with the two additional forces that operate at the smallest levels of matter.

The **weak force** holds the nucleus of the atom together. When you think about it, present in the nucleus of every atom are particles (protons) that have like charges (+), and therefore that must repel one another. So a nuclear glue-like force holds the protons together as a unit — the nucleus. But if something comes along to break the nucleus apart — like a subatomic particle in a particle accelerator — some of that weak force is released as energy. That is the basic principle upon which atomic reactors and atomic bombs operate. In those two cases, a supply of neutrons is exposed to atomic nuclei so that collisions will occur.

The **strong force** holds the *quarks* together to form the subatomic particles — the protons and neutrons. We encountered the subject of quarks when we attempted to define black holes in space. Recall that quarks are not particles of matter, but rather units of energy. At the finest level of matter, there is no matter — matter appears to be ultimately made of energy units. Theory tells us that there are 6 kinds (particle physicists refer to them as *flavors*) of quarks, and it is the different arrangements of these basic six energy units that form the particles of matter we call neutrons and protons.

Those 6 flavors are called "up", "down", "strange", "charmed", "truth", and "beauty". Each of these flavors in turn can have one of three properties called colors, often called "red", "blue", and "green". Therefore there are 18 subtypes of quarks. And to round out what must seem to be an excursion through a science-fiction story, each subtype has its opposite pair — the anti-subtype. So there is an *anti-strange-red quark* hanging out in your body somewhere! There are (finally?) 36 subtypes of quarks. You can appreciate why particle physicists are busy! Recent work with accelerators have confirmed the existence of five of the six quarks, and unconfirmed evidence for the sixth (the "truth" quark).

Specific mathematical equations express the manner and conditions under which each of these four forces operates. So there are four equations — one expressing the characteristics of each of the known forces. Einstein was convinced that the universe is so beautifully organized that there must be a single equation which incorporates within it all four of the equations. At the time of his death in 1956, he was searching for it. Some of the calculations he was working on at the time are still on the blackboard behind his desk. Other scientists have since taken up the task. Notable amongst them is Steven Hawkings of Cambridge University in England, famous for his book "*A Brief History of Time.*"

Return to Particle Accelerators One pregnant direction of research is that involving particle accelerators like that of SLAC, where beams of electrons are accelerated to speeds very close to that of light, and then caused to collide with a target of atoms. Of course, most of an atom consists of the space between the nucleus and the electron shells, so most of the electrons go through the target without hitting a nucleus. But a few electrons manage to find nuclei and collide with them, and those are called events. The fragments of the colliding particles are carefully studied, because they tell us about the underlying makeup of all matter in the universe, and therefore the very stuff that the entire universe grew out of.

More recent experiments involve the acceleration of protons in one ring going in one direction, and the acceleration of antiprotons in an adjacent ring going in the opposite direction. At an opportune moment, the two beams are caused to collide, creating the constituents of the protons and antiprotons which can then be studied. The point of all this is that the conditions under which these events occur are not unlike the conditions that existed close to the moment of the Big Bang.

Subatomic particles didn't exist at the moment of the Big Bang — so gravity, electromagnetism, the weak and strong forces did not exist separately at that moment. They were one. There was only one force that governed the behavior of everything in the universe. Only as the universe expanded and energy evolved into matter did the separate forces in a sense split out or "distill" out of the original one force. It is rather like what happens when water turns from

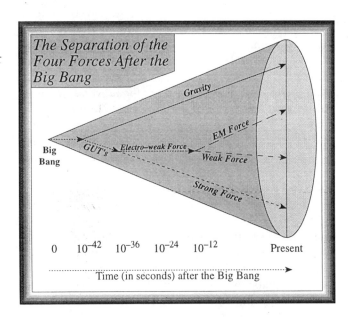

Figure 15–13 The study of quantum physics attempts to explain how the 4 forces that govern the universe today evolved out of a single force that was "born" into the universe at the time of the Big Bang. As energy evolved out of matter and the matter arranged itself into larger structures, new forces "evolved" out of the original single force. Research in modern nuclear accelerators recreate the conditions that would have existed in the early universe when these "phase changes" were occurring.

CHAPTER 15 / The Origin of the Universe   **329**

a liquid into a solid, or a liquid into a gas. This is what chemists call a *phase change*. The splitting of the forces of nature as the conditions in the universe changed can be thought of in the same way.

This was the life work of Einstein — to find a set of equations that describe that one force. The Force. The universe of matter and forces that we see today distilled from The Force: the strong force emerged to unit quarks to form protons and neutrons — then the weak force to hold the protons and neutrons together in nuclei — then electromagnetism to hold electrons in orbit about the nuclei — and finally gravity to pull collections of atoms together to form stars and planets. All so that we might evolve the awareness and curiosity with which to probe the simplicity and beauty of how it all got started? Stated in that manner, humankind is the universe's way of being aware of itself.

Cause of the Big Bang Experiments conducted at various particle accelerators are revealing what Einstein suspected — at higher and higher temperatures, the four forces converge to become one. In 1979, the Nobel Prize in Physics was awarded to a team of particle physicists for confirming the existence of the *electroweak force* — a combination of the electromagnetic and weak forces. This force no longer operates in the universe (under natural conditions). It operated under conditions that existed when the universe was very young. To detect it, therefore, the team of physicists had to recreate the conditions of that time period. And that is only possible in particle accelerators.

Having accomplished that feat, the team (and other teams at other accelerators as well) are attempting to recreate the conditions under which there were only two forces operating in the universe — gravity and the *GUT* force. Actually, the searched-for force — the combination of the strong and electroweak forces — has not as yet been named. We refer to it as the **GUT** force because the work being conducted in that direction is called the **G**rand-**U**nified **T**heories.

The evidence is not complete by any means. There are a lot of unanswered questions. But the pattern is clear — *the nature of matter at the subatomic level is thoroughly compatible with the Big Bang theory.* If only we were able to experiment with matter and/or energy in such a way that we could find out where the universe itself came from! We have traced its history as far back as the creation of the quarks from which everything in the universe has evolved. But from just where did the quarks come?

Quite familiar to atomic physicists are what are known as **virtual particles** — particles that appear from nowhere, exist for a brief moment, and then decay again to "nothingness." For example, we learned from our study of Sun in Chapter 7 that a positron — a positively-charged electron — appears momentarily from nowhere to be a participant in the proton-proton chain that is responsible for sunlight and starlight. In that particular case, it does not return to nothingness — it finds an electron with which to unite to create a gamma-ray photon that escapes from Sun's surface as an ultraviolet photon to cause your suntan (Figure 15-14).

Virtual particles are electron-positron pairs that appear spontaneously from nothingness, exist for a brief moment, and then return to nothingness — giving off energy in the process. Physicists refer to this process as **pair production**. Imagine, if you will, the entire universe having an origin analogous to this phenomenon. There existed 20 billion years ago an ocean of nothingness out of which a superdense pair of virtual particles spontaneously appeared, existed for a brief moment, then returned to nothingness. In the act of mutual annihilation, the superdense pair became the quarks which then proceeded to form matter into its myriad of sizes, shapes, and behaviors. This is called the theory of **vacuum genesis** — also known as the **Big Foam** theory.

Think of it this way. When we use the word "nothingness," you may think the word means "nothing." But of course it is only that the word masks our ignorance about the conditions that existed before the universe began. "Nothing" is the word we choose to use to name something that we don't yet understand. When we do, we'll give it a name.

And what do we mean by "began" anyway? We mean time began, of course. But then what was the universe "doing" before time "began"? It wasn't "doing" anything, since "doing" implies that time was going on. But if time didn't exist until <u>after</u> the Big Bang occurred, then "nothing" was "doing" nothing before the Big Bang. What was God doing before He "created" the universe (in Christian theology)? Nothing. And perhaps everything? You see, in order to deal with the beginning of the universe, we must put on a different type of thinking cap. We humans are limited to thinking in terms of time. So to speak of a possible reality (God, vacuum genesis, etc.) that doesn't include time is limited by the extent of our imaginations.

The Big Foam Theory So try to imagine an endless foam of space-time "bubbles" — forever appearing and disappearing. The "bubbles" come into existence for a brief moment — like virtual particles in particle accelerators — and then vanish into timeless nothingness. But once in a great while — although the probability is extremely small that it will happen — one of the "bubbles" rises from the space-time foam, expands rapidly, and creates time and matter where there was none before.

All of the events and processes covered in earlier Chapters then proceeded up to this very moment. Even if the probability is extremely small that one of the "bubbles" will escape the Foam, it only has to happen once. And here we are. Could there be other such fortunate (?) "bubbles"? And if so, do we have anything in common with them?

Amazingly enough, the Big Foam theory is not incompatible with the known laws of physics. The chances of such an event occurring are admittedly extremely small —

but again, it only had to happen once. As yet, there is no good theory to account for all aspects of such an initial state of matter and energy, but particle physicists are presently working in that direction.

Scientists possess an inherent faith that the picture gets clearer and sharper as they work with particle accelerators in examining matter and energy at the finest level. It seems ironic that in order to understand the universe at its grandest and largest scale, we must study it at its finest and smallest scale — in order to explain the entire universe of galaxies, we must study the behavior of the tiniest quarks.

Assumptions of Cosmologists

If this discussion of cosmology has led you to conclusions that are incompatible with common sense or your personal beliefs, you might want to identify any assumptions we have been making that might be faulty. One obvious assumption we use is that the laws of physics with which we are familiar and which we used to successfully land men on Moon have been operating consistently throughout the universe in space and time since the Big Bang.

Beauty and the Universe For example, we assume that by studying the radiation received from quasars, we are able to interpret the atomic processes taking place in those distant objects — we assume that atomic processes operating billions of years ago are identical to those operating today. We have no proof of this one way or the another. We have a built-in faith that it is so. Why? Because we have inherited (in our culture, at least) the ancient Greek belief that the universe is rational and knowable — there is a certain beauty and simplicity to it.

We see that beauty and simplicity all around us in the preciseness with which laws of physics and chemistry operate and with which we can explain what we observe. There would be a consistency to this beauty and simplicity if in addition the laws of physics and chemistry are constant throughout the history of the universe. But didn't we discover in Chapter 4 that the Greeks also had a belief and faith in the simplicity and beauty of the circle — that the orbits of planets and stars must therefore conform to that shape? And likewise, might our faith in what appears to be beauty and simplicity in the laws of physics and chemistry be eventually shaken by new discoveries? Of course.

How — then — can we hold onto that same faith as regards the laws of physics and chemistry? Because it is a starting point from which we conduct our research and experiments. We must begin somewhere. Eventually we will — in the act of investigating — uncover evidence that our assumptions were too narrow, that there is a larger simplicity and beauty that exists beyond what we currently know. We replaced circles with ellipses — which are no less beautiful, although perhaps less simple. We might, likewise, discover that the laws of physics <u>have</u> changed with time — but in an orderly and predictable way.

Testing Assumptions If this seems rather abstract, let us consider a concrete example — the universe is expanding. The Big Bang theory begins with this fact. If the universe is not expanding, then there was no Big Bang. But how do we know the universe is expanding? Because we assume that the redshifts of the galaxies are due to the Doppler effect. Why do we assume that the redshifts are due to the Doppler effect? Because we use it in hundreds of practical applications, both in space and here on Earth (have you been caught in a police radar trap recently?). The Doppler effect is presently the only way to explain the redshifts. Simply put, we just don't have a better explanation for the redshifts observed in the galaxies. We make assumptions based upon the known, having faith that continued experimentation and observation will tell us if we are right or wrong.

Modified Steady State Theory A good example of assumption-testing is the proposed *Modified Steady State theory* that incorporates the new discoveries that contradicted the old Steady State theory. In 1971, the former champion of the Steady State theory, Sir Fred Hoyle — having abandoned his theory because of the discovery of

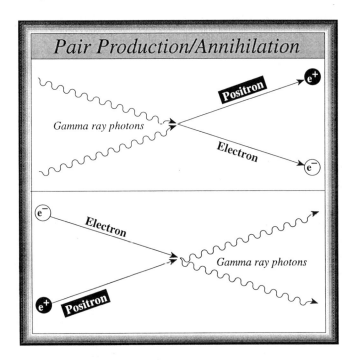

Figure 15–14 Working with nuclear accelerators, particle physicists have long recognized that at the sub–atomic level, particles of matter spontaneously appear from "nowhere" and return to "nowhere" during brief moments of time. If the same principle for the appearance of these virtual particles operates at the larger level of the universe in its entirety, then we have an explanation for how the universe came out of "nowhere."

quasars and the microwave background radiation — proposed a new idea. He suggested that the masses of subatomic particles such as protons and electrons change with time — that the very ruler with which we measure the behavior of radiation changes in length as the universe ages. This proposal in essence suggests that the laws of physics and chemistry are not constant with time.

If, for example, the masses of the electron and proton decrease as the universe ages, then there is weaker and weaker bonding between the two oppositely charged particles. That weaker bonding in turn would cause shorter and shorter wavelengths of photons to be emitted by the atoms making up the aging universe. Therefore, the radiation from the more distant galaxies is redshifted not because the galaxies are moving rapidly away from us, but because they are younger and made of less-massive atoms that the nearby galaxies. According to this model, the universe is neither expanding nor contracting — it is a static loaf of raisin bread. The Big Bang is a myth based upon an erroneous assumption.

In spite of the reasonableness of this theory, astronomers are not yet prepared to abandon the Big Bang and the immense research and experimental evidence accumulated so far to explain it. Hoyle's theory is neither supported by a larger well-developed theory nor by the wide range of observations that the Big Bang theory is. "What is the origin of the microwave background radiation if there was no Big Bang," I am tempted to ask Hoyle.

Someone may eventually find an explanation for the background radiation within the framework of Hoyle's theory. But until then, the Big Bang picture is more complete. The tendency to believe in a scientific theory is not so much a matter of it successfully describing a single phenomenon or making a single correct prediction, but of the theory's ability to weave diverse phenomena into an intricate quilt-work pattern of understanding.

A Shrinking Universe? In addition to questioning whether subatomic particles themselves change and evolve as Hoyle does, one might also question the very measuring sticks with which we determine the distances and motions of objects in the universe. Take the raisin bread loaf, for example. Instead of the entire loaf expanding — taking the raisins along with it — how do we know that it isn't rather the raisins that are shrinking, making it appear as if other raisins are moving further and further away from our raisin?

In other words, the universe need not expand for us to detect the redshifts of the galaxies as we do. Since we have no evidence to the contrary, we simply find it convenient to assume that rulers we use on Earth are equally useful in the larger universe. And we simply do not know if that assumption is valid.

Now don't get down on scientists for working in this fashion. You and I and the government institutions we elect to structure our lives operate in the same way. We assume that technical progress is a desirable thing — to be adopted by all countries on Earth. We assume that a market economy is the best manner in which to conduct business. We assume that all people are innocent before being proved guilty. We assume that women are as capable as men in performing the tasks necessary to run our society. We haven't always assumed those things. We have adopted those assumptions because we found through painful experience that the alternatives did not meet the needs of people bent on survival and happiness.

Future of the Universe

To conclude this discussion of cosmology, let us imagine the possible futures the universe can have — extrapolated from the cosmological model we accept. If Steady Staters are correct, then business proceeds as usual — star birth, life, and death. Matter in the form of hydrogen atoms that is spontaneously "born" into the universe is an eternal source of fuel for stoking the fires within stars. The Steady State theory (Type A universe) predicts an eternal universe, past and future.

If, however, the expanding universe was "born" in a Big Bang explosion, but lacks the mass needed to reverse the expansion (Type B universe), then stellar evolution gradually peters out. The remaining hydrogen gas created at the time of the Big Bang is eventually exhausted — having been fused into metals which make up the dead corpses of black dwarfs, neutron stars, and black holes. Galaxies drift further and further apart as the universe expands into the infinity of space, with eventually nothing left to shine. The ever-expanding universe theory of the Big Bang predicts that the universe will slowly run down and be extinguished as it expands forever.

If the missing mass is found, then the model we are left holding predicts that the universe will expand to some limit before reversing direction, eventually fall back on itself, galaxies will collide, stars will run into one another in a blaze of light, and all matter and energy will return to an identical state of near-infinite density from which the universe originally came — the *Big Crunch*. This is the Type C universe of finite matter in finite space — a universe that cannot expand forever.

Perhaps the universe at that point of infinite density just turns inside out — and all of the processes discussed in the book repeat themselves in similar fashion on the "other side." Or perhaps the universe rebounds from that point, and gives other life-forms a chance to figure out how it all happened. The universe merely goes from Big Bang to expansion to contraction to Big Bang to expansion and so on — forever. We are presently on one of the expansion legs of an **oscillating universe**. That should keep your mind active for awhile.

Summary/Conclusion

Perhaps the most obvious connection between the topic of cosmology and the search for life elsewhere is the importance of keeping an open mind when dealing with either. The evidence is few, and not clear at that. What is important as we continue to accumulate new data is to keep our file of assumptions in the active mode. If we care to speculate with the evidence that is currently available, however, it appears that the universe itself evolved from some preexisting state. So life seems to fit into a much broader pattern than what our immediate surroundings would indicate.

LEARNING OBJECTIVES: *Now that you have studied this Chapter, you should be able to:*

1. State the basic assumptions upon which astronomers form cosmological models.
2. Describe the three possible universes in which we live in terms of an intergalactic traveler.
3. Explain how the three types of universe are related to the shape of space.
4. Describe the Big Bang theory, explaining its assumptions and its ability to explain fundamental observations.
5. Describe the Steady State theory, explaining its assumptions and its ability to explain fundamental observations.
6. Compare and contrast the Big Bang and Steady State theories.
7. Explain how the detection of the microwave background radiation supports the Big Bang theory, and contradicts the Steady State theory.
8. Outline the theoretical process of chemical element formation in the Big Bang theory, and offer evidence from previous chapters that supports it.
9. Explain how the determination of the mass of the universe can tell us something about the future of the universe.
10. Describe how experiments in particle accelerators are providing clues about the origin of the matter at the moment of the Big Bang.
11. Discuss the assumptions that cosmologists use while formulating theories about the origin of the universe, and give one example of an assumption which, if invalid, would lead to a revision of the Big Bang theory.
12. Define and use in a complete sentence each of the following **NEW TERMS**:

Big Crunch 320
Big Foam theory 330
Closed universe 317
Cosmic Background Explorer (COBE) 325
Cosmic background radiation 324
Cosmology 315
Fermi Lab 323
Grand Unified Theory (GUT) 330
Hyperbolic paraboloid 318
Microwave background radiation 324
Missing mass 322
Open universe 317

Oscillating universe 332
Pair production 330
Perfect cosmological principle (PCP) 321
Steady state theory 321
Strong force 329
Time of decoupling 323
Type A universe 316
Type B universe 317
Type C universe 317
Vacuum genesis 330
Virtual particles 330
Weak force 329

★★★★★★★★★★★★★★

Section 4

Evolution of the Planets

Chapter 16

The Origin of the Solar System

Figure 16–1 Comets are as much members of our solar system as are the planets, and indeed provide important clues toward understanding its origin and evolution. This is Comet West, as seen in 1976. It came close enough to Sun to be visible to the naked eye. (Lick Observatory)

CENTRAL THEME: What objects are in our solar system, and how do their chacteristics and behaviors suggest that they evolved with Sun?

Stars and clouds of gas and dust appear to make up at least 99.9% of the universe. They are gravitationally bound within the large-scale building blocks of the universe, the galaxies. The evidence from astronomical studies suggests that stars form out of the clouds of gas and dust, live for awhile, and die. The galaxies themselves seem to have evolved from the behavior of radiation and matter shortly after the Big Bang. Everything we have studied so far appears to have evolved out of something before it. Without stars, there would be no light, no galaxies, no luminous clouds of gas and dust, no life even.

So I have up to this point concentrated on the origin and life cycles of the stars. One of the subthemes I tried to emphasize along the way is that everything happening out there appears to have either <u>encouraged</u> or at the least <u>allowed</u> for the flourishing of life here on our planet. Stars are essential to life, but life cannot evolve or survive on the surface of a star. So although the amount of material in the universe that is inside of planets and satellites (moons) and debris is quite small and indeed insignificant from a mathematical point of view, it is certainly worthy of our interest and study.

Exploration of the Solar System

There are practical as well as academic reasons for our interest in exploring and understanding the contents of our family of planets, satellites, and debris. We are obviously curious about its origin and evolution – how Earth got here in the first place, for example. It would be nice to fit a theory of the solar system's origin into the broader theory of stellar evolution, simply to round out that Chapter of our *Book of the Evolving Universe*. Are planets – in other words – common by-products of the evolution of stars, or are they consequences of rather unusual and chance combinations of events that occurred before or after Sun evolved from a collapsing nebula?

If the evidence suggests that planets, satellites, comets and asteroids are the direct result of events commonly associated with star birth, then it is logical to assume that most stars (if not all) possess systems of planets as well. And, if that is the case, the next question is obvious – "Might similar events leading up to intelligent life have occurred elsewhere? Might there be life out there on any of those planets?" Is it possible that we are but a single civilization living within a larger family of millions (or billions) of other civilizations, just in our galaxy alone? Sobering thought.

On the other hand, if we find that the solar system is the product of a unique set of uncommon circumstances that occurred in association with or independently of Sun's origin, then we'd conclude that we are freaks in the universe – there are very few (if any) civilizations elsewhere. We are alone in the universe. Sobering thought. Regardless of the outcome of our studies, we should stand in awe – if we are common and mundane, how exciting it will be to contact other civilizations! If we are alone in the universe, so much more important should be our efforts to preserve this unique experiment called life.

Comparing Earth with other Worlds To be sure, the preservation of our planet is of great concern to all of us. Aside from any academic interest we have in determining the origin of our solar system, there are some lessons we may learn while studying other environments in the solar system that can be applied to our own planet. Ever since NASA launched the satellite *Mariner 4* in the year 1962, the first space vehicle to explore the environment of another planet (Venus), the field of **planetary geology** has grown by leaps and bounds.

Textbooks used in college geology classes usually contain a Chapter on the subject of planetary geology. The reason should be obvious – we understand things best when we compare them with something with which we are already familiar. So we understand our own planet much better as we compare it with other planets. I can only hint at the benefits that may accumulate as a result of such comparisons, because we have only been active in this matter of studying solar system contents in detail for a couple of decades. There are several areas of study that may bear fruit.

Earth's Climate The study of Earth's climate (**meteorology**) is made difficult by the large number of variables that influence it – Earth's rotation, the tilt of Earth's axis, the storage of sunlight in the oceans and land masses, the reflection of sunlight off vegetation along the equator. It is difficult to know which factor is more influential than any other at any particular moment. Do you have confidence in weather predictions presented during the evening news?

But scientists are gradually piecing together parts of the puzzle by having a few natural laboratories in the solar system. Venus has a very thick atmosphere, but rotates slowly. Mars rotates at approximately the same rate as Earth, but has no oceans to affect its thin atmosphere. Jupiter consists entirely of atmosphere, and rotates rapidly. These planets are models with which we are able to compare Earth.

We are plagued by uncertainties in Earth's climate, the recent droughts in Africa being a prime example. If we are able to obtain accurate warnings of such trends, we might well be able to minimize the human suffering that accompanies such climatic changes. And there are also the effects that we humans are imposing on our atmosphere through industrialization – the pouring of sulphur dioxide, carbon dioxide, and fluorocarbons into the air we breathe.

To what extent are these artificial additives influencing our climate? To what extent is the cutting of the tropical rain forests along the equator in the name of progress affecting the growing of crops in the mid-Western states, the breadbasket of America (and to some extent, the world)?

Venus, whose atmosphere contains sulphur dioxide and lots of carbon dioxide, may provide a good test case for any theory we develop regarding the future of our atmosphere.

Space Habitation While President of the United States, Ronald Reagan announced a U.S. goal of establishing a permanently-manned space station in orbit around Earth by the year 1994. Achieving that objective by that date has been frustrated by the *Challenger* disaster and a sagging economy, yet there is little doubt that it will eventually become a reality. The **Space Station Project Alpha** is currently being reevaluated in light of ongoing debate over budget. It will most likely be scaled back to achieve more with less (Figure 16-2).

And what happens after that? Will we be content to have a single station, rotating crews to and from Earth at one-year intervals? Or will we expand our presence into space even further, mining the surface material of Moon and/or the smaller asteroids for the basic materials needed for space colonies? Is there a basic difference between that scheme and one of mining iron ore in Minnesota, processing it into iron ingots, and shipping the ingots to Japan where they are molded into a automobile bound for a dealership in California? Is it a different in kind, or merely different in amount?

In broader terms, are we as a species destined to remain here on the surface of this planet? Or will our future Space Shuttles, like the covered wagons that crossed the Great Plains to carry our ancestors to the wide-open spaces of California, carry us to new living environments in space? The South Pole is not a particularly comfortable location to live, yet people live there throughout the year at various military and research stations.

What I am saying, of course, is that the knowledge astronomers accumulate about the planets, satellites, asteroids, and general environments of the solar system is used in planning for such possibilities. The emerging picture goes something like this — it is not a matter of whether we can expand into space but whether we want to. I will elaborate on this question in more detail in Chapter 21, after we survey the contents of the solar system with a view toward finding suitable targets for human exploration.

Solar System Survey – Contents

Most of the mass of the solar system is in Sun. And of that mass that isn't in Sun (0.15%), most is in the planet Jupiter. Reflecting on the distribution of matter in the solar system in Figure 16-3, one can easily feel a degree of humility about our own importance or significance — at least in terms of size. Yet there are objects above and beyond those with which we are most familiar, the planets and moons (technically called **satellites**). And aside from the major pieces of debris in the solar system — comets and asteroids — there are billions of tons of microscopic particles in orbit around Sun.

Planets Outside of Sun itself, the planets are the major components of the solar system (Figure 16-4). Jupiter alone contains two-and-a-half times more material than all of the rest of the planets, satellites, asteroids, and comets put together. Since the characteristics of the planets are directly related to their respective distances from Sun, we will tour the solar system from the inside out — from the nearest planets to Sun to the most distant. So you may want to memorize their names in order of their distances to Sun, as you may have done in elementary school:

> "My Very Elderly Mother Just Served Us Nine Pizzas."
> Mercury Venus Earth Mars Jupiter Saturn Uranus Neptune Pluto

Moons There are at least 61 moons or satellites in the solar system. You notice that throughout the book I use the word "satellite" to refer to bodies orbiting planets. Planets revolve around Sun — satellites revolve around planets. Moon, therefore, is Earth's natural satellite. We humans have placed artificial satellites in orbit around Earth as well as a few of the planets, so we need to be clear when we refer to a satellite as to whether we mean a "natural" or "artificial" satellite.

It is tempting to think of the satellites and debris of the solar system as minor components and therefore of little interest to astronomers, but don't entertain that notion for long. Some of the satellites are actually more interesting than a couple of the planets — at least they have features that reveal a more dynamic and dramatic past history. Pre-

Figure 16–2 Continued work on the International Space Station Alpha *indicates humankinds drive to explore the new vistas of space. Since it is designed to be not only an Earth–monitoring station but a way–station for exploration of other planets and beyond, it is the next logical step into space.* (NASA)

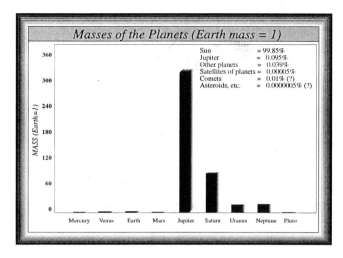

Figure 16–3 This diagram illustrates the breakdown of matter in the solar system as regards where it is located. Sun certainly has the lion's share.

sumedly, the satellites formed in orbits around the planets just as the planets formed in orbits around Sun.

Comets As every school child knows, Pluto is the most distant object in the solar system—right? Wrong. Surrounding Sun and planets in a huge spherical cloud (called the **Oort Cloud**), whose radius is half the distance to the nearest star, are countless chunks of icy material called **comets**. At that distance, they are unobservable, since they are at most a few miles in size. A comet is observable only when it leaves the cloud and enters an orbit that brings it close enough to Sun to develop a long "tail." The tail may then reflect enough sunlight to be observed from Earth's surface.

Unfortunately, although an average of 10 to 15 comets pass around Sun each year, seldom is one bright enough to be seen without the use of a telescope. That is why—in many people's minds—comets are not thought of as being members of the solar system. In fact, a good percentage of people think of them as objects that flash across the sky, rather like meteors.

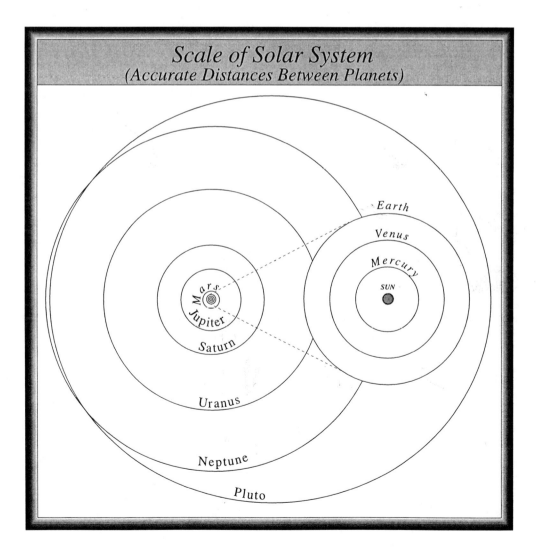

Figure 16–4 The solar system to scale with respect to distances from Sun and one another. Mostly the volume of the solar system consists of space. It is not entirely empty—solar wind particles, gas and dust particles and of course radiation travel throughout.

CHAPTER 16 / The Origin of the Solar System

✮✮✮✮✮✮✮✮✮✮✮✮✮✮

Figure 16–5 Comets are as much members of the solar system as planets. But they are normally placed too far away from Sun to be visible to the naked eye. comet Bennett, shown here as it appeared in 1970, was an exception. (Lick Observatory)

For many school children, only Halley's comet is associated with the solar system (Figure 16-5). The other comets come from that vast "void." This is because Halley's comet is a returning comet — one that returns to the inner portion of the solar system every 76 years. For that reason, astronomers refer to it as a **periodic comet**. Periodic comets are common, but few ever approach close enough to Sun to develop large enough tails to be visible to the naked eye. Comets represent an important clue as to the solar system's origin. In addition to their sometimes sudden arrival and appearance in the inner solar system, they also show evidence of disintegration and disappearance.

Asteroid Belt Located generally between the planets Mars and Jupiter is planet number 10 — the planet that failed to form. Or at least that is one way of theorizing the presence of the asteroids there. It makes sense. The Titius–Bode rule suggests that there should be a planet at that distance from Sun. And the proximity of Jupiter's immense gravitational field would certainly explain why the "missing" planet was unable to form from the debris orbiting the early Sun.

The name asteroid means "starlike" because of its dot-like appearance in a telescope. A photographer can easily determine if a suspected dot is a star or an asteroid by taking a long time-exposure while the telescope to which a camera is attached is set to track the stars. If an asteroid is in the field of view of the telescope, it leaves a streak on the photographic plate, since it moves at a different rate and in a different direction than Earth's spin.

In fact, this is just the method used to discover new asteroids (yes, you too can discover an asteroid!). Some 5,000 asteroids have been detected and numbered (or named), and it is thought that perhaps there are an additional 100,000 or so yet to be sighted. The reason is simple — they are mostly small bodies, since even the most powerful telescope on Earth reveals each as a dot. If they were sizeable, they would show up as disks on photographic plates.

Solar System Survey – Appearances

Movements Let us imagine viewing the solar system from a point far above Earth's North Pole (Figure 16-4). Over a period of time we observe a very obvious pattern in the movements of almost all the contents. The predominant motion is *counterclockwise*. All of the planets and asteroids revolve in a counterclockwise direction around Sun. Sun rotates in a counterclockwise direction, as do all the planets except *Venus* and *Uranus*. The asteroids — at least those few that are large enough to be observed in detail — rotate as well, but there is no obvious pattern or general rule.

Some of the satellites of the solar system deviate from the norm as well, either by virtue of backward rotation or backward revolution or both. One of Saturn's outer satellites — *Phoebe* by name — is one of them. It revolves in a clockwise direction — opposite to that of Saturn's rotation. We say this satellite has **retrograde revolution**.

A satellite of Neptune – *Triton* – also exhibits retrograde revolution. The ring systems of Jupiter, Saturn, Uranus and Neptune all revolve in the usual counterclockwise direction. In general, however, with so few exceptions to the counterclockwise direction of motion in the solar system, it would seem that this pattern is an important clue as to its origin.

The planets and asteroids are in well-defined elliptical orbits, as Kepler first determined in 1606. But you notice in Figure 16-4 that the shapes of these elliptical orbits are not all the same — some are more elongated than others. For example, the orbit of the planet Pluto is rather squashed in shape, causing it to cross the orbit of Neptune in two locations.

You can now understand why Pluto is not always the most distant planet in the solar system! In fact, at the moment, it isn't. Pluto moved inside of the orbit of Neptune in 1979, and will remain there until 1999. Knowing this, you might wonder if Pluto and Neptune will ever collide, since their orbits cross one another. The probability of such an event is extremely small.

Using basic laws of motion, astronomers have com-

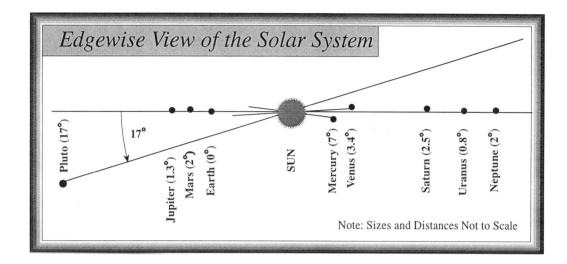

Figure 16-6 *This diagram illustrates the appearance of the solar system from a point above Earth's north pole (bottom), and from along the edge. The direction of revolution for all planets is the same. And mostly the solar system is flattened. Pluto's orbit is the only one that is obviously unique.*

puted the movements of the two planets' positions backward and forward through time, and have concluded they have never been at the same point of intersection at the same time within the 5-billion-year history of the solar system, nor will they ever be within the expected life of Sun. This conclusion is made less surprising by another unusual feature of Pluto's orbit — its large angle of inclination with respect to the ecliptic. To visualize that, zip to a different viewing position, looking at the solar system from along the edge, along the place of the ecliptic (Figure 16-6).

Flattened Shape We performed this very feat back in Chapter 3, except at that time we viewed the movements of the planets from Earth's surface. We observed the movements of the planets generally along the ecliptic, movement with respect to the background of the stars. Now I am asking you to view the solar system from some distant position outside of its boundaries. From this position along the edge, another pattern is obvious — the flattened shape of the solar system.

Again, most of the planets are very close to the ecliptic. But Pluto stands out like a sore thumb — its orbit takes the planet from 17° above the ecliptic to 17° below the ecliptic. So in order for a collision to occur between Pluto and Neptune, not only would they both have to be at the point where their orbits cross, but Pluto's steep orbit would just have to be cutting across Neptune's flattened orbit at the same time.

As before, I use the satellite system of Saturn to illustrate the general orientations of orbits of satellites in the solar system (Figure 16-7). Two of its satellites, *Iapetus* and *Phoebe*, move along orbits which cause them to move alternately far above and far below Saturn's <u>equatorial</u> plane. Their steep-angled orbits are rather like that of Pluto's around Sun.

Now this is rather interesting, because we expect the satellites of a particular planet to orbit along the equatorial plane of the planet, not just along an arbitrary orbital plane. In other words, we have reason to believe that the satellite system of a planet should behave like a miniature solar system (recall that the planets revolve generally above Sun's equatorial plane). The fact that some satellites in the solar system deviate from this expected pattern demands explanation.

Presumably, the satellites formed in orbits around the planets just as the planets formed in orbits around Sun. Certainly diagrams of the satellite systems support that hypothesis. The outer eight satellites of Jupiter, like *Iapetus* and *Phoebe*, move in irregular orbits inclined at varying angles to Jupiter's equatorial plane. Astronomers therefore categorize two types of satellite systems in the solar system — the **regular satellites** and the **irregular satellites**.

Irregular satellites are those whose orbits are either retrograde or are inclined at a steep angle with respect to the equatorial plane of the parent planet. Using these criteria, the irregular satellites of the solar system are the eight outermost satellites of Jupiter, the two outermost satellites of Saturn, the two outermost satellites of Neptune, and Earth's Moon (see Figure 18-16). All others are regular. Our interest in classifying satellites according to this characteristic is that it provides clues to their formation early in the history of the solar system.

Solar System Survey – Distances

In order to express the distances between objects in the solar system, I could try to impress you with large numbers — the average distance between Sun and Earth is 93 million miles, for example. But for me personally, I like to think of those distances in terms of the time required to go from one of those objects to another.

Now of course, that means we have to think in terms of speed. And so I do. The speed of light is one unit of measurement. And another is the speed at which satellites (personnel and un-personnel alike) travel as they leave Earth to visit components of the solar system. We learned earlier that light requires 8 minutes to travel between Sun

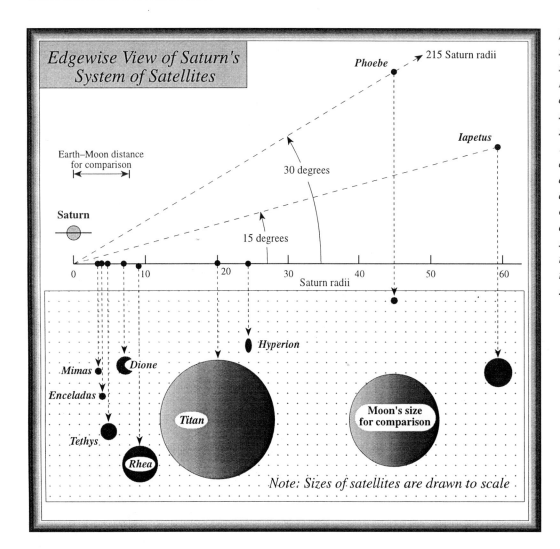

Figure 16–7 *Saturn's system of satellites is a solar system in minature. Its appearance edgewise, as in this diagram, is similar to that of the solar system in Figure 16–6. The satellites that deviate from the normal are irregular satellites and are probably captured asteroids. Some of the satellites are shown to scale in size so that you can appreciate the range of sizes in the solar system.*

and Earth. Radio signals sent to Earth from a satellite in the vicinity of Neptune require about 4 hours to make the journey. Sending astronauts to Moon required a couple of days. To send a team to the surface of Mars will require about 9 months.

These distances really don't tell us anything directly about the architecture of the solar system or even hint at its origin. It is when we compare the distances one to another that an interesting relationship appears. Study the progression of numbers in Figure 16–8. You notice that we can obtain the approximate distances of the planets away from Sun in terms of Earth's distance (the AU) simply by playing around with an arbitrary arithmetic progression. Or is it arbitrary?

Actually, this "rule" for the solar system is referred to as the **Titius-Bode rule** for the two Eighteenth Century scientists who first pointed it out in the year 1772 (Figure 16–8). In fact, it came as no great surprise when in 1801 the first asteroid was discovered because it conformed exactly to the "predictions" of the Titius-Bode rule. There is no scientific basis for the rule. That is, there is no reason to believe that the planets had to form in orbit around Sun at their present distances from Sun. Astronomers know of no principles of physics that deal with that aspect of the solar system's formation. But it certainly is curious. Perhaps there is some yet-to-be discovered principle of physics that operates under conditions of a collapsing cloud of gas and dust that dictates the spacing between globs of matter. And the *Titius-Bode rule* is just an approximation of that principle.

Solar System Survey – Chemistry

We would lack rather important clues about the origin of the solar system if we failed to consider the chemistry of its various components. This is quite obvious when we study a scaled-down diagram of the planets according to their sizes. In Figure 16–9, all of the planets are placed according to their sizes relative to that of Sun. The comets (without their tails) and asteroids are so small on this scale that I have chosen to ignore them. Although a few of the satellites of the solar system are quite impressive, two of them actually being larger than the planet Mercury, I have also left them out since I will discuss them in detail in Chapter 19.

A casual glance at Figure 16–9 suggests that the solar

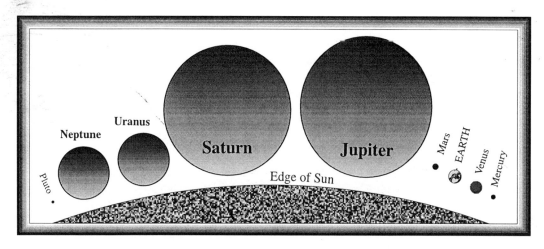

Titius-Bode Rule for the Distances of Planets from Sun

Planet	Column 1		Column 2		Column 3		Column 4		Bode Numbers	Actual Distances (AU's)
Mercury	0		4		4		10		0.4	0.39
Venus	3		4		7		10		0.7	0.72
Earth	6		4		10		10		1.0	1.0
Mars	12	+	4	=	16	÷	10	=	1.6	1.52
-	24		4		28		10		2.8	-
Jupiter	48		4		52		10		5.2	5.20
Saturn	96		4		100		10		10.0	9.54
Uranus	192		4		196		10		19.6	19.18
Neptune	384		4		388		10		38.8	30.06
Pluto	768		4		772		10		77.2	39.44

Figure 16–8 A simple arithmetic scheme approximates the distances of the planets from Sun. Known as Bode's rule, it may in some crude fashion illustrate a subtle mechanism that determined the distances to the evolving planets when the solar system was quite young.

Figure 16–9 The planets of the solar system scaled according to size.

system consists of two groups of planets. The **terrestrial** (Earth-like) planets are close to Sun, are small, have very few (if any) satellites, and chemically consist mostly of metals (elements heavier than helium). The **Jovian** (Jupiter-like) planets are distant from Sun, are large, have many satellites, and chemically consist mostly of hydrogen, helium, and a few other lighter elements.

Densities of the Planets In fact, there is an interesting and definite pattern in the chemical makeup of the planets if they are compared according to their average densities — mass per unit volume. Those with the smallest densities obviously consist of the lightest chemical elements, and vice versa. The chart in Figure 16-10 reveals the pattern – *the closer a planet is to Sun, the higher is its density.*

At least it is true for planets out to Saturn. Initially, one would think that if the entire solar system collapsed from an interstellar cloud of gas and dust, all of the planets should be similar and consist of the average composition of that original cloud. A simple experiment explains why that should not be the case.

Imagine yourself on each of the planets in turn, looking up at Sun in the sky. Figure 16-11 represents the sizes of Sun as seen from each of the planets. Use your familiarity with Sun in the sky as a comparison. Notice the dramatic change in the apparent size of Sun as you hop from the surface of Mars to that of Jupiter. Since the amount of

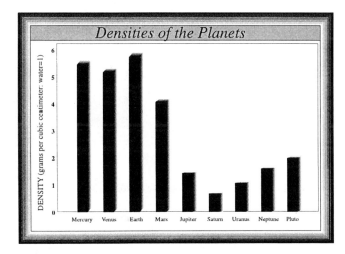

Figure 16–10 For the terrestrial planets, at least, there is signifcance to the pattern of their densities with respect to their distances from Sun. If we compensate for Earth's increased density due to size, we find that the closer a planet is to Sun, the greater its density.

CHAPTER 16 / The Origin of the Solar System **343**

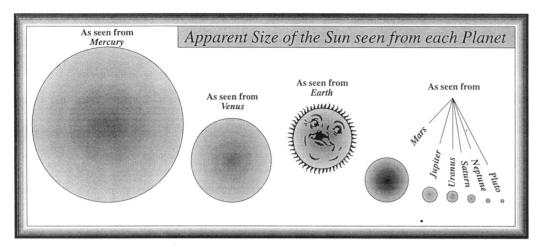

Figure 16–11 Much of the differences between the planets in the solar system can be easily understood when we consider the size of Sun in the sky as seen from their surfaces. Obviously, the larger Sun appears in the sky the hotter it is on the surface.

sunlight falling on a planet, which determines its temperature, depends on the inverse-square law, planets close to Sun are expected to be warmer than those distantly located.

Assuming that all of the planets began forming with the same average chemical makeup, it is easy to imagine why they have ended up so differently. It is rather easily demonstrated by boiling some water from your tap in a clean container. If you boil all of the water away, you notice a thin film of residue on the bottom of the container. This scum consists of the minerals and impurities that are in the water but that could not be vaporized and boiled away by the limited heat output of your stove.

During the process of refining crude oil, the oil is heated to higher and higher temperatures. Each of the vaporized substances that boils away at a particular temperature is collected as some type of product. This is called **fractional distillation**. A similar process must have occurred during the early formation period of the solar system.

The terrestrial planets had most of their lighter chemical elements boiled away, leaving behind those not easily vaporized — the heavy elements beneath our feet. On the other hand, not only did the Jovian planets retain their original chemical elements, thereby making them chemically like Sun, but they also collected some of the lighter elements being blown outward from the terrestrial planets by solar radiation and the solar wind. The large masses of the Jovian planets attest to this. This means, of course, that the solar system began with more mass than it presently has. Based upon the condensation sequence (Figure 16-12), the estimated ratios between the original and present planetary masses are as shown in Figure 16-13.

Solar System Debris: Comets

Comets as Messengers of Doom? Historically speaking, comets have aroused fear and anxiety in people, and have been interpreted as omens of doom, disaster, and terrible things to come. When Halley's comet appeared in 1910, many people feared for their lives when they learned that Earth would be passing through the comet's tail known to consist of the poisonous gas *cyanogen*. Their fears were abetted when some enterprising people offered relief by selling them "comet pills"!

Don't laugh, now. Both in 1973 — when comet *Kohoutek* encountered Sun — and 1986 — when comet Halley was last here — religious organizations predicted that the two comets were here to announce the "Second-Coming of Christ." They predicted great catastrophes for Earth! Members of the organizations distributed leaflets around classrooms at a famous West Coast community college just as the two comets mentioned were making their appearances.

The Condensation Sequence
(Temperatures at which Substances Condense)

Temperature (degrees F)	Condensate	Planet (Estimated temperature at formation: degrees F)
2,240	Metal oxides	Mercury (2,060)
1,880	Metallic oxide, nickel	
1,700	Silicates	
1,340	Feldspars	Venus (1,160)
764	Troilite (FeS)	Earth (620)
		Mars (350)
–144	H_2O ice	Jovian (–144)
–190	Ammonia-water ice	
–243	Methane-water ice	
–342	Argon-neon ice	Pluto (–342)

Figure 16–12 Experimenting in the laboratory, astronomers can calculate which chemicals will "boil away" and which will remain behind in a cloud of debris surrounding the birthing Sun. The condensation sequence, then, verifies our theory that the differences between the planets has much to do with distance from Sun. The innermost planets are very dense because the lighter substances were "boiled" away.

| Masses of Nebular Matter Required to form Planets ||||
Planet	Assumed composition	Present mass (kg)	Ratio of required to present mass
Mercury	Silicates, Iron	3×10^{23}	410
Venus	Silicates, Iron	5×10^{24}	380
Earth	Silicates, Iron	6×10^{24}	380
Mars	Silicates, Iron	6×10^{23}	370
Asteroids	Silicates, Iron, (ices?)	$\sim 1 \times 10^{21}$	250
Jupiter	Hydrogen, ices	2×10^{27}	10
Saturn	Hydrogen, ices	6×10^{26}	16
Uranus	Methane, ammonia, water, ices	9×10^{25}	67
Neptune	Methane, ammonia, water, ices	1×10^{26}	64
Pluto	Ices, methane, (silicates?)	7×10^{23}	75
Comets	Methane, ammonia, water, ices	$>1 \times 10^{27}$	5

Figure 16–13 According to the mechanism of fractional distillation, the planets began to form around Sun when they were considerably more massive than they are today. The lighter elements were boiled away. Depending upon distance from Sun, a ratio between starting and finishing mass can be calculated for each planet.

If you were to read the brochure associated with Halley's comet, you would notice that it provides evidence of catastrophes have occurred on Earth with each passage of the comet. One such disaster is recorded on the famous Bayeaux tapestry, on which is depicted the sequence of events that led up to the defeat of King Harold by William the Conqueror in the year 1066. Halley's comet appeared over the battlefield, and (apparently) evoked enough fear amongst Harold's soldiers as to affect their morale and to lead to their defeat (Figure 16-14). Of course what was bad for Harold was good for William, so it difficult to understand how one could argue that the comet caused some type of catastrophe. Museums throughout Europe possess numerous drawings and paintings that reflect the fear of comets appearing in the sky.

Fear of comets goes back at least to the time of medieval Europe, at which time the prevailing belief about the universe was that it consisted of two separate hemispheres — the universe above our heads and Earth beneath our feet. This was a major theme inherited from Aristotle and elaborated on by the Church. Objects in the sky were eternal and unchangeable — their behaviors repetitive and predictable. Objects on Earth were the very opposite. So when a comet appeared, it had elements of both worlds — it came from the sky, but it came without warning.

Comets were totally unpredictable as regards arrival in the sky. So the idea developed that since the sky was the domain of God, the comet was obviously sent by Him/Her as a message to people on Earth. The obvious message was the one mentioned in both the Book of Matthew (24: 29-30) and the Book of Luke (21: 11) — the Second Coming of Christ will be preceded by "signs in the sky." Because comets arrived from the sky without warning, they met that criteria perfectly.

There are two misconceptions about comets that relate to this drama. First of all, comets do not "streak" across the sky. They are seen for weeks and even months. So when we learn that Halley's comet appeared to the armies of Harold and William, there was plenty of time for fear to set in — probably for months.

Second of all, Halley's comet was not "discovered" by Edmond Halley — he just happened to be the first person to propose that at least one comet returns to the sky with regularity approximately every 76 years. He was a contemporary and friend of Sir Issac Newton, and he fully appreciated the power of Newton's laws of motion. Halley studied the historical records of comet appearances and noticed one particular regularity — the orbits of three unusually bright comets (in the years 1531, 1607, and 1682) were quite similar. He suggested that all three were one and the same comet.

Knowing the 76-year period required to orbit Sun, he used Kepler's third law to determine the distance of the comet from Sun — he found it to be 18 AU. But to be as bright as it was, Halley knew that it had to be close to Sun on a highly "squashed" elliptical orbit. That meant that the comet would — at greatest distance from Sun — reach almost as far out as the planet Uranus. This type of orbit had never been observed before for any object in the sky. Confident of his theory, he predicted the comet's return in 1758. Unfortunately, he died before he could confirm his theory with observation, but the comet was named after him when it did arrive as predicted — 16 years after his death.

Origin of Comets The arrival of a comet is not as unusual as one would think from the amount of attention that the media gives to them. But very few ever get bright enough to be visible to the naked eye. So astronomers have all the fun as far as observing comets is concerned. And most are onetime visitors to the inner portion of the solar system. We observe them once — when they are close to Sun — and then they vanish into the void of space, never to be seen again.

These particular comets travel on open, parabolic orbits rather than closed, elliptical orbits. Occasionally, when a comet makes its brief run to the inner part of the

★★★★★★★★★★★★★★

Figure 16-14 The Bayeaux tapestry depicts a major battle between rival armies in the year 1066, during the same year Halley's comet made an appearance. Its appearance over the battlefield—depicted in this particular section of the huge tapestry—was interpreted as a sign of bad luck for the loser of the battle. (Tom Bullock)

solar system, it may pass close enough to one of the larger planets (Jupiter, for example) to be gravitational disturbed and captured permanently within the inner solar system. Henceforth it is a *periodic comet* (Figure 16-15).

Halley's comet is a periodic comet. Its 76-year orbit brings it inside of Earth's orbit around Sun, and outside as far as Uranus. Conforming to the principle of Kepler's second law (refer back to Figure 4-13), it moves considerably faster when it is near perihelion than when it is traveling near **aphelion** (most distance point from Sun). Thus we see it for only a few brief months during its long orbital period (Figure 16-17).

Notice, however, that its orbit is not in the ecliptic plane – it moves within those regions of the solar system that the planets never get to visit. This feature of Halley's comet is not unusual. In fact, the observation that comets approach Sun from every conceivable angle – both direct and retrograde – is good evidence that comets come from a reservoir somewhere.

Solar radiation and solar wind particles create tails in a comet as it swings close to Sun, but that eventually spells doom for the comet. The tails are permanently lost to the solid "head" of the comet after it passes around Sun. The material that is "boiled" off the head to form the tail during close approach does not get pulled back into the solid head.

But neither does it just drift off into space. It remains in the original orbit around Sun, separated from the head. It is just no longer concentrated within the larger body of the comet. A small fraction (~1%) of the comet is lost to the length of the orbit at each passage around Sun. This fact suggests that comets are not eternal, that they gradually deteriorate.

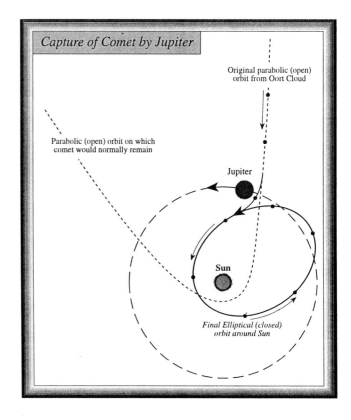

Figure 16-15 Comets "invade" the inner solar system all the time, but only if they are gravitationally disturbed by a giant planet do they become periodic or returning comets.

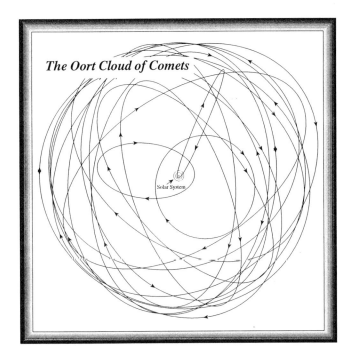

Figure 16–16 Comets are believed to originate in a vast cloud of comets that circle Sun at a distance half that to the nearest star. Their orbits must be completely random as far as direction and angle of inclination to the ecliptic plane are concerned. This is called the Oort Cloud.

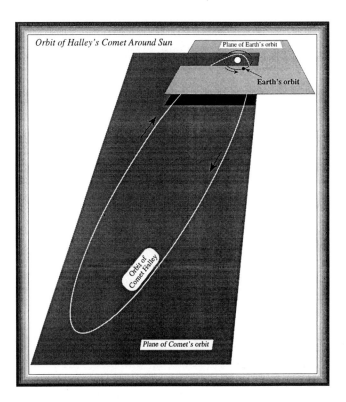

Figure 16–17 The calculated orbit for Halley's comet reveals that it is on a very squashed elliptical orbit that is highly inclined to the flattened plane of the solar system. In fact, its orbit is retrograde when compared to the orbits of the planets.

Therefore, a fresh supply of comets must be available in a region somewhere out beyond Sun. How else can we explain the discovery of 10 to 15 new comets every year! To date, some 1,000 comets have been observed and catalogued. Their orbits suggest that they originate in (come from) a spherical cloud that surrounds Sun at a distance equal to about one-half the distance to the nearest star (approximately 50,000 AUs). The existence of such a cloud would also explain why comets approach the solar system at steep as well as shallow angles with respect to the ecliptic, as well as in retrograde motion — as in the case of Halley's comet. The astronomer who first proposed this hypothesis was Jan Oort (Netherlands). Hence the name *Oort Cloud* is given to the hypothetical cloud from which comets are thought to originate (Figure 16-16). The Cloud cannot be detected from Earth. The small sizes of comets — less than 10 miles, in most cases — makes them entirely undetectable beyond 10 AU's or so.

Why, you are asking yourself, would a comet leave the Oort Cloud to approach the inner solar system in the first place? Any object located at a distance of 50,000 AUs from Sun is in an unstable region, inasmuch as passing stars could easily perturb its orbit around Sun. And that is exactly what astronomers propose happens to cause some comets to leave the Oort Cloud to begin their excursions into the inner regions of the solar system. Once they swing around Sun, they return to the Oort Cloud never to return, unless they happened to encounter a major planet while close to Sun and are captured to become a periodic comet such as Halley's.

This traditional explanation for the existence of periodic comets is currently being challenged by a newer theory that proposes that they actually come from a belt of comets that formed closer to Sun just beyond the orbit of Neptune. Computer simulations of comets passing by one of the massive planets revealed that they would not be captured into the orbits we observe. This additional reservoir of cometary objects is called the **Kuiper belt**, named after the astronomer who first proposed the theory in 1951 (Figure 16-18). It is thought that the Kuiper belt is disk-shaped and that comets contained therein share the general east-to-west rotation of the solar system.

Prior to the upgrading of the HST, astronomers were unable to detect objects in the belt simply because of their great distances from Sun (and therefore their inherent dimnesses). But in 1995, the HST detected 30 faint, comet-like objects that have the necessary characteristics to qualify them as Kuiper belt comets. Based upon their distribution in space, astronomers estimate that 1 to 10 billion of these objects inhabit a flat disk beyond the orbit of Neptune. Pluto is also embedded in the disk, which makes it the largest inhabitant of the disk. We will soon find that the Kuiper belt theory for the origin of periodic comets also provides a better explanation for the origin of a few

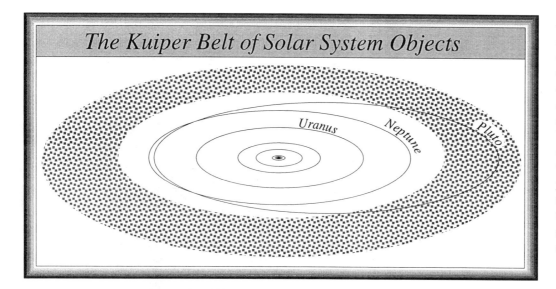

Figure 16–18 The Kuiper Belt is proposed to explain the presence of icy dwarfs discovered between the orbits of Neptune and Pluto and out beyond Pluto's orbit. This may explain the origin of periodic comets whose periods are 20 years and less, since they, unlike those from the Oort Cloud, are mostly confined to the ecliptic plane.

strangers in the outer solar system: *Triton*, an icy satellite of Neptune, Pluto, and Pluto's single satellite *Charon*.

Comet Structure The head of a comet consists of an icy planetesimal (the **nucleus**) surrounded by gases (the **coma**) (Figure 16-19). The nucleus is a glob of ice and dust that accumulated from the solar nebula some 4.5 billion years ago. It is thought to be a sample of the basic building blocks out of which the planets and satellites first formed. In that sense, the nucleus of Halley's comet is the Rosetta Stone of solar system astronomy. Knowing its structure and chemical composition allows astronomers to understand the processes by which the solar system formed from an interstellar cloud. Up until the time the *Giotto* satellite encountered this peanut-shaped sample in 1986 (and obtained close-up photographs), astronomers had been limited to Earthbound telescopes for observing comets. The resolutions of even the best telescopes, however, did not allow for close-up study of the nuclei of any comet. Besides, when comets are closest to Sun — and therefore easiest to observe — the solid nucleus is engulfed in the very dust and (mostly) gas that sunlight has "boiled" off its surface. That gas and dust prohibits telescopic observation of the nucleus within. This cloud is the coma, which consists mostly of neutral gas atoms, but also some ionized gas atoms.

Surrounding the head (nucleus + coma) at a great distance is a hydrogen cloud a million miles or so in diameter. This cloud probably forms from the breakup of water molecules in the nucleus by solar ultraviolet radiation. If the nucleus of a comet is the size of a peanut, then its coma is about 1,000 feet wide. And the greatest extent of the hydrogen cloud is about 20 miles across. While a comet can be huge in dimension, especially in length, the amount of matter in the extended tail and hydrogen cloud is quite small (Figure 16-19).

Comet Tails During most of its orbit around Sun — certainly during the time it remains in the Oort Cloud — a comet consists of nothing more than the solid nucleus. The small percentage of comets that experience close passage around Sun are subjected to the conditions necessary for creating tails and comas. By the time a comet arrives at Mars' distance from Sun, solar radiation has caused some of the ices in the nucleus to become gases. Since dust is mixed in with the ices, it is released along with the gases. Earlier, I used the word "boiling" to describe this process. But it is not "boiling" in the sense that we are used to thinking of it.

The process of a solid material going directly to the gaseous state with the application of energy (heat) is called **sublimation**, something you have undoubtedly observed yourself. If you have ever observed dry ice (frozen carbon dioxide) at room temperature, you probably noticed that the solid material was evaporating — going directly from the solid state to the gaseous state without going through the liquid state. This is the very process that occurs on the nucleus of comets as solar radiation energy hits it. Carbon dioxide is a major component of the icy nucleus.

Two solar mechanisms now act on the gas and dust that accumulates around the solid nucleus. Solar radiation, in addition to sublimating some of the material in the nucleus, acts on the dust particles by *pushing* them away from the nucleus. Radiation consists of bundles of energy, and anything that absorbs energy moves in a direction away from the source of energy. Sunlight falling on the dust particles pushes them away from the nucleus in the direction opposite to that of Sun. This explains the observation that tails of comets always point away from Sun.

In a similar manner, the solar wind particles (mostly electrons and protons) interact with the gas ions in the coma and "push" them into a gas tail behind the solid nucleus. Just why the gas tails of comets always point away

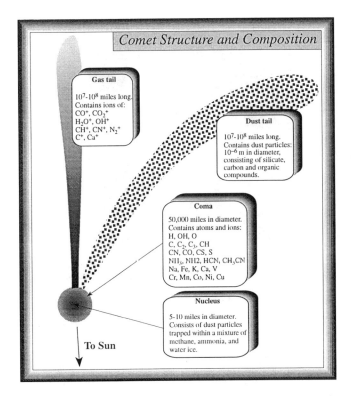

Figure 16–19 This diagram illustrates a model of a typical comet, showing structure and chemical composition. Note the direction to Sun.

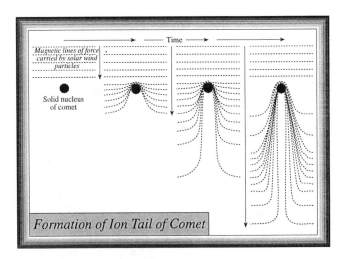

Figure 16–20 The gas (ion) tail of a comet is formed by the interaction between the charged particles in the solar wind and ions in the coma of the comet.

from Sun was not understood until the nature of the solar wind was understood. Astronomers now understand the behavior of charged particles when they encounter other charged particles — such as occurs in the auroras — and can therefore explain the behavior of the gas tails of comets (Figure 16-20). Since like-charged particles repulse one another, the charged particles that make up the solar wind repulse ("push") the charged particles (ions) in the coma. Therefore comets frequently display two tails, an **ion tail** (electrically-charged particles) and a **dust tail** (Figure 16-21).

Chemistry of Comets Earlier we learned that determining the chemical composition of a cloud of gas-emitting photons is fairly straight forward (emission spectrum). So too is determining the chemical composition of a cloud of hot gas located between a source of radiation and ourselves (absorption spectrum). But planets reflect light.

The analysis of sunlight reflected off Mars, for example, reveals the absorption spectrum of Sun's atmosphere plus absorption lines due to the photons selectively absorbed by minerals and/or gases on the surface of Mars. To some extent, it is possible for astronomers to obtain an analysis of the chemicals present on reflected surfaces.

Since some of a comet's structure consists of dust particles that have solid surfaces, some of the light that defines the shape of a comet is reflected off the dust particles. Comets also emit radiation when they are close to Sun, since their gases are heated up by solar radiation and are thereby caused to emit wavelengths of radiation according to the gases present. Astronomers estimate that approximately 15% of the light by which we observe comets is emitted light, and the other 85% is reflected sunlight. Comets reveal their chemical makeup through emission spectra (Figure 16-22).

Such studies reveal that a comet is a frozen snowball of mostly water ice and carbon dioxide (dry ice), with a sprinkle of meteoric dust thrown in. It is this dust that is eventually responsible for meteors in our nighttime skies. So a commonly heard description of a comet is that it is a "dirty snowball." Considering what we know about the abundances of the chemical elements in the universe, the chemistry of comets is about what we would expect if they indeed are samples of the material from which the solar system formed.

The frozen nucleus of comet Halley consists of the gases H_2O (80%), CO_2, CO (carbon monoxide), methane, ammonia, and meteoric material similar to that we find inside of meteorites that fall to Earth's surface. The surface of the nucleus is covered with a very black layer of that dust, reflecting less than 4% of Sun's radiation. Analysis of the dust grains shows two major groups. The organic grains, which consist of the light elements (hydrogen, carbon, oxygen, and nitrogen) contrast with those containing the mineral-forming elements (iron, silicon, calcium, sodium). There is also evidence of hydrocarbons (combinations of hydrogen and carbon atoms) in the comet.

Some scientists report evidence that Earth is bombarded by about 20 house-sized comets every minute. Microwave observations of the normally dry upper atmosphere have detected brief water vapor "puffs" that appear to be the tracks of the comets as they enter the atmosphere. With a consistency of fluffy snowballs, the ice vaporizes as it is heated up by friction with the atmosphere.

CHAPTER 16 / The Origin of the Solar System   **349**

Figure 16–21 (above) This classic photograph of Halley's comet was taken at Cerro Tololo in Chile on March 6, 1986. Make note of the gas (right) and dust (left) tails. (National Optical Astronomy Observatories)

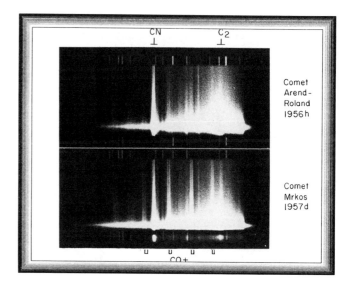

Figure 16–22 The chemistry of comets is determined by the spectra of the feeble light emitted as atoms in the gas tail are excited by sunlight. (Lick Observatory)

The water vapor eventually finds its way into the water cycle on Earth's surface. If this frequency (10 million per year) has been consistent over the 4.6-billion-year history of Earth, then comets could account for all of the water in our oceans and polar ice caps. It could even have played a critical role in the evolution of life on Earth. Remember that when you next go for a swim!

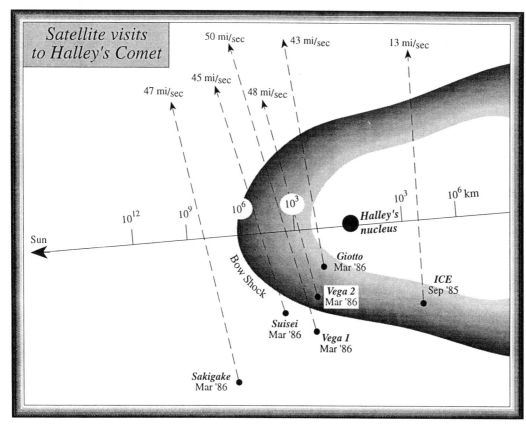

Figure 16–23 An armada of satellites from several countries took advantage of Halley's comet return in 1985–86 to study a comet up close for the first time. This diagram shows their trajectories and distances from the cometary nucleus.

On the other hand, comets might have offered a threat to life on Earth as well. I will discuss a theory that attempts to link comets with the demise of the great reptiles later in the Chapter. But according to a recent computer analysis run by an astronomer, Earth has been saved by an excessive number of cometary collisions by the presence of Jupiter in the solar system. Although that massive planet can certainly capture comets and retain them in the inner solar system, by and large it generally hurls them back out into space, thereby preventing them from doing an excessive amount of damage. So it seems that perhaps life on Earth has always been teetering between threats by comets and assists from comets.

Visiting Halley's Comet Aside from the small percentage of people who retain ancient superstitions about comets, the return of Halley's comet every 76 years is always an occasion for great excitement. Planetariums that normally have sparse attendance at public shows are suddenly overwhelmed with people wanting to know how and when to see the comet. It was no different when it last appeared in 1985-86. Commemorative stamps from countries around the world drew public attention to it. Tee-shirts, buttons, coffee mugs and of course viewing instruments of every description were also used to stimulate the world's economy. Many people travelled to destinations in the southern hemisphere in order to get a better view of it there.

Inasmuch as the Oort Cloud is believed to be the unadulterated remains of the original interstellar cloud from which the solar system formed, it stands to reason that astronomers are interested not only in the orbits and behavior of comets, but their chemical compositions as well. After all, knowledge of their chemistries could assist in our understanding the conditions that are necessary for the formation of planetary systems in general. Although there is no reason to believe that Halley's comet is unique amongst the comets, because its recent swing around Sun allowed scientists to engage it with the widest range of sophisticated technologies ever trained on a comet, Halley's comet is now our model for comets in general.

A worldwide organization (*International Halley Watch*) was established in 1979 to coordinate the numerous observations that would be made from different geographical locations on Earth's surface. Because the orbit of Halley's comet is inclined 162 degrees with respect to the solar system's ecliptic plane (Figure 16-17), and because the comet moves in a retrograde manner, observers at different locations had better views of the comet at different times during its closest approach to Sun.

In addition to the ground-based observations, three space agencies — the European Space Agency (ESA), the Soviet-led consortium *Intercosmos*, and the Japanese Institute of Space and Astronautical Science (ISAS) — sent spacecraft to intercept the comet. The ESA sent **Giotto**, Intercosmos sent *Vegas 1* and *2*, and Japan sent *Suiei* and *Sakigake*. The United States, with the most advanced and

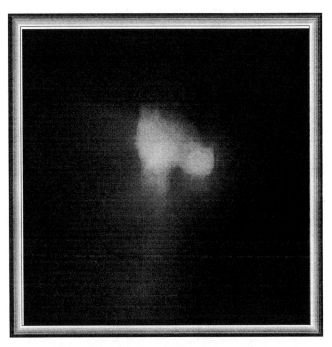

Figure 16–24 *The Giotto satellite obtained this first–ever photograph of the actual solid portion (nucleus) of a comet, in this case Halley's comet. Notice the jets of gases sublimating from the solid portion. Those gases eventually were influenced by the solar wind to form the gas tail of the comet. (NASA)*

sophisticated space program in the world, sent nothing. Giotto was launched so as to study the nucleus of the comet close-up (within 350 miles!), whereas *Sakigake* was programmed to fly through the outer reaches of the comet's environment by a few hundred thousand miles. And, after encountering another comet for which it had been designed to study, NASA's *International Cometary Explorer* (ICE) was rerouted to the vicinity of Halley's comet to make observations from 20 million miles away (Figure 16-23).

In order to maximize the return of scientific information from the various spacecraft, all four space agencies joined into a communication network called the **Pathfinder Project**. Targeting a probe to engage a comet's nucleus that is hidden by gas and dust is a difficult problem. After the *Vegas* encountered the comet, Soviet scientists provided ESA with a more accurate position of the nucleus so that *Giotto* could fly within 375 miles of the nucleus of the comet. NASA's Deep Space Network of radio telescopes supported the effort by tracking the *Vegas* for the then Soviet Union. I am sharing with you some of the administrative details of scientific investigation so that you can appreciate the primary interest scientists have in obtaining knowledge about the universe above and beyond political, cultural, and language barriers.

No doubt the most exciting result from the *Giotto* mission, perhaps from the entire *Pathfinder* project, were the close-up photographs of jets of gases and dust sublimat-

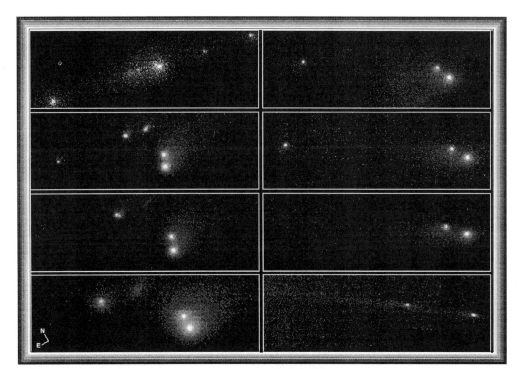

Figure 16–25 This series of eight HST "snapshots" shows the evolution of the P-Q complex, also called the "gang of four" region, of comet Shoemaker-Levy 9. The frames reveal changes in the comet during the 12 months before colliding with Jupiter. The sequence shows that the relative separations of the various cometary fragments, thought to range in size from about 500 meters to almost 3 miles across, changed dramatically over this period. The apparent separation of Q1 and Q2 was only about 680 miles in July 1993 and increased to 17,400 miles by July 1994. (NASA/STScI)

ing from the nucleus (Figure 16-24). Estimates are that the gas production taking place was about 30 tons per second during the encounter! For an object that is 10 miles by 5 miles by 4 miles in size, that is no great loss in mass. But it means the eventual doom of the comet. Once gas and dust is forced away from the nucleus to form the tail, it is lost to the nucleus forever. The gas and dust particles continue to move in their original orbit, but gradually drift further away from the nucleus through the interaction with solar radiation and solar wind particles. We must conclude, therefore, that the tail of a comet is mostly nothing. It may reach a length of millions of miles, but it is not substantial enough to affect an object (like Earth) that passes through it.

Even after comet Halley's closest approach to Sun in 1986, observers continue to track its retreat into the deep freeze of the solar system. In 1991, five years after its swing by Earth, an immense dust cloud erupted from the comet's surface, making it 300 hundred times brighter than it was supposed to be at that distance from Sun. During a routine check on the progress of the famous comet as it travelled out into the fringes of the solar system, astronomers noticed that it had sprouted a shiny dust cloud that extended for about 180,000 miles away from the nucleus. Since solar energy is thought to cause such outbursts, Halley's behavior at this great distance of 1 billion miles was rather surprising.

Giotto Visits Comet Grigg–Skjellerup Because of the great expense of building, launching, tracking, and processing the data returned by satellites, astronomers attempt to get as much mileage out of a given satellite as possible. So it was with the European satellite *Giotto* after it encountered comet *Halley* in 1986. After traveling for another six years, Giotto encountered its second target, comet *Grigg–Skjellerup*, in 1992. This comet approaches the inner solar system every five years, in contrast with comet Halley's 76-year period. The much larger, relatively younger, nucleus of *Halley* contains enough icy material to present a stunning celestial display, complete with bright, spectacular tail. On the other hand, frequent trips around Sun have taken their toll on *Grigg–Skjellerup*. Much of its icy nucleus has been lost to former tails of gas and dust material that has gradually dissipated throughout the elliptical orbit of the comet. It is much smaller and older in appearance than *Halley*. Its nucleus is somewhere between 200 meters and 2 kilometers across, and is surrounded by only a small coma of gas and dust. Most of the comet's icy materials have long since sublimated away.

Comet Shoemaker–Levy 9's Collision With Jupiter
During a routine search for new comets, an astronomer in 1993 noticed a "squashed" comet on a photograph taken at the Palomar Observatory. Observing it in greater detail, astronomers realized that the comet wasn't squashed at all, but that it consisted of individual fragments—21 in all, and orbiting the planet Jupiter rather than Sun. Researchers tracked the comet's orbit back in time and realized that an extraordinary event had occurred. The comet had skimmed a mere 12,000 miles above Jupiter's cloud tops during the previous year, and the gravitational forces on the solid body had shattered it into more than 20 large fragments and thousands of smaller ones (Figure 16-25).

Tracking it even further back in time, they calculated that Jupiter had captured the comet about 100 years ago. Named after the codiscoverers, comet **Shoemaker–Levy 9** (its fragments, of course) slammed into Jupiter's atmosphere in 1994, creating an opportunity for astronomers to gather valuable information about the dynamics and composition of that intriguing atmosphere. It was the opportunity of a lifetime for astronomers, inasmuch as such an event had never been witnessed before. Modern telecommunication systems such as the Internet allowed scientists as well as nonscientists to view the event as it happened or shortly thereafter. The twenty-one pieces of the comet, strung out into what appeared in telescopes like a "string of pearls", slammed into the atmosphere of Jupiter at a speed of 135,000 miles per hour. Since the fragments were up to 3 miles in diameter, there was an immense release of energy (up to 6 million megatons per fragment) during the collisions that occurred over a six-day period.

To astronomers, this was an opportunity to learn something not only about the behavior and composition of comets, but about the composition and dynamics of the atmosphere of Jupiter as well. Using the full resources of Earthbound telescopes as well as the HST and the Jupiter-bound satellite *Galileo*, astronomers accumulated an avalanche of astronomical imagery that will require at least a decade to fully analyze and digest. Since the impacts occurred on the nightside of Jupiter, only *Galileo* had a direct view of the moments of collision. The other instruments had to wait for Jupiter's rapid rotation to bring the impact sites into Earthbound view. At the moment of impact, Galileo observed 13,000-degree Fahrenheit fireballs, followed by rising columns of superheated jovian air that took the form of dark plumes larger than Earth in Jupiter's atmosphere.

Presumedly, each comet fragment was vaporized in the impact and its constituent molecules broken apart. Much of this material was detected in the dark plumes in the form of sulfur compounds, carbon monoxide and water — common constituents of comets. But astronomers are still uncertain as to what material from the depths of the jovian atmosphere were added to the plumes, or even how deep the penetrations took place before the comet fragments were entirely vaporized. Keep tuned as astronomers unravel the data and advance theory based upon it.

Now Comes Comet Hale-Bopp During July of 1995, two amateur astronomers independently discovered a new comet in the constellation of Sagittarius (Figure 16-26). Named after the codiscoverers, Comet **Hale-Bopp** promises to be the brightest comet to appear in Earth's sky in 20 years and, as such, will be a major event for astronomers and the general public alike until the Fall of 1997. Presently approaching its perihelion date of April 1997 (when it will be at its brightest at $m_v = -1.7$), the comet will be visible with the naked eye by August 1996.

With the observational data already available, astronomers have already concluded that Comet *Hale-Bopp* has some unusual characteristics. It is much larger than most comets that venture close to Sun. The envelope of dust that surrounds the solid nucleus measures well over 1.5 million miles, making it larger than Sun's volume. The solid nucleus, measuring dozens of miles — larger than the average comet — has already been observed squirting jets of sublimated gases through cracks in its surface as Sun's influence increases during the comet's gradual approach. It will approach to within 85 million miles to Sun at time of perihelion.

All indications are, therefore, that comet *Hale-Bopp* will develop quite a spectacular tail as it is a combination of solar energy and solar wind particles act on the solid nucleus and convert some of it to the gas and dust that will make up an increasingly long tail. Just how long and bright the tail will be to Earthbound observers can only be estimated at the moment. But all indications are that the prospects are good for a treat for all who make the effort to observe the comet.

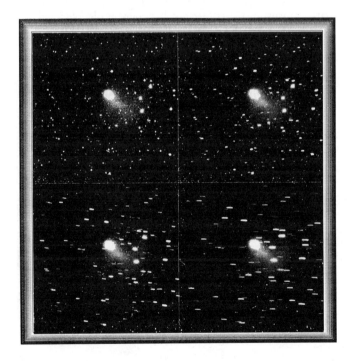

Figure 16–26 Comet Hale–Bopp *is rather typical of the sudden and unexpected appearance of a comet. First discovered independently in 1995 by backyard astronomers Alan Hale in New Mexico and Thomas Bopp in Arizona, it will put on its brightest show during April 1997. At that time, it is predicted that it will have an apparent visual magnitude of (–1.7).* (NASA/STScI)

★★★★★★★★★★★★★★

Solar System Debris – Meteoroids

Less than 200 years ago, the idea that meteorites — or anything else — could come to Earth from outer space was considered totally preposterous and unscientific. Thomas Jefferson, scientist and United States President, upon hearing that two Yale University professors had recovered and weighed 300 pounds of meteorites that had fallen at Weston, Connecticut, in December of 1706, declared that it was easier to assume that the professors were lying than that stones could fall from the sky.

During the past few years, there has been increased interest in — and indeed excitement over — the minor components of the solar system such as asteroids, comets, and natural satellites. Much of that interest stems from the simple fact that the *Voyager* probes provided stunning photographs of features on worlds orbiting the Jovian planets that no one had ever suspected.

But as far as the asteroids and comets are concerned, the excitement is over the increasing evidence that collisions played a much more important role in the formation and evolution of the solar system than had been suspected. And much of this evidence comes from a renewed interest in our own planet — Earth.

Meteors You are probably more familiar with debris in the solar system than you think. Those occasional streaks or flashes of light that you observe in the sky, what you used to call *shooting* or *falling* stars before you learned otherwise, are believed to result from particles of cometary debris colliding with Earth's atmosphere. Actually, it would be better to think of it more as a matter of Earth running into or colliding with the particles, since Earth orbits Sun and runs into the particles in the process. Called **meteors**, these light phenomena are caused by the energy of motion of a sand-sized particle being transformed into radiation.

Friction between the particle (traveling between 10 and 20 miles per second) and the molecules making up the atmosphere causes the particle to slow down, and the atoms and molecules to speed up (higher temperature). This in turn causes more collisions between molecules, which causes electrons to jump to higher energy levels, which

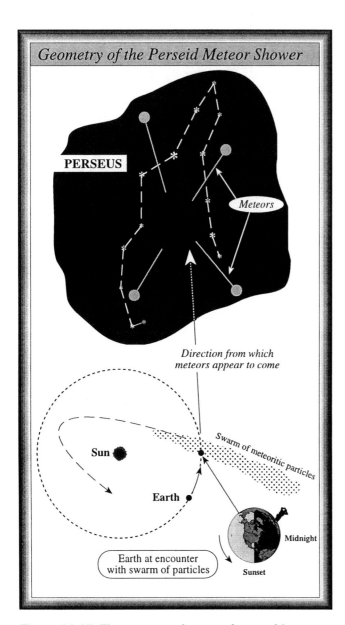

Figure 16–27 The geometry of meteor showers. Meteors are best seen after midnight, since that is the leading "face" of Earth as it orbits Sun. Hence that is the "face" that runs into the meteoritic debris.

Some Prominent Meteor Showers

Shower	Date of Maximum	Associated Comet	EXPECTED HOURLY RATE
Quadrantids	January 3	??	40
Lyrids	April 21	Comet 1861 I (Thatcher)	10
Eta Aquarids	May 4	Comet Halley	20
Beta Taurid	June 30	Comet Encke	25
Delta Aquarids	July 30	??	20
Perseids	August 11	Comet 1862 III (Swift-Tuttle)	50
Draconids	October 9	Comet Giacobini-Zinner	up to 500
Orionids	October 21	Comet Halley	30
Taurids	November 7	Encke	10
Leonids	November 16	Comet 1866 I (Tuttle)	up to 100,000
Geminids	December 13	(Asteroid *Phaethon* ?)	50
Ursids	December 22	Comet Tuttle	30

Figure 16–28 A list of the major meteor showers is provided so that you can practice your sky-viewing as you conduct a study of the frequency of meteors visible from your location.

354 SECTION 4 / Evolution of the Planets

Figure 16–29 A set of photographs of the Leonid *meteor shower of 1966, a peak year for this active yearly shower. The Leonids make their appearance—and take their name—from a point in the constellation Leo. It is difficult to take photographs such as this, because they are too sudden to catch by hitting the shutter on the camera, and not usually frequent enough to leave the shutter open long enough to "capture" one. And there is always the matter of city lights and/or moonlight. (National Optical Astronomy Observatories)*

results in electrons dropping to lower energy levels, which results in emitted visible photons. If it seems difficult to imagine a single grain of sand causing such a bright streak of light in the sky, try your hand at a simple mathematical calculation for the amount of energy contained in an object traveling 20 miles per second.

If you are surprised at the speed of the particles themselves, keep in mind that Earth's average speed in circling Sun is about 18 miles per second, and, in a sense, Earth is running into the particles rather than vice versa. Evidence of this claim is easily gathered by a careful observer like yourself. Spend an entire clear night outside away from the lights of the city – lie down on a soft cushion, get comfortable with a clipboard in hand, and record by time every meteor you see during the night.

You will find that the greatest majority occur during the hours after midnight. This is simply due to the fact that the side of Earth that faces its direction of motion around Sun is the "after midnight" face. Meteors you observe prior to midnight are those that Earth's gravity pulls inward and toward the opposite side (Figure 16-27).

Meteor Showers Some comets have long since disintegrated to the point that they are no longer detectable directly, the only evidence of their having been here at all being their remains that continue to orbit Sun. This debris is not uniformly distributed throughout the orbit of the original comet, however. When Earth passes through the orbit during its yearly orbit around Sun it can encounter a particularly large number of grains (Figure 16-27). The result is a spectacular display of meteors – **a meteor shower**. Since Earth's orbit is well defined, meteor showers occur with yearly precision.

Figure 16-28 is a list of the common meteor showers you can witness. Astronomers refer to the shower according to the name of the constellation in which it occurs (Figure 16-29). By referring to a constellation we are indicating the direction in the sky toward which you must look in order to observe the meteors. Look again at the sketches I suggested you make as you observe meteors. Your sketches should reflect a certain geometry amongst the streaks of light. Notice that they appear to come from a common point in the direction of the constellation for which they are named. Since Earth is running into the swarm of particles, the meteors appear to travel away from that point toward which Earth is headed. The situation is much like driving through a tunnel – the sides of the tunnel appear to move outward and away from the end of the tunnel toward which we move.

Meteor showers usually excite enough interest to be mentioned during the evening news. Just how spectacular they will be is not easily predictable, since that depends solely on where in the orbit the greatest groupings of grains are located. Historically, there have been meteor showers so spectacular that people feared that the entire sky was falling! Imagine being outside as an average of 1500 meteors *per minute* streak across the sky? That is about 30 per second! This was the case on the night of November 13, 1833, when people along the Atlantic seaboard were treated to the display for *three hours* in the constellation of Leo (the *Leonid* meteor shower).

The subsequent (at 33-year intervals) displays of 1866, 1899, and 1932 were progressively weaker. On November 17, 1966, however, a fairly spectacular meteor shower was observed in the southwestern part of the United States. So we will have to wait until 1999 for the next display due to that particular disintegrated comet. That would be quite a nice way to begin the celebrations certain to occur as we approach the next century. Meteors are not very bright, however, unless viewed from a dark location away from city lights. So you'll have to plan out your viewing location in advance.

Certainly the richest, most reliable and most easily-observed – because of the time of year – meteor shower is the *Perseids*, which takes place each August 11-13. The Perseid's meteor swarm of particles is orbiting through space in the same path as its parent comet, *Swift-Tuttle*, named for the two 19th Century American astronomers who discovered it in 1862. The discovery occurred in the midst of the United States' Civil War, and the proximity of the comet to Earth was heralded by spectacular showers of meteors observed by soldiers on both sides for several nights. Sky-watchers reported that "sparks of fire" flashed 400 to 500 times an hour! The comet itself was brighter than the North Star Polaris and had an impressive 30-degree tail.

It was thought that Comet Swift-Tuttle traveled in a highly elliptical orbit that lasted 120 years, and so its return was eagerly awaited in 1982, but to no avail. Astronomers finally spotted the comet in September 1992 on its way back into the inner solar system, and its period of return has now been calculated to be 131 years.

Cometary Dust So what eventually happens to those grains as they plunge into the atmosphere and slow down? Referred to now as **micrometeorites**, they slow down and

Figure 16–30 The remote manipulator arm of the Space Shuttle suspends the Long Duration Exposure Facility *prior to releasing it to space. It remained in orbit for 6 years before being returned to Earth by another Shuttle crew. Among other tasks, it collected data about the concentrations, directions of motions, sizes and masses of micrometeoritic dust in Earth's vicinity. (NASA)*

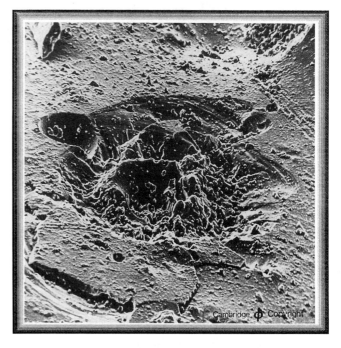

Figure 16–31 A photograph taken with a scanning electron microscope (800 magnification) of a rock returned from Tranquility Base on Moon reveals a microimpact crater caused by infalling dust. Lacking an atmosphere to slow dust grains down, Moon's surface is continuously "bathed" by dust. (NASA)

float to Earth's surface, there to mix and mingle with the rest of the surface material. They are in the air we breathe, the food we eat, the water we drink, and, of course, in vacuum cleaner bags. Since oceans cover most of Earth's surface, much of it ends up on the ocean floors as a thin layer of silt. When the grains strike the surface material of Moon, they leave evidence of their speeds in the form of microimpact craters (Figure 16-31).

Estimates based upon the data returned from the **Long Duration Exposure Facility** (LDEF)) (Figure 16-30) are that about 40,000 tons of this cosmic debris are added to Earth <u>daily</u>, and eventually become integrated into Earth's life support system. We are made up not only of *star stuff* but *comet stuff* as well. In a very real sense, the solar system is still forming, inasmuch as the planets (and satellites) are still collecting leftover material within the solar system and adding it to their existing masses. Prior to encountering Earth's atmosphere, these particles are called **meteoroids**. They are collected by high altitude balloons, airplanes, and even satellites that have been send beyond Earth's atmosphere. So we are quite familiar with their sizes and chemical compositions. The LDEF determined that the greatest number of particles encountered during its 6-year duration in Earth orbit were about 200 micrometers in diameter (0.004 inch).

Fireballs Hopefully, during the meteor-observing activity I suggested you perform, you witnessed a spectacular streak of light in the sky that astronomers call a **bolide** or **fireball**. At the time, you probably thought of it as a particularly bright meteor. But I am sure that its sudden appearance will be an unforgettable experience, especially if you happened to be in a dark, remote location. *Fireballs* can light up the entire sky, as well as the surrounding landscape. For a few seconds, it can seem like daytime. Occasionally a muffled sound can be heard coming from the direction of the bolide — especially if the meteoroid responsible for the event breaks into pieces when it hits Earth's atmosphere.

Although the physical process responsible for the emitted light of a bolide is the same as for meteors, the *size* and *origin of the particle* responsible for the light is quite different. Whereas the meteoroids responsible for meteors weigh less than a gram, those responsible for bolides are the sizes of walnuts or larger.

The solar system contains meteoroids of various sizes, ranging from those like grains of sand up to the largest chunk called *Ceres*, which is about 500 miles in diameter. You are aware that there is a certain overlapping of terms here. Keep in mind that words are like brushes — words convey ideas in the mind, brush-strokes form a picture that conveys ideas or feelings. When does a particle from a dead comet become a meteoroid? When does an asteroid become a meteoroid? An answer that allows us to get on with the subject of the solar system's formation is "...when it gets close enough to Earth to interact with it." — for example by colliding with the atmosphere to cause a meteor or bolide.

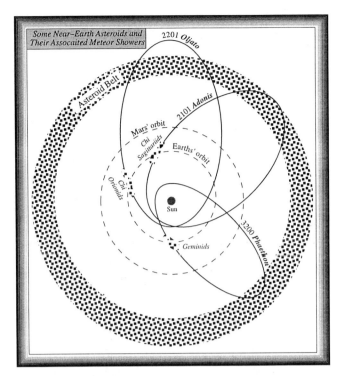

Figure 16–32 Although most asteroids orbit Sun within a broad band between the orbits of Mars and Jupiter, there is a good percentage that have orbits that take them close to and far away from Sun.

The point is that the small meteoroids appear to come from dead or dying comets, whereas the larger ones appear to originate in the **asteroid belt**, the belt of rocks that orbits Sun. *Ceres* is the largest member of that belt. Occasionally the observation of a bolide allows investigators to recover pieces of the object that caused it — a **meteorite**. *If a meteoroid survives its passage through Earth's atmosphere while causing a meteor or bolide – and is later found on Earth – it is called a meteorite.* Most meteorite finds are accidental, however, and not the result of someone observing their fall through the atmosphere.

Solar System Debris – Asteroids

The asteroids currently are receiving considerable interest amongst astronomers. There are a few good reasons why that is so, not the least of which is the concern that Earth may be a target for a devastating collision with one sometime in the foreseeable future. On a positive note, the push of humanity into space may create a demand for the mining of raw materials for the construction of space habitats on Moon, Mars, or even in orbit around Earth or Sun. The asteroids would be natural targets for the invasion of mining operations so that we don't have to pollute Earth itself to obtain the necessary materials. From a strictly scientific point of view, however, the interest in asteroids stems from their role in the solar system's formation.

CHAPTER 16 / The Origin of the Solar System

★★★★★★★★★★★★★★

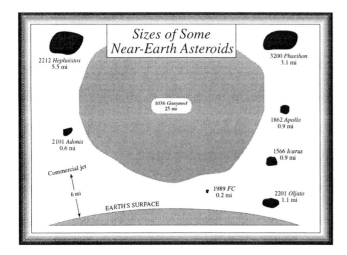

Figure 16–33 The sizes of some asteroids are compared to objects in the solar system with which you are familiar.

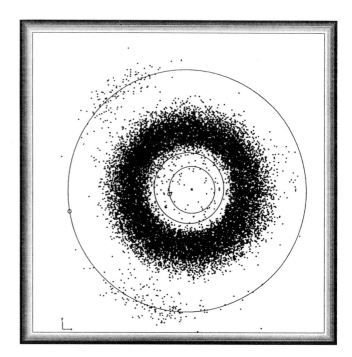

Orbits and Motions All asteroids revolve around Sun in the same direction — the counterclockwise direction that holds true for just about everything in the solar system. Most asteroids orbit in the plane of Earth's orbit (the ecliptic plane), the average inclination being about 10 degrees. A few are more than 25 degrees, and the orbit of *2102 Tantalus* is inclined the greatest of all — 64 degrees.

The shapes of their orbits are, on the average, about the same as the planets. 90% are in the main belt between Mars and Jupiter, but the remaining 10% are scattered both near and far (Figure 16-32). Two asteroids wander out beyond Jupiter: *944 Hidalgo* and *2060 Chiron*. The average spacing between asteroids is about 3 million miles — 12 times greater than Moon's distance from Earth! It is not surprising, therefore, that not one of the four satellites (*Pioneers 10* and *11*, *Voyagers 1* and *2*) sent out to explore the outer regions of the solar system managed to survive passage through the belt without a single collision.

Sizes of Asteroids With the exception of the few asteroids that have been photographed up close by the *Galileo* satellite, astronomers must use indirect methods for determining the sizes of asteroids. The most accurate method involves determining when one will pass in front of a star. Then an observer accurately times the interval during which the star is hidden (occulted) by the asteroid. Kepler's laws of motion are used to determine how fast the asteroid is traveling, and its size can then be calculated. In this method, the asteroid is not necessarily observed directly — the star merely blinks out when one edge of the asteroid first occults the star, then blinks back on when the other edge passes the star. The occultation typically lasts only a few seconds. We can even get a rough idea of its shape if several observations of the occultation are made from different locations on Earth.

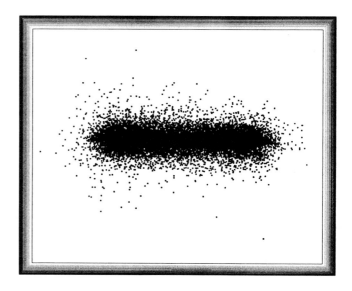

Figure 16–34 With painstaking effort at the telescope, an astronomer recently published his plot of 21,644 asteroids detected up until April 1, 1993. The bottom view is the plot as seen along the ecliptic plane of the solar system. The top view is the plot as seen from above the plane. The location of Earth is indicated by the small circle on the inner–most orbit ring in the top view. (Edward Bowell, Lowell Observatory)

Radar-imaging is a second, less-accurate, method. And it can be applied only to very close-by asteroids. Nevertheless, using this method, astronomers determined that asteroid 1989PB is about 400 meters in size and is shaped like a dumbbell. Finally, astronomers compare the amount of infrared radiation being emitted by an asteroid to the amount of reflected light, and that allows them to calculate the size of the asteroid. The amount of solar radiation falling onto the asteroid is known because its distance from Sun is known, so the difference between the amount absorbed and the amount reflected has something to do with the size. Most asteroids reflect very little light — 3 to 5 percent. That makes them about as dark as coal. Others may reflect 15% or so and even as high as 60%. The vast majority seem to be primitive bodies composed of silicates mixed with dark organic carbon compounds.

Using such techniques, we have determined that asteroids are all quite small compared to planets (Figure 16-33). That doesn't make them unimportant. The largest is Ceres, which is about 500 miles across. Ceres, Pallas, Vesta and Juno — the "Big Four" asteroids — account for more than half of the combined mass of all known asteroids. "Known" asteroids? Well, we have been able to calculate rather accurate orbits for about 5,000 asteroids, and we know that there are another 10,000 we could track if we wanted. There are probably tens or hundreds of thousands of smaller ones that are invisible from Earth. Astronomers have reason to believe there are 100,000 objects whose diameters are one kilometer or less. Nevertheless, their combined mass is still considerably less than that of our Moon.

Origin of Asteroids Where do asteroids come from? Location, orbital behavior, and chemical composition provide the information that allows for a theory. Some time ago, scientists entertained the theory that they are debris from a planet that once orbited between Mars and Jupiter, and that for some reason exploded or broke into pieces of different sizes. With advances in our knowledge of the processes accompanying the formation of stars in general and the formation of the solar system in particular, it is now theorized that asteroids represent the debris left over from the formation of the solar system that failed to collect together to form a planet.

The reason that the debris did not collect together to make another planet whose name you'd have to remember — in contrast to the planets that obviously did form — is explained by the proximity of the debris to massive Jupiter. That planet's gravity disturbed the rocks' orbits sufficiently that they were never able to pull one another together enough to collect into a single object.

This theory becomes more believable when one learns that Jupiter continues to influence the orbits of both comets and asteroids to this very day. Before considering the nature of the pieces of asteroids that occasionally fall out of the sky, let's look closely at the behavior of meteoroids in the asteroid belt, from which meteorites are thought to originate in the first place. The placement of the asteroid belt in Figure 16-32 was only a general one. The diagrams in Figure 19-34 are much more accurate in that it shows how widely scattered the asteroids are in the belt. It is assumed that they were once in a narrower band, but have been dispersed by Jupiter over the past 5 billion years.

The general rule is that asteroids circle Sun between the orbits of Mars and Jupiter. But as I mentioned earlier, there are many exceptions. The asteroid Icarus actually goes closer to Sun than does Mercury. So what is the nearest object to Sun? And in 1977 an object named **Chiron** was discovered to be orbiting Sun between the orbits of Saturn and Uranus. At first, astronomers didn't know if Chiron was perhaps the first discovered member of an outer belt of asteroids, or if it was an object that represented a much-suspected connection between asteroids and comets. It is now considered to be a comet from the Kuiper disk of comets generally located beyond the orbit of Neptune. Perhaps this comet has been perturbed by one of the giant planets into an orbit that brings it closer to Sun than the others. It may well be that many (if not all) periodic comets originate in the Kuiper belt.

Both asteroids and comets represent samples of the cloud of small bodies once filling the entire solar system and from which the planets and satellites are believed to have formed. Over the past couple of years, Chiron has been getting brighter than expected. After close observation, astronomers concluded this extra brightness comes from an extended dust atmosphere, indicating that Chiron is a very large comet rather than an asteroid. It is about 110 miles in diameter, or ten to twenty times as large as Halley's comet. This makes Chiron rather unusual as a comet, since the solid part of a comet is not typically that large. This may be due to the fact that it has escaped alteration by solar radiation because of its great distance from Sun.

The importance of the discovery and study of Chiron is that the solar system contains a variety of objects orbiting Sun in a variety of orbital types, and the orbit itself does not tell us what the object is. Comets and asteroids may mingle with one another as far as orbits are concerned, but they have very different compositions. All of the asteroids added together would form an object only about one-thousandth the mass of Earth. Ceres alone constitutes about one-fifth of that total mass. So even though it is common to refer to the asteroid belt as the "planet that failed," it certainly would not have been much of a planet if had successfully formed into a planet.

Families of Asteroids In 1917, a Japanese astronomer noted that a handful of asteroids shared similar orbital properties such as distance from Sun and inclination and eccentricity of orbits. Since he referred to them as "families" of asteroids, they have become known as **Hirayama families**. He believed that a "parent" asteroid had broken into two or more "children" asteroids, and that the latter — although now separate from parent — still "remember" the orbit of the parent. Shortly after the event that caused the fragmentation, the pieces were in a small clump orbiting Sun, different relative speeds and the presence of Jupiter spread them out to their present orbits.

The three most common groupings are known as the *Koronis*, *Eos*, and *Themis* families. Of the 5,000 or so numbered asteroids, about 860 belong to families. 600 alone belong to these three large families. A piece of evidence that supports the fragmentation theory is that the member asteroids within each family are frequently physically similar — just as you would expect if they had a common parent. It is easy to calculate the frequency of collisions between asteroids if we know their average size and the space volume they occupy. Calculations suggest that an individual 3-mile asteroid will experience one collision every one billion years. Thus the interval between collisions for all asteroids is about 100,000 years.

Colliding Asteroids During the 5 billion years that have elapsed since the planets and asteroids and comets first formed, there has been plenty of time for Jupiter to gravitationally disturb the asteroids out of their original circular orbits around Sun. Today, the orbits of asteroids are distributed rather randomly. In addition, because they are so easily disturbed in their orbits, innumerable collisions have occurred between asteroids.

If two colliding asteroids are quite unequal in size, the smaller one causes craters to form on the larger one. If their sizes are approximately equal, however, a collision could result in fragmentation of both. It is these fragments that are believed to be responsible for the *bolides*, and that we occasionally recover as meteorites as Earth's orbit crosses the orbit of a chunk of asteroidal material.

This hypothesis for the origin of meteorites is supported by close-up photographs of the two satellites of Mars by two U.S. probes sent to Mars in 1976 (Figure 16-35) and the Soviet *Phobos* probe in 1989. Both satellites are quite small — **Phobos** is about 17 miles long and 12 miles wide, and **Deimos** is about 18 by 10 miles. Both reveal numerous craters, and *Phobos* has a system of parallel grooves a few hundred yards wide and perhaps 30 feet deep. These grooves may have been caused by whatever impact caused the crater *Stickney*, which appears so prominently on Phobos. The darkness of the surface material of these satellites reminds us of the material typically found in meteorites, so there is strong suspicion that these two errant asteroids were captured by Mars early in the solar system's history.

Impacts of Asteroids You undoubtedly have the impression that astronomers believe that the planets and satellites formed from a cloud of rocks and chunks of ice as gravity pulled them together. But if that is true for Earth as well,

Figure 16–35 Asteroid Gaspra (top) is compared with the two Martian satellites Phobos (lower left) and Deimos (lower right). The similarity in sizes and surface brightnesses and features suggests that Mars captured these two wandering asteroids long ago. (NASA)

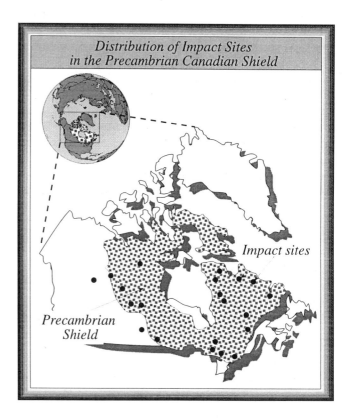

Figure 16–36 A map of the locations of remnants of impact craters in the oldest exposed section of Earth's surface in northern Canada. This finding conforms to the pattern found on Moon that large impacts occurred early in the history of the solar system, and presently only smaller asteroids are available to collide with planets or satellites.

Figure 16–37 This view of the gibbous moon reveals striking evidence of the violence that marked Moon's formation and evolution. Note especially the "rays" of ejected matter away from some of the larger impact features. (Lick Observatory)

Actually, there are a few craters found on Earth's surface, but they are the result of rather recent collisions, not those that occurred during the final stages of its early formation process. Erosion, tectonic, and other geological processes have removed all traces of the earliest craters. The best preserved of all craters on Earth is that located just outside Winslow, Arizona. The meteorite that created the **Barringer Meteorite Crater** must have been about 160 feet in diameter and weighed about 30,000 tons to have produced the 3/4-mile diameter, 500-foot deep crater. It is believed to have occurred some 50,000 years ago. Pieces of the meteorite — which naturally fragmented into thousands of pieces upon impact — are found in museums around the world (Figure 16-38). The site of this crater is well worth visiting.

Other craters are currently being detected by using orbiting satellites and studying core samples obtained by deep drilling methods. Some are located deep beneath the surface of Earth, and are discovered by analyzing the types of minerals found in those core samples. Erosion has erased most of the crater walls and filled in the centers of the surface craters. In a few cases — Sudbury in Canada, Nördlingen in Germany, and Mora in Sweden — towns have been established and flourished inside the filled craters.

then where is evidence of the craters caused by the final chunks that fell to Earth's surface? Well, Earth is not a good place to begin a search for craters, although geologists have located the remnants of many large craters in the far north of Canada at which location the oldest exposed portion of Earth's crust is exposed (Figure 16 – 36). But the best evidence is right in front of us, up in the sky — at least when Moon is visible.

Even through a small-sized telescope or pair of binoculars, the battered surface of Moon reminds us of what it must have been like 5 billion years ago when the solar system was forming — regardless of how Moon was acquired by Earth. One of the more impressive features to be seen on Moon through a modest-sized telescope is the crater named *Copernicus*. Running in a radial direction away from the crater are bright **rays** (Figure 16-37). These consist of the lighter (in color) material ejected outwards in all directions at the time of the collision that created *Copernicus*. The lighter material fell back down on top of the darker, more ancient surface material.

Cratered surfaces such as that of Moon are common in the solar system. Those planets whose surfaces consist of liquid (like Jupiter), or those that have substantial atmospheres (like Earth) lack evidence of the early violence that accompanied the formation process. Being confined to Earth's surface for most of human history has limited our imaginations toward explaining Earth's origin.

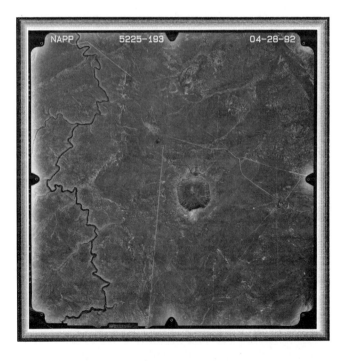

Figure 16–38 This photograph of the Barringer Meteor Crater in Arizona was taken by an orbiting satellite 500 miles above Earth's surface. Look carefully around this best-preserved Earth impact crater for stresses in Earth's crust created at the time of impact. (U.S. Geological Survey, EROS Data Center, Phone 605/594-6589, NAPP Project, Roll 5225, Frame 193)

Since there are plenty of asteroids having the potential for hitting Earth, you won't be surprised to learn that there is a good chance that a collision with one will occur soon. Soon? Well — soon in geological terms at least. Geologists have discovered 139 craters left by comets and asteroids. Some of the impacts must have been devastating enough to kill much of the life then present on the planet. Here is a short list of the most recent impact craters on Earth:

• Meteor Crater:	Arizona, U.S.A.	50,000 years ago
• Lonar:	Maharashtra, India	52,000 years ago
• Wolf Creek:	Western Australia	300,000 years ago
• Kara-Kul:	Russia	10 million years ago
• Ries:	Germany	15 million years ago
• Kamensk and Gusev:	Russia	65 million years ago
• Manicouagan:	Quebec, Canada	212 million years ago
• Clearwater Lakes:	Quebec, Canada	290 million years ago

Current estimates are that an asteroid whose size is about that of a football field will hit Earth every 10,000 years. One close to six miles in size will hit once every 100 million years. The estimated energy associated with a collision of this latter magnitude is equivalent to 1 billion atomic bombs. Since Earth's surface is mostly water, the greatest probability exists for a collision at sea. That would make quite a splash! Keep your surfboards ready!

In 1972, an object weighing approximately 1,000 tons was detected by Air Force reconnaissance satellites as it just missed Earth. Instead, it skipped off the outer atmosphere and returned to its orbit around Sun. And what about smaller collisions? As far as we know, only one person has ever been hit by a meteorite. In 1956, Mrs. Hewlett Hodges of Sylacauga, Alabama, was hit by an 8-pound meteorite after it tore through the roof of her home and bounced off the sofa on which she was sitting. She suffered only a large bruise on her leg.

In October 1992, a brilliant fireball was observed over much of the eastern United States in the middle of the daytime. A 27-pound meteorite punched a hole through the trunk of 18-year-old Michelle Knapp's old red 1980 Chevy Malibu in Peekskill, New York, and gouged out a 3-inch deep crater in the ground below. An enterprising group of rock-hounds reportedly paid her $50,000 for meteorite *and* damaged car. It was the type of meteorite that is most commonly found on Earth. Since the fireball occurred during the high school football season, there were several video recordings of the event. Astronomers are using the tapes in order to determine the direction from which the meteoroid came, and thereby determine the location within the asteroid from which it originated.

A chain of elongated depressions in the Argentine pampas has been interpreted to be the result of the explosion of a 500- to 1000-foot meteorite as it grazed Earth less than 10,000 years ago. The energy released at the time of the explosion is calculated to have been equivalent to 350 million tons of TNT. Laboratory tests show that the object approached Earth at an angle of less than 7 degrees. As it collided with the dense lower layers of Earth's atmosphere, it exploded and gouged out the depressions.

Spacewatch, a program designed to detect small asteroids passing near Earth, has counted about 100 times more near-Earth asteroids than astronomers had previously estimated. The study concludes that as many as 50 asteroids smaller than the size of a football field pass within Moon's distance from Earth each <u>day</u>. Even the U.S. Congress expressed concern over the results of the survey. A hearing took place before that body on March 24, 1993 (*The Threat of Large Earth-Orbit-Crossing Asteroids*).

Earth-Approaching Asteroids There is great interest amongst astronomers to learn more about those asteroids whose orbits bring them close to Earth. Estimates are that there are some 1000 to 2000 in the quarter-of-a-mile size range. There are three groups of Earth-approaching asteroids:

- The *Atens* orbit Sun within Earth's orbit
- The *Apollos* cross Earth's orbit
- The *Amors* cross Mars' orbit

Their orbits are unstable, and therefore will eventually impact a terrestrial planet, impact a terrestrial satellite, or be ejected from the inner solar system by one of the planets. The time scale for one of these events is on the order of 100 million years or so. This fact is rather interesting, because of its rather obvious association with evidence of such collisions on Earth in the past.

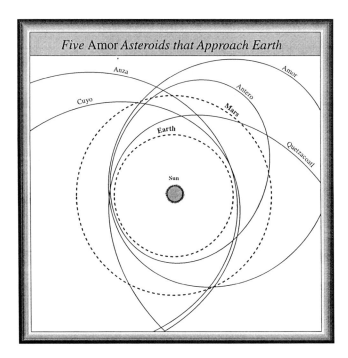

Figure 16–39 The Amors *family is a group of asteroids that cross Mars' orbit while approaching Earth.*

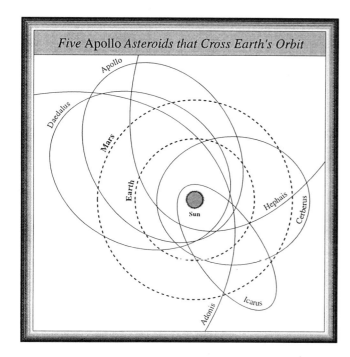

Figure 16-40 The Apollo *family is a group of asteroids that orbit Sun in elliptical orbits such that they cross Earth's orbit, creating the unlikely but eventual possibility for a collision.*

The *Planet-Crossing Asteroid Survey* (PCAS, for short) is a continuing program to search for small solar system objects that pass near Earth (Figures 16-39 and 16-40). So far, at least 163 **Near-Earth asteroids** have been detected and recorded. They range in size from a few hundred yards across to a few exceeding 10 to 15 miles. Most are in the one- to two-mile range. The largest is asteroid *1627 Ivar*, about 5 miles across. Assuming that humanity pursues its interest in space travel, about a dozen of these are easier to reach — in terms of energy — than Moon. We could prospect these asteroids robotically. They are perhaps the best candidates for missions using solar sails powered by the most efficient energy source known — sunlight. And, if we are serious about personned missions to explore the entire solar system, missions to asteroids would be excellent tests for more ambitious voyages to Mars and beyond. A few examples of these asteroids and what we know about them follows.

- In January 1991, Earth had a fairly close call with a newly discovered asteroid, called *1991AQ*. It zipped past within 5 million miles of Earth (20 times more distant than Moon). There have been closer calls, however. The most recent was *1989FC*, which came within 470,000 miles — about twice the distance of Moon from Earth.

- In terms of our interest in mining asteroids for the purpose of building habitats either in space or on the surface of Moon or Mars, there are three asteroids that travel in orbits that bring them close to Earth's orbit, so that a spacecraft could reach one with a minimal amount of fuel. These are *1989ML*, *1982DB* and *1982XB*.

- Asteroid *1986DA* comes no closer than about 20 million miles to Earth, but there is some interest in its composition — estimates are that it contains about 10,000 tons each of gold and platinum, worth some $90 billion and $1 trillion respectively on today's commodity market. Based upon spectroscopic studies of its surface material, astronomers suspect that it came from the interior of a much larger object that existed billions of years ago and that was broken apart by a massive collision.

- Asteroid **951 Gaspra** was imaged by the *Galileo* space probe in November 1991. It is a small (10 miles by 5 miles) main-belt asteroid, a member of the Flora family (Figure 16-35). This is the closest approach yet to such an object by a human machine. The photograph in the Color Plate section was taken when the probe was 10,000 miles from *Gaspra*. Careful examination of this and other photographs suggest that Gaspra might consist of two or more large lumps of rock held loosely together by their own feeble gravity. From every angle, the asteroid looks lumpy, rather the kind of shape you would get if you stuck two lumps of clay into a peanut shape. Based upon crater counts on its surface, astronomers believe that it is only 200 million years old.

Figure 16-41 The asteroid Ida *and its smaller companion* Dactyl *were the first discovered double asteroid system. They were imaged by the* Galileo *satellite.* (NASA)

CHAPTER 16 / The Origin of the Solar System **363**

- Thought to be from the same family as Gaspra, asteroid **243 Ida** was the second asteroid encountered by the *Galileo* space probe (August 1993). Looking much like Moon's surface, it is a heavily-cratered rock about 25 miles long, and irregular in shape (Figure 16-41). Its large size and extent of cratering suggests that it is older than *Gaspra*, perhaps 2 billion years. It is actually a **double asteroid** system – *Ida* has a small companion orbiting at a distance of 60 miles. Named *Dactyl*, it is only 2,300 feet in diameter. This is not the first pair of asteroids to be detected (at least two others have been imaged by radar), but this is the first of which high resolution images are available. Double asteroid systems are interesting, inasmuch as at least three of the 28 largest well-preserved craters on Earth are doublets. This connection may suggest that double asteroid systems may be very common.

Asteroids and Dinosaurs

I hope the foregoing has left you with the impression that colliding with an asteroid is not an event that seriously threatens our existence. It is far more dangerous for you to get into your automobile and drive to a lecture about asteroidal collisions. Having said that, let me take you on an exciting adventure story to illustrate our interest in the subject of asteroidal collisions and near-Earth asteroids. In 1980, Nobel Prize Laureate (physics) Luis Alvarez and his geologist son Walter shocked the scientific world by announcing the results of work that suggested that dinosaurs became extinct as the result of a collision between Earth and a large rock.

Disappearance of Life-forms The Alvarezes found a thin clay layer at a site in the Apennine Mountains of Italy that had all the characteristics of material that would have been laid down as the result of such a collision. Prior to coming to rest on Earth, however, this material would have blocked out sunlight for weeks or even months and caused Earth to cool. Acid rain and wildfires would also have ravaged Earth's surface after the collision, and all of these factors would have caused plants to die, as well as many of the animals that ate them.

Even before the Alvarezes had proposed their theory, there had been strong agreement within the scientific community that 65 million years ago there occurred a rather sudden and dramatic extinction of life forms on Earth. Approximately 60% of all animal species disappeared, the dinosaurs being the ones most commonly referred to. Although there has been considerable debate over exactly what it was that caused the extinction, there is impressive evidence that it was related to a sudden climatic change. The thickness of the clay layer itself suggested a short time interval for the deposition of the material, and hence implied a sudden change in the climate.

Evidence of Fallout The reason that the layer is believed to be fallout from a cosmic collision is because it is rich in the chemical *iridium*, which is rare in Earth rocks but abundant in meteorites from the asteroid belt. Although it is rare in Earth's crust, it is more plentiful in the mantle. Modern samples taken from the Kilauea volcano in Hawaii show an enormous iridium enrichment in deep-mantle lavas of about 100,000 times over other Hawaiian lavas.

Evidence for the timing of the collision coinciding with the extinction of the dinosaurs (and other species) derives from the discovery that the clay layer is found between two geologic layers – the Cretaceous layer below contains fossils of dinosaurs (and other species), while the Tertiary layer above contains little evidence of those species, but lots of fossils of evolving mammals. It is thus easy in any case to relate the clay layer to extinction.

The boundary between the two layers is known as the **K-T boundary** *(the* K stands for Cretaceous, the T for Tertiary). The same clay layer has now been found in dozens of sites around the world. At one site in Montana, 33 dinosaur species that were present 100 million years ago gradually decline to 13 species just below the *iridium* layer. They gradually disappear over the next 500,000 years.

Extraterrestrial amino acids discovered at the K-T boundary also provide convincing evidence of the impact theory of mass extinction. These particular amino acids are extremely rare on Earth, but are among the major amino acids found in carbon-bearing meteorites and are likely to be found in comets and asteroids. However, the amount of organic compounds found in them is far too low to be a source of life on Earth.

Estimate of Crater Location and Size The change in climate, according to some scientists, could have been caused by a comet or six-mile-wide asteroid traveling through space at a speed of 9 miles per second, and hitting Earth with a force equivalent to a 10 million megatons of TNT. It would have ejected a vast quantity of *iridium* dust up into the atmosphere, blocking out much of sunlight. The dust would eventually have settled back down to Earth's surface, depositing a one-half inch layer of clay worldwide. Scientists began looking for such preserved layers of clay, since they should contain a small percentage of the pulverized meteoritic matter, as well. Besides the layer found at Gubio, Italy, others have been uncovered at Stevens Clink and Lake Ackerman in South Australia.

According to the amount of material found in the clay layer, scientists estimated that the crater left behind by the collision should be about 100 miles across. Geologists identified 139 impact craters scattered around Earth. But the thickness of the clay layer found in the different sites around the world suggested that the collision occurred somewhere in North America. Coarse rocky debris, thought to be deposits carried by giant tidal waves spread out by the impact, were found in several sites along the U.S. Gulf Coast. Then an 18-inch layer of debris full of small blobs of glassy rock (**tektites**) which form from melted rock ejected during an impact was discovered in Haiti.

Crater Located? When scientists working for Pemex — Mexico's national oil company — reported magnetic and gravitational anomalies in the region of the northeastern corner of Mexico's Yucatan Peninsula in 1990, they dug a ring of sinkholes in the region to detect the perimeter of what is the largest known impact crater on Earth — 110 to 190 miles in diameter, located one-half mile beneath Earth's surface under the town of *Chicxulub* (pronounced CHEEK-shoe-lube).

Interestingly enough, this is a Mayan name meaning "*tail of the devil.*" Earth scientists have concluded that this crater-like structure dates from the exact time of the extinction of the dinosaurs 65 million years ago. The dating of the crater itself yields the same results as impact debris found around the Caribbean/Gulf of Mexico region, suggesting a connection between the two. Scientists used a radiometric technique that relies on the radioactive decay of *potassium-40* to *argon-40* over millions of years. Overwhelming evidence has been gathered that points to Chicxulub as ground zero of this event: shocked quartz (fractured crystals indicative of high-velocity impact), tektites, impact-melted rock hundreds of meters thick that has been dated to 65 million years ago, and tsunami wave deposits. All of these have been found in or around the crater.

Periodic Extinctions? Some **Paleontologists** — scientists who study Earth's fossil record — point to evidence that not only was there a dramatic extinction of life 60 million years ago, but that there have been similar extinction events before and since. When the number of species that died off is plotted against time — assuming the methods to date the fossils is accurate, of course) — there appears to be a pattern to those extinctions (Figure 16-42). They appear to occur at intervals of 28 million years or so. In addition to that interesting pattern of periodic life extinctions, a plotting of the number of impact craters with time suggests a similar pattern (Figure 16-43). Now that is a very interesting idea, because **periodic extinctions** would suggest some mechanism at work that is associated with motions within the solar system. While the geological evidence for the extinctions is being pondered and debated by scientists, an astronomer at the University of California is putting an ingenious theory to test.

If, indeed, there are periodic extinctions every 28 million years as a result of an asteroid or comet slamming into Earth, one naturally attempts to explain why they prefer to hit only at those time intervals. The astronomer — Allan Mueller — proposes that Sun may in fact a member of a binary star system. The second star — which he dubbed the **Nemesis** star — is in a very elongated orbit around the Sun-Nemesis barycenter, and is currently close to aphelion (greatest distance from Sun).

Every 28 million years, while passing through the Oort cloud on its approach to the inner solar system, it disturbs some of the orbits of cometary bodies enough to cause them to hurl in toward Sun. One or more collide with Earth to cause the damage we uncover in the form of the K-T boundary or fossils in general. This theory explains what we observe in terms of the periodic nature of the extinctions,

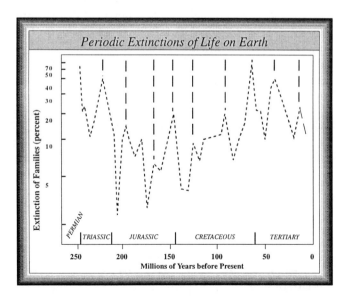

Figure 16-42 Although not agreed upon by all scientists, a group of paleontologists have plotted the extent of life form extinctions by date, and have found a pattern that suggests a periodic mechanism.

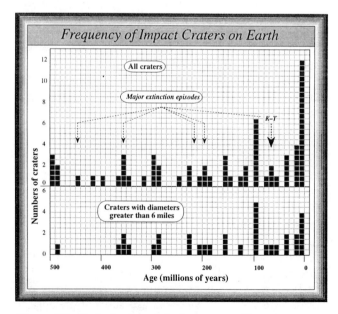

Figure 16-43 A plotting of the frequency of impact craters on Earth by date reveals a pattern that closely resembles the pattern of major life extinctions determined by paleontologists.

but so far the search for Nemesis has had negative results. Stay tuned.

The Rebuttal The scientific community is not unanimous in its acceptance of the Nemesis theory, nor of periodic extinctions in general. They are hotly-contested theories, made so by the inherent difficulty of accurately identifying and dating fossil remains of the "victims." Paleontologists for example point out that the dinosaurs died out over a period of 100,000 to 500,000 years — hardly what one would call a sudden event. Some geologists calculate that the dust cloud required to deposit the thickness of the K-T boundary layer was a thousand times greater than that created by the volcanic eruption of Krakatoa in 1883.

There is evidence that the eruption of the volcano Toba 75,000 years ago caused a cloud four hundred times that of Krakatoa, and yet there is no evidence of extinctions taking place. The shellfish and plankton that were extinguished along with the dinosaurs also died out over a very long period of time. And why, some paleontologists ask, did the most common of the dinosaurs — triceratops — die out before the iridium layer was deposited?

Finally, if the dust cloud wiped out the vegetation upon which the extinguished 60 percent of all life forms depended, how did the remaining 40 percent manage to survive — including birds, mammals, amphibians, some reptiles, and a large host of plankton, shellfish, and other marine life? Well, you get the idea. The evidence is circumstantial, at best. But that's how science works. More observations will be made, evidence will accumulate, and the theory revised. Stay tuned.

Tunguska Event

There is also the case of the gigantic explosion that rocked a remote portion (near the **Tunguska** River) of Siberia in 1908, an event that continues to receive occasional attention in the media. If an identical event occurred in a populated area today, people would probably associate it with an atomic explosion. Seismic waves were registered 600 miles from the site, observatories in England 2,000 miles away detected unusual air pressure changes, and observers 300 miles away reported a deafening sound and a bright cloud in the sky. A few reindeer herders less than 50 miles away had the tents in which they were sleeping blown into the air. There was so much dust ejected up into the atmosphere that at night, sunlight reflecting off the dust allowed people to read newspapers at midnight throughout parts of Europe and Siberia!

Because of the political events occurring in the Soviet Union at the time, Soviet scientists did not conduct an investigation at the site of the explosion until almost 20 years later — in 1927. Although they observed that trees were knocked over by the blast out to a distance of 20 miles from the center of the explosion, they found no crater and no meteorites. Most investigators have theorized that the object causing the blast was a small (100 yards or so) crumbly asteroid that disintegrated in Earth's atmosphere — or a similar-sized icy comet fragment. Fortunately, modern surveillance satellites are capable of warning us of the approach of such objects today, at least to the extent that we would not misinterpret it as an incoming enemy nuclear-tipped missile.

In 1993, a major scientific study of all of the available evidence concluded that the culprit for the event was a stony meteoroid 100 feet in diameter or so. Stony meteorites are the most common to hit Earth. A cometary nucleus or carbonaceous meteoroid were ruled out because they would have exploded or broken apart too high in Earth's atmosphere to account for the reported blast. On the other hand, an iron-rich meteoroid would have exploded so close to Earth's surface that surely craters would have resulted from the blast.

Mission to an Asteroid The *Comet Rendezvous/Asteroid Flyby* (CRAF, for short) mission was scheduled for launch in February 1996. It was targeted to fly in tandem with the comet *Tempel 2*, then to fly by the asteroid *449 Hamburga* in January 1998, and finally in August of the year 2000 to accompany periodic Comet *Kopff* on its inward journey to Sun. The satellite's resolution was scheduled to be from 3 feet to 2 inches. During its rendezvous with Comet *Kopff*, it would have sent a penetrator into the 6-mile frozen nucleus, scrape a sample from the surface material, and test it in a chamber. Budget cuts of 36% were made in 1991, and the program was completely cancelled early in 1993.

A new mission was considered and carried through to launch in early 1996. The target is near-Earth asteroid *433 Eros*, to be engaged in the year 1999 by the **N**ear **E**arth **A**steroid **R**endezvous (**NEAR**, for short) satellite (Figure 16-49). NEAR will approach within 300 miles of the 25-mile-long asteroid, and then be gravitationally captured to take up orbit around it for a period of one year. This requires very delicate maneuvering, inasmuch as Eros is only about twice the size of Manhattan. The mission will allow astronomers to accumulate considerably more information than during the usual pass-by trajectory (e.g., *Gaspra* and *Ida*). Resolution of the cameras aboard the satellite will allow for objects ten feet and larger to be imaged. A possible bonus for NEAR is a close encounter (800 miles or so) with the main asteroid belt object *253 Mathilde* on the way to *Eros*. With a diameter of 38 miles, this would be the largest asteroid ever visited by a satellite. In addition, it is believed to be composed of a slightly different set of minerals than either *Gaspra* or *Ida* or even *Eros*. So the scientific return would be significant toward understanding the raw material from which the solar system evolved. For the first time ever, humans will hopefully learn what an asteroid has to say about our mutual origins. In particular, they would like a sample of that material before it passes through Earth's atmosphere

Figure 16–44 A close-up view of a stoney iron meteorite weighing about 40 pounds, which was found in Antarctica by a team of scientists searching for meteorites. (NASA)

Some Famous Meteorite Falls			
Name	Fall/Find	Classification	Total Mass
Ahnighito	(Greenland)	Iron	31 tons
Allende	Mexico, Feb 8, 1969	Carbonaceous	5 tons
ALHA 81005	(Antarctic)	Achondrite	31 g
Canyon Diablo	(Arizona)	Iron	30 tons
Chassigny	France, Oct 3, 1815	SNC	9 lbs
EETA 79001	(Antarctic)	SNC	5 lbs
Hoba	(Namibia)	Iron	60 tons
Innisfree	Alberta, Feb 5, 1977	Chondrite	9 lbs
L'Aigle	France, Apr 26, 1803	Chondrite	80 lbs
Lost City	Oklahoma, Jan 3, 1970	Chondrite	40 lbs
Murchison'Nahkla	Australia, Sep 28, 1969	Carbonaceous	220 lbs
Nahkla	Egypt, June 28, 1911	SNC	90 lbs
Norton County	Kansas, Feb 18, 1948	Achondrite	2 tons
Orgueil	France, May 14, 1864	Carbonaceous	70 lbs
Pribram	Czechoslovakia, Apr 7, 1959	Chondrite	13 lbs
Shergotty	India, Aug 25, 1865	SNC	11 lbs
Sikhote–Alin	Siberia, Feb 12, 1947	Iron	23 tons
Sioux County	Nebraska, Aug 8, 1933	Achondrite	9 lbs

Figure 16–45 A list of some of the more famous meteorite "falls" and their masses and types provides a crude survey of the most common types of space debris encounter Earth. The SNC meteorites appear to come from the planet Mars, perhaps ejected off that planet at the time of an impact with a huge asteroid early in its history. The ejected chuck of material has just recently arrived at Earth.

or collides with Moon. To obtain samples of an asteroid is a much more challenging task. But astronomers are encouraging the development of a satellite named *Nereus* that would return a sample from an asteroid early in the next century.

Meteorites

Aside from the few hundred pounds of rocks returned from Earth's Moon — and material available from the surface of Earth itself — we have only samples of the asteroids in the form of meteorites as a basis for directly knowing the chemistry of objects in space. This fact makes them extremely valuable to astronomers. With the continued opening up of Antarctica for scientific research, teams of scientists search for meteorites that have been preserved in the vast open frozen terrain there. Hundreds of pounds have been found and examined, giving them a good cross-section of the types that strike Earth (Figure 16-44). There have been many famous meteorite "falls" in populated areas, however, although no one has ever been seriously injured by one (Figure 16-45).

In light of the obvious role that asteroids played in the shaping of the inner solar system, it is not difficult to understand why some astronomers specialize in the study of these samples of the raw material from which the solar system formed. Of special interest to the astronomer is the chemical composition of meteorites, inasmuch as chemistry tells us of the processes that the material has gone through.

Types of Meteorites Analysis in the laboratory allows astronomers to separate meteorites into three distinct categories – *irons*, *stones*, and *stony-irons* (Figure 16-46). Those most commonly <u>found</u>, and therefore the ones you have probably had the opportunity to handle, are the metallic meteorites, or **irons**. They are quite heavy for their sizes (high densities), inasmuch as they consist mostly of iron and nickel — two quite heavy chemical elements.

Those that most commonly <u>fall</u>, however, are the stony meteorites, or **stones**. Because their chemical composition is not unlike rocks found on Earth's surface, they are more difficult to identify and more easily broken down by erosional forces such as rain and wind. The reason we believe that stones fall more often is that they are the type generally found subsequent to someone observing a *bolide*. They contain high percentages of silicates, common in the rocks lining your garden.

Meteorites in the third group — the **stony-irons** — contain a mixture of the two. They appear to be the remnants of larger bodies that were once melted so that the heavier metals and lighter rocks separated into different layers. What is rather interesting about all of this is that the three types of meteorites conform pretty much to the pattern of rock types in our planet and (presumedly) other terrestrial planets, as well. That is, the interior of Earth is thought (through seismic studies) to consist of an iron-nickel core, surrounded by less dense minerals toward the surface.

A special class of the *stones* is interesting because of its connection to the question of life. **Carbonaceous meteorites** are relatively rich in carbon and water, and contain very little iron or nickel. They are physically very weak, capable of being crushed between the fingers. So erosional forces have destroyed those that fell long ago. In 1969, one fell

Types of Meteorites

Meteorite Type	Percentage of Falls*	Remarks
Stony		
Carbonaceous chondrite	5.7	**Oldest**, least altered material from early solar system
Chondrite	80.0	**Most common** type. Named for millimeter-size spherical, glassy silicate inclusions called *chondrules*
Achondrite	7.1	**Similar to terrestrial rocks.** *Chondrules* have been detroyed by a heating process
Iron	5.7	Nickel-iron material, **similar to Earth's core**. Most dense of all meteorites. Cut, polished and eched samples reveal crystal structure
Stony-iron	1.5	**Composite** of stony and iron sections
Total	100	

* Note: Because the "stony" meteorites most closely resemble terrestrial rocks, they attract less attention, and therefore less commonly found. The percentages of "finds" by type are: STONY (26%), IRON (66%), and STONY-IRON (8%).

Figure 16–46 The categories of types of meteorites that scientists have studied are shown in this diagram, together with characteristics that are used to determine possible origins.

near the small town of Murchison, Australia, a few months after a similar one had fallen near the village of Allende, Mexico.

Samples of both the **Murchison** and **Allende** meteorites were recovered and rushed to NASA's moonrock laboratory in Texas before they could become contaminated by Earth's environment. Extracting core samples from them, scientists found sixteen amino acids, some of which are identical to those known to make up living organisms on Earth. Headlines in the newspaper at the time stated that scientists had discovered that the universe out there was a "Life Factory!"

The important point is, of course, that although the universe outside of Earth is made up mostly of the simple elements hydrogen and helium, there appear to be the necessary environments in space within which simple elements can combine together to form more complex arrangements of atoms. And so perhaps when we consider the evolution of life on Earth from the basic chemicals present on the early Earth, we may want to keep in mind that many of those chemicals necessary for life could have been stowaways aboard the meteorites (and even comets) raining down on the primitive Earth.

Formation of Meteorites When the early Earth was in a molten state, the heavier substances sank to the center, forcing the lighter minerals to float to the surface. In-falling rocks of all sizes, up to hundreds of miles in diameter, colliding and breaking up into smaller chunks — but eventually sticking together by gravitational forces — was a source of the heat that has kept Earth's core hot for billions of years. Five billion years later, the surface is cool and hardened, but the core is still in a semi-molten state due to the high temperature preserved from Earth's violent beginning. This process of a planet evolving so that it is layered according to differences in temperature and/or chemistry is called **differentiation**.

So the irons are samples of the core of a terrestrial planet, whereas the stones are samples of the outer crust. The stony-irons must be from the region between those extremes. Am I suggesting that a planet once lived where the asteroids now reside? No — it is not believed that the solar system is missing a planet. But most likely much larger asteroids existed in the solar system when it was young, and frequent collisions during the intervening 5 billion years have gradually ground down the sizes so that today *Ceres* reigns supreme. According to this scenario, the stones represent the outer mantle of the larger asteroids, and the irons are samples from their cores. There is good evidence to support this model (Figure 16-47).

If an iron meteorite is sliced in half, polished, and etched with acid, one can see a crystal pattern named after the scientist who first discovered it, the **Widmanstätten lines**. When molten rock cools at different rates, the crystal patterns of molecules vary in size and shape. In essence, astronomers can determine the rate at which this class of

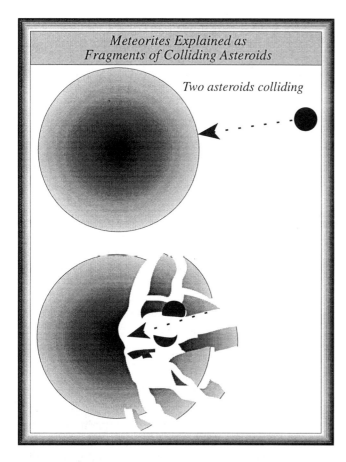

Figure 16–47 When asteroids collide with one another, fragments from different depths are scattered randomly. Since the types of material at different depths are chemically different, scientists can estimate the origin of a particular meteorite by its composition.

pounds were made by living organisms, only that they contain complicated carbon-based molecules. The organic molecules appear to have been produced at a late stage in the cooling of the solar nebula, the cloud of gas and dust from which Sun and planets condensed.

Complex molecules such as these have been produced in the laboratory by reactions between carbon monoxide and hydrogen helped along (catalyzed) by dust particles. With ammonia added, these reactions produce amino acids and other compounds of biological importance. The solar nebula would have contained all these gases and dust in abundance. What I am presenting is a possible scenario for the chemical processes that may have occurred on the early Earth that eventually led to life. The discovery of these chemical-bearing rocks from space suggest that the basic building blocks for life might be scattered around the solar system in liberal quantities.

There is general agreement that *carbonaceous chondrites* are probably fragments of the dark, C-class asteroids. *Ceres*, by far the most massive asteroid, falls into this category. Such C-type asteroids predominate in the outer portion of the asteroid belt. Thus they appear to have extended farther out originally, when the solar system was quite young. After Jupiter formed, most of these outlying objects were drawn into elongated orbits, causing them to collide with the planets and satellites. Thus the newly-formed Earth is likely to have been enriched by organic material from the asteroid belt, perhaps assisting in the beginnings of life.

meteorites cooled — it was rather <u>slow</u>. Now at the distance of the asteroid belt from Sun, the cooling of a small baseball-sized meteorite would have been rather rapid, considering the minus 150°F or so temperature there.

Conclusion — the irons must have originated in the interior of a larger asteroid, the material surrounding it acting as insulation to prevent heat from escaping rapidly. But as asteroids continued to collide, they were whittled down into smaller and smaller pieces (Figure 16-47). Some of the smaller pieces ended up on trajectories that crossed Earth's orbit to become the fiery objects plunging to Earth's surface and eventually found as irons. So the general theme of the formation of the solar system remains the same — remnant material in the solar nebula collected together to form planets and satellites, but some of it failed to collect together and that material now makes up the asteroid belt.

Life-Bearing Meteorites About 5 percent of all meteorites found on Earth are known as **carbonaceous chondrites**, so named for the richness of carbon found in them. Traces of organic compounds have been found in many of them. The word organic does not imply that the com-

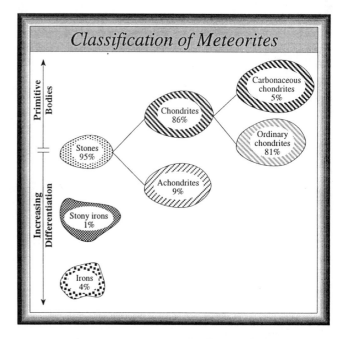

Figure 16–48 The general classification system for meteorites assists scientists in understanding the early history of the solar system in terms of chemical composition and manner in which material around birthing Sun collected together to form planets and satellites.

CHAPTER 16 / The Origin of the Solar System **369**

The asteroid *Vesta* provides evidence that it is a volcanic asteroid. With an albedo of 30%, it is the brightest of the main belt asteroids, and appears to be covered with basalt — a volcanic rock. Some 30 meteorites have been found on Earth that have a very similar composition to that of Vesta. In addition, their spectra match that of Vesta. It cannot have originated on Mars, Moon, or Earth. Evidently there were some very interesting events that occurred during the early solar system's history.

Gaps in the Asteroid Belt One particularly empty gap in the asteroid belt is theorized to be the source of some of the meteorite impacts on Earth. The gap in question is where a particle in the belt revolves three times around Sun for every single circuit that Jupiter completes. To explain the number of these meteorites found on Earth, scientists theorize that particles are shot out of the gap after spending only about a million years orbiting there. How does the ejection process work? Jupiter's perturbations produce a chaotic zone in the gap, and particles wandering in from elsewhere arrive with many different initial conditions. So they end up with completely different dynamical states after invading the gap. Some small percentage of those particles is bound to be flung out and sent toward Mars or Earth. One out of five leaving that particular **Kirkwood gap**, according to theory, is hurled to Earth. This prediction matches very closely the observed pattern of how ordinary chondrites actually fall.

Missiles from Mars There is another rather special class of meteorites — the **SNC meteorites.** All of these have a similar chemical composition, and therefore are believed to have a common origin. Most astronomers believe that they were ejected from the surface of Mars 180 million years ago when it was hit by a large asteroid. Recently, astronomers found small traces (~0.4%) of water in many of them, suggesting that Mars had water present on its surface long ago. Meteorites that fall to Earth certainly absorb some water from their surroundings, but the water found in these meteorites contains different amounts of heavy oxygen isotopes than water on Earth. The fact that we find these particular meteorites on Earth adds more evidence to our theory that the planets formed by collecting the debris orbiting the early Sun.

Extrasolar Planetary Systems

I am sure it has occurred to you long before this to wonder if there is sufficient evidence to support the theory that planetary systems are common throughout the galaxy and the universe in general. That is a very important question, especially as regards the possibility of life elsewhere. We don't think that the nature of life would allow for life in any form (that our chemistry and physics allows us to entertain as being possible) to either develop or live on the surface of a star.

So a crucial factor in considering the existence of civilizations elsewhere is that of the possible existence of planets elsewhere. Up until as recently as 1995, the opinion amongst astronomers was that the evidence for the existence of extrasolar planets was at best circumstantial, but taken as a whole, it convinced most that planetary systems were common throughout the universe. The sheer number of stars in the universe, the commonness of Sun as a star, and the seemingly easy explanation of how our planets arrived on the scene along with the birthing Sun: these were the factors that were most convincing. But the task of actually detecting other planets seemed elusive and fraught with technical problems.

First, consider the problem that we face. Imagine yourself an extraterrestrial, approaching our solar system from a distant star system. From a great distance, Sun far outshines any of our planets, inasmuch as planets only reflect light from Sun, not emit any light of their own. The task would be like trying to see a firefly perched on the rim of a searchlight. Only in getting rather close to the solar system would any of the planets become visible, and only then if you have the means to block out the blinding light

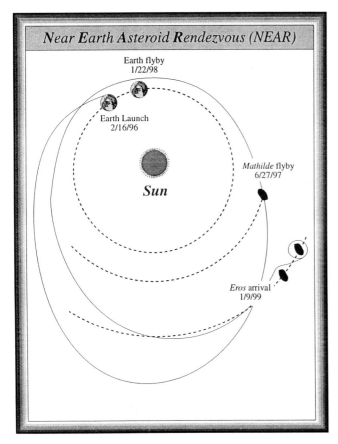

Figure 16–49 The satellite NEAR *is presently on the way to rendezvous with asteroid Eros in 1999. In addition to taking close-up photographs, it will go into orbit around the asteroid for one year. This will provide spectacular photos.*

Heat Flows in the Solar System	
Object	Measured heat flow (Joules/m^2-sec)
Sun	6.6×10^7
Earth	0.055
Moon	0.03
Jupiter	7.6
Saturn	2.8
Uranus	0.18
Neptune	0.28

Figure 16–50 Measured heat flows from the surfaces of objects in the solar system reveals different responsible mechanisms for different objects. Jupiter and Saturn, for example, appear to give off energy as they slowly collapse.

of Sun and look carefully to the side to catch the feeble reflected light of one of the planets. And what then would be the easiest thing to see? Jupiter. That is our solar system from a distance — Sun, Jupiter, and some debris.

Stars vs Planets Jupiter contains two and a half times more material than all of the planets, satellites, and debris put together. In a sense, we live in a binary star system in which one of the stars — Jupiter — failed. What I mean by "failed" is that Jupiter could have become a star in partnership (a binary star system) with Sun, but something went wrong. From a chemical point of view, Jupiter is composed mostly of the same chemicals of which Sun is composed — hydrogen and helium. But since Jupiter's mass is insufficient to create the high temperatures and pressures necessary at the core for the proton–proton chain to get started, it does not shine. Scientists estimate that if Jupiter contained ten times more mass than it has (the calculated critical value is 0.08 solar mass), it would be shining as a feeble red dwarf star. But there is no place within the solar system that such material is available, except for Sun itself. And it doesn't seem likely that Jupiter can get additional mass from Sun or anywhere else.

The point is that there is no fine line of distinction between a feeble star and a massive planet. Recall an earlier Chapter (Figure 9-27) in which the subject of brown dwarf stars arose as I discussed the process of stellar birth. Astronomers now have evidence (in the form of the companion of *Gliese 229*) that starlike/planet-like objects exist. Astronomers arbitrarily define a star as an object in which the proton–proton chain takes place in the core, sending out photons of radiation. But viewed from the outside, how would one be able to tell? Perhaps, you say, by measuring the radiation coming off the surface of the object. Well, that gets a little complicated. You see, Jupiter actually gives off more energy than it receives from Sun, but at infrared wavelengths, not at all wavelengths of the spectrum (Figure 16-50). Our interpretation is that Jupiter, being a large ball of gas and liquid, is slowly collapsing, releasing infrared radiation in the process. This collapse is simply a consequence of the aging of Jupiter. It has been collapsing since its formation 5 billion years ago. It has always been collapsing.

But our task here is not to determine whether an object is a star or a planet, but to emphasize the fact that there is no clear distinction between the two as far as formation is concerned. The commonness of binary star systems detected close to us suggests that solar systems are common. In some cases, a nebula collapses and fragments into two large components that form into two stars orbiting a common center of gravity (binary star system), and in some cases a nebula collapses and fragments into a large component and one or more small components. The large component evolves into a star, and the smaller component(s) into a star or stars.

The Search for Extrasolar Planetary Systems

Infrared Evidence of Planets In 1983, the Infrared Astronomy Satellite (IRAS), while mapping the concentrations of dust along the Milky Way band where stars are believed to be forming, discovered that some stars give off much more infrared radiation than expected for stars of their temperatures. *Vega* (alpha *Lyra*) was one of those stars. It is one of the brightest stars in the sky, and also one of the nearest to us. Its excess infrared radiation comes from a region extending out to approximately 80 AU's from the star itself. *Fomalhaut* (alpha *Piscis Austrini*) and epsilon *Eridani* are additional stars displaying patterns of excess infrared radiation. Astronomers reason that the infrared radiation comes from rings of particles in orbit around *Fomalhaut*, *Vega*, and epsilon *Eridani* — material that may eventually collect together to form planets. At the moment, the ring particles are only a fraction of an inch in size, hardly what one would call planets. But gravity is expected to gradually pull that material together.

A more direct detection of disk material around a star has been made in the case of the star **beta Pictoris**. It was first photographed from Earth by an ingenious device that blocks out starlight so that ring material can be detected. The HST more recently imaged the star and its associated disk, avoiding the limitations of atmospheric seeing (Figure 16-51). Estimates are that about 200 times as much mass as Earth is present in the form of solid material that extends no more than one billion miles from the star.

The disk is tilted almost edge-on to Earth. The fact that the disk is so thin increases the probability that comet-sized or larger bodies have formed through accretion in the disk. When this discovery was first made, the news media widely reported it as the discovery of planets. But there is a long

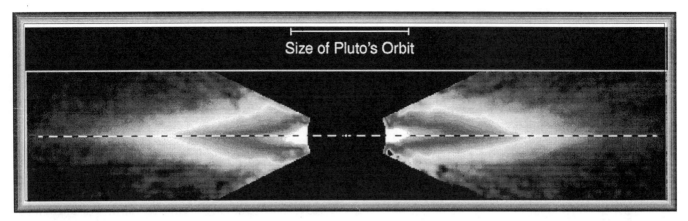

Figure 16–51 HST image of the star beta Pictoris *reveals rings of dust in orbit around the star, perhaps in the process of accreting into planets and satellites. The very slight "warping" of the disk away from the orbital plane (dashed line) may be due to purturbations caused by presence of planets close to the disk material.* (NASA/ESA)

way to go from a ring of small particles to planet. Yet a recent photograph provided by the HST reveals that the ring is bowed slightly away from the orbital plane, suggesting that perhaps the presence of planets in the vicinity are responsible for the "warping" of the disk. Intriguing.

The HST has added 15 stars to the list of stars with extended disks of dust. It discovered the disks around newly-formed stars in the Orion Nebula located 1,500 light-years away. The disks are visible to the HST because they are illuminated by the hottest stars in the nebula, and some of them are seen in silhouette against the emission clouds behind the disks. Each protoplanet appears as a thick disk with a hole in the middle where the cool star is located. Radiation from nearby hot stars "boils off" material from the disk's surface at an estimated rate of about one-half the mass of Earth per year. This material, according to computer models, will eventually be blown back into space by stellar winds of radiation and subatomic particles streaming from nearby stars. Based upon this estimate of the erosion rate of the disk material, a protoplanets initial mass would be at least 15 times that of the planet Jupiter.

Detection by Proper Motion Studies In Chapter 10, while discussing the death of stars, I proposed a method (*astrometric binary*) by which it is theoretically possible to detect invisible black dwarf stars orbiting around stars visible to us (Figure 16-52). As a matter of review, Newton's first law states that an object in motion continues moving in a straight-line direction unless acted upon by an outside force. A single star orbiting the galactic nucleus therefore appears to change positions along a straight line as plotted on three or more photographic plates taken at intervals of several years.

This change of position of a star relative to the backdrop of much more distant stars is the same proper motion responsible for the changing shapes of the constellations. Proper motion is the result of the differential motion of the stars in orbit around the galactic nucleus. Its effect only shows up on photographs taken of the nearest of stars, since changes of position of more distant stars does not show up over periods of even several years. Future generations will be able to compare their photographs of the sky with those being taken today to detect motions of more distant stars. In fact, any extraterrestrial civilization within about 20 light-years distance from us, using instruments equally as sensitive as ours, could deduce the presence of planets around Sun through studying the proper motion of Sun against the backdrop of more distant stars. Knowing the masses and distances of the planets on our solar system from Sun, it is easy to calculate how that proper motion would appear on their photographic plates (Figure 16-53).

The star found to have the greatest rate of proper motion of any known star is named *Barnard's star*, located in the constellation of *Ophiuchus*. It is at a distance of six light-years away from us. The amount of change in position it exhibits on a photographic plate in one year is 10 seconds of arc (a second of arc is one-sixtieth of a minute of arc, which is one-sixtieth of a degree of arc, which is one-360th of a circle). This small angle amounts to the thickness of a nickel seen at a distance of 100 yards. On every available clear night during the period 1937 to 1962, an astronomer at Sproul Observatory in Pennsylvania took a photograph of Barnard's star against the backdrop of more distant stars. Peter van de Camp then accurately measured the slight change in position of Barnard's star relative to the seemingly fixed stars around it, and plotted its resulting movement over that 25-year period.

The movement revealed on the plot was not along a straight line. Some unseen companion object(s) must be orbiting the common barycenter of the two (or more) objects, thereby causing the proper motion of the visible star to appear to "wiggle" (Figure 16-52). As I mentioned during our investigation of binary star systems, the amount

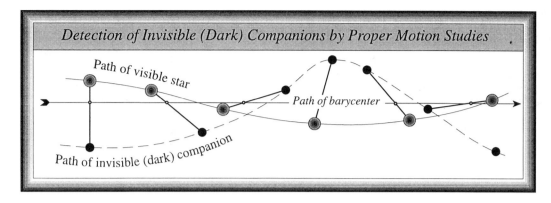

Figure 16–52 *The plotting of a star's position with respect to stars in the vicinity over a period of time allows astronomers to detect variations that would imply the presence of an orbiting invisible object such as a planet.*

of the wiggle in the proper motion of the visible component in the system is related to the mass of the unseen dark companion. The greater the dark companion's mass, the greater the wiggle in the proper motion of the visible star. Based upon the measured amount of wiggle for Barnard's star, Peter van de Camp calculated that there are two dark objects in orbit around Barnard's star, and that their masses must be just slightly less than that of Jupiter (0.7 Jupiter mass, and 0.5 Jupiter mass, to be exact). These objects cannot be stars, for reasons already explained.

Unfortunately, subsequent to van de Camp's study of Barnard's star, two additional surveys were performed with different telescopes and no wiggle was observed. At the moment, there is no consensus as to why it was that van de Camp's study revealed a positive result. When you realize that the measured wiggle on the original photographs was less than the thickness of the paper on which this is written, you can appreciate the difficulties of this particular technique, and the chances of error in using the measuring equipment itself. In any case, all is not lost. The technique is valid. We just need to perform the study under conditions in which equipment error and limitations due to Earth's atmosphere are minimized. Although the HST was not designed with this kind of study in mind, some astronomer maay find a way of using it for that purpose. Then we won't have to compensate for at least the blurredness of the star's image due to Earth's atmosphere.

Detection by Doppler Effect Studies This method takes advantage of the Doppler effect, so useful in so many astronomical applications. Consider a star's motion relative to Earth, either going away from or coming toward us. When the presumed orbiting planet is between its star and us, its slight gravitational pull causes the star to speed up (if it is approaching us), or slow down (if it is receding from us). Conversely, when the planet is on the opposite side of the star from us, it causes the star to slow down (if it is coming toward us), or to speed up (if it is going away from us). In any case, by detecting the ever-so-slight alternate shifts in spectral lines as the planet orbits the star, we can determine the mass of the object doing the pulling (Figure 16-54). Ingenious, eh!

Using the same technique at radio rather than at visible wavelengths, in 1992 radio astronomers detected Doppler variations in the radio signals from a pulsar lying some 1500 light-years from Earth. They used the Arecibo radio telescope in Puerto Rico to take precise timing measurements from the pulsar **PSR 1257+12**. This is one of just 21 known stellar objects, called **millisecond pulsars**, that spin thousands of times faster than typical stars, broadcasting powerful radio pulses as they evolve. Astronomers noticed a tiny "wobble" that could be caused by two orbiting planet-sized bodies. The size of the "wobble"

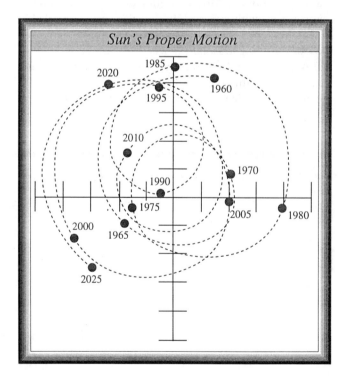

Figure 16–53 *The varying position of Sun against the backdrop of more distant stars (proper motion) as seen from a distant star would reveal the presence of Jupiter orbiting Sun. This technique, applied to other stars, might eventually allow astronomers to detect planets orbiting distant stars.*

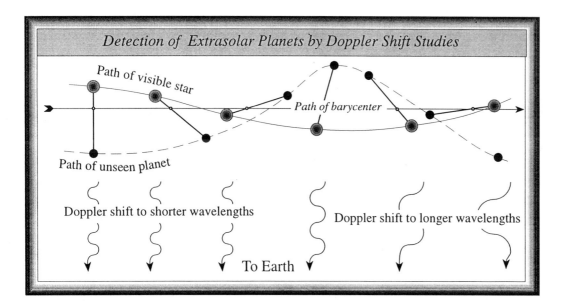

Figure 16–54 Plotting the proper motion of a star over a long period of time allows astronomers to detect the existence of any dark companion(s) the star may have in orbit around it.

suggests that they are 2.8 and 3.4 times the mass of Earth, and are 0.5 and 0.4 AU away from the pulsar. They move in almost circular orbits with periods of 98 and 67 days, matching the timing of the fluctuations. (Figure 16-55). In 1994, further observations not only confirmed the previous findings, but revealed the presence of a third body, with mass comparable to Moon, orbiting only 0.2 AU from the neutron star, with a period of 25 days. These remarkable observations provide the very first convincing evidence for the existence of extrasolar planets in the universe.

Although the long search for confirmation that our solar system cannot possibly be unique in the immensity of the universe ended in success, the discovery of planets orbiting a neutron star is puzzling. It is highly unlikely that the planets are native to the neutron star and therefore formed in the same way as ours. Any planets orbiting the star that exploded and collapsed to neutron star stage would most certainly have been destroyed during the supernova explosion. Consequently, astronomers are quite uncertain as to how the planets came into being. Perhaps they condensed from some of the debris ejected outwards during the explosion, a process that would be difficult to verify by observations currently available to astronomers. In order to truly establish our un-uniqueness in the universe, we must locate planets that formed as a natural by-product of star formation–like ours. It is rather ironic that

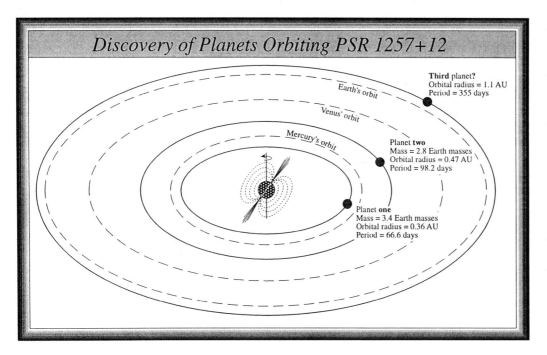

Figure 16–55 The first "discovery" of extrasolar planets was surprising since they orbit a pulsar, the remnant of a collapsed star. How the planets survived the violence that is thought to accompany the formation of pulsars is anyone's guess. But the study of the changes in pulsation of the pulsar suggest the presence of planets nevertheless.

the first planets discovered beyond our solar system should turn out to be orbiting a dead star, and have very little in common with Earth.

Discovery of Planets Around Sun-Like Stars The year 1995 may well go down in the annals of science as a banner year for astronomy. If so, certainly the discovery of planets orbiting the star **51 Pegasi** will be central to that distinction. *51 Pegasi*, located between 55 and 60 light-years from Sun, appears to be a G3-type star nearing the end of its hydrogen-burning life cycle. In October 1995, a Swiss astronomer reported detecting periodic changes in the star's spectrum indicative of an invisible companion's gravitational influence on the parent star itself. This landmark discovery was confirmed ten days later by a more sensitive instrument at Lick Observatory near San Jose, California.

Based upon the pattern of shifts in the spectral lines of the 5.5-magnitude Sun-like star *51 Pegasi*, astronomers estimate that the planet is about half as massive as Jupiter, and orbits the star once every 4 days. Kepler's laws of motion immediately tell us that the planet is a mere 5 million miles from the star (compared to Mercury's 30 million miles), placing it within the star's corona! Calculations establish its surface temperature as something on the order of 1000 degrees Celsius. These conclusions raise more questions than answers as regards the nature of planetary systems beyond our own. Just how planets can form that close to their host stars is difficult to understand. In any case, the discovery amazed the astronomical community, and has sent many astronomers back to the drawing boards to work out new models of planetary formation.

Right on the heels of announcing the discovery of the *51 Pegasi* planet (now called *51 Pegasi B*), the same astronomers announced in early 1996 that they detected additional planets around the stars **47 Ursae Majoris** and **70 Virginis**. These two planets, although still strange by solar system standards, are nevertheless located at distances from their host stars that provide their "surfaces" with more reasonable temperatures. Both are more massive than Jupiter, but they orbit their stars much closer than Jupiter orbits Sun. The stars themselves are similar to Sun in terms of size, mass, and temperature (Figure 16-56). Calculations suggest that *47 Ursae Majoris B*, having an orbit comparable to the solar system's asteroid belt between Mars and Jupiter, has a "surface" temperature of about –112 degrees F. *70 Virginis B* orbits its host star at approximately Mercury's distance from Sun, providing its "surface" with a temperature of 185 degrees F — about the temperature of lukewarm coffee. The codiscoverers nicknamed the planet "Goldilocks" because the temperature is "just right" for liquid water

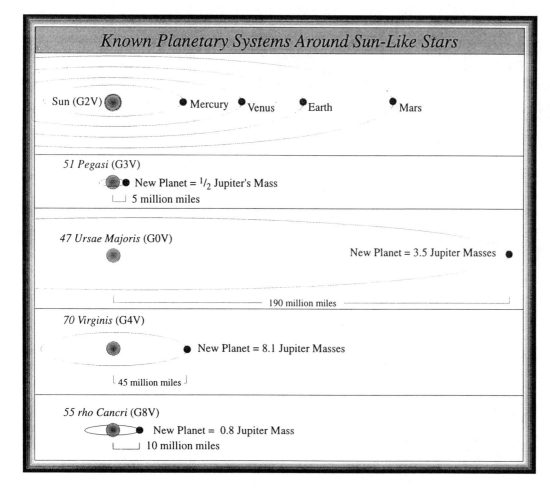

Figure 16–56 The most recent detection of planets around Sun-like stars astonished the scientific and nonscientific worlds alike. In this diagram, the four planetary systems are compared to our own solar system. Knowing the distances between the planets and their parent stars, astronomers can estimate conditions such as temperature and even chemical composition.

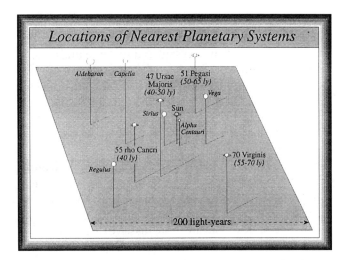

Figure 16-57 The locations of the three Sun-like stars that possess planets, with respect to our solar system.

(Figure 16-57). Another planet was discovered by the team in April 1996, suggesting that the new technology being used (the most powerful spectrograph in the world) has ushered in a "planet-a-month" era. **55 rho Cancri** is also a Sun-like star, but its planet is situated quite close to the star itself, similar to that of *51 Pegasi*. The planet's temperature is at least 700 degrees F, making it much too hot to support life as we know it.

But the significance of the predictions about *47 Ursae Majoris B* goes deeper as astronomers, biologists, and chemists begin to model the conditions that might exist on such a planet, and the chemical arrangements that might evolve under the condition calculated to exist there. That is not to say there is a strong possibility of life existing anywhere within any of these newly-discovered environments, but certainly organic molecules — the basic building blocks of life — could form under the conditions that most likely exist at least around *47 Ursae Majoris B*. There is still some reason to suspect the existence of such molecules in the atmosphere of Jupiter, in our own solar system. And even if the planets themselves are unlikely sites for life, it's entirely conceivable that they possess satellites with atmospheres that make surface conditions more moderate, allowing life to flourish. Since all of the giant planets in our own solar system possess a family of satellites and even a set of rings, it seems reasonable to assume that many of the planets discovered around other stars will be similarly endowed.

The most exciting spin-off of these recent discoveries, however, stems from a course survey of our own system of planets and the inherent limitations of the instrument that detected the four new planets. The sensitivity of the spectrograph that was used to detect the planets is presently limited to those that are at least as massive as Jupiter. In our solar system, there are 5 planets that are much less massive than Jupiter, and therefore would be undetectable by someone using our same spectrograph at a great distance from Sun. So, isn't it reasonable to assume that where we find little ones, there may well be big ones! So you can rest assured there is at least one astronomer that is at this very moment working in his/her optics laboratories devising a more powerful instrument so that he/she will be the first to find the very first Earth-sized planet. Stay tuned.

Are There Other Planets? In light of the foregoing, we currently conclude that there is convincing direct evidence that planets exist around other stars, and that increasingly sophisticated techniques will result in the detection of more in the near future, possibly before this Book is off the presses. Now we look at the question from another viewpoint — from our position in the solar system. Are the properties (movements, chemical makeup, locations, temperatures, etc.) of the objects in our solar system compatible with a theory that suggests they formed along with Sun? Patterns within the properties and characteristics of objects in the solar system provide important clues.

Summary/Conclusion

The surface of Earth certainly does not show dramatic evidence of its formation from a cloud of gas and dust in accordance with our general theories developed in Chapter 9. But fortunately we have access to most of the rest of the solar system via telescope and artificial satellites that provide a wealth of information about the objects in the solar system. The general characteristics of the solar system meet the general requirements for a natural formation. The predominant motions of all the planets and asteroids attest to a collapse theory. But it is the study of Moon rocks, meteorites, and comets that provide the best clues to the material that went into the making of the early planets. And as if that were not enough to convince us that the formation of planets is a part of the process of star formation, detection of planets around *51 Pegasi*, *70 Virginis*, and *47 Ursa Majoris* offers hope that extraterrestrials have stable environments within which to evolve just as we do.

LEARNING OBJECTIVES: *Now that you have studied this Chapter, you should be able to:*

1. Identify at least two advantages to our being to explain the origin of the solar system and its contents.
2. Describe the evidence currently available that suggests the presence of planets around other stars.
3. Sketch the general appearance of the solar system and its contents—both top-wise and edgewise views.
4. Describe and compare the properties of the debris of the solar system: comets, asteroids, meteoroids, and meteorites.
5. Explain how the properties of the comets, asteroids, meteoroids, and meteorites offer clues as to the manner in which the solar system formed.
6. Describe the mechanism of accretion as it pertains to the formation of the solar system.
7. Describe the chemical condensation process, and how it accounts for the radical difference between the terrestrial and Jovian planets.
8. Describe the general scenario for the formation of the solar system, and evaluate how well it explains the physical and chemical features of the solar system.
9. Define and use in a complete sentence each of the following **NEW TERMS**:

Allende meteorite 367
Aphelion 346
Asteroid 357
Asteroid belt 357
Barringer Meteorite Crater 361
Beta Pictoris 371
Bolide 357
Carbonaceous chrondrite 368
Carbonaceous meteorite 367
Chiron 359
Coma 348
Comet 339
CRAF 352
Deimos 360
Differentiation 368
Double Asteroid system 364
Dust tail 349
Fireball 357
Fractional distillation 344
Gaspra 363
Giotto 351
Hale-Bopp 353
Hirayama families 359
Ida 363
International Space Station alpha 338
Ion tail 349
Irons 367
Irregular satellite 341
Jovian 343
Kirkwood Gaps 370
K-T boundary 364
Kuiper Belt 347
Long Duration Exposure Facility (LDEF) 357
Meteor 354
Meteor shower 355
Meteoroid 357

Meteorite 357
Meteorology 337
Micrometeorite 357
Millisecond pulsar 373
Murchison meteorite 367
NEAR (Near Earth Asteroid Rendezvous) 366
Near-Earth asteroid 362
Nemesis star 365
Nucleus 348
Oort cloud 339
Paleontologist 365
Pathfinder Project 351
Periodic comet 340
Periodic extinctions 365
Phobos 360
Planetary geology 337
PSR 1257+12 374
Rays 361
Regular satellite 341
Retrograde revolution 340
Satellite 338
Shoemaker-Levy 9 353
SNC meteorites 370
Stones 367
Stony-iron 367
Sublimation 348
Tektites 365
Terrestrial 342
Titius-Bode rule 342
Tunguska Event 366
Widmanstätten lines 368
51 Pegasi 375
55 rho Cancri 375
47 Ursae Majoris 375
70 Virginis 375

CHAPTER 16 / The Origin of the Solar System 377

Chapter 17

Exploring the Earth–Like Planets

Figure 17–1 The terrestrial planets have been targets for several probes from Earth. This is an artist's conception of Ishtar Terra, the highest and most dramatic continent–sized highland region on Venus, based on topography measurements by the Pioneer Venus Orbiter spacecraft. The dense atmosphere of Venus prevents direct imaging of its surface. (NASA)

CENTRAL THEME: What are the physical features of the terrestrial planets, and how do they argue for a natural origin of the planets themselves and the solar system in general?

*I*t usually comes as a surprise to people when they learn of the number of artificial satellites that have been sent out to explore the environment of space around Sun, and to visit the planets and satellites of the solar system. Planetary exploration began in the early 1960s, riding on the coattails of President Kennedy's announced goal of landing a man on Moon by the end of that decade. The decision to spend $24 billion to return to Earth some 850 pounds of Moon rocks and reams of photographs was not made solely on the basis of its scientific merits. The United States could just have easily sent an un-personned rover to accomplish the same purpose, just as the Soviet Union did in 1970.

Competition in space travel has its merits, as long as the goal is that of obtaining knowledge (science) and not military advantage. Both the former Soviet Union and the United States have been active in planetary exploration, so there is a huge body of data available on both sides. By and large, that information travels freely between scientists, regardless of the country of origin. In other words, planetary scientists are generally more eager to unlock the secrets of the solar system than to argue international politics.

It might seem obvious why anyone would want close-up images of the distant planets. But from the perspective of the scientist, the value of photographic coverage of planets by satellite is that it allows them an opportunity to formulate meaningful questions (Figure 17-1). Scientists, in other words, were not wise enough to know in advance what questions to ask about the planets, other than the most obvious and therefore most trivial.

So as I take you on an imaginary tour of the planets of the solar system, keep in mind that the evidence upon which most of our theories is based is youthful. The data was provided by the numerous artificial satellites that have visited the planets only within the past three decades. Therefore, our explanations of what we observe are subject to change, and indeed change from month to month as new interpretations of the data are offered. In order to more fully appreciate this fact, you may want to refer to Figure 17-2 frequently during the tour.

General Factors in Planet-Building

Up until the planetary probes of the 1960s, astronomers had only a vague understanding of the processes involved in planet-building. We could study Earth at close hand, but we really didn't know which features were peculiar to Earth and which were the result of processes that operated throughout the early solar system. What has become increasingly clear during the past few years is the important role that collisions played in the formation of planets and satellites as well. This fact in itself is extremely important in attempting to quantify the number of life-sustaining planets that might exist in the universe. If life gets started as a consequence of some freak collision between rare objects around an evolving star, then we may well be the only ones around.

Heat in the Interior The key to understanding the forces that shape a planet's surface is its heat budget. On a dynamic planet such as Earth or Venus, heat is held in its interior, left over from its formation or generated by the decay of radioactive elements. The heat gradually flows outward to the surface and escapes through the crust. This heat flow provides a driving force for volcanic eruptions, fractured features and folded rocks. The internal heat also determines the structure of the interior of a planet. In particular, astronomers are interested in obtaining answers to two questions about the interior of a planet or satellite: (1) is there a molten core? And (2) is the mantle layered? The answers to these questions have implications as far as the possibility of life developing on the surface is concerned. I will establish that connection as I proceed through this and following Chapters.

Satellites to the Terrestrial Planets

Spacecraft	Launch	Arrival	Remarks
Mercury			
Mariner 10	11/73	3/74	Trajectory allowed three working flybys
		9/74	
		3/75	
Venus			
Mariner 2	8/62	12/62	Flyby
Venera 4	6/67	10/67	Atmosphere probe
Mariner 5	6/67	10/67	Flyby
Venera 5	1/69	5/69	Atmosphere probe
Venera 6	1/69	5/69	Atmosphere probe
Venera 7	8/70	12/70	Lander, 23 minutes of data from surface
Venera 8	3/72	7/72	Lander (50 minutes)
Mariner 10	11/73	21/74	Flyby
Venera 9	6/75	10/75	Orbiter and lander
Venera 10	6/75	10/75	Orbiter and lander
Pioneer/Venus			
Orbiter	5/78	12/78	Operated until late 1992
Multiprobe	8/78	12/78	Five atmospheric probes
Venera 11	9/78	12/78	Flyby and lander
Venera 12	9/78	12/78	Flyby and lander
Venera 13	10/81	3/82	Lander
Venera 14	11/81	3/82	Lander
Venera 15	6/83	10/83	Orbiter with imaging radar
Venera 16	6/83	10/83	Orbiter with imaging radar
VEGA 1	12/84	6/85	Lander and balloon atmospheric probe
VEGA 2	12/84	6/85	Lander and balloon atmospheric probe
Magellan	5/89	8/90	High-resolution radar mapping mission
Galileo	10/89	2/90	Flyby on way to Jupiter (arrival 12/95)
Mars			
Mariner 4	11/64	7/65	Flyby
Mariner 6	2/69	7/69	Flyby
Mariner 7	3/69	8/69	Flyby
Mariner 9	5/71	11/71	Orbiter (ceased functioning in 1972)
Mars 2	5/71	11/71	Orbiter and lander (no data from lander)
Mars 3	5/71	12/71	Orbiter and lander (no data from lander)
Mars 5	7/73	2/74	Orbiter and lander (lander failed at touchdown)
Mars 7	8/73	3/74	Orbiter (lander missed planet)
Mars 6	8/73	3/74	Orbiter and lander (lander crashed)
Viking 1	8/75	6/76	Orbiter (ceased functioning in 1980)
		7/76	Lander (ceased operating in 1982)
Viking 2	9/75	8/76	Orbiter (ceased functioning in 1978)
		9/76	Lander (ceased functioning in 1980)
Phobos 2	7/88	6/89	Lost contact with Earth and failed in 1989, after gathering data for two months
Mars Observer	9/92	?/93	Lost contact with Earth and failed in 1993, shortly before arrival at Mars

Figure 17-2 The terrestrial planets have been targets for many a visiting probe from Earth. The United States led the way to Mars, whereas the former Soviet Union concentrated on Venus. The data returned by these satellites will be discussed throughout this Chapter.

★★★★★★★★★★★★★★

The interior dynamics of a planet are also linked to its atmosphere (and therefore life). Volcanic eruptions are one of the mechanisms that can add gases to an existing atmosphere (Earth and possibly Venus), or one that is solely responsible for an atmosphere (Jupiter's satellite Io). On Earth, the fact that surface material is recycled into the interior causes carbon dioxide to be removed from the surface and formed into carbonate rocks — thereby preventing it from being released into the atmosphere.

Craters and Dating Planetary Surfaces Astronomers are presently in possession of samples of material from three objects in the solar for the purpose of dating the solar system — meteorites, Moon rocks, and Earth rocks. The 4.5 billion-year age of the solar system that I have used throughout the Book comes directly from those samples. Actually, the oldest rocks on Earth fall somewhat short of that figure. That fact must no doubt be due to the erosional forces that are so active on our planet — forces that essentially erase much of our past history. Since **radioactive dating** is the method by which such ages are determined, I digress for a moment to explain the process.

Some chemical substances are naturally unstable, the heavy nuclei within the atoms breaking into two or more lighter elements. But not all of the atoms break apart at once. Nature controls the rate of the decay process — each radioactive substance is unique in that it has a specific rate of decay. That is, over a given period of time, a specific

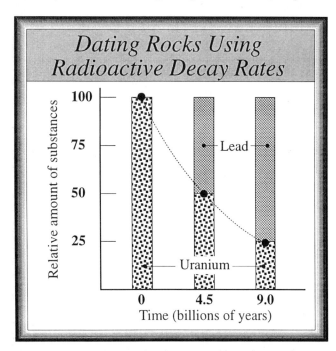

Figure 17–3 Scientists use radioactive dating in order to determine the absolute age of an object. But a sample of that object must first be examined in the laboratory. So other than for Earth, Moon and meteorites, we presently are very limited in knowledge about most of the contents of the solar system.

Figure 17–4 Astronomers use crater density as the basis for determining relative ages of different planetary or satellite surfaces. Where craters are scarce, some mechanism is responsible for erasing the craters that were presumed once there Use that logic in estimating the relative ages of different portions of Moon's surface. (Lick Observatory)

percentage of the atoms of the "parent" material decays into lighter products. The rate of decay for a particular element is expressed as the **half-life** of that element (Figure 17-3).

The *half-life* of a particular substance is the time required for half the original sample to disintegrate. For example, the half-life of Uranium is 4.5 billion years. A ten-pound chunk of uranium that was collected by Earth as a component of its planet-building process will today — 4.5 billion years later — be found as a five-pound chunk of lead embedded in five pounds of helium nuclei. During that 4.5-billion-year period, energy escaped each time an atom disintegrated. That energy, amongst other things, participated in the heating up of the material surrounding it. The radioactive dating technique provides absolute ages for the samples. Thus we know that the oldest regions on Moon are about 4.6 billion years old.

Carbon-14 is another radioactive chemical substance commonly used to date ages of materials on Earth, especially those associated with early human remains. The half-life of carbon-14 is about 6000 years. The by-products of that disintegration are electrons and nitrogen.

Lacking samples of material from Mercury, Venus, and Mars with which to date those planets, scientists use **crater density** as a measurement of relative age of a given surface. We simply assume that the greater the density of craters for a given region, the longer that region has been exposed to space debris. If a region lacks extensive cratering, we assume that some eroding process has been or is at work. In looking at Moon (Figure 17-4), for example, it is readily apparent that the large dark areas are younger than the surrounding heavily-cratered areas. This technique necessarily provides relative dates. We can say that one region appears to be twice as old as another. But combining the relative dating technique with the absolute dating technique, and applying some basic assumptions regarding similarities of crater-types, allows us to create models for the evolution of surfaces of solid objects in the solar system.

Planet Earth

I promised earlier that I would begin our tour of the solar system from the planet closest to Sun and work outward, eventually ending up on tiny Pluto. Allow me to deviate from that itinerary slightly in order to develop a vocabulary that we can use during the rest of the tour. I begin with Earth. Since you have some degree of knowledge of Earth's features, it is convenient to begin with what we already know.

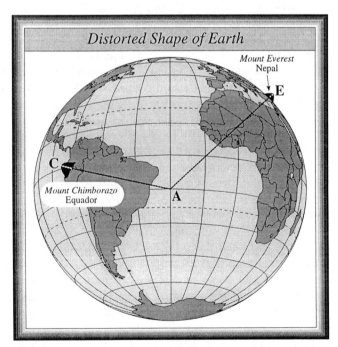

Figure 17-5 Earth has been distorted out of perfect roundness by a combination of rapid rotation and Moon's differential gravitational attraction. So simple geometry tells us that the point on Earth's surface that is most distant from the mathematical center of Earth is not Mt. Everest, but Mt. Chimborazo.

Then as we explore the other planets and satellites, we will have a basis for comparison. At the same time, of course, our attempts to explain what we observe on other bodies in the solar system are based upon what we claim to know about our own planet. So to begin our tour with Earth is not to emphasize its importance in the solar system but merely to apply what we know about it to other planets and satellites.

Shape Ask any school child what the shape of Earth is, and he/she is bound to reply that it is "round." As a general description, that is certainly true. But nothing in space is perfectly round. And the extent to which an object is "out-of-round" is a vital clue to the internal structure of that object, and also a clue as to its past history. Earth is no exception, although it is certainly not obvious when viewed from a distance — say from Moon.

From that vantage point, to the human eye, Earth appears as a perfectly round ball. To we who live on its surface, even Earth's 80-mile thick atmosphere seems like a terribly substantial and significant aspect of our planet. But compared to the 8,000-mile diameter solid portion, it is insignificant. Comparatively speaking, our atmosphere is less than the thickness of the skin on a Fuji apple.

The exact shape of Earth is determined through the use of orbiting artificial satellites. If Earth were perfectly round, the orbits of satellites would conform to exact ellipses. But since the distribution of Earth's mass is not uniform, the orbits of satellites we place above its surface are continuously changing. Having the ability to accurately track satellites from stations around the world — one reason that the United States tries to maintain friendly relationships with countries around the globe — we can determine that distribution of mass by knowing the extent to which the orbits of the satellites change over time. It turns out that Earth is slightly pear-shaped, with a bulge around the equator, making the distance around the equator slightly greater than the distance from pole to pole (25 miles). The very slight depressions in Earth's shape are along the northern hemisphere, making the southern hemisphere more "bulgy" than the northern hemisphere.

From a geocentric point of view, one could argue that Mt. Everest is not the highest mountain on Earth. Since Mt. Chimborazo (Ecuador) is closer than Mt. Everest (Nepal) to the equator, and is therefore located on that **equatorial bulge**, the summit of Chimborazo is actually farther from the mathematical center of Earth (Figure 17-5)! Of course, we usually use mean sea level as the datum from which heights of mountains are measured. In that case, Everest (29,028 feet) wins out over Chimborazo (20,561 feet). But what if Earth lost its oceans? Then we would probably look for a mountain standing highest above the surrounding plain. Using that criteria, Mauna Loa on the Big Island of Hawaii would outclass all others.

Rotation and Tilt Although Earth is not the fastest

CHAPTER 17 / Exploring the Earth-like Planets   **381**

★★★★★★★★★★★★★★★

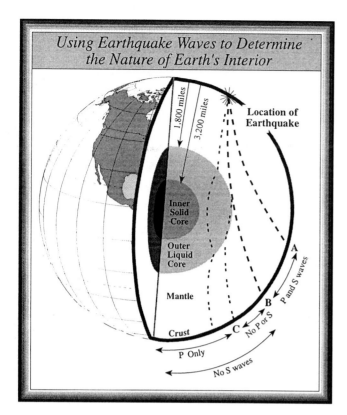

Figure 17-6 Earth's interior in cross-section illustrates what geologists know about the different consistencies of materials making up Earth. The detection of seismic waves of different patterns tell us what types of material the waves had to travel through to get to the stations positioned around the globe.

rotating of the planets (Jupiter and Saturn rotate much faster), its rotation is responsible for its non-spherical deformity. During the past 5 billion years, semi-molten material within Earth's interior has flowed from higher latitudes toward the equator to form the equatorial bulge. This is a simple consequence of centrifugal force, a force you are certainly familiar with when you ride a roller coaster or make a sudden left-hand turn in your car. This means, of course, that Earth is far from being solid.

Actually, we encountered this idea back in Chapter 3 when we learned that the very surface upon which we stand rises about an inch every time Moon crosses the meridian twice a day. The flow of material within Earth occurs because the interior is still hot. Still? When Earth formed from a pre-solar nebula of material collecting together, tremendous amounts of energy were released as material collided and stuck together. In addition, some of that material contained radioactive elements which, in the act of disintegrating, release the heat that keeps Earth's interior in a semi-molten state. Figure 17-6 illustrates how we verify this to be the case.

Earth's rapid spin may have been caused by collisions with Mars-sized planetesimals early in its formation. Computer models of the conditions in the early solar system suggest that Earth would be rotating every 200 hours instead of the 24 hours it takes. Both Mercury and Venus rotate at the expected slower rates. The fact that current theories for the origin of Moon favor the collision of a similar-size object with Earth strengths this explanation for Earth's rapid rotation. Once in orbit around Earth, Moon began to play an important role toward insuring a stable environment on Earth's surface. Moon may even have played a pivotal role in allowing life to develop and flourish.

The tilt of Earth's axis, as we learned in Chapter 3, is responsible for seasons. Scientists propose that Moon acts as a kind of gravitational gyroscope to stabilize the 23-degree tilt. Without Moon, the tilt could be as much as 85 degrees. Such a radical tilt would be catastrophic for life on Earth. A change of a mere 1.3-degrees in its tilt — according to some scientists — would result in ice ages. A tilt greater than 54 degrees would give the equator less sunshine than the poles, with resulting long seasonal patterns.

Seismic Waves Although people who live in California might have an opposite opinion, it turns out that Earthquakes have some value. **Seismic stations** have been set up all over the world to monitor waves generated whenever an Earthquake occurs. Sometimes geologists even use artificially-induced explosions to create the waves for the purpose of studying the cross-sectional layers of rock beneath Earth's surface. Waves generated by either of these methods are of two types — **pressure** and **shear**.

Each travels at a different speed, and each speed depends upon such factors such material density, compressibility, and rigidity. But what is important is that the shear waves do not move through liquids, whereas the pressure waves do. When an Earthquake occurs, waves of both types are sent from the point of origin (**epicenter**) out in all directions through Earth. As they encounter the interfaces between layers of different densities, they refract or reflect, or, in the case of the shear waves, are absorbed.

By timing the receipt of the waves at the various stations according to strength and type, geophysicists deduce a cross-section of Earth that looks like Figure 17-6. The hot dense inner core is thought to be composed of iron and nickel, material similar to that found in the iron meteorites. The outer core surrounds the inner core and exhibits the properties of molten iron and nickel, a consistency that allows atoms to slide around one another and strip away electrons in the process. The flow of the resulting charged particles within Earth's interior (a process called the **dynamo effect**) creates the magnetic field around Earth. It is this magnetic field that is responsible for influencing your compass and for capturing the charged particles from Sun (*solar wind*) that create the *auroras* discussed in Chapter 7.

The envelope that surrounds the outer core is the **mantle**, which consists mostly of solid rock containing combinations of such elements as magnesium, iron, silicon and oxygen. Lying atop the mantle is a thin skin on which we live, the **crust**. The crust also consists of the elements

Earth's Tectonic Plates and Their Directions of Motion

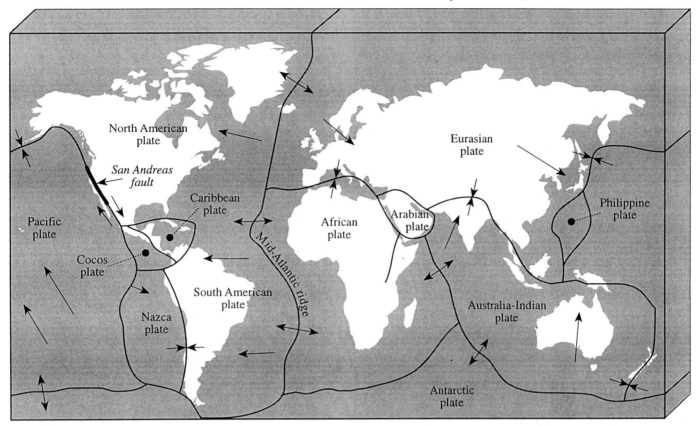

Figure 17-7 Earth's surface is broken up into "plates" that are constantly shifting and readjusting to a certain about of movement within the semi-plastic nature of the planet. When these "plates" rub against one another are the most diverse and interesting landforms such as mountain ranges.

magnesium, iron, silicon, and oxygen, although in different combinations than in the mantle. There are also compounds of sodium and potassium in this rather rigid zone that extends about 50 miles below Earth's surface on the continents, but only about 5 miles below the ocean floor. Gaining access to Earth's mantle, were we interested in doing so, would be easiest by drilling down through the ocean floor.

Crustal Movement What is most important in this discussion of Earth's interior at the moment is that region sandwiched between the crust and mantle — a 100-mile thick zone of partially molten material known as the **asthenosphere**. This zone flows under pressure much like asphalt or wet cornstarch. Earth's crust essentially floats on a thin layer of semi-molten material. Due to movement of the crust during Earth's long history, the crust is cracked in places so that it is presently divided it into about 12 pieces, or **plates**. I say presently because there is good evidence that the crust has had different patterns of cracks at different times in its past. There is clear evidence that the African plate is in the process of splitting into two or more plates.

These moving plates are frequently referred to as **continental plates**, although their boundaries are not to be confused with those of the popularly known land continents.

There is some similarity between the plates and the continents, as you see in Figure 17-7. But of course the continents themselves are so identified only because they happen to stand tall above the oceans that cover some 67% of Earth's surface. The plates move relative to one another quite slowly because material from within the semi-molten asthenosphere slowly "oozes" up through cracks between plates, forcing the plates to move away from one another along that crack. The term **fault line** is preferred over "crack." The motion of the crustal plates is known as **plate tectonics**, but more popularly called **continental drift**.

Much of your knowledge of geography and geology is probably related to the effects of this movement, since it not only influences the types of landforms along the cracks, but the movement itself is occasionally felt as Earthquakes. Actually, there are several ways that the plates move with respect to one another, and it is these different types of boundaries that are responsible for the different types of landforms.

CHAPTER 17 / Exploring the Earth-like Planets **383**

Lateral Fault Now let us do a brief bit of geography to excite your curiosity about our evolving planet. Certainly everyone is familiar with the state of affairs in California as to the commonness of Earthquakes. California, in fact, lies astride the separation or crack between two plates (Pacific Plate and North American Plate in Figure 17-7). As the *Pacific Ocean Plate* moves laterally toward the northwest relative to the *North American Plate*, Los Angeles, on the Pacific Plate, is gradually approaching San Francisco, on the North American Plate, at a snail's pace of 1 inch per year. Of course, that is only an average. There may be only gradual slippage for several years, and then one year a sudden slippage of several inches or feet — an Earthquake.

The severity of an Earthquake is an indication of the amount of slippage, as if pressures due to internal forces gradually build up until they overwhelm the force of friction between adjacent plates. Geologists are able to determine the slippage that has occurred between plates over time by comparing and dating rocks found on opposite sides of fault lines. To look at a map of the recorded Earthquakes that have occurred over a period of time is to look at the boundaries of the continental plates themselves.

You might also look at a list of volcanic eruptions that have occurred in the past, because while plates slip against one another, friction causes heat which in turn melts the rocky material which in turn releases gases which may push the molten material up and out. The Cascade Range of mountains in the Pacific Northwest, of which Mt. St. Helens is a member, is a good example of volcanic activity associated with plate movement. Just think of a volcanic explosion you have heard about — Mt. Etna, Mt. Vesuvius, Krakatoa, Fuji, Mt. Lassen. Each is associated with a plate boundary.

Hawaiian Islands You certainly are familiar with the Hawaiian Islands, and perhaps with media attention given to the occasional volcanic activity on the Big Island of Hawaii. Although they are certainly volcanic in origin, they are not located at or even close to a plate boundary. So what's the deal? Well, there is another mechanism that can be responsible for volcanoes. Particularly hot regions within the asthenosphere, called **magma chambers**, may occasionally force molten material up through weak spots in the middle of a plate and cause it to ooze out to form a

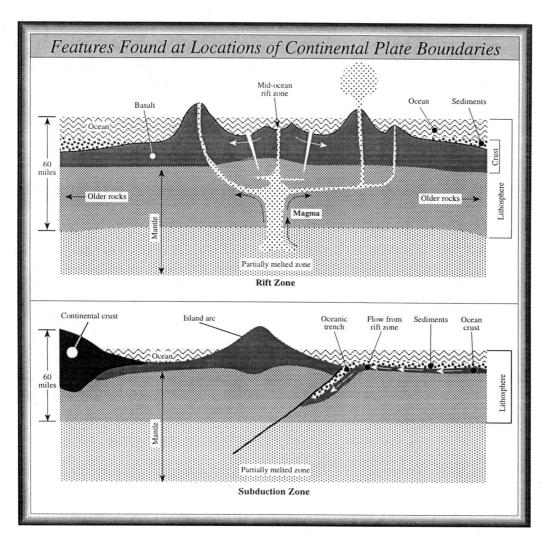

Figure 17–8 This diagram illustrates the various landforms that result when material from within Earth "oozes" to the surface (upper part of diagram), or when two "plates" run into one another head–on and one is forced up as the other subducts back into Earth's interior.

shield volcano. This happens most often where the crust is thin, as on the ocean bottoms. Because the material is only slowly pushed up and out, the sides of shield volcanoes are more gradually sloped than volcanoes associated with plate boundaries. The volcanoes located along plate boundaries are generally eruptive, frequently catastrophically so.

As highly liquid lava oozes up, it flows down the sides of the shield volcano and builds up material as it cools, slows, and hardens. This additional material adds to the size of the volcano. As I mentioned previously, if the oceans were removed from Earth, the Hawaiian Islands would be the highest mountains on Earth. There is a similar shield volcano on the planet Mars — called *Olympus Mons* — which has built up so much lava on its top and sides that it presently rises 78,000 feet above the surrounding plain. That makes it over three times higher than Mt. Everest above sea level.

The Hawaiian Islands would be that high too, if it were not for the fact that the Pacific Ocean Plate on which the Islands "ride" is gradually moving north-westward, carrying the Islands (volcanoes) past the magma chamber embedded in the beneath. Fresh molten material is gradually pushing up through the crust to the southeast of the present Hawaiian Islands, which will in turn create new volcanoes (and new islands for additional condominiums!). Geologists have already mapped (and named) the slowly-rising seamount **— Loihi**. It is presently 3,000 feet beneath the present ocean level.

This scenario is also obvious if we look at the configuration of Earth's crust to the northeast of the present Hawaiian Islands. There is a chain of submerged, extinct volcanoes that long ago were situated over the magma chamber, but plate movement has caused them to move farther and farther away from the chamber that is embedded in the asthenosphere. Mars lacks crustal movement due to the greater thickness of the crust, and therefore the molten material continues to ooze up to form an increasingly higher and higher mound.

Mountains, Ocean Trenches, and Mid-Ocean Ridges
Have you fantasized taking a cruise through the Mediterranean? When you eventually do, you will cruise through waters that have filled the depression caused by the separation of the Arabian and Eurasian plates. If you decide to hike in the majestic Himalayas, you will walk along the crease formed by the Indian Plate jamming against the Eurasian Plate (bottom portion of Figure 17-8), causing the plates to "buckle" upward where they meet.

Similarly, a jaunt through the mighty Andes mountain range that runs along the west coast of South America is an excursion along the fold created by the South American Plate pushing against the Nazca Plate. In this case, however, instead of the two plates buckling upward, the Nazca Plate is **subducting** ("leading under") beneath the South American plate, pushing up its edge in the process.

Finally, an example of a mid-ocean ridge through which new material is oozing from below is the Mid-Atlantic ridge (Figure 17-7). This region is not so obvious to the tourist, since it lies on the floor of the Atlantic, and therefore inaccessible to the average traveler. But it was extensively studied and photographed by deep-diving submersibles like *Alvin* in the late 1970s.

Appearing along the ridge are pillow-like formations of newly solidified lava, having just recently been pushed up from the asthenosphere below. And most convincingly, rock samples taken from either side of the ridge at increasing distance from the ridge reveal a definite pattern — the closer to the ridge the sample is located, the younger it is.

Figure 17-9 The satellite Lageos *was launched in 1981 to determine the rates at which Earth's "plates" are moving with respect to one another. Geophysicists placed aboard for future generations a plaque on which are the projected locations of land masses in the past (200 million years ago) as well as into the future (10 million years). (NASA)*

✯✯✯✯✯✯✯✯✯✯✯✯✯✯

Figure 17–10 The non–uniform distribution of various species of animals, such as the Komodo dragon of Indonesia, is explained by the theory of plate tectonics. (Tom Bullock)

That is just the result we expect to obtain if our hypothesis of plate movement is true.

If you want to visit a site where plate separation is just beginning to develop, where a new plate is in the process of forming, try an African safari in the country of Kenya. In fact, it is along just such a split in the African plate in Kenya that fossils of early humans were uncovered by anthropologists such as the Leakeys. Continental drift is assisting our studies of early humans.

In October 1992, the **Laser Geodynamic Satellite** (**LAGEOS II**, for short) was placed into Earth orbit for the purpose of monitoring the movement of crustal plates of Earth's surface, and to detect wobbles in Earth's rotation. Covering the surface of *Lageos* are 426 retroreflectors, which reflect any laser beam aimed at it directly back to Earth. Thirty laser stations distributed around Earth measure the time that the laser beam requires to make the round-trip journey, which in turn allows each station to determine its own position relative to the other stations to within a fraction of an inch. Any crustal movement between stations is detected and measured (Figure 17-9). Detection of wobbles in Earth's rotation allow geophysicists to detect the movement of material deep in Earth's core. A third *Lageos* satellite was launched in 1994.

Pangaea Hypothesis Based upon what geologists presently know about the rate and direction of motion of each continental plate, a map of Earth's surface as it must have looked 200 million years ago reveals that there was a single land mass. We call it **Pangaea**. It is tempting at this point, of course, to pose a question that is similar to one we raised while discussing the Big Bang theory — "From whence came *Pangaea?*" Well, it is generally agreed that this supercontinent came from some previously drifting plates that came together, recoiled, and which are now gradually separating. So it is analogous to the pulsating or oscillating Big Bang theory — alternate expansion and contraction, Big Bang, expansion and contraction, etc.

The theory of Pangaea has important consequences as regards the distribution of animal species around the world, as studied by biologists. Because as the plates moved apart from the supercontinent Pangaea and were subsequently separated by the ocean's waters, animal species common to the plates when they were together were left to evolve separately, gradually developing dissimilar features.

For example, an isolated island of Indonesia is inhabited by an animal whose ancestors date back to the age of the great reptiles 100 million years ago, at which time the plates were closer together and even connected in some cases. The Komodo dragon is found no where else in the world (Figure 17-10). Apparently, when the continental plates separated, the ancestors of the Komodo dragon on other land masses died off, leaving only those on Komodo Island to carry on. The environmental conditions that existed there were conducive to their survival. Elsewhere, they were not.

Magnetic Reversals In addition to gathering evidence showing that Sun's magnetic poles reverse polarity every 11 years, scientists have gathered evidence that Earth's magnetic field reverses as well, although over much longer time intervals. Were you to be holding a compass at the time of a reversal, its needle would simply rotate through 180 degrees.

The most recent reversal occurred about 700,000 years ago, but a more typical interval is 200,000 years (Figure 7-24). In geologic terms, they occur rapidly, the poles going end to end in a few thousand years or less. Just why reversals occur is not understood, but a recent theory connects them to the same events that may have caused the extinction of the dinosaurs some 65 million years ago.

The theory goes something like this — Sun is a member of a binary star system. The other star, called *Nemesis*, having a period of revolution around Sun of about 26 million years, is presently located close to aphelion, at its greatest distance from Sun. After passing perihelion and entering the outer portions of the solar system, the star gravitationally disturbs the orbits of some asteroids and/or comets, some of which then collide with Earth at speeds of 20 miles per second. Dust from the impacts blocks sunlight for months or even years, leading rapidly to an ice age. Species that survive only within a narrow range of such parameters as temperature, die off when the temperatures drop. This is the *Nemesis Star* theory discussed in the last Chapter.

Because of the rapid buildup of glaciers and ice caps

near the poles at the expense of water in the oceans, the rotation of Earth speeds up (conservation of angular momentum). Researchers calculate that dropping the level of the oceans 30 feet and piling it up as ice near the poles would speed Earth's rotation by one part in a million. But this redistribution of mass on the surface causes the rigid crust to speed up slightly compared to the softer interior. In 3,000 years or so, the crust has made one extra rotation compared to the interior, thus mixing up the magnetic field. Eventually, the inside of Earth speeds up to match the surface, and the dynamo reorganizes itself. The entire scenario reads much like that of Sun's 11-year magnetic field cycle.

Future of Earth's Surface

If we project continental drift backward in time, then we should be able to predict the future face of Earth. As the North American Plate continues to separate from the African and Eurasian plates, the Atlantic and Indian Oceans will enlarge while the Pacific Ocean will shrink. The African Plate's northward movement against the Eurasian Plate will eventually make Mediterranean cruises a luxury of the past (refer back to Figure 17-9).

And Los Angeles, along with a sliver of the State of California, will be cast adrift at sea, destined to eventually subduct under another plate whose boundary is along the Aleutian Island Trench. Los Angeles will then be ground up and recycled somewhere else in place and time. At an average rate of slippage of one inch per year along the boundary of any particular plate, these changes in the surface configuration of Earth seem extreme. But of course we are talking in terms of enormous periods of time.

For example, in 50 million years, movement amounts to about 800 miles. And Earth is 5 billion years old, or 100 times older than that. Time, in fact, has consistently been an obstacle to our appreciation of or even acceptance of theories regarding Earth's origin. We live for such a short period of time compared to the time intervals required for large-scale changes to occur on Earth's surface.

The same is true regarding debates over the origin of life on Earth. The scientific theory of evolution says that over a 5-billion-year period of time, chemical reactions in some stinky pools of primordial ooze resulted in self-replicating molecules (molecules that create copies of themselves), which eventually directed the development of a structure that became aware of itself, you. That seems so simplistic and improbable.

But how can we condense 5 billion years of chemical-physical experimentation into one sentence? How can we, in the space of the 10 seconds required to read this sentence, fully comprehend and appreciate the kinds of things that chemistry and physics can do in 5 billion years of time? I am not arguing for evolution. Remember — "The thrill is in the chase, not in the kill!" What I am saying is that time limits our understanding and appreciation of nature's workings, and that is certainly true as regards the dynamics of continental drift.

I began discussing continental drift by presenting evidence for the semi-molten nature of Earth's outer core, mantle, and asthenosphere — kept molten by heat escaping from the interior. That heat is escaping because the behavior of energy flow is from high energy to low energy. Heat from Earth's interior is released to the coldness of surrounding space. That isn't too obvious to us since we live within the protective blanket of the atmosphere which traps and holds much of the heat close to the bosom of Earth's surface (the **Greenhouse Effect**). If the atmosphere were not present, the average temperature on Earth would be about 20°F. Everything would be frozen solid!

The point that I want to make is that Earth is gradually cooling off, losing heat to its surroundings, just as a hot potato loses heat to its surroundings when placed in the refrigerator. *Ergo ipso facto* ("Therefore, it goes without saying"), this gradual cooling from heat loss will eventually lead to that time when the asthenosphere flows less freely, and continental drift grinds to a halt. Large-scale changes associated with such drift (mountain-building, volcanic eruptions, etc.) will cease. The study of geology will turn dull. But by then, estimated to be some 2 billion years in the future, geologists will have their hands full trying to explain discoveries made by astronauts who have landed on distant planets in the Andromeda galaxy.

The Earth as a Planet

The subject of continental drift is interesting not only because it relates to the environment in which we live out our lives, but because it offers a starting point in our understanding of the internal structure of other planets and the satellites of our solar system. If we find evidence of plate boundaries on other planets or satellites, we presume molten interiors for them. Or, if we find that a planet or satellite has a magnetic field, we presume a molten interior for it, and therefore that it is still evolving.

Earth's Surface Even a casual glance at the surfaces of the planets and satellites (like Moon) close to us suggests that the task to understand the interiors is not easy, however. That is because, by and large, their surfaces are dominated by **impact cratering**, that phase of planet- and satellite-building that ended about 3 billion years ago.

Estimates are that only about 10% of the terrain features of Earth can be associated with the last stages of the era of impact cratering, with the remaining 90% being due to what is called **heat tectonic activity**. The outflow of heat from Earth's interior, resulting in mountain-building and the other side effects of plate movement, together with erosion by wind, water, and life processes, have been dominant on our planet ever since.

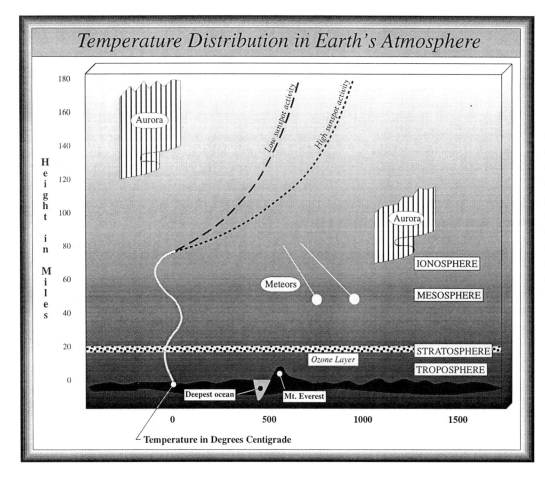

Figure 17-11 A cross-section of Earth's atmosphere illustrates the altitudes at which various phenomena occur. Note also the pattern of temperature change with altitude, and the effect Sun has on those temperature differences.

Atmosphere Other than the presence of life, what one usually associates with the uniqueness of Earth is its atmosphere. I have already mentioned its importance to us not only in terms of our survival (trapping Sun's heat, filtering out damaging radiation and charged particles from Sun), but also its role in shaping the environment through erosion.

So the atmosphere plays an extremely important role for us beyond the obvious one of supplying the oxygen necessary for us to breathe. It has no particular thickness — it just gets less and less dense with distance from Earth's surface. Its cross section up to about 180 miles is shown in Figure 17-11. Some of the atmospheric phenomena observable from the surface is included to further your appreciation of its dynamic nature. Chemically, it remains rather uniform, there being a delicate balance between the external (solar radiation) and internal (life processes, etc.) forces that attempt to modify its composition. For example, each time you leave an iron tool outside to rust, you remove some oxygen from the atmosphere as iron oxide forms on the tool.

Origin of Oxygen It is the presence of oxygen in our atmosphere, however, that notably sets our planet apart from all others. We know of no other object in the solar system that has molecular oxygen in its atmosphere. Therefore, if we are at all to propose an acceptable theory of the evolution of life on Earth that is based on average conditions in the universe, we must account for this strange gas that we depend upon. The oxygen we breathe is almost entirely a result of photosynthesis, produced primarily by the green plants in the oceans and to a much lesser extent by land plants. Therefore, if oxygen is a by-product of life

Gases Added to Earth's Atmosphere by Volcanic Outgassing
(percent composition by weight)

Gas	Observed from eruptions that tap the mantle (Hawaii)	Observed from continental fumaroles and geysers
H_2O	58	99.4
CO_2	23	0.33
CL_2	0.1	0.12
N_2	6	0.05
S_2	12	0.03
Others	0.9	0.07
Total	100	100

Figure 17-12 There is a difference in the chemistries of the gases released from the deep interior of Earth and those released from the crust. This tells geologists something about the chemistries of Earth's interior, and therefore about its evolution.

processes, it seems reasonable that the atmosphere of Earth has evolved just as has its surface.

It is generally believed that the early Earth's atmosphere contained no oxygen at all, and that it became oxygen-rich only after life got a foothold and began to change the primitive atmosphere. This is not difficult to understand if you keep in mind that the nebula from which Earth formed was chemically identical to that of Sun. After all, the nebula from which Earth is believed to have formed was a fragment of the solar nebula.

So the chemicals contained in the early Earth's atmosphere were no doubt combinations of those very chemicals making up Sun today — carbon dioxide (carbon and oxygen), water vapor (hydrogen and oxygen), and ammonia (hydrogen and nitrogen). This means, of course, that the earliest of life forms would have evolved in an environment quite different from that which exists today. Today's atmosphere would be lethal to the earliest life-forms, as theirs would be to us today.

To support this theory, there is a class of organisms that live on Earth today whose members are poisoned by oxygen. They survive by living in environments out of contact with the atmosphere, such as oceanic muds and some soils. Biologists are of the opinion that these **anaerobes** are the survivors of those earliest life forms that evolved prior to the atmosphere acquiring oxygen status. They survived by going underground, so to speak. The only other life-forms that survived are those that gradually learned to adapt to the very poisonous by-product exhaled by the planets — oxygen. So we have evolved to use that poisonous substance for burning our food.

You see, oxygen is quite a reactive substance, and quite readily combines with many different atoms to form oxides. Many of the varied colors you observe in exposed rocks are due to the presence of such oxides, which indicates that humans and rocks compete with one another for obtaining and removing oxygen from the atmosphere. If the resupply of oxygen back into the atmosphere were cut off for some reason, after several tens of thousands of years there would be none left.

Oceans and Atmosphere But, you say, from whence came that original atmosphere within which the first life forms developed. And what was its composition if not oxygen-rich? No doubt there was significant **outgassing** from the interior of Earth in the process of it forming. Heat trapped within its interior dissociated minerals to release gases through early volcanoes and smaller **fumaroles**. Earth's gravity is sufficiently strong to hold onto these gases, even though the early solar radiation attempted to boil it away (Figure 17-12).

Moon was not so fortunate. Its small gravity (17% that of Earth) allowed for outgassing to be lost to interplanetary space, perhaps even to be captured by more massive bodies such as Jupiter and Saturn. Mars has a gravitational field 38% that of Earth, which allowed it to retain only about 1% of its outgassed gases. The intense heat on the surface of an object closer to Sun, such as Mercury, was sufficient to boil away all outgassed chemicals — even though the gravity itself is sufficient to retain a thin atmosphere.

A study of the gases exhaled from volcanoes today gives us clues as to what the composition of our early atmosphere must have been. They are water vapor, carbon dioxide, nitrogen, and small amounts of methane, ammonia, and sulfur compounds. Keep these gases in mind, as we will encounter them frequently as we cruise the solar system. So the early atmosphere of Earth must have been rich in carbon dioxide, just as those of present-day Mars and Venus.

As the water vapor fell out of Earth's early skies to form the beginnings of oceans, the carbon dioxide was easily absorbed by the water. Anyone familiar with carbonated drinks knows how easy it is to do. It is a very inexpensive process. But the oceans could not absorb all of the available carbon dioxide, so minerals within the seafloor combined with it to form limestone and sand and other mineral sediments. This left the atmosphere rich in nitrogen, to be eventually enriched with oxygen after life got a start.

Conditions Favorable to Atmospheres

We are not generally aware of how important Earth's atmosphere is until we are deprived of it (like swimming under water) or we are made aware of treats to it by modern civilization. It is rather like the fish who suddenly finds itself dangling from a fisher-person's hook. We take it for granted because we are surrounded by it. If we expect to find life anywhere else in the solar system, or the conditions that could have made life possible, certainly we place atmosphere at the top of the list of requirements. The discovery of an atmosphere elsewhere in the solar system is reason for great excitement amongst scientists. In is curious, therefore, that two planets side by side in the solar system — Earth and Venus — are so different with respect to their atmospheres.

Lacking liquid water on its surface because of its closeness to Sun, Venus followed a different course. There was no other mechanism for removing the carbon dioxide outgassed from the interior, so the atmospheric gases trapped more and more of Sun's radiation, releasing additional carbon dioxide from the surface minerals.

Mass and Temperature Today, the atmosphere of Venus is so thick that we are unable to observe the surface directly, in spite of the fact that it is the nearest planet to Earth. So proximity to Sun and mass of the object determine how much of the original outgassed material is retained. But there is one additional factor, one that will allow you to better understand the profound differences between the terrestrial and Jovian planets.

Presumedly, the composition of the original solar

nebula from which the planets and satellites formed was similar to that of the interstellar medium itself. Earlier we found that the nebulae consist mostly of hydrogen and helium, with a smattering of the heavier chemical elements. So we expect that the rings of material that surrounded the proto-Sun shared that same relative distribution of elements. In other words, why aren't the atmospheres of all the planets pretty much the same?

Weights of Chemicals The additional factor that determined what chemicals a given planet or satellite kept was the weight of the chemical present in the early outgassed atmosphere. The closer to Sun, the higher the surface temperature, and the more easily a given chemical will boil away. By boil away, I mean that the gas molecules absorb enough energy to give them escape velocity from the objects surface, just as giving a rocketship enough energy causes it to escape from Earth. Therefore, a planet loses its lighter molecules more easily than its heavier molecules because the lighter ones travel faster than the heavy ones.

When we apply all three of these factors (temperature, gravity, molecular weight) to the terrestrial planets, we find that none of them could have held onto hydrogen or helium in any significant amounts. But because of their masses and distances from Sun, Earth and Venus were able to hold onto their thickish atmospheres.

Because of its small mass and great distance from Sun, Mars is just barely able to hold onto water vapor, methane, and ammonia, but because it is a heavier molecule, carbon dioxide is more easily retained. In fact, some of the original water vapor on Mars was probably broken by solar radiation into its two components — hydrogen and oxygen. The hydrogen escaped to space and the oxygen combined with the iron in the surface material to form iron oxide, or rust. That is what gives Mars its predominant red color today.

Moon

Earth's satellite — Moon — is odd. Whereas the rule of thumb in the solar system is that big planets have big satellites, and small planets have small or no satellites at all, Earth possesses a relatively large satellite. Because of this, Earth-Moon system is frequently referred to as a double-planet system, even though Moon is not quite as large in diameter as the United States is from coast to coast. Its mass is only about 1% that of Earth.

In addition, whereas another rule of thumb in the solar system is that the orbits of satellites are in the equatorial planes of their parent planets (refer back to Figure 16-7), Moon orbits Earth very close (within 5°) to the plane of the ecliptic (refer back to Figure 3-17). It is tempting, for these two reasons, to theorize that Moon formed not along side Earth, but came from elsewhere in the solar system. If it formed as a secondary condensation with Earth, we expect its orbit to be along the equatorial plane.

After spending some $25 billion to bring moon-rocks back to Earth, we thought that we would have the riddle solved. But as of this date, there is no consensus as to its origin. The evidence is ambiguous, just the kind of stuff that spurs scientists on to further study. But before engaging in the controversy, let us look at the general features of this our closest neighbor in space. Perhaps there are important clues to be found in those features.

Surface Appearance Anyone who has even casually looked at Moon is aware of observing dark blemishes amidst a whitish background. Galileo was the first to give the dark areas a name, not because he was the first to see them, of course, but because his first views of Moon through his telescope provided a much more detailed look at them. He called the dark areas **maria** (seas), because they

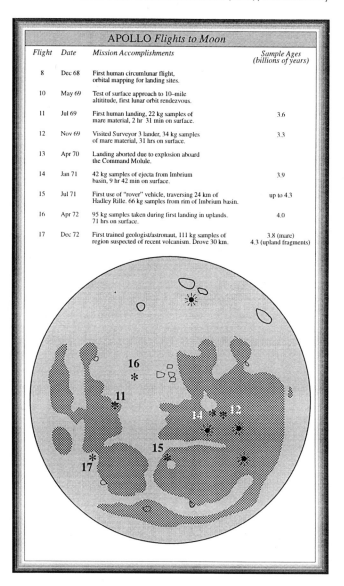

Figure 17–13 The general landing sites of the Apollo astronauts are shown in this figure, together with some of their accomplishments.

Figure 17–14 *The large circular dark regions on Moon, visible to the naked eye, are* maria*. They are large solidified lava fields that formed as Moon was cooling, and liquid lava flowed upward through cracks caused by giant impact.* (Lick Observatory)

Figure 17–15 *This close–up photograph of Moon clearly reveals evidence of massive collisions between Moon and debris in the early solar system. The ejected material (rays) was thrown out in all directions, as can be seen around the crater Copernicus.* (Lick Observatory)

reminded him of dried-up ocean beds (Figure 17-14).

The rest of the cratered surface he called **terrae** because they reminded him of the mountainous regions on Earth. The *maria* are dark because they reflect less light from Sun. The first astronauts to land on Moon in 1969 landed in one of these "seas" — the **Sea of Tranquility** — because of its relative smoothness. To land in one of the rougher areas may have threatened the mission (Figure 17-13).

The maria are covered with layers of lava quite similar to those found around terrestrial volcanoes. Radioactive dating of samples of the lava returned to Earth reveal them to be about 3 billion years old, quite young compared to the 4.5-billion-year age of samples returned from the cratered **highlands** associated with the rest of Moon's surface. Yes, it was discovered that the maria are low-laying areas, whereas the *terrae* are higher, mountainous areas.

Cratering The lunar mountains tend to concentrate along the edges of the rather circular maria, suggesting that they are the result not of moving plates buckled upward like Earth, but rather the remains of the ridges of great craters caused by meteorites colliding with Moon early in its history. The violent collisions caused cracks to form in the floors of the resulting craters, and molten rock from Moon's interior oozed up through crustal cracks to flood the crater bottoms. The lava gradually cooled to become the present-day maria. This scenario explains the youthfulness of the maria and the old ages of the mountains.

Another observation that supports this theory is the cross-sectional shape of the mountain ranges around the maria. They rise more steeply on the side facing the maria than on the opposite side, just as we would expect were they the remains of such a collision. Actually, there is a specialized field of physics that deals with things hitting other things, called **collisional physics** (how would you like to spend your workday throwing objects at various targets?). The various types of craters found throughout the solar system can be duplicated in the laboratory by varying such factors as material content, temperature, size, speed at impact, etc.

Seen through large telescopes, some 30,000 craters are visible on Moon's surface, ranging in size from that of a football field up to 200 miles in diameter. This does not include the maria, which range in size from 200 to 700 miles in diameter. Can you imagine observing a collision that leaves a 700-mile crater! Not all of the craters are of the same age, however. This is obvious when you carefully study close-up photographs of Moon's surface — craters are seen on top of craters, which are seen on top of other craters, etc.

The general rule is that smaller craters lie on top of larger craters, causing the outline of the original craters to be quite obscure. But there are some quite youthful impact craters, made obvious by the streaks of ejected material that fell back down to Moon's surface to form *rays* (Figure 17-15). To view these craters and their rays close-up through

★★★★★★★★★★★★★★

Figure 17–16 Astronaut James Irwin gives a military salute at the Apollo 15 landing site at the Hadley–Apennine landing site. Hadley Delta in the background rises about 13,000 feet above the plain. The base of the mountain is about 3 miles away. (NASA)

Figure 17–17 One of the lunar samples returned from Moon is examined in a clean room. It is nicknamed "Big Bertha" because of its basketball size. (NASA)

a telescope is a fascinating experience. It seems as if the collision just occurred! Because the ray material lies atop many smaller craters believed to have formed more recently than the larger ones, the small craters are estimated to be only some 500 million years old (Figure 17-16).

In spite of the fact that Moon appears quite brilliantly white in the sky, even the lighter terrae reflect only about 7% of the light that falls on them from Sun. In general, Moon is made up of the same chemical elements as is Earth, but in slightly different proportions. This alone would suggest that it evolved from the same cloud of material as did Earth and at approximately the same distance from Sun. But as I mentioned earlier, there are other factors that must also be considered in proposing an origin theory.

To be more specific, Moon contains less iron that Earth. But it contains more of the substances that cannot be melted easily, such as calcium, aluminum, and titanium (keep this latter substance in mind – it may be the very substance needed for building habitats on Moon and elsewhere in the solar system). The chemicals that we associate with life processes, such as water and organic (carbon-based) compounds, are completely lacking in Moon's surface material, although almost 50% of Moon consists of oxygen locked up in combination with other elements. Thus what we can generally say is that Moon's chemistry is closer to that of Earth's mantle, but, taken as a whole, quite dissimilar (Figure 17-17).

Rotation From Earth's surface, we see only one side of Moon. It is locked into what we call **synchronous rotation** around Earth. This is not unusual — most of the satellites in the solar system that orbit close to their parent planet behave in this fashion. What happened early in its history, no doubt, is that Earth's gravity very slightly distorted Moon's shape into that of an egg, and then gravitationally

Figure 17–18 The backside of Moon, seen here in a photograph taken by the lunar orbiter, has very few maria. Apparently Moon's crust on the backside is thicker than on the side facing Earth. (NASA)

locked onto the elongated part of that egg. Therefore its period of rotation is equal to its period of revolution. Sunrise to sunset and sunset to sunrise each last about two weeks on Moon.

Since it cannot be seen from Earth, the far or backside of Moon was first photographed and mapped by a Soviet satellite. What is surprising is that there are no sizeable maria on the far side — it consists almost entirely of craters. In total, the cratered highlands cover about 83% of the lunar surface, whereas the maria cover only about 17%.

There are 16 maria on the near side, 4 on the far side. Lacking sizeable maria, the far side has no extensive mountain ranges (Figure 17-18). Astronauts did not explore any of the regions on Moon's backside, because that would have placed them out of direct contact with ground control in Houston, something any of us would request were we to be sent to Moon ourselves. Presumedly, the age of the far-side material is similar to that of the 4.5-billion-year age of the near-side terrae.

Accurate tracking of orbiting satellites reveals that the crust on the near side is thinner than that on the far side, allowing for deeper penetration of those early giant meteorites. This would explain why the backside of Moon has so few maria. The molten lava, in trying to make its way to the surface, took the path of least resistance, the shortest route where the crust was thinnest, the side of Moon that faces Earth.

Scientists believe that the nonuniform crust thickness is a result of the differential gravitation attraction that Earth has on Moon, much as Moon's effect on Earth's tides. The denser material tried to sink to the center of Moon, while the lighter crust material floated to the top. At the same time, however, for the near side of Moon at least, Earth had a stronger attraction for the more dense material, and subsequently prevented a thick crust from forming.

Interior In addition to collecting samples from Moon's surface, the astronauts left seismographs behind in order to obtain data with which to understand Moon's interior. It was reasonable to assume that Moon was still settling to some extent, since its interior was thought to be warmer than its exterior. But also, Earth must exert a large tidal strain on Moon that contributes to a certain amount of shifting of material within Moon. Seismic events (Moonquakes) confirm this suspicion, although they reveal different patterns of waves than those in Earth.

After a Moonquake, the waves build up gradually and then take several hours to subside, much like striking a cathedral bell. They also originate much deeper within Moon than Earthquakes, about 500 miles or so below the surface in contrast to about 50 miles for Earth. This fact suggests that Moon possesses a thick shell of strong, rigid rock. Just as with Earth, the quakes occur at the plastic base of the rigid shell, under which is the semi-plastic asthenosphere. But the seismic waves change speed at a depth of 40 miles or so, suggesting that the hard shell is differentiated into a crust and a mantle.

The energy associated with Moonquake activity is less than one ten-billionth that of Earth. Since Moon is so much smaller than Earth, it is reasonable to assume that it cooled off much more rapidly than did Earth, and therefore lacks a sufficiently molten core to allow for significant seismic activity. This is verified by heat-flow experiments conducted by astronauts on Moon's surface. An amount that is only one-third that of Earth's heat flow is lost through Moon's crust. There is virtually no magnetic field around Moon — only about one ten-thousandth that of Earth. If, indeed, strengths of magnetic fields are measurements of the extent to which bodies have molten interiors, then the very slight measurement for Moon confirms the results of the seismic studies that suggest a rather solid interior for Moon.

Origin of Moon

So with the information that is now available to scientists, what can we say about Moon's origin? Well, it certainly appears obvious that it formed in the same manner as that of Earth, accreting from planetesimals that were part of the contracting material of the pre-solar nebula.

Studies of moon-rocks returned to Earth reveal that much of the original crust was destroyed by the flooding of the basins of huge craters, and by slow modification by meteoritic impacts that continue up to the present day. But as to where Moon formed, we are still quite unsure. Prior to the lunar landings of the late 1960s and early 1970s, there existed three theories to explain Moon's origin.

Moon Flew off Earth Because Moon is our nearest neighbor in space, and so familiar to us in the sky, it has always been easy to entertain the idea that Earth and Moon somehow evolved together. One popular theory proposed that when Earth was youthful and in a molten state, it was spinning quite rapidly (recall that Earth is slowing down as a result of tidal friction). Centrifugal force and instabilities in the molten material might have caused a "blob" of the molten material to fly off and remain in orbit around Earth, gradually cooling off to become the present-day Moon. Hence the name — **Earth fission** or *"daughter"* theory.

The scar left on Earth by escaped Moon was subsequently filled by water, and is today the Pacific Ocean. This was therefore referred to as the **Pacific Ocean Basin theory**, first proposed by George, the son of Charles Darwin. He also suggested that perhaps bits and pieces from this ejected piece fell back to Earth's surface to form the Pacific Coast Range of mountains.

However, calculations reveal that the total energy content of Earth-Moon system does not allow for an early Earth rotating the once-every-2.6 hours necessary to eject a piece of Earth into orbit. Also in contradiction to this theory are geological studies of Moon-rocks and rocks that ring the Pacific Ocean and the conclusion that they do not match one another at all.

Earth Captured Moon A more recent theory, popular during the 1960s, was that Earth might have gravitationally "captured" a meandering planet that formed elsewhere in the solar system. Such a **capture theory** does not require that the material compositions of Earth and Moon be similar, but it does require that the energy lost by Moon in slowing down during capture must have been absorbed by Earth (*conservation of angular momentum*). Such a huge amount of energy would surely have created geological features on Earth that have yet to be found (Have you ever tried to "capture" a 280-pound fullback running at full speed toward you?). Besides, the chances of such an encounter, in light of the tremendous distances between bodies in the solar system, are very slight — especially one that has just the right parameters to allow for capture.

Earth and Moon Formed Together The serious objections to the foregoing theories led scientists to hover around the **co-formation theory** — Earth and Moon formed together from a common "blob" of material encircling the proto Sun. They thus formed as a "double-planet" system, orbiting around a common center of gravity (barycenter). The fact that Earth and Moon have such different chemical properties, however, is a serious impediment to accepting the possibility that both bodies evolved from the same "blob." In entertaining this theory, however, scientists maintain that it is easier to live with the dissimilarities in chemical makeup than to live with the difficulties associated with the fission and capture theories.

Moon was Knocked off Earth Recently a competitor theory arose, sort of a hybrid of the capture and fission theories, although its general acceptance may require another visit to Moon. Astronomers call it the **collisional-ejection** (or **impact**) **theory**, since it proposes that a Mars-size object collided with the early Earth. The dislodged material from Earth created an Earth-orbiting cloud of debris that eventually collected by gravity together to become Moon. This dramatic scenario for Moon's origin is also referred to as the **Big Whack theory**.

Using a super-computer, scientists have created a reenactment of the proposed collision and resulting Earth-Moon system, working on the assumption that it was more of a glancing blow than a head-on collision. Although most of the features of Moon and the differences between Earth and Moon are explained by this theory, much work needs to be done to fill in the gaps. Assuming that Earth had an iron core at the time of the collision, the glancing blow would have resulted in a chemical similarity between Earth's mantle and Moon. This is exactly what we find.

This question of Moon's origin is an excellent example not only of the methods of science, but also the notion that to a scientist, "The thrill is in the chase, not in the kill!"

Venus

You may well feel some affection for the planet Venus because of its occasional brilliant appearance in the sky as the Evening Star (Figure 17-19). You may have a fondness for Venus as well because it is the same size as Earth and the nearest planet to us. The fact that it was named after the Roman goddess of beauty suggests that you are not alone — other civilizations paid homage to her as well. The Aztecs of central America were fond enough of Venus to go to great pains to integrate the rising and setting points of Venus into the alignment and construction of various temples in the Yucatan of present-day Mexico — notably at the site of Uxmal.

But as regards Venus being referred to as our Sister or Twin planet, the kinship ends abruptly after their sizes are compared. There will certainly be no "giant step for Mankind" on the surface of Venus for quite some time. Even if we could brave the thick clouds of sulfuric acid that blanket the planet and the tremendous pressures the thick atmosphere creates on the surface, we'd be fried by temperatures that are similar to those in self-cleaning ovens.

Atmosphere Venus was first visited by the satellite

Figure 17-19 The over-exposed crescent moon (upper-right), Venus (left of and lower than Moon), and Mercury (left of and lower still than Moon) can be occasionally seen during the evening or morning twilight hours. On this photograph, you should be able to sketch in the approximate position of the ecliptic. (Tom Bullock)

Figure 17–20 Our first good close–up look at Venus by Mariner 10 revealed a completely cloud-covered planet, disappointing some astronomers who had hoped to get a peak at the surface. (NASA)

Mariner 2 in 1962. It was the very first planet to be visited by an alien space vehicle — as far as we know. Why Venus? Well, its closeness to Earth was certainly an important reason. But another has to do with the most prominent feature of Venus — its terribly thick atmosphere (90 times thicker than Earth's). You are well aware of it already — although only in an indirect way. Aside from Sun and Moon, Venus *can* be the brightest object in the sky. In fact, on a moonless night when Venus is at crescent phase, Venus can cast your shadow on the ground around you. You can observe a "Venus shadow" caused by the light of Venus!

This is possible only because the thick atmosphere of Venus reflects some 78% of the sunlight that falls on it — in contrast to the 39% reflected by Earth's atmosphere (Figure 17-20). The percentage of light that reflects off an object is called that object's **albedo**. Determining or calculating the albedo of an object in the solar system is very important to astronomers because it provides clues about the chemistry of the chemical(s) responsible for reflecting

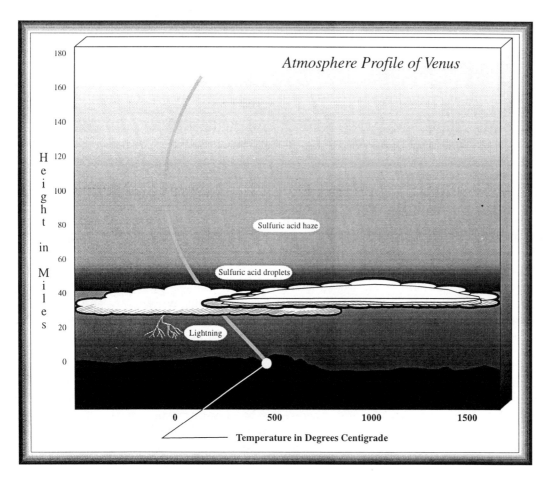

Figure 17–21 A model of the atmospheric profile of Venus, obtained from data returned to Earth by numerous probes that have descended to the surface of Venus and registered conditions during the descent.

CHAPTER 17 / Exploring the Earth–like Planets

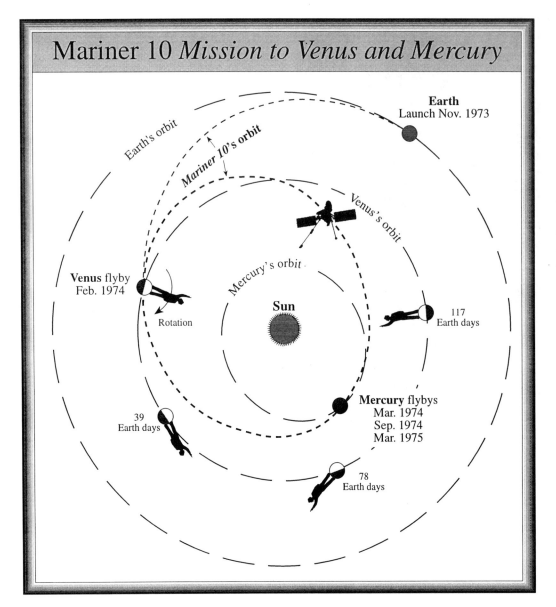

Figure 17–22 Mariner 10's trajectory allowed astronomers to kill two birds with one stone. Both Mercury and Venus were visited and imaged by the satellite. This diagram also illustrates the unusual rotation of Venus, and the length of time for an observer on the surface to go from sunrise to sunrise again.

the light in the first place. Coal dust reflects less light than does ice, for example. So the brightness of Venus in the sky is a result of both small distance from Earth and highly-reflective atmosphere (Figure 17-21).

Getting to Venus Whereas a one-way trip to Moon in a modern rocket ship takes about 2 days, the journey to Venus requires 6 months. Erase from your mind any image of rocket ships of the Flash Gordon era, shown sputtering out exhaust continuously as they journey around the solar system. That is an enormously inefficient way to get a space vehicle from one planet to another. It is slower – but more fuel-efficient – to send satellites on **least-energy orbits**. This type of orbit allows them to move as miniature planets controlled by the gravity of Sun. Once launched from Earth and given an initial "shove" to get it away from Earth's gravitational influence, the satellite obeys Kepler's laws of planetary motion (Figure 17-22).

Most of the energy needed to get the satellite from Earth to its destination is obtained from Sun, not from rockets strapped to it. In most cases, in order to provide sufficient energy to cause the satellite to visit more than one object, we take advantage of the gravity of a planet to swing the probe around and out beyond the planet to head toward another planet at a faster speed than when it approached the first planet. Because the probe has to steal a bit of the rotational energy of the planet as it swings around the planet, astronomers call this maneuver the **slingshot effect**.

Satellite Visits The high albedo of Venus may be great for a romantic evening under the stars, but the thick atmosphere plays havoc with attempts to discover what the surface of Venus is like. Observed through even the best of telescopes, Venus appears bland and blurry, simply because the atmosphere hides from our view any secrets of the

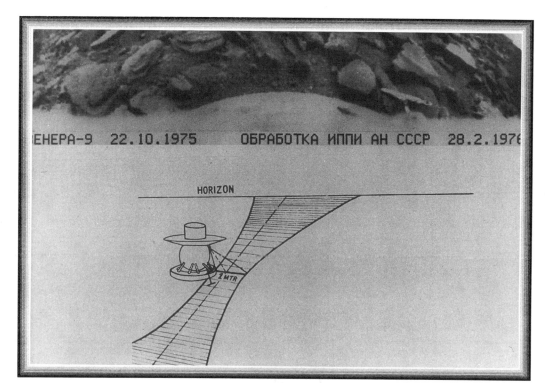

Figure 17-23 One of the images taken by the Soviet Venera satellites shows an extreme wide angle view of the surface of Venus. The horizon is severely curved, so that the surface stretches off into the distance in the upper-right corner of the photograph. (NASA)

surface. Since the *Mariner 2* mission to Venus in 1962, over 20 artificial satellites have visited it. Sixteen alone were sent by the then Soviet Union. Four of the Soviet **Venera** satellites, in fact, actually landed on the surface of Venus. Once on the surface, they took photographs and studied rock compositions before succumbing to the hellish conditions that exist there — 900°F in temperature and a pressure equivalent to that existing 1/2 mile beneath the ocean's surface.

Veneras 13 and *14* analyzed the chemical composition of the rocks at their respective landing sites. The rocks look very much like a rare type of volcanic rock on Earth, except with the important difference of containing a larger amount of sulfur. This finding indicates that sulfur participates in chemical weathering on Venus — the rocks and soil removing the element from the atmosphere. Were it not for the active volcanoes, there would be no sulfur or sulfur compounds in the atmosphere today, having long ago been absorbed into the surface material.

Later satellites dropped atmospheric probes that slowly descended to the surface while taking measurements of conditions at different heights above the surface. The information regarding temperature, pressure, etc. was then sent to an orbiter, which in turn relayed it to waiting radio telescopes on Earth. One of the most successful of these was the U.S. *Pioneer/Venus Orbiter and Multi-probe* program of 1978.

So we now have a wealth of information about the conditions within the atmosphere — its composition and temperature and pressure — and of course those on the surface as well. But detailed information about the surface features themselves is difficult to come by, inasmuch as the clouds prevent any broad view or photographic coverage. The photographs taken on the surface by the landers revealed what appeared to be vast lava flows broken up into angular fragments, suggesting a rather young-ish surface (Figure 17-23).

Radar Coverage So astronomers, intent on solving the mystery of what the broad surface features of Venus are like, developed a unique technique of "feeling" the surface features of Venus by using radar. Radio telescopes are able not only to collect electromagnetic radiation, but to transmit it as well. Transmitting radio energy instead of collecting it is accomplished simply by replacing the receiver with a transmitter.

Beginning in 1965, short bursts of radar energy from the Arecibo radio telescope were sent to different areas of that side of Venus facing Earth. A few minutes after each burst, a portion of the energy returned to the telescope at which its strength, component wavelengths, and time of receipt were accurately recorded. Since the thick atmosphere of Venus is unable to block or scatter radar energy, it is presumed that the returned signal reflected off the surface of Venus.

The *strength* of the reflected signal is a measurement of the roughness of the terrain off of which it reflected. When a pulse hits the side of a steep mountain, less energy is reflected back to Earth than when a pulse hits a level plain. The *time interval* between transmitted and reflected pulses reveals the heights of terrain features. A pulse hitting the top of a peak returns to Earth before a pulse hitting a low-laying

★★★★★★★★★★★★★★★★

Figure 17–24 These photographs are radar images that show the complete surface of Venus. Obtained by data from the Magellan orbiter as well as the Pioneer–Venus Orbiter, the lights and darks have been computer–enhanced in order to emphasize terrain differences. (NASA)

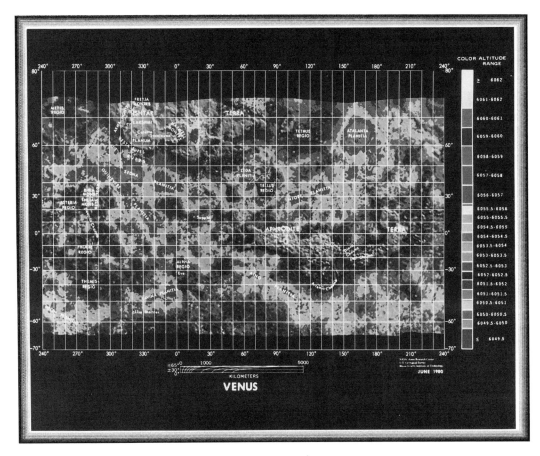

Figure 17–25 A topographical map of the surface of Venus has been compiled by combining all of the data received from the various satellites and radar images obtained by Earth–based telescopes. (NASA)

398 SECTION 4 / Evolution of the Planets

plain. The curvature of the planet is considered while making the calculations in this experiment (Figure 17-25).

The **Pioneer/Venus** spacecraft mentioned earlier was actually an orbiting radar telescope, and it provided more complete coverage of the surface of Venus than the Arecibo radio telescope. But its resolution of 25 miles was significantly less than that of the earthbound radio telescope — which is about 2.5 miles. *Pioneer Venus* was not, however, able to study the polar regions of Venus. In 1984, the Soviet *Venera 15* and *16* radar orbiters provided us with the best resolution yet, about 3/4 mile.

The Magellan Mission to Venus As you can readily see as you examine the data returned by successive probes to Venus, scientists have been striving for ever-better resolution of the images returned to Earth. So you won't be surprised to learn that the satellite that is currently orbiting Venus is superior to all of its successors. The **Magellan Venus Orbiting Radar** satellite, launched in May 1989 from the Space Shuttle *Atlantis*, arrived in August 1990 to begin its 5-year radar-mapping mission with a resolution capability of 100 meters. *Magellan* has provided us with a more detailed and a more complete coverage of Venus than we have of our own ocean floors. It took just one day for the sophisticated probe to map 84 percent of the planet's surface. That is, one Venus day — which is equal to 243 Earth-days (Figure 17-22)!

In September 1992, the last phase of its mission began — mapping the gravity of the Venusian surface. Changes in the spacecraft's altitude during each orbit tells astronomers about variations in the gravitational field around Venus over a region from 40° North latitude to 30° South. This allows astronomers to study the crust of Venus to determine what the surface layers are made of. Venus reveals no tectonic activity like that occurring on Earth. And since it is tectonic activity that is responsible for Earth's mountain ranges, we naturally wonder what forces are responsible for creating the mountains of Venus.

When the gravity-mapping mission ended in 1993, *Magellan* was turned off. It would have been possible to extend the mission and obtain even higher resolution photographs of the surface by "aero-braking" the satellite — slow it down by causing it to collide with the upper atmosphere, and consequently bring it to a mere 110 miles from the surface. But funds to extend the mission were not made available. So what have these latest images told us about Venus?

Having by now photographed 97% of the surface of Venus — with about 30% of it mapped with stereo 3-D imaging techniques - *Magellan* reveals a landscape dominated by volcanic features, faults, and impact craters. Huge areas of the surface show evidence of multiple periods of lava flooding with flows lying on top of previous flows. This is not difficult to understand, since the planet has some 430 volcanoes 12 miles or more in diameter, and tens of thousands of smaller ones (Figure 17-24).

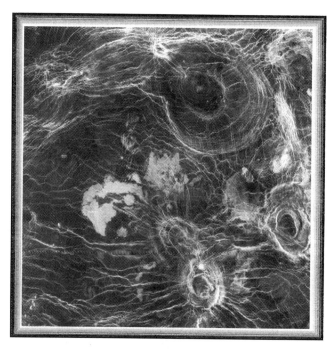

Figure 17-26 This extremely detailed radar image provided by Magellan *reveals features on Venus that are found no where else in the solar system except Earth. The presence of arachnoids indicate the upward movement of molten material sometime in the past.* (NASA)

Surface Features-Water Prior to *Magellan*, there were a few scientists who thought that we would find evidence of ancient shorelines or long-dry riverbeds on the surface of Venus. Their optimism was no doubt a result of our surprise finding of dry riverbeds on Mars and vast amounts of water frozen in the soil. But no such evidence was revealed in the images sent to us by the satellite. At least there is no evidence of water having been on the surface within the past 500 million years — the estimated age of most of the surface regions. The surface lacks small impact craters, no doubt because small meteoroids burn up in the dense atmosphere.

If some of the older-looking regions had small craters, that would suggest that the atmospheric shield was lowered in the past (the atmosphere was thinner), but that is not the case. So if the atmosphere has been at its present thickness for a long time, it is not possible for water to have existed on the surface. That reasoning leads us to two possible hypotheses: (1) Venus formed water-rich with an early ocean, but lost it through the high surface temperature created by the dense atmosphere, or (2) Venus formed dry and the small amount we do detect is in a steady state — that which is lost to space is replenished by volcanic activity.

Surface Features-Volcanism Certainly volcanism is a dominant process on the surface of Venus. There is no other way of explaining the vast assortment of features imaged by *Magellan*. The most intriguing of these are the sinuous channels, which average a mile wide and 500-1000 feet deep. The longest is 4,200 miles long, longer than Earth's longest river — the Nile. It cannot have been carved by liquid water. It was most likely liquid rock — lava. But most lavas become sluggish and "freeze" after flowing several hundred miles at the most. So we are curious about the composition of the lavas of Venus.

The photograph in Figure 17-26 reveals features found nowhere else in the solar system. They are called **arachnoid** formations because they resemble spiders (*arachnids*) and cobwebs. Found by the fine resolution of the *Magellan* radar camera, arachnoids are circular to ovoid in shape with concentric rings and intricate outward-extending fractures. Those shown in the Figure 17-26 are located at 40 degrees N latitude, 18 degrees E longitude, and range in size from 31 to 143 miles in diameter.

The bright lines, which extend outward for many miles, are fractures that may have been created when molten rock upwelled from the planet's interior, causing the surface to crack. Cooling and retreating of the magma resulted in collapse of the center. The bright patches in the center of the image are lava flows, suggesting volcanic activity. Some of the fractures cut across the lava flows, suggesting that the fractures are younger than the flows themselves. So by studying the relations between different structures, scientists are able to determine the relative ages of various geological formations.

Smaller-sized arachnoids, called **pancakes** because of their appearances, are believed to be small dome volcanoes. They are caused by lava oozing through Earth's crust and cooling as it spreads in a circular pattern. They are scattered around California, the easiest to observe located adjacent to Mono Lake on the eastern side of the Sierra Mountains.

The difference between Earth's volcanic landscapes and those of Venus is that the forces of erosion dominate the surface here, erasing or at least modifying evidence of the activity that formed the features. Volcanic features that form on Earth are soon covered by plants and eroded by wind and water. On Venus, a lava structure remains in its virgin condition until another geologic force alters it.

The *Venera* spacecraft first detected large circular features up to 700 miles in diameter that have been given the name of **coronae**. They are clearly not craters, since they lack the sharp ridges that accompany features caused by colliding asteroids. Rather than being depressions in the surface of Venus, they appear to be slightly domed and are bounded by concentric ridges. It is easy to conclude that the

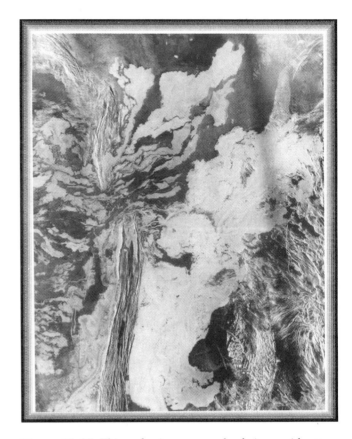

Figure 17-27 This radar image reveals obvious evidence of lava flows on the surface of Venus. It is not known if lava still flows today, although circumstantial evidence suggests that it does. (NASA)

Figure 17-28 (right) This radar image reveals three large impact craters on the surface of Venus. Their diameters range from 23 to 30 miles in diameter, and can be seen located in a region of fractured plains. The craters show many features typical of meteorite impact craters, including rough terrain, ejecta material, terraced inner walls and central peaks. (NASA)

coronae are bulges caused by plumes of molten rock below the crust. In addition, it seems plausible that the arachnoids are the precursors to coronae, that the latter have simply been the result of longer periods of upwelling of molten material and are consequently larger than the arachnoids.

The photograph in Figure 17-27 is of the *Lada* region of Venus, centered at 47 degrees S latitude, 25 degrees E longitude, and covers an area approximately 340 by 390 miles. The radar-bright and-dark lava flows reveal eastward movement of the lava, breaching a ridge belt (the elongated feature left of center) that runs north-south. After pushing through the ridge belt, these lava flows are deposited over approximately 38,000 square miles of the Venusian surface. The lava deposits appear as bright, swirling features in the radar image. The source of the lava flows, a caldera named Ammavaru, lies about 200 miles west of this scene.

The photograph in Figure 17-28 reveals evidence of massive collisions of meteorites on the surface of Venus. Smaller asteroids no doubt break into even smaller fragments when they collide with the thick atmosphere, but the larger ones get through. *Magellan* detected features on the surface that are caused by shock waves that strike the surface when some of the smaller asteroids hit the atmosphere and break up, but do not leave visible traces of the tiny pieces that strike the surface.

Surface Features-Tectonics Related to volcanism is the process of tectonics — faulting, fracturing, and folding of surface rocks. On Earth, as you learned, plate tectonics is at work modifying its surface. Seafloors cover 65 percent of our total surface, while the remaining 35 percent is covered by continents that rise — on the average — 2 miles above sea level. Venus, on the other hand, is rather flat. Figure 17-24 is a relief map of Venus similar to those you commonly see of Earth. 70 percent of the surface area of Venus consists of gently rolling, open plains, 20% consists of lowlands, and 10% consists of highlands.

The highland areas, which resemble Earth's continents in size at least, are found almost exclusively in two large plateaus that rise a few miles above the rolling plains. **Ishtar Terra**, which is about the size of the United States, contains at the eastern end the highest feature on Venus — **Maxwell Montes** (Figure 17-29). It rises almost 7 miles above the plains, higher than Mt. Everest in Nepal — which is only 5.5 miles high. **Aphrodite Terra**, located in the southern hemisphere, is the second highest plateau. But what it lacks in height it makes up for in size — it is twice as large as *Ishtar Terra*, and has significantly rougher terrain. The crust appears buckled and fractured, suggesting large compressive forces at work.

Two smaller elevated features, *Alpha Regio* and *Beta Regio*, appear to be a pair of volcanoes. And finally, there is a giant canyon almost 1000 miles long, 3 miles deep, and 250 miles wide. Thus the difference in elevation between the highest and lowest spots on Venus is about the same as

Figure 17-29 This is an artist's conception of Ishtar Terra, a huge up-lifted plateau on Venus. At its highest point above the surrounding terrain (in the vicinity of the state of Pennsylvania on the sketched-in map of the U.S.), it is higher than the highest point on Earth. (NASA)

Earth. Scarring the flank of Maxwell Montes is a 1.5-mile deep impact crater named *Cleopatra*. Almost all features on Venus are named for women — Maxwell Montes, Alpha Regio and Beta Regio are the exceptions ("regio" means region). Ishtar Terra contains some of the most dramatic terrain to be found anywhere in the solar system. Craters survive on Venus for perhaps 500 million years or so because there is no water and very little wind erosion.

Extensive fault-line networks cover the planet, probably the result of the same crustal flexing that produces plate tectonics on Earth. But on Venus the surface temperature is sufficient to weaken the rock, which cracks just about everywhere. This prevents the formation of major plates and large quake faults like the San Andreas Fault in California.

Surface Features — Volcanoes? There is considerable evidence that there are active volcanoes on Venus, although there is also considerable debate as to the interpretation of that evidence. Aboard both the *Pioneer/Venus* and *Venera* satellites were instruments capable of detecting lightning discharges from within the thick atmosphere of Venus. On Earth, it is quite common for lightning discharges to accompany volcanic eruptions. They occurred during the 1980 eruption of Mount St. Helens in Washington, for example. Discharges appeared around Beta Regio and two other suspected volcanoes.

There is further evidence — this time coming from studies of the atmosphere of Venus itself. As mentioned earlier, carbon dioxide dominates the contents of the atmosphere — 96% of it is CO_2. Nitrogen makes up 3.5%, and the remaining small percentage is accounted for by water vapor and oxygen. But all of these gases are transparent to visible light. What actually hides the surface from our view are dense clouds of sulfuric acid, formed when sulfur dioxide combines with water vapor in the company of sunlight. But sulfur dioxide itself is a common by-product of volcanic eruptions.

Comparisons of the data received from *Pioneer Venus* in 1978-79 with that taken earlier with Earthbound telescopes indicated a dramatic increase in the amount of concentrated sulfuric acid haze above the clouds during the interval between measurements. The most probable explanation for this sudden increase is volcanic eruption.

Volcanic gases do not change the concentration of sulfur dioxide in the entire atmosphere in a few years. If that were so, we ourselves would be threatened by eruptions occurring on Earth. But it is possible, according to scientists, that the convection currents created by the release of enormous amounts of heat during eruptions carry the discharged sulfur dioxide to the uppermost levels of the atmosphere, there to be detected by our instruments.

It is possible that Venusian volcanism occurs on a grand scale. Heat from Earth's interior prefers to escape through the rift zones between plates, primarily through the mid-oceanic ridges. Lacking such rift zones on a global scale, Venus releases internal heat through specific hot spots, similar to the Hawaiian Islands. Without local heat outlets, the crust of Venus would melt and be unable to support sizeable continent-sized plateaus known to exist.

Therefore, Venus may have a small number of these hot spots above which tremendous volcanic eruptions occur, accompanied by electrical discharges within the dense atmosphere just above. In 1982, instruments aboard the Soviet *Venera*s 13 and 14 detected not only a decrease in the amount of electrical activity on Venus — presumed to be associated with volcanic activity — but also the disappearance of the sulfuric acid haze and lowering of the sulfur dioxide concentration in the upper atmosphere. The detection of methane in the atmosphere is compatible with models of gases released by volcanic eruptions. Now that Magellan has completed three mapping cycles of the surface of Venus, scientists are comparing photographs taken of the same region during each of the three passes in order to detect changes (like signs of volcanic eruptions) that might have occurred on the surface during the intervals between passes.

Surface History of Venus Astronomers and geologists attempt to explain the surface features of Venus according to what we understand of similar features on Earth. Heat generated by radioactive decay in the interior of the planet creates pockets of molten rock (magma), which is occasionally pushed toward the surface. It may break through the surface to create a shield volcano or, if it stops just short of the surface, push the crustal material up into a dome. The temperature, composition, and gas content of the molten rock determine whether the magma has the flow characteristics of motor oil or toothpaste. These, in turn, determine the size and shape of the resulting volcanic formations on the surface.

In addition to volcanic activity on both Earth and Venus, tectonic forces leave behind evidence of their behavior. The primary force on Earth, as you learned earlier, involves lateral movements of crustal plates, creating high mountains such as the Andes and deep-ocean trenches. Venus, on the other hand, shows evidence of more localized tectonics — smaller rift-like features. Whereas upwelling magma on Earth often times emerges from the mantle through mid-ocean ridges, it forms highlands such as Alpha and Beta Regio on Venus.

Hotspot volcanoes and coronae also form over these magma upwellings, much like the Hawaiian Islands on Earth. Overall, the plate tectonics on Earth is largely a horizontal process, while on Venus it is largely vertical. For this reason, scientists refer to it as **blob tectonics** to contrast it with the plate tectonics operating on Earth. Volcanoes and highland regions form over mantle upwellings, while regions that are sinking may form mountain belts due to the thickening and compression in the crust.

Surface Features–Determining Age Recall that astronomers use crater density to determine the relative ages of regions on a planet's surface. But that is not possible for Venus — its low crater count prevents us from using that method, while at the same time telling us that the surface is quite young. Astronomers offer two interpretations of that observation: (1) a great global catastrophe occurred 500 million years ago, during which extensive lava floods erased all evidence of the ancient surface. The 900 craters visible on the surface today are the result of random hits during the past 500 million years. And (2) there is an equilibrium between small volcanic events that flood and bury craters and the rate at which crater form.

One surprise about the craters on Venus is their rather pristine appearance. On Earth, those not destroyed by plate tectonics are weathered by water, wind, and volcanic processes. There is no water on Venus, but many craters are partly buried by younger lava flows. One would think that the thick atmosphere would rapidly erode newly-formed craters away, but the probes that parachuted to the surface reported that the clouds in the upper atmosphere move 60 times as fast as the atmosphere at the surface. So there seems to be very little wind erosion on Venus.

Magnetic Field? Now for a problem with Venus. Even before it was visited by alien probes from Earth, scientists suspected that it should possess a magnetic field comparable in strength to that of Earth. After all, since the two planets are similar in size, and presumedly accreted from the same pre-solar nebula, they should have lost heat at about the same rate.

In other words, the internal structure of Venus should be quite similar to that of Earth — a semi-molten outer core and so on. And the dynamo effect should be generating a magnetic field comparable in strength to that of Earth. However, none of the satellites sent to Venus so far have detected such a field. Instead, it is quite weak. So weak, in fact, that solar-wind particles penetrate to the very surface of the planet without being trapped in radiation belts like those surrounding Earth. It would be lethal to stand unprotected on the surface of Venus — even if it were possible to shield oneself from the searing heat.

What I am trying to emphasize is the importance of a magnetic field to the presence of life on a planet's surface. As if the high temperature, high pressure, and acid environment of the Venusian surface were not lethal enough, the solar wind particles would discourage any life-forms from setting up shop on Venus. There are, however, a couple of explanations for the lack of a substantial magnetic field around Venus. It rotates quite slowly, completing one rotation in 243 days. This, coupled with less of a fluid-like core, might result in a weak magnetic field.

All of this new knowledge supplied by *Magellan* has been of great interest and excitement to scientists who specialize in the study of mountains on Earth. The lack of liquid water and strong winds on Venus mean that the surface features forming on the planet are left intact between periods of volcanic or tectonic activity. Erosion by rain, wind, glaciers, as well as the freeze-thaw cycle has stripped away evidence of the earliest mountain chains on Earth, leaving only their cores. Thanks to the images provided by Magellan, we can get a hint at what Earth's surface would be like if conditions on the early Earth had been slightly different, or if Earth had been slightly closer to Sun.

Origin of Atmosphere I previously discussed the Venusian atmosphere while stressing the reasons for the uniqueness of Earth's watery surface environment. Volcanic outgassing during the early histories of Venus, Earth, and Mars should have resulted in the release of carbon dioxide, nitrogen, and water vapor in approximately the same proportions as we observe in terrestrial volcanic eruptions today. On Earth, the water vapor condensed to form oceans, the nitrogen remained in the gaseous state, and the carbon dioxide combined with surface material to form the extensive deposits of limestone found around the world.

Since Venus receives about twice the amount of solar energy as does Earth, the water vapor as well as the carbon dioxide remained in the gaseous state. The carbon dioxide trapped infrared radiation by the *greenhouse effect*, causing the surface temperature to rise, which in turn caused the surface material to break down and release more carbon dioxide, to trap more infrared radiation, to raise the temperature even further, to release more carbon dioxide to ... and so forth. This spiral of cause-effect to increase the temperature on a planet is called the **runaway greenhouse effect**.

Meanwhile, solar ultraviolet radiation broke apart the water vapor molecules into hydrogen and oxygen atoms. The lighter hydrogen atoms were boiled away to space, while the heavier oxygen atoms combined with crustal rocks to form oxides. A smaller amount of the water vapor combined with the sulfur dioxide to form the sulfuric acid clouds.

Hopefully, you are beginning to appreciate not only how special our planet is, but also the circumstances that provide for that special-ness. Here are two planets of similar size and chemical composition, and which even have a similar location within the solar system, but which are dramatically different. Those differences result from differences in their distances from Sun, rates of rotation, and internal compositions. We naturally wonder just how narrow the constraints must be in order for life in some form to evolve and flourish. This is one of the thoughts that spin off from our studies of other worlds like Venus.

Mercury

Just about everything we know about Mercury originates with the three flybys of the satellite **Mariner 10** in 1974 and 1975 (refer back to Figure 17-22). Venus was the first to be imaged by the cameras aboard the satellite. Then as it swung by Venus, it stole a bit of the rotational energy of the planet in order to get closer to Sun to intersect the orbit of Mercury. The parameters of *Mariner 10*'s orbit around Sun were established so that the satellite would return to the same spot in Mercury's orbit on two additional occasions so that we could obtain data from three encounters with Venus for the price of a single satellite.

Observing Mercury in the Sky As seen from Earth, this innermost planet, like Venus, never appears to be positioned very far away from Sun. Thus observers are limited to seeing Mercury either just after sunset or just prior to sunrise (Figure 17-30). During those times, however, there is always a certain amount of residual light (twilight) from Sun that interferes with our ability to separate out the faint dot that is Mercury. This is especially the case when viewing takes place from the vicinity of a metropolitan area.

Technically speaking, either Venus or Mercury can be called the **Evening Star** or **Morning Star**, even though neither is a star. Very few people, for reasons explained above, have knowingly seen Mercury. Anyone who claims to have seen the Evening Star is probably referring to the planet Venus, the brighter and more distant of the two from Sun. Venus is not always in the evening sky, however, nor even in the morning sky. So the concept of the Evening Star is somewhat like that of ghosts — many may speak of them, but few ever see them.

General Features *Mariner 10* imaged only 45 percent of the surface at an average resolution of 1 km, and less than 1 percent at resolutions between 100 to 500 meters.

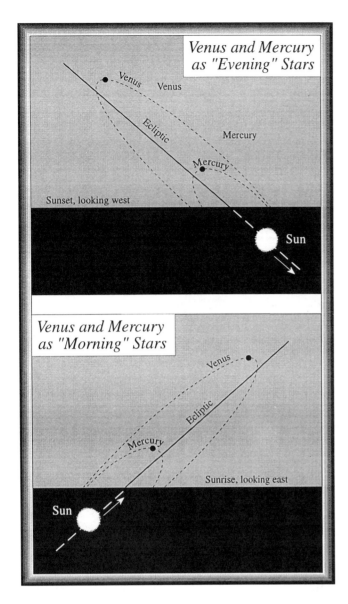

Figure 17–31 This is a composite photograph of all the separate photographs taken by Mariner 10 *during all three passes by Mercury. On the outside, at least, Mercury looks very much like Moon.* (NASA)

Figure 17–30 The planets Venus or Mercury may appear as either the "evening star" or "morning star," depending upon which side of Sun they are on.

Unfortunately, the satellite's orbit was such that it had to image the same illuminated face of Mercury on all three encounters. This coverage and resolution is about the same as our views of Moon prior to the time of spaceflight.

There were very few surprises for astronomers when the first photographs of Mercury arrived from *Mariner 10*. As we expected, the surface is heavily cratered like Moon. It is quite hot (800° F) during the daytime and freezing cold (–300° F) at nighttime. Because Mercury has virtually no atmosphere, once Sun goes down, there is no insulation or blanket to retain the heat that is stored in the surface material. So the temperature drops rapidly and dramatically.

Since the planet rotates quite slowly — once every 59 days — the period of time during which Sun is either in the sky or below the horizon is about 88 days. Thus one can imagine that somewhere along the terminator between daytime and nighttime, the temperature is just about right for human survival. But since Mercury slowly rotates, it is certainly life on the move — there is no mercy for those who dart ahead or fall behind.

Surface Features At first glance, the craters on Mercury's surface appear strikingly similar to those on Moon. But upon closer examination, there are some subtle differences that are due to a slightly different surface evolution than Moon and other terrestrial planets. After all, it is the closest planet to Sun, and it therefore formed where the largest collection of sizeable planetesimals with which to collide orbited the early Sun, not to mention the immense solar radiation that strikes the surface. As far as comparing its surface to that of Moon, its greater mass can account for the differences we observe (Figure 17–31).

Like Moon, there are maria on Mercury, but they are significantly smaller than those on Moon. There appear to be far fewer smaller craters on Mercury than on Moon. There is extensive overlapping of lunar craters, but the craters on Mercury are often separated by smooth plains, giving the surface a speckled appearance. There are rays

Figure 17-32 This closeup view of the surface of Mercury reveals a pitted appearance much like that of Moon. (NASA)

associated with many of the Mercurian craters, although its greater gravitational field prevents the dispersed ray material from being thrown as far as on Moon.

The largest surface feature viewed by *Mariner 10* is the **Caloris basin**, an enormous impact crater 800 miles in diameter. It is filled and surrounded by smooth plains resembling the lunar maria. Unlike the lunar maria, however, the floor of the basin has closely spaced ridges and fractures that are arranged in both concentric and radial patterns. These patterns were probably caused by alternate lowering and rising of the basin floor. On the opposite side of Mercury, directly opposite *Caloris basin*, is peculiar terrain consisting of hills and valleys. This region is thought to be the result of the seismic waves caused by the Caloris impact to converge to a focus on the opposite side of Mercury.

The one visible surface feature that distinguishes Mercury from any other object in the solar system is that of the **lobate scarps**. These are cliff-like features, roundish in shape, one or two miles high and hundreds of miles long, often running right through craters (Figure 17-32). These are thought to be the result of a slight shrinkage of Mercury as it cooled early in its early history, similar to what happens when a plum is left in Sunlight to dry to become a prune.

In 1991, while Mercury's axis of rotation was tilted slightly toward Earth, radar observations made from Earth revealed that a small region around the north pole was radar bright. This suggests that there may be deposits of ice frozen into the soil there. Despite the fact that Mercury's proximity to Sun causes it to experience a dramatic temperature change between day and night, the poles may remain cold enough for ice to remain in the soil.

Interior Mercury's density of 5.44 times the density of water is greater than that of any planet or satellite in the solar system except Earth. But Earth's larger size compresses its internal material to a higher density (5.52) than it would have if it were smaller. When we mathematically remove the additional pressures caused by size on both Earth's and Mercury's interiors, and recalculate their densities, that of Earth is only 4.4 whereas that of Mercury is now 5.3. This means that Mercury's interior consists of an enormous fraction of iron — its metallic core must be about 75 percent of its diameter, or 42 percent of its volume. In contrast, Earth's iron core is only 54 percent of its diameter, or 16 percent of its volume.

Mariner 10 detected a magnetic field strength around Mercury of about 1% that of Earth. So astronomers have reason to believe that Mercury's iron core is partially molten. In order for such a small planet to retain a partly molten core over the 4.5 billion year history of the solar system, however, requires the presence of an additional lighter element that would lower the melting point. Otherwise, the interior would have solidified long ago. It is rather like putting an antifreeze in the radiator of your automobile. Sulfur is the suspected chemical. The problem is that the presence of sulfur does not work well into models of the condensation sequence of the birth of planets. A return visit to Mercury could clear up these confusions. However, there are no plans to return to the planet again, so what we know today may appear in astronomy textbooks for years to come.

Return to Mercury? It is surprising that so little attention has been given to Mercury since the 1974-75 flybys of *Mariner 10*. Mercury may hold the key to understanding the birth and early evolution of all the terrestrial planets. It formed in the hottest part of the solar system, and current models of the chemical condensation process have a difficult time explaining the large percentage of iron in Mercury's core. And if we can't explain Mercury's birth, then it is difficult to explain those of the other terrestrial planets.

Mars

Mars is an entirely different story than that of Mercury. It reveals some tantalizing clues to an extremely interesting past history. In fact, although there is no direct evidence of there ever having been life on its surface, when what has been learned about it is looked at in its broadest perspective, speculation about life continues. But, I am getting ahead of the story. The fascination with Mars as a possible abode for life goes back long ago.

When you stop to think of it, Mars is the only planet that consistently elicits remarks, jokes, and even media attention about extraterrestrial life. Go to a costume party dressed as an alien creature, and you will undoubtedly be accused of being a "Martian." In fact, the word "Martian" is loaded with cultural — at least American — overtones, much as are the words "Jack Daniels." If you don't believe me, try this experiment during the next lull in conversation at a gathering of your friends. Do something strange and then say – "I'm a Mercurian (or Venusian or Neptunian or Saturnian)!" Compare the responses you get then with those you get when you say "I'm a Martian!"

Early Telescopic Observations So just where did we get this cultural bias? Let's go back to the earliest observations of planet number four from Sun. Even with the telescopes in use as early as the 1700s, Mars was observed to be similar to Earth in a few important respects. White splotches appeared to wax and wane during the two Earth years required for Mars to circle Sun. These were interpreted to be polar ice caps similar to those on Earth, which of course also wax and wane with the seasons. Our seasons, we learned early in the book, are due to the axis of Earth being tipped 23.5° with respect to its orbital plane.

Well, Mars is tipped by almost exactly the same amount: 24°. This provides an immediate conclusion that Mars has seasons. With the development of larger telescopes, astronomers resolved darkish features on the Martian surface that provided astronomers with a basis for both speculation and for specific data about the planet. By measuring the length of time it took for a given dark spot to make a complete circuit around behind the planet and return to its original position, astronomers determined the Martian rotation rate — the day-night cycle. It turned out to be almost exactly 24 hours, identical to Earth's rotation period.

Seasonal Cycles The speculation about these spots was far more exciting to some astronomers than the rate of rotation. The dark splotches were not permanent features. They lasted long enough for the purpose of measuring the rotation rate of Mars, but their overall patterns gradually changed during the course of the two Earth years required for the planet to orbit Sun. Figure 17-33 shows two photographs taken at different times of the Martian year. What is quite noticeable is not only the waxing and waning of the polar ice caps and the changing pattern of the dark areas, but the relationship between the two.

As the polar ice cap recedes, the equatorial areas appear to get darker and darker. As the polar ice cap grows again, the equatorial areas regain their former brightness. This is

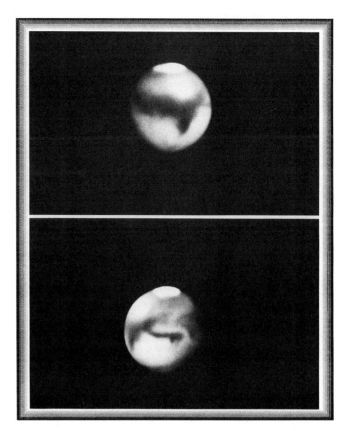

Figure 17–33 These two Earth–based telescopic photographs of Mars were taken at different times of the planets orbit around Sun, and therefore reveal the seasonal changes that were interpreted not too long ago as indications that life existed there. (Lick Observatory)

called the **wave of darkening**. Since the early observers of this phenomenon lived prior to the invention of color photographic film, the assignment of color to a celestial object was strictly a matter of personal judgment. To the naked eye, however, Mars appears an obvious red color. The early observers interpreted the dark areas of Mars to be greenish in color compared to the reddish color of the general surface. So what they noticed was that as the polar ice caps receded, greenish areas began to grow on the Martian surface. As the ice caps grew, the greenish areas returned to their reddish appearance again. Might they be observing the growing and decaying of vegetation with the seasonal cycles of Mars?

Martian Atmosphere Looking along the edge of Mars through a telescope, it was also quite obvious that the planet possessed a thin atmosphere. Instead of appearing sharp and distinct as does Moon's edge, that of Mars was fuzzy. This is also the case with Earth (from an orbiting space vehicle, for example), Venus, Jupiter, and Saturn. Once in a great while, Mars passed in front of (to hide = **occulted**) a bright-ish star. By observing the manner in which the starlight was diminished and then extinguished at the edge

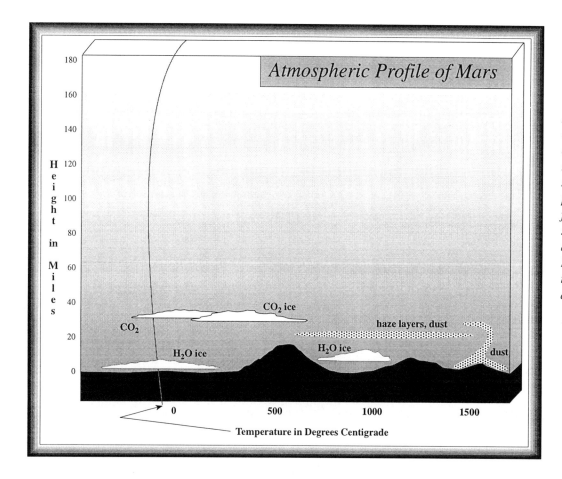

Figure 17–34 *Even the earliest telescopic views of Mars suggested that it had an atmosphere, since the edges of the planet appeared slightly fuzzy. This diagram shows a model of the atmospheric profile of Mars, obtained from all the data accumulated to date.*

of the planet, an approximate density (thickness) of the atmosphere was determined. So early in the drama Mars was known to have at least a thin atmosphere (Figure 17-34).

Before I tell you what all of this is leading to, I would like to explain why the history of Martian observations is so important to science in general and astronomy in particular. It is a classic tale of how a theory is only as good as the assumptions upon which it is based, and how important it is to hold onto theories only tentatively when the data is in short supply.

Sometimes our "wants" to believe something get in the way of the assumptions used in arriving at the belief, and we fail to see that we are treading on weak terrain. In other words, we believe lots of things without having sufficient evidence to support the belief because we have some other reason for wanting to maintain the belief. In the Martian drama, there was an additional human element that crystallized the popular belief that Mars was inhabited by intelligent "beings."

So far, you see, arguments for the existence of life on Mars were based solely on its similarity to Earth in terms of observed features and behavior. There was certainly no evidence for vast oceans or even lakes on Mars. Were there such bodies of water on the surface, the strongly reflected sunlight could easily be observed. But, you say, perhaps the retreating (melting) ice cap leaves liquid water behind that percolates into the soil, thereby feeding the dormant plant seeds that eventually sprout into healthy plants during the springtime!

Actually, this very effect is what you would observe from a space station hovering above Earth's north polar region. As the snow/ice cap recedes northward from Michigan through Canada, it leaves behind a carpet of greenery. In the case of Mars, all one need assume is that the polar caps are made of water ice and that the greenish color of the wave of darkening indicates the presence of plant life. But the Martians that created fear in the hearts of many — especially during the broadcast of *The War of the Worlds* in 1938 — were not person-eating plants. They were thought to be more intelligent than plants, and intelligent enough to visit Earth in their advanced spaceships and with their advanced weapons! So how did that idea arise?

Reports of Canals In 1877, the Italian astronomer Giovanni Schiaparelli was visually studying Mars through a telescope when suddenly Earth's atmosphere stood still and the seeing conditions were superior. During these few moments of time, he observed unusually distinct features on the Martian surface. Recall that Earth's atmosphere offers a distinct disadvantage to Earthbound astronomers, inasmuch as it is in constant motion, and thereby causes blurry images either visually or photographically. For a brief moment, Schiaparelli was able to observe under #10 seeing conditions. What he thought he saw he described as thin dark lines running across the Martian surface.

Schiaparelli was not able to photograph them – the superior seeing conditions did not last long enough. After pondering what he had just seen, he wrote down in his journal the Italian word *canali* to describe the lines. This word is freely translated as channel or depression. The rest is history. When the media in the United States heard of the report, they simply dropped the *i* from *canali* and reported the discovery of "canals" on Mars, a word we associate with the efforts of intelligent creatures.

Percival Lowell, an American diplomat-turned-astronomer, was so intrigued by Mars and the discovery of canals that he made it his lifelong ambition to study it. He even built an observatory in Flagstaff, Arizona, for that purpose. Soon other astronomers claimed seeing the thin dark lines during good seeing conditions as well. The Martian "canals" fever spread. Lowell collected these observations and formulated a complete map of the Martian network of canals, and attempted to explain their patterns.

He theorized that the Martian atmosphere was getting thinner and thinner, and there was no longer sufficient air pressure to hold bodies of liquid water on the surface. The "Martians," dying of thirst, were forced into a last-ditch effort to conserve the melted water of the ice caps by building the elaborate network of canals to carry the water to the Martian cities. As springtime approached and the ice cap began to melt, water would flow from the polar to the equatorial region where the "cites" were located. Noting that where the dark lines crossed over one another there were particularly large dark splotches, Lowell theorized that the Martians would have built the canals to intersect at the locations of their cities in order to make the task of tapping the water easier (Figure 17-35).

Skeptical astronomers pointed out the fact that the resolving power of the existing telescopes was insufficient to see features as small as a canal at such great distances. But Lowell responded by suggesting that the observed dark lines included not only the width of the canals but the fields of irrigated crops that lined both sides of the canals! Well, of course, the Martians would certainly be intelligent enough to extract water from the viaducts in order to irrigate the fields of green vegetation on either side.

And so very gradually it became acceptable to speak publicly about life on Mars. Lowell popularized the theory by talking to anyone who would listen. He wrote articles for popular journals of the day, including *Ladies Home Journal*. By 1938, the possibility of intelligent life on Mars seemed so reasonable that few were surprised that people panicked during the *War of the Worlds* broadcast. There was another fear present in peoples' minds during that autumn radio program as well – Hitler's sword-rattling in Europe.

Invasion of Mars Instead of Earth being invaded by Martians, it was a case of Earthlings invading Mars. In 1965, the truth about cities and canals on Mars became obvious. Twenty-two close-up photographs were televised back to Earth from *Mariner 4*. They were not terribly detailed photographs, but they were sufficient to reveal Mars as a bleak and barren planet, much more like Moon than Earth. Two additional satellites, *Mariners* 6 and **7**, flew by in 1969, and in 1971 **Mariner 9**, orbiting the planet, sent back more than 7,000 photographs (Figure 17-36).

In 1976, while two **Viking** landers descended to the Martian surface to photograph and conduct experiments with soil samples, two orbiters refined our understanding of this intriguing planet. There were many surprises, especially the photographic evidence for extensive flowing water at some time in the past. Before the Soviet *Phobos 2* mission to the planet Mars and its largest satellite Phobos ended suddenly and unexpectedly in March 1989, it managed to return some important data on the Red planet and its Moon.

Before providing a summary of the exciting discoveries of these various missions, let's consider the phenomena whose interpretation had led to the belief that intelligent life existed on Mars in the first place. What did Schiaparelli and Lowell see if they were not canals and adjacent fields of crops?

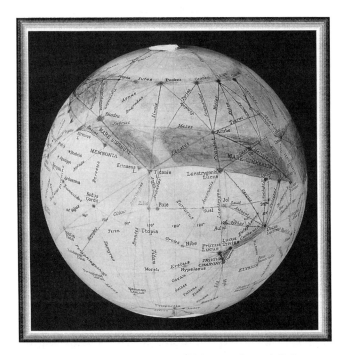

Figure 17-35 *This is the map of Martian "canals" that Lowell and his team of astronomers claimed to observe on the surface. Prior to the time of high-speed photographic film, it was impossible to take photographs that could reveal what they were looking at.* (Lowell Observatory)

Polar Caps While it is true that the Martian ice caps wax and wane with the seasons, they consist not of water ice but of carbon dioxide ice – commonly called "dry ice." Even so, the atmospheric pressure is so low — about 1% that of Earth's — that neither carbon dioxide nor water can exist there in the liquid state. Instead, the carbon dioxide goes directly from the solid state to the gaseous state, just as you observe it behaving on Earth. Even were there canals on Mars, no water would flow, at least for long.

To be accurate, there are vast amounts of water ice on Mars, lying beneath the layers of carbon dioxide ice observed at the polar caps. Estimates are that it exists in layers perhaps several hundreds of feet thick, enough to force visitors to wade in water up to their knees if it melted and spread evenly over the entire surface of the planet. There is also evidence that additional water is frozen into the Martian soil as **permafrost**.

Wave of Darkening What had been interpreted as the growth and decay of vegetation in the appearance of the wave of darkening turned out to be the result of seasonal wind patterns transporting fine grains of dust from one region of the Martian surface to another. Although the carbon dioxide atmosphere is quite thin, unequal heating of the Martian surface causes the thin atmosphere to move rapidly enough to cause massive dust storms. These storms occasionally engulf the entire planet and last for months. When the dust is blown away from rough, rocky terrain, the regions look darker than when the dust covers the coarse rock. So the seasonal winds cause regions to alternately change from dark to light and back to dark again.

What about the greenish color that was associated with the wave of darkening? Apparently, when a person looks at a dark splotch on a red background, the dark splotch appears greenish in color because they are complementary colors. So the greenish color claimed for the canals was a result of the "communications center" of the human brain making a poor judgement — or at least sending an improper message to the "consciousness" part of the brain. And the canals? Well, those were the most interesting of all, and again, the result of a breakdown in the manner in which the human brain processes information fed to it by the senses.

When Lowell's map of the Martian canals was compared to a map compiled from photographs returned by the *Mariner* series of planetary probes, there was nothing to be

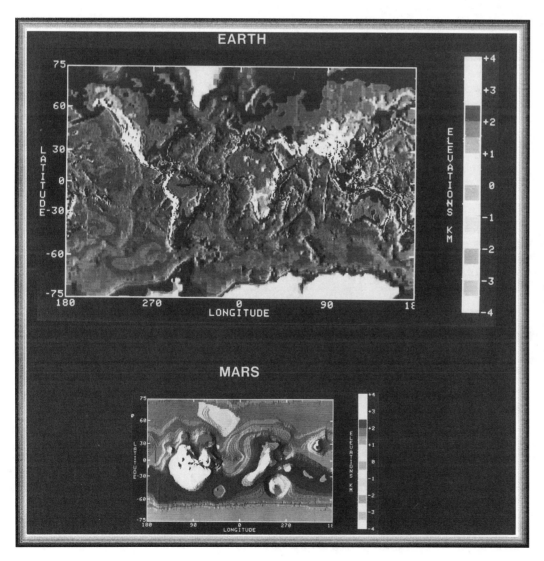

Figure 17–36 The surfaces of Earth and Mars are compared in these shaded relief maps of the two planets. The data for that of Mars was obtained from the tens of thousands of photographs taken by the several satellites that have visited the planet. The two maps are to scale. The total land mass on Mars is about equal to that on Earth. (NASA)

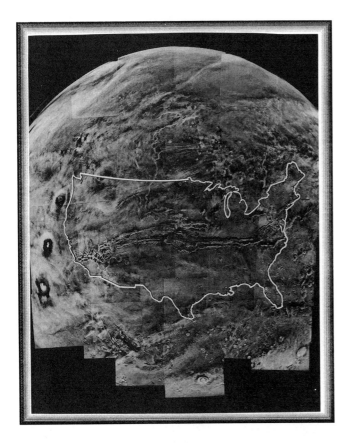

Figure 17–37 This composite photograph of Mars shows some of the dramatic features found there. The largest canyon in the solar system runs the length of the superimposed outline of the U.S. The giant volcano Olympus Mons *lies just to the left of the State of Oregon. (NASA)*

found on the map that corresponded to the lines on his map. The surface is heavily cratered in places, although the presence of even a thin atmosphere has been responsible for greater erosion than on either Mercury or Moon. Apparently the early observers, working close to the limits of the resolving power of their telescopes, observed patterns of terrain features like craters, but consciously perceived them as connected features.

This is actually a well-known characteristic of the human eye-brain system. We call it an **optical illusion** (Figure 17-38). Now keep in mind that eyes do not see — they record the receipt of photons. Absorption of photons on the retina of the eye causes electrical impulses to travel to the brain, which then interprets the impulses. The brain likes order, filling in gaps when they exist between data points. This is especially true when the seeing conditions are marginal. So optical illusions explain the origin of the belief that canals existed on Mars. So alas, no canals, no vegetation, no Martians. So naturally you wonder if anything exciting was discovered if Martians were not.

Volcanoes and Canyons To geologists attempting to reconstruct the past history of Mars, the photos taken by both the *Viking* orbiters and the two landers are fascinating. Look at the photograph in Figure 17-39, which is of a giant volcano over 90,000 feet above the surrounding plain — about the size of New Mexico in total area. Although not an active volcano, it is nevertheless quite young. In addition to having very few impact craters on its sides (indicating youth), it and several other huge volcanoes have flooded a large part of the surface with newly-formed lava. **Olympus Mons**, like the Hawaiian Islands, is a shield-type volcano (refer back to the section on the Hawaiian Islands earlier in this Chapter).

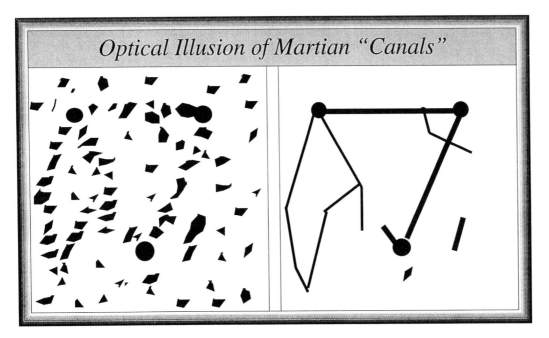

Figure 17–38 To understand the Martian "canals" illusion, try this experiment. Stare at the diagram on the left while squinting your eyes quite tightly (almost closed). Compare what you see with what you see in the diagram on the right.

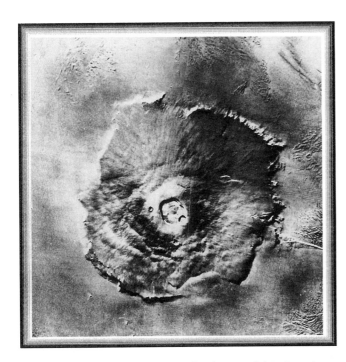

Figure 17–39 This is a spectacular image of the gigantic volcano Olympus Mons, seen from directly above by the camera aboard the Viking Orbiter. *It would cover the entire State of Oregon if placed on Earth.* (NASA)

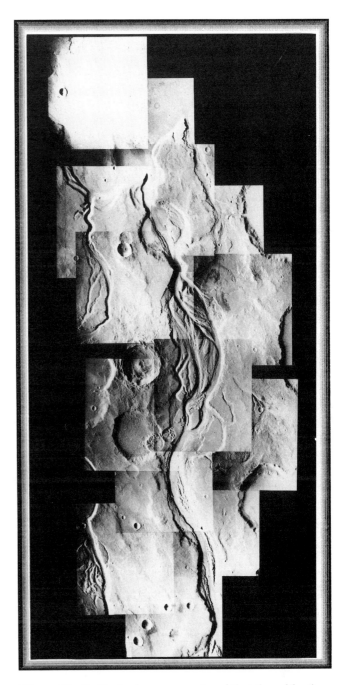

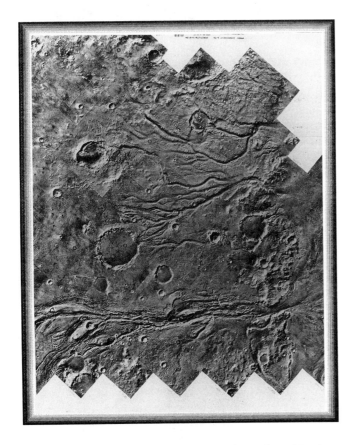

Figure 17–41 Geologists are convinced that these Martian surface features are the result of flowing water sometime in the past. (NASA)

As if possessing the largest volcano (and highest mountain, as far as that is concerned) in the solar system were not enough, Mars also has the distinction of having a "Grand-er" Canyon. **Valles Marineris** is about as long as the United States from coast to coast, is about 50 miles wide, and about 20,000 feet deep. Visual inspection of its features suggest that it is a result of vertical faulting, not the result of horizontal faulting associated with plate tectonics. Its origin is probably associated with the volcanoes lying at one end of the great canyon (Figure 17-37).

Figure 17–40 Geologists are convinced that these Martian surface features are the result of flowing water sometime in the past. (NASA)

CHAPTER 17 / Exploring the Earth–like Planets

★★★★★★★★★★★★★★

Figure 17–42 *The pattern of terraces at one of the Martian poles, imaged by the* Viking Orbiter. *(NASA)*

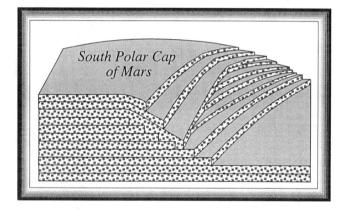

Figure 17–43 *This diagram illustrates the manner in which deposits at the Martian poles may be in a pattern like this due to repetitive freeze–thaw cycles. This is an attempt to explain what we observe in the photograph taken by the* Viking Orbiter *shown below.*

The magnetic field of Mars is quite weak, so there is probably no molten core. Because of its small size, Mars cooled rapidly. As you see in the geological maps of Mars in Figure 17-36, the planet is divided almost equally into a southern hemisphere of old cratered terrain and a northern hemisphere of younger volcanic terrain

Water on Mars As I mentioned earlier, it was the discovery of dry river beds on Mars that most excited scientists. It was certainly an unexpected discovery, because the present-day atmosphere does not allow water in the liquid state to exist. Estimates are that the atmosphere would have to be some 10 to 50 times more dense for water not to evaporate immediately. Yet the photographs are unmistakable — only water could have carved the features visible in photographs like those of Figures 17-40 and 17-41.

Apparently Mars is presently experiencing an ice age, during which the temperatures are consistently too low to allow the frozen carbon dioxide and water ice to <u>completely</u> change to the gaseous state. Summer daytime temperatures at the equator only occasionally rise above freezing. At night and during the winter days temperatures slip to hundreds of degrees below zero. But precession is at work at Mars as well as at Earth. Mars orbits Sun in an elliptical path, moving alternately closer to and farther away from Sun. Presently, the north polar cap is tilted away from Sun when Mars is closest to Sun, so that the ice cap never completely melts to release water into the atmosphere.

Figure 17–44 *The first invader of the surface of Mars—the* Viking Lander. *It was designed to search for evidence of life, but found none.* (NASA)

Figure 17–45 *The first photograph of the Martian surface taken from the surface, taken by the* Viking Lander *of its own foot and the surface material of Mars.* (NASA)

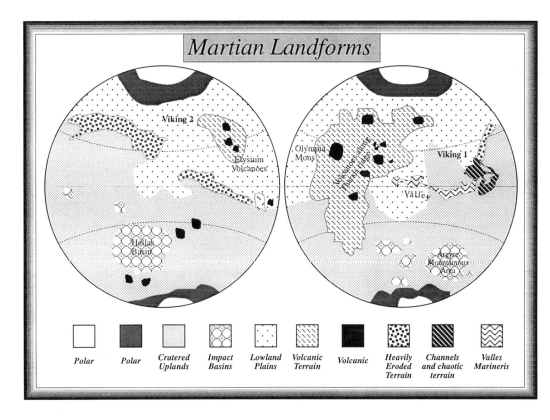

Figure 17–47 A general geological map of Mars, showing the major terrain types. Note the locations of the Viking landers.

Figure 17–46 A photographic view of the Martian surface taken by the Viking Lander. *Is this where you would like to spend a vacation?* (NASA)

Precession will gradually change that. It may be that Mars goes through a succession of freezing and thawing cycles along with the precessional cycle (Figure 17-42).

This is the best explanation for the layered terrain found near the poles (Figure 17-43). Photographs taken by the *Viking* orbiter reveal frozen ice on a vast network of terraces, each of which is deposited (according to the theory) during a particular precession cycle. If so, and there is sufficient water ice in the ice cap and soil to be released into the atmosphere, perhaps the rains will come again and the rivers will once again flow. Unfortunately, we might have to wait 175,000 years or so to witness this event.

Viking Landers Although the *Mariner* and *Viking* photographs gave the intelligent Martians a hasty funeral, the discovery of dry river beds on the Martian surface nudged scientists toward theorizing that life might have evolved in the watery environment long ago. Perhaps some of the life forms learned to survive by adaptation as conditions changed and the water froze again. Or, perhaps the two landers would find only their skeletal remains. But the *Viking* landers found in the soil neither signs of biological activity nor the organic compounds that we associate with living organisms (Figure 17-44).

Figure 17–48 The photograph of the crater nicknamed "slurpy" reveals obvious evidence of the presence of water in the soil of Mars. Some of the heat released upon impact with an asteroid melts ice and throws it and rock outward. (NASA)

The instruments aboard the landers were designed to detect specific characteristics of terrestrial life – for example, chemical exchanges between life forms and the surrounding environment (metabolism). It may well be that Martian life is indeed present, but behaves in ways not detectable by the *Viking* instruments. The landers sampled only within several feet of their landing positions – they did not wander around the surface and sample various regions (Figures 17-45 and 17-46). But they nevertheless filled in important gaps in our survey of the surface of Mars (Figure 17-47).

History of Life (?) on Mars Mars has roughly the same chemical composition as Earth, and hence a goodly supply of the biogenic elements. We know that liquid water once flowed on Mars, and we believe that the presence of liquid water was the crucial factor in the origin of life on Earth (Figure 17-48). Ultraviolet radiation from Sun, a possible energy source for the first chemical reactions leading to life, falls on the martian surface in nearly the same intensity as it did on the early Earth, prior to the life-driven formation of the ozone layer.

Evidence uncovered by our spacecraft suggests that about 3.8 to 3.5 billion years ago was the time water carved out the martian channels. This just happens to be the very time that life seems to have arisen on Earth. The earliest fossils yet found on Earth are about 3.5 billion years old, and these cyano-bacteria (blue-green algae) life-forms were already using photosynthesis to power their activities.

The first living things were undoubtedly more primitive and probably arose millions of years earlier, perhaps during the same time in which water flowed on Mars.

Studies conducted in the permanently frozen permafrost regions of Russia indicate that organic material is preserved – and organisms are still viable – after being frozen for 3 million years at 10-14 degrees Fahrenheit. At the even lower temperatures on Mars, about minus 94 degrees Fahrenheit in the permafrost regions, we expect survival over even longer periods to be possible. Scientists believe that clement conditions on Mars deteriorated about 3.5 billion years ago, although there may have been occasional periods of warming since then.

As Mars cooled and its water was immobilized into permafrost, any organic material and microbial life would have been incorporated and frozen into the sediments. The *Viking* life science experiments suggest that any organic material remaining on the surface has long since been destroyed by photochemically produced oxidants. With the surface of Mars barren of organic material, subsurface permafrost deposits of organics would be an important clue to the biological history of early Mars. Material several meters below the surface would be shielded from cosmic radiation and solar ultraviolet radiation. The martian climate of the past 3 billion years, while unsuited for life, has been ideal for preserving frozen samples.

Mars does not seem to have plate tectonics, which on Earth continually moves continents, recycles the crust and resurfaces the planet. The low martian erosion and burial rates, estimated at about 1 meter per billion years, would not hide possible fossil-bearing deposits beyond our reach. Thus the permafrost regions of Mars remain high on the list of targets for future exobiological investigations of Mars.

The very best place to go today to get Mars-like conditions on Earth is Antarctica. There are areas on the frozen continent where there is no ice or snow, such as the McMurdo Dry Valleys. The annual average temperature is minus 4 degrees Fahrenheit – in the summer, temperatures barely climb above freezing. The region receives less precipitation than the Gobi Desert. A major habitat for life in this extreme region is below the perennial ice cover of the lakes found on the valley floors. Here, beneath 13 to 16 feet of ice, algae, diatoms and other microbial life-forms thrive in a liquid water environment buffered from the cold, dry conditions above.

Studies of life-forms on Earth in those harsh environments that resemble the surface of Mars, taken together with our analysis of the surface conditions on Mars itself, continue to give scientists a glimmer of hope (hypothesis) that there is still much to be learned about life on Mars. If it is found there, either in its deceased, frozen, or living state, surely that would be the most important discovery ever made. If, on the other hand, no signs of life are found anywhere on Mars, we will be a step closer to understanding the constraints within which life develops and evolves. We will be closer to knowing our origins in either case.

The Human Invasion of Mars The human exploration of Mars is the ultimate goal of the **S**pace **E**xploration **I**nitiative (**SEI**, for short). In preparation for that exciting adventure, there is an international effort to send further probes to Mars. One of the first, the US Mars *Observer*, was launched in 1992, but failed shortly before it was to arrive at the planet. Duplicates of many of the instruments in the payload of the Mars *Observer* will be aboard the **Mars Global Surveyor**, due for launch in November 1996. In early 1998, it will begin its two-year mapping mission 250 miles above the martian surface. In addition to providing photographic coverage at resolutions down to 10 feet, it will act as a relay station for data collected by the Russian **Mars '96** Mission that also launches in November 1996. This satellite includes a large orbiter, two penetrators and two small meteorological stations. The orbiter carries 12 different instruments to study the martian surface and atmosphere, plus six instruments to measure the solar wind at Mars. The 220-pound penetrators will be released from the orbiter and penetrate to a maximum depth of 20 feet into the martian surface. Once buried, instruments aboard the penetrators will measure seismic activity, local magnetic fields, and temperatures beneath the surface.

The Jet Propulsion Laboratory has been working on a robot vehicle designed to navigate across the surface of Mars. Nicknamed "*Sojourner*," the rover has stereoscopic vision in order to avoid obstacles that it might encounter (like a large boulder or deep crater). During trial tests, it successfully picked its way along a 110-foot long course in a dry riverbed near JPL's laboratory in Pasadena, not unlike the terrain known to exist on Mars. This is the first time an autonomous vehicle has ever navigated rugged terrain without the aid of a human pilot. To perform this feat, *Sojourner* had to sense the terrain and make a map of it, two meters at a time. Whenever it encountered an obstacle, it devised a safe path around it.

Sojourner will be aboard **Mars Pathfinder**, due for launch in December 1996. The rover is small (22 pounds), solar-powered, and will operate for up to one year within view of the lander (50 feet or so). The lander cameras will work together with the rover imaging system to reveal martian geologic processes and surface-atmospheric interactions at a scale of inches to yards — comparable to the resolution of the *Viking* cameras. The rover's primary mission is to perform experiments designed to provide information to improve future planetary rovers. The Mission (*Pathfinder*) takes its name from (hopefully) a series of missions to follow every two and a half years, when the positions of Earth and Mars in their orbits allow for quick passage between them (*least energy orbits*). Mars *Global Surveyor 2* will launch in early 1998, and will include part of the Mars *Observer* payload and two imaging cameras.

Astronomers would like very much to conduct a more extensive mission of the Martian surface, and NASA is considering a project whereby a robot rover on the Martian surface would collect samples from a variety of locations

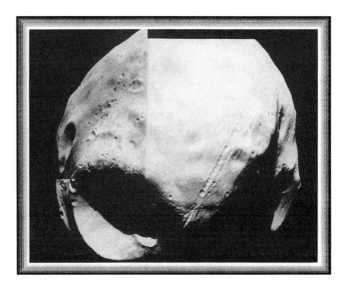

Figure 17–49 *One of the small satellites of Mars,* Phobos, *reveals evidence of a massive impact during its time in the asteroid belt. The fact that it is so small suggests that it was captured by Mars.* (NASA)

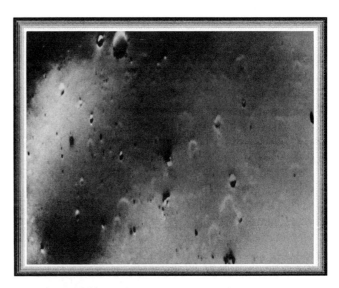

Figure 17–50 *The Martian satellite* Deimos, *showing evidence of half-buried craters on the surface.* (NASA)

and return them to Earth. This is identical to the technique the former Soviet Union used to obtain lunar samples without sending humans to the lunar surface. The goal of the mission would not be to determine the presence or absence of life, but to better understand the nature of the Martian environment and its evolution over time.

A *Mars Network* is being considered by NASA for the 1998 to 2003 launch windows — with up to five launches of four landers each – and a Mars *Sample Return* and *Phobos*

Sample Return mission planned for the year 2005. As a part of the overall SEI program, several other missions are being considered for the future, such as a flyby of Pluto, a mission to a near-Earth asteroid (*Nereus*), and a probe into Sun in 2003. Scheduled for launch in 1999, *Stardust* will return samples of interstellar dust after encountering comet *Wild-2* in January 2004. Return of the samples will occur in January 2006. Then President Bush targeted the year 2019 for the first human landing on Mars.

Japan has set its sights on Mars and Moon, most likely because of the interest of both the United States and Russia in establishing outposts on both bodies. *Lunar-A* is scheduled for launch to Moon in 1997, following the success of the probe *Hiten* that was launched in 1990 and which released a subsatellite into orbit around Moon. *Lunar-A* will fire three penetrators into the lunar soil at 650 mph, allowing for 3- to 10-foot penetration. Inside of the penetrators will be seismometers and heat probes. The Mars probe, *Planet-B*, will be launched in 1998, and will orbit 150 miles above the Martian surface as it studies the interaction of the solar wind with the atmosphere of Mars.

Tiny Satellites of Mars

I mentioned the two satellites of Mars – *Phobos* and *Deimos* – while discussing asteroids. I did so because of their similarity to asteroids. In fact, they were no doubt captured by Mars long ago as they were gravitationally disturbed by Jupiter and wandered close enough to Mars to fall into its "gravity well." We had hoped that the *Phobos* probe launched by the Soviet Union in 1989 would have provided the necessary information to verify or refute this theory, but unfortunately the probe stopped transmitting shortly before arrival at the satellite.

Above and beyond appearance, the theory is easy to adopt in light of the fact that the orbit of Mars lies along the inner edge of the asteroid belt, and therefore has a ringside seat to passing asteroids. A detailed understanding of these satellites could prove of practical advantage to us as well as filling out our knowledge of asteroids. Extensive space habitation requires that we find suitable and easily-obtained raw materials for building the habitat structures. We may want to set up mining operations on *Phobos* and *Deimos*.

Surface Features Close-up photographs of Phobos, obtained by *Mariner* and *Viking* probes as they approached Mars, reveal an impressive crater called *Stickney* (Figure 17-49. Radiating outward from the crater are long parallel grooves about 500 feet wide and 80 feet deep. Most likely these deep grooves formed in conjunction with whatever collision caused *Stickney*.

Deimos has no grooves, nor even prominent craters. In fact, Deimos looks rather smooth because of a thick layer of dust covering its surface (Figure 17-50). The dust no doubt fills small craters and other irregularities on its surface. There is every reason to believe that it – like Phobos – experienced collisions in its past, so there is surely evidence of fractures buried beneath the dust. The presence of a layer of dust is somewhat of a problem to explain, however.

Both of these satellites are quite small – Phobos is about 17 by 12 miles, and Deimos is about 10 by 7 miles. That explains, of course, why they are oblong and not spherical in shape. The gravity must be extremely weak on both satellites, and this fact should have caused dust to escape at the time it was (presumedly) created during collisions with other objects. Why Deimos has been able to retain its dust is a mystery.

Think of it this way – a high-jumper who can clear a 6-foot pole on Earth could clear a 1.7-mile pole on Phobos! The manner in which Phobos affected the *Viking* probe as it swept past in 1976 revealed that the density of the satellite is very close to that of carbonaceous meteorites. This, together with the dark gray color of the satellites, suggests that they originated in the outer portions of the asteroid belt. If that be so, then the collisional features on their surfaces – if studied at close range – could tell us something of the behavior of objects within the asteroid belt.

Formation of the Solar System

Now that we have explored the characteristics of the terrestrial planets, we are in a position to propose a more detailed and comprehensive theory for their origin and evolution (Figure 17-51). Keep in mind that a natural theory for its formation is relatively modern, inasmuch as detailed knowledge of their interiors, surfaces, and atmospheres has only recently been acquired. So although the general scenario is rather clear, astronomers continue to search for additional clues (Figure 17-52). So don't be surprised if new discoveries cause major revisions.

Mass-Surface Relationship A Rule of Thumb for the solar system is: *the more massive the object, the younger and more active its surface. The less massive the object, the older and less active its surface.* Let's consider this rule for a few objects. There are four bodies intermediate in size between Earth and Mercury: Venus, Ganymede (satellite of Jupiter), Titan (satellite of Saturn) and Mars.

- Mercury is only 40% more massive than Moon. They both cooled rapidly. Their magma oceans allowed some of the lighter minerals to float to the surface and form primordial crusts which later broke and cracked as lava oozed up from the interior.

- On more massive Venus, basaltic volcanism dominated and covered much of the planet with lava flows. Some small, continent-sized concentrations of granitic rock may have formed, but there was insufficient subsurface activity to allow for full-fledged plate tectonics.

On Earth, remnants of the primeval crust must have been erased by the basalts that erupted to form the seafloor crusts, and the plate tectonics that broke up and recycled the ancient surface. Earth's crust remains hot up to the present day, allowing for the slow movement of the crustal plates. The mixing up of the internal material and surface material allowed for low-density minerals to accumulate on the surface as the continents (Figure 17-53).

Surface features vs planet size Between 4.5 billion and 4 billion years ago, there appeared intense early bombardment. This was the great sweep-up of interplanetary debris left over from the formation of the planets themselves. The surfaces of the terrestrial planets reveal the competition between the production of internally derived features and their destruction by external bombardment.

Let's begin with the assumption that all of the planets began with magma oceans and hot interiors, results of the dynamics of accreting matter and the release of energy by radioactive materials deep within the planets' interiors. Volcanic mountains, lava flows and fault cliffs would form on those planets that remain hot and active and whose lithospheres are not too thick.

A planet that is small cools rapidly and forms a thick

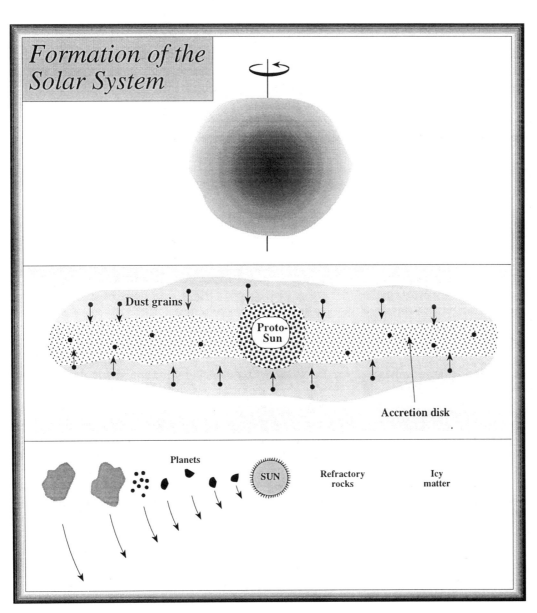

Figure 17–51 This figure illustrates the favored explanation for the origin of the solar system some 4.5 billion years ago. Its sequence of events best fit the observations and laws of physics and chemistry that govern the universe.

CHAPTER 17 / Exploring the Earth–like Planets

> ## Characteristics of the Solar System to be Explained by a Theory of Origin
>
> 1. All the planets' orbits lie generally in a flat plane.
> 2. Sun's rotational equator lies nearly in this plane.
> 3. The planets and Sun all revolve in the same west-to-east direction.
> 4. All planetary orbits are nearly circular.
> 5. Planets have much more angular momentum than does Sun.
> 6. Meteorites contain inclusions of minerals that formed from grains that condensed at different temperatures.
> 7. Planets differ in chemical composition, correlating roughly with distance from Sun.
> 8. Meteorites differ in detailed chemical and geologic properties from all known terrestrial and lunar rocks.
> 9. The distances between the planets follow a pattern (Titius-Bode rule).
> 10. All closely studied planets and satellites reveal impact craters caused by objects up to 60 miles in diameter.
> 11. Most planets and asteroids rotate with similar periods of about 5 to 10 hours, except for those in which tidal forces have slowed them down.
> 12. Except for Venus, Uranus, and Pluto, all planets have direct rotations with axial tilts less then 30 degrees.
> 13. As a group, comets' orbits define a large, spherical swarm around the solar system.
> 14. Major planet–satellite systems resemble the solar system on a smaller scale.

Figure 17–52 This is a list of the general characteristics of the solar system that scientists must keep in mind as they propose theories to explain its origin.

lithosphere quickly, perhaps in less than 500 million years. Surface area per volume for a small sphere is greater than surface area per volume for a larger sphere, so that a small sphere at the same initial temperature will cool off sooner than the larger sphere. So the formation of volcanic and tectonic features stops after 500 million years, and yet the intense bombardment continues for another few hundred million years. This bombardment erases the constructional features of the first 500 million years and leaves craters atop the older features.

On a slightly larger object, such as Moon, a few late-stage lava flows might manage to slip through the lithosphere onto the surface after the 1 billion year early intense bombardment stage had stopped. These lava flows would be visible today as the dark plains on Moon. On a still larger object, such as Mercury, the surface would take even longer to solidify, and such long-term cooling could cause the surface to "wrinkle" as it shrinks. On still larger objects, like Earth, the interior still remains hot to the present day, allowing material from the interior to penetrate through the lithosphere to the surface in the form of faults, mountain ranges, volcanos, and so forth (Figure 17-54).

Solar System Collisions Between 4 and 3.8 billion years ago, after the surfaces of the terrestrial planets had solidified, the remaining planetesimals in the inner solar system appear to have been responsible for a major intense bombardment. The outer solar system also experienced a period of major bombardment, but it is unclear if it occurred at the same time as that of the inner solar system. This era is called the **Late Heavy Bombardment**.

On Moon, objects up to 100 miles across crashed onto the surface to form at least eight huge, ringed basins more than 150 miles across. The Orientale Basin, over 500 miles across, was created during this period. In fact, the planetesimals themselves could have formed Moon. According to one theory, Moon was formed 3.8 billion years ago when one of the Mars-sized planetesimals, traveling along a similar path as that of Earth, struck Earth at low velocity.

A computer model of the event reveals an impact and a resulting plume of gas expelled from the point of impact. A disk of molten and gaseous debris settled into orbit around Earth. According to recent computer models of the collision, much of the material formed into a slowly rotating bar of material after the impact. The iron from the impacting object was in the half of the bar nearest Earth and the lighter rock was in the outer half. The iron particles streamed back into Earth, penetrating through the mantle to the core. Moon formed primarily from debris from the impacting body.

Evidence of such an impact may be apparent in the chemical composition of Earth's mantle. Since the pre-impacted early Earth was in a molten or semi-molten state, heavy substances such as gold and platinum should have sunk to Earth's iron core. But the mantle presently contains 100 to 1 million times more of these substances than it should. Moon, on the other hand, lacks these chemical elements. So Moon may reflect the pre-impact mantles of Earth and the impacting bodies.

Early Atmospheres The Rules of Thumb for planetary atmospheres are: (1) The higher the temperature, the greater the average speed of atoms/molecules of air, (2) The less the mass of the atom/molecule of air, the higher the speed at a given temperature, and (3) The more massive the planet, the higher the speed needed for an atom/molecule to escape into space.

Planetary atmospheres could also have been, to some extent at least, created by impacts. All of the inner planets no doubt had much thicker atmospheres in the past than they do now. But surely the early solar wind must have stripped away many of the gases from the inner solar system bodies, especially the lighter gases like hydrogen and helium. The outer planets would have been protected by their extreme distances from Sun.

The planet interiors were filled with hot rock and gases. An impact by a planetesimal or meteorite could easily have ruptured the young crusts and released gases such as carbon monoxide, carbon dioxide, nitrogen and water vapor. Such icy planetesimals from beyond the orbit of Uranus may have been flung into the inner solar system by Jupiter. These icy bodies would have struck the inner planets, melting or vaporizing on impact and contributing water and other volatile compounds to the planet's atmosphere and surface.

Crater Counts and Ages Once the surfaces of the inner planets and their satellites had solidified, they began to keep a record of further impacts, clues to their times and natures. The early history of the inner solar system bombardment has been largely deduced through the study of Moon (the *Apollo* missions). There is no atmosphere or geologic activity available to change the craters, so Moon is huge fossil containing a comprehensive chronological record of inner solar system bombardment.

The basic method of reconstructing the record of bombardment is to count the craters in a given area and their relative sizes. If there are more craters in one area than another, the first must be the older. The assumption is that the longer a surface is exposed to bombardment, the more impacts it will show. In actuality, it is a bit more difficult than that, because craters can be erased completely or modified by internal activities such as lava flows or crustal movements, and by external activities such as further direct bombardments or material ejected from nearby collisions. In either case, the initial crater densities can be affected.

Application of this method and reasoning to Moon have led scientists to the conclusion that its surface can be divided up into two general regions or epochs: the young, dark maria of ancient lava fields with very few craters; and the older, light-colored highlands with crater upon crater. The highlands have twenty times more craters of a given size than the maria. That doesn't mean that the highlands are twenty times older than the maria, however, because the rate of bombardment has not been constant.

This is known to be true for Moon because we have had a chance to go there and date the rocks collected from those

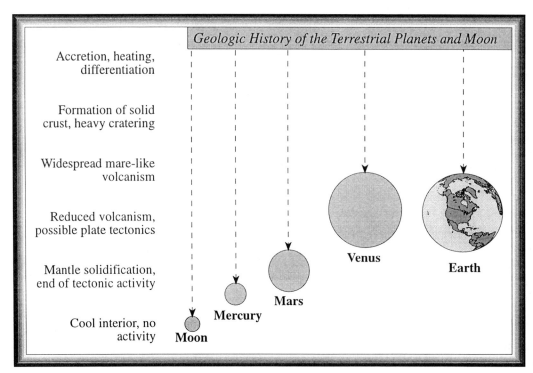

Figure 17–53 The planets had different evolutionary histories primarily due to their different masses. This concept is illustrated by comparing the inner planets side by side according to size and state of internal development.

Surface Material on Various Bodies in the Solar System

Object	Diameter (in kilometers)	Surface material
Saturn ring bodies	<0.1	H_2O ice
Deimos	12	Dark, carbonaceous silicates
Phobos	22	Dark, carbonaceous silicates
J7 Elara	80	Dark, carbonaceous silicates (+ H_2O ice?)
J6 Himalia	180	Dark, carbonaceous silicates (+ H_2O ice?)
S9 Phoebe	220	Dark, carbonaceous silicates (+ H_2O ice?)
J5 Amalthea	270 x 155	Dark, reddish (carbonaceous and sulfur?) soil
S2 Enceladus	500	H_2O ice (virtually pure)
1 Ceres	914	Dark carbonaceous hydrated silicates
S3 Tethys	1,048	H_2O ice
S4 Dione	1,120	H_2O ice
U1 Ariel	1,160	H_2O ice + dust?
P1 Charon	1,190	H_2O ice
S8 Iapetus	1,440	Dark, carbonaceous silicates (leading side); H_2O ice (trailing side)
U4 Oberon	1,520	H_2O ice + dust?
S5 Rhea	1,530	H_2O ice
U3 Titania	1,580	H_2O ice + dust?
Pluto	2,300	CH_4 ice
N1 Triton	2,700	N_2 ice + CH_4 ice
J2 Europa	3,130	H_2O ice
Moon	3,476	Basaltic soil and rock
J1 Io	3,630	Sulfur compounds
J4 Callisto	4,840	Dark (carbonaceous?) soil + H_2O ice
Mercury	4,878	Basaltic soil and rock
S6 Titan	5,150	? (obscured by clouds, differentiated?)
J3 Ganymede	5,280	H_2O ice and (carbonaceous?) soil
Mars	6,787	Basaltic, chemically weathered soil and rock
Venus	12,104	Basaltic and some granitic soil and rocks
Earth	12,756	H_2O liquid; basaltic and granitic soil; rock
Neptune	49,528	H_2 liquid?
Uranus	51,118	H_2 liquid?
Saturn	120,660	H_2 liquid?
Jupiter	142,800	H_2 liquid?

Figure 17–54 A survey of the materials on the surfaces of objects in the solar system assists astronomers in understanding the conditions under which the objects formed in the early history of the solar system.

different regions. The samples from the maria reveal ages of 3.5 billion years, whereas those from the highlands are about 4 billion years old. Furthermore, the craters in the two regions were formed by different types of objects. Apparently Moon went from the Late Heavy Bombardment era to one of very few impacts in a little more than a billion years.

In the asteroid belt, planetesimals did not form into a planet because of the gravitational influence of Jupiter, which caused them to collide with one another. The mass of the belt today—about 2% that of Moon—is much smaller than its original mass. The belt has been depleted by collisions that have left distinct clusters of the *Hirayama* families of asteroids, which travel in the same orbit.

Large planetesimals, scattered by Jupiter into this region, gravitationally perturbed asteroids into planet-crossing or escape-bound trajectories. So there has been considerable mixing up of the material in the belt, and it is unclear whether the present-day material originally formed in the belt or was moved there from elsewhere. For example, some of them, especially those with eccentric orbits, may well be dead comets, their ice and gas shells stripped away by repeated encounters with Sun.

Early Era of Bombardment Astronomers would like to know what kind of objects were responsible for the period of Heavy Late Bombardment. Was it a hail of comets, or simply the sweeping up of the remaining planetary debris? And was it unique to specific regions of the solar system, or was it solar system-wide? It is generally agreed that surface features on all of the members of the solar system reveal evidence of an early era of heavy bombardment. For inner solar system, astronomers assume that a similar population of objects were responsible for the shaping of surface features of the planets and satellites, based upon analysis of the crater densities/sizes on Moon, Mars, and Mercury.

But that same approach cannot be used for the outer solar system, since we do not yet have any absolute ages for any of the objects there (i.e., we do not have any samples from those bodies). Many scientists assume that this era occurred at the same time for both inner and outer solar system. But we don't know that for sure. We have only one sure measurement — the age of the Late Heavy Bombardment on Moon. Because the surface features of Mars and Mercury are so similar to those on Moon, it is assumed that the period during which they occurred was the same as for Moon. The dust of the debate will only settle when we are able to obtain surface samples from Mars and Mercury.

When one evaluates the crater statistics for objects in the outer solar system, they don't look the same as those for the inner solar system. These statistics have to be taken for the satellites, since the planets themselves — being gaseous in nature — do not reveal evidence of impact craters. *Voyager 2* images of the satellites of Jupiter, Saturn, and Uranus, for example, reveal evidence for different populations of impacting objects at different times. This leads to the conclusion that the objects responsible for the different crater statistics had different origins.

Perhaps there were two families of objects involved. Objects in the inner solar system were confined to the inner solar system and were in heliocentric orbits. But in the outer solar system, craters were formed by different populations but they were not in heliocentric orbits. Rather, they were in orbits around the large planets (planetocentric), possibly remnants left over from the formation of the satellites rather than the planets. Some astronomers cite the example of Saturn's rings to support the idea of planetocentric impactors. But the lack of any examples that are still present today is offered as a rebuttal to the idea.

Once the era of heavy bombardment ended, comets took over as the prime source for impacts in the solar system. In fact, some scientists believe that evidence points to the comets as a likely source of the raw ingredients that were needed for life to get started on Earth. This concept, called **panspermia**, proposes that life could be spread from one planet (or even one star) to another by the movement of material.

Chaos in the Solar System During the past couple of decades, scientists have developed the new discipline of **chaos theory** in an attempt to explain some of the major

Future Missions to Planets, Asteroids, or Comets

Year/Mission	Country	Remarks
1996		
Mars *Global Surveyor*	NASA	Nov '96 launch of first satellite in the Mars Surveyor program, carrying many of the *Mars Observer* payload
Mars '96	Russia	Nov '96 launch of large orbiter, two small stations and two penetrators
Mars *Pathfinder*	NASA	Dec '96 launch of small landers to explore Mars
1997		
Lunar Prospector	NASA	Jun '97 launch of lunar polar orbiter that will map the surface composition of Moon
Cassini/Huygens	NASA/ESA	Oct '97 launch of Saturn orbiter and Titan atmospheric probe (arrival 2004, probe entry 2005)
Lunar-A	Japan	A lunar orbiter that will send three penetrators to the lunar surface
1998		
Mars *Global Surveyor 2*	NASA	Jan '98 launch of part of *Mars Observer* payload and two imaging cameras
Planet-B	Japan	Aug '98 launch of Mars orbiter that will study the interaction of the solar wind with the atmosphere
Clementine 2	US Air Force	Launch to study three Earth-crossing asteroids close-up, including penetrators
New Millennium	NASA	Interplanetary flight test of miniature components for future planetary missions
1999		
Stardust	NASA	Feb '99 launch of Discovery mission to comet Wild-2 (arrive 2004) to return samples of interstellar dust (2006)
Mars Surveyor Lander	NASA	A lightweight lander targeted to a near polar latitude
2001		
Mars Surveyor/*Mars '01*	NASA/Russia	Satellite that will include an orbiter and a lander, followed by a rover (marsokhod)
Pluto Express	NASA	First satellite to visit Pluto
2002		
Nereus Sample Return	Japan	Visit to near-Earth asteroid, with sample return in 2002
2003		
Rosetta	ESA	Jan '03 launch of satellite to visit asteroid and rendezvous with comet, with science station deployed on comet's nucleus
Intermarsnet	NASA/ESA	ESA will supply an orbiter for a network of US Mars Surveyor landers
Solar Probe	NASA	Under development
2005		
Mars Sample Return and Phobos Sample Return	NASA/Russia	Under development as part of *Fire and Ice*, an American/Russian initiative

Figure 17–55 Future missions to objects in the solar system.

problems in astronomy as well as other sciences that deal with movements of large numbers of particles. Just how much of a role chaos has played in forming the solar system is an open question. Did the present system settle down to its configuration in its first few million years, or has it gradually evolved over the past 5 billion.

Determining the orbits of two objects in close vicinity to one another (like binary stars) is quite easy to do. Any more than two is extremely complicated because of their constantly changing positions. The 3-body problem is formally unsolvable — although computers can get close. There are limits to computer power, however. Chaos theory had great strides in 1963 when weather phenomena were first studied. What Lorenz of MIT found was that only a very slight change in initial conditions caused drastic changes over long periods of time. This sensitive dependence on initial conditions is a hallmark of chaotic systems. As one varies the initial properties only slightly, the changing outcomes diverge from each other with exponential rapidity. The application of chaos theory to a couple of solar system bodies is illustrative:

- *Hyperion*, a satellite of Saturn, is an irregularly shaped blob that appears to be wobbling and spinning at erratic rates as it moves in its elongated orbit. It may have gone from no rotation at all to one every 10 days in as few as two trips around Saturn. In addition, its spin axis varies both in space orientation and its physical location on the satellite. Why had it not settled down into a synchronous orbit of 21 days? Other observations have revealed a 13-day period, but further show something different.

- Just before the *Voyager 2* flyby of Neptune, a proposed history for Triton was advanced. Coming from an independent orbit around Sun, it was captured by Neptune when it collided with one of the satellites already circling the planet. Another scenario is that Triton's initial orbit was highly eccentric, but tidal forces made it perfectly circular within a few hundred million years. In the process, it cannibalized most of the planet's regular satellites. Only those closer than 5 planet-radii would have survived, perturbed into inclined orbit by interactions with Triton.

Summary/Conclusion

A brief tour of the solar system provides us with convincing evidence that the planets were participants in the formation of Sun from a collapsing cloud of gas, dust, rock and ice. The planets close to Sun have retained much of the history of the early violence that accompanied their formations. Although many of the details are missing that would explain how the planets got to be so different, it is easy to explain most of their differences in terms of their proximity to Sun.

Although there is absolutely no evidence of the existence of life anywhere in the solar system, there are some intriguing places we would like to explore more in more detail — especially Mars. If nothing else, we are learning a little more about the constraints within which life gets started. So by comparing the features of other planets with those of Earth, we are in a better position to evaluate how common life may be elsewhere. Future missions to the objects in the solar system will certainly help us to refine our knowledge about not only our own solar system's formation, but those around other stars as well (Figure 17-55).

LEARNING OBJECTIVES: *Now that you have studied this Chapter, you should be able to:*

1. Name the satellites that have visited the planets of the solar system, and list the advantages of exploring them that way.
2. Explain how a model of Earth's interior structure is obtained, and how that model argues for Earth having formed through accretion.
3. Explain how the semi-molten interior of Earth results in its varied surface features through the mechanism of continental drift.
4. Describe a model to explain Earth's present atmosphere and oceans.
5. Describe the major surface features of Moon, and offer a theory to explain their origins.
6. Describe each theory proposed to explain Moon's origin, and explain the features of Moon that create uncertainty about its origin.
7. Describe the observed features of Venus, and the difficulty of studying them with telescopes.
8. Explain the features of Venus discovered through the use of radar and satellites that have landed on the surface.
9. Offer arguments for the existence of active volcanoes on Venus.
10. Explain how it was that Earth and Venus, similar in many respects, ended up so differently if they formed from the same cloud of gas and dust.
11. Describe the observed features of Mercury, and explain how they support the general theory for the formation of the solar system.
12. Describe the observed features on Mars that led to the claim that intelligent life existed on that planet.
13. Describe how the discovery of dry riverbeds on Mars argues for the possible existence of life on Mars.
14. Describe the major findings of the Viking Landers, and their limitations as regards the question of life on Mars.
15. Define and use in a complete sentence each of the following **NEW TERMS**:

Albedo 395
Anaerobes 389
Aphrodite terra 401
Arachnoid formation 400
Asthenosphere 385
Big Whack theory 394
Blob tectonics 402
Caloris basin 405
Capture theory 394
Chaos theory 421
Co-formation theory 394
Collisional-ejection theory 394
Collisional physics 391
Continental drift 383
Continental plates 383
Coronae 400
Crater density 381
Crust 383
Dynamo effect 382
Earth-fission theory 393
Epicenter 382
Equatorial bulge 381
Evening star 403
Fault line 383
Fumaroles 389
Greenhouse effect 387

Half-life 380
Heat tectonic activity 387
Highlands 391
Impact cratering 387
Ishtar Terra 401
LAGEOS satellite 386
Late Heavy Bombardment 402
Least-energy orbits 396
Lobate scarps 405
Loihi 385
Magellan satellite 382
Magma chamber 384
Mantle 382
Maria 390
Mariner 9 mission 408
Mariner 10 mission 403
Mars Global Surveyor
Mars Pathfinder
Mars '96
Maxwell Montes 401
Morning star 403
Occult 406
Olympus Mons 410
Optical illusions 410
Out-gassing 389
Pacific Ocean Basin theory 393

Pancake formation 400
Pangaea 386
Panspermia 421
Permafrost 409
Pioneer-Venus mission 399
Planetary Society 415
Plates 383
Plate tectonics 383
Pressure waves 382
Radioactive dating 380
Runaway greenhouse effect 403
Sea of Tranquility 391
Seismic stations 382
Shear waves 382
Shield volcano 385
Sling-shot effect 396
Space Exploration Initiative (SEI) 415
Subduction 385
Synchronous rotation 392
Terrae 391
Valles Marineris 411
Venera 397
Viking 408
Wave of darkening 406

Chapter 18

Exploring the Gas Giants Jupiter and Saturn

Figure 18–1 The HST imaged the sudden appearance of a huge storm in Saturn's atmosphere in 1994. Such storms are frequent occurrences, and reveal the dynamic nature of the gaseous planets. (NASA/STScI)

CENTRAL THEME: How do Jupiter and Saturn differ from the terrestrial planets, and how are those differences explained by the theory that all planets formed around Sun together?

*I*n order to accurately understand the Jovian planets that we are about to encounter in the outer reaches of the solar system, we must put on completely different thinking caps from the ones we used while exploring the terrestrial planets. These consist of the lighter chemical substances, similar to those in Sun, but not at the high temperatures as in Sun. In fact, except for those buried deep within the planets, the gases exist mostly at temperatures hundreds of degrees below zero. Remember that their positions in the deep freeze of the solar system have allowed them to retain the general makeup of the nebula from which they formed.

Voyager Mission

Aside from the landing of astronauts on the lunar surface in 1969, surely the greatest achievements of the U.S. space program of the 1960s and 1970s were the twin space probes, **Voyager 1** and **2** (Figure 18-2). Launched in 1977, they performed almost flawlessly as they visited and photographed and measured the environments of all the Jovian planets — Jupiter, Saturn, Uranus, and Neptune. *Voyager 1* visited only the planets Jupiter and Saturn, but in doing so, provided the information necessary for accurately determining the masses of the two planets.

That information, in turn, allowed scientists to make minor mid-course corrections to the trajectory of *Voyager 2* so that it approached the two planets at the right angle for the "slingshot" effect to fling it out to Uranus and Neptune. Besides that, the advantage of sending two spacecraft through the asteroid belt for the first time is obvious — scientists didn't know how concentrated the asteroids are in that region of the solar system, and they didn't want to take the chance by sending a single spacecraft.

The data returned to Earth by these robot cameras is being examined and analyzed even today, so voluminous was the quantity. It is fair to say that most of our knowledge of these gas giants stems from the information supplied by *Voyager 1* and *2*. The entire *Voyager* mission cost just less than $1 billion — which amounts to about 20 cents per year for each American citizen. The *Voyager* spacecraft is one of the most complex robots ever constructed. It has 65,000 parts, and the computer system alone has as many circuits as 2,000 color TV sets.

The resolution of the television camera would allow one to read an automobile license plate at a distance of one mile. Thirty-eight different antennas on four continents were required to receive and amplify the extremely faint signals the spacecraft returned to Earth. For example, the strength of the signal by the time it reached us from the planet Neptune was about twenty billion times less than the power required to operate a digital watch!

The intensity of sunlight at Neptune is about a thousand times less than that at Earth. Taking a photograph in the vicinity of the planet, therefore, was rather like taking a photograph of the contents of your closet with all of the lights turned off. The exposures required for a photograph were necessarily long — the moving spacecraft actually had to rotate backwards to compensate for its movement during each exposure. This wouldn't seem like a great challenge in itself, except when you realize that all of the commands for taking the photos and rotating the spacecraft had to be calculated and programmed in advance. New instructions took four hours to reach the spacecraft! In spite of all of these technical challenges, the two spacecraft performed almost flawlessly and returned spectacular images of the gas giants.

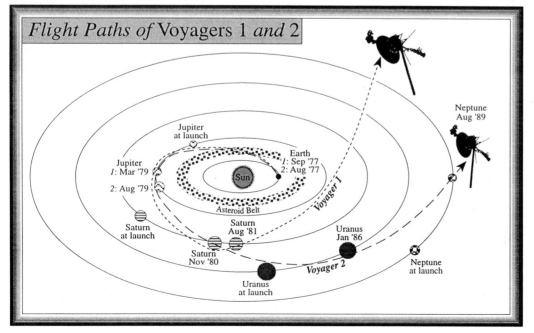

Figure 18–2 The two Voyager *satellites are the most recent and successful missions to explore the giant planets. They returned tens of thousands of images to astronomers, who are still sorting through them to better understand our solar system.*

CHAPTER 18 / Exploring the Gas Giants Jupiter and Saturn

★★★★★★★★★★★★★★★

Figure 18-3 The solar system's four largest planets are compared with Earth in this photomontage. Unlike Earth, these giant worlds are composed primarily of hydrogen and helium with lesser amounts of ammonia, methane, and other gases. (NASA)

Jupiter

Jupiter is a fascinating planet to observe through even a smallish telescope. Even though it is 5 times farther from Sun than Earth, its huge size (11 times the diameter of Earth) and reflective clouds cause it to appear big and bright. To the naked eye, it can be the fourth brightest object in the sky behind Sun, Moon, and Venus. Of all the planets, Jupiter is the only one whose atmospheric details are easily seen and studied in smallish telescopes, like the ones that you would use at a star party (Figure 18-3).

There is also something fascinating about being able to observe satellites circling another planet, and again, Jupiter is the only planet that allows us to easily observe this phenomena. The four Galilean satellites can be seen lined up on either side of the planet itself (unless one or more is momentarily in front of or behind the planet). If you observe them from one night to the next, you can easily notice their movements around Jupiter). With its 16 (there may be more) satellites, Jupiter is like a miniature solar system itself, although of course it doesn't shine. As we will see later in the Chapter, however, Jupiter continues to have significant influences on its nearest satellites, a fact that is obvious when we examine their surfaces.

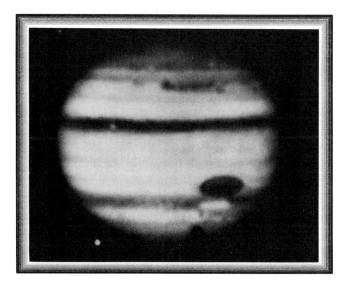

Figure 18-4 A ground-based photograph taken with the 200-inch telescope at Palomar reveals the cloud structure of Jupiter quite clearly, as well as the Great Red Spot for which Jupiter is easily remembered. (NASA)

Atmosphere From an observational point of view, the most interesting feature of Jupiter is its continuously changing cloud patterns. This is not obvious through a telescope, since the changes occur slowly over time (Figure 18-4). But four probes have visited the planet and provided thousands of photographs and data that provide clues toward explaining Jupiter's features. In order to depict the changing cloud patterns, large numbers of photographs taken over a long time interval have been arranged in sequence and then looked at sequentially as if it were a movie or one of those old Walt Disney flip comic books. The movie shows quite vividly the behavior of the atmosphere, very much like the speeded-up video segments used to show the patterns of Earth's clouds during the evening TV weather broadcast.

In order to interpret these time-lapse motion studies of Jupiter's atmosphere, keep in mind that Jupiter consists almost entirely of atmosphere. As far as we know, there is no solid surface in the sense that you and I think of surface. Although in the case of Earth, there is chemical interchange between the gases in the atmosphere and the solid (and liquid) surface material, there is an abrupt difference in density between the two. That difference in density is what we usually mean when we refer to the "surface" of a planet (Figure 18-5).

There is certainly solid material deep within the interior of Jupiter, but you would not suddenly arrive at it if you were to fall into the atmosphere. The density of the gases — and consequently the pressure — would get greater and greater as you penetrated deeper and deeper into the atmosphere. What started off as a thin gas would seem to gradually become the consistency of liquid and in turn the consistency of solid material. If you were simply falling, you

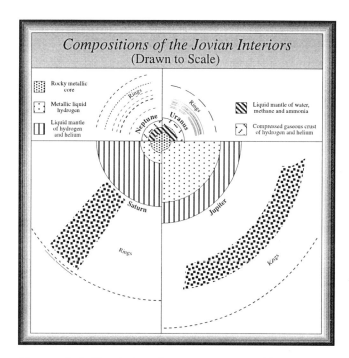

Figure 18–5 The Jovian planets are compared by size, cross–sectional internal composition, and extent of rings. You might find it useful to refer back to this diagram as the Jovian planets are discussed each in turn.

would go only to that depth at which the volume of your body displaces a volume of Jovian gases whose weight equals the weight of your body. This is **Archimedes's Principle** at work. It is the same principle under which I can claim that drowning in "quicksand" is impossible. That is, *"An object displaces a volume of fluid whose weight is equal to the weight of the object itself."* The gases of which the outer planets are made are in about the same abundance percentage wise as Sun (Figure 18-6). In light of earlier discussion about the condensation sequence, and how light gases in the terrestrial planets were "boiled" away because of their proximity to Sun, it is easy to understand why the outer planets do consist of gases.

At Jupiter's core, the great pressures due to the overlying material crushes the hydrogen and helium of which it is mostly composed into what we would — if we possessed a sample — call *metallic liquid* hydrogen and helium. But there is an important difference between that substance and what you would think hydrogen and helium metal is. The atoms are ionized — electrons have been stripped away from atomic nuclei by the great pressures and temperatures.

As a result, Jupiter has a magnetic field 10 times stronger than that of Earth. Surrounding this core of plasma is probably a liquid mantle of hydrogen and helium which is in turn surrounded by the clouds of ammonia and methane that we observe. There are probably no abrupt boundaries between these layers. There is no surface upon which to land or stand.

Cloud Pattern and Rotation It seems strange that the largest planet in the solar system is also the most rapidly spinning planet in the solar system (once every 10 hours). This feature is obvious through a telescope, although not because you can actually see it spin. Rather, Jupiter appears flattened at the poles and "bulgy" along the equator, as if something squashed it from top to bottom. That is due, of course, to the liquid nature of the interior. Even more so than Earth, Jupiter forms an equatorial bulge as its rapid spinning throws the material away from the spin axis.

Looking at Jupiter carefully through a telescope, you observe that the atmosphere consists of alternate light- and dark-colored cloud bands running parallel to the equator (Figure 18-7). The computer-generated movie mentioned earlier reveals that these cloud bands move at different rates, and constantly change in color and intensity (Figure 18-8). The darker bands (called **belts**) generally appear brown or red, although they may shade into blue-green. The lighter bands (called **zones**) are yellow-white. This is probably due to the formation and dissolution of clouds of varying chemical compositions at different altitudes (and therefore at different temperatures). Thus the belt-zone circulation is related to the high- and low-pressure areas we see on weather maps of Earth. The belts are the lower, descending bands of gases, whereas the zones are the higher, ascending bands of gases (Figure 18-9). It is the average rate of movement of the belts and zones that we refer to as the rotation period of Jupiter, inasmuch as it doesn't spin as a solid object.

Although evidence suggests that the cloud bands have been constant in latitude and speed for at least the past 100 years, the movie clearly shows that dramatic changes occur

Atmospheric Compositions of Jupiter and Saturn		
Percentage by Mass (Estimated)		
Gas	Jupiter	Saturn
H_2 (Hydrogen)	79	88
He (helium)	19	11
Ne (neon)	1?	?
H_2O (water)	Trace	?
NH_3 (ammonia)	0.5?	0.2
Ar (argon)	0.3?	?
CH_4 (methane)	0.2?	0.6
C_2H_6 (ethane)	Trace	0.02
PH_3 (phosphine)	Trace	Trace
C_2H_2 (acetylene)	Trace	Trace

Figure 18–6 A list of the most commonly–detected chemicals in the atmospheres of Jupiter and Saturn reveal the two planets being very similar. Notice that hydrogen and helium account for over 98 percent of the masses of the two planets.

Figure 18–7 *The relative motions of features in Jupiter's atmosphere are revealed in these remarkable cylindrical projection photographs taken by* Voyager 1 *(top) and* Voyager 2 *(bottom). Notice that the GRS has moved westward and the white oval features eastward during the time between the acquisition of the images. Also, significant changes are evident in the recirculating flow east of the GRS, in the disturbed region west of the GRS, and is seen in the brightening of material spreading into the equatorial region from the more southerly latitudes. (NASA)*

at the interfaces of adjacent bands that move at different speeds. Even in still photographs like that of Figure 18-4, the turbulence between adjacent bands is obvious. There are small (about the size of Earth) eddies and ovals that form along these lines of shear. They can be thought of as immense cyclones formed by colliding cloud bands.

The most impressive of these cyclones is the phenomena for which Jupiter is best known, the **Great Red Spot**. Oval in shape, three times the size of Earth, it is a rather permanent feature in the atmosphere. Although it has been observed continuously since its discovery 300 years ago, it does change in size and degree of prominence.

Ovals in Jupiter's southern hemisphere, including the Great Red Spot, circulate counterclockwise. Those in the northern hemisphere circulate in the opposite direction, a fact which makes such features similar to the high-pressure cells in Earth's atmosphere. The red color of the spot itself and of some of the bands is not known, although the most likely candidate is sulfur.

It is not clear why the atmosphere of Jupiter behaves the way it does. Astronomers create computer models to explain the observed phenomena, but additional data is needed to refine our understanding (Figure 18-10). One thing is for sure, however. Whereas Earth's atmosphere is driven from the outside — as sunlight falls (mostly) along the equator to form the basic wind patterns — Jupiter's atmosphere is driven from the inside as heat from its slow contraction gradually leaks out into space.

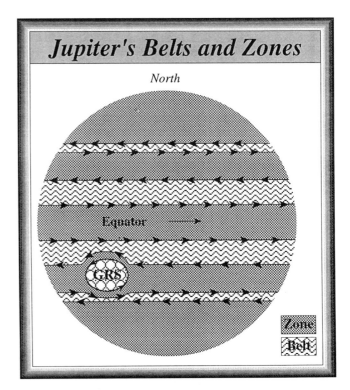

Figure 18–8 *The stripped appearance of Jupiter's atmosphere is the result of horizontally-moving currents of gases that move at different rates, as illustrated in this diagram. The GRS is caught between two of these currents, which cause it to rotate in a counterclockwise direction.*

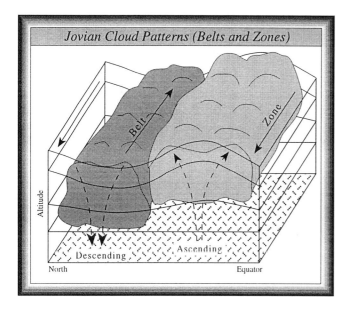

Figure 18-9 *Jupiter's belts and zones are at different heights, suggesting that there is vertical as well as horizontal movement between them. This also accounts for their different brightnesses, since the descending belts are at lower temperatures.*

Yes, it is believed that Jupiter is slowly shrinking. This is not something directly observable — measurement of the electromagnetic energy given off by Jupiter indicates that it releases three times more energy than it receives from Sun. Something is going on inside of the planet that allows for the release of energy. Slow collapse can explain the excess energy measured on Earth. In a sense, Jupiter is still forming as it very slowly collapses. So Jupiter is much more like Sun in its properties than it is like Earth.

Radio Emission Jupiter is like Sun in another respect. The giant planet is second only to Sun in the amount of radio energy it emits. Its powerful magnetic field traps solar wind particles which, when disturbed by the innermost of the Galilean satellites, **Io**, emit radio energy bursts. As is the case of Earth, trapped ionized particles also interact with Jupiter's atmosphere to form impressive auroras. In 1992, the Hubble Space Telescope photographed a Jovian aurora. This offers strong evidence that *Io* plays a major role in aurorae. The aurora photographed by the HST lies in the northern polar region, near where magnetic field lines passing through the orbit of *Io* enter Jupiter's atmosphere.

In addition to photographing these colorful displays, the *Voyager* space probes photographed what resemble immense lightning discharges within the clouds. These are easily seen in the photograph taken of the nighttime side of Jupiter, Figure 18-11. Actually, the probes were pro-

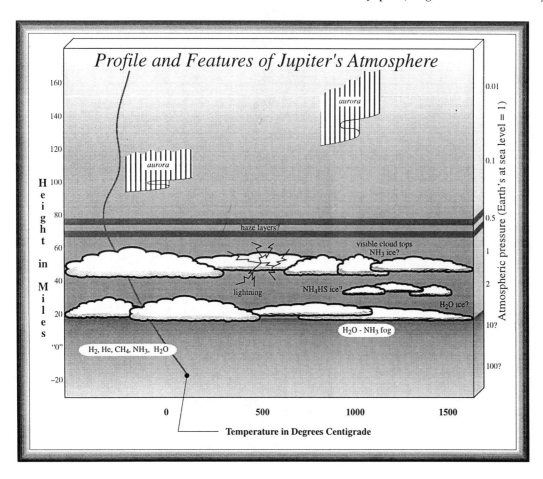

Figure 18-10 *A model of Jupiter's atmosphere profile, showing the heights at which the various phenomena we observe occur. Note especially the locations of the lightning and auroras.*

CHAPTER 18 / Exploring the Gas Giants Jupiter and Saturn

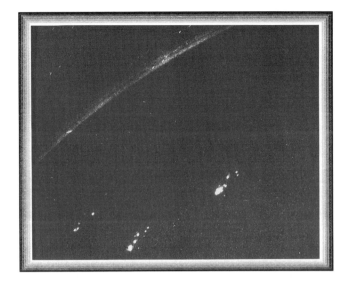

Figure 18–11 Lightning and auroras were imaged by the Voyager *satellite as it passed behind to observe the nightime side of Jupiter.* (NASA)

grammed to fly through the radiation belts in order to measure their strengths and locations. In the process, they received dosages of radiation that were hundreds of times more lethal than the maximum dose for humans. Humans intent on exploring Jupiter should take note of that fact.

Collision of Comet Shoemaker–Levy 9 In Chapter 16, while surveying the contents of the solar system, I mentioned that a once-in-a-lifetime event occurred in late 1995 that turned every major telescope on Earth and in space toward the planet Jupiter. The collision of the 21 fragments of Comet *Shoemaker–Levy 9* with the atmosphere of Jupiter gave astronomers the opportunity to study almost in real time an event that must be similar to those that helped shape the early solar system. It was hoped that observations of the collisions would shed light on not only the composition and dynamics of Jupiter's outer atmosphere, but on the composition of comets themselves. Unfortunately, all 21 collisions occurred on the backside of the giant planet, so that Earthbound observatories had to wait for the impact sites to rotate into view. Fortunately, Jupiter rotates so rapidly that details of the sites were still

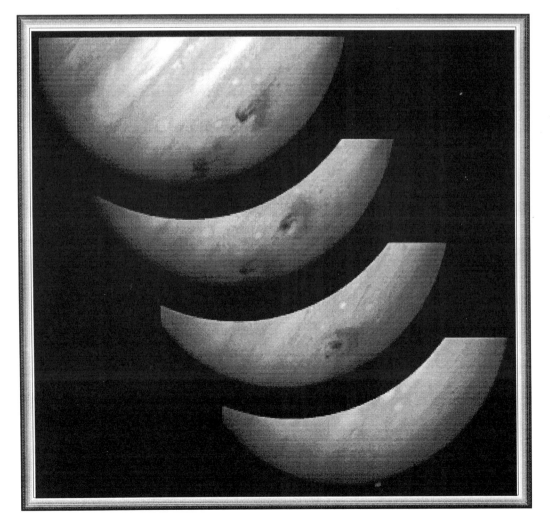

Figure 18–12 This mosaic of HST images shows the evolution of the G impact site on Jupiter. The images from upper left to lower right show: (a) the impact plume about 5 minutes after the impact; (b) the fresh impact site 1.5 hours after impact; (c) the impact site after evolution by the winds of Jupiter, along with the L impact (right), taken 3 days after the G impact and 1.3 days after the L impact; and (d) further evolution of the G and L sites due to winds and an additional impact (S) in the G vicinity, taken 5 days after the G impact. (NASA/STScI)

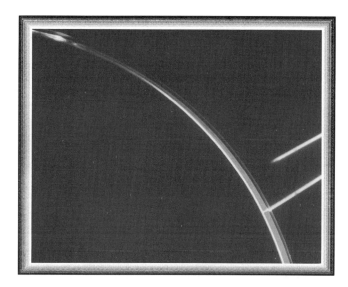

Figure 18–13 The Voyager *satellites provided a complete surprise to scientists when they imaged the existence of a very thin ring system around Jupiter. The rings are backlit in this image since it was taken while the satellite was passing behind the planet.* (NASA)

visible by the time they were viewable from Earth (Figure 18-12).

There was, however, one observatory that was in a favorable position to witness the impacts directly: the *Galileo* satellite, enroute itself to Jupiter for its own collision with the planet (actually, a probe that it carried was released and plunged into the atmosphere). I will discuss that mission shortly. Observations concluded that:

- Comet fragments were traveling 40 miles per second when they hit Jupiter's atmosphere,

- *Galileo* observed hot fireballs produced by the plunging comet fragments,

- Plumes rising from the impact sites contained a mixture of material from the comet and Jupiter's clouds,

- HST measured the heights of all the plumes and found them to be about 2,000 miles above Jupiter's upper cloud deck, and

- The hot spots observed at the impact sites by Earth-bound infrared telescopes arose from debris from the plumes falling back onto the upper atmosphere and heating the gases there.

The observations, in general, allowed (and continue to allow) astronomers to fine tune their models of Jupiter's atmospheric content and dynamics. There are, however, some unanswered questions that will require further study of the data to resolve:

- What were the sizes of the fragments that hit Jupiter? (Estimates range from one-half to 2 miles in diameter)

- How deep into Jupiter's atmosphere did the fragments plunge prior to disintegrating and becoming gas and dust components of the atmosphere?

- Did the dark material that formed above the impact sites come from the comet or from Jupiter itself?

- Why did all of the plumes rise to about the same height, but the hot spots that formed in the vicinity have different sizes?

Rings Space probes are much like Christmas presents — they sometimes provide surprises. As *Voyager 1* passed behind Jupiter, it photographed a faint ring system in its equatorial plane. This discovery took astronomers quite by surprise, since the rings of Saturn were thought to the result of a moon-like object wandering too close to Saturn and being pulled apart by tidal forces into the rings we see. For that to have happened twice in the same solar system was thought to be highly unlikely.

Compared to Saturn's ring system, however, the ring system of Jupiter is not very impressive. As *Voyager 2* passed behind Jupiter four months after *Voyager 1*, it took photographs of the ring backlit by Sun. Small particles are more efficient than large particles in scattering light forward — you are aware of that every time you are blinded by sunlight scattered by dirt particles on your automobile windshield as you drive "into" Sun. Knowing what size particles are necessary to scatter the amount of light detected by *Voyager 2*, scientists obtained an estimate of the particle sizes.

The pictures of the backlit rings reveal three different elements of the ring system — the main ring (Figure 18-13) is about 4,000 miles wide and is made of very small particles, about 1/10,000th of an inch across. A diffuse "sheet" extends away from the main ring, and a faint "halo" doughnut-shaped ring 12,000 miles in diameter circles Jupiter far above the cloud tops. The ring system is quite thin, less than 20 miles thick.

Laws of dynamics tell us that these particles could not remain in stable orbits around Jupiter for long. So the suspicion is that they are not permanent members of the ring. There must be some agent responsible for resupplying particles to the ring as other particles spiral into Jupiter. Perhaps Jupiter collects them from interplanetary space as it sweeps around Sun, and as Sun orbits the galactic nucleus. It could also be matter ejected off of the satellite Io

Figure 18–14 The satellite Galileo *is launched from the Space Shuttle, with Earth as a backdrop.* (NASA)

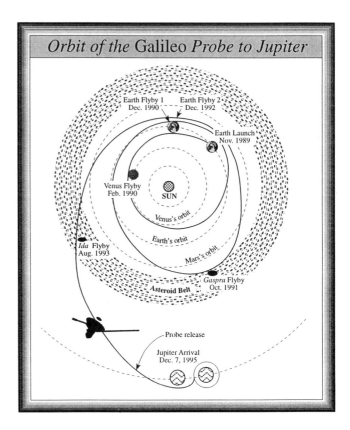

Figure 18–15 The probe Galileo *visited Jupiter to follow up on the success of the* Voyager *Mission. During its long route to the planet,* Galileo *obtained some remarkable photographs, such as the first close-up photos of asteroids* Gaspra, Ida, *and* Dactyl *(Refer to Figures 1–17 and 16–41).*

as a result of the active volcanoes on its surface. Further observations and computer-modeling are necessary.

Galileo Visits Jupiter The **Galileo** satellite was launched from the Space Shuttle in 1989 (Figure 18-14). It did not have sufficient velocity at time of launch to reach Jupiter directly, so it picked up some additional energy with three gravity-assists by Venus (one visit) and Earth (two visits). It arrived at Jupiter and began returning data in late 1995 (Figure 18-15). Five months prior to arrival a probe separated from the orbiter and flew on toward Jupiter by itself. On December 7, 1895, the probe became humankind's first entry into the atmosphere of an outer planet.

It hit the atmosphere at 115,000 mph, a speed that would allow you to travel from San Francisco to New York in 90 seconds. While the orbiter relayed data back to Earth, the probe plunged about 400 miles through the planet's brilliantly colored cloud layers into the hot, dense atmosphere below. Eventually its instruments were fried by radiation. But during the descent, reams of data were collected by the probe, allowing us to model more accurately the inner-workings of this intriguing planet, and even providing clues as to its early formation process. Some of the important conclusions obtained are:

- Jupiter's atmosphere is much more Sun-like in composition than astronomers had expected,

- The probe's atmospheric entry site lacked the expected three layers of clouds, and what thin clouds it did encounter contained no water (contrary to what the *Voyager* satellites and comet *Shoemaker-Levy 9* encounters indicated),

- Lightning is rare, and seems to have produced no complex organic molecules,

- The atmosphere is much denser and windier than astronomers had guessed, and the winds did not diminish as the probe descended deeper into the atmosphere, and

- There is a radiation belt extending from Jupiter's filmy dusty ring down almost to its cloud tops.

The widowed orbiter is presently performing a loop-the-loop trajectory that will eventually take it past the large Galilean satellites **Io**, **Europa** (Dec '96, Feb '97, Nov '97), **Ganymede** (Jun '96, Sep '96, Apr '97, May '97) and **Callisto** (Nov '96, Jun '97, Sep '97). The current goal is to obtain about 50,000 to 100,000 images during the encounter. Astronomers are anticipating some more stunning photographs of these interesting satellites, and of course new discoveries and new questions. Even during its long voyage to Jupiter, *Galileo* was busy at work. In October,

Characteristics of the Satellites of the Solar System

Planet	Satellite	Discoverer	Inclination of Orbit to Planet's Equator	Inclination of Orbit to Planet's Orbit Around the Sun	Orbital Eccentricity
Earth	**Moon**	?	23.5	5	0.05
Mars	Phobos	A. Hall (1877)	1	1	0.02
	Deimos	A. Hall (1877)	2	3	0
Jupiter	Metis	*Voyager* (1979)	0	0	0
	Adrastea	*Voyager* (1979)	0	0	0
	Amalthea	Bernard (1892)	0.5	0.5	0
	Thebe	*Voyager* (1979)	0	1	0
	Io	Galileo (1610)	0	0.3	0
	Europa	Galileo (1610)	0.5	0.5	0
	Ganymede	Galileo (1610)	0	0.2	0
	Callisto	Galileo (1610)	0	0.25	0
	Leda	Kowal (1974)	26.7	26.7	0.15
	Himalia	Perrine (1904)	27.6	27.6	0.16
	Lysithea	Nicholson (1938)	29	29	0.12
	Elara	Perrine (1904)	28	25	0.21
	Ananke	Nicholson (1951)	33	147	0.17
	Carme	Nicholson (1938)	16	163	0.21
	Pasiphae	Melotte (1908)	35	147	0.40
	Sinope	Nicholson (1914)	27	156	0.28
Saturn	Pan	*Voyager* (1991)	0	0	0
	Atlas	*Voyager* (1980)	0	0	0
	Prometheus	*Voyager* (1980)	0	0	0
	Pandora	*Voyager* (1980)	0	0	0
	Epimetheus	*Voyager* (1980)	0	0.3	0
	Janus	*Voyager* (1980)	0	0.1	0
	Mimas	Hershel (1789)	1.5	1.5	0.02
	Enceladus	Hershel (1789)	0	0	0
	Tethys	Cassini (1671)	1	1	0
	Calypso	*Voyager* (1980)	0	1	0
	Telesto	*Voyager* (1980)	0	1	0
	Dione	Cassini (1684)	0	0	0
	Helene	*Voyager* (1980)	0	0	0
	Rhea	Cassini (1672)	0	0	0
	Titan	Huygens (1655)	0	0	0
	Hyperion	Bond (1848)	0	0	0.1
	Iapetus	Cassini (1671)	14.7	14.7	0.03
	Phoebe	Pickering (1898)	30	150	0.16
Uranus	Cordelia	*Voyager 2* (1986)	0	0	0
	Ophelia	*Voyager 2* (1986)	0	0	0
	Bianca	*Voyager 2* (1986)	0	0	0
	Cressida	*Voyager 2* (1986)	0	0	0
	Desdemona	*Voyager 2* (1986)	0	0	0
	Juliet	*Voyager 2* (1986)	0	0	0
	Portia	*Voyager 2* (1986)	0	0	0
	Rosalind	*Voyager 2* (1986)	0	0	0
	Belinda	*Voyager 2* (1986)	0	0	0
	Puck	*Voyager 2* (1986)	0	0	0
	Miranda	Kuiper (1948)	3	3	0.02
	Ariel	Lassell (1851)	4	0	0
	Umbriel	Lassell (1851)	0	0	0
	Titania	Hershel (1787)	0	0	0
	Oberon	Hershel (1787)	0	0	0
Neptune	Naiad	*Voyager 2* (1989)	4	0	0
	Thalassa	*Voyager 2* (1989)	0	0	0
	Despina	*Voyager 2* (1989)	0	0	0
	Galatea	*Voyager 2* (1989)	0	0	0
	Larissa	*Voyager 2* (1989)	0	0	0
	Proteus	*Voyager 2* (1989)	0	0	0
	Triton	Lassell (1846)	23	160	0
	Nereid	Kuiper (1949)	29	28	0.76
Pluto	**Charon**	Christy (1978)	17	120	0

* Note: Boldface data suggest that satellite was captured by parent planet.

Figure 18–16 This is a complete listing of the known satellites of the solar system, together with some of their basic characteristics. Note especially any deviations from a pattern in the data. That usually means that the satellite had an origin different from the others. For example, under orbital eccentricity (the larger the number, the less circular the orbit) the outer satellites of both Jupiter and Saturn are probably captured asteroids. The same deviation in the pattern of inclination of orbit holds true for those satellites as well.

1991, it flew close to the main-belt asteroid *Gaspra* and returned data (and photos) to Earth (see Figure 1-17). Later it passed by and imaged the double asteroid system *Ida-Dactyl (Figure* 16-41).

These were the very first detailed close-up photographs of an asteroid. In December 1990, *Galileo* swooped in toward Earth and picked up enough speed to slingshot it toward another encounter with Earth in December 1992, and then on for its rendezvous with Jupiter in 1995. As it approached and then departed from Earth, it continuously snapped images of our planet and our Moon. This long-distance view gave astronomers a glimpse of the Moon's far side, which has not been photographed since the *Apollo* missions.

But camera technology has improved considerably since the 1970s. The difference between images taken by the cameras aboard *Apollos* and those aboard *Galileo* are incredible — they have been likened to the difference between a snapshot taken by a Brownie camera and that taken by a high-tech camera with a 400-color spectrum. That is another reason that astronomers are so anxious to receive images of the Galilean satellites that the spacecraft will send back beginning in summer 1996. *Galileo's* outer-space perspective has allowed astronomers to compile a "movie" of Earth as it rotates, creating a "never-before-seen" spectacle.

Satellites of the Outer Solar System

There is probably a tendency for us to consider the planets as having more importance and hence greater interest than their satellites. My personal experience is that young school children and even adults have a fairly good ability to name the planets and their increasing distances from Sun. But few can name even a single satellite other than Earth's Moon, let alone the name of its associated parent planet. And yet not only are at two of them actually larger than the planet Mercury, but several have surface features that are more interesting from a scientific point of view than those of some of the planets.

Figures 18-16 and 18-17 provide some of the basic characteristics of the satellites. You may find it useful to refer to them as you read descriptions of each in this and the following Chapter. Note especially how certain characteristics appear to go together. For example, the outer eight satellites of Jupiter are all small, located far from Jupiter, have steep orbital inclinations, and are in rather elliptical orbits around their parent planet. These patterns provide the major clue for their origin as captured asteroids. Look for other patterns in the Figures that may provide clues as

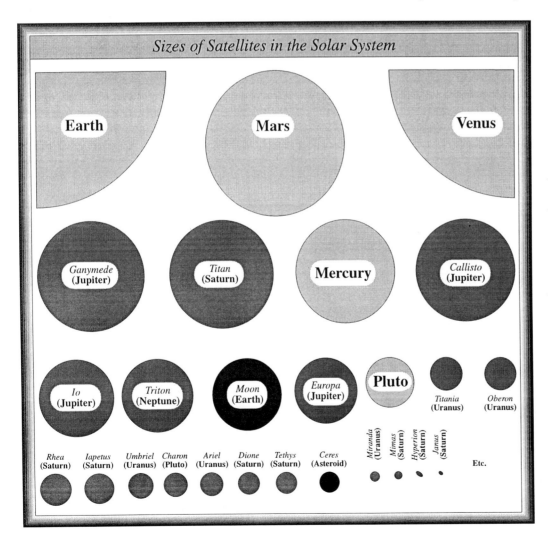

Figure 18–17 This chart of the relative sizes of the major satellites in the solar system, together with the list of their characteristics (Figure 18–16), will prove useful in your appreciating the assortment of minor bodies in the solar system.

to origin.

Out of the 65 known satellites in the solar system, only seven are large. The "Big Seven" are Moon, Jupiter's four Galilean satellites, Saturn's *Titan*, and Neptune's *Triton*. Titan and Triton are the most distant of the Big Seven, and therefore may provide clues about the processes that occurred to make the solar system. Those two may also help us to understand the nature of that most distant and as yet unexplored world, Pluto.

Saturn's Titan is especially interesting because it has an atmosphere much more substantial than that of Mars and slightly more substantial than Earth's. There is even speculation that life could be present in the liquid environment that may exist on the surface of that satellite. Neptune's Triton is most unusual by virtue of its retrograde revolution around its parent planet. Of the 61 known satellites, only six revolve clockwise — and Triton is the only large one to do so. On dynamical grounds alone, it could not possibly have formed in the normal way (the manner in which other satellites formed).

There are at least two reasons why astronomers are interested in the features and characteristics of satellites. One is that they provide clues to the origin of the solar system, which of course is important to the question of solar systems existing around other stars. But the unusual features of some of the satellites provide us with a broader understanding of the forces and geological processes that are possible in the universe, including Earth. Before you begin this tour of the satellites of the solar system, you may want to clear your mind of any lingering "planet chauvinism" and respect each satellite as a world unto itself.

Jupiter's Family of Satellites

There are 16 known satellites orbiting Jupiter, although only the 8 nearest to the planet are probably native to it. Since the other 8 move in irregular orbits inclined at varying angles to Jupiter's equatorial plane, they are probably captured asteroids (Figure 18-18). The most interesting are the four *Galilean* (those first observed by Galileo) satellites, and *Amalthea*, the nearest satellite to Jupiter. All five are caught in *synchronous rotation* by Jupiter's immense gravitational field, meaning (like Earth's Moon) they always present one face toward Jupiter. By and large, these are icy worlds made of frozen mixtures of water ice and carbon-based dirt (Figure 18-19). I begin the tour with the outermost of the *Galilean* satellites and work inward so that you can readily appreciate the past and continued effect Jupiter has on its family of satellites (Figure 18-20).

Callisto The satellite *Callisto* is uniformly and heavily cratered (Figure 18-21). In fact, there are no smooth areas on Callisto — its surface of wall-to-wall craters makes it the most heavily-cratered object in the solar system. Apparently it cooled rapidly, preventing any internal processes

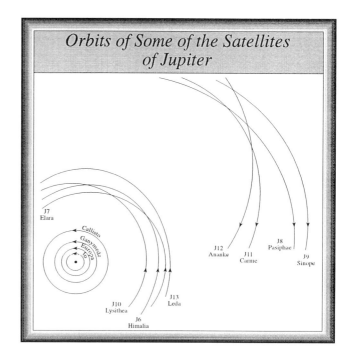

Figure 18–18 Compare this correctly–scaled diagram of the orbits of the major satellites of Jupiter with the list of their characteristics (Figure 18–16), and it becomes obvious why it is so easy to entertain a capture theory for the outer satellites.

from erasing evidence of the early collisions that occurred while forming. But the cratered surface of Callisto appears significantly different than that of Moon or Mercury or even Mars. There is very little vertical relief associated with the craters — basically because ice, a principal component of these outer worlds, is quite weak and unable to support high crater walls. This fact is applicable to many of the satellites in the outer solar system. Callisto has the lowest density of the Galilean satellites, suggesting the presence of a large percentage of water.

The largest impact feature in the solar system, called **Valhalla**, is also found on Callisto (Figure 18-21). It cannot really be called a crater, since at the time of impact the surface ice melted, flowed back into the depression created by the collision, and then refroze. Today we see only the compressional waves frozen into the surface ice, causing it to look like a giant bull's-eye. It is much like throwing a pebble in a pond of water, and the pond suddenly freezing to preserve the ripples moving outward from the point of impact. Within the 6- to 12-mile range of craters on Callisto are found chains of craters, suggesting that comet breakup and collision has occurred on satellites as well as planets. Chains of craters are also found on Jupiter's largest satellite, Ganymede.

Ganymede The satellite *Ganymede* is the largest in the solar system, actually larger than the planet Mercury. Its density suggests that it is about 50 percent water or ice and the rest rock. Photographs taken by the two *Voyager* probes reveal craters that remind us of Moon's surface (Figure 18-22). Like Callisto, these craters are in an icy surface. Ganymede, however, being somewhat larger, cooled more slowly than Callisto and consequently suffered surface modification from interior forces.

Comparing photographs of Callisto and Ganymede, you notice that the impact craters appear to be lighter than the surrounding crater-less regions. In fact, the rays leading away from the impact craters appear lighter as well, suggesting that water-ice lies just beneath a thin coating of dark material. During asteroidal collisions, the dark coating is penetrated and the water-ice is thrown upward and outward. Fresh water-ice is sprayed out in all directions and deposited on the surface over the dark material.

Ganymede has two rather distinct terrain types. The dark terrain is less heavily cratered, and possesses extensive grooves that are sets of ridges and troughs whose widths are a few miles in extent and whose depths are 1,000 feet or so. These grooves wander for thousands of miles, intersecting in intriguing patterns. Occasionally, they are offset in a manner similar to fault lines on Earth (Figure 18-23). These features suggest that as the satellite slowly cooled and solidified, material oozed out of the interior, causing the slippage of dark crustal blocks of ice which in turn caused the grooves to form. Although significantly different from Earth's tectonic activity, this is the only other known case of the process of block slippage in the solar system.

Europa Moving a step closer to Jupiter, we arrive at a satellite – *Europa* – that is still being influenced by the presence of its parent planet (Figure 18-24). In strong contrast to both Callisto and Ganymede, Europa reveals a surface almost completely void of craters. Rather, it has a chaotic system of light and dark stripes that crisscross in no obvious pattern. These stripes are typically several miles wide and some are thousands of miles long. And yet Europa is probably the smoothest object in the entire solar system. It is the brightest of the Galilean satellites, suggesting a high percentage of highly-reflective ice on its surface.

The stripes appear to be cracks in the icy crust that have been subsequently filled by liquid material pushed up from below. Since the temperature conditions reach a maximum of -220°F at this distance from Sun, however, the liquid quickly freezes to become the stripes we observe. Thus there could be a liquid ocean beneath a frozen crust on Europa. To keep the interior of Europa warm enough to account for such an ocean even to this day requires a source of heat.

Although located at a distance from Sun at which the temperature is about 100 K, Europa may still experience substantial heating. One source is simply the decay of radioactive elements in its interior (as in the case Earth).

Figure 18-19 This is a composite photograph of Jupiter and its four Galilean satellites, shown according to their order of distance from the parent planet. (NASA)

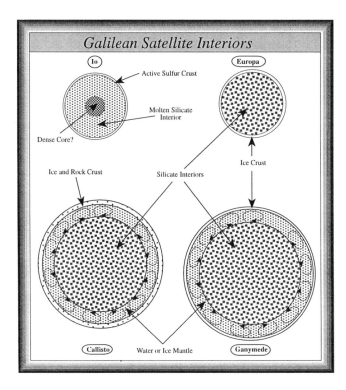

Figure 18-20 The interiors of the four Galileo satellites are modeled according to what chemicals were detected on the surfaces by the Voyager satellites, and the densities determined by knowing their masses and volumes.

Another possible heat source for maintaining Europa's liquid mantle is called **tidal dissipation**. In principle, it is similar to the one-inch *land tides* caused in Earth's surface material as Moon crosses overhead (Chapter 3). Europa is caught in a gravitational tug-of-war between Jupiter and the satellite Io. This results in a certain amount of friction between particles of matter that make up the satellite, with the subsequent release of heat.

Since Europa's orbit around Jupiter is not perfectly circular, nearby satellites must be continuously distorting it and causing Europa to move alternately closer to and farther away from Jupiter. The enormous gravity of the giant planet itself distorts Europa's shape, causing a tidal bulge that varies in size as Europa's distance from Jupiter varies. It is similar to the method you use to break a coat hanger in half by rapidly bending it back and forth. While the liquid ocean is estimated to be tens of miles in depth, the solid crust is only a couple of miles in thickness and therefore very susceptible to fracturing (Figure 18-25).

Once this model was developed, it was natural to speculate on the possibility of life in that ocean. But it is sheer speculation that arises out of the model (the movie *2010* portrayed Europa as the home base for extraterrestrials). Spacecraft exploration has a way of making theorists' hopes of such findings evaporate in the face of better data. The two *Voyagers* simply didn't examine Europa closely enough to resolve the issue. The *Galileo* spacecraft, having cameras with better resolution, just might. For example, if liquid water is exposed to Europa's surface when new fractures form in the ice cover, the water should boil vigorously and create geyser-like clouds of vapor. These precesses, if occurring, could easily be detected by *Galileo*.

If the model *is* correct, then the temperature and pressure in the ocean might just be favorable enough for simple life forms like algae to live there, since some have been found to live at the bottom of permanently ice-covered lakes in the Antarctic. It may be that during the first few years after a crack forms in the crust of Europa, enough sunlight penetrates down through the crack to support the photosynthesis necessary for algae to survive. Until the *Galileo* probe arrives at Jupiter in 1995 and takes an extended look at Europa, however, we won't know for sure that an ocean exists and therefore that algae-Europeans are even able to exist there.

The latest surprise found on Europa occurred in 1995 when astronomers analyzing data supplied by the HST concluded that it has an extremely tenuous atmosphere of molecular oxygen. This makes Europa the first satellite found to have an oxygen atmosphere, and only the third such solar system object beyond Earth (Mars and Venus have traces of molecular oxygen in their atmospheres). Now don't think that astronauts going there will be able to survive on the surface without a space suit. The atmospheric pressure on the surface is barely *one hundred billionth* that of Earth. And it doesn't mean that life might have evolved on Europa either. The temperature measured

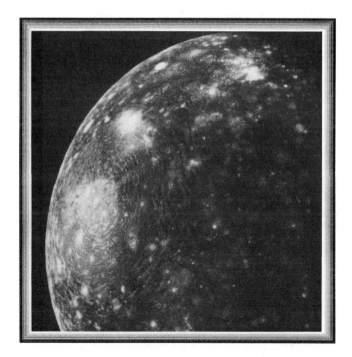

Figure 18–21 The satellite Callisto *is the most heavily cratered object in the solar system. Notice the bright areas where fresh water ice has been thrown outward at the time of impact.* (NASA)

Figure 18–22 The largest satellite in the solar system, Ganymede, *shows evidence of large blocks of frozen ice-rock mixture having been pushed around on the surface.* (NASA)

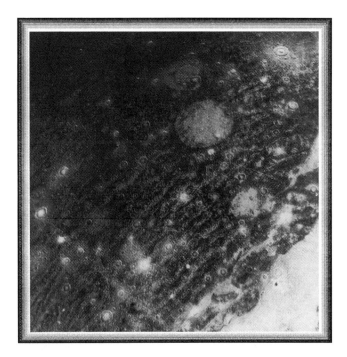

Figure 18–23 An extreme close–up of Ganymede *reveals a wealth of detail about the tortured terrain on that now–frozen satellite. When it was young and forming, however, there must have been a considerable amount of shifting around of huge globs of molten material.* (NASA)

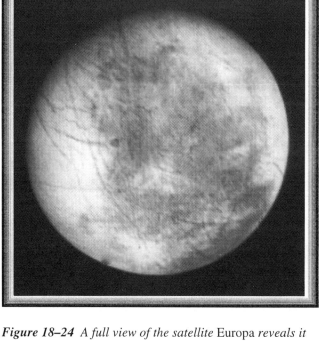

Figure 18–24 A full view of the satellite Europa *reveals it to be different than the more distant Callisto and Ganymede — there are no craters. The surface, therefore, must be young.* (NASA)

on the surface is –230 degrees F, too cold for life as we know it to survive.

The thin oxygen atmosphere is the result of purely non–biological processes. Europa's icy surface is exposed to sunlight and is impacted by dust and charged particles in the solar wind. Together, these processes cause the frozen water ice on the surface to produce water vapor as well as fragments of water molecules. The relatively lightweight

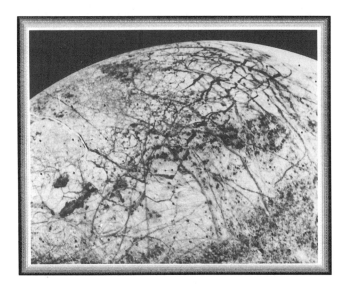

Figure 18–25 With this extreme close–up of the surface of Europa, *we see revealed the ridges of frozen material that may have originated in a sub–surface ocean deep within the satellite.* (NASA)

Figure 18–26 A close–up of the limb of Io *reveals one of the eight volcanoes erupting. There are no craters on the satellite's surface, since it is constantly being reworked and covered over by sulfur and sulfur compounds ejected from the volcanoes.* (NASA)

hydrogen gas escapes into space, while the heavier oxygen molecules accumulate to form an atmosphere that extends 125 miles above the surface.

Io Color photographs of the innermost *Galilean* satellite, Io, make it appear more like a poorly-made pizza than a satellite in the solar system. Acre for acre, it is the most volcanic object in the solar system. This distinction was not earned until *Voyager 1* photographed 8 active volcanoes on Io's surface as it cruised through Jupiter's satellite system in 1978 (Figure 18-26). Four months later, when *Voyager 2* cruised through, 7 were still erupting. Plumes of ejected material were observed as far as 175 miles above the satellite's surface. It is not surprising, therefore, to learn that there are no impact craters on Io's surface, since the active volcanoes are continuously reworking the surface and erasing evidence of past asteroidal collisions.

Unlike volcanic activity associated with Earth, however, sulfur appears to be the predominant ingredient of the ejected material on Io. Thus compounds of sulfur no doubt account for the myriad of colors seen on Io's surface. The scientists who first studied the images of the active volcanoes theorized that the source of heat required to melt the subsurface material and cause it to be ejected outward as volcanic eruptions was the same *tidal dissipation* that accounts for the observed stripes on Europa. Since Io is closer to Jupiter than Europa, the tidal stress is therefore greater — with a corresponding greater amount of thermal energy being released (Figure 18-27).

Now, 15 years after the *Voyager* probes first detected Io's volcanoes, the original speculations on potential causes of volcanism are still the most plausible. The basic idea is that Io is slightly deformed due to strong tidal interaction primarily with Jupiter and Europa, the next satellite out. The tidal pulling distorts Io into an oblate shape, and then allows it to relax as its position relative to Europa changes.

This repeated flexing of Io's outer shell generates considerable internal heat that, over time, mobilizes most of the volatile materials near the surface including water — with sulfur and its compounds being left behind. This strong tidal heating is probably generating various kinds of volcanism simultaneously on Io: surface flows and ponds of quickly cooling liquid sulfur; surface flows of much hotter silicates; and the spectacular, primarily sulfur dioxide driven umbrella-shaped plumes so familiar from the *Voyager* flyby pictures.

Although the solid sulfur compounds ejected from the volcanoes fall back down to cover the surface of Io, much of the gas (thought to be sulfur dioxide) is dissociated and ionized by the charged solar wind particles caught in Jupiter's powerful magnetic field. The sulfur ions are then distributed into a doughnut-shaped swirl that lies roughly in the orbit of Io around Jupiter.

As Io orbits Jupiter inside this swirl of charged particles, it runs into the very particles it ejected from its interior. This collision causes bursts of radio energy to be emitted. It was not until the *Voyager* probes revealed the existence of the volcanoes and the swirls of ionized particles that such radio bursts coming from the vicinity of Jupiter

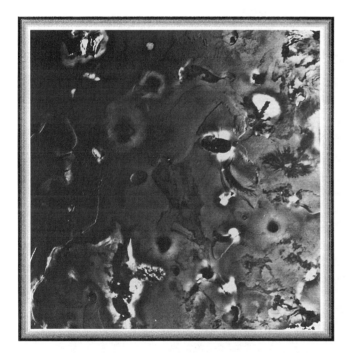

Figure 18–27 An extreme close–up of the surface of Io reveals the fine detail of the surface that is constantly being reworked and covered over by sulfur and sulfur compounds ejected from the volcanoes. (NASA)

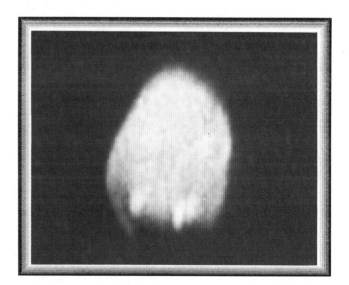

Figure 18–28 When seen in color, the satellite Amalthea looks like Io, suggesting that some of the sulfur ejected out of the volcanoes of Io goes into orbit around Jupiter, and this small satellites collects it as it orbits the planet. (NASA)

were explained. Now they are. Another mystery in the solar system solved.

Amalthea The satellite **Amalthea** is really quite puny in size compared to the Galilean satellites. It measures about 170 by 100 miles (Figure 18-28), and is another example of an object that is not massive enough to have pulled itself into a spherical shape as it cooled and solidified. The reason that I mention the existence of this satellite at all is that its reddish surface color reminds us of the reddish color of the sulfur covering much of Io's surface. Based on this observation, astronomers theorize that in the process of orbiting Jupiter within the swirl of sulfur ions ejected from Io's interior, Amalthea sweeps up enough of the reddish sulfur material to form a thin layer on its surface.

Here is another strange event in the solar system — a satellite sharing some of its material with a neighboring satellite. These kinds of situations are constant reminders to us that objects do not exist by and within themselves in the universe. There are subtle, if not always recognizable, influences that objects exert on one another. The *Voyager* satellites discovered three previously unknown satellites of Jupiter, making the total of known satellites 16. There may be others that are simply too small to have been imaged by the *Voyagers*, so don't be surprised if the *Galileo* satellite discovers additional ones when it arrives in 1995.

Saturn

Saturn has to be one of the most impressive celestial objects to observe in a telescope, if not *the* most impressive. Of course, it is the ring system that makes it so. In just about all other respects, Saturn is similar to Jupiter. It is the second largest planet behind Jupiter, 90 times more massive than Earth, but contained within a volume 830 times that of Earth. That means Saturn has an especially low density — less than that of water. Saturn would float in a large bathtub, although once the water is removed a ring remains around the tub. It rotates just slightly slower than does Jupiter, so a glance through a telescope reveals a flattened shape similar to that of Jupiter (Figure 18-29).

General Features Saturn's internal composition is thought to be quite similar to that of Jupiter's. But because of its greater distance from Sun (and corresponding lower temperature), Saturn's cloud bands are much more subdued and certainly less spectacular. Close-up photographs returned by the *Voyager* probes reveal the presence of oval-shaped circulation cells similar to those in the atmosphere of Jupiter (Figure 18-30). Like Jupiter, Saturn emits more energy than it receives from Sun, suggesting that it too is slowly collapsing. A model of the profile of its atmosphere (Figure 18-31), when compared to that of Jupiter (Figure 18-10), suggests that the two atmospheres in general are

Figure 18-29 This is a remarkable composite photograph of Saturn and some of its major satellites, as imaged by the Voyager satellites. At the conclusion of the Chapter, you should be able to identify a few of the satellites by name. (NASA)

Figure 18–30 (right) A close–up image of Saturn's atmosphere reveals patterns that appear very similar to those in Jupiter's atmosphere. (NASA)

quite similar.

In September 1990, amateur astronomers discovered a white "spot" near Saturn's equator. Over the next few days planetary astronomers saw the spot grow from an area the diameter of three Earths to a storm that nearly encircled the giant planet, with ammonia clouds billowing 150 miles high. So in November NASA turned the HST onto Saturn and obtained pictures of the storm (Figure 18-1). At the moment, there is no explanation for the sudden outburst. One planetary scientist from California Institute of Technology offered the idea that Saturn just sort of "burped."

Unlike Jupiter, Saturn has seasons, although certainly not like what we experience here on Earth. Its axis of rotation is tilted 27 degrees with respect to its orbital plane, allowing us to observe the rings alternately from above and below during its 29-year orbital period around Sun. That also means that twice during the same period, as seen from Earth, the rings appear edge-on. And since the rings are quite thin — 20 miles at the most — they nearly disappear from view.

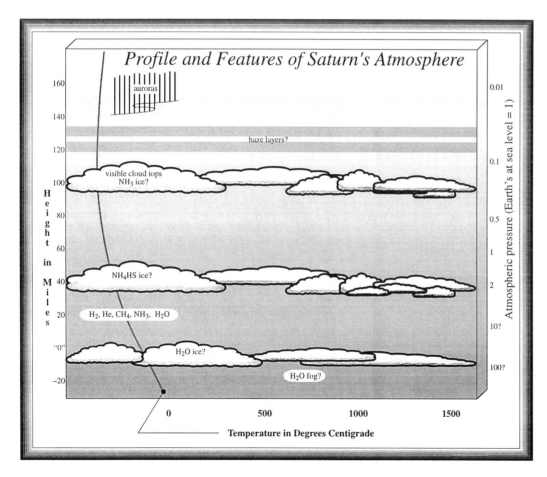

Figure 18–31 A model of the profile of Saturn's atmosphere. Compare it with the profile of Jupiter's atmosphere (Figure 18–10).

CHAPTER 18 / Exploring the Gas Giants Jupiter and Saturn   441

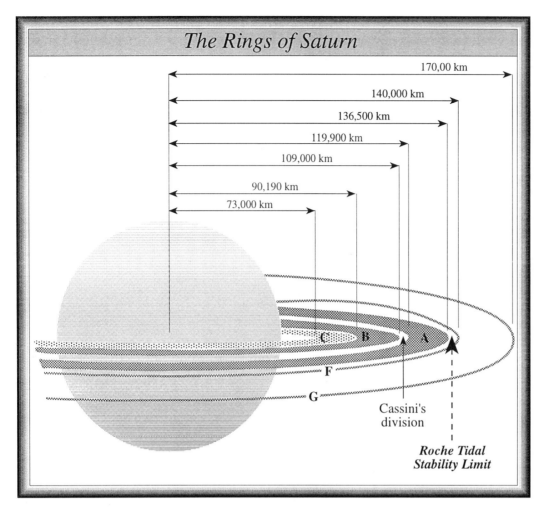

Figure 18–32 The ring system of Saturn is illustrated in this diagram, which shows the placement of the rings to scale in terms of distance from Saturn's outer atmosphere. Notice the location of Cassini's division and the Roche Limit.

Structure of the Rings When Galileo first trained his small telescope on Saturn, he observed what seemed to him to be two moons next to the planet — one on either side. Since he was not entirely free of the notion that objects in the sky had to be spherical in shape as well as behavior, he misinterpreted what he observed. After Kepler destroyed the addiction to perfect circles in his elliptical-orbit theory, another astronomer — Christiaan Huygens — correctly interpreted them as concentric rings around the planet. But he thought they were solid and flat.

Even through a small telescope, the rings reveal detail. Three concentric rings have been observed for centuries — from outside to inside they are designated the *A-*, *B-*, and *C-rings*. Through a small telescope, you should see a separation between the A-ring and B-ring, called **Cassini's division** (Figure 18-32). The two *Voyager* probes found an additional *D-ring* even closer to Saturn, and thin *E-*, *F-*, and *G-rings* outside the bright rings.

The reason that astronomers entertain the theory that the rings are the result of the collision of a moon-like object with the tidal forces of Saturn is because all three major rings (A-, B-, and C-rings) are located within **Roche's limit**.

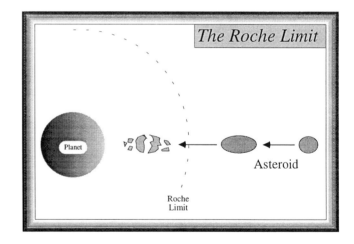

Figure 18–33 The observation that all of the major rings systems in the solar system are within the Roche Limit suggests that asteroids that invade the inner reaches of a massive planet's domain may be pulled apart by gravity. That is one explanation for the presence of the rings in the first place.

Figure 18–34 A close–up image of the ring structure of Saturn reveals an incredibly complex system, far more so than suggested by ground–based images. The rings actually consist of smaller "ringlets" whose formation is still being debated. (NASA)

Figure 18–35 The "spokes" of dark, dusty material suspended above the ring plane by electrostatic charges are shown in this close–up of the rings. They rotate along with the ring material, but do not conform to Kepler's laws of motion. (NASA)

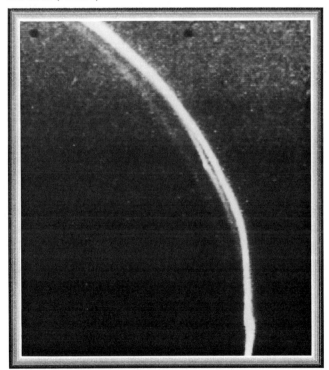

Figure 18–36 The F–ring or "kinky" ring of Saturn shows evidence that some of the smaller satellites gravitationally affect the rings. These nearby satellites are called "shepherding" satellites. (NASA)

This is the mathematical distance within which Saturn's gravity would pull apart any sizeable object (Figure 18-33). Roche's limit is nothing more than an extension of the mechanism responsible for Earth's tides. Actually, it applies to any two objects that are in close proximity to one another, and so we will encounter its application in many a situation in the solar system.

When the high-resolution photographs from the *Voyager* probes were carefully studied, it became obvious that the major rings themselves are made up of hundreds or thousands of narrower **ringlets**, making the entire ring system appear like a close-up view of a phonograph record. Even Cassini's division was found to have a hundred or so ringlets within what appears to be empty space (Figure 18-34).

The easiest way to explain this discovery, as well as the characteristics of the entire Jovian ring system, is to assume that the rings did not form along with the planet, but are constantly changing. That is to say — which is in keeping with the general theme of this book — that ring systems also go through a life cycle of creation, development, and decay.

Ring Formation The mechanism for this process is a familiar one — collisions. Assume that two large particles are orbiting a planet at approximately the same distance from the planet, and that they collide. After the collision, the faster-moving object is moving faster, and the slower one even slower. Energy lost by one is gained by the other: But if an object orbiting a planet gains energy, it spirals further away from the planet.

So we explain the width of a ring system according to the time it has taken to spread out since the time of the original collision. "What original collision?" you ask? Well, here is our present scenario (hypothesis) of what happened during the history of the ring systems in the solar system.

A small satellite orbits close to a planet, having formed as a by-product of the cloud that collapsed to become the planet itself. A passing meteoroid or comet creams the satellite, causing fragments to spread. First a narrow ring forms, then as the fragments themselves collide, the ring broadens out. As this broadening continues, the visibility of the rings decreases along with the sizes of the particles making up the rings. Eventually, only a thin, transparent ring system like that of Jupiter remains. Even the small particles disappear as moons within the ring itself (called **ringmoons**) sweep up or gravitationally "throw" the dust into the atmosphere of the planet.

A ring system that forms from a very massive satellite breaking up might well out-last the solar system. Perhaps the broad rings of Saturn are such ancient rings. But they may be the exception. The rings of the other planets arise as old comets and other space debris collide with the ringmoons. The rings of Jupiter, Uranus, and Neptune are a passing stage, to be replaced by newer satellites at some time in the future. According to this theory, therefore, planetary rings might have been more common longer ago, and even Earth may have had a ring system.

The F-ring of Saturn, located just outside the A-ring, is not strictly circular. It has knots, braids, and twists, and astronomers refer to it as the **kinky ring**. Two small satellites, called **shepherding satellites**, are responsible for the strange behavior within the F-ring, and even the narrowness of the ring itself. One of them orbits just inside the F-ring, while the other orbits just outside. Their gravitational influences prevent any particles from wandering away from the ring and, at the same time, influences the shape of the ring (Figure 18-36).

Composition of Rings So we conclude that the material making up the Saturnian rings varies in size from a few microns (one micron is one-millionth of an inch) to 50 feet or so. The most common-sized particles are distributed differently throughout the ring system. Analysis of infrared data returned by the *Voyager* probes tells us that the particles are made up entirely of water-ice or solid particles covered with water-ice.

Spokes One final surprise. *Voyager* photographed dark streaks crossing the broad rings in a radial direction, and their appearances earned them the name **spokes**. They are not permanent features, but gradually fade away as Keplerian motion causes the inner portions to separate from the outer portions (Figure 18-35). Then new spokes appear in their places. They seem to consist of fine dust-like particles suspended by **electrostatic forces** above the ring plane, much like the static electricity you experience on dry days when you drag your feet across a carpet.

Satellites of Saturn

Of Saturn's 22 known satellites, the innermost 15 form a regular system that probably formed along with the planet itself. The outer two, **Phoebe** and **Iapetus**, have orbits that are highly inclined with respect to Saturn's equatorial plane (refer back to Figure 16-7). And *Phoebe*'s motion around Saturn is retrograde. As I mentioned before, we do not really know how to explain the origin of such *irregular* satellites, except to hypothesize that they were captured long ago when there was a greater number of solar nebula fragments circling the newly-forming Sun. Those not captured are today found as comets or asteroids. Jupiter's family of satellites includes four large ones and twelve small ones, whereas Saturn's family consists of one large satellite, four intermediate satellites, and twelve small satellites. For Saturn as for Jupiter, most satellites have one side always facing their parent planet. This gives rise to some interesting differences between the leading and trailing sides of a couple of the satellites.

Titan The satellite **Titan**, once thought to be the largest satellite in the solar system, is the only satellite known to

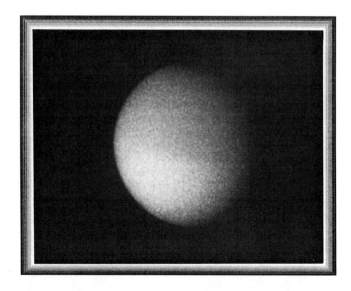

Figure 18-37 Saturn's satellite Titan *doesn't look very impressive in this* Voyager *photograph, but the fact that it has a thick atmosphere containing an assortment of hydrocarbons suggests that it may have liquid oceans on its surface. Very interesting chemical reactions could go on in an environment like that. (NASA)*

have a substantial atmosphere (Figure 18-37). Actually, if the thickness of the atmosphere is included in your measurement, it is the largest satellite. Otherwise, Ganymede claims the honor. Composed mostly of nitrogen (85%), argon (12%), and methane (3%), the atmosphere is dense enough to have a surface pressure 60% greater than that at sea level on Earth.

The existence of this atmospheric pressure is important because it is vital to the possibility that liquid exists on the surface — and liquid on the surface is vital to the possibility of life on the surface (we believe). It isn't, by the way, a matter of life needing something to drink — although that may well be true, too. In order for life to have evolved in the first place, there must have been a mechanism by which simple chemical arrangements came together to form more complex arrangements. After all, any life form, no matter how primitive, is a highly complex chemical arrangement.

Titan has an atmosphere rich with organic molecules, the carbon-based molecules out of which life is thought to have evolved. Its atmosphere is mainly nitrogen and methane (CH_4). In the upper atmosphere, these molecules are broken apart by electrons caught in Saturn's magnetosphere. The fragments (C, H, and N) recombine into more complex molecules and gradually settle down to the satellite's surface. Experiments have been performed in the laboratory to simulate the conditions known to exist on the surface, and these lead to the conclusion that — over geologic time — at least 330 feet of organic muck could have fallen onto the surface as a result of such processes.

The methane clouds and smog completely concealed the surface from *Voyager*'s cameras. We have not a single image of a single surface feature on Titan. The smog

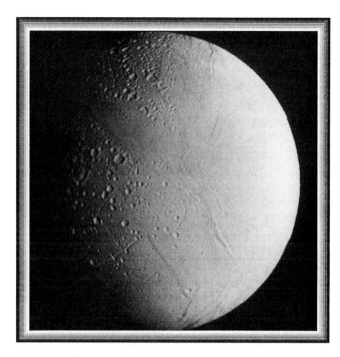

Figure 18–40 Saturn's satellite Enceledus. (NASA)

consists of various arrangements of hydrogen and carbon atoms as sunlight acts on the atmospheric methane. It is very similar to the process by which sunlight acts on gases produced by automobiles and industrial processes to form the smog that plagues many of our large cities. *Voyager*'s instruments detected many such **hydrocarbons** in Titan's atmosphere, including *ethane* (C_2H_6) *acetylene* (C_2H_2), *ethylene* (C_2H_4), and *propane* (C_3H_8).

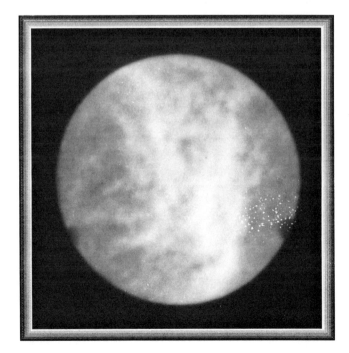

Figure 18–38 Saturn's satellite Rhea. (NASA)

Figure 18–39 Saturn's satellite Dione. (NASA)

Obviously, one would not want to light a match on Titan's surface. But the surface may not consist of solid material. Rather, it could be an ocean of liquid methane, at least in the warmer regions. So methane on Titan could play the same role as water does on Earth, forming methane clouds and methane polar caps. Obviously, in order for life to have evolved, it also required a supply of organic constituents readily available in the environment, and it is those that we are interested in learning about. This is not to say that Titan is a promising spot for life to be found. The temperature at the surface is minus 290 degrees Fahrenheit, far below the freezing point of water.

Besides organic molecules, life appears to require liquid water or some type of liquid within which molecules can come together to form more complex molecules. What all of this means is that Titan could be a tremendously interesting place for our understanding of the prebiological processes that may have operated on Earth before life got started. Understanding Titan's chemistry could help us do this. This is one of the motives behind the **Cassini** mission to Saturn, and the **Huygens** instrument probe that *Cassini* will drop into Titan's atmosphere in June 2005 (Figure 18-44).

Currently there are no *Viking*-type missions planned for either Titan or Europa. One of the lessons we learned from the *Viking* project to Mars is that it is extremely difficult to explore a world for microscopic life before the geology and surface chemistry of the world are reasonably well understood. If we are lucky, the *Galileo* mission will tell us whether there is any basis for getting excited as far as possible environments for life on Europa are concerned. After *Galileo* arrives at Jupiter in 1995, we will be in a much better position to make decisions about further explorations to suspected life-supporting environments like Titan.

Intermediate-Sized Satellites The Voyager satellites found that the other satellites of Saturn have densities between 1.0 and 1.5, suggesting that they are mostly water ice with some rock thrown in for good measure. The four intermediate-sized satellites of Saturn, whose diameters are over 1000 kilometers, probably consist of a rocky-ice mixture much like cometary nuclei. They are so cold that the ice is as rigid as rock and therefore retain evidence of crater impacts.

Iapetus is strange in that its trailing half is highly reflective, whereas the other half reflects about as much light as coal dust — ten times darker than the trailing half. The dark hemisphere is the leading face of the satellite as it orbits Saturn, so evidently the dark surface material is collected as it orbits Saturn or it is the result of a collision with another object. There is a large circular feature, *Cassini Regio*, that is probably an impact crater outlined by dark material. The density of Iapetus suggests that it is the only satellite containing an interior partly made of methane ice in addition to water ice. Water ice covers the bright trailing side. Thus the dark material covering the leading face may be a hydrocarbon formed by sunlight reacting on some of the methane that wells up from within the satellite. Spectra of the dark material suggest that it is similar to material making up some meteorites and asteroids.

Rhea (Figure 18-38) and **Dione** (Figure 18-39) both

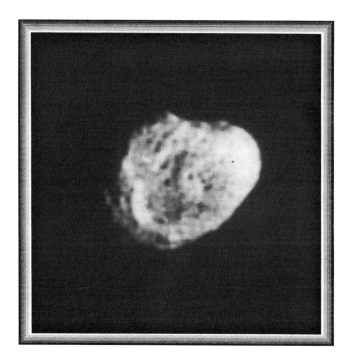

Figure 18–41 Saturn's satellite Mimas. *(NASA)*

Figure 18–42 Saturn's satellite Hyperion. *(NASA)*

have heavily cratered leading sides, and trailing sides that are marked by wispy-looking patterns that may be valleys in the icy surfaces. Or perhaps they are fresh deposits of water-ice that have outgassed from their interiors. Some of Dione's craters have rays of ejected debris surrounding them. Some of Rhea's craters have sharp rims, suggesting they are recent, whereas some craters have subdued rims as a result of long term erosion from small impacts.

Tethys, on the other hand, is heavily cratered throughout. In fact, it has a huge circular feature called Odysseus that is one-half the diameter of the satellite. The crater is unlike those on the rocky planets and Moon, however. It has been flattened by the flow of softer ice, probably early in its history when the interior was warmer. Tethys also has huge valleys in the icy surface, larger than any such feature on any satellite in the solar system. One, *Ithaca Chasma*, extends almost two thirds of the way around the satellite. It is several miles deep, and is either a fault or the result of the expansion of Tethys as its early warm interior froze solid.

Tiny Satellites Only three of the small satellites have features worthy of mention:

Enceladus (Figure 18-40), the most highly reflective satellite in the family of 22, has rather large, smooth areas free of craters. Evidently, recent collisions have melted the water-ice on the surface, which then flowed to submerge the older features, and finally froze again. Or, as an alternative explanation, Enceladus is caught in a gravitational tug-of-war like Jupiter's Io and Europa. There are also long sets of linear grooves on the surface similar to those on Ganymede, and are probably the result of similar geologic faults.

Mimas has a huge impact crater (relative to its small diameter of 250 miles) called *Herschel* that makes it look like an eyeball in space (Figure 18-41). The crater has a raised rim and a peak at the center, very typical of Moon craters and craters on the terrestrial planets in general. There is a canyon that goes halfway around Mimas, and may have been caused at the time of the collision. It is not quite clear how Mimas managed to avoid being broken apart when the collision responsible for the crater and canyon occurred.

Hyperion, as revealed by the cameras aboard the *Voyager* satellites, looks like a hamburger or a hockey puck. In spite of its small size, it reveals lots of craters on its surface (Figure 18-42).

Newly-Discovered Satellites Astronomers have announced the discovery of at least two, and possibly as many as four, new moons orbiting the giant planet Saturn (Figure 18-43). This discovery was based upon HST images taken in May, 1995, when Saturn's rings were tilted edge-on to Earth. Two of the satellites seen by Hubble are in orbits similar to those of Atlas and Prometheus, a pair of moons discovered in 1980 by the *Voyager 1* spacecraft. If all four satellites are new, then the total number of known moons

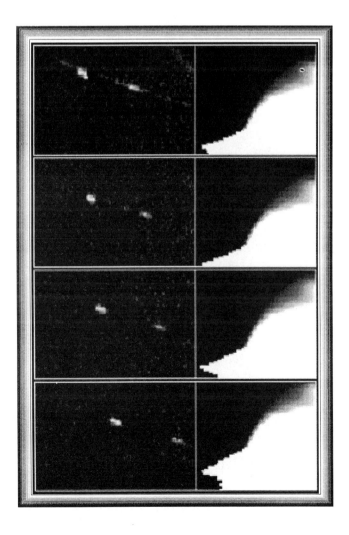

Figure 18-43 (left) This four-picture sequence (spanning 30 minutes) shows one of four new satellites discovered by the HST, in images taken of Saturn in 1995, when Saturn's rings were tilted edge-on to Earth. Identified as S/1995 S3, the satellite appears as an elongated white spot near the center of each image. It lies just outside Saturn's outermost "F" ring and is no bigger than about 15 miles across. The brighter object to the left is the satellite Epimetheus, *which was discovered during the ring-plane crossing of 1966. Both satellites change position from frame to frame because they are orbiting the planet. Saturn appears as a bright white disk at far right, and the edge-on rings extend diagonally to the upper left. The long observing times account for Saturn's bright, overexposed appearance. (NASA/STScI)*

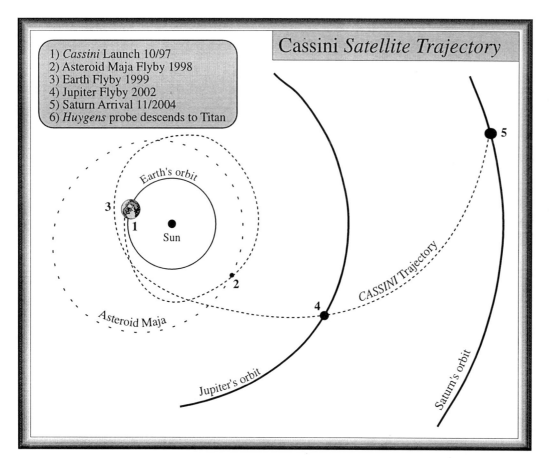

Figure 18–44 *This diagram illustrates the Cassini mission scheduled for launch in 1997. Considering the advances in electronic and computer technologies since the Voyagers were built, the photographs returned of Saturn and its satellites, especially Titan, will be stunning.*

orbiting Saturn will grow from 18 to 22. Two of the new moons (called 1995 S1 and S2) lie inside Saturn's thin, eccentric "F" ring; a third moon (S3) lies just outside the F ring; and the fourth moon (S4) is 3,700 miles (6,000 kilometers) beyond the F ring.

The moons are no bigger than about 45 miles across. Hubble's sharp view is ideal for detecting faint new satellites that have previously gone unseen. The astronomers identified the new moons by first processing the Hubble pictures to remove residual light from the ring edge, and noting the locations of Saturn's known satellites. After this was done, the researchers saw four objects moving from frame to frame that did not correspond with any of the known satellites.

Saturn ring plane crossings happen only once every 15 years, and historically have given astronomers an opportunity to discover new satellites that are normally lost in the glare of the planet's bright ring system. Astronomers discovered 13 of Saturn's moons during ring-plane crossings from 1655 to 1980. Other satellites were identified during the *Voyager* spacecraft flybys of Saturn in the early 1980s. The next such event that will be visible from Earth won't occur until the year 2038. (Saturn will be edge-on in 2009 and 2025, but will be too close to the Sun to be observed from Earth.)

Future Missions to Saturn If all goes well, the *Cassini Saturn orbiter* satellite, presently planned for launch in October 1997, will arrive at Saturn in November 2004 to begin a 4-year orbital tour of the ringed planet and its 22 satellites. The orbiter is being built by NASA and the European Space Agency (ESA) is building the probe. The instruments aboard the satellite have continued to increase in number as scientists develop more and more sophisticated measurement devices. So *Cassini* must take the long path through the inner solar system to obtain "gravity-assists" from Venus (two visits), Earth (one visit), and Jupiter (one visit) to build up the velocity it needs to reach Saturn (Figure 18-44).

After its first Earth gravity-assist in 1998, it will pass by and photograph asteroid *Maja* in 1999. The mission at Saturn is designed so that *Cassini* will orbit the planet 36 times while studying the composition and dynamics of the planet's atmosphere and rings. It will also carry the probe *Huygens* that will plunge to the surface of cloud-covered Titan — the only satellite in the solar system that possesses a dense atmosphere. Current theory is that a thick atmosphere is essential for the chemical processes necessary for life to even get started. On the surface of Titan, there may be some interesting chemistry going on. Our hope would be to find some of the same chemicals from which we think the first life-forms on Earth evolved.

LEARNING OBJECTIVES: *Now that you have studied this Chapter, you should be able to:*

1. Compare and contrast the Jovian and terrestrial planets, and explain their major differences.
2. Compare and contrast the interior of the Earth with those of Jupiter and Saturn.
3. Describe the behavior of the cloud patterns of Jupiter and Saturn in terms of their rapid rotations.
4. Explain why planetary rings must be temporary features.
5. Compare the rings of Saturn and Jupiter, and offer an explanation for their differences.
6. Compare and contrast the surface features of the Galilean moons of Jupiter, and explain how their distances from Jupiter explain their differences.
7. Describe the features of the largest moon of Saturn, Titan, and explain the unusual amount of interest astronomers have in it.
8. Define and use in a complete sentence each of the following **NEW TERMS**:

Amalthea 439
Archimede's principle 427
Belts 427
Callisto 432
Cassini's division 442
Cassini Saturn Orbiter 446
Dione 346
Electrostatic forces 449
Enceladus 347
Europa 432
Galileo Mission 431
Ganymede 432
Great Red Spot 428
Huygens probe 446
Hydrocarbons 445
Hyperion 447
Iapetus 444

Io 429
Kinky ring 444
Mimas 447
Phoebe 444
Rhea 446
Ringlets 442
Ringmoons 444
Roche's limit 442
Shepherding satellites 444
Spokes 444
Titan 444
Tethys 346
Tidal dissipation 437
Valhalla 435
Voyager 1 and 2 satellites 425
Zones 427

Chapter 19

Exploring Uranus, Neptune, and Pluto

Figure 19–1 This has to be one of the most dramatic photographs of any object imaged by either of the Voyager satellites. This is of the satellite Triton, revealing all of the dramatic surface features for which it is famous. Notice especially the dark streaks across the surface which are geysers of dust, gas and nitrogen ice. (NASA)

CENTRAL THEME: How are Uranus, Neptune, and Pluto different from the other planets, and how are those differences explained by the theory that all planets formed around Sun together?

Human societies have used the seven days in a week calendar from earliest recorded time. The origin of that practice, as I discussed in Chapter 4, was the fact that seven objects appear to move against the backdrop of stars in the sky. Thus early peoples thought they must be of particular significance, and even honored them by assigning them the names of their gods — from which the seven days of the week are derived (Figure 3-14).

It must have come as a shock to some, therefore, when in 1781 the English astronomer William Herschel reported the discovery of an eighth object, planet number seven from Sun. Actually, Uranus had been recorded as a star on several sky maps at least during the hundred years prior to Herschel's announcement, but no one had singled it out as anything other than a star. That is both because it is quite dim and because it moves so slowly around Sun (its revolution period is 84 years).

General Features of Uranus

After *Voyager 2* passed by and photographed Jupiter and Saturn, it went on to give us our first good close-up photographs and accurate data on Uranus. Prior to that 1986 fly-by, we had only tantalizing clues as to its appearance and character. That it was a cold and forbidding planet we had no doubt, since the temperature at its distance from Sun is about -325°F.

Actually, from a dark location away from the lights of a city, a dedicated and persistent observer can see Uranus with the naked eye. But since it is close to the threshold of visibility, you need to refer to accurate star charts to verify the find — even through smallish telescopes, Uranus appears as a faint dot like the stars around it. Having a high albedo, it was long known to have a thick cloud cover.

The planets Uranus and Neptune are similar in several respects, just as the case of Jupiter and Saturn. Both are about the same size and mass. They are four times greater in diameter than Earth, and 15 times more massive. They are similar to Jupiter and Saturn in that they have no solid surfaces as such, although they have higher proportions of heavier elements mixed in with the hydrogen and helium of which they are mostly composed.

The latter fact must be true because measurements of their densities gives us values of 1.2 and 1.6 grams/cm^3, for Uranus and Neptune respectively. Astronomers believe that methane (CH_4) is the additional major component in their atmospheres. Thus the percentage of carbon is about 20 times higher in Uranus and Neptune than in Sun. This fact is supported by the hypothesis that these planets formed from a core of rocky and icy planetesimals.

Rotation of Uranus The most notable feature of Uranus is its strange rotational tilt. All of the planets we have encountered up to this point rotate such that their axes of rotation are roughly at right angles to their orbital planes. But the rotational axis of Uranus lies only 8 degrees from its orbital plane (Figure 19-2). That fact has important implications not only as regards other features of Uranus (such as seasonal patterns), but as regards the conditions within which Uranus formed in the early solar system. We are about to see that the unusual axial tilt combined with the strange features found on one of the satellites of Uranus may provide the clues to prove that a massive collision occurred as Uranus was forming.

Atmosphere Telescopically, Uranus appears greenish blue. This is due to the presence of methane in its atmosphere. Methane absorbs the red and orange wavelengths that arrive from Sun, leaving only the green and blue wavelengths to be re-emitted to space. The fact that the planet is tilted 8 degrees with respect to its orbital plane would lead you to suspect seasonal changes in the extreme.

But being so far away from Sun, and because of the effective mixing of the atmospheric gases due to its 17-hour rotation rate, Uranus has a surprisingly high degree of temperature uniformity in the atmosphere at different locations and at different times. It is estimated that the temperature varies only about 5°F from one season to another. Keep in mind that Uranus orbits Sun once every 84 years, and therefore each pole spends about 42 years in sunlight.

Computer-enhanced photographs reveal a slight banded appearance in the atmosphere of Uranus and cloud circulation patterns that resemble those of both Jupiter and Saturn. And what do these clouds conceal underneath? Well, we don't exactly know, of course. But mathematical models allow for a gradually increasing density of the gases with decreasing depth.

In fact, one prediction is that there is an enormous liquid ocean beneath the clouds, although not with a well-defined surface like the Earth's oceans. One reason for this prediction is the presence of a substantial magnetic field around Uranus — about as strong as the Earth's. This observation requires that Uranus have a liquid-conducting medium within the planet, and water is a strong candidate (Refer back to Figure 18-5).

Magnetic Axis Alignment But there is something quite odd about the magnetic field of Uranus. In spite of the unusual tilt of its axis of rotation, we would still expect its magnetic axis to line up closely with the rotational axis. That expectation springs from what we understand about the *dynamo effect* responsible for magnetic fields of planets.

For sake of comparison, the Earth's magnetic axis is tilted about 12 degrees with respect to its rotational axis, Jupiter's about 10 degrees, and Saturn's is exactly lined up. But the magnetic axis of Uranus is tilted about 60 degrees (Figure 19-3)! As if that didn't make Uranus odd enough, the magnetic axis is offset from the center of the planet by about 5,000 miles, whereas the Earth's is offset by a mere 300 miles.

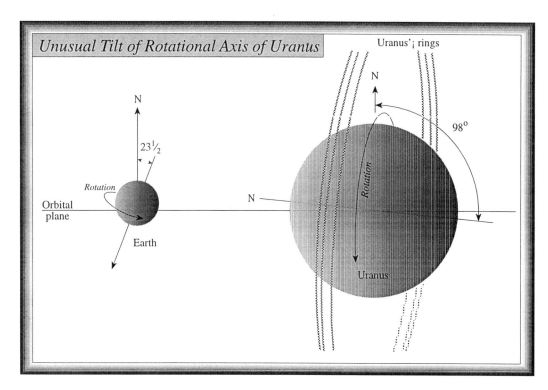

Figure 19-2 The most notable feature of Uranus is the unusually large tilt of its axis of rotation with respect to its orbital plane. This makes for unusual seasons on Uranus, although it is always cold compared to Earth anyway. This strange tilt may be due to a major collision between Uranus and a large object during the time it was forming.

How might we explain these peculiarities? (1) Perhaps astronomers need to reevaluate the dynamo effect theory, and enlarge it to include these strange features of Uranus. One suggestion is that the dynamo effect occurs in a thin electrically conducting shell outside the core of the planet, rather than deep in the core as in Jupiter and Earth. Or (2) it may simply be that we are observing Uranus in the midst of a reversal of the polarity of its magnetic field, as happens in the Earth every 200,000 years.

Finally, (3) there may be a connection between the strange orientation of the magnetic axis and the tilt of the rotational axis. There is reason to believe that a catastrophic collision occurred between Uranus and another object while Uranus was forming, and that the events surrounding the collision could be responsible for these present-day irregularities (Figure 19-4).

The magnetic field of Uranus is responsible for some curious features on the surfaces of a few of its satellites. Its magnetic field, like the Earth's and Jupiter's, traps the charged particles of the solar wind. The Earth's Moon orbits outside the radiation belts that contain those particles. But in the case of Uranus, the five major satellites (*Miranda, Ariel, Umbriel, Titania,* and *Oberon*) orbit within the belts. As they orbit their parent planet, the satellites collide with and trap some of the charged particles. Their surfaces, covered primarily with methane ice, darken in the process, making the satellites among the darkest in the solar system.

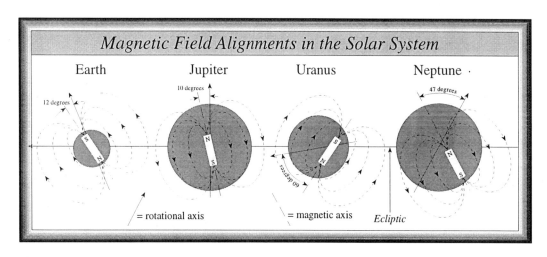

Figure 19-3 A comparison of the alignments of the magnetic fields of the planets reveals some unusual features. Especially notable is the great difference between rotational and magnetic axes (Uranus, Neptune), the large off-settedness of the center of the magnetic field (Uranus, Neptune), and the reversed polarity (Earth, Uranus).

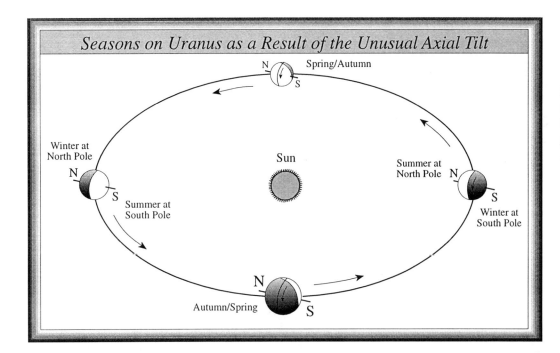

Figure 19–4 *The tilt of the rotational axis of Uranus as it orbits Sun gives the planet unusual seasonal variations.*

Ring Discovery In 1977, a group of astronomers working aboard the *Kuiper Airborne Observatory* (Chapter 5) discovered a ring system around Uranus. They were studying the behavior of the light of a star that Uranus was about to occult. But long before Uranus was close enough to occult the star, the starlight dimmed off and on nine times, and repeated that behavior after Uranus completed its passage in front of the star. The most reasonable interpretation of this observation was to theorize that Uranus possessed nine rather thin rings.

Voyager 2 and Uranus Finally Meet

Voyager 2 encountered Uranus on January 24, 1986. Since sunlight was only 1/370th as bright as it is at Earth, exposures by Voyager's cameras had to be long. Technicians at the **J**et **P**ropulsion **L**aboratory (**JPL**, for short) in Pasadena, California, developed a technique of turning the satellite very slightly during each exposure to compensate for *Voyager's* rapid movement past the planet. The radio signals used to return the photographs and other data to *JPL* required almost 3 hours to reach Earth.

Atmosphere *Voyager 2* got to within 50,000 miles (one-fifth the Earth-Moon distance) of Uranus. But even from that distance the planet appeared bland and uninteresting. This came as no great surprise, given the great distance from Sun (and therefore low temperatures) inhibits chemical reactions. That is why we put food items we want to remain unchanged in the freezer. Infrared studies provide a temperature measurement of 58 K for the outer layers of the atmosphere. There is no evidence of an internal heat source as is the case for the other Jovian planets. The infrared images reveal a dark polar cap, although it is best interpreted as high-level photochemical haze (Figure 19-5).

A few low-contrast bands were imaged in the atmosphere, especially at mid-latitudes. Tracking the patterns within the bands showed that the winds circulate east-west rather than north-south. This is certainly the case for Jupiter and Saturn, and to a lesser extent on other planets with atmospheres. The fact that the axis of Uranus is tilted so that the equator is almost at right angles to the orbital plane might suggest that atmospheric circulation would be from pole to pole. But the opposite discovery for Uranus illustrates how important rotation is for weather on a planet, as opposed to whether Sun is overhead, heating the pole more than the equator.

A strange ultraviolet emission was discovered in the atmosphere by the detectors aboard *Voyager 2*. Called the **electroglow**, it occurs on the sunlit side of the planet some 1000 miles above cloud-top level. The most reasonable explanation for the *electroglow* is that sunlight splits up the hydrogen molecules into protons and electrons, which then collide with other hydrogen molecules to cause the glow. Astronomers have since discovered a very faint electroglow on Jupiter, Saturn, and Titan.

Rings After astronomers aboard NASA's flying observatory discovered rings around Uranus, ground-based astronomers went to work to image them with their telescopes. Success was not long in coming (Figure 19-6). They were detected in infrared photographs in much the same way as the dust rings around *beta Pictoris* were first detected. But it is readily apparent that the finest images come from a satellite that went close to Uranus.

Images of the rings taken as *Voyager 2* approached the

CHAPTER 19 / Exploring Uranus, Neptune, and Pluto

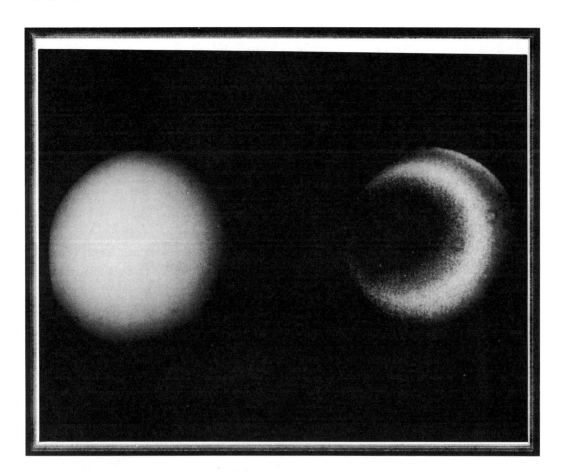

Figure 19–5 These are two images of Uranus provided by Voyager 2's cameras. The image on the left is a straightforward photograph. The one on the right has been computer–enhanced in order to show temperature variations within the upper atmosphere of the planet. The "bull's–eye" appearance results from the fact that the image is of the side facing Sun. Uranus is rather bland compared to Jupiter and Saturn. (NASA)

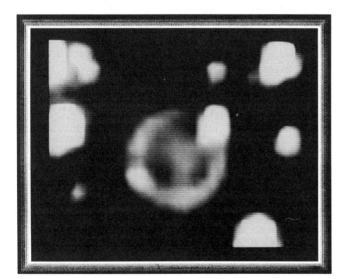

Figure 19–6 This infrared image is the first photograph of the rings of Uranus with a ground–based telescope. Since the planet is so much brighter than the rings, the rings are impossible to detect at visible wavelengths. By using infrared film, astronomers detect the energy that is absorbed and then re–emitted by the dust particles that make up the rings. (National Optical Astronomy Observatories)

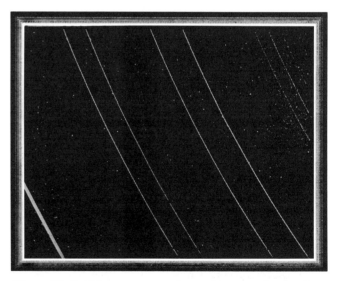

Figure 19–7 This computer–enhanced image taken by the Voyager 2 spacecraft reveals the thin rings of Uranus. (NASA)

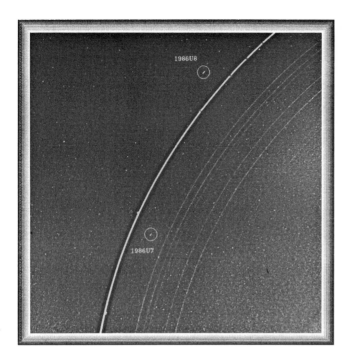

Figure 19-8 This computer-enhanced image taken by the Voyager 2 *spacecraft reveals not only some of the thin rings of Uranus, but also two new satellites.* (NASA)

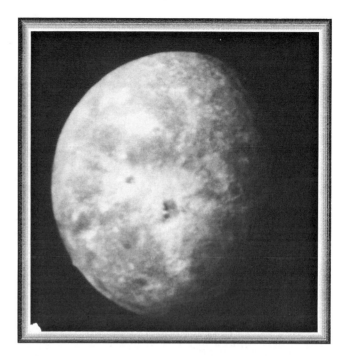

Figure 19-9 A Voyager 2 *photograph of the satellite* Oberon. (NASA)

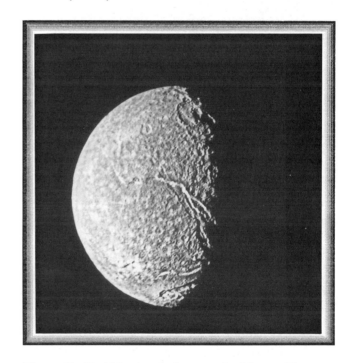

Figure 19-10 A Voyager 2 *photograph of the satellite* Titania. (NASA)

Uranian system revealed that they consisted, like Jupiter's rings, of very dark material—much darker than the material in Saturn's rings (only 2 percent of the light falling on the ring material is reflected). The images also revealed two additional faint rings, one a narrow ring like the previously-known 9 rings, and a broad ring (Figure 19-7). That brings the total to 11 rings. Viewed from behind the planet, the rings appeared much broader, no doubt because of the presence of large amounts of small-sized particles that scatter sunlight forward (Figure 19-7).

But there appears to be much less dust than in the cases of Jupiter and Saturn. Less than 0.001 percent of the particles in Uranus's rings are micron-sized, in comparison with a few percent for Saturn's rings and 50 percent for Jupiter's rings. Therefore, there must be some mechanism for removing the dust from the rings of Uranus. Perhaps an extended atmosphere of Uranus is responsible.

The sizes of the ring material could be estimated by observing the behavior of the radio signals from *Voyager*'s transmitter as it passed behind the rings after the satellite imaged the planet itself. The strength of both the 3.6-cm and 13-cm wavelength signals lessened as the rings occulted the satellite, suggesting that the ring particles must be at least many centimeters in size. Recall that objects absorb wavelengths of radiation equal to or smaller than their diameters.

Because the rings are rather narrow, with large empty spaces between them, there is suspicion that shepherding satellites orbit Uranus within the rings and confine the ring material. This is similar to the case of Saturn's F-ring. As with the ring system of Jupiter, we cannot offer a good explanation how the Uranian ring system maintains itself so close to the strong gravitational field of the planet except by assuming that the ring system is young, and is therefore gradually decaying.

CHAPTER 19 / Exploring Uranus, Neptune, and Pluto

★★★★★★★★★★★★★★★★★

The fact that all of the rings orbit Uranus above its equator suggest that they could not have originated along with the planet if indeed it was "knocked" on its side by a massive collision. The rings could not, under such circumstances, have retained their original orbits. Besides, there is no mechanism for holding the ring material in orbit for over 100 million years or so. Therefore, the rings we observe today must have formed relatively recently, perhaps by a collision between a ringmoon and an asteroid or comet.

Satellites of Uranus

When *Voyager 2* visited Uranus in early 1986, only 5 satellites were known to exist around the planet. All 5 were known to move in nearly circular orbits that lie close to the equatorial plane of Uranus — in the same plane as the thin rings but at much greater distances from the planet. All 10 of the new satellites found by *Voyager 2* lie within the orbits of the 5 previously known satellites. Only one, nicknamed "Puck," was large enough (100 miles) to offer much information. The 5 largest previously-known satellites stole the show (Figure 18-16).

General Features All of the 15 known satellites rotate synchronously as they orbit Uranus. That is, one side of the satellite always faces the parent planet — like Moon. Since Sun was shining almost directly down on their south poles as *Voyager* flew through the system, we obtained images only of their southern hemispheres. We have no idea what the northern hemispheres are like. In general, their albedos are lower than those of the Saturnian satellites, except for *Phoebe* and *Iapetus*. Carbon in the form of soot and graphite could explain the darkness of these satellites. Carbon is plentiful in the solar system.

Gravitational disturbances of *Voyager*'s orbit as it passed through the satellite system of Uranus allowed astronomers to determine the masses and therefore densities of the satellites. They are about 1.5 times the density of water, higher than those of Saturn's satellites. The densities do not vary in any meaningful way with distance form Uranus. They are probably composed of varying mixtures of rock and ices of water, ammonia, and methane.

All of the satellites have craters on them. The largest are probably remnants of the collisions of planetesimals or other debris during the formation of the solar system over 4 billion years ago. The smaller craters are no doubt the result of secondary debris of collisions between satellites and impacts with smaller satellites directly. Any craters occurring at the present time probably result from impact with short-period comets, captured into orbits that carry them close to Uranus.

The five previously-known satellites can be remembered by their initials: **MAUTO** (**M**iranda, **A**riel, **U**mbriel, **T**itania, **O**beron).

Oberon and *Titania* are similar in size — about 1000 miles across. Both have albedos of 20 to 30 percent, and appear grey.

- **Oberon** (Figure 19-9) is heavily cratered with large craters (greater than 60 miles across). A few are bright-rayed craters with darkish material partly covering their bottoms, probably material that oozed up from *Oberon*'s interior. There is a fault running across most of the southern hemisphere, suggesting early tectonic activity during which movements of the crust were caused by internal forces and heating. Beyond the limb of the satellite, a mountain is seen sticking up 7 miles above the limb, but Voyager was not able to look around the

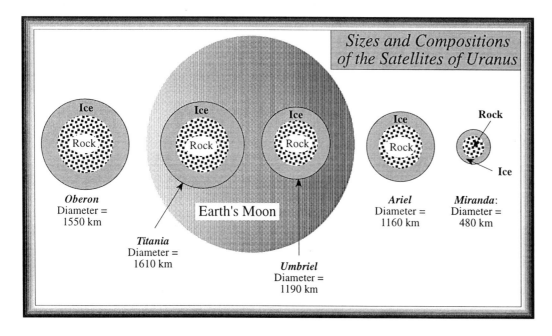

Figure 19-11 The major satellites of Uranus are compared by size and internal composition in this diagram. Moon is placed for comparison sake.

456 SECTION 4 / Evolution of the Planets

limb to see its base. So we have no idea how much taller it is than the calculated 12 miles.

- **Titania** (Figure 19-10) displays few large craters, but many small ones. One large impact feature is 200 miles across. Therefore, early tectonic forces must have erased the large craters that surely existed there. This idea is supported by the finding of satellite-wide fault systems or scarps 1 to 3 miles high on *Titania*. In some cases, the scarps split craters right in half, suggesting that they are rather young features. There are a few bright deposits on some of the scarps, again suggesting their youthfulness.

Umbriel and *Ariel* are similar in size (700 miles in diameter) and density, but they have remarkably different appearances. Their past histories must, therefore, have been quite different.

- **Umbriel** is the darkest of Uranus's satellites. It appears to be covered with large, old craters, but lacks high contrast between bright and dark areas on the surface. This uniform dark appearance suggests that *Umbriel* avoided large-scale collisions that would have been violent enough to break through the dark crust to reveal the bright water-ice beneath. It has the oldest surface of any of the satellites — its appearance is similar to those of the ancient cratered highlands of the inner planets.

  Appearing close to *Umbriel*'s limb is a large bright ring feature. It is a high-albedo deposit that covers the floor of a 25-mile diameter crater, called *Wunda*. It is no doubt a young-ish feature. It is difficult to explain why Umbriel is so uniformly dark. The fact that the other satellites do not have so much dark material on their surfaces suggests that it is peculiar to *Umbriel*. Perhaps the dark material was ejected from the interior by volcanism, or there was a massive collision that spread the dark material around the entire satellite. Both of these proposals lack evidence.

- **Ariel** (Figure 19-11) wins the prize for fault systems in the Uranian system. Some are over six miles deep and hundreds of miles long, all on a satellite that is only 700 miles in diameter. Thus *Ariel* appears to have experienced large crustal movements early in its history. Since it has the brightest surface of the Uranian satellites, with smooth areas overlapping craters, it is easy to imagine that *Ariel* renewed its surface in the distant past as melted water and rock oozed to the surface.

  In one location, a half-buried crater can be seen. Sinuous valleys similar to those on Moon are visible, some with parallel ridges alongside. These features are similar to lava channels and ridges found on Earth and other inner planets. But it is too cold to water to melt on this tiny satellite. A mixture of ammonia and water ice has been proposed as the mechanism that caused these channels. It may even be that tidal effects participated in their formation (Figure 19-12).

Miranda (Figure 19-13) is in a class of its own. Half the size

Figure 19-12 A Voyager 2 *photograph of the satellite* Ariel. (NASA)

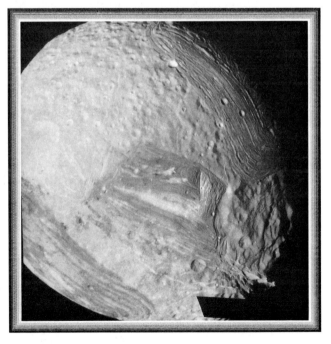

Figure 19-13 A Voyager 2 *photograph of the overall surface features of the satellite* Miranda. (NASA)

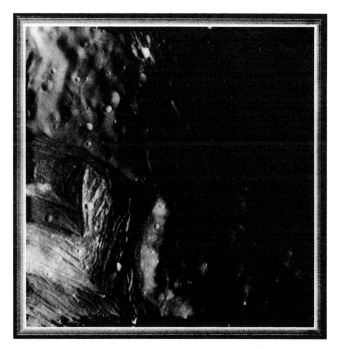

Figure 19–14 Close–up photograph of the feature called the "chevron" on the satellite Miranda. (NASA)

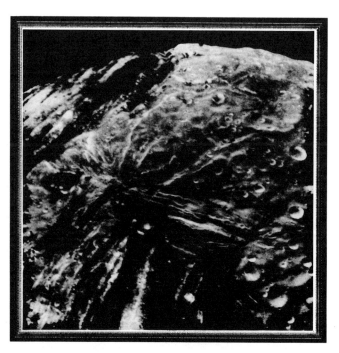

Figure 19–15 Close–up photograph of 3–mile–high ice cliffs (along the edge) on the satellite Miranda. (NASA)

of Ariel and Umbriel, a third the size of Titania and Oberon, no one expected anything dramatic to be found on this satellite prior to *Voyager*'s arrival.

- **Miranda** Of all the new discoveries that *Voyager 2* made at Uranus, it was surely the spectacular photographs of *Miranda* that stunned the scientists who received the photographs. On this single, small (300 miles in diameter) satellite are found just about all of the strangest features of the solar system. It is a bizarre hybrid consisting of the valleys and layered deposits of Mars, the grooved terrain of Ganymede, and the lobate scarps of Mercury. Some of the scarps reach 3 miles in height, much higher than the walls of our Grand Canyon.

  A bright region, called the "chevron" because of its shape, strongly argues for some internal process operating on the satellite (Figure 19-14). The sharp angles evident in the pattern of surface material could not be produced by an external event. A part of Miranda's surface is covered with undulating cratered plains and the other part is covered with terrain that reveals parallel dark bands with a few bright bands. These remind us of the grooved regions on Jupiter's satellite Ganymede. All of these features argue for some type of internal heat source that would be responsible for the upwelling of material onto the surface to form the patterns we now observe.

  But it could also be that Miranda formed from different globs of material. Calculations of the probability of collisions in the Uranian system suggest that Miranda may well have suffered impacts large enough to break it apart several times, only to reform as gravity pulled the pieces back together again. This theory explains why there are such distinct boundaries setting apart such features as the "chevron" from the surrounding regions. Some of the energy of the collision(s) would have gone into heating the interior, resulting in the strange surface features as material oozed up from the interior (Figure 19-15).

10 New Satellites *Voyager 2* discovered 10 new satellites orbiting Uranus. They are all in nearly circular orbits and all but one orbit between the outermost ring, the *epsilon ring*, and Miranda. Their albedos are all quite low – 5 to 7 percent, which is lower than not only Saturn's smaller satellites but also the major satellites of Uranus. Thus the conditions under which they formed must have been quite different than those for Saturn's smaller satellites.

Neptune

Uranus was accidentally discovered by Sir William Herschel in 1781 as he was observing stars in hopes of detecting parallax. After calculating an orbit for Uranus based upon the laws of planetary motion devised by Kepler and Newton, mathematicians in 1840 discovered that Uranus was not moving exactly in accordance with that calculated orbit. They predicted that the presence of another planet, somewhat more distant from Sun, would account for the slight discrepancy, and even where that other planet would have to be in order to cause the discrepancy.

Sure enough, Neptune was finally discovered in 1846 within 1 degree (two full-moon widths) of the predicted position (Figure 19-16). You and I cannot fully appreciate the effect of such a discovery at the time, since we take laws of motion for granted. But it was a sensation, finally putting to rest the lingering belief that the heavens were not for people on Earth to describe and explain.

Although Neptune was discovered over 200 years after Galileo revolutionized our thinking about the nature of the universe beyond the Earth, people at the time commonly believed it was presumptuous for humans to assume they could apply the rules governing matter on Earth to explain and predict behavior of objects in space. Well, the discovery of Neptune destroyed that reluctance once and for all, although there are people today who are unwilling to assume that the laws of biology are universal just as the laws of physics are universal.

General Features Up until *Voyager 2*'s encounter with Neptune during the summer of 1989, astronomers had only the vaguest knowledge of the planet. Through Earth-based telescopes, Neptune appears very nearly like a star — a tiny, featureless disk. Although Neptune seemed like a twin to Uranus in size and color, the two planets appeared to differ significantly in other characteristics. Whereas Uranus is tilted 98 degrees with respect to its orbital plane, Neptune is tilted only 29 degrees.

Neptune was known to receive less than one-half sunlight Uranus receives, yet it is about the same temperature. Like both Jupiter and Saturn, this suggests that Neptune has a heat source in its interior, perhaps from gradual contraction. Whereas the atmosphere of Uranus is rather uniform and bland in appearance, Neptune's atmosphere was observed to be quite variable even from the Earth. Features within the atmosphere were tracked for hours and even days.

Astronomers had also attempted to detect a ring system around Neptune by monitoring the apparent brightnesses of stars as the planet moved in front of them. Most of those occultation studies suggested a thin ring system, but one that was not a complete circle around the planet. It appeared to consist of at least three partial segments, or arcs, suggesting youthfulness. *Voyager 2* did find three arcs, but they are clustered together in the same orbit as part of a thin ring that encircles the planet.

Magnetic Field Eight days before arrival at Neptune, *Voyager 2* detected short bursts of radio waves coming from the planet's vicinity. This was first evidence that a magnetosphere populated by charged particles lay ahead of the spacecraft. Since the bursts occurred at intervals of 16 hours, it is concluded that Neptune's interior must be rotating at the same 16-hour rate. Recall that a magnetic field arises from electric currents generated as the result of a fluid material rotating around a solid, metallic core. The magnetic field presents an obstacle to the ionized solar wind particles flowing outward through interplanetary space, and creates a shock front, or bow shock, in the wind upstream from the planet.

As *Voyager 2* passed through this region, it detected the presence of the magnetic field in a manner much more gradual than was expected. When the data was examined

Figure 19–16 Prior to the arrival of Voyager 2 *at the Neptune system, this was about the best photograph astronomers had of Neptune. The dot to which the arrow is pointing is the satellite* Triton. *(Lick Observatory)*

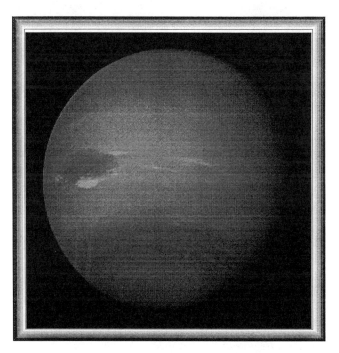

Figure 19–17 The planet Neptune, as imaged by Voyager 2. Notice the Great Dark Spot. (NASA)

carefully, it was discovered that the planet is strange in a manner similar to that of Uranus — the magnetic axis is tipped 47 degrees with respect to the planet's spin axis (that of Uranus is 59 degrees). Consequently the entire magnetic field gyrates wildly as Neptune spins. *Voyager 2* happened to enter the bow shock environment when the magnetic pole was pointed only 20 degrees from Sun.

So it had entered the magnetosphere through a funnel-shape entrance over the magnetic pole where the plasma can penetrate deeply before being repelled. As if this was not strange enough, it was discovered that the source of the magnetic field is not in Neptune's core itself, as one would expect, but some 55 percent of the way out toward its surface! That of Uranus is similarly offset — by about 30 percent (Figure 19-3).

This discovery is causing big problems for magnetospheric physicists (yes, there are scientists who specialize in that field of study!). In the case of Uranus, you recall, astronomers attempt to explain the off-setted-ness of the rotational axis of the planet by invoking a possible collision early in the formation history of the planet. But there is no evidence to support a similar collision theory in the case of Neptune, aside from the strange magnetic field tilt.

It would be strange indeed had two almost identical collisions occurred to two adjacent planets. Although collisions were certainly the most likely processes by which the planets formed in the early solar system, two collisions that would result in the same effect on two adjacent planets would be highly improbable. Evidently, whatever is driving Neptune's magnetic field arises in a circulating fluid outside a presumably solid core.

Cloud patterns Earth-based studies had prepared astronomers to expect ever-changing patterns in the atmosphere of Neptune. Specifically, it was well known that the winds change velocity with latitude, causing any patterns to change as rapidly as cars changing lanes on a busy highway. Some winds have been clocked at 1400 mph with respect to the interior. Although the largest component (75 percent) of Neptune's atmosphere is hydrogen gas, its predominate bluish-green color is due to the presence of methane (CH_4), which makes up only about 1 percent. This gas absorbs red light, allowing the blue and green wavelengths from Sun to be reflected back out into space. Helium makes up about 25 percent of the atmosphere.

Certainly the most impressive feature in the full-length photographs of Neptune is the huge, dark, oval-shape area dubbed the **G**reat **D**ark **S**pot (**GDS**, for short). It is located at nearly the same 20 degree South latitude as Jupiter's Great Red Spot, and is proportionally the same size with respect to the size of the planet. Unlike the slow-moving GRS, however, the GDS drifts rapidly westward at more than 300 meters per second with respect to the deep interior of the planet. In the process, it appears to roll like a ball between the wind zones on either side. It changes size and shape as it sloshes around. It appears to make one complete rotation in a counterclockwise direction every 16 days, implying that the Earth-size feature is an enormous high-pressure area (Figures 19-17 and 19-18).

A second dark feature, located further south at about 55 degrees latitude, is known as D2. Scientists are intrigued by this feature because a bright cloud blossomed in its center as *Voyager 2* approached the planet. Another bright patch — actually a collection of several streaks at 42 degrees South — is called "Scooter" because it moves around the planet faster than many other cloud features. Infrared radiation studies by *Voyager 2* instruments indicates that the outer atmosphere has a temperature of minus 213 degrees F. It is not yet understood why Neptune gives off 2.7 times more energy than it receives from Sun.

Neptune has two main cloud decks in its upper atmosphere. Condensed crystals of methane ice form a partial blanket, and some 30 miles lower down is a continuous and very opaque cloud that may consist of ammonia (NH_3) or hydrogen sulfide (H_2S) ice. It also has a thin, high-altitude haze, which consists of hydrocarbons created after the dissociation of methane molecules by sunlight. Eventually these by-products collect into particles large enough to rain down into lower, warmer levels of the atmosphere. There they react with hydrogen to form methane, which is in turn carried upward by convective currents to begin the cycle over again.

The planet has a distinct pattern of dark and light bands, and the GDS is embedded in the brightest of these.

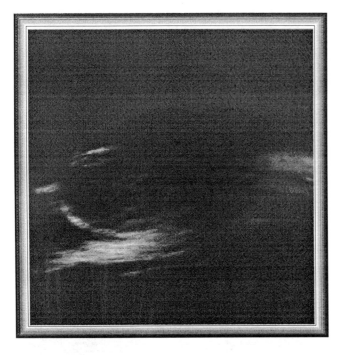

Figure 19-18 The Great Dark Spot in the atmosphere of Neptune. Its position with respect to Neptune's equator is almost identical to that of the Great Red Spot with respect to Jupiter's equator. (NASA)

In contrast to Jupiter and Saturn, however, these bands are not symmetric with respect to the equator. Only a faint hint of aurora emissions was detected by *Voyager 2* UV instruments, unlike the strong emissions on the other giant planets.

Red Spot/Dark Spot The duration of the cyclonic storms on both Jupiter and Neptune, in contrast to those with which we are familiar on Earth, is due to the inherent different natures of the interiors of these gas giants. Earth has a solid surface with protrusions (mountains) and indentations (valleys). There is friction between the surface and the atmosphere, as well as temperature differences — both of which assist in storm-decaying motions.

In comparison, the Jovian planets are completely fluid objects with no solid surfaces (except at the very cores). As a result, there is no friction between the atmosphere and a surface, and there are no horizontal temperature differences under the clouds to affect atmospheric motions. Their atmospheres behave more like huge oceans of gas than atmospheres as we know them on Earth. In these outer planets, once a storm system develops, it decays only through the viscous interactions that take place within the swirling gases. In some cases these features last for years, decades and possibly centuries.

Figure 19–19 These two 591-second exposures of the rings of Neptune were taken by the Voyager 2 camera when it was 175,000 miles from the planet. Visible in the image is the inner faint ring and the faint band which extends smoothly from the 33,000 mile ring to roughly halfway between the two bright rings. These long exposures were taken while the rings were back-lighted by Sun. This viewing angle enhances the visibility of dust and allows fainter, dusty parts of the ring to be seen. The bright glare in the center is due to over-exposure of the crescent of Neptune. Numerous bright stars are evident in the background. (NASA)

Neptune's Rings Astronomers had suspected the existence of rings around Neptune even before *Voyager's* arrival in 1989. In fact, ground-based occultation studies had suggested the existence not of complete rings, but rather of *ring arcs* — incomplete rings, that is. As our record-setting satellite approached the planet, it became clear that they were complete rings after all. The material in one of the rings is quite clumpy, and that Earthbound observation had led to the incorrect conclusion about ring arcs.

There are two distinct outer rings, a faint inner ring, and a broad "plateau" ring that extends halfway between the two outer rings. The rings showed up in backlighted images taken by *Voyager* after it had passed Neptune (Figure 19-19). The material forming the rings revolves in the same direction as Neptune rotates, and are in almost perfectly circular orbits close to Neptune's equator. The fact that the rings are much brighter when backlighted again tells us that there is much small-sized dust particles in them. And since small particles gradually settle out of rings, the rings of Neptune are young-ish. This again suggests that there is a mechanism for renewal of ring material — perhaps collisions between satellites, between asteroids and satellites, or between comets and satellites.

Satellites of Neptune

The only two satellites known to orbit Neptune prior to the arrival of *Voyager 2* — **Nereid** and **Triton** — are odd. *Nereid* follows the most eccentric orbit of any satellite in the solar system, and *Triton* is the only Moon-sized satellite in the solar system that orbits its parent planet in a retrograde direction (Figure 19-1). Furthermore, both satellites travel in orbits highly inclined to the planet's equator. Is it possible that Neptune, too, experienced a catastrophic collision during its formation period?

Triton Even before the arrival of *Voyager* at *Triton*, there was great interest in the satellite, not only because of its strange orbit around Neptune, but because occultation studies had revealed the presence of a thin atmosphere. But unlike the case of Saturn's *Titan*, *Triton's* atmosphere was calculated to be thin enough to allow *Voyager's* cameras to get images of the surface. And the scientists were not disappointed in the least as the photographs began to arrive. Uranus's *Miranda* and *Triton* have to compete with one another for the distinction of having the weirdest terrain in the solar system.

Since the density of *Triton* is about twice that of water, it is probably 70 percent rock and 30 percent water ice. It is more dense than any Jovian-planet satellite except *Io* and *Europa*. Much of the region imaged by *Voyager* was near the south polar cap. It seems to be covered with seasonal ice — probably nitrogen. The measurement temperature was 37 K. The ice appears slightly reddish, no doubt the result of organic compounds formed as solar ultraviolet radiation

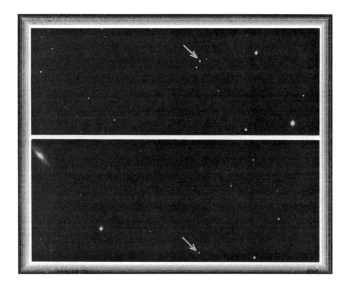

Figure 19-20 Two photographs of the planet Pluto were taken at high magnification at a one-day interval. Since it is so far away from Sun, it has not made a complete orbit around Sun during the time that its presence has been known. (Lick Observatory)

and ions caught in the magnetic field act on methane in the atmosphere and on the surface.

Nitrogen and methane were the only chemicals detected on the surface, but the observed craters and cliffs would slump if they were made of only those materials, so astronomers suspect that water ice must be a major component. Most of the craters are theorized to have been created during collisions with comets over the past 4 billion years.

Giant faults cross Triton's surface (Figure 19-1). About 50 dark streaks are seen aligned parallel to one another. Analyzing *Voyager 2* photographs taken at different angles of Triton's surface lead astronomers to the conclusion that five-mile-high black plumes are the result of geysers on the surface. An alternative explanation, offered to explain the fact that they remain narrow as they rise, is that they are really swirling funnels of dust, gas and nitrogen ice. This would make them similar to the *dust devils* that are found in windy deserts on Earth and Mars.

Much of the trailing hemisphere of Triton is so puckered that it is called **cantaloupe terrain.** This terrain contains 20-mile diameter depressions that are crisscrossed by ridges. It has few impact crater, suggesting a youthful surface. Some regions are broad, flat basins that are probably old impact basins. The ones seen in the images have been extensively modified by flooding, melting, faulting, and collapse.

Triton's highly-inclined (157 degrees), retrograde orbit suggests that long ago this satellite was an interplanetary traveler that ventured too near Neptune and was captured by the planet's gravity. However, the satellite had to encounter some form of resistance in order to lose enough energy to end up in permanent orbit around Neptune.

Two possibilities are that it collided with an existing satellite or that it was slowed down as it passed through a dense dust-and-gas cloud that surrounded the planet as it was forming. In either case, Triton's orbit would have been very elliptical at first. But over a billion years it became circular through tidal interaction with Neptune or through drag created by the nebula that surrounded Neptune. In the process, Triton became hot enough to differentiate, the dense material ending up in the core and the lighter, more volatile material like water ending up in the crust and mantle.

Astronomers recently discovered the presence of carbon-monoxide and carbon-dioxide ices on the surface of Neptune's large satellite Triton. *Voyager 2* had detected the presence of methane ice on Triton, but the discovery of these two ices strengthens the idea that the satellite is a captured object and the idea that comets, icy satellites such as Ganymede and Callisto, and the material that formed the cores of the Jovian planets have a common heritage. A new infrared spectrometer on the UK Infrared Telescope on Mauna Kea was used in making the discovery.

Nereid The other previously-known satellite of Neptune, *Nereid*, was not in an opportune position to be well imaged by Voyager. Its orbit is inclined by 29 degrees, and is highly elliptical.

Six New Satellites All six of the newly discovered satellites orbit Neptune in circular orbits in the same direction as Neptune rotates. In addition, all orbit within 5 degrees of Neptune's equator. They, like Triton, are locked in synchronous rotation so that one side always faces their parent planet. The number of craters found on the larger satellites suggest that the smaller satellites could not possibly have survived the 4 billion year history of the solar system. They must, therefore, have originated more recently, either by fragmentation by collision of larger bodies or by capture from somewhere else in the solar system. Most of them are located closer to Neptune than Neptune's rings. Since objects cannot grow by accretion within the Roche limit, the existence of the satellites in that location also argues for a collision and/or a capture origin.

Pluto

Pluto is a stranger in the solar system. Astronomers don't know whether to call it a Jovian or a terrestrial planet. It is located amongst the Jovian planets, but its size is that of a terrestrial planet. It, like Neptune, was discovered as a result of prediction — the discovery of Neptune failed to completely satisfy the criteria needed to explain the orbital motion of Uranus. From 1905 until its discovery in 1930, Pluto was sought for at the very observatory founded by Percival Lowell for the purpose of studying the canals of Mars.

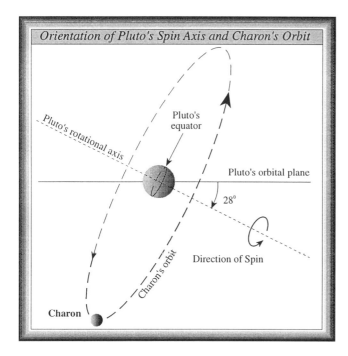

Figure 19–21 *This diagram illustrates the orbital characteristics of the Pluto–Charon system as determined by careful studies through ground–based telescopes. Since Pluto and Charon are close to having the same size, this is almost what one would call a double–planet system rather than a planet–satellite system.*

When Pluto was discovered, it appeared as a faint dot on a photographic plate (Figure 19-20). It still does, except for the very largest telescopes using sophisticated light-detection devices and computer programs to enhance the data. Unfortunately, Pluto was not at a point in its 249-year orbit to make it an available target for either *Voyager 1* or *Voyager 2*, so we have very little direct knowledge of the planet. The most recent upgrading of our knowledge comes from two sources: the Hubble Space Telescope, and the observed behavior of Pluto's only satellite **Charon**.

HST Studies of Pluto Pluto was the first solar system object to be imaged by the Hubble Space Telescope. Thanks to the ESA's Faint Object Camera aboard the Telescope, we now have the very first picture ever of Pluto and *Charon* seen as distinct objects. This was also the first long-duration HST exposure ever taken of a moving target. The telescope had to be carefully programmed in order to avoid smearing the images as a result of the combined motions of Pluto, Earth (both of which orbit Sun at different rates) and the HST (which orbits Earth). Pluto is currently near its closest approach to Earth in its 249-year journey around Sun.

The HST traced the motions of Pluto and *Charon* to determine their masses. Pluto is 12 times the mass of Charon. Charon is only 740 miles in diameter, which is half the size of Pluto. But that makes it the largest satellite of any planet in the solar system (relative to the size of the parent planet). Earth's Moon is second. So it would not be incorrect to refer to Pluto and Charon as a double-planet system. They perform a spiral dance together as they orbit around a common center of gravity. They rotate/revolve

Figure 19–22 *Slight differences in the amount of light reflected off the surface of Pluto suggests that there are some dust deposits. This seems reasonable, since just about every other solid object in the solar system has some amounts of dark material concentrated at different locations. The light to dark shades on this map of Pluto represent the areas of highest and lowest reflectivity of surface features. (NASA/STScI)*

CHAPTER 19 / Exploring Uranus, Neptune, and Pluto

★★★★★★★★★★★★★★★★★

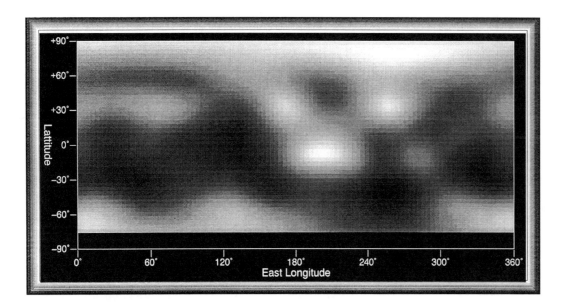

Figure 19–23 The planet Pluto's surface, as imaged by the HST. (NASA/STScI)

once every 6 ⅓ days, always keeping the same face toward one another (Figure 19-21).

Knowing the masses and sizes of the two bodies allows us to calculate their densities. Pluto is one-third the density of Earth, which means that it consists of one-half ice and one-half rock. *Charon's* density is one-sixth that of Earth, suggesting that it consists almost entirely of water ice. Further HST observations of these frigid worlds (minus 419 degrees Fahrenheit) will be conducted to assist scientists in understanding their nature and origin.

Pluto, An Escaped Moon of Neptune? Actually, there may be a connection between the unusual orbits of the satellites of Neptune and the existence of planet number 9 from Sun — Pluto. As you recall from Chapter 16, the orbit of Pluto is the most elliptical in shape of any planet, and actually crosses the orbit of Neptune at two positions. But that statement is rather misleading, since the crossing points must be viewed in three dimensions, not just in two. When one considers that Pluto's orbital plane is tilted some 17 degrees with respect to the ecliptic, it is obvious that there is an extremely small likelihood that Neptune and Pluto will be at the same point at the same time. Nevertheless, the unusual geometry of Pluto's orbit has caused many to theorize that Pluto may be an escaped satellite of Neptune.

In spite of the fact that mathematicians have run the two planets' orbits back in time in their computers, only to find that they have never been in the same place at the same time in the 5-billion-year history of the solar system, the *escaped-satellite hypothesis* was alive and well as late as 1978. In that year, a satellite was discovered orbiting Pluto (Figure 18-21). Since Pluto was named after the Greek god of the underworld, the satellite was named *Charon* after the boatman who rowed passengers across the river Styx to that underworld.

The discovery of a satellite in orbit around Pluto pretty much rules out the escaped-satellite theory, since that would mean that before Neptune lost a satellite, it possessed a satellite that itself possessed a satellite. Even though there is no physical reason why such an arrangement is not possible, it would definitely be an exception in the solar system. We know of no such case amongst the 61 satellites. And scientists are always leery of exceptions. So for the moment — at least — we assume that Pluto is a naturally-evolved planet.

Determining Properties of Pluto and Charon Shortly after *Charon's* discovery, scientists calculated that Pluto and its satellite were about to go through a succession of mutual eclipses during the period 1985-1990. Much like the case of an eclipsing binary star, these eclipses would allow for the determination of some basic properties of both objects. Since eclipses like this happen only once every 124 years, many astronomers dropped other projects they were engaged in to milk the situation for all that it was worth.

Astronomers had a spectrum of the single dot that had turned out to be Pluto <u>plus</u> its satellite, so the challenge now was to obtain a spectrum of just Pluto while Charon was being eclipsed. By subtracting the spectrum of just Pluto from that of Pluto+*Charon*, we could obtain a spectrum of just *Charon*. *Charon* appears to be covered primarily with rigid water ice — at a temperature of minus 360 degrees. Pluto itself consists of methane ice, nitrogen ice, and carbon monoxide ice.

The ices of Pluto are much more "mushy" than those of *Charon*, and are therefore too structurally weak to hold cliffs and scarps. The south pole of Pluto is dazzlingly bright, suggesting the presence of frozen frost. But this would imply a youthful surface, since dark dust particles seem to coat just about everything else out in that region of the solar system. Perhaps there are seasons on Pluto, a result of the elliptical shape of its orbit.

Atmosphere on Pluto? When Pluto passed in front of a

star in 1988, astronomers were surprised to find that the star did not go off and back on in a sudden fashion as would be expected if it had a solid surface. An evaluation of the behavior of the starlight led astronomers to the conclusion that Pluto has an extremely thin (one one-hundred thousandth that of Earth) atmosphere of nitrogen compounds and gaseous nitrogen (Figure 19-23). Now this is very interesting, because the gravity of Pluto is insufficient to hold such an atmosphere close to the surface. It probably extends out to envelop *Charon*! This atmosphere probably waxes and wanes with the seasons on Pluto.

Origin of Pluto and Charon The similarity in sizes and the differences in densities of Pluto and *Charon* strongly suggest that they were not born together. In fact, they may both be related to another icy world, Neptune's largest satellite *Triton*. Both Pluto and Uranus have dramatically tilted axes of rotation, and the orbits of Neptune's two outer satellites are quite disturbed. These features suggest that collisions occurred in this region of the solar system long ago.

The discovery of objects that appear to belong to the theorized *Kuiper belt* may provide the missing link toward our understanding of the origin of Pluto and Charon as well as of the unusual characteristics of some of the planets and/or their satellites. It may well be that Pluto, *Charon*, and *Triton* are the remains of a large number of small, icy objects that once roamed the outer solar system. Called **Plutons** or **icy dwarfs**, these objects would have been more common in the early history of the solar system, and hence collisions (explaining the axial tilts of Uranus and Pluto) and capture (explaining how Pluto obtained *Charon* and Neptune obtained *Triton* and *Nereid*) would have been more commonplace. At the moment, this theory is rather speculative, and will require further investigation of not only the outer planets but the *Kuiper belt* members as well.

Pluto Express Satellite There is serious consideration currently being given to launching a satellite to Pluto in order to round out our knowledge of the major objects in the solar system. Inasmuch as it has never been observed up close by a probe, it will remain an intriguing target until we do send something there, especially since we have the technology to do so anytime we want. The proposed mission, called the **Pluto Express**, must be launched prior to the end of this century in order to maximize the scientific knowledge that we could obtain from the planet. The reason for this has to do with Pluto's strange orbit around Sun.

Since its orbit is so eccentric (out of round, elliptical in shape), Pluto varies in its distance from Sun — from a minimum of 2.8 billion miles to a maximum of 4.6 billion miles. You may recall that Neptune is presently the most distant planet in the solar system, Pluto having moved inside the orbit of Neptune in 1979 and destined to remain there until 1999. Scientists theorize that when Pluto is closest to Sun, the slightly higher surface temperature causes methane and nitrogen to bubble out of the surface material to create an atmosphere. As it gradually moves away from Sun, the gases refreeze again and fall back to the surface. In this sense, Pluto appears to act somewhat like a comet.

The plan, then, is to get a probe to Pluto before it gets so far away from Sun that the atmosphere freezes and drops to the surface. In other words, astronomers want to obtain information about Pluto's atmosphere. If we don't it this time, we'll have to wait another 249 years until Pluto is again at closest approach to Sun. Since the probe requires 6 to 8 years to even get to Pluto, the planning and construction stages must be speeded along — something that government bureaucracy is not prone to do. Actually, the Fast Flyby mission will — assuming that all goes well — consist of two probes that will rendezvous with Pluto within a year of one another. In that way, astronomers will be able to detect any changes that occur within the time interval.

Beyond Pluto

One often hears references to a planet beyond Pluto, usually referred to as **Planet X**. Although there is no physical reason why such a planet or even several planets cannot exist out beyond Pluto, extensive searches have found none. If any planets do exist between the orbit of Pluto and the Oort cloud of comets, they must be too small to reflect enough light to have been detected yet. There are no doubt small objects out there orbiting Sun, but because of their small sizes we will probably think of and refer to them as asteroids or comets.

The Future of the Voyager Satellites

After passing Neptune in 1989, the planetary probe *Voyager 2* became the **Voyager Interstellar Mission (VIM,** for short), expected to operate well into the next century. *Voyager 1* has already completed the solar system phase of its mission, and it too is on its way out of the solar system. Just how long the spacecraft can continue to function depends on several things, one of the most important being the availability of electrical energy. Aboard the satellites, radioisotope thermoelectric generators (RTG's) convert heat from the radioactive decay of plutonium directly into electricity, but this can't go on indefinitely. There is only so much plutonium.

Current predictions claim that the RTG's can power basic spacecraft operations until about the year 2025. Full instrument operations should be possible through 2000. However, in addition to the requirement that the *Voyagers* be able to send information, there must be a receiver on Earth sensitive enough to detect the increasingly weak signal. Two options are available — we can build larger tracking antennas or instruct the satellites to send the data back more slowly. At the slowest rate possible, the satellites can be tracked with 34-meter antennas until the year 2015, or by 70-meter antennas until beyond 2030 (assuming the electrical supply holds out that long).

Another consideration is the necessity that the spacecraft remain "locked" on Sun for the purpose of pointing the antenna toward the Earth. As the distance increases, Sun-sensing instruments will have difficulty identifying our star amongst the billions of stars in the galaxy. Eventually, when the sensors lose the Earth, the *Voyagers* will tumble out of control. Analysis of the sensors indicates that this will not happen until about the year 2030.

Assuming that all goes well, what exactly will the *Voyagers* tell us about the outer reaches of the solar system? Although the spectacular photographs of the Jovian planets stole the show, the *Voyagers* actually returned a considerable amount of valuable data about Sun while it was cruising from one planet to another. Instruments aboard have been studying the solar wind and the intense radio bursts that accompany solar flares. Taken as a whole, the observations have provided fundamental information on the physics and dynamics of Sun's magnetosphere, the extended region influenced by Sun's magnetism. But we'd like to know how far out into space this region extends.

After detecting and passing this "boundary" of Sun's influence in space, called the **heliopause**, we expect that the instruments will begin to record what might be called the "interstellar wind." After all, stars are suns too, so it is assumed that they emit a steady stream of charged particles just as Sun does. There is reason to believe that there are some interesting physics that goes on at the boundary where solar and interstellar winds collide. There is also a cosmic ray instrument aboard each of the satellites. These very high energy particles (misnamed rays) can originate from supernova explosions, galactic magnetic fields, and other high-energy events. So keep tuned. You haven't heard the last from the *Voyagers* yet! Here is *Voyager 2*'s predicted itinerary:

- 2,012 AD   crosses the heliopause into the interstellar medium.
- 8,571 AD   closest approach to Barnard's Star, which may have a planetary companion.
- 20,319 AD   closest approach to Proxima Centauri, star nearest to our Sun.
- 20,629 AD   closest approach to Alpha Centauri.
- 23,274 AD   closest approach to Lalande 21185.
- 26,262 AD   enters the Oort Cloud, from which comets originate.
- 28,635 AD   leaves the Oort Cloud and so exits the solar system.
- 129,084 AD closest approach to Sirius, the brightest star visible from the Earth.

LEARNING OBJECTIVES: *Now that you have studied this Chapter, you should be able to:*

1. Compare and contrast the general characteristics of Uranus, Neptune and Pluto with the other Jovian and terrestrial planets, and explain their major differences.
2. Compare and contrast the interior of the Earth with those of Uranus, Neptune, and Pluto.
3. Describe the behavior of the cloud patterns of Uranus, Neptune, and Pluto in terms of their distances from Sun, interior compositions, and rotations.
4. Compare the rings of Uranus and Neptune with those of Jupiter and Saturn, and offer an explanation for their differences.
5. Describe the features of Uranus that make it unusual, and offer a theory that explains those features.
6. Explain the features of Pluto that make it a stranger in the solar system, and explain why a theory to account for that unusualness is difficult to find.
7. Compare and contrast the surface features of the satellites of Uranus, Neptune, and Pluto, and explain how we can account for differences between them.
8. Describe the features of the small moon Miranda, and explain how they may be explained by the same theory that explains the unusual features of its parent planet Uranus.
9. List the planets and satellites in the solar system that offer the best chance for finding life, and explain why that it so in terms of the conditions of each object.
10. Define and use in a complete sentence each of the following **NEW TERMS**:

Ariel 457	**Icy dwarfs** 465	**Pluto Fast Flyby** 465
Cantalope terrain 462	**MAUTO** 456	**Plutons** 465
Charon 463	**Miranda** 458	**Titania** 457
Electroglow 453	**Nereid** 461	**Triton** 461
Great Dark Spot (GDS) 460	**Oberon** 456	**Umbriel** 457
Heliopause 466	**Planet X** 465	**Voyager Interstellar Mission** 465

Section 5

Evolution of Life in the Universe

Chapter 20

Is Anyone Out There?

Figure 20–1 An average galaxy in the universe like M33 contains approximately 100 billion stars, and perhaps an equal number of planets gathering light and heat as they orbit those stars. Can we be the only conscious beings in the entire universe of galaxies to appreciate that understanding? (Lick Observatory)

CENTRAL THEME: **Why are we conducting searches for radio signals from other civilizations, what are our chances of hearing from someone, and how do we decide what to say and how to say it?**

Ponder this chart for a moment before proceeding to the next page:

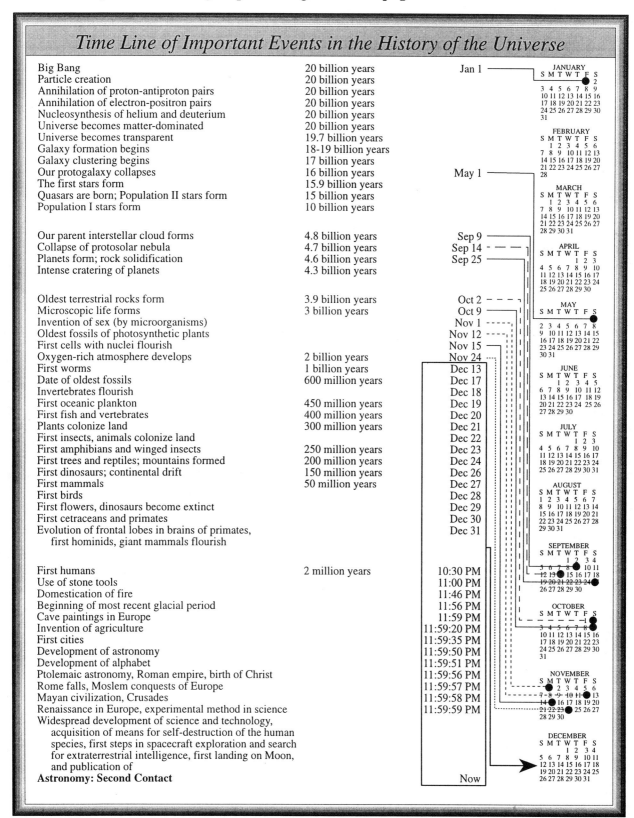

Figure 20-2 *The 20-billion year history of the universe is scaled down to the period of one year. Some of the important events covered in this Book are shown at the corresponding time of the year they occurred. As you see, human presence is a rather recent occurrence.*

★★★★★★★★★★★★★★★★★★

Reflecting on the general theme that ties the preceding Chapters together — evolution — it is difficult to avoid the conclusion that events that occurred on Earth can very likely happen elsewhere as well. There just may be some other "guys" out there with whom we can communicate. I am sure this thought crossed your mind even before you picked up this Book for the first time. And it is not just because Hollywood has provided us with thrilling episodes of galactic battles with high-tech space weaponry.

People have been wondering about life beyond Earth throughout recorded history — probably from that very first moment from which consciousness evolved (Figure 20-1). A reality that is difficult to escape is that there is life here — somehow we got here. If we got here, then perhaps "They" got there. It is really that simple. What has brought the subject out into the open, considered by scientists as well as philosophers, is the vast amount of knowledge scientists have accumulated about what is going on out there as well as here. This Book has been a summary of the major findings and conclusions about how the material objects of the universe got here. Now we consider what seem to be special material objects of the universe — you and I.

Time Scale of the Universe If the major themes I have developed throughout the Book are essentially correct — that is, stars, galaxies, the universe, and planets evolve — then we reach the major conclusion that we humans are recent in the time-scale of the universe. Using the ages of planets, stars, galaxies, and the universe determined by scientists, let's compress the entire 20-billion year history of the universe into a one-year calendar as in Figure 20-2. Spend a few moments studying it carefully, and then reflect on the recency of human presence in the universe in terms of some important historical events.

The striking feature of this chart is the enormous time span within which the events have occurred. That fact leads scientists to conclude that if there <u>are</u> some folks out there living around other stars, they are most likely either very, very, very, very, very much smarter than we are, or terribly dumber. In other words, the chances of us running into (or being visited by) a civilization that is at approximately our level of technological (and social?) development is rather unlikely. What are your chances of running into someone on the street that has your same birthdate?

That doesn't mean that we are going to be successful in our search for other civilizations, even if "they" are out there. But if we don't try, we'll never know what the success factor is. Perhaps it is all for curiosity's sake anyway. In order to convince taxpayers that their money is being wisely spent, however, those engaged in the SETI Program have developed a line of reasoning that argues for the search. I outline that rationale by considering the questions "When should we search?", "Why should we search?", "Who do we search for?", "How do we conduct the search?", and "What do we say or expect them to say?".

"When" Should We Conduct the Search?

We can gain some perspective on the problem of contacting other civilizations by looking briefly at some previous attempts to communicate with "them" out there. As we contemplate the reasonableness of the attempts, we should be asking ourselves what assumptions they were using in hoping for success. Perhaps we can then better evaluate our own current assumptions as we engage in the search.

Old Attempts to Signal Our Existence In the 1820s, the famous mathematician *Karl Gauss* proposed that a 10 mile-wide forest of pine trees be grown in the pattern of a triangle, the lengths of whose sides would express the Pythagorean theorem (Figure 20-3). That is, the ratios of the lengths of the three sides of the triangle would be 3 to 4 to 5 ($3^2 + 4^2 = 5^2$). It was presumed that a distant civilization capable of observing the surface of Earth would be able to see this huge work of intelligence.

Similarly — in 1840 — the director of the Vienna Observatory, Joseph von Littrow, proposed that a ditch 20 miles in diameter be dug in the Sahara desert of Africa, filled with water, and then kerosene floated on top. By lighting the flammable fuel at night, we could create a flaming ring that would advertise the presence of intelligence (?) on our planet (Figure 20-3).

Another imaginative scheme proposed building a giant, steerable, concave mirror with a predetermined focal length that would focus reflected sunlight onto the surface of the planet Mars. The mirror would essentially act as a giant magnifying glass, concentrating Sun's rays onto the vast sand deserts thought to cover the Martian surface. By carefully guiding the mirror, we could scrawl a huge message across the deserts by fusing the sand with the concentrated sunlight. This would be the solar system's first billboard in space! Presumedly we would be careful not to scrawl the message through the presumed Martian city!

Although none of these three proposals was ever carried through to completion, there were common underlying assumptions in each:

The ET's (extraterrestrials) were:

- thought to be nearby, most likely within the solar system itself.
- believed to be scientific and/or mathematical in their thinking.
- capable of recognizing a triangle or a blazing circle or scribbling in the form of patterns as obvious signs of intelligence.
- Although it may seem too obvious to even mention, there was also the implied assumption that "They" would want to respond to our attempts to advertise our presence in space. That is, it was assumed that They would be interested in responding to our attempts at communication.

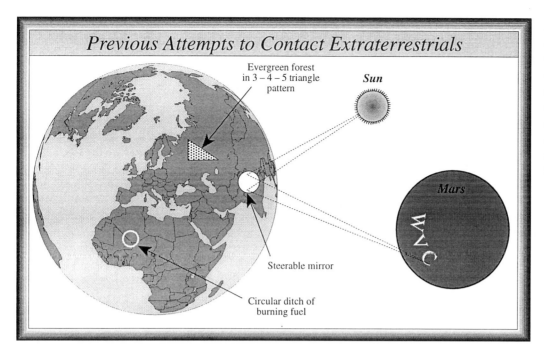

Figure 20–3 Prior to the 20th Century, the attempts to contact extraterrestrials was based on the assumption that "They" were close to us—within the solar system. This diagram illustrates a few of the novel ideas.

More Recent Attempts In the very first Chapter of the book I presented a list of the more recent attempts to detect signals from distant civilizations. In order to emphasize the international scope of the interest in the search, I repeat that list below:

Attempts to Contact Extraterrestrials

- 1960 USA 2 stars
- 1963 USSR 1 quasar
- 1966 Australia 1 galaxy
- 1968–69 USSR 12 stars
- 1970–72 France 10 nearest stars
- 1972–76 USA 674 stars
- 1972–present USSR All–sky search
- 1974–76 Canada 70 stars
- 1974 USA 1 star cluster
- 1975–79 Netherlands 50 star fields
- 1977 USA 200 stars
- 1977–present Germany 3 stars
- 1978 USA 185 stars
- 1978 USA/Australia 25 star clusters
- 1979 Australia Close Sun–like stars
- 1979–81 USA 200 stars
- 1981 Netherlands Center of Milky Way
- 1982 Canada Center of Milky Way
- 1985–95 USA Project *META*
- 1992–94 USA NASA's *HRMS*
- 1990–present Argentina Project *META II*
- 1994–present USA (Arecibo) Project *SERENDIP*
- 1995–present USA Project *Phoenix*
- 1995–present USA Project *BETA*

What has changed between the 19th Century and the modern era is that the search program is a serious scientific enterprise, one that costs the taxpayer money and engages the minds of some of our brightest scientists.

So we might ask ourselves — as we have for topics throughout the book — what the assumptions are upon which these searches are based. What do we expect "them" to say, for example? What language do we expect "them" to use? Why should we even assume "they" will say anything at all — much less want to say anything to us? What do we even have in mind when we think of a "they"? Beyond those questions, we might want to reflect upon the desirability of even trying contacting them in the first place.

Objections to Searching After all, there is no foregone conclusion that "they" will be peaceful — if they find out that we are here, perhaps they will want to come here and do something horrible, like eat us! Even if we don't expect "them" to actually come here, what would be the social effects if we merely <u>knew</u> "they" were out there? Especially if they were significantly more advanced than we. How would our political and religious leaders react to a communication coming from a civilization millions of years more advanced than we are, even if they didn't intend to run for political office or take over our churches, mosques, temples, and synagogues? What do we mean by "advanced," anyway?

The question that comes up most frequently when this topic is discussed, however, is this — "Why should we waste money and time in an attempt to communicate with distant civilizations when we obviously haven't learned yet how to communicate well with one another?" Or put the question in a another way — "Why search for intelligent life else-

CHAPTER 20 / Is Anybody Out There? **471**

where when we are not convinced yet that there is intelligent life here?" Good questions.

But regardless of how you personally feel or what you personally believe about the matter of attempting to contact others in space, it is really quite too late to do anything about it. We have already announced ourselves to the universe – <u>intentionally</u> and <u>unintentionally</u>. Aside from leaving some debris scattered around the solar system in the form of satellites and satellite parts that might be found by extraterrestrials on tour through this region of the galaxy, we have been *unintentionally* beaming radio and radar signals out into space – mostly in the form of radio and TV broadcasts. Only one *intentional* message has ever been sent out – sent with the most powerful radio telescope in the world in 1974. I will discuss the message in detail later in the Chapter. The point is – these artifacts of our civilization are already out beyond Earth penetrating deeper into the cosmos. We can't get them back. So "someone" may eventually get them – or already have them!

Divine Creation of Life? Another frequent objection to searching for life elsewhere is that there is no one out there to contact because life was divinely created on Earth and nowhere else. Evidence in the direction of this argument is not based on any specific reference in sacred writings or documents (the Holy Bible is not specific on this point), but on the apparent uniqueness of life and the improbability that it happened solely by chance — by chemicals coming together to bond into brain cells.

In fact, one of the arguments frequently used by creationists is that scientific evidence points clearly to the fact that the universe is running down — matter is converted into stellar energy that disperses throughout the universe and is thereby participating in the increasing disorderliness of the universe. This is the argument based on the **Third Law of Thermodynamics**-the universe tends toward disorder, not order. Since life is the ultimate state of order in the universe, so the argument goes, there needs to have been a creation event to navigate life against the natural current of disorder.

Scientists agree that within any *closed* system there is an inevitable trend toward disorder — perpetual motion machines are impossible for this reason. But (1) we don't know that the universe is a closed system, and (2) within the vastness of a system as large as the universe there can be local events that temporarily go against the current toward disorder. As regards (1) – if the universe is constantly being replenished with fresh supplies of hydrogen gas (steady state theory), it is not a closed system.

As regards (2) – certainly stars are more orderly systems than the clouds of gas and dust from which they emerged, and yet they too eventually become disorderly. So if gravity can temporarily pull together gas and dust particles to form stars, why is it difficult to imagine electromagnetic forces pulling chemicals together to form life-sustaining arrangements? It is a difference in complexity and probability, not a difference in principle.

Complexity and probability are certainly serious considerations as regards the possibility of life evolving elsewhere, and should definitely not be swept under the rug and ignored. But one of our biggest problems is that we have only one kind of life to study, and therefore we have no point of comparison in establishing what those words mean as far as life is concerned. I <u>feel</u> complex and improbable when I compare myself to the rock in my garden. But I have no way of calibrating that feeling.

Perhaps in the universe as a whole, life is about as common and mundane as rocks are on Earth's surface. Until we find life elsewhere, or find life lacking everywhere but here, we are limited to what is immediately around us. We are rather like the early civilizations that were able to argue for their Earth-centeredness because of the limit to what they were able to explore around them. Will we be led out of our life-centeredness as we search for signs of life elsewhere?

Now don't take me wrong. There is just as likely the possibility that we will find that we are absolutely alone in the universe — that there is no one out there to talk to. We are it. We are one of a kind. We are unique. We are freaks in the universe. Even were that to be so, do you think that such a realization would shake us to action to preserve that uniqueness? Would that realization cause us to more fully appreciate the one experiment that gave the universe an opportunity to be aware of itself? Would we see mass migrations back into churches, mosques, temples, and synagogue?

If you find yourself in the creationist camp or leaning toward its *a priori* assumption that life is only to found on Earth, then the value of the reminder of this Chapter is that it helps you understand and maybe even appreciate the contrary opinion and the line of reasoning that leads them to that opinion. In either case, the discussion that follows is not an argument for the search, but simply an outline of the factors that are considered in determining if the search has reasonable grounds for success.

Peaceful or Aggressive? There is another belief camp that argues against the SETI Program as well. Those of this persuasion argue that "they" are not only out there, but "here" as well. They claim that there is already sufficient evidence of their "visits" that we should devote ourselves to opening contacts with the "ones" who are already here (Figures 20-4 and 20-5). Assuming that "they" are attempting to contact us at this very moment, perhaps we should ponder the question of what their intent is of contacting us — be they hostile or peaceful. Unfortunately, Hollywood film producers and science fiction authors have biased our thinking. We commonly think of ET's as ugly and menacing creatures, anxious to enslave us or in the very least put us in zoos as mere cosmic curiosities.

If we entertain the opinion that their favored method of contact is to physically come here in their advanced space

Sightings of Aerial Phenomena at First Described as Unidentified for the Year 1963								
Explanation of the Phenomenon								
Month	Astronomical[a]	Aircraft	Balloon	Insufficient	Other[b]	Satellite	Unidentified	Pending	Totals
January	4	4	1	5	1	2	0	0	17
February	5	3	1	3	3	2	0	0	17
March	8	2	1	4	7	7	0	0	29
April	4	7	1	4	2	8	0	0	26
May	3	6	3	3	6	0	2	0	23
June	3	10	6	12	4	24	1	0	60
July	9	9	6	3	3	11	1	0	42
August	11	15	4	6	7	4	2	1	50
September	12	7	1	4	3	12	2	0	41
October	10	6	0	6	3	6	4	0	35
November	8	2	0	2	6	2	0	1	21
December	5	1	1	3	3	3	3	2	21
Totals	82	72	25	55	48	81	15	4	382

Figure 20-4 This breakdown of the evaluation of UFO sightings for a single year by members of the "Project Bluebook" team suggest that people are not well-informed about atmospheric phenomena, nor are they astute observers of the sky. Of course, one can insist that the team members were part of a "cover-up" to prevent the public from knowing the truth about aliens. A further breakdown of that category of sightings called astronomical is shown in Figure 20-5.

Cases Explained by Astronomical Phenomena				
Month	Meteors	Stars & Planets	Other	Totals
January	3	1	0	4
February	3	2	0	5
March	8	0	0	8
April	3	0	1	4
May	3	0	0	3
June	1	1	1	3
July	5	4	0	9
August	4	5	2	11
September	7	4	1	12
October	7	3	0	10
November	6	2	0	8
December	5	0	0	5
Totals	55	22	5	82

Figure 20-5 Again, a breakdown of the phenomena reported as UFO's and explained as astronomical objects or phenomena suggests that people are often times mistakened in their explanation of events in the sky. Those reports listed under "other" included such things as hallucinations, hoaxes, and even a hole in the ground.

ships, it hardly seems to their advantage to expend the energies required to go from one star to another in the galaxy just to obtain slaves or curiosities. That is not very energy-efficient. Besides, if they still harbor aggressive or destructive traits like those that lead to enslavement and zoos, then they have very likely already misused the vast amounts of energy available to destroy themselves (or been destroyed by another hostile civilization with those same traits).

Furthermore, if they are seeking less-advanced civilizations to enslave, how is it that they avoided being enslaved by an even superior civilization long before now? Is it possible that there an hierarchy of civilizations within the galaxy, each biding the will of its Master, with one top-dog civilization calling the shots and directing the enslavement of backward civilizations? It seems more reasonable that a civilization even the slightest bit more advanced than ours will be peaceful and accept us into its fold as long as we, too, throw off the yokes of violence, destruction, aggression, and enslavement.

This is usually the point at which someone responds — "Well, isn't there sufficient evidence already that we are being watched and/or visited by ET's? Haven't there been enough UFO reports to convince us that more advanced beings are monitoring our activities on Earth, perhaps waiting for us to change our attitudes toward our planetary responsibilities before we are allowed into the Galactic Federation? Well, even if that <u>is</u> the case, the fact of the matter is that the world's population by and large doesn't seem to be affected by such reports sufficiently so as to choose peace as its primary goal.

Many people claim that the ET's do not allow themselves to be detected because they want us to learn on our own. They do not want to interfere with our "free will." I have two responses to that suggestion or claim. One, if we assume that they are already here, but do not present themselves because we have not yet demonstrated ourselves worthy of contact, then we might as well go about the business of finding new methods toward becoming worthy — including attempts to detect signals from other civilizations who are "planet potatoes" and attempt to contact other civilizations from the comforts of their own planet.

Secondly, the argument that "they" know what is best for us and can even straighten us out if they wanted — but don't want to interfere — is not intellectually satisfying. It smacks too much of the theological idea that God can do anything — that He knows what bad things we are about to do — that He can interfere at anytime — but He doesn't want to interfere with our "free will."

Then there is the conspiracy theory — "they" <u>have</u> landed and contacted "chosen" members of our civilization, but the government doesn't want us to freak out so they are holding the information from us. Or the "chosen" members are common citizens like you and I who now have a message for us — just buy their book ($80) or attend their workshop ($800) and you too can become one of the Chosen. I think that at this point we should invoke one of the criteria by which scientists operate — **extraordinary claims demand extraordinary evidence.**

★★★★★★★★★★★★★★★★

"Why" Should We Conduct the Search?

Aside from basic curiosity, I can think of a couple of arguments for engaging in the search for extraterrestrials. I like to think that I have a long and prosperous life ahead of me. There are no certainties in life, however, and so I also consider the possibility that serious illness, injury, or even death will catch me by surprise. But there are others who depend upon me for their own hoped-for long and prosperous lives. So I have chosen to give up some of my immediate needs and wants in order to pay for medical and life insurance policies that provide for them in case I succumb to early illness or death. At the same time I attempt to maintain my health through physical exercise and thoughtful diet. But there are still uncertainties.

Should we perhaps be thinking of our planet in a similar way? After all, we humans have stewardship of a planet that is also capable of getting sick. You are certainly aware of the many challenges that humankind faces today — challenges that involve threats not only to the very health of our planet, but to the very survival of life itself. One need only reflect on current events to get the impression that humankind is out of control of the forces that trigger nuclear warfare or ecological disaster. I won't depress you with a list of the world's concerns — you can just as easily keep a list of your own as you watch the evening news.

So doesn't it seem reasonable that we should be making an effort to insure the survival of the human species <u>in the universe</u> in case our efforts to preserve life on Earth fail? Shouldn't we be exploring the means by which to get a few living representatives of our planet off Earth to establish permanent habitats elsewhere in space? Might that realization actually be — at the deepest levels of our psyches — the driving force behind the efforts to get humans out into space by the space programs undertaken by various nations?

And might it also be possible that extraterrestrial civilizations with whom we can communicate already know the solutions to our problems, since they confronted and overcame the same hurdles long ago? Perhaps this is another criteria we could use to define an "advanced civilization" — one that has mastered space travel to the extent that some of their members have learned to travel to, survive and live in other environments in space. By virtue of their being more advanced, they have learned how to minimize environmental threats to their survival.

What I am trying to say is that perhaps our attempts to survive on Earth should include — but not be limited to — *SETI* and *Space Habitation* programs. The SETI program is the theme of the current Chapter, whereas the latter is the theme of Chapter 21.

"Who" Should We Conduct the Search For?

Space is too vast to simply turn on our telescopes and scan the sky for obvious signs of intelligent life. Besides, if it were that simple, we would have already had picked up some evidence in the act of conducting astronomical research. So there must be a search strategy — we must decide what the "theys" are like that we hope to contact. The assumptions we use are not too different from those we use in our current SETI programs.

Perhaps 50 or 100 years from now people will laugh at our current attempts just as you probably laughed at the proposals of centuries past. The argument against these assumptions appears in the form of a questions that goes something like this — "But you are talking about life as <u>we</u> know it! It isn't valid or even reasonable for us to talk about how ET's think or talk or act, since we are limited to imagine them only in ways common to our own experiences!"

I have two responses to that argument. Firstly, while it is true that we can talk and think only about life as we know it, we may never be able to talk about life as "we don't know it" if we simply wait around until "it" comes to us. At least if we state our assumptions clearly, proceed ahead in our search, we are prepared to modify our assumptions if we encounter evidence that doesn't fit the existing theory. In other words, since we are not quite clear what we are really searching for, we begin with what we know and try to extend that knowledge into realms with which we are unfamiliar.

Secondly, if you object to searching for life as we know it, then please tell us something about life as we <u>don't</u> know it so can change our search strategies! Our search for life on Mars using the *Viking* landers used the assumption that the general characteristics of life on Mars would be similar to those of life on Earth. We may, as we explore Mars in closer detail with robot probes like the *Mars Observer*, discover that life forms on Mars operate according to significantly different strategies than life on Earth.

But there is another important consideration involved in the question of searching for signs of intelligence elsewhere. Scientists have come a long way in understanding the survival strategies of all species of life on Earth. If, in our SETI program, we use the principle of the *assumption of mediocrity* that we have used throughout the book, then it seems reasonable that life forms throughout the universe evolve using similar principles. The principles of physics and chemistry appear to be universal, so why not those of biology as well! So we come full circle back to Galileo's challenge to the Church — can we assume that the principles determining the behavior of matter in our local region of space also apply to distant environments?

Evolutionary Strategy The modern theory of the evolution of life on Earth has a few general assumptions, one of which is that life forms at every level of complexity and sophistication survive by competing for the available resources necessary to their survival. It is tempting to think of *food* when we think of resources necessary to survival, and that is probably okay if we generalize the word *food* to mean *energy*. For example, trees may compete with one another for sunlight because they need that resource for survival.

Basically, the argument for life throughout the universe possessing the same general features as life on Earth goes like this — if life everywhere in the universe evolves as a by-product of stellar and planetary evolution, adapting to environmental constraints such as abundances and distribution of chemicals, temperature, types of stellar radiation falling on the planet, etc., then its heritage will be similar to ours. They must have — in the process of evolving — competed for available resources.

What is it that we spend most of our time doing, but engaging in activities that directly or directly assist in maintaining our survival. We eat, sleep, and reproduce in order to survive as a species. Think about all the things you do during a particular day. How many of them are for the sheer pleasure of doing them? And can't we claim that even those activities done for sheer pleasure are also fundamental to our survival? They make us happy, and happy organisms have a greater chance of survival than unhappy ones.

It would almost seem that life revolves around reproduction as the dominant theme. Now listen carefully. I am not talking about sex. Without the ability to reproduce, a species comes to an end — another deadend road in the limitless number of possibilities for life forms. But in order to reproduce, the members of a species must be capable of finding and approaching the mate. That requires a detection system and an energy system. We humans see not only for the purpose of finding a mate but for the purpose of finding food so that we can maintain the integrity of our bodies while searching for a mate.

I suspect that you may be taking exception to what I am saying. You are thinking: "What about love?" How can one explain such a complex and pervasive human trait as love in terms of survival?" Well, Somerset Maugham once wrote that "Love is only the dirty trick played on us by nature to achieve continuation of the species." I am not claiming that the purpose of existence or the *Meaning of Life* is to reproduce. Certainly there is a prevalent theme in every human culture generally categorized as spiritual — the belief that material existence is a mere prelude to or consequence of an "existence" outside of and unconfined by time and space. But that is certainly not the primary motivation for living day to day. If it were, we wouldn't be so uncomfortable about approaching death.

Isn't it interesting, though, that in spite of the variety of belief systems throughout the world that preach life after death, there is not a single one that encourages or practices early departure (death) in order to more quickly arrive at a "timeless" and more Perfect existence! In the Judeo-Christian tradition, the emphasis is on the doing of good works on Earth as the necessary preliminary to *better things to come*. To intentionally cause early departure to that more Perfect World is considered a sin.

No matter how one looks at it, we are left with the same conclusion — we compete within an intricate economical and political system for the available resources of our planet so that we can further our survival. Just how much we need in order to survive is, of course, a personal decision. Food and shelter are the absolute minimum. But we in modern cities demand much more.

The point I am trying to make is that if "others" out there evolve in the same way as we apparently did — competing for limited resources in a finite volume of space — then perhaps we can assume that "they" too associate progress (or technology) with the acquisition of more and more energy resources. Look at any comparison of the countries of the world in terms of their standing in the world community, and you find that it is expressed in terms of their levels of technological development. At the same time, look at the energy consumption of each of those countries. There is, as you would expect, a direct relationship between the two.

The United States, with only 6% of the world's population, consumes 32% of the world's energy. Is it surprising that Americans have a high standard of living? Someone defined work as *the moving of a piece of Earth from one place to another, or telling someone else to do it*. Initially, the energy used by a civilization is that which is locally available and most directly usable — burning wood or coal, water- or wind-powered devices, or simple gravity (dropping a rock, for example).

Over time, some of the local sources of energy are used to create modes of transportation that allow remote energy deposits to be tapped and utilized. With ever-increasing surpluses of energy, components of Earth's surface are heated and cooked to extract metals which are used to build larger and faster transportation systems to tap greater amounts of energy, and so on. Eventually, we tackle the ultimate energy source: Sun. Having discovered the means by which Sun provides the light and heat that nourishes life on Earth, we possess an unlimited source of energy for insuring our survival. By duplicating Sun's fusion process, we tap the energy bound up in the very atoms of which everything is made.

Levels of Advancement The Soviet astrophysicist N. S. Kardashev proposed that civilizations anywhere in the universe will go through similar steps in arriving at the Atomic Age. These steps are the logical outcome of the search for explanations for the manner in which nature operates. Scientists are attempting to simply mimic nature through the process of what we call "science."

Kardashev proposed 3 categories for civilizations, according to the availability of energy (Figure 20-6):

- Type I: A level close to contemporary terrestrial civilization with an energy capability equivalent to the amount of solar radiation arriving at Earth's surface (10^{16} watts).

- Type II : A civilization capable of capturing and utilizing the entire radiation output of its star (4×10^{26} watts).
- Type III : A civilization capable of capturing and utilizing the entire radiation output of the Milky Way galaxy (4×10^{37} watts).

Those that learn to recreate their star's energy source on the surface of their home planet are referred the **Type I** civilizations. We are on the threshold (hopefully) of becoming such a civilization. Although we have learned to duplicate the mechanism by which Sun shines (fusion) in hydrogen bombs, we have not yet learned how to use the energy derived therefrom creatively. Hydrogen bombs release their energy in one single swoop rather than slowly — a firecracker rather than a generator.

Research is currently being conducted in several countries around the world in an attempt to imitate the fusion process and bleed the energy off slowly, but success is not yet in sight. There is not even a consensus amongst scientists that it will <u>ever</u> be possible to do so economically. Kardashev assumes that civilizations are eventually successful, and that the next logical step is to harness the entire energy output of the star around which their planet orbits.

Just think how much energy in the form of sunlight is being wasted at this very moment — only an extremely small percentage of it is actually intercepted by the surfaces of the planets and other solar system debris. The remaining sunlight goes out into space, hardly ever to be intercepted by anything. Mostly, space is empty of matter. So doesn't it seem logical that civilizations initially install solar collectors on the roofs of dwellings.

Then, as property gets too expensive to build larger collectors to meet the increasing demand for energy (progress), they place them in orbit around their planet. Finally, they build a stellar collector that surrounds the entire star. By this time, the civilization, now a **Type II**, has left their planet behind to set up housekeeping on the inside surface of this sphere or ring that surrounds the star (Figure 20-7). This is called a **Dyson sphere** after the Princeton physicist who proposed the idea. Life within such a civilization is described by Larry Niven in his sci-fi book "Ringworld."

So a **Type III**? That is a civilization that goes on to tap the energy of all the stars in the galaxy in which they live! Farfetched, to be sure. But it seems to be an inevitable consequence of our appetites for energy and knowledge. Certainly there are other possible scenarios. We don't know what new developments may lead us along different paths. But again, until we know of other alternatives and/or whether any particular path is better than any other, we feel free to speculate about the possibilities.

"How" Should We Conduct the Search?

So there may be highly technical civilizations out there, civilizations that may be millions or billions of years more technically advanced than we are. Why do we even assume they will return our calls, let alone send out messages on

Characteristics of Extraterrestrial Civilizations

Level	TYPE I	TYPE II	TYPE III
Characteristics	Planetary Society	Stellar System Society	Galactic Civilization
	Developed Technology • understand laws of physics • space technology • nuclear technology • electromagnetic communications	Construction of space habitats "Dyson Sphere" as an ultimate limit Search for intelligent life in space	Interstellar communication/travel Very long societal lifetimes (10^8 to 10^9 years) Effectively "the immortals", for planning purposes
	Initiation of spaceflight, interplanetary travel, settlement of space	Long societal lifetimes (10^3 to 10^5 years) Initiation of Interstellar travel/colonization	Energy resouces of the entire galaxy (10^{11} - 10^{12} stars are commanded)
	Early attempts at interstellar communication Starting to push planetary resource limits	Ultimately all radiant energy output of native star is utilized	
Manifestations	Intentional or unintentional electromagnetic emissions, especially radio waves	Electromagnetic • radio waves • optical lasers • X-rays • Gamma rays Gravity Waves Mass Transfer • probes • panspermia • stellar ark	Feats of astroengineering Exotic forms of interstellar and intergalactic communication • neutrinos • tachyons? • Waves "I"?

Figure 20–6
Kardashev's classification of civilizations in the universe according to the amounts of energies at their disposal. Thinking about what types of technologies more advanced civilizations might possess allows us to develop strategies for detecting their presence in space.

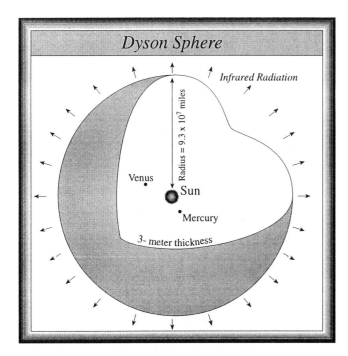

Figure 20–7 The Dyson sphere is based upon the assumption that any advancing civilization will eventually learn how to use the energy of their parent star to disassemble a planet and construct a sphere around the star to tap the entire energy output of the star. "They" could even live on the inner surface of the sphere.

their own initiative? Just think about our own civilization's track record as we approach the *Type I* category.

Send or Receive? While unlocking the secrets of the atom, we also unleashed the primitive parts of our reptilian brains that cause us to use that energy to compete for Earth's limited resources. The claim that someone else would have done it if we hadn't is not the point. All developed countries are caught up in the struggle to harness the destructive power of nuclear weapons so that they can maximize their piece of the resource pie.

Is it possible that civilizations throughout the cosmos encounter the same dilemma as they seek greater reservoirs of energy? Anything can be used either for constructive purposes or for destructive purposes. In that sense, we seem to be at a crossroads. Either we learn to live with the bomb while cutting back on its production and distribution as we learn to tolerate one another's differences, or we continue along the present path until someone accidently (or purposely) rubs the bottle to release the genie of destruction.

Conceivably civilizations that choose the constructive path are not limited in their lifespans. Once they discard war and violence as methods of resolving differences, there is no limit to their technological growth save the limits imposed by nature itself. There may be natural disasters that wipe out even the most advanced of civilizations. But at least the civilization doesn't set its own limits.

So if there are some of those very ancient (and wise) civilizations around, might they want to share the painful lessons they learned as they approached the point of too much accumulated energy that could have led to self-destruction? As offspring of the same galaxy of stars, might they feel some kinship with distant, fledgling, immature civilizations (like ours) stuck at the crossroads, not knowing in which direction to go?

They need not even know specifically of our existence — they need know only that the evolution of life (and intelligence) is as inevitable as the formation of stars from clouds of gas and dust, and that others struggle through puberty just as they did. They may even know that some fail. *Shakespeare* did not specifically know that his writings would assist me in my growing-up years. He no doubt felt that the lessons learned in his own life might be of some help to others — either then or now or in the future. Perhaps at this very moment there are similar messages being beamed from some stars in the galaxy, being absorbed by the rafters of the roof over your head, instead of by the dish of a radio telescope pointed in their direction! We won't know if we don't try.

Time Limitation So assuming "they" are out there, that "they" are interested in communicating with others, and that we need not fear being enslaved by "them," how do we go about establishing contact? What are our choices — do we send out messages, and then wait for a response? Or do we sit around passively with our instruments aimed at the sky waiting for the alarm to go off? At this stage of our technological development, there seems to be only one thing that we can do — listen. There are two reasons why I say listen, and not send — one quite obvious and the other less so. As far as we know, nothing goes faster than the speed of light. The nearest stars to us are about 5 light-years away. That means a response from a civilization living around one of those stars cannot be expected for at least 10 years after we send a message — somewhat less than immediate gratification.

What are the probabilities that the very nearest stars are even inhabited? And if they are, why haven't astronomers detected some sign of their presence already? An organized search program costs money — money that mostly comes out of public coffers (do you know of any private citizen who is interested in putting up a million dollars for such a program?). Politicians control the purse strings of public monies, and, unless there is a loud clamor of excitement from taxpayers (you and I), they usually hesitate to vote for programs that do not have immediate benefits. By "immediate" I mean, of course, within the elected term of the politicians.

And what do I mean by "listen" you say? Certainly not sound, since sound travels neither very rapidly nor at all in the vacuum of space. Scientists have pondered this question in depth, and created a list of potential carriers of extraterrestrial information (Figure 20-8), including the

CHAPTER 20 / Is Anybody Out There?   **477**

★★★★★★★★★★★★★★★★★★

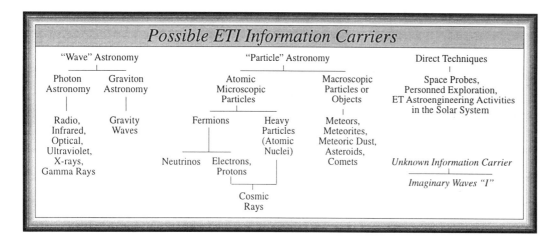

Figure 20–8 Before conducting a search for ET intelligence, we must first consider how that intelligence disseminates information into space. Based upon what we know about the universe, scientists speculate that any of those listed in this chart could be carriers of information.

imaginary "I" wave that has not yet been discovered. In principle, at least, every form of energy and every particle in the universe could be used as a vehicle for carrying information from one star system to another. It would seriously tax the resources of astronomical facilities to investigate each and every one, so astronomers have decided that our own experience with investigative techniques may be shared with the Others. (1) Electromagnetic energy travels at the speed limit of the universe (speed of light), and therefore requires the shortest possible time to get from one location to another. (2) Of all types of EM energy, only radio and visible radiation penetrate through Earth's atmosphere without significant absorption. To send a message at visible wavelengths would require that the receiving civilization be able to separate the feeble beam of visible radiation from the overwhelming visible energy output of Sun – a difficult task if conducted at any great distance form Sun.

Radio telescopes are our preferred instrument for establishing contact with any civilization in space. A radio telescope must focus on a specific point in space – it doesn't collect energy from wide regions of space. It is not a matter of scanning the sky in broad sweeps. A search program requires a systematic search from one star to another. So the vast distances between the stars in the galaxy would seem to prohibit (or seriously restrict) two-way communication between civilizations. Yes, I know that some of you are objecting to my lack of imagination in limiting us to the speed of light. So would you please step forward and tell me of your discovery of something that goes faster, and how we might use it for communicating with others?

While we are waiting for such new discoveries – if and when they are made – we work on the assumption that electromagnetic energy is the most efficient means by which information can be transmitted across the vast reaches of space. That is the manner in which stars send us information regarding their chemical makeups, temperatures, and so forth. Certainly scientists will be interested in any new discoveries having to do with anything that travels faster than the speed of light, but if we are to act now, we are restricted to what is known at the moment.

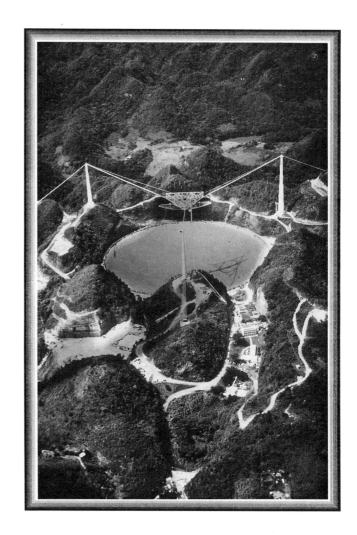

Figure 20–9 The world's largest radio telescope at Arecibo, Puerto Rico, is capable of having a dialogue with a twin telescope just like itself that is located anywhere within the Milky Way galaxy. This capability has only recently been acquired. (National Astronomy and Ionosphere Center operated by Cornell University under contract with the National Science Foundation)

Limitation of Knowledge The less obvious reason why we might choose not to send out signals to other civilizations is that there is really nothing much we can say to "them" that they don't already know, except that we are here (a fact that they might know already, for reasons I explain shortly). In other words, there is a *very small chance that there are any civilizations out there with whom we can communicate that are dumber than we are!*

Admittedly, there may be zillions of planet's on which there are life-forms in various stages of evolution, but the only ones capable of receiving signals we send must be at least at our level of development. Remember that we have already outlined the assumptions as to what "development" means — we have only recently acquired the means by which to detect signals from other civilizations.

Assuming that such signals are sent at radio wavelengths (I will discuss this assumption shortly), we have possessed the means by which to detect radio signals from space only within the last 50 years, since the development of the radar and radio telescopes during the Second World War. So the chances of our signals being received by a civilization that is at the same level of development as us are enormously small.

Think of it this way — the 100-billion or so stars in our galaxy range in age from 0 to 15 billion years. Assuming that the 5-billion-year history of our solar system is typical of the time required for any civilization to evolve to the point of being capable of receiving radio signals, then only stars that are at least 5 billion years old can be orbited by such civilizations. So the range of ages for such technical civilizations is 0-10 billion years beyond that of our own!

The chances are significantly greater that a message sent by us will land in the lap of a vastly superior civilization, one that is thousands or even millions of years more advanced than us. What could we possibly say to such a civilization? Perhaps a feeble but honest cry of "Help!" Chances are that we stand to gain more from what it tells us than the other way around.

But if we choose not to send, why should we assume that another civilization will be sending? Precisely for the reason I mentioned earlier — having gone through puberty itself in learning how to deal with vast surpluses of energy on a finite planet, "it" wants to share its lessons with us. Perhaps it is for the good of the galaxy that it do so. Otherwise, we may extend our violent natures out into the cosmos to threaten the very stability of what might be a vast network of peaceful, advanced intelligences.

So why, then, did we choose to send an intentional message out into space in 1974? And why have there been no other messages sent, even out of simple curiosity? When astronomers completed the upgrading of the surface of the huge radio telescope at Arecibo in Puerto Rico in 1974, giving it the capability to participate in a two-way conversation with its hypothetical identical twin on the opposite side of the Milky Way galaxy, they decided to use the dedication ceremony as a forum to send a *message* to Earthlings by sending a signal into space. That is, they decided it was time for people on Earth to be aware that we are at a level of technology that allows us to announce our presence to the rest of the galaxy. They wanted all of the peoples of the world to participate in dialogue to determine if we should proceed in that direction. It is not the role of a scientist to make decisions for the rest of the planet — especially if those decisions can have such a fundamental impact on the future of life on the planet. What if, for example, Someone receives the message we sent and comes here in a battlecruiser and annihilates us? Won't the scientists who sent the message feel somewhat guilty about their actions?

But since the thrust of the message-sending effort was to alert the peoples of Earth about our newly-acquired technological skill, astronomers decided to direct the radio beam carrying the message not to a nearby Sun-like star, but rather to a very distant (10,000 light-years) group of old, M-type stars — the globular cluster in Hercules. M-type stars, you may recall, lack the heavy elements upon which life as we know it depends. Since the signal has travelled a mere 22 light-years since it was sent, the soonest that we can expect a reply is some 19,978 years from now! Unless, of course, They (the recipients) are so advanced that They can skip around the galaxy at warp speeds, and even have the ability to detect the message before it gets too far away from Earth. If that is the case, I argue, then They must already know that we are here, and consequently need not worry about having sent the message in the first place.

Eavesdropping Earlier in the Chapter, I suggested that ET civilizations may already know that we are here. That would seem to be a contradiction, in light of what I said earlier about the chances that ET's are already here. But if "they," too, are in a "listening" mode, and radio is also their preferred method of communication, then they may at this very moment be detecting some of the radio energy that leaked away from Earth's surface long ago. The earliest radio broadcasts began on a sizeable scale in the 1920s. Those transmissions were not sent outward in a straight line (unidirectional), but outward in all directions (omnidirectional).

Even though the inverse-square law tells us that the signal decreases in strength with increasing distance, it is possible that they are capable of detecting such weak signals across interstellar distances. In fact we might, during our current search program, eavesdrop on the cultural transmissions of another civilization!

Just reflect on that for a moment. Signals from our earliest radio broadcasts are at this very moment reaching stars located about 70 light-years from us, having already passed the thousands of stars located within a sphere of that radius. Stars 50 light-years from us are just now receiving the earliest TV signals.

Stars 25 light-years from us are just now receiving TV

coverage of the Viet Nam war. Stars just 25 light-years away are just now listening to early Beatles' music. 10 light-years, *General Hospital*. 15 light-years, *Watergate* coverage. Those are signs of our culture distributed to the masses of stars. That could be what "they" know of us. And we wonder why "they" don't attempt to contact us! It is a sign of their intelligence that we haven't yet been contacted.

Perhaps "they" have given up on us? Does that mean "they" had no similar difficulties in growing up? If so, then perhaps we are freaks in the universe — a bad mutation, a poorly-formulated experiment. Then perhaps we don't deserve to survive — survival of the fittest, you know. But if all civilizations go through a similar awkward stage in the process of maturing, then it seems reasonable that they would want to assist in our growth by sending out information important to our continued survival. Perhaps they were helped by another more advanced civilization, and we in turn will assist others after we mature. But will we really listen to what they have to say?

"What" Kind of Signal Do We Search For?

Hopefully, you are presently in an optimistic mood about the future of Earth. Perhaps you are even willing to invest in a strategy to detect signs of intelligence in outer space. Assuming that the most logical approach at the moment is to listen for signals beamed our way, to what frequency do we tune our receivers? How can we be sure that the wavelength to which we tune our telescopes is that on which "they" are sending their messages? Let us look at the problem from the largest perspective. Here we are, located under an ocean of air contemplating the receipt of an electromagnetic signal. Immediately we must rule out the search for signals at wavelengths outside the atmospheric windows discussed in Chapter 5.

"Tuning In" Indeed, other civilizations may be sending messages at X-ray wavelengths, or ultraviolet wavelengths, or even gamma-ray wavelengths. But since those types of radiation are by and large absorbed by our atmosphere, there would be no sense at all in trying to detect them, at least from Earth's surface. Mostly the types of radiation that freely penetrate Earth's atmosphere are visible and radio wavelengths. Presently, then, those are the two intervals of electromagnetic wavelengths that are capable of carrying a message from outer space through our atmosphere to our waiting detectors on the surface.

The visible wavelengths we immediately rule out simply because of our inability to separate out the intelligent signal from the visible light from the star itself. Across interstellar distances, radiation emitted from a planet orbiting a star will appear to be coming from the star itself — the separation distance will be so small compared to the distance of the star from us. The visible radiation reflected from a planet orbiting the star will be overwhelmed by the visible radiation emitted by the star itself — it would be like trying to detect the presence of a firefly sitting on the rim of a searchlight.

But stars are radio-quiet — that is, they do not emit great quantities of radio radiation. So a radio message from a civilization living on a planet orbiting a star could be detected at Earth's surface. The largest radio telescope on Earth, located in Puerto Rico, is capable of detecting signals sent by a clone located on the far side of the galaxy, at a distance of 100,000 light-years (Figure 20-9). So the technology necessary to detect radio signals from any civilization within our galaxy is already available to us and in place. A signal sent out by any civilization with a telescope comparable to our largest could be detected by the Arecibo telescope.

Choice of Wavelengths Still, there are an infinite number of wavelengths of radio energy to which we can tune our receivers. Do we just scan that entire range of wavelengths while focused on each star, much as we scan the FM radio band when we search for our favorite radio program? In view of the number of stars that we expect we will have to listen to before success is achieved, that approach would be incredibly time-consuming. Wouldn't it be easier if we assumed that the ET's are making every effort to ensure that their message is easily received and understood?

That seems like a good assumption. Why else would they be sending the message, especially since they must assume that it will probably be received by some terribly backward civilizations? So we must ask ourselves if there is a particular wavelength that an advanced civilization would assume another civilization with radio telescopes would certainly know about. Fortunately, nature has provided us with a natural wavelength at which messages might be exchanged between civilizations — 21 centimeters. Recall that the most abundant atom hydrogen announces its presence in space by emitting the 21-centimeter wavelength photon. What could be more obvious? Send a message at the very wavelength emitted by the most common substance in the universe!

So it is that most all of the searches conducted to date have tuned either to that wavelength, or one quite close to it — the 18-centimeter wavelength. This particular wavelength is emitted by another common constituent of the interstellar medium, the **hydroxyl molecule**. It consists of a hydrogen atom and oxygen atom bonded together. When a hydroxyl molecule (OH^-) and hydrogen atom (H^+) join together, the result is a water molecule (H_2O) — the liquid of life (as we know it).

Now that is very interesting. It just so happens that within the intervals of wavelengths of the electromagnetic spectrum shorter than 18-centimeters and longer than 21-centimeters, there exists static and background cosmic "noise" that interferes with our search for interstellar signals. It is within the interval of 18 to 21-centimeters that optimum conditions exist for detecting a message. Since water may be a necessary ingredient for life throughout the universe, this presents another argument for the use of these wavelengths for interstellar communication. Out of respect for the importance that water plays not only in our lives but the lives of life forms throughout Earth, this interval of wavelengths between 18 and 21-centimeters is called the **water hole** (see Figure 20-10). Perhaps galactic neighbors meet at the *water hole*.

Recognition of Message I'm certain that something has been bothering you every time I have mentioned the word "message." How will we possibly be able to recognize a message from an alien civilization anyway? How can we expect them to know our language, or any language at all! Naturally we don't presume they have learned English, or French, Spanish, or even Esperanto. If they send messages, they know and use mathematics — the language in which nature reveals how it (nature) works. That is the only language (but important one) that we assume they use when sending messages, and so for us as well.

No, we don't think that the message will be a series of numbers and rows and rows of equations. Rather, we assume that they will use the language of mathematics to send pictures. After all, one picture is worth a thousand words. To get an idea of the language and assumptions we use when we consider the matter of the content of a message, look at a copy of the message engraved on gold plates attached to the two *Pioneer* satellites that were sent past Jupiter and Saturn in the early 1970's. Similar plaques were attached to the support struts of the *Voyager* satellites that visited not only Jupiter and Saturn, but one of which

went on to provide tens of thousands of images of Uranus and Neptune. All four satellites are at various stages of leaving the solar system. Since they were designed to eventually leave the solar system entirely, it was deemed important to include a greeting card just in case they eventually fell into the hands of aliens. Can you make any sense of it (Figure 20-12)? Spend a few minutes deciphering it. If you need help, you'll find the intended solution in the caption.

But now for the serious stuff. It will take thousands of years for the Pioneers and Voyagers to get even close to the nearest stars to Sun. And they are not even programmed to visit other star systems. So the chance that intelligent beings will ever see them again is highly unlikely. As I mentioned earlier, we humans have sent only one intentional message. In Figure 20-14 is the raw data of the message we sent in 1974 via the Arecibo telescope. The method of sending a message like that works something like this — a button is attached to the transmitter circuit of a radio telescope. Each time the button is pressed and released, the circuit is momentarily closed and a pulse of 21-centimeter radio energy leaves the telescope in the direction of the target star at the speed of light. Someone at the receiving end records that pulse of energy as a blip on a piece of moving graph paper (or whatever equivalent is used in "their" radio telescope).

If a series of pulses of radio energy separated by quiet gaps are sent, and the pattern of the pulses and gaps according to strength of signal is plotted on a graph against time, it should look something like the uppermost portion of Figure 20-13. Hopefully the receiving civilization will recognize the data received by their radio telescope (as you do!) as being organized in **binary language**. Binary lan-

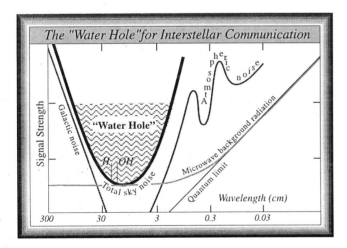

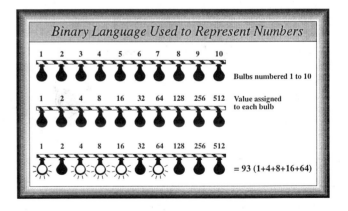

Figure 20–10 That portion of the electromagnetic spectrum that has the least amount of "static" or "noise" that would drown out a weak interstellar message from another civilization just happens to be located at the wavelengths associated with the two ingredients of water. Thus astronomers refer to the "dip" in the curve of radio "chatter" between the two wavelengths as the water hole.

Figure 20–11 It is convenient to use a row of lightbulbs to represent the idea of binary numbers. We assign the numbers 1, 2, 4, 8, 16, etc. to the bulbs in the row. If a given bulb is lit, then the number that bulb represents is being expressed. If two or more bulbs are lit, then the number being expressed is the total of the numbers represented by the lit bulbs.

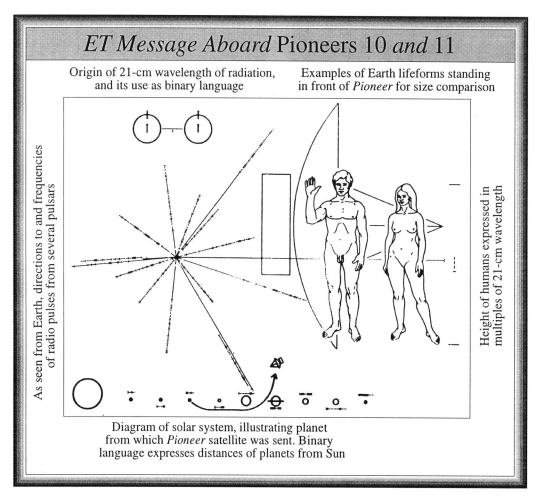

Figure 20–12 What is obvious to you and I would not necessarily be obvious to aliens, and vice versa. Being significantly different from us in physical appearance, "they" may wonder what the figures are on the right side of this plaque on Pioneers 10 and 11. The center–top is an illustration of the hydrogen atom in its two spin states, between which the 21–centimeter wavelength is emitted. The tic–mark between the two hydrogen atoms indicates the 21–centimeter length, which then can be used as a standard "ruler" throughout the rest of the message. The lines running outward from a point represent the directions to known pulsars from Earth. The tic–marks along the lines represent the pulsation rates of the pulsars in units of 21–centimeters. If "they" identify pulsars by pulsation rates, "they" will be able to calculate our location in the galaxy relative to them. The solar system and its contents is represented along the bottom, along with the distances of the planets from Sun. The curved line shows the planet from which the satellite was launched, and a diagram of the satellite stands behind mom and dad so that "they" know the sizes of humans.

guage is the very basis of modern computers. It is such an efficient system of expressing information that any civilization calling itself advanced must certainly use it or have once used it while evolving to their present level of development.

Binary Language Think of the five fingers of your hand representing five light bulbs — the bulb lit when the finger is up, unlit when curled up in your palm. Those five fingers (or bulbs) can be used to represent 33 numbers, or 33 pieces of information. With no bulbs lit, the closed fist represents the number 0. Nothing (no light bulbs on) is a piece of information. The first bulb in the sequence lit up represents the number 1. If only the second is lit, the number 2 is represented.

The efficiency of this scheme reveals itself when you realize that in order to represent the number 3, you simply light up bulbs 1 and 2. As the number of bulbs increases, so also does the amount of information they can store and express. Specifically, notice that for each digit added, twice the previous amount of information can be stored — six bulbs allow for the expression of up to 65 pieces of information, seven bulbs up to 129 pieces of information, and so on (Figure 20-11).

No wonder enormous amounts of information are stored in a small computer chip. Those chips consist of miniature light bulbs that individually are either on or off, and the pattern of on's and off's expresses the information being expressed. In this case, we want to use the binary

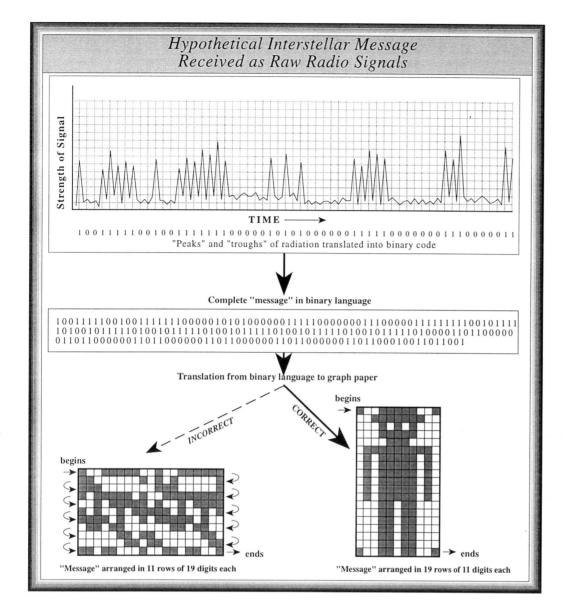

Figure 20–13 This diagram illustrates the sequence of steps in going from the raw radio energy data to the finished message. Compare the "peaks" and "troughs" in the graph of radiation with the pattern of ones and zeros, as well as with the filled-in squares that form the message.

language to send a picture. So we need to tell the receiver of the message two important clues — the dimensions of the picture, and which bulbs to turn on and which to leave off. The latter is rather straightforward — the "blips" in the radio signal represent the "on's", and the spaces between the "blips" represent the "off's." Actually, it doesn't really matter if they do it the opposite way — the content of the picture is the same, since the picture is simply the contrast between the on's and the off's (Figure 20-13).

Creating the Picture But how will "they" determine the dimensions of the picture? What we send and what they receive in their radio telescope is a series of pulses separated by patterns of blank spaces. So the received message might look something like that in Figure 20-15. Of course, we would not send the data making up the picture just once, hoping they just happen to be listening at that particular moment. If they begin receiving and recording the message during the middle of the transmission, and we send the transmission only once, then of course they will not get the full message. In order to create the "picture" from the raw data, the entire sequence of pulses and gaps must be recorded. So after sending the complete message for the first time, we wait a few minutes, repeat it — wait, repeat, wait, repeat, and so on.

Let's consider the message we sent out, the raw data of which is in Figure 20-14. They receive a string of blips and spaces from our transmission. Hopefully, it should occur to them — assuming they do not misinterpret it as the product of a process associated with the behavior of stars or such — that it is binary language. If so, they will count the total pieces of information contained in the string of pulses. In the message we sent, that number is 1,679. Again, if they think mathematically as we do, then someone will recog-

CHAPTER 20 / Is Anybody Out There? 483

★★★★★★★★★★★★★★★★

```
"Text" of ET Message Sent in 1974
Begin
→ 0000001010101000000000001010000001010
  0000001001000100010001001011001010101
  0101010101001001000000000000000000000
  0000000000000001100000000000000000000
  1101000000000000000001101000000000000
  0000000010101000000000000000000111110
  0000000000000000000000000000000110000
  1110001100001100010000000000000110010
  0001101000110001100001101011111011111
  0111110111110000000000000000000000000
  0100000000000000001000000000000000000
  0000000000010000000000000000011111100
  0000000000111100000000000000000000000
  0011000011000011100011000100000001000
  0000001000011010000110001110011010111
  1101111011111011110000000000000000000
  0000000001000001100000000001000000000
  0011000000000000000001000001100000000
  1111100000110000001111100000000000110
  0000000000001000000001000000001000001
  0000011000000010000000110000110000000
  1000000000011000100001100000000000000
  0110011000000000000011000100001100000
  0000110000110000001000000001000000100
  0000010000010000001100000000010001000
  0000011000000001000100000000010000000
  1000001000000010000000100000000100000
  0000001100000000011000000001100000000
  0100011101011000000000001000000010000
  0000000001000011111000000000000001000
  0101110100101101100000010011100100111
  1111011100001110000110111000000000010
  1000011101100100000010100000111111100
  1000000101000011000000100000110110000
  0000000000000000000000000000000011100
  0001000000000000011101010001010101010
  0100111000000000101010100000000000000
  0010100000000000001111100000000000000
  0001111111110000000000001110000000111
  0000000001100000000000110000000110100
  0000000101100000110011000000011001100
  0010001010000010100010000100010010001
  0010001000000010001010001000000000000
  0100001000010000000000001000000000100
  0000000000000010010100000000000111100 11
  11101001111000 → End
```

Figure 20–14 The "pulses" of 21-centimeter energy and spaces between "pulses" that were transmitted into space in 1974 are represented here by a series of zeros and ones, respectively. This is simply translating the data of the radio signal into binary language, a process that allows any information to be digitized and transmitted from one place to another. This is identical to the manner in which CD players operate.

nize 1,679 as the product of two prime numbers, 23 and 73.

So the picture is 23 units by 73 units, or 73 units by 23 units. In other words, they can use 23 units in the first line before going to the second of the 73 lines of the picture, or they can use 73 units in the first line before starting the second of 23 lines. There is a 50–50 chance that their first attempt will be successful, but in something as important as this, no one is going to consider the failed attempt to have been a waste of time. Someone whimsically playing with the received data might add up the individual numbers of the 1,679 total (1+6+7+9 = 23) to arrive at the number 23. That clue is not only one of the two prime numbers, but the horizontal measurement of the message as well. That clue will not necessarily be added to each and every message, but only serves to illustrate how "chance" may play an integral role in the sending/receiving of interstellar messages.

You see, when we sent out the message in 1974, we wanted to make it as easy as possible for "them" to decipher it correctly. So a specific total of data points was selected in order to increase the probability of their success. After determining the dimensions of the picture, their next task is simply to transfer the binary language onto a piece of graph paper (or spreadsheet program), filling in the squares that represent the "blips" or the spaces between the "blips," and leaving the opposite blank. The solution to the message sent out in 1974 is shown in Figure 20-15. Spend some time interpreting it, identifying the binary language contained within the message itself.

So our message arrives and is deciphered. But will "they" understand it? That we don't know. Certainly if "they" have gotten as far as getting the picture, they will spend a lot of time attempting to interpret it. The mathematical and scientific language will no doubt be readily understood. I can imagine our receiving a message, and the scientific community making a copy available to each and every member of the human race. And a reward of $1 million to the person with the best interpretation! Do you think that such a reward would be a strong enough incentive for some rather creative solutions?

Effects of Contact

Here is one of the scenarios that scientists entertain about the effects that a single intelligent signal from a civilization might have on us. At first, it would have front-page coverage, the leadoff item on the evening news. We would all be talking about it, reflecting on its importance and possible consequences. Then gradually, as the more immediate demands of life demanded attention — washing the car, emptying the garbage, planning for the summer's vacation, studying for final examinations as the more immediate demands of life require our attention — wash the car, empty the garbage, plan for next summer's vacation, study for final examinations — we would lose interest.

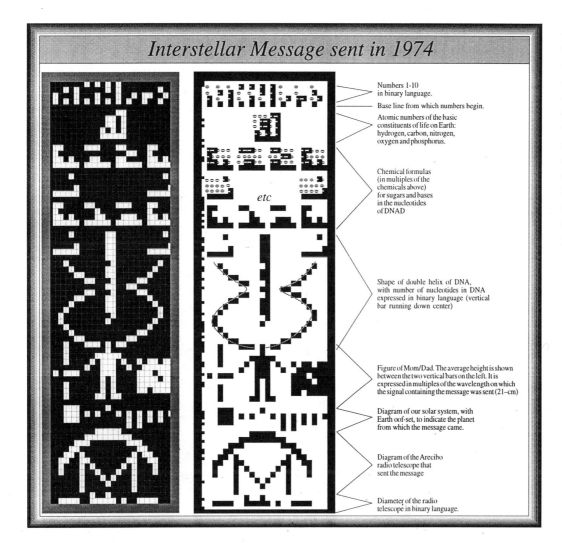

Figure 20–15 *The correct decipherment of the message sent to the globular cluster in Hercules (M13) in 1974. A portion of the binary code of zeros and ones is included in the interpretation diagram on the right in order to illustrate the association between the binary code and the forming of the picture.*

Weeks later, a two-inch column in Section F of the Sunday newspaper announces some scientist's speculation about the star from which the message came, and the varied interpretations of the content of the message. Meanwhile, scholars are busy at work attempting to decipher the language used, contemplating the psychology and sociology of the senders, and perhaps deciding how to structure the language and content of a proposed response.

Gradually, the awareness of the presence of *others* creeps into our consciousnesses, and becomes one of the components of the tapestry of our belief and value systems, from which spring our actions and deeds. In other words, the effect of the "contact" might not be sudden and dramatic. It might well be more like high school graduation — terribly important at the time — but only gradually understood and appreciated as we encounter life on our own.

Probability of Successful Contact

So assuming we decide — for the moment at least — only to listen, what is the probability of success? Success? Why are we interested in a probability of success? You might argue that until we are contacted, we don't know that there is anyone there. So there is no basis for determining probability of success. Well, that is one way to look at it. But each of us has a "gut feeling" about a subjective probability. And certainly if Congress authorizes the use of funds to establish a search program, it requires that an evaluation of the return on its investment be made. A part of that evaluation will certainly be a subsection entitled "probability of success." So we state what we know, what assumptions we make, and proceed.

Mathematical Treatment of the Probability Frank Drake, presently teaching at the University of California at Santa Cruz, was the first scientist to direct a radio telescope

CHAPTER 20 / Is Anybody Out There? **485**

Estimate of the Number of ET Civilizations in the Galaxy

$N_c = N_s \times f_s \times f_p \times f_L \times f_c \times f_i \times f_{ET}$

Factor		Pessimistic	Optimistic
N_S =	Number of stars in the Galaxy	2×10^{11}	2×10^{11}
f_s =	Fraction of stars that are suitable	1 of 100	1 of 2
f_P =	Fraction of stars possessing planets	1 of 100	1 of 1
f_L =	Fraction of those planets whose surfaces have habitable conditions, and that remain in the life zone of the parent star for at least 5 billion years	1 of 100	1 of 1
f_c =	Fraction of those habitable planets on which life does evolve and flourish	1 of 100	1 of 1
f_i =	Fraction of those life-bearing planets on which life evolves to an "intelligent" level	1 of 100	1 of 1
f_{ET} =	Fraction of those planets containing intelligent "beings" who endure and want to communicate	1 of 10 million	1 of 10
N_C =	Product of factors in each column: Number of stars with planets having intelligent life	2×10^{-6}	10^{10}
Implication: Distance to nearest civilization		10^7 **light-years**	**15 light-years**

Figure 20–16 The Drake equation attempts to provide an estimate of the number of technical civilizations in our galaxy with whom we could communicate. Having such a number, astronomers can calculate the average spacing between civilizations so that the probability of contact can be estimated. By its very nature, the equation is based upon assumptions, and everyone has his/her own assumptions, and therefore estimates based upon the equation vary amongst scientists and non–scientists alike.

to the sky and scan three stars for intelligent signals in 1960. Prior to initiating the search, he proposed an equation to determine such a probability of success. He didn't want to waste his time in a project whose outcome was to be a failure at the offset. Called the **Drake equation**, it allows us to estimate the total number of intelligent civilizations within the galaxy with whom we are able to establish contact. Having calculated that number and the volume of space within our galaxy, we can easily determine the average expected separation between civilizations.

For example, if we determine that there are 5 civilizations in our entire galaxy of 100 billion stars, then the average separation is about 20,000 light-years. If the calculated number is more like a billion civilizations, then the average separation is about 50 light-years. The greater the number, the closer to us will another civilization be located, and the *fewer the stars we will have to study before we hit upon an inhabited one*. There are fewer stars close to us than there are at greater distances.

The Drake equation is shown in Figure 20-16. It consists of a number of factors (assumptions). Frank Drake assigned a value to each factor and arrived at a total number of civilizations in our Galaxy of 10^{10} stars (using optimistic values). Certainly not all of the factors are based upon scientific criteria. For most, there is insufficient data upon which to base a value that we can all agree on. Feel free to use your own values if you don't like his.

Actually, it doesn't really matter what values you assign to each factor. There is an additional factor yet to be considered that far outweighs the importance of all the others. Before discussing the key factor in determining the density of civilizations in space, let me make a few comments about each of the factors, and try my own values:

- f_s – You have learned that not all stars are suitable for the arising of life and intelligence. O- through F2-type stars do not remain on the main sequence for the 5 billion years necessary for life to evolve to our level of development. K5- through M-type stars do not have the metals necessary for the evolution of life and technology. That leaves a narrow range (F2 through K5) of main sequence stars potentially populated. Approximately **20%** of the stars in our galaxy fall within this range.

- f_p – Although there is no hard evidence that planet's exist around other stars, current evidence strongly suggests that they do. The prevalence of binary star systems and the evidence for our own solar system's formation strongly support this conclusion. Not every star will be endowed with planets, but a healthy percentage will. I venture to guess that **10%** of all stars will have a system of planets.

- f_L — Of the numerous planet's and satellites in our solar system, only one appears to harbor life. The basic constraints for life to have evolved in the solar system are temperature and chemistry, which of course are based upon distance from Sun. We also learned that distance from Sun also determines the chemical environment on a planet's surface.

If we can use our solar system as an example of a typical planetary system anywhere in space, then only those planets within a certain interval of distance away from their star will have the minimum requirements of proper temperature and chemistry. For our system there is only one planet — but then, that is all that is needed. Other stars may have none within that **life zone**, while some may have several. I assume that every planetary system will have at least one planet located within the life zone (Figure 20-17). So every solar system — 100% — will have a suitable planet.

f_c — We do not know, however, the minimum environmental requirements for the arising of life on a planet even if the temperature and chemical environment are suitable. We are here. But we don't know if we would be here if Earth rotated once every 48 hours instead of its present 24 hours, or if the axis of rotation were tilted by 5 degrees rather than 23.5 degrees. We don't know if it could happen again, even if the very same conditions that existed on early Earth exist on another planet somewhere else.

Perhaps the probability of life getting started is so extremely slight that it has only occurred once. We are cosmic freaks. We are the only ones who are conscious of the fact that there is a universe out there — the only ones wondering if we are alone. As I have said before, scientists do not like to assume special-ness about any part of the universe. Unless there are natural laws to the contrary, we assume that what can happen once in one corner of space and time can happen again in another corner of space and time.

Naturally if you are of the opinion that life arose only once in the entire universe, either because you believe that it was the result of a combination of extremely unlikely events or of a Divine Plan (**creationism**), you need go no further in considering the other factors in the Drake equation. Working on the assumption that a Supreme Deity created life only once, there is obviously no one out there with whom to communicate. For the rest of you — even though the likelihood may not seem very great — I will be rather conservative and assume that on a mere **1%** of the planets within the life zones of other stars has life actually evolved.

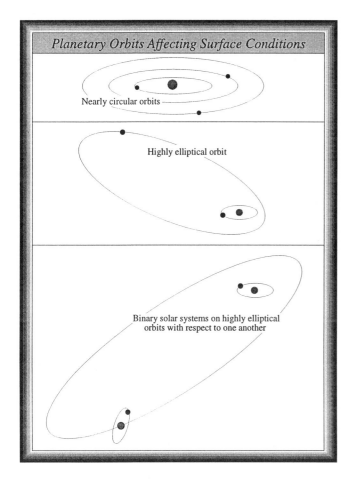

Figure 20-17 As an example of the range of possibilities that our planet could have suffered, and conditions that could have prevented the development of life here. Planets orbiting stars may have many different orientations, especially if they are in binary systems—which make up over 50 percent of all stars. Our solar system may be quite unique as far as the arrangement of planets is concerned. We won't know until we understand the conditions in other planetary systems.

- f_i — So life arises rather routinely in the universe — but is it always going to evolve into an intelligent species? What do we mean by "intelligent" anyway? A traditional definition of intelligence is the ability to use tools. Now, however, we know that such animals as chimpanzees and even birds use primitive tools. Another definition is that involving the use of language, but evidence mounts that chimps and dolphins communicate by using a language.

Even honeybees use a language of dance to communicate.

So with respect to communicating across interstellar distances, we choose to use the criteria of radio technology for intelligence. After all, that is the method we are using to listen, and if they aren't sending in that same form then we probably won't hear from them. Unless, of course, they know something about traditional methods that we don't. Since the laws of physics appear to work everywhere, radio communication is possible everywhere. "They" do not have to be carbon-based. "They" don't have to be living on a solid planet like Earth. "They" could be plasma beings that live in interstellar space.

In a certain sense, anything that is alive is intelligent. If it is alive, it has survived through the use of intelligence. Only those species (like dinosaurs) that are no longer with us lack intelligence. Intelligence has survival advantage because it allows for the more efficient gathering of resources necessary for survival. Increased intelligence means having a wider range of control over the environment so that changes within that environment do not threaten the long-term survival of that intelligence. Humans are more intelligent than the Bald Eagle because it cannot adapt to the changing conditions (destruction) that we impose on the environment within which it lives and upon which it depends for survival.

We humans, through the use of technology, adapt ourselves to a great diversity of environments that under normal circumstances would threaten our survival (e.g., living at the South Pole). Building radio telescopes for the purpose of listening for signals from ET's could therefore be considered a part of that increased intelligence. Something "they" tell us may help us improve our survival strategies on Earth. For the purposes of our argument, at least, we must take intelligence to mean the capability to engage in interstellar communication.

If they have not advanced to that level, then they will obviously not be communicating across light-years of space. We can imagine lots of intelligent life forms that don't build radio telescopes.
But do all evolving life forms inevitably reach a level of intelligence that allows for radio communication? Not necessarily so. What about a planet that is completely covered with oceans, where are no land creatures? Under the ocean's surface, there is no driving force behind the development of arms and hands and fingers with which to manipulate the electronic components necessary for the building and operating of radio telescopes.
What is necessary for survival in the oceans is a set of strong fins to escape predators or some efficient disguise so that they can't see you. Fingers are not really necessary under water for survival. But they come in quite handy when you need to climb a tree rapidly in order to escape that bear! The amount of gray matter in our brains needed to instruct our fingers to turn this page is phenomenal.
The consensus amongst anthropologists is that when our prehuman ancestors left the oceans to live on land, they developed hand-eye coordination in order to compete with the large land animals. Climbing trees — in particular — required delicate manipulative skills, which in turn required increased brain size to handle the complex task of coordinating hand and eye. Those life forms that lacked the coordination necessary to get them up the tree fast enough became someone else's meal, and hence they failed to survive. So again, I will be conservative and assume that only **10%** of all intelligent species reach at least a level of technology comparable to ours.

Having assigned numerical figures for each of the factors, we can now estimate the total number of intelligent, communicative civilizations that have evolved and developed within our galaxy of 100 billion stars:

- $10^{11} \times 20\% \times 10\% \times 1\% \times 10\% \times 10\% = 200{,}000$

Limitations of Estimate If I had been more optimistic on any of the factors, the total would obviously have been greater — more pessimistic, less. What does this number 200,000 actually mean, even if we take it as a reliable estimate? For one thing, it means that the average distance between civilizations is approximately 800 light-years. And it also means that — statistically speaking — we have to examine 1,000 stars before finding one from which intelligent signals are coming.

But now we consider the final factor — the one that far outweighs the other factors in importance. The factor is that of longevity — how long does the average civilization last? Those 200,000 civilizations evolved during the 15-billion-year lifetime of our galaxy. How do we know they are still around and sending out messages? When we consider the ways in which life on our planet — advanced or otherwise — could be extinguished, it seems plausible that has occurred to some, if not all, of those civilizations in the Galaxy. A nuclear holocaust, biological disaster from some mutant organism against which we have no defense, ecological disaster caused by human stupidity in changing the climate through pollution and industrialization, or even a cosmic disaster such as a collision with a major asteroid or comet — these are possibilities we face.

So the question remains — how long do civilizations continue to communicate after reaching the communication stage of development? Do they typically self-destruct once they harness the atom, using that knowledge for aggression rather than for further acquisition of knowledge? If so, then maybe there is no one out there sending messages, in spite of the fact that all of the other factors are busy at work on the planet — life evolving, developing higher and higher levels of intelligence, and even developing space programs. But soon after harnessing the atom, they destroy their planet and all higher forms of life. In other words, intelligence has survival advantage until it develops nuclear weapons, at which time it loses that survival advantage.

Well, the point of this discussion of the Drake equation is this — even if the scientific community is able to assure us that technical civilizations routinely evolve to the level of interstellar communication throughout the Galaxy, there is no assurance that they are sending out messages. As long as we survive, we are optimistic that other civilizations have abilities to survive as well. Once we destroy ourselves, however, we will have better data to use in evaluating the probabilities of others having done the same thing. To that extent, the fact that we have not yet found conclusive evidence that others are out there, we are in a position to be pessimistic about our own future. It certainly would be comforting and encouraging to know that even <u>one</u> other civilization has managed to survive the long-range challenges to managing a planet.

So the number of civilizations you personally estimate to be out there depends ultimately not so much on your knowledge of astronomy, but on your degree of optimism about the ability of civilizations to survive past a certain level of development (f_{ET} in Figure 20-12). And your degree of optimism, of course, can change from moment to moment — perhaps it depends on how often you read the morning newspaper or watch the evening news.

- f_{ET} — Even assuming that intelligent civilizations evolve to levels of technology surpassing that of humans, should we assume they will necessarily be interested in sending out signals to the cosmos, notwithstanding the arguments I presented earlier? Maybe not. Perhaps there exists an *every-civilization-for-itself* attitude out there. Or perhaps shortly after developing the means by which to communicate to other civilizations, they turn that very technology to destructive ends such as warfare or environmental depletion. To be on the safe side, I assume that only **10%** of the evolved technological civilizations actually engage in programs to transmit messages both because they have learned to live in peace with others and because they have learned how to live with limited resources (or have learned how to tap the rest of the universe for unlimited resources). Therefore, the total number of civilizations that are still around and communicating is

10% × 200,000 = <u>20,000</u>

There are 20,000 guys out there waiting for our signals, or sending us signals. That is 20,000 out of 100,000,000,000 stars in our galaxy. Statistically-speaking, we must examine 5,000,000 stars before we successfully detect a message being sent our way. And why would we want to assume that we are worthy of a message anyway?

Attempts to Contact Others

Project Sentinel In 1983, Harvard University and the Smithsonian Astrophysical Observatory assumed responsibility for Project **Sentinel** at the 85-foot radio antenna at Harvard University. At the time, the receiver had 131,000 radio channels. Astronomers wanted to have at least 8 million channels, so Steven Spielberg offered the necessary money and in 1985 he threw the switch that turned on the new system, called **META** (*Megachannel Extraterrestrial Assay*). A duplicate of this system called **META II** is installed in Argentina so that the entire sky can be surveyed for signals from distant civilizations.

Figure 20–18 This radio telescope was the first to conduct long-term searches for signals from distant civilizations as a part of the SETI Program. (Harvard–Smithsonian Center for Astrophysics)

Publically-Funded Search On October 12, 1992, NASA began observations with its own SETI system, which had fewer channels than META, but covered a much larger frequency range — the entire microwave region from 1 to 10 gigaherz. Tests conducted with a prototype of the system detected faint signals from the old *Pioneer 10* spacecraft that is presently out beyond the orbit of Pluto. 500 years after Columbus landed in the New World — the **SETI HRMS** (*High-Resolution Microwave Survey*) was the most comprehensive attempt ever undertaken to answer the question of whether signals are being sent our way. Conducted at a cost of $10 million a year, it employed some of the world's most sensitive radio telescopes and antennas belonging to NASA's Deep Space Network. The initial observing period was to last until the year 2001, and would have probed 10 billion times more search space than the sum of all previous searches.

It employed two complementary strategies — a *Targeted Search* and a *Sky Survey*, which was conducted by Ames Research Center in Mountain View, California. The Targeted Search used the largest radio telescope in the world (1,000-foot) at Arecibo, Puerto Rico, and focused on 1,000 solar-like stars within 100 light-years at frequencies between 1 and 3 gigahertz. It scanned the radio sky with a 14-million-channel system.

The *Sky Survey*, conducted by the Jet Propulsion Laboratory in southern California, used the 112-foot radio antennas of the Deep Space Network at Goldstone, California, and began performing a high-resolution search of the entire celestial sphere at frequencies from 1 to 10 GHz. The high-tech receivers were capable of tuning in on more than 2 million channels at once, and would eventually have been enlarged to handle 32 million channels.

Unfortunately, Congress decided to drop the program in 1994 not so much as a cost-saving measure, but because of a growing public demand for smaller government and lower taxes. Subsequently, the HRMS Program has been reborn as Project **Phoenix**, with funding coming from private sources such as the *Planetary Society*. Since the Jet Propulsion Laboratory is associated with NASA, however, only the *Targeted Search* will continue. Meanwhile, radio astronomers at the University of California at Berkeley have developed a method of using a detector to ride "piggyback" which listens in the background as radio telescopes (such as Arecibo) are involved in other projects. This is Project **Serendip**.

The latest addition to SETI went on-line in October 1996, using the 84-foot radio telescope near Harvard University (Figure 20-18). Called **BETA** (**B**illion-channel **E**xtra**T**errestrial **A**ssay), it takes advantage of the ever-increasing power of computer chips to look at a quarter of a billion channels (wavelengths) within the waterhole. It is funded by hi-tech firms such as Advanced MicroDevices, Fluke, Hewlett-Packard, Intel, and Xilinx.

Several years from now, the Russians plan to launch a 10-meter radio telescope called **Radioastron** into Earth orbit. It will primarily eavesdrop on galaxies, quasars and other astronomical objects. But when a suspicious object is found by an astronomer, *Radioastron* will examine it carefully to see whether it has the characteristics of a radio signal from another civilization. Other SETI programs currently in session are — a pulsar-aided search at Nancay, France; the Algonquin System in Canada; the Russian *Aelita* and *Zodiac* Projects; and Project *Telescope* at CSIRO in Australia. The Russians have also been searching for pulses a civilization might transmit by laser, anything from 1 pulse every 1,000 seconds to 10 million per second. Using their giant 6-meter telescope, they have found 20 objects so far. But all seem to be natural objects.

And so *Provision One* of our "*Insurance Policy for the Long-Term Survival of the Human Species in the Universe*" is in force and is operational. At this very moment, radio telescopes and their associated computers are analyzing the "blips" on radio energy charts, looking for indications of intelligent signals from a distant civilization. Somewhere, scattered amongst the volumes of data on one of those charts, may be encoded advice designed to teach us how to manage our planet's resources and to guide our behavior toward all living creatures in the universe — large and small.

As we proceed with these first (probably) clumsy attempts, we should keep our imaginations open and working. For example, a really advanced civilization might not be confined to a single planet. So we should not confine our searches just to stars, but to the entire sky as we do in META and in NASA's All-Sky survey. If a "being" could live on a billion-ton comet, for example, 100 million watts of starlight could be tapped wherever the comet happened to be in the Galaxy. In fact, comets could be used by extraterrestrials as space vehicles. They could use the vehicle for water, fuel and the chemical building blocks of food. Over the centuries, they could transfer from one comet to another. And what are they looking for, you ask? Perhaps they erect "field stations" at interesting places in the Galaxy — black holes, pulsars, globular clusters and even active galactic nuclei.

ET's: a Case Story During the past 15 years or so, there have been reports of patterns in fields of crops such as wheat and corn. They have appeared mostly in fields of southern England, and to a lesser extent in the fields of 20 other countries. They have been given the name Crop Circles, primarily because of the shape in which the wheat or corn is usually trampled down. Some people have established careers of investigating such occurrences, and of writing books for a public fascinated with claims that no earthly cause could be responsible for their complexity and features.

No fewer than 35 Britons claim to be experts on the phenomena. One book – *Circular Evidence* – sold more than 50,000 copies. A new scientific discipline, called *cerelogy*, emerged. A group of scientists formed the *Circles Effect Research Unit* to study them, and argued that they were the result of unusual wind patterns. A group of Japanese scientists argued that a form of ball lightning could be the culprit. They created crop-like circular patterns both in the laboratory and on a computer programmed to simulate ball lightning. In 1991, two British landscape painters stepped forward and revealed that for the past 13 years they had been sneaking around southern England at night, creating as many as 25 new circles each growing season. Although this does not account for all of the circles reported, no doubt other people, inspired by newspaper accounts, copied the idea. Their method was simple.

After making a scale drawing of the design they wished to create, the two men proceeded to a field with their equipment — a 4-foot wooden plank, a ball of string and a baseball cap with wire threaded through the visor as a sighting device. One stood at the center of the intended site holding one end of the string. The other tied the end of the string to the plank, which he held horizontally at knee level, and circled around the center pushing down the wheat or corn as he went. Asked why they finally stepped forward after so many years, they stated that they were concerned over efforts on the part of some researchers to obtain government funding for their work on circles.

LEARNING OBJECTIVES: *Now that you have studied this Chapter, you should be able to:*

1. Describe the basic assumptions upon which the belief that intelligent life exists out there is based.
2. Describe the 3 phases of advancement that a civilization might go through, and the consequences of that in terms of our search strategies.
3. Explain why the general scientific opinion is that intelligent civilizations will be peaceful, and therefore will be attempting to send signals out to us.
4. Explain why it is that the search programs so far have been passive (listening) programs, not sending programs.
5. Explain the value of the Drake equation, and briefly describe each factor used in calculating the number of civilizations in the Galaxy.
6. Explain how it is that the average life-span of a civilization is the most uncertain factor in the Drake equation toward determining an accurate figure.
7. Explain why it is that astronomers have settled on the 21-centimeter wavelength for the chosen signal for listening.
8. Explain how mathematical language can be use to send information to another civilization, and state the assumptions upon which that conclusion has been reached.
9. Decipher a message written in mathematical (binary) language.
10. Briefly describe the evidence that scientists use to support the theory that life arose on the early Earth through the process of evolution.
11. Define and use in a sentence each of the following **NEW TERMS**:

Binary language 481
Creationism 487
Drake equation 485
Dyson Sphere 476
High Resolution Microwave Survey (HRMS) 489
Hydroxyl molecule 480
Life zone 486
Project BETA 490
Project META 489

Project Phoenix 490
Project Sentinel 489
Project Serendip 490
Radioastron 490
Third law of thermodynamics 472
Type I civilization 476
Type II civilization 476
Type III civilization 476
Water hole 481

Chapter 21

The Future of Life in the Universe

Figure 21–1 The first person to set foot on an extraterrestrial object: Moon. (NASA)

CENTRAL THEME: Is it reasonable to assume that humans will expand their presence into the universe by building space habitats? What are some of the anticipated discoveries in astronomy in the future?

This final chapter is not so much about astronomy as it is about the practical uses to which astronomical knowledge might be put. Although what frequently happens is that practical application of knowledge leads to new methods of obtaining additional knowledge. So we can't predict accurately in advance just what new discoveries will occur as we apply our astronomical to something practical — for example, building space habitats in orbit around Earth. Astronomers pursue their work through a motivation to understand nature's workings rather than a need to create new products for the consumption of society. In effect, knowledge is the product of science, but it is free to the consumer.

It should be obvious by now that the process of obtaining and applying knowledge is not an intentional, smooth, and uninterrupted flow of human thought and activity. Columbus didn't really know where he was going. He believed that the ocean covered only one-seventh of the globe, the rest being dry land. That was the conclusion of orthodox Christian authorities. His voyages of discovery eventually led Europeans to the knowledge that their certainties about geography had been false. This was the great significance of his voyage — the discovery of European man's ignorance of the world.

Now there is a difference between inventors and "questors". Thomas Edison was not interested in knowledge. He was interested in payoff. Questors don't apply the rules of cost-benefit analysis. Space exploration is the obvious parallel today (Figure 21-1). The pace of investigation and discovery is different today. It took centuries for the full impact of Columbus's encounter with the West to develop. The explosion of the atom changed the world in the blink of an eye. With this speed comes unpredictability. Progress cannot be carried out in a cost-free manner. With advancements come problems. The Garden of Eden must have been a very boring place. But it is not my intent to direct your choices as to how our knowledge about space should be applied, or even that it should be applied — but to raise questions which will hopefully broaden our perspective on those choices.

Planetary Survival In the previous Chapter, I suggested that humankind might want to consider strategies for insuring the survival of life and Earth and in space. The general theme of that Chapter was that if we were to establish contact with an advanced civilization, we might learn some new techniques toward that end. *Provision One* of our **Cosmic Insurance Policy** therefore supposes that "Someone" else will provide the cure for healing Earth and install the Preventive Maintenance Program that will allow us to inhabit the planet for billions of years into the future.

But even with that in place, according to our understanding of the life cycle of stars, Earth will eventually experience a natural death when Sun dies some six or eight billion years from now. So *Provision One*, although in force for a long period of time, is not a truly long-term policy against the death of the human species.

Provision Two of the Policy is. Besides the intelligent use of Earth's resources and striving for international peace, we might insure humanity's survival by a vigorous Space Program — one whose ultimate goal is to launch some members of our species out into space and to give them the wherewithal to survive there forever. It would be a modern-day Noah's Ark. Beneath all of the hoopla surrounding the *Apollo* flights to Moon was the realization that we were taking a first step toward that goal (Figure 21-2).

Provision Two, unlike *One*, is not founded on a pessimistic view of humans — it simply recognizes that everything in the universe of matter is temporary. In a sense, it is the ultimate optimistic view of humanity and life on Earth in general. It states that although human bodies come and go, the wisdom of one generation is passed on to succeeding generations so that the human species gradually learns how to leave Earth permanently and to spread throughout the universe to live forever. In doing so, we transcend the universe of matter that comes and goes. Perhaps this uniquely human capacity for transcendence allows for the universe to become a kind of "Heaven" — a place of timeless existence and bliss!

The father of the modern rocket ships that carry humans out into space for momentary glances back at our fragile planet Earth is considered to be the Russian rocket pioneer *Tsiolkovsky*. Considered to be a madman by the Russian government at the time, he made a remark that sums up the attitude of many people concerned about what we are doing to our planet: "Earth is the cradle of the mind, but one cannot live in the cradle forever." Just ponder that thought for a moment.

Is there any fundamental difference between early colonists braving the hardships of crossing the Atlantic Ocean and the North American continent to settle and develop the fertile West Coast, and astronauts setting up

Figure 21-2 This plaque was left on Moon. (NASA)

permanent, self-sustaining habitats on Mars? Is it a difference in kind, or simply a difference in degree? What will the solar system be like 29,672 years from now?

Will humans still be confined to Earth, or will most of the planets and satellites have permanent stations inhabited perhaps by humans who are no longer exactly human as we understand that term today — the rigors and challenges of space habitation having caused biological changes in the voyagers themselves. Do you share the opinion offered by some that it is not a question of possibilities, but a question of will? That is, isn't it more a matter of whether we want such a future rather than whether such a future is technically possible?

Now I doubt that anyone argues that we should tear apart Earth in our haste to provide humankind with permanent settlements in space. We are at the same time working diligently to preserve our precious environment, not only through direct action to prevent ill effects due to chemical spills, leakage of radiation from atomic power plants, and so on — but also working for changes in attitudes toward and within our educational, political, religious, and personal institutions (relationships).

In spite of all the efforts made to date — in spite of our awareness that we must work even harder in that direction — there is no guarantee that our efforts will succeed and be rewarded with utopia on Earth. And I am not suggesting that the threat to our survival comes only from within the ecosystem. Ignoring the possibility that some ET's may cruise down here and find that we are tasty enough for a huge banquet, there is also the matter of asteroidal or cometary collisions with Earth — or even Sun unpredictably going into supernova stage. We are infants as regards the amount of knowledge we have about such possibilities. The only comfort we seem to possess is the knowledge (or at least belief) that we have been here for a very long time, and that we have somehow managed to survive.

Limits to Growth? It is rather interesting that much of the current interest in the idea of space habitats began with a single article written in a *Physics Today* magazine (1964) by Gerard O'Neill, then a physicist at Princeton University. I say interesting because it is common to hear people claim that a single person cannot make a difference in society — that the private citizen is powerless to effect change in the direction that our society and world goes. And yet here was a simple yet elegant idea, worked out in quantitative terms and explained in precise detail.

The initial impetus for his work was the conclusions of the *Club of Rome* report of the 1960s, which said that the only viable future for humanity was to give up the idea of economic and political progress — in the sense of greater political freedom — and accept absolute physical limits on the capability of Earth to support a population. But in such a steady-state society, can freedom of thought be permitted? The Club concluded No — because the slightest diversion from the steady-state would result in the world blowing up. The line of reasoning went something like this:

- All societies run on energy. The most advanced use about 100 times as much per capita as the average resident of India. But the living standard is about 50 times as high.

Figure 21–3 One of the Shuttle astronauts works in the cargo bay of the space vehicle to prepare a satellite for launch. Such missions have become routine, and argue for the building for ever-larger units in space. (NASA)

Figure 21–4 One of the Shuttle astronauts floats in space outside the Space Shuttle, kept alive by a life-support system. Notice the thin fuzzy "film" that coats the surface of Earth. That thin film is the life-support system for life on the planet. (NASA)

- We have not yet figured out a good way of having an affluent use of energy without substantial harm to the environment.

- At our present rate of energy consumption, by the year 2050 we will need 63,000 nuclear reactors in operation all over the world.

- Solar energy intercepted on the surface of Earth is not an answer because of the day-night cycle, clouds, and the arrival of the energy at a funny angle most of the time.

- Sun is the one perfect source of energy in the solar system, and has been steady for 5 billion years.

- It takes twice as much energy to get to the surface of Moon as to get to a high orbit over Moon. One power satellite would put out as much as 10 nuclear plants on Earth. We could put one up built out of lunar material in 20 years if we wanted.

NASA conducted three studies of space habitats in 1975, 1976, and 1977. These examined the issues of closed-cycle life support, radiation shielding, habitat design and construction, economics and logistics, and lunar and asteroid mining. The initial reason for the colonization of space would be to use the resources of space to provide for the needs of our home planet. Since 1977, most of the research has been carried out by the nonprofit *Space Studies Institute* of Princeton, New Jersey, whose President is Gerald O'Neill.

A habitat has 3 essential characteristics: (1) a tightly closed-cycle ecological system capable of replenishing the air, water and food with only trace elements from outside the system; (2) have enough radiation-shielding so that indefinite stays are possible; and (3) provide sufficient artificial gravity to permit the inhabitants to reside on a permanent basis without bone-calcium loss or other effects due to prolonged exposure to microgravity.

Space stations such as NASA's *Skylab* and the Russian's *Salyut* and *Mir* have flown in low Earth orbits and were therefore protected by Earth's magnetic field. The astronauts aboard the *Apollo* flights were protected by the short time durations of the flights. A privately-funded United States organization called *Space Biospheres Ventures* has surpassed the previous Soviet-dominated art of Closed Environmental Life Support Systems (*CELSS*) by constructing the **Biosphere II**.

Located near Oracle, Arizona, *Biosphere II* is nestled a 3-acre mini-world dedicated to investigate human abilities to create and live in a 100% recycled environs. Since 1984, a community of scientists, engineers, adventurers and dreamers prepared for Closure Day (December 1990)

Figure 21–5 The International Space Station's design has not yet been finalized, but this is one of the possibilities. It will consist basically of a few trusses to which will be attached various living and working and utilities modules. The idea is to allow for future expansion, somewhat like the construction of a suburban area. (NASA)

when eight "biospherians" (bionauts) entered the tropical interior and closed the air lock for two years. Inside, they lived and worked in different artificially-created climatic zones.

The chief motivation behind *Biosphere II* is to develop a biologically-based technology for creating self-sufficient space habitats. There is no way that the battleship-like steel-and-concrete structure could — or would — be assembled on Mars. Miniaturization is not part of the experiment. *Biosphere II* is a test of concept, a bold leap into the unknown. No endeavor on this scale could fail to influence our eventual approach to inhabiting space. The immense resources invested in Biosphere 2 — $100 million or so — comes from a Texas millionaire.

International Space Station Alpha

Actually, it is probably too late to debate the merits and demerits of a program to inhabit space. The basic framework for the venture has already been established. As early as 1984, President Reagan, in his State of the Union address, committed the United States to establish a permanently personned Space Station in orbit around Earth by 1994. Over the years since then, the design of the space station — named the **International Space Station** — has been scaled down significantly — partially because of budget cuts, but also because of the *Challenger* disaster and the state of the economy in general.

As presently conceived, it will be assembled by a succession of 23 shuttle flights, and astronauts will undertake extravehicular activity (EVA) to build and maintain it. The crew capacity will be 4, and the target date for "permanent manned capability" is September 1999. The launch of the first element is planned for 1997. Europe and Japan are building modules to attach to the space station.

It seems to me that the words of Neil Armstrong, upon first setting foot on Moon in 1969, reflect the same vision of humankind's future in space — "That is one small step for man, one giant step for mankind." The Russian's too seem bent on rushing humans into space. They are, in fact, much further along in the testing of long-duration effects on humans living in the weightlessness of space. They have had astronauts living and working under such conditions for almost a entire year, a duration the United States has not even approached.

A space station is certainly a far cry from establishing a permanent, self-sustaining habitat either in orbit around Earth or Sun or on the surface of a planet or satellite itself. But of course a space station is just a step in the direction — just as a manned landing on Moon can now be seen as a prelude to more extensive ventures out into the cosmos (Figure 21-5).

President Bush proposed the next step beyond that of

Figure 21–6 The Shuttle's role in the construction of Alpha *is to act as the cargo vessel for delivering parts and personnel to Earth orbit. Since the environment there is gravity-free, there is no need for the massive and heavy structures on Earth. New types of lightweight but strong materials are being developed for this purpose. (NASA)*

Space Station *International Space Station*. He targeted the year 2019 as that year in which to land an expedition of humans on the planet Mars. *International Space Station will* be the jumping off point for that venture. Figure 21-9 offers a long-range program to populate the cosmos. The program is actually a very detailed flow diagram from present to the future, based upon available technologies and those that can be reasonably anticipated.

Construction Materials In order for anything to be constructed anywhere, there are 3 requirements — raw materials, energy, and manipulative devices for rearranging the raw materials into useful structures. Initially, the raw materials necessary for building space habitats will come from Earth. Solar collectors aboard the Space Shuttle provide much of the energy needed to run the electronics and operate the tools. Eventually, Earth will cease serving the function of providing those requirements, except for those materials and substances not found in the asteroids or in Moon material. The raw materials and energy necessary for constructing large-scale structures outside Earth's environment will eventually be "mined" in space.

Much of the energy (and hence cost) consumed in launching a Shuttle vehicle is for the purpose of getting the mass of the payload (including the unspent fuel itself) away from the gravity of Earth and through the friction-causing atmosphere. That is less of a problem in getting raw materials off Moon, where the gravity is one-sixth that of Earth — and even less of a problem in getting raw materials from near-Earth asteroids.

Raw materials obtained from Moon and asteroids might then be used to construct enormous **Solar Power Stations** placed in orbit around Earth, or on Moon's surface. These power stations could in turn be used for processing larger amounts of raw materials, for making even larger power stations, and so on. The recycling of scarce resources — such as water — will of course be routine. The same will be true of the air needed for breathing by the humans present, although no doubt robots will perform the lion's share of the mundane chores required. Mining the surface material of Moon, for example, should not require long-term human presence. Automated assembly lines, patterned after those of the automobile industry in which robotics play the key role, will be common.

Mass-driver electromagnetic launchers have been developed at MIT and Princeton University. Because of the one-sixth gravity there, the length of the accelerator on Moon can be reduced from 5 miles to 500 feet. SSI studies have shown that the cost of a lunar-sourced power satellite is only 3 percent of the cost of the same satellite built from Earth-launched materials. In 1991, a workshop of the *International Astronautical Federation* at a conference on solar-power satellites suggested a vigorous program of international experimentation on the use of space resources

Figure 21–7 Among other things, Alpha *would act as a way–station for astronauts going to Moon. A special lunar shuttle craft would go back and forth between* Freedom *and Moon. The lunar shuttle is seen in this artist's drawing.* (NASA)

to provide energy to Earth.

Food? Well, production of that too will eventually be a self-sustaining activity. After all, what is needed for plants to grow besides seed, sunlight, and chemical nourishment? Soil is an important element on Earth because it physically supports plants while they collect sunlight, at the same time providing a reservoir and canal network for nutrients to reach the roots of the plants. Water is simply the medium within which the nutrients move in getting to the roots. There are many alternate strategies for achieving the same result in space, and several are currently being tested in laboratories around the world for their possible application in space.

Robots in Space Robotics have not only become routine in the automobile manufacturing process, but we are on the threshold of their expansion into other technological endeavors. The next generation of "smart" computers — in conjunction with robotics — will certainly displace humans from many repetitive and mundane tasks. Although some fear that robots will take over the planet, robots will certainly allow us to more fully enjoy the pleasures of the mind as long as we determine in advance exactly what their limits should be.

That is not to say that space exploration should or will be conducted by intelligent machines, at least in the short run. Robots lack the flexibility necessary to accomplish everything that needs to be done in space. A good screwdriver-bearing astronaut cannot be replaced by a robot. On one of the Shuttle flights, a photographic experiment was saved when an astronaut performed a delicate repair on a jammed film drive. On the other hand, during that same flight, a crystal-growing experiment was ruined when another astronaut accidentally kicked the "off" switch.

The price tag for maintaining the human expertise in space is astronomical — currently it is about $1,000 per pound of material delivered to low-Earth orbit. Much of that cost goes toward life support and redundant safety systems for the astronauts. It is comparable to the price tag for a new television set that includes room-and-board for a TV repair person! So it stands to reason that robotics will be utilized whenever possible.

The *International Space Station* is not an end in itself, but a doorway to further space exploration. In fact, just about any place that we consider exploring in space will require a space station — a stopover point between Earth and the destination. The next step will certainly be the building of habitats on Moon or on Mars. Prior to the breakup of the Soviet Union, it was the opinion of many that the Russians would attempt a personned mission to Mars sometime in the 1990s.

Although recent events in that country certainly preclude such an ambitious accomplishment during this decade, it is probably only a matter of time that they do so once they get back on their feet. Successful accomplishment of that feat would certainly stimulate the United States' space program — there is certainly nothing wrong with a little competition.

Terraforming Planets Setting up shop on the surfaces of planets and/or satellites within the solar system may not just be a matter of adapting to the unique environment of the chosen site, but may also of consist of changing the environment to suit the needs of the visitors. This is not

Figure 21–8 Alpha would also act as a way-station for astronauts going to Mars and beyond. The landing of people on Mars will most surely occur within your lifetime, a result of international effort more than that of national pride of a single country. That is because of the expense involved. (NASA)

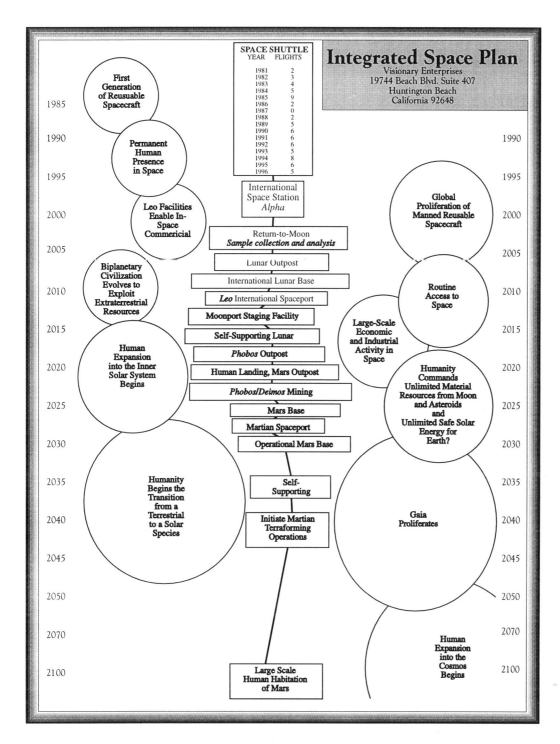

Figure 21-9 Here is a plan for the future of human presence in space. It is not a matter of feasibility, but one of will. Esssentially, if we want to do it we will. (Visionary Enterprises)

unlike the human presence on Earth, although the scale of the endeavor elsewhere could be much greater. There has been considerable thought, for example, given to the idea of injecting blue-green algae into the atmosphere of Venus to change the hellish conditions on that planet to more Earth-like conditions. This type of endeavor of rearranging the environments of other planets to suit human presence is called **planetary engineering** (or **terraforming**).

Consider Venus. The first thing that we would want to do would be to reduce the amount of carbon dioxide in the atmosphere and to produce limited amounts of oxygen. It is, after all, the greenhouse effect of the CO_2 that makes the planet's surface temperature extremely hot — about 860 degrees Fahrenheit. But we must remember that Venus is also closer to Sun. Unfortunately, the process of converting carbon dioxide to oxygen requires more than just green plants — it also requires water.

One of the biggest problems in terraforming Venus, therefore, would be in providing the water needed for photosynthesis. Photosynthesis is, of course, the process by

CHAPTER 21 / The Future of Life in the Universe **499**

which green plants take CO_2 from the atmosphere and convert it to the energy they need to grow. They then exhale O_2 back into the atmosphere. This process is the source of most of the O_2 in Earth's atmosphere.

Venus has no water on its surface and very little in its atmosphere. Because of the large amount of additional water required, it would be impractical to import it from Earth. Some scientists have proposed that it might be possible to divert the orbits of large numbers of comets to make them collide with Venus. Because comets consist largely of water ice, this solution could — in theory at least — solve the water shortage. But we'd still have to worry about other problems, such as making sure that the oxygen did not recombine with organic carbon to reform into carbon dioxide and water. All in all, when all of these difficulties are considered, most scientists believe that Mars is a more promising target for the purpose of terraforming.

Interstellar Travel Human departure for the stars is not a likelihood that is just around the corner. Think of it this way. Using state-of-the-art rocket technology allows us to send space vehicles about 30 miles per second in leaving the solar system. That is what the probe *Pioneer 10* is presently doing. Assume that some brave astronaut is a stowaway aboard, and that the probe is headed for the nearest star beyond Sun (which it isn't). Moving at the rate of 30 miles per second, he/she should plan to arrive in about 40,000 years! Visiting another solar system will certainly not be routine — at least not for the time being.

But an interesting thought arises here. If our goal is, in fact, to insure the survival of our species in the cosmos, is there any reason to be concerned about the length of time required to "get" somewhere? Does our representative have to be an adult, fully grown? Does he/she even have to be conceived yet? Why not send frozen eggs and sperm into space, to be defrosted upon arrival — and then caused to fertilize within an artificial womb?

Science fiction stuff, right? But it may be less a matter of whether or not it is currently feasible than a matter of whether or not we want to do it. After all, we find ourselves stranded on Spaceship Earth without an owner's manual. We are gradually adding pages to that manual started by the earliest humans, but it is far from clear what the long-term goal is of our presence on Earth.

Astronomy in the Future

One of my major goals throughout the book was to impress you not only with the vast amount of knowledge that is currently available to us, but also with the recency of most of that knowledge. This latter emphasis has, I believe, important implications — much of what we (claim to) know is probably wrong, or is at least incomplete. If the history of astronomy tells us nothing else, it tells us that new tools and techniques for investigating radiation and particles from events in space will expand our understanding of the cosmos. The search just goes on and on. Just what those new discoveries will be is impossible to predict, but the following are some areas bound to be affected by any new discoveries.

Solar System Although we have amassed a tremendous amount of data about the solar system since *Sputnik* was launched from Earth's surface in 1957, there are many questions yet to be answered. The major question continues to be — "How did it form?" While the general theory of a contracting cloud of gas and dust from the interstellar medium is compatible with the wide range of observational and experimental evidence currently available, a full description of the entire process is still a matter for theoretical conjecture. Additional probes and landers sent to the planets, asteroids, comets, and satellites of the solar system will be necessary to provide the detailed answers.

The minor components of the solar system — the asteroids and comet's — are largely mysterious objects with respect to their participation in the overall scheme of its origin and evolution. We would like to know, for example, the compositions of the various groups of asteroids and comet's. For example, are comets representative samples of the interstellar medium?

Sun, our nearest star, confronts us with major questions. Is our model of its internal composition and structure correct? How does it contribute to the magnetic field and its related effects such as sunspots and flares? Why do we detect so few neutrinos from Sun's core? And what is the relationship between solar cycles of activity and changes in climatic conditions and the ozone layer on Earth?

Stars Using our Sun as a starting point, the theory of stellar evolution is quite capable of explaining the vast range of types of stars observed in space. But we don't know if any of them possess planets comparable to Earth. Or — if they do — how they are distributed according to spectral types. What kinds of stars, in other words, are predisposed toward planetary system status?

Spiraling off from this question is one that haunts just about everyone — is there intelligent life elsewhere? And leading off from that question is the obvious one of just how life got here on Earth in the first place? Will we be able to identify through spectral analysis the existence of interstellar molecules so complex as to suggest that they are primitive life forms?

As to the various stages through which stars evolve, we are quite uncertain about the specific details of their progress through those stages. In what manner do stars in particular ranges of mass reach particular end conditions, and what nuclear reactions actually occur in the process? During the initial phases of stellar birth, what exactly goes on in those dark clouds such as Orion? What is the role of magnetic fields during star formation? How old are the oldest stars?

Interstellar Matter Molecules of ever-increasing complexity are being detected in the cool dark interstellar clouds. How do they form in such hostile environments, and what role do they play in the formation of stars? What is the chemistry of the interstellar gas in the various regions of our galaxy and other galaxies as well? Studies to answer this latter question may provide a clearer understanding of the chemical evolution of galaxies.

Galaxies The questions to be resolved here are enormous, basically because most galaxies are too distant to be studied in detail. How did they form in the first place, and what factors went into their evolving into different sizes and shapes? Do they have different ages as well? How do peculiar galaxies get to be? If, in fact, the universe began with a big bang, how is it that the universe formed into its present "lumpy" distribution of matter? Are there perhaps faint, relatively undeveloped galaxies in the apparently empty voids between the "lumps"?

Much observational work is needed on active galaxies and quasars. What exactly accounts for their enormous releases of energy, and subsequent ejections of matter into space? Are the quasars as far away as their redshifts imply? What exactly are they? And what is the relationship between galaxies and quasars? Might quasars represent the death throes of an earlier generation of galaxies?

The Universe Here the possibilities seem endless. What are the boundaries or constraints of that which we call the "universe"? Are there additional constraints which would allow us to conjecture the existence of other "universes" using different combinations of those constraints? If the universe is indeed expanding, will it expand indefinitely into the future, or will it eventually contract back onto itself? And if so, when? If indeed a Big Bang occurred, will particle accelerators allow us to creep up to the very conditions present at the moment of creation, so that we can peek into the "void" of what the universe was doing before the Big Bang?

Conclusion These are certainly not trivial questions, by any stretch of the imagination. And I certainly hope that you do not feel cheated if the answers to these questions are not forthcoming during your lifetime. The chances are, in fact, that as we open up new methods of exploring the universe, the questions themselves will change. Future astronomy textbooks will have a different set of "Questions for future astronomers." That is your homework during your next lesson in astronomy in **Astronomy — Third Contact**.

I hope that I have inspired you to want to keep informed of the latest developments in astronomy and be sensitive to the changing nature of the questions being asked. Not only does the universe and its contents evolve, but scientific knowledge itself evolves. I hope that you share in that evolution during your future **Contact** with astronomy.

LEARNING OBJECTIVES: *Now that you have studied this Chapter, you should be able to:*

1. Argue for the construction of space habitats as an extension of our space program.
2. Describe where and with what materials using what energy source those space habitats could be built and operated.
3. Explain why interstellar space travel is not feasible at this time.
4. Describe some of the future discoveries that are expected to be made in astronomy, and how they will fit into current theories about the universe.
5. Define and use in a complete sentence each of the following **NEW TERMS**:

Biosphere II 495
Cosmic Insurance Policy 493
International Space Station Alpha 496
Planetary engineering 480
Solar Power Station 479
Terraforming 499

Glossary

Figure Glossary–1 Earth as viewed from space. (NASA)

CENTRAL THEME: What new words have I added to my volcabulary, and on what page of the Book can I find them first used?

Absolute visual magnitude [Mv] (167) The apparent visual magnitude an object would have it if were located at a distance of 32.6 light-years from Earth. A measurement of the energy output of an object.

Absolute zero (14) The lowest temperature that something can get. Atomic motion ceases at this temperature. In the Kelvin system, zero degrees.

Absorption graph (132) A graph on which is plotted the amount of absorption that occurs by specific wavelength.

Absorption nebula (194) Interstellar gas and/or dust observed in silhouette against a background of luminous gas, or detected indirectly by absorption lines in spectra.

Absorption spectrum (128) A spectrum that reveals the nature of the absorbing material through the study of absorption lines.

Accretion disk (238) The whirling ring of material that collects around a compact object such as a white dwarf, neutron star, or black hole as a result of strong gravitational forces.

Active galaxies (297) Galaxies that exhibit an unusually high output of energy at various wavelengths, often–times coming from the nucleus.

Active optics (107) A new development in building telescopes in which the mirrors are actively shapened by electronics so that the radiation falls on the detector at a point.

Advanced X–ray Astrophysics Facility (AXAF) (118) Satellite to be launched in 1999 that will extend the previous studies of sources of X–rays in the sky. It will be 100 times more powerful than previous satellites.

Age of Aquarius (48) In astrology, the belief that the slippage of the vernal equinox from Pisces into the constellation of Aquarius will have a profound effect on humankind.

Age of Moon (48) The number of days since new Moon.

Age of Reason (80) That period of history around the Renaissance during which people learned to rely increasingly on human reason rather than faith in authorities.

Albedo (395) The fraction of light that reflects off an object, a function of the composition of the reflecting material.

Albert Einstein (85) That scientist/mathematician who revolutionized our thinking about the universe, most notably by showing that gravity is the warping of space rather than the attraction between objects.

Alchemists (78) Predecessors to modern chemists, having had as their goal the turning of the 4 fundamental elements into gold, the quintessence.

Algol (178) The best studied eclipsing binary star, located in the constellation of Perseus.

Allende meteorite (367) A meteorite that fell to Earth that was found to contain numerous kinds of amino acids.

Amalthea (439) The nearest to the planet itself of the 16 known satellites of Jupiter.

Anaeobes (389) Tiny organisms on Earth that are unable to survive where oxygen is present, suggesting that Earth's atmosphere has not always contained oxygen.

Ambient temperature (217) The prevailing temperature surrounding an object under study.

Andromeda galaxy (5, 277) The nearest galaxy to the Milky Way that is about as large as our galaxy.

Angstrom unit (129) A very small unit in which wavelengths of radiation are expressed.

Annular eclipse (50) A type of solar eclipse during which a bright ring of the solar photosphere can be observed around the new moon.

Antiparticles (159) Subatomic particles created artificially in accelerators and which have the opposite properties (such as electrical charge) of matter of which Earth is made.

Aphelion (346) That point in the orbit of an object around Sun when it is most distant from Sun.

Aphrodite Terra (401) One of the two large plateaus detected on the surface of Venus through the use of radar.

Apparent infrared magnitude (94) A scale of numbers used to express the apparent brightnesses of objects in the sky in terms of the amount of infrared photons we receive from each.

Apparent visual magnitude (m_v) (28) A scale of numbers used to express the apparent visual magnitude of objects in the sky in terms of the amount of visible photons we receive from each.

Arachnoid formation (400) Formation found on Venus that suggests the upwelling of magma from the interior.

Archimede's principle (427) A principle of physics that an object loses as much weight as the weight of the volume of fluid it displaces.

Arecibo radio telescope (113) The largest radio telescope in the world (1000 feet in diameter), noted for important discoveries as well for having sent only intentional radio message into space.

Ariel (457) A satellite of Uranus.

Aristotle (69) An early Greek scholar whose ideas in the realms of physics and metaphysics were adopted by the Church as dogma as they applied to the relationship of humans to the cosmos.

Asteroid (357) A chunk of rock that orbits Sun.

Asteroid belt (357) A belt of asteroids that generally lies between Mars and Jupiter.

Asthenosphere (383) That semi-plastic layer of material beneath Earth's crust that is responsible for the movement of crustal plates (continental drift).

Astrolabe (81) An ancient device developed by the Arabs in order to locate direction to Mecca by relating positions of objects in the sky to local time.
Astrology (24) The ancient practice that is based upon the belief that the positions of celestial objects influences the behavior of people on Earth.
Astrometric binary system (218) A system of two stars orbiting a common center of gravity, deduced by the study of the gravitational effect that the unseen star has on the proper motion of the visible companion star.
Astronomical unit [AU] (8) The average distance between Earth and sun, about 8 light-minutes.
Atmospheric window (97) Those portions of the EM spectrum that are not absorbed by Earth's atmosphere, and are therefore detected at Earth's surface.
Atomic number (125) The number of protons in the nucleus of an atom.
Aurora Australius (149) The auroras as seen from southern latitudes, caused by solar wind particles interacting with air molecules high in Earth's atmosphere.
Aurora Borealis (149) The auroras as seen from northern latitudes.
Autumnal equinox (41) The position of Sun on the celestial sphere as it crosses the celestial equator on its way southward, or the date of that occurrence (about September 23).

Babcock model (153) The preferred theory that attempts to explain Sunspot cycle and related phenomena.
Bailey's Beads (54) A phenomenon seen during a total solar eclipse, at which time the last bit of sunlight peaks through depressions on Moon's surface.
Balmer lines (129) The prominent spectral lines that result from electrons moving between the second and higher energy levels of the hydrogen atom.
Barred spiral galaxy (287) A spiral galaxy in which the spiral arms take off not from the nucleus, but from the ends of bars of stars bulging out from the nucleus.
Barringer Meteorite Crater (361) A well-preserved crater in Arizona created 50,000 years ago during Earth's collision with a meteoroid.
Barycenter (175) The center of gravity of two or more objects that are gravitationally bound to one another.
Belt (427) The cooler, descending brown/red bands of gases in Jupiter's atmosphere.
BETA (490) The most recent addition to the SETI program, a search for extraterrestrial signals using computers that can analyze a billion channels at once.
Beta Pictoris (371) A nearby star whose infrared portrait reveals a ring of dust encircling it, suggesting an eventual planetary system.

Big Bang Theory (138) The theory, based on observational and theoretical evidence, that all space, matter, and time began with an explosion about 20 billion years ago.
Big Crunch (320) Term to describe the universe coming back together again, if and when it reverses its present expansion.
Big Foam theory (330) A theory to explain what the universe was doing before the Big Bang, and how the universe could have come out of that condition.
Big Whack theory (394) The theory that the universe, after reversing its present expansion, will collapse into another dense state and explode again.
Binary language (481) The mathematical system used in modern computers to store and process information, and the assumed method by which intelligent civilizations exchange information.
Binary star (4) A star that is gravitationally bound to another star, and around which it orbits.
Binary star system (173) Two stars gravitationally bound together, revolving around a common barycenter.
Biosphere II (495) A 3-acre mini-world in Arizona dedicated to investigate human abilities to create and live in a 100% recycled environs. The chief motivation is to develop a biologically-based technology for creating self-sufficient space habitats.
Black dwarf star (217) The hypothetical object that remains when a white dwarf star cools off completely.
Black holes (240) Regions wherein space has been infinitely warped by the collapse of massive stars, and from which nothing, not even light, can escape.
Blob tectonics (402) Proposed geological process on Venus to explain the surface features. Semi-molten material collides rather than solid plates.
Bok globules (199) Small, compact absorption nebulae seen against the background of glowing gas clouds, and believed to be the sites of star formation.
Bolide (357) A particularly bright meteor, the result of a larger meteoroid hitting and "burning up" in Earth's atmosphere.
Brahe, Tycho (82) Danish astronomer of the 1500s, noted for the precision of his work in plotting positions of celestial objects, from which Kepler was able to deduce the elliptical nature of planetary orbits.
Brown dwarf (206) A not very massive star that is on the border line between being a star and a planet, thus not radiating much light.

Callisto (432) One of the 4 Jovian satellites, noted for its densely cratered surface.
Caloris basin (405) A large impact feature on Mercury.

Cantaloupe terrain (462) Unusual sculptured terrain on Neptune's satellite Triton, marked by crisscrossing ridges and depressions, and showing evidence of youthfulness.
Capture theory (394) One of the theories to explain how Earth obtained Moon.
Carbon detonation (219) That phase in the death of a star during which carbon in the collapsing core gets hot enough to fuse and "ignite."
Carbonaceous chrondrite (368) A type of meteorite that has small inclusions of carbon in it.
Carbonaceous meteorite (367) A general class of meteorite that contains a high percentage of carbon.
Cassini **Saturn Orbiter** (446) A proposed mission to send a satellite to the planet Saturn, during which time the probe *Huygens* will be dropped into the thick atmosphere of the satellite Titan.
Cassini's division (442) The apparent "gap" between the A- and B-rings of Saturn, easily observed in amateur telescopes.
Celestial equator (30) The projection of Earth's equator onto the celestial sphere, dividing the sky into northern and southern hemispheres.
Celestial (North, South) poles (36) The two points on the celestial sphere toward which Earth's rotational axis points.
Celestial sphere (22) That imaginary sphere in the sky against which the positions and movements of celestial objects can be plotted.
Cepheid variable stars (256) A class of dying stars whose changes in luminosity are related to their average luminosities, and which can thereby be used as distance indicators.
Chandrasekhar limit (228) The upper mass limit for a star to die without collapsing to become a neutron star or black hole.
Chaos theory (421) A newly–evolved science that attempts to explain the behavior of movements of objects in space by rules of chaos, such as asteroids and rings.
Charged–couple device (CCD) (109) A plate of small detectors that records the receipt of photons digitally. The modern replacement for the photographic plate.
Charon (463) Pluto's only satellite.
Chemical element (13) One of the 92 naturally–occurring basic substances in the universe. Specified by the number of protons in its nucleus.
Chiron (359) A small object orbiting Sun out near Uranus, perhaps a wayward asteroid or a comet, perhaps a member of the Kuiper Belt.
Chromatic aberration (105) An optical shortcoming of refracting telescopes, in which different wavelength photons are focused to different points within the telescope, causing fuzziness in the image.
Chromosphere (55) That layer of Sun just above the visible photosphere, seen as a reddish color during a total solar eclipse.
Circumpolar (37) Stars and/or constellations seen from a particular latitude that move around the celestial pole, and are above the horizon throughout the year.
Clock (81) A device used to tell the angular distance between Sun and the meridian, expressing that in units of time.
Closed universe (317) A universe in which space does not go on forever, but bends back on itself.
Co-formation theory (394) A theory proposing Moon's origin as being the result of a collapsing cloud of matter along with and close to Earth.
Collecting power (103) The major criterion upon which the worth of a telescope is expressed, a function of the diameter of the objective lens or mirror.
Collisional–ejection theory (394) A theory proposing that Moon formed as a result of a collision between Earth and a planet-sized object, ejecting material that condensed into Moon.
Collisional physics (391) That branch of study involving the collision of objects in an attempt to explain the variety of types of impact features observed in the solar system.
Color (93) The human brain's expression for the eye's receipt of particular wavelengths.
Coma cluster of galaxies (280) A nearby cluster of galaxies.
Coma–Sculptor cloud (279) A nearby group of galaxies.
Coma (348) A large gaseous envelope surrounding the head of a comet.
Comet (339) Ice-rock remnant of the solar nebula which occasionally approaches close enough to Sun to develop a tail and thereby be observed.
Compact object (215) An object that is extremely dense, such as a white dwarf, neutron star or black hole.
Compact sources (298) A type of radio galaxy whose nucleus is responsible for the emission of strong radio energy.
Conic sections (83) The shapes (circle, ellipse, parabola, hyperbola) obtained when a cone is sliced in various ways. These are the shapes of orbits of moving objects in the universe.
Conservation of Angular Momentum (59) A principle of physics which says that a collapsing object must spin faster and faster.
Constellations of the zodiac (24) The 12 constellations that lie along the ecliptic, and through which the apparent Sun moves as Earth orbits Sun.
Contact binary system (238) A binary star system in which the stars are so close together that mass of gasses pass back and forth between them.

Continental drift (383) The phenomenon that "plates" constituting Earth's crust move relative to one another as material within the asthenosphere oozes up through the crust.

Continental plates (383) Those sections of Earth's crust that move with respect to one another.

Continuous spectrum (104, 127) The presence of all wavelengths of the visible spectrum spread out in the colors of the rainbow.

Convection currents (148) The flow of material from a hotter to a cooler region. For example, the plasma bringing heat from the interior of Sun up to its surface.

Copernican system (72) The model of the solar system proposed by Copernicus that placed Sun at the center, thereby replacing the Ptolemaic or earth-centered system.

Copernicus (72) The scholar of the 1500s who sparked a major revolution by proposing a sun-centered universe.

Corona (55) Sun's outer atmosphere, easily observed during a total solar eclipse.

Coronae (400) Domed features found on Venus that appear to be the result of molten material that tried to push up through the surface.

Coronagraph (152) An instrument used in conjunction with a telescope for artificially eclipsing Sun, allowing for the study of Sun's corona.

Coronal holes (152) Regions of Sun's corona within which magnetic lines of force are disconnected from the photosphere, allowing solar wind particles to escape.

Cosmic Background Explorer (COBE) (118, 325) Satellite orbiting Earth that measured and detected the radiation left over from the Big Bang explosion.

Cosmic background radiation (324) Radiation in the form of microwaves that are left over from the moment of the Big Bang, and which was first detected in 1967.

Cosmic Insurance Policy (493) The attempt to prolong the life of and/or insure the survival of humankind by engaging in SETI programs and building habitats beyond Earth.

Cosmic strings (295) A theory to propose why the galaxies are distributed in space the way they are.

Cosmology (68, 315) The study of the origin and large-scale structure of the universe.

Coulomb barrier (213) The region around identically charged, subatomic particles that resists further approach of the particles.

CRAF (352) A proposed mission to an asteroid.

Crab nebula (227) The remains of a star that exploded in the year 1054.

Crater density (381) The number of craters per unit area, used to establish relative dates for regions on the surfaces of solid planets and satellites and asteroids.

Creationism (487) The belief that the creation (and operation) of people and the entire universe was the result of a higher Intelligence, rather than the working out of physical principles of physics and chemistry.

Crust (383) That layer of Earth's surface material that is broken into plates that move with respect to one another.

Cygnus X–1 (249) A likely candidate for a black hole in our galaxy.

Dark matter (292) Matter in the form of *machos* and/or *wimps* thought to constitute most of the universe.

Dark matter particle theory (294) The proposal that the "missing mass" of the universe consists of *machos* and/or *wimps* in intergalactic space, proposed to account for unpredictable motions of galaxies in clusters.

Dark nebula (195) Clouds of dust in interstellar space observed against the background of luminous clouds of gas.

Declination (30) One of the two angles used to express the position of an object on the celestial sphere. It is the angular distance between the object and the celestial equator.

Deferent (71) In the Ptolemaic system, the orbit of the epicycle around Earth at the center.

Deimos (360) One of the two tiny satellites of Mars.

Density (9) Measurement of mass per unit volume of an object.

Density wave theory (269) The proposal that disturbances within the disk of a spiral galaxy cause the collapse of clouds of gas and dust to form stars.

Deuterium (156) Heavy hydrogen atom, in which there is a neutron in the nucleus with the proton.

Diamond Ring effect (54) A phenomenon seen during a total solar eclipse, at which time the very last bit of light leaks through a single depression on Moon's surface.

Differentiation (368) The layering of different densities of substances as a planet cooled, the heavier material sinking and the lighter floating to the top.

Differential rotation (142) The flow of currents of material making up a non-solid object at different rates; for example, in Sun and Jupiter.

Dione (446) A satellite of Saturn.

Distance indicator (283) A well-understood object located in a distant galaxy that can be used for determining the distance to that galaxy.

Doppler effect (137) The effect that different observers detect the spread-out or compressed waves of sound or light emitted by a moving object, according to their placement with respect to that object.

Double asteroid system (364) Two asteroids gravitationally revolving around a common center of gravity, such as *Ida* and *Dactyl*.

Drake equation (485) A mathematical method used to estimate the number of intelligent civilizations that exist within our galaxy with whom we might be able to communicate.

Dust tail (349) That part of a comet's tail that is influenced by solar radiation.

Dwarf elliptical galaxy (279) The most common type of galaxy, a small, symmetrically shaped group of old stars.

Dynamo effect (382) The theory to account for magnetic fields around celestial objects such as Earth, the result of a solid object rotating differentially with respect to semi-molten material within.

Dyson sphere (476) The hypothetical construction of a Type II civilization, whereby they have dismantled a planet and constructed a sphere around their star, collecting the entire energy output of the star.

Earth–fission theory (393) The proposal that Moon formed as a result of the rapidly spinning earth ejecting a glob of material that condensed to become Moon.

Eclipsing binary systems (178) Two stars of a binary system whose orbital plane is lined up with Earth, deduced from the changing behavior of the apparent visual magnitude as a result of the stars alternately eclipsing one another.

Ecliptic (39) The apparent path of Sun's movement on the celestial sphere as Earth revolves around Sun.

Einstein Orbiting X–ray Observatory (232) A satellite designed to detect sources of X-ray energy, with which the first black hole was discovered.

Electroglow (453) A phenomenon seen in the atmosphere of Neptune by *Voyager 2*. Its cause is not well understood.

Electromagnetic force (123) One of the 4 forces governing the behavior of matter in the universe, holding electrons within the energy levels of atoms due to their opposite charges with respect to the protons in the nuclei.

Electromagnetic energy/spectrum (91) The pattern of energies in the form of photons emitted by various processes in the universe, distinguished according to wavelength, frequency, or energy content.

Electron (123) A negatively charged particle that is one of the three major subatomic particles making up atoms.

Electrostatic forces (444) The forces between identically charged particles (repulsion) or oppositely charged particles (attraction), resulting in such effects as sparks while walking across carpets.

Elliptical-orbit law (82) The proposal first made by Kepler that the planets move in elliptical orbits around Sun, not perfectly circular ones.

Emission nebula (194) A cloud of interstellar gas that emits photons as a result of absorbing high-energy radiation from nearby hot stars.

Emission spectrum (129) The spectrum of lines obtained when the emitted photons of different wavelengths are recorded on photographic film.

Enceladus (447) One of the satellites of Saturn.

Energy (12) The ability of something to do work.

Energy curve (98) A plot of the energy output of an object according to wavelength.

Epicenter (382) The point at which an earthquake occurs.

Epicycle (71) In the Ptolemaic system, the orbit of the Sun, Moon and planets. The center of the epicycle was itself in orbit around Earth.

Equal areas law (82) Kepler's second law of planetary motion, which says that an object orbiting another object sweeps out equal areas in equal intervals of time. This explains why Earth moves faster in its orbit around Sun when it is closest to Sun.

Equatorial bulge (381) The greater circumference around the equator of objects like Earth and Jupiter as a result of rapid rotation throwing material outward by centrifugal force.

Equinox (41) Equal. Days and nights are equal in length when Sun crosses the celestial equator at the times of the vernal and autumnal equinoxes.

Escape velocity (241) The minimum velocity to which an object (like a spaceship) must be accelerated in order to escape the gravitational field of another object (like Earth).

Europa (432) One of the Galilean satellites of Jupiter, noted for its cracked surface of frozen ice.

Evaporating Gaseous Globules [EGGs] (204) Dense globules of gas and dust that remain behind as photo–evaporation caused by nearby hot stars clears away less–dense clouds of gas and dust.

Evening star (403) The planet Venus or Mercury seen just after sunset just above the place where Sun set, occurring when either planet is positioned most eastward from Sun.

Event horizon (245) The theoretical boundary of warped space-time in the vicinity of a black hole within which nothing can escape (escape velocity = infinity).

Evolution (16) The act of something changing into something else.

Evolutionary track (202) On an H–R diagram, the plot of the positions a star would have as it evolves.

Excited (124) The condition of an atom when an electron is in an orbit of higher energy content than that of the ground state.

Extended sources (298) Locations of radio-emitting regions on either side of what appears to be a normal elliptical galaxy.

Extreme Ultraviolet Explorer (EUVE) (117) An artificial satellite launched into Earth orbit in 1992, designed to detect objects that emit the shorter wavelengths of ultraviolet radiation.

Eyepiece (104) A device inserted at the focus of a telescope for the purpose of magnifying the size of the image of the object being observed.

Faculae (156) Bright eruptions that occur on Sun at time of sunspot maximum.

Fahrenheit (131) A scale of expressing temperature.

"Falling" star (23) A meteor, the result of a dust particle burning up in Earth's atmosphere.

Fault line (383) The boundary between two continental plates, where slippage may occur.

Federal Communication Commission (96) The Federal agency responsible for allocating specific wavelengths of the electromagnetic spectrum to users.

Fermi lab (324) The largest and most powerful particle accelerator in the United States, used to investigate the nature of subatomic particles.

Filaments/voids (295) Terms used to describe the largest–scale structure of galaxies in space.

Filaments (144) Dark, wavy-looking features observed on the face of Sun, believed to be solar prominences seen face-on rather than edgewise.

Fireball (357) Another term for a bolide, a particularly bright meteor.

Fission (160) In atomic physics, the breaking of heavier elements into lighter ones.

Focal length (104) The distance between the objective lens or mirror of a telescope and the point at which the photons are focused, upon which the magnification of the telescope is based.

Focus (82) Mathematically, that point within a conic section shape that is the center of motion for objects moving along that shape.

Foucault pendulum (36) An instrument that proves Earth is spinning. The pendulum swings back and forth relative to stars, but Earth turns underneath it.

Fractional distillation (344) The process by which a mixture of chemicals is separated into its component chemicals according to weight in the act of heat being applied, the lightest being boiled away first. This attempts to explain why the planets closest to Sun contain the highest proportion of heavy chemicals.

Frequency (92) For electromagnetic energy as well as for sound, the number of crests of emitted waves that pass a point in a given interval of time.

Fumeroles (389) Regions of the surface of a planet (or satellite) through which gases escape from the interior of the planet, having been released from heated material.

Fusion (160) The method by which two lighter chemical elements join together to form a heavier element, releasing (or absorbing) energy in the process. The basic method by which stars emit radiation.

Galactic cannibalism (289) The act of one galaxy gravitationally capturing and "digesting" another galaxy.

Galactic (open) cluster (191, 260) Another term for an open cluster of stars, most often found along the flattened plane of spiral galaxies, and containing exclusively young stars.

Galactic plane (19) The flattened portion of a galaxy. The band of the Milky Way lies along the galactic plane of our galaxy.

Galaxy (5) A large collection of stars, bound together by their mutual gravitation.

Galileo (68) The father of the modern scientific method, the Renaissance scholar who confronted the Church with the radical theory of Copernicus, who suggested that Sun, and not Earth, stood at the center of the universe.

Galileo **mission** (431) An artificial satellite launched in 1991 bound for Jupiter. A component of the satellite will descend into the thick clouds to determine their composition and behavior.

Gamma rays (96) The shortest and therefore most energetic type of radiation.

Gamma Ray Observatory (GRO) (116) An artificial satellite to be launched in 1990 and placed in earth orbit for the purpose of detecting sources of gamma rays in space.

Ganymede (432) One of the Galilean satellites of Jupiter, noted for its crisscross pattern of folded terrain features.

Gaspra (363) An asteroid imaged up close by the *Galileo* satellite.

Geocentric system (71) Earth-centered system associated with Aristotle and Ptolemy.

Geomagnetic storm (149) At time of solar outbursts, increased solar wind particles cause magnetic disturbances in Earth's upper atmosphere, including the auroras.

Giotto (351) The ESA satellite that imaged the solid nucleus of Halley's comet up close.

Gliese 229B (207) The best candidate for a brown dwarf star, orbiting the star *Gliese 229* in the constellation of *Lepus*.

Glitch (232) A change in the rotation rate of a neutron star, revealing itself as a change in the pulsation rate of the pulsar we measure.

Globular cluster (256) A large group of old M-type stars gravitationally bound together, generally located in the halo of a galaxy, and containing virtually no gas or dust.

Golden Age of Astronomy (98) The current era of astronomical research, whose accomplishments are principally due to the use of artificial satellites in orbit above Earth's atmosphere.

Grand Unified Theory [GUT] (330) The proposal that three forces of nature (electromagnetism, strong and weak nuclear forces) were united as one when the universe was a mere fraction of a second old, at which time the temperature was immensely high. The search for this unity occurs in experiments conducted in atomic accelerators.

Granulation (147) The splotchy appearance of the photosphere of Sun, the result of ascending and descending convection currents bringing heat to the surface from the interior.

Granule (147) One of the dark or light splotches observed on Sun's photosphere, being the top of an ascending or descending current.

Gravitational lens effect (307) The cause of the double image of a quasar on a photographic plate as a result of a massive galaxy distorting space-time, suggesting the extreme distance of the quasar.

Gravity (12) One of the four forces of nature, the result of large collections of atoms warping space–time.

Gravity wave (250) The predicted disturbance of space when compact objects collide or swallow stars. Detection would prove the existence of black holes.

Great Dark Spot (GDS) (460) A feature found in the atmosphere of Neptune by *Voyager 2*. It is similar to Jupiter's Red Spot, and probably due to similar process.

Great Observatories Program (118) U.S. program to cover most of the electromagnetic spectrum with Earth–orbiting satellites. The Hubble Space Telescope is one of them.

Great Red Spot (GRS) (428) A feature observed in the upper cloud deck of Jupiter for centuries, no doubt the result of a cyclic storm caused by the planet's rapid rotation.

Greenhouse effect (387) The mechanism by which a substance (for example, carbon dioxide in the atmosphere of Venus) allows for the passage of visible radiation but blocks that of infrared, causing the buildup of a high temperature.

Ground state (124) The lowest energy state of an atom that an electron can occupy.

GS 2000+25 (249) One of the best candidates for a black hole in the Galaxy.

H I region (194) Gaseous regions in space in which atoms have been excited by radiation from nearby hot stars.

H II region (194) Gaseous regions in space in which atoms have been ionized by radiation from nearby hot stars.

Half–life (380) The period of time required for half the mass of a radioactive substance to disintegrate into other elements.

Hale–Bopp comet (353) A newly–discovered comet that became visible to the naked eye during summer 1996.

Heat (91) The popular term for infrared radiation.

Heat tectonic activity (387) The mechanism by which an object such as a forming planet or satellite will, as it cools to the surrounding temperature, exhibit surface features such as wrinkling of the crust, fumeroles, faulting, and the like.

Heliocentric system (72) Sun-centered system proposed by Copernicus and defended by Galileo, currently the accepted model of the solar system.

Heliopause (466) The boundary of Sun's influence in space, and where the influence of other stars begins to be detected (stellar winds, especially).

Helioseismology (148) The study of the large–scale motions of plasma on Sun's surface.

Helium flash (219) The condition within the collapsing core of a dying star, at which time the temperature of the helium "ash" reaches the point at which time atoms fuse together to form carbon, releasing enormous amounts of energy.

Herbig–Haro objects (200) Dark splotches found embedded in clouds of gas and dust and believed to be newly–formed stars in the act of expelling gas and dust away from them.

Highlands (391) The mountainous regions of Moon, found mostly around the perimeters of the maria, the result of the pushing up of material as a result of the collisions that created the maria during Moon's formation.

High Resolution Microwave Survey (HRMS) (489) A Congressionally–funded program to search for microwave signals from distant civilizations, terminated in 1994 after only two year due to budget concerns.

Hipparchus (28) Early Greek astronomer who organized the celestial sphere, marked the positions of the celestial equator and ecliptic, and developed a system of expressing the brightnesses of stars still used today.

Hirayama families (359) Families of asteroids that have similar orbits around Sun.

H-R diagram (170) A plot of stars according to their surface temperatures (spectral types) and energy outputs (luminosities or absolute visual magnitudes), used to predict stellar evolution and determine the ages of star clusters.

Hubble constant (284) The value of the rate of expansion of the universe, taken from the plot of galaxies on Hubble's diagram.

Hubble Space Telescope (HST) (98) One of the Earth–orbiting telescopes of the *Great Observatories Program* that provides high resolution images of objects in space. Fully operational in 1994, it has provided theory–challenging images and data.

Hubble's law (284) The observation that the more distant a cluster of galaxies is located from Earth, the faster is its moving away from us, the interpretation being that the universe is expanding.

***Huygens* probe** (446) A probe that will be carried aboard the *Cassini* mission to Saturn, and which will drop into the dense atmosphere of the satellite Titan.

Hydrocarbons (445) Molecules consisting of atoms of hydrogen and carbon, important in astronomy because of their close association with the chemistry of life.

Hydrogen (97) The most common chemical in the universe, also the simplest and lightest.

Hydrogen-alpha filter (144) A device attached to the eyepiece of a telescope to enable the observer to view those features on Sun such as prominences that emit radiation primarily at a wavelength associated with hydrogen.

Hydrostatic equilibrium (181) The gravity-pressure balance within a star that accounts for its long lifetime on the main sequence of the H-R diagram.

Hydroxyl molecule (480) One of the two chemical components of water, and which emits an 18–cm wavelength of radiation to signal its presence in space.

Hyperbolic paraboloid (318) The shape of the curvature of space if space is infinite in extent, associated with the Big Bang, ever-expanding universe, and the Steady State universe.

Hyperion (447) A small, irregularly–shaped satellite of Saturn.

Hypothesis (87) In the scientific method, a single statement that requires testing.

Iapetus (444) One of the satellites of Saturn.

Icy dwarfs (also **Plutons**) (465) Presumed small, icy objects that comprise the Kuiper Belt, and which may have caused the unusual axial tilt of Uranus and be responsible for the origin of Triton, Nereid, and even Pluto itself.

Ida (363) Asteroid imaged by the satellite *Galileo* as it passed by on the way to Jupiter. It is a double asteroid system with *Dactyl*.

Impact cratering (387) The process believed to be responsible for the observed craters on objects in the solar system, the result of the accretion of material in the pre-solar nebula during the formation period of the solar system.

Implosion (239) The opposite of explosion. When a star explodes, it also implodes inward.

Infrared radiation (91) That portion of the EM spectrum adjacent to visible light, and often called *heat* energy. Longer wavelength and less energy than visible radiation.

Infrared Astronomical Satellite [IRAS] (116) An artificial satellite launched into earth orbit in 1983 in order to detect those objects emitting radiation at infrared wavelengths, such as cool stars and regions within which stars are believed to be forming.

Interferometer (106) The integrated use of two or more telescopes located at some distance from one another to achieve higher resolution than if the telescopes were used independently.

International Space Station *alpha* (338, 496) The proposed (and currently being worked on) station in orbit around Earth that will be permanently manned, and be a launching pad for distant exploration of space. Should be in orbit by the turn of the Century.

International Ultraviolet Explorer [IUE] (117) Launched in 1978, this satellite has operated continuously while imaging sources of UV radiation in space.

Interstellar (95) The region between stars, occupied for example by clouds of gas and dust.

Interstellar clouds (194) Clouds of gas and/or dust that occupy the regions between the stars in a galaxy. Also called nebulae.

Interstellar medium (194) Referring to any of the material or processes in the regions between the stars.

Interstellar reddening (193) The process by which distant objects will appear redder than they should due to intervening dust clouds scattering the blue photons traveling from those objects, but allowing the red photons to arrive uninterrupted.

Inverse–square law (167) A mathematical description of the magnitude of the force of gravity or the intensity of radiation with increasing distance between objects (for gravity) or between the source of radiation and detector (for radiation).

Io (429) One of the Galilean satellites of Jupiter, noted for its active volcanoes.

Ion tail (349) That gaseous portion of a comet's tail that is influenced by the solar wind.

Ionized (124) The condition of an atom after it has lost one or more electrons, leaving behind a net electric charge due to the excess protons in its nucleus.

Irons (367) The most commonly found type of meteorite, consisting of a mixture of iron and nickel, and believed to be a fragment of the deep interior of a broken-up asteroid.

Irregular galaxy (279) Those collections of stars that have no particular shape or symmetry, and found to contain a high proportion of gas, dust, and young stars.

Irregular satellite (341) Satellite whose orbit is retrograde or is inclined at a steep angle with respect to the equatorial plane of its parent planet.

Ishtar Terra (401) The largest of the two uplifted plateaus on the surface of Venus detected by radar, and which contains the highest feature on the planet.

Jovian (343) Associated with Jupiter by virtue of size, composition, behavior, etc.

Keck telescope (106) Presently the largest telescope in the world, located on Mauna Kea in Hawaii.

Kepler, Johannes (82) Renaissance scholar who destroyed the medieval notion of a universe dominated by perfect circle through the discovery that the planets revolve around Sun in elliptical orbits.

Kepler's 3 laws of planetary motion (82) The laws of motions that govern the motion of planets around Sun, and indeed the motion of any object around another object.

Kinky ring (444) The thin, outer F-ring of Saturn's system of rings, gravitationally distorted by the proximity of two small satellites.

Kirkwood gap (370) A gap within the asteroid belt, believed to be the source of many meteorites found on Earth.

K–T boundary (364) Geological layer of material found in various locations on Earth that may be the fallout from a collision with an asteroid or comet.

Kuiper Airborne Observatory (101) A high–flying jet aircraft with an infrared telescope aboard operates above the radiation–absorbing layers of Earth's atmosphere, and images sources of infrared in space.

Kuiper Belt (347) A belt of comets orbiting Sun between the orbits of Neptune and Pluto and extending far out beyond Pluto, believed to be the source of comets orbiting around Sun in less than 20 years.

LAGEOS **satellite** (386) A geophysical satellite that measures very accurately the slippage of continental plates with respect to one another.

Land of the Midnight Sun (42) Those locations on Earth at which an observer sees Sun in the sky during a 24–hour period.

Late Heavy Bombardment (418) Based upon crater studies around the solar system, there appears to have been a period of intense collection of asteroidal material onto surfaces of planets and satellites even after that of the initial formation period.

Latitude (30) One of the two angles used to express the position of someone or something on Earth's surface. It is the angular measurement of that person or thing from the equator.

Least–energy orbits (396) Those orbits on which artificial satellites are sent to other planets, since by making them behave according to Sun's gravity just as the planets do, a minimum expenditure of energy is necessary.

Life zone (486) That volume of space around a star within which the temperature conditions that are necessary (according to earth standards) for life to flourish and be sustained exist.

Light curve (178) A plot of the apparent visual magnitude of an object with time, used to detect eclipsing binary stars, Cepheid variable stars, pulsars, etc.

Light–year (7) The distance that light travels in one year.

Lighthouse effect (231) The explanation for the behavior of a neutron star in being detected as a pulsar, from which we observe a flash of radiation each time a beam of emitted photons from the rapidly spinning star lines up with Earth.

Lobate scarps (405) Long, arch-like cliffs photographed on the surface of Mercury, believed to be the result of the shrinking of the planet early in its cooling-off period.

Local Group (7, 279) The group of 20 or so nearby galaxies to which the Milky Way and Andromeda galaxies are gravitationally bound.

Local hypothesis (305) The assumption that quasars may be close to the Milky Way galaxy instead of being the most distant objects in the universe, used in an attempt to avoid having to explain the enormous energy outputs of quasars if they are so distant.

Lohihi (385) The slowly–rising seamount located to the southeast of the Big Island of Hawaii, the result of motion of the Pacific Ocean plate over molten material rising from a magma chamber within Earth.

Long Duration Exposure Facility [LDEF] (357) Satellite that remained in orbit around Earth for 6 years, collecting micrometeoritic dust in Earth's vicinity, returned to Earth by Space Shuttle.

Longitude (30) Along with latitude, used to express the position of a person on Earth's surface. It is the angular measurement of that person East or West from Greenwich, England.

Luminosity (28, 168) An expression for the total energy that an object emits.

Luminosity class (172) One of five classes into which a star is placed according to it position on the H–R diagram.

machos (293) Proposed theory of matter in the universe to explain the behavior of galaxies moving within clusters of galaxies.

Macrocosm (78) The larger world outside people, the things we see directly around us.

Magellan **Venus Orbiting Satellite** (399) Currently orbiting Venus, this satellite is providing complete and detailed images of the surface of the cloud–covered planet.

Glossary/Index 511

Magellanic Clouds (279) The nearest galaxies to the Milky Way, named clouds because of their appearance.
Magma chamber (384) Hot regions within Earth's asthenosphere, origin of molten material that flows to surface to form shield volcanoes.
Magnetic force lines (146) The lines of force that run between opposite ends of a magnet, made obvious when iron filings are sprinkled near the magnet. Important in the understanding of solar activity and pulsars, for example.
Main sequence stars (170) Stars in that period of evolution during which they are in hydrostatic equilibrium, steadily emitting energy as hydrogen is fused together to become helium.
Mantle (382) The outer portion of a planet, satellite, or asteroid. That region of asteroids from which the stony type of meteorite is believed to originate.
Maria [singular = mare] (390) Those darkish regions easily seen on the near-side of Moon, believed to be vast lava plains caused by the upwelling of molten material subsequent to major impacts of accreting material during its formation.
***Mariner 9* mission** (408) Early satellite sent to Mars, providing complete coverage of the martian surface, revealing the network of dry riverbeds.
***Mariner 10* mission** (403) Early mission to photograph both Mercury and Venus. The only satellite to visit Mercury.
Mars Global Surveyor (415) US satellite departing Earth in November 1996, carrying many of the instruments lost when the *US Mars Observer* failed upon reaching Mars in 1993.
Mars Pathfinder (415) US satellite departing Earth in December 1996, carrying a lander and rover for testing development of future larger rovers, aside from various scientific experiments.
Mars '96 (415) Russian satellite launches from Earth in November 1996, carrying two small landers that will monitor climatic changes, and two penetrators that will penetrate up to 20 feet into the martian surface to study subsurface conditions.
Mass (8) The measurement of the amount of matter making up an object, basically the sum total of the atoms and/or molecules making up that object.
Massive black hole (273) A black hole formed not by a star collapsing, but by massive clouds of gas just after the Big Bang. Detected at the centers of many galaxies, including our own.
Maunder Butterfly diagram (143) A plot of the number of sunspots observed as a function of time and location on Sun's surface with respect to its equator, revealing Sunspot cycle of approximately 11 years.

Maunder minimum period (153) That historical period between 1645 and 1715 during which there was a notable lack of sunspots and related activity, accompanied by particularly cold winters in Europe.
MAUTO (456) Abbreviation for the five known satellites of Uranus prior to arrival of the *Voyager* satellite: Miranda, Ariel, Umbriel, Titania, Oberon.
Maxwell Montes (401) The highest point on Venus, a part of Ishtar Terra.
Meridian (39) An imaginary line running directly over an observer's head between the north and south points on the horizon, cutting the celestial sphere into an east half and a west half.
Metals (263) A term to designate those atoms heavier than that of helium, inasmuch as it is believed that it is only within the interiors of dying stars that such atoms are formed.
Meteorite (357) A meteoroid that has survived its' collision with Earth's atmosphere and fallen to the surface, believed to be a fragment of a broken-up asteroid.
Meteoroid (357) Chunk of matter in interplanetary space, into which Earth continuously runs. Believed to be debris left over from the interstellar cloud from which the solar system condensed.
Meteor (popularly called "**shooting**" or "**falling**" **star**) (23, 354) The observed phenomenon of a streak in the nighttime sky, the result of Earth running into a tiny meteoroid.
Meteor shower (355) The appearance of an unusually large number of meteors in the sky, believed to be the result of Earth encountering some of the scattered remains of a disintegrated or disintegrating comet.
Meteorology (337) The study of Earth's atmosphere.
META search program (489) An ongoing search for radio signals from distant civilizations, conducted at observatories throughout the world.
Microcosm (78) The world of things not seen directly, the world inside of people.
Micrometeorite (357) Extremely small meteorite found on Earth's surface, having been too small to vaporize in the atmosphere at the time of colliding with it.
Microwaves (97) A component of the radio region of the EM spectrum, in the shorter wavelength region.
Microwave background radiation (324) That radiation accidently detected in 1965 as microwaves that for theoretical reasons should be the radiation left over from the very explosion that created the universe 20 billion years ago.
Milky Way (5) The galaxy of 100 billion stars of which Sun is a member. Also, the band of white seen in the sky, our view of the edgewise portion of the galaxy.

Millisecond pulsar (373) Pulses of radio radiation emitted by rapidly spinning neutron star, spinning thousands of times faster than typical stars.
Mimas (447) One of the small satellites of Saturn, noted for the appearance of a sizeable crater making it look like the "death star" in the movie Star Wars.
Miranda (458) A satellite of Uranus, appearing to have formed in some kind of violent manner, perhaps related to whatever caused the planet to be tipped on its side.
Missing mass (322) The calculation that the amount of matter (and its equivalent in the form of energy) in the known universe is insufficient to account for its expansion slowing down and reversing, resulting in collapse.
Missionary galaxy (290) A galaxy that collides with and gravitationally absorbs another galaxy.
Model (78) In the scientific method, a description of a natural phenomenon.
Molecular cloud (195) Cloud in which radio telescopes detect a vast assortment of complex molecules. These are usually associated with the sites of active star birthing.
Moon illusion (57) The optical illusion that makes Moon close to the horizon appear larger than when it is higher in the sky.
Morning star (403) The planet Venus or Mercury observed in the eastern sky just above where Sun is about to rise, occurring when the planet is furthest west of Sun in its orbit around Sun.
Motion energy (12) The energy an object has by virtue of its motion.
Multi-Mirrored Telescope [MMT] (106) A telescope located in Arizona utilizing the new principle of several mirrors focusing radiation to a single point, thereby avoiding the problem of distortion due to the sheer weight of a large mirror causing it to sag.
Multiple star system (173) Three or more stars that orbit around a common center of gravity. At least one-half of the stars in the sky are in this category.
Murchinson meteorite (367) A meteorite that fell to Earth and which contained a large number of amino acids.

NASA (101) The National Aeronautics and Space Administration, the space agency of the United States.
National Science Foundation (107) The organization that generally oversees the progress of scientific achievement in the United States.
Natural law (87) In the scientific method, a theory that just about everyone accepts as being true, by virtue of it having been tested successfully over and over again.
Neap tides (58) The lowest range of tidal variation between high and low tides, occurring at quarter moon phases.

Near Earth Asteroid Rendezvous satellite (366) Launched in early 1996, the satellite will visit the asteroid *Eros* in 1999, there to go into orbit for one year while studying the asteroid.
Near–Earth asteroids (362) Asteroids whose orbits bring them close to Earth.
Nebula [plural = nebulae] (5) A cloud of gas and/or dust in interstellar space that is sufficiently dense to be observed either by virtue of emitted radiation, reflected radiation, or absorbed radiation from some source behind it.
Nemesis star (365) Hypothetical star that is, with Sun, member of a binary star system. Could be responsible for the periodic collision of comets with Earth, explaining periodic extinctions of life forms on Earth.
Nereid (461) One of the two known satellites of Neptune, noted for its highly elliptical orbit.
Neutrino (160) A tiny, massless (?) bundle of energy generated in the cores of stars, important in their contribution to the deaths of some stars as supernovae.
Neutrons (125) One of the three fundamental particles making up the atom, neutral in charge and contributing only mass to the atom.
Neutron star (229) One of the 3 possible end products of the death of a star, a dense concentration of neutrons resulting from the collapse of the original star to the point that the protons and electrons combine to form neutrons. Observed as a pulsar.
Newtonian focus (109) That viewing position of a reflecting telescope located close to the end of the tube into which the light of objects being studied enters.
Newton, Sir Isaac (84) English scientist of the 1600's who completed the Copernican revolution by explaining why the planets orbit Sun, and how the same force of gravity operating on earth causes the observed motions in the solar system.
Node (47) As applied to Moon, that point in time in its orbit around Earth when it crosses the ecliptic, that being one of the conditions necessary for an eclipse to occur.
Normal spiral galaxy (287) A galaxy whose spiral arms connect with the round shape of its nucleus.
North star (37) That star toward which Earth's rotational axis approximately points. Presently it is Polaris.
Northern (Southern) lights (149) Popular terms for the *Aurora Borealis* and *Aurora Australius*, respectively.
Novae (236) The outbursts observed from a close-binary star system in which material is being gravitationally pulled off one star and onto the accretion disk of the neighboring compact object, resulting in sudden releases of radiation.

Nuclear fusion (12) The process during which elements collide and join together to form a heavier element.

Nucleosynthesis (219) The process occurring inside of stars in which lighter elements are fused together to form heavier ones, gradually increasing the percentage of heavy elements in the universe.

Nucleus (123, 348) The center of an atom, containing at least one proton and, except for hydrogen, neutrons. The solid portion of a comet, located within the coma.

Oberon (456) One of the satellites of Uranus.

O.B.A.F.G.K.M. (133) A sequence of letters used to express the surface temperatures of stars, obtained by studying the spectra of stars and therefore referred to as the spectral sequence.

Object (114) Something in space that is detectable by the visible light it emits.

Objective lens (102) That lens (or mirror) of a telescope that first collects the radiation from objects being studied, and that causes it to be focused to a point. The feature of a telescope that determines its collecting power.

Occultation/Occult (301) The condition of an object covering up or passing in front of another object, for example during a solar eclipse or the manner in which the rings of Uranus were discovered.

Olympus Mons (410) The gigantic Martian volcano, the tallest feature in the solar system.

Oort cloud (339) The proposed but unobserved cloud of icy-rocky chunks of material surrounding Sun at a great distance, used to explain the origin of comets.

Open universe (317) A universe in which space goes on forever, represented by the hyperbolic paraboloid.

Optical binary (175) Two stars that appear close to one another on the celestial sphere, but are not gravitationally bound in orbit around a common center of gravity and are actually at a great distance from one another.

Optical Illusions (410) A condition that led to the theory that canals existed on Mars. When the human brain is presented with disorganized information from the eye, it will organize it into something with which it is familiar. In the case of Mars, disorganized surface features such as craters were organized as straight lines in the minds of early observers.

Optics (104) The study of how radiation is affected by lenses and mirrors.

Orion nebula (194) A prominent emission nebula located in the constellation of Orion, and a much-studied site of active star formation

Oscillating universe (332) One of the 2 versions of the Big Bang theory, in which the presently-expanding universe eventually slows down and collapses, only to rebound and create another expanding universe which in turn slows down, collapses, and so on to infinity.

Outgassing (389) The releasing of gases from the interior of a planet or satellite.

Ozone layer (97) A layer of the oxygen molecule ozone in Earth's upper atmosphere responsible for absorbing ultraviolet radiation.

Pacific Ocean Basin theory (393) The now-discarded proposal that the Pacific Ocean Basin is the scar remaining after the rapidly-spinning earth ejected some material into orbit that eventually formed into Moon.

Pair production (330) The process in which electron–positron pairs appear briefly from nowhere, then return to nothingness while emitting energy.

Paleontologist (365) One who studies fossils.

Pancake formation (400) Formations found on Venus that suggest molten material tried to push through the surface, but instead cooled and collapsed slightly.

Pangaea (386) The hypothetical supercontinent that existed 200 million years ago, and which broke up into pieces that have gradually separated out to their present positions as plates on Earth's surface.

Panspermia (421) The proposed idea that complex molecules can be pushed by starlight from one region of space to another.

Parallax (76) The apparent change in the position of a star on the celestial sphere as a result of Earth orbiting Sun, used to determine the distance to that star.

Parsec (167) A unit of distance, defined such that it is equal to 3.26 light–years.

Partial lunar eclipse (51) That type of eclipse during which Moon does not completely enter Earth's shadow.

Partial solar eclipse (51) That type of eclipse during which Sun is only partially blocked by Moon.

Particle accelerators (159) Mile-long instruments in which subatomic particles are smashed into atoms at speeds close to the speed of light in an effort to understand the nature of the subatomic particles and the forces at work within them.

Pathfinder project (351) The multiple–satellite mission to image and study Halley's comet up close in 1985–86.

Peculiar galaxies (297) Galaxies that appear distorted and which usually emit huge amounts of energy, the causes of which are not very clear.

Penumbra (56) That portion of a shadow cast by an object, within which an observer would see only a portion of the light source blocked out by the object.

Penumbral lunar eclipse (56) Partial shadow. A penumbral eclipse occurs when the full moon moves through that part of the shadow cast by Earth in which only a part of Sun is being blocked.

Perfect Cosmological Principle [PCP] (321) The assumption used amongst Steady Staters that the universe is uniform in the distribution of matter in time as well as in space: In other words, the average density of the universe is unchanged over time.

Perigee (49) For an object orbiting Earth, that point in time when it is at closest approach to Earth. This is one of the conditions necessary for a total solar eclipse to occur.

Perihelion (41) That time during the year (approximately January 4th) when Earth is closest to Sun.

Period-luminosity relationship (257) That relationship found to be true of some variable stars, for example Cepheids. The longer the period of time between points of maximum brightness, the greater the average luminosity of that star. This relationship can be used as a distance indicator.

Periodic comet (340) A comet that has been diverted in its orbit from the Oort Cloud around Sun, probably in the act of passing close to Jupiter, and consequently caught in the inner solar system to return according to a predicted schedule like Halley's comet.

Periodic extinctions (365) According to the fossil record, large percentages of life-forms have been extinguished at periodic intervals during Earth's history. The Nemesis star theory attempts to explain this.

Permafrost (409) Frozen soil, found at extreme latitudes on earth and throughout (it is assumed) Mars.

Permitted orbits (124) The unique pattern of energy levels associated with each of the 92 chemical elements, between which electrons are able to move in absorbing and emitting photons.

Phase change (295) The change of a substance from one state to another, as when ice melts or water turns to steam.

Phase of moon (45) The shape of the visible Moon, determined by angle between Sun, Earth and Moon.

Phobos (360) One of the two small satellites of Mars.

Phobos 2 (408) A Soviet satellite that ceased functioning when it arrived at the Martian satellite Phobos.

Phoebe (444) The outermost of Saturn's 17 known satellites.

Photo–evaporation (204) Process by which ultraviolet radiation from nearby hot stars "pushes" gas and dust away, revealing denser clouds of gas and dust within.

Photometer (28) An instrument which, when attached to a telescope, measures the number of photons arriving from the object being studied. Used for the purpose of determining the apparent visual magnitude of an object.

Photons (29) Bundles of energy that make up radiation, different types of which are classified according to wavelength, frequency, or energy content.

Photosphere (55) That layer of Sun observed with the human eye, on which such phenomena as sunspots occur.

Pioneer–Venus **mission** (399) The first satellite to image Venus by radar as it orbited the planet.

Pitch (137) In sound, the measurement of wavelength.

Plage (148) A hot spot on Sun's surface, showing up as white in photographs.

Planet X (465) A presumed planet located beyond Pluto, but as yet never detected.

Planetary engineering (also **terraforming**) (499) The concept that another planet could be made livable for humans through massive alterations of the atmosphere and/or surface.

Planetary geology (216) The study of the geology of other objects in the solar system, using the processes on Earth as a comparison.

Planetary nebula (216) That observed stage in the death of a not-so-massive star at which time the outer expanded layers form a bubble-like cloud around the collapsed core, seen as a white dwarf star. Called planetary because they appeared as planets to early observers.

Plasma (147) A fluid of charged particles, such as that of which Sun and (presumedly) stars are composed.

Plate tectonics (383) The theory that explains phenomena such as earthquakes, volcanoes, mountain-building, etc. as being the result of Earth's crust being broken into plates, which "rub" against each other as semi-molten material oozes up from within Earth.

Plates (383) Portions of Earth's crust that collide or slip with respect to one another, causing the dominant surface features we observe.

Pleiades (191) A group of bluish stars prominently seen in the winter and spring skies, and often-times referred to as the Seven Sisters. They are an excellent example of an open or galactic cluster, containing young stars and lots of gas and dust.

Pluto Fast Flyby (465) NASA's proposed mission to explore Pluto, as yet the only planet that has not been visited by a satellite from Earth.

Plutons (also **Icy dwarfs**) (465) Presumed small, icy objects that comprise the Kuiper Belt, and which may have caused the unusual axial tilt of Uranus and be responsible for the origin of Triton, Nereid, and even Pluto itself. (465) Presumed small, icy objects that comprise the Kuiper Belt, and which may have caused the unusual axial tilt of Uranus and be responsible for the origin of Triton, Nereid, and even Pluto itself.
Polaris (37) Presently our North Star, the star that is closest to the projection of Earth's rotational axis onto the sky.
Polarity (146) The condition of having a positive or negative charge, as in the case of the opposite ends of a battery or magnet.
Poor cluster of galaxies (7) A small (less than a hundred or so) group of galaxies gravitationally bound together, and presumedly which formed together.
Population types (263) A classification of stars in our galaxy in terms of their locations with respect to the shape of the galaxy, their spectral types, their ages, and their chemistries and movements.
Population I extreme (264) Very young stars located in the spiral arms and having a high percentage of metals.
Population II extreme (264) Very old stars found in the globular clusters, and quite poor in metal content.
Population I intermediate (264) Young-ish stars found in the disk of the galaxy, and containing moderate proportions of the metals.
Population II intermediate (264) Old-ish stars located in the nucleus of the galaxy, and relatively poor in metal content.
Positron (159) A negatively charged electron, an antiparticle, created at the core of a star during the fusion process, but emitting a gamma ray photon when it mutually annihilates an electron.
Powers of ten (5) A mathematical system of expressing large numbers by indicating the numbers of zeros after the decimal point by a superscript of that number above the number 10.
PPL 15 (206) The first brown dwarf star discovered by the Keck telescope in the star cluster *Pleiades*.
Precession (47) The slowly changing direction toward which the rotational axis of Earth (and Mars) points as it orbits Sun. The cause of the slippage of the vernal equinox through the various constellations of the zodiac.
Pressure waves (382) The seismic waves that transmit through molten material within Earth.
Prime focus (109) The favored position of an observer (or other detector) at the eyepiece of a reflecting telescope, since at that position there are the fewest number of mirrors between the object and the person.

Prism (93) A triangular piece of glass that breaks "white" light up into the continuous spectrum of wavelengths seen as a rainbow of colors.
Project *Phoenix* (490) Rebirth of the cancelled *HRMS Program* of SETI, with funding coming from private sources such as the *Planetary Society*.
Project *Sentinel* (489) One of the earliest attempts to detect signals from distant civilizations. More sophisticated systems have replaced it.
Project *Serendip* (490) Radio astronomers at the University of California at Berkeley use a SETI detector to ride "piggyback" as radio telescopes (such as Arecibo) are involved in other projects.
Prominence (144) Loops of gases extending above the edge of Sun when observed in profile, associated with magnetic lines of force twisted up away from the photosphere and into the corona.
Proper motion (22, 218) The apparent motion of a star relative to the backdrop of more distant (and therefore more stationary) stars on the celestial sphere, the result of the galaxy rotating differentially.
Proplyds (204) Disk–shaped globules of gas and dust that have the appearance of protoplanetary disks, first discovered by the HST in the Orion nebula.
Proton (123) One of the 3 fundamental subatomic particles making up an atom, having a positive charge and residing in the nucleus.
Proton-proton chain (158) The series of subatomic events occurring within the core of a star that is responsible for the energy emitted by the star, basically the conversion of hydrogen into helium.
Protostar (199) That stage in the evolution of a star during which it is still collapsing from a cloud of gas and dust, having not yet achieved a temperature high enough to initiate the proton-proton chain at the very center. The position of a star on the H-R diagram prior to its arrival on the main sequence.
Proxigean tide (59) Exceptionally high ocean tides, the result of the combined effects of Sun and Moon alignment together with Earth at perihelion.
PSR 1257+12 (374) A pulsar around which the first extrasolar planets were discovered.
Ptolemaic model/system (70) Earth-centered model of the universe that survived for over 12 centuries, having been consolidated into a concise theory by Ptolemy.
Ptolemy (70) The scholar of Alexandria who in 140 a.d. incorporated the ideas of Aristotle into a concise earth-centered theory to explain the motions and behavior of celestial objects.
Pulsar (229) The phenomenon we observe suggesting the presence of a rapidly spinning neutron star. It consists of precisely timed pulses of radiation, oftentimes observed coming from the center of a supernova remnant.

Quantum mechanics (240) The field of study of the particles and behavior of particles within the nucleus of the atom. Especially it studies the conditions that are thought to have been present at the time of the Big Bang.

Quarks (239) Theoretical massless particles that make up all of the subatomic particles, and which are therefore the basic energy units of which all things are made.

Quasars (297) Presumedly the most distant objects in the universe, appearing as stars on photographic plates, but moving at enormous velocities away from us, as determined by the Doppler effect. The source of their enormous energy output is not clearly understood.

Quasi-stellar radio source (297) The term originally assigned to the objects we now call quasars, inasmuch as they emit enormous amounts of radio radiation.

Quintessence (78) In the geocentric system adopted by the early Church, the fifth elementary substance, believed to be eternal and incorruptible, of which all celestial objects are made.

Radiation pressure (181) Since photons are bundles of energy, when they are absorbed by matter, some force is applied to that matter. Sunlight hitting your body exerts a very slight force in pushing you away from Sun.

Radio galaxies (298) Galaxies that are notable by virtue of the enormous amounts of radio energy that they emit, either from their cores or from locations to either side of their cores.

Radio interferometer (113) Two or more radio telescopes, located miles or hundreds of miles apart, used in conjunction with one another in order to improve the resolving power of a single telescope.

Radio lobes (299) Those regions on either side of a visible object from which radio energy is detected.

Radio radiation (96) The longest wavelength radiation, also the weakest.

Radioactive dating (380) The method of dating a material that contains a radioactive substance.

Radioactive decay (13) The process during which one element naturally breaks up into two or more lighter elements, releasing radiation in the process.

Radioastron (490) A Russian program to place a radio telescope in orbit around Earth for the purpose of listening for signals from distant civilizations.

Radiograph (273) A radio plot of a region in space, the equivalent of a photograph, but imaged with a radio telescope.

Rays (361) Streaks of lighter material radiating outward from impact craters such as those on Moon, the result of material ejected outward at the time of the impacts.

Red giant (171) A star in the death stage, having expanded to an enormous size and cooled off in the process of hydrogen-fusing ceasing at the core.

Reflecting telescopes (105) That type of telescope employing the use of a mirror in collecting radiation and focusing it to a point.

Reflection nebula (191) A bluish-appearing nebula, made so by the presence of dust grains scattering blue photons more efficiently than red photons coming from hot, young stars in the vicinity.

Refracting telescopes (102) That type of telescope employing the use of a glass lens through which radiation must pass in being focused to a point.

Refraction (112) The slight bending of light as it passes from one medium into another, a principle that allows astronomers to break the light from some distant object into its component wavelengths in passing through a prism.

Regular satellite (341) A satellites whose orbit is prograde and in the plane of its parent planet's equator, suggesting that it formed with the planet.

Renaissance (68) That historical period (14th to 16th centuries) marked by the rebirth of learning in the arts and sciences, during which time the events leading to the methods of modern science began.

Resolution (103) That feature of a telescope (or any optical device) that enables a viewer to see two separate objects at a distance as two separate objects.

Retrograde motion/revolution (71, 340) The reversal of the usual direction of motion of an object. Some satellites revolve in a direction around their planets that deviates from the norm. Also, the usual direction of planetary motion on the celestial sphere is from west to east, but planets appear to reverse direction when Earth is overtaking them.

Revolution (36) The motion of one object around another object, for example Moon around Earth.

Rhea (446) One of the satellites of Saturn.

Rich cluster of galaxies (5) A large (over a hundred or so) group of galaxies gravitationally bound together, and presumed to have formed together.

Right ascension (30) With declination, used to express the position of an object on the celestial sphere. It is the angular measurement along the celestial equator of the object's distance from the vernal equinox.

Ringlets (442) The pattern of thousands of smaller rings that make up the main rings of the planet Saturn, and whose origin are poorly understood.

Ringmoons (444) Small moons found inside of the rings of some of the Jovian planets, and believed to influence the shapes of the rings.

Roche's limit (442) The distance from a sizeable object (such as Earth) within which the gravitational tidal forces would pull another object apart.

Roche lobe (238) The region around an object

within which the gravity of that object dominates, anything approaching within that region being affected.

Rosat X–ray telescope (116) The largest and most precise X–ray telescope placed in Earth orbit in 1990, a team effort by the United States, Germany, and Great Britain.

Rotation (35) The spinning of an object around an axis. The rotation of Earth results in day and night.

Rotation curve (291) A plot of the speed at which stars revolve around the center of the galaxy with respect to their distances from that center.

Rotational axis (36) The imaginary line around which an object rotates.

Runaway Greenhouse effect (403) The proposed explanation for the unusually high temperature at the surface of Venus, the result of increased carbon dioxide in the atmosphere trapping more sunlight, which in turn caused the further release of more carbon dioxide, which trapped more sunlight, and so on.

Sagittarius A (273) Designation of that source of violent activity at the center of our galaxy.

Saint Thomas Aquinas (70) Religious theologian who integrated the work of Ptolemy with that of Christian doctrine to create a fusion between the Earth–centered universe and role of humans on Earth.

Satellite (338) A body that orbits a planet. It is commonly called a moon. If placed there as scientific package, called an artificial satellite.

Schwarzschild radius (245) The radius of the event horizon around a black hole within which light cannot escape.

Scientific model (87) A physical, mathematical, or other type of representation used to illustrate a working or process of nature.

Sea of Tranquility (391) One of the larger mare on Moon, the site of the first personned landing in 1969.

Seeing (100) Term used to describe the state of the atmosphere during an evening of viewing the sky, basically an expression of just how still it is. Good seeing means good observing.

Seismic stations (382) Detectors installed around Earth's surface in order to monitor waves emitted at the time of earthquakes, partially in order to determine the interior composition of Earth.

Selection effect (287) When counting galaxies in space, we mistakenly count too many elliptical galaxies because a very distant spiral galaxy will look like an elliptical galaxy.

Separation distance (176) The actual distance between two stars orbiting a common center of gravity, the value of which must be known in order to calculate the masses of the two stars.

Seyfert galaxy (297) A normal-looking spiral galaxy, but whose nucleus emits an unusually large amount of radiation, suggesting some type of violent event occurring there.

Sextant (38) A device used to measure the angular distance of an object in the sky above the horizon. Use for navigational purposes.

Shear waves (382) Those seismic waves that do not travel through molten or liquid material within Earth.

Shepherding satellites (444) Small satellites of the planet Saturn that orbit the planet quite close to the outermost F-ring, and which gravitationally influence the behavior of the ring.

Shield volcano (385) A volcano that forms as molten material from a magma chamber pushes up through Earth's crust, building up a large, gently-sloping mound of lava. The Hawaiian Islands are a good example.

Shoemaker–Levy 9 (353) A comet that was gravitationally torn apart by Jupiter in 1992, its pieces colliding with the planet in late 1995.

"Shooting" star (20) See meteor and "falling" star.

Singularity (240) A mathematically undefined point at which some value goes to infinity. In the case of a black hole, the very center of the collapsed star exhibits infinite gravity.

Sirius (26) The brightest star in the sky, outside of Sun. It is located in the constellation of Canis Major.

Size (8) The dimension of an object, expressed as length, diameter, radius, volume, etc.

Slingshot effect (396) In sending artificial satellites to other planets, the use of the gravity of one planet to propel the satellite out to another planet at a higher velocity, thereby shortening the travel time.

Slope of curve (284) In Hubble's diagram, the rate of motion of a galaxy divided by its distance is the slope of Hubble's law.

SNC meteorite (370) A meteorite whose composition suggests that it came from the planet Mars, ejected outward, perhaps, by a collision with a larger asteroid.

Solar constant (160) A mathematical value expressing the energy output of Sun.

Solar flares (148) Sudden outbursts of radiation and high-speed particles in the vicinity of sunspots, believed to be the result of the sudden release of energy stored up in the magnetic lines of force as the lines suddenly break.

Solar Max (152) A satellite that investigated Sun extensively and provided important data about Sun.

Solar–neutrino experiment (161) The counting of the number of neutrinos detected at Earth in an effort to determine the rate at which the reaction at Sun's core is proceeding.

Solar power station (497) Proposed stations built in space by space colonists that would collect solar energy and beam it down to an energy–hungry planet.

Solar prominence (54) Loops of gases extending above the edge of Sun when observed in profile, associated with magnetic lines of force twisted up away from the photosphere and into the corona.

Solar system (4) Sun, together with the system of planets, satellites, asteroids, comets, and general debris that are in orbit around it.

Solar wind (149) High-speed electrons and protons and helium nuclei that escape from Sun during solar storm activity, and which are responsible for the auroras in Earth's atmosphere.

Source (114) The location in the sky from which radiation is coming, but nothing can be seen in visible light.

Space Exploration Initiative (415) The United States space program that has as one of its goal to land people on Mars early in the next century.

Space Infrared Telescope Facility [SIRTF] (118) A new generation of satellites to study celestial objects at various wavelengths will be launched in the 1990s. The SIRTF will detect objects that emit radiation at infrared wavelengths.

Space Shuttle (98) The foundation of the U.S. Space Program, this reusable space vehicle delivers payloads into Earth or interplanetary orbits, or allows astronauts to conduct scientific experiments under weightless conditions.

Spectral sequence (133) A system of letters (OBAFGKM) used to express the surface temperatures of stars, obtained through the study of the spectra of those stars.

Spectral type (133) For a particular object, the letter assigned to it according to its surface temperature and luminosity class.

Spectroscope (127) A device which, when attached at the eyepiece of a telescope, separates the radiation from some distant object into its component wavelengths, allowing for a determination of important characteristics about that object.

Spectroscopist (133) A scientist who specializes in determining the characteristics of objects in space by reading their spectra.

Spectroscopic binary system (177) A binary star system detected through the periodic splitting of spectral lines as a result of the Doppler effect operating while the two stars orbit a common center of gravity.

Spectroscopic parallax (181) A method of determining the distance to a star by knowing its location on the H–R diagram.

Spectrum (129) A photograph of the collection of wavelengths of radiation that an object emits or absorbs.

Spicules (148) Small, flame–like structures on Sun's surface.

Spin–casting (108) The shaping of the mirror for a reflecting telescope by spinning it as the molten glass cools.

Spiral galaxy (287) A galaxy like the Milky Way that has young stars in patterns of arms that stretch out from the nucleus.

Spokes (444) Dark, linear-like features observed in close-up photographs of the rings of Saturn, and which do not behave according to Kepler's equal areas law.

Spring (vernal) equinox (41) That moment of time (about March 21st) during Earth's orbit around Sun when Sun appears to cross the celestial equator on its way northward, at which days and nights are equal in length.

Spring tide (58) High tide, the result of Sun and Moon in alignment with Earth.

Stanford Linear Accelerator Center [SLAC] (159) One of the major atomic accelerators being used to explore the inner world of the atom in an attempt to recreate the conditions that existed shortly after the moment of the Big Bang.

Star cluster (5) A group of stars gravitationally bound together, and presumed to have formed together.

Star party (104) An organized gathering of amateur astronomers and interested persons for the purpose of observing celestial objects with eyeballs and telescopes.

Steady State theory (321) The cosmological theory that the universe always has been and always will be, there having been no beginning as in the Big Bang theory.

Stone (367) The most common type of meteorite that falls to Earth's surface, looking very much like a common garden rock, believed to have come from the mantle of a broken-up asteroid.

Stony-irons (367) A type of meteorite that is a mixture of the nickel-iron and stone-type meteorites.

Strong force (329) One of the 4 forces that govern all processes in the universe, the strong force being that which holds quarks together to form the subatomic particles such as protons and neutrons.

Subduction (385) On Earth, location of one continental plate pushing against and slipping underneath an adjacent plate, pushing up mountain ranges in the process.

Sublimation (348) The process by which a solid turns directly into the gaseous state without going through the liquid state, as in the case of comets.

Summer solstice (41) That time of the year (about June 21st) when Sun is located at its greatest distance from the celestial equator, so that the day is longest and the night is shortest.

Sunspots (141) Cooler regions on Sun's surface, believed to be the result of broken magnetic lines of force preventing the upward flow of energy from within Sun.

Supercluster of galaxies (7) Clusters of clusters of galaxies, simply the result of matter gravitationally attracting other assemblages of matter.

Supernova (211) One of the 3 possible deaths of a star, in which material is ejected explosively out into interstellar space, leaving behind either a neutron star or a black hole.

Supernova remnants (226) The remains of a supernova explosion that have been ejected out into space.

Synchronous rotation (392) The condition of one side of a satellite always facing the planet around which it revolves, having been captured into that condition by gravitational tidal forces. Earth's moon is a prime example.

Synchrotron radiation (230) A type of radiation emitted when high-speed electrons from some energetic event are caught in a strong magnetic field.

SN1987A (233) A supernova that occurred in the Large Magellanic Cloud in 1987, and which gave astronomers an excellent opportunity to learn more about supernovae.

T Tauri variable (203) A type of star whose characteristics and behavior suggest that it is just about to leave the protostar stage of evolution and become a main sequence star.

Teide 1 (207) One of the first brown dwarf stars discovered, located in the star cluster *Pleiades*.

Tektites (365) Glassy rocks believed to be formed during high-speed impacts of asteroids with Earth.

Temperature (14, 131) The measurement of the average speed at which atoms and/or molecules in a medium are moving.

Terrae (391) That portion of Moon's surface that is cratered, as opposed to the marias that are smooth, flat regions.

Terraforming (also **Planetary engineering**) (499) The concept that another planet could be made livable for humans through massive alterations of the atmosphere and/or surface.

Terrestrial (342) Pertaining to Earth, or earth-like in size and composition.

Tethys (446) One of the satellites of Saturn.

Theory (87) In the scientific method, a system of rules and principles that is capable of being applied to a wide variety of experiences in the "real" world.

Third Law of Thermodynamics (472) A principle of physics that says that systems gradually run down, that the universe tends toward disorder. It is used by Creationists to argue that humans (as orderly systems) could not have evolved on Earth without a higher Intelligence.

Tidal dissipation (437) The principle causing an internal energy source in the Galilean satellite Io leading to active volcanoes. Jupiter's gravity distorts the shape of the satellite during its revolution, like bending a coat hangar back and forth.

Time of decoupling (323) That time (about 100,000 years) after the Big Bang at which radiation no longer disintegrated matter that was forming, allowing for matter to condense into the stars that today make up the universe.

Time–lapse photography (151) Photographs taken of an object at intervals in order to track changes in atmospheric movement, etc.

Titan (444) A satellite of Saturn, the only one in the solar system known to have a substantial atmosphere, and thus a surface on which prebiological chemicals may be found.

Titania (457) A small satellite of Uranus.

Titius–Bode rule (342)

Total eclipse (50) Event that occurs when Moon completely blocks out Sun (solar), or Moon completely enters Earth's shadow (lunar).

Total lunar eclipse (56) Event when Earth's reddish shadow falls onto Moon, turning it red.

Tracking (111) The feature of a telescope mounting that allows an observer to move the telescope precisely at the same rate that Earth rotates but in the opposite direction, so that the image of the object under study remains steady in the eyepiece or on the photographic plate or other detector.

Trigonometric parallax (165) An accurate method of determining the distance to a nearby star by photographing its position relative to other stars at 6–month intervals.

Triton (461) The large satellite of Neptune, known for the presence of active geysers and young–ish terrain (cantaloupe terrain).

Tunguska event (366) A massive explosion in Siberia might have been the result of a small comet colliding with Earth.

Turnoff point (261) That position on the main sequence of the H-R diagram at which stars in a star cluster are beginning to leave on their way to red giant stage. Used to determine the age of star clusters.

Type A universe (316) A universe whose space has zero curvature, in which radiation travels in straight lines.

Type B universe (317) A universe with negative curvature, in which radiation travels along the shape of a hyperbolic paraboloid, representing an infinitely expanding universe.

Type C universe (317) A universe with positive curvature, in which radiation travels along the shape of a sphere, representing an expanding universe that will eventually collapse back onto itself.

Type I Civilization (476) A civilization that has learned to duplicate the process by which its star shines on the surface of its planet.

Type II Civilization (476) A civilization that has learned to harness the entire energy output of its star, perhaps by enclosing it with a large sphere of energy-collecting devices.

Type III Civilization (476) A civilization that has learned to harness the entire energy output of the galaxy in which it lives by combining the harnessed energy of each of the stars.

Ultraviolet radiation (95) Radiation adjacent to visible light, shorter wavelength, higher energy.

Ulysees **mission** (150) A satellite currently exploring the north and south poles of Sun, having used the slingshot effect to be thrown out of the ecliptic plane.

Umbra (50) The shadow cast behind any object being illuminated by Sun, from inside of which Sun would be invisible.

Umbriel (457) One of the satellites of Uranus.

V404 Cygni (249) One of the best candidates for a black hole in the Galaxy.

Vacuum genesis (330) In the Big Bang theory, an attempt to explain how the universe came into existence from nowhere (a vacuum).

Valhalla (435) The largest impact feature in the solar system, found on the Galilean satellite Callisto.

Valles Marineris (411) A huge canyon found on the surface of Mars, as long as the United States is wide.

Vega (48) One of the brightest stars in the sky, located in the constellation of Lyra.

Venera (397) The Soviet satellites that landed on the hot surface of Venus to obtain the only photographs of that surface.

Vernal equinox (30) The location of Sun on the celestial sphere when it moves from below the celestial equator to above, on or about March 21st. One of the two intersections between the ecliptic and celestial equator.

Very Large Array [VLA] (113) A large collection of radio telescopes located in New Mexico used as a radio interferometer.

Very Long Baseline Array [VLBA] (114) A proposed enlargement of the VLA to include radio telescopes separated by thousands of miles in an attempt to improve the resolving power of radio telescopes.

Viking (408) Unpersonned satellite mission to Mars that took place in 1976, at which time 2 landers photographed and examined soil samples.

Virgo supercluster of galaxies (280) The large collection of clusters of galaxies of which the Milky Way is a member.

Virtual particles (330) In particle accelerator research, particles that appear from nowhere, only to return to nothingness in a very brief moment.

Visible light (91) That portion of the EM spectrum to which human eyes are sensitive.

Visual binary system (175) A binary star system in which both stars can be observed, and which are observed to be moving around a common center of gravity.

Voyager **satellites** (425) The highly successful pair of satellites that returned tens of thousands of images of the planets Jupiter, Saturn, Uranus and Neptune and their satellites.

Voyager **Interstellar Mission** (465) The mission of the *Voyager* satellites to return a limited amount of data as they head out of the solar system.

Waning moon (45) Moon is declining in brightness, occurring between full and new phases.

Waterhole (481) The interval of wavelengths of the electromagnetic spectrum within which communication between intelligent civilizations might take place.

Wavelength (92) With regard to different types of radiation, the measured distance between crests of the waves that constitute radiation.

Wave of darkening (406) The observed changing pattern of dark features on Mars that led to speculation about plant growth on that planet, but now thought to be the result of vast dust storms.

Waxing moon (45) Moon is growing in brightness, occurring between new and full.

Weak force (329) One of the 4 fundamental forces governing the universe, this one operating within the atomic nucleus to hold the protons and neutrons together.

Weight (58) The gravitational force that one object (usually Earth) has on another object.

White dwarfs (171) The most common type of stellar death, being the collapsed, dense core of a star after it has gone through red giant stage.

Widmanstatten lines (368) Patterns in the crystalline structure of iron meteorites that tell us that they cooled slowly during their origin in the asteroid belt.

Wimps (293) A theory proposed to explain in what form the "missing" mass of the universe is in.

Winter solstice (41) That point in time (about December 23rd for northern hemisphere) at which Sun reaches lowest position below celestial equator, and the time of the longest nights.

Worldview (65) The collective belief system of a particular culture about such matters as its relationship to the physical universe in which it lives. Religion is a world view.

Worm hole (249) A theoretical subway system from one region of space to another, created by the formation of a black hole.

X–rays (96) High–energy, short wavelength portion of the EM spectrum.

Zeeman effect (144) A phenomenon observed in the lines of a spectrum taken of a region of Sun's surface where a strong magnetic field exists, causing the lines to split.

Zenith (38) That point on the celestial sphere directly above the head of an observer.

Zones (427) The rising, warmer yellow–white bands of gases in Jupiter's atmosphere.

21–centimeter radiation (112) The radiation that hydrogen atoms emit in space. It is the preferred choice of wavelengths for interstellar communication.

51 Pegasi (375) Sun–like star around which planet was recently discovered.

55 rho Cancri (375) Sun–like star around which planet was recently discovered.

47 Ursae Majoris (375) Sun–like star around which planet was recently discovered.

70 Virginis (375) Sun–like star around which planet was recently discovered.